POLYMER *and* POLYMER-HYBRID NANOPARTICLES

From Synthesis to Biomedical Applications

POLYMER *and* POLYMER-HYBRID NANOPARTICLES

From Synthesis to Biomedical Applications

Stanislav Rangelov
Stergios Pispas

CRC Press is an imprint of the
Taylor & Francis Group, an **informa** business

CRC Press
Taylor & Francis Group
6000 Broken Sound Parkway NW, Suite 300
Boca Raton, FL 33487-2742

First issued in paperback 2019

CRC Press is an imprint of Taylor & Francis Group, an Informa business

ISBN-13: 978-1-4398-6907-9 (hbk)
ISBN-13: 978-0-367-37956-8 (pbk)

Visit the Taylor & Francis Web site at
http://www.taylorandfrancis.com

and the CRC Press Web site at
http://www.crcpress.com

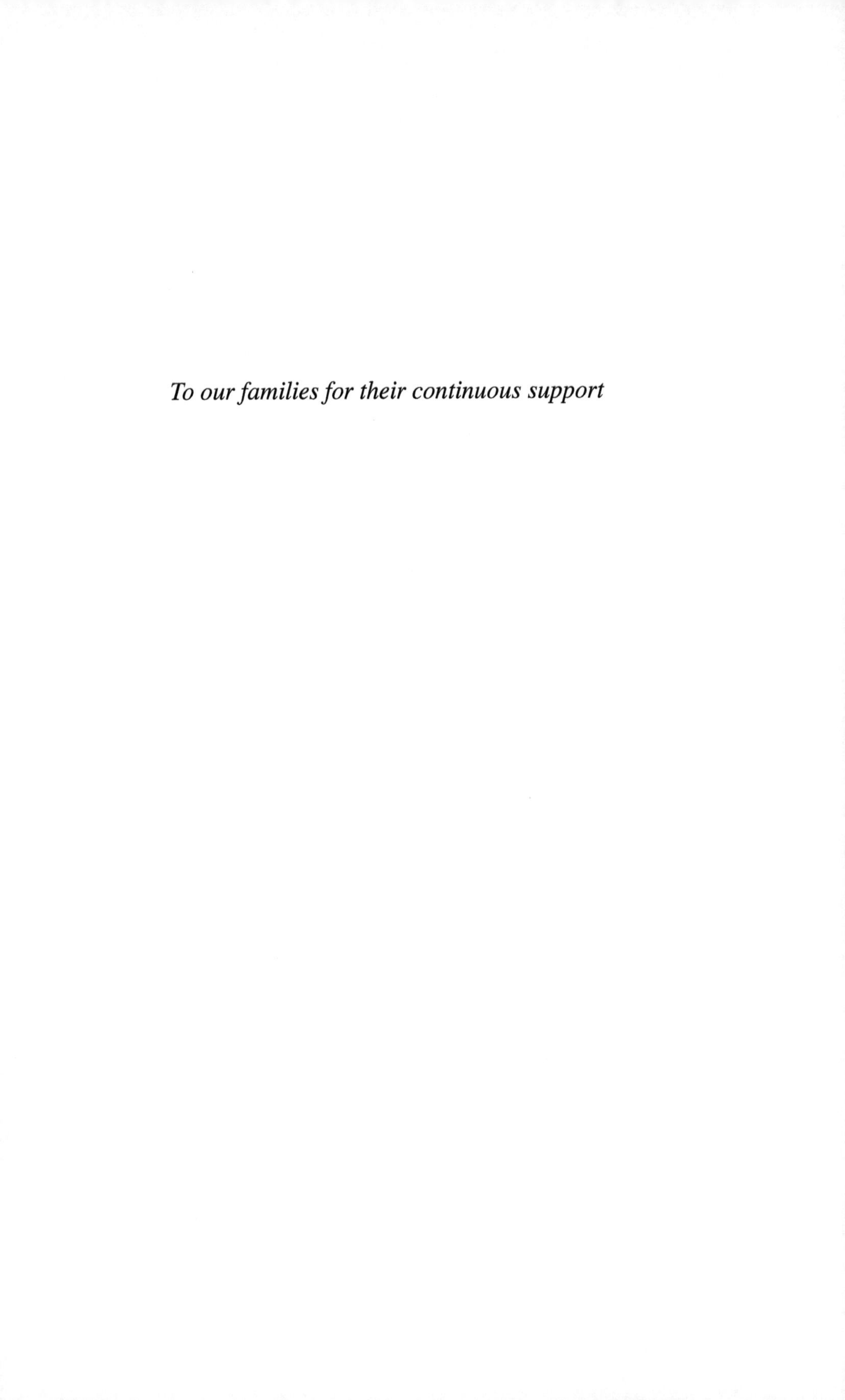

To our families for their continuous support

Contents

Preface

In recent times, polymeric and hybrid nanoparticles have attracted increased scientific attention, both in terms of basic research as well as in terms of commercial applications. Current abilities of polymer and materials chemistry allow for the development of various polymeric and polymer-containing hybrid nanoparticles that employ a range of preparation schemes, starting from conventional synthetic organic/polymer chemistry and reaching up to supramolecular chemistry and self-assembly schemes. The variety of nanoparticle morphologies and properties can be fine-tuned to a great extent. Applied preparation protocols and utilization possibilities from material schemes up to biomedical applications are increasing rapidly. The construction of a great variety of polymer-based nanoparticles and colloidal assemblies is also possible due to the availability of a large number of building blocks with diverse properties and the possibility of utilizing several assembly principles (hydrophobic and electrostatic interactions, hydrogen bonding, etc.). The results so far promise a variety of applications of nanostructures in social and economic fields, including bionanotechnology and medicine.

Research in these fields is ever growing and developing. While a detailed account of all relevant work is rather impossible, this book condenses the relevant research into a comprehensive reference and covers topics ranging from synthetic procedures and macromolecular design to possible biomedical applications of nanoparticles and materials based on original and unique polymers. By detailing the synthetic procedures, techniques for characterization and analysis, properties, and behavior in selective solvents and dispersions, the reader is provided with a well-rounded picture of objects as different as simple polymeric micelles and complex hybrid polymer–based nanostructures.

Each chapter contains sufficient background and introductory information, summarizing some generalities on the discussed nanosystems. Examples of representative works of experts in the respective fields are included as well, which, together with a comprehensive/focused discussion, should help create a stable basis for readers' own work, interests, or needs.

The chapters pass on to the reader the expertise and knowledge gained so far on the following:

1. Designed synthesis of functional polymers (mainly block copolymers)
2. Construction of block copolymer micellar (i.e., core–corona type) and nonmicellar (e.g., vesicular and bilayer structures) self-assembled structures and ways to control their morphology, functionality, and responsivity to changes of the physicochemical parameters of the environment
3. Construction of organic–organic hybrid nanosized particles and principles that control the formation, structure, and properties of macromolecular complexes between nonionic copolymers, lipid–polymer complexes (e.g., long-circulating liposomes and bicontinuous structures), complexes formed

by electrostatic interactions between block polyelectrolytes and low molecular weight surfactants, as well as functional polyelectrolytes, including proteins and DNA

4. Construction of organic–inorganic hybrid nanoparticles and nanoassemblies based on block copolymers/polyelectrolytes and inorganic building blocks like metal, semiconductor, and magnetic nanoparticles, as well as other inorganic nanostructures, that is, fullerenes, carbon nanotubes, and graphene

Authors

Stanislav Rangelov studied chemistry at the University of Chemical Technology and Metallurgy of Sofia, Bulgaria. He obtained his PhD from the Institute of Polymers, Bulgarian Academy of Sciences, in 1997. Following postdoctoral stays in the Czech Republic (Institute of Macromolecular Chemistry) and Sweden (University of Uppsala), he rejoined the Institute of Polymers in 2003. He has served as associate professor since 2005 and as full professor since 2011 in the same institute. Presently, he serves as head of the Laboratory of Polymerization Processes and chair of the Scientific Council of the Institute of Polymers. His research interests include controlled polymerization processes, self-assembly of amphiphilic copolymers, and polymer and polymer–hybrid nanosized particles.

Stergios Pispas studied chemistry at the University of Athens, Greece, and obtained his PhD in 1994. In 1994–1995, he served as a postdoctoral fellow at The University of Alabama at Birmingham, Birmingham, Alabama. Since 2009, he has served as a senior researcher at TPCI-NHRF. He also served as editor of the *European Physical Journal E* (2003–2012). He has received the American Institute of Chemists Foundation Award (1995) and the ACS A. K. Doolittle Award (2003). His current research focuses on the synthesis of functional block copolymers and polyelectrolytes, as well as the study of complex, self-organized, "hybrid" nanosystems based on polymers and surfactants, biomacromolecules, and inorganic nanomaterials.

List of Abbreviations

AFM	atomic force microscopy
AOT	sodium bis(2-ethylhexyl)sulfosuccinate
ATRP	atom transfer radical polymerization
BO	butylene oxide (unit)
C3M	complex coacervate core micelles
CAC	critical aggregation concentration
CLSM	confocal laser scanning microscopy
CMC	critical micelle (or micellization) concentration
CMT	critical micellization temperature
CP	cloud point
Cryo-ET	cryo-electron tomography
Cryo-TEM	cryogenic transmission electron microscopy
CTAB	hexadecyltrimethylammonium bromide (cetyltrimethylammonium bromide)
CTAC	hexadecyltrimethylammonium chloride (cetyltrimethylammonium chloride)
DDAB	didodecyldimethylammonium bromide
DDGG	didodecyl glycidyl ether
DDP	1,3-didodecyl propane-2-ol
DLS	dynamic light scattering
DMF	*N,N*-dimethylformamide
DMSO	dimethyl sulfoxide
DOPE	dioleoylphosphatidylethanolamine
DOX	doxorubicin
DP	degree of polymerization
DPPC	dipalmitoyl phosphatidylcholine
DSC	differential scanning calorimetry
DSPC	distearoylphosphatidylcholine
EO	ethylene oxide (unit)
EPC	egg phosphatidylcholine
EPR	enhanced permeation and retention
G	glycidol
GMO	glyceryl monooleate
GPC	gel permeation chromatography
(I)PEC	(inter)polyelectrolyte complexes
LCST	lower critical solution temperature
LGP	linear polyglycidol-polypropylene oxide copolymers
MRI	magnetic resonance imaging
mRNA	messenger RNA
Nagg	aggregation number
PAA	poly(acrylic acid)

PB	polybutadiene
PBLG	poly(γ-benzyl L-glutamate)
PBO	poly(butylene oxide)
PBS	phosphate buffer saline
PCL	poly(ε-caprolactone)
PDEAEMA	poly(*N*,*N*-diethylaminoethyl methacrylate)
PDMA	poly(*N*,*N*-dimethylacrylamide)
PDMAEMA	poly(*N*,*N*-dimethylaminoethyl methacrylate)
PDMS	poly(dimethylsiloxane)
PDPA	poly(2-(diisopropylamino)ethyl methacrylate)
PDT	photodynamic therapy
PEE	poly(ethyl ethylene)
PEEGE	poly(ethoxyethyl glycidyl ether)
PEG	poly(ethylene glycol)
PEI	polyethyleneimine
PEO	poly(ethylene oxide)
PG	polyglycidol
PIC	polyion complex
PLA	polylactide
PLGA	poly(lactic-*co*-glycolic acid)
PMA	poly(methyl acrylate)
PMAA	poly(methacrylic acid)
PMMA	poly(methyl methacrylate)
PMOXA	poly(2-methyloxazoline)
PMPC	poly(2-(methacryloyloxy)ethyl-phosphorylcholine)
PNIPAM	poly(*N*-isopropylacryl amide)
PO	propylene oxide (unit)
PODA	poly(*n*-octadecyl acrylate)
PPO	poly(propylene oxide)
PS	polystyrene
PVA	poly(vinyl alcohol)
PVCL	poly(*N*-vinylcaprolactam)
ROS	reactive oxygen species
SANS	small angle neutron scattering
SAXS	small angle x-ray scattering
SDS	sodium dodecyl sulfate
SEM	scanning electron microscopy
siRNA	small interfering RNA
SLS	static light scattering
SO	styrene oxide
SSSR	super strong segregation regime
TEM	transmission electron microscopy
THF	tetrahydrofuran

1 Polymer Synthesis

1.1 INTRODUCTION

In recent years, polymeric nanoparticles have attracted considerable interest of the polymer science community, both in terms of the basic science and from the applications points of view.

"Polymeric nanoparticle" should be regarded as a general term that describes polymeric/macromolecular entities of nanoscale dimensions suspended in a solvent medium. Since the scope of this book is focused on polymeric nanoparticles with relation and potential applications mainly to the biomedical field, we mostly consider nanoparticles in water. The detailed structure of polymeric nanoparticles can be varied widely and mainly depends on the chemical or physicochemical process followed to prepare them. They can be actual polymeric substances, for example, nanosized macromolecules of various architectures (linear, branched, etc.) or cross-linked hydrophobic/hydrophilic/amphiphilic polymeric chains suspended in a solvent medium (polymeric colloids), or particles that have been formed through noncovalent self-assembly of exclusively preexisting macromolecules and from macromolecules and other tectonic blocks (of low or higher mass). These assembled nanoparticles are formed through the action of solvation forces and several kinds of interactions, like van der Waals forces, solvophobic or electrostatic interactions, and hydrogen bond formation. Regardless of the type of nanoparticle structure and formation mechanism, the actual chemical nature of the nanoparticle (in the interior or maybe more importantly on the surface) is of great importance and dictates nanoparticle properties and uses. Therefore, chemical synthesis routes can lead directly or indirectly to the creation of polymeric nanoparticles. For example, polymeric colloidal nanoparticles can be directly prepared through emulsion polymerization techniques. On the other hand, solution polymerization of monomers leads to the synthesis of homopolymers, polyelectrolytes, and block copolymers that can be utilized in the formation of polymer-based nanoparticles via routes that will be exemplified in the following chapters.

In this chapter, we focus on the chemical synthesis of macromolecules that can be utilized for the production of functional and smart polymeric nanoparticles that present interest from the biomedical point of view in virtue of their physicochemical properties. We examine the so-called living/controlled polymerization mechanisms that can lead to synthetic polymers of well-defined molecular structure, that is, tunable molecular weights and compositions, low molecular weight, and compositional heterogeneity and controlled macromolecular architecture. We also discuss the preparation of polymeric nanoparticles via (mini)emulsion polymerization methodologies concentrating on purely polymeric structures. Some relevant examples of syntheses are also presented for clarity and for better understanding of the synthetic strategies available up to date. The presentation is by no means exhaustive but gives a good view of the basic concepts and design principles, as well as the advancements in the field of polymer synthesis.

1.2 GENERALITIES ON LIVING/CONTROLLED POLYMERIZATIONS

There are a number of polymerization mechanisms that allow the synthesis of homopolymers and block copolymers of various chemical structure and architecture having well-defined molecular characteristics under certain experimental conditions (Hadjichristidis et al. 2003). The most important of them, based on the criteria of wide applicability on a large number of functional and important monomers and of the precision control of molecular architecture, are presented briefly in the following sections, together with some examples of synthesis of block copolymers, which can be utilized in the formation of self-assembled polymeric nanoparticles.

1.2.1 Anionic Polymerization

Anionic polymerization is the oldest polymerization methodology that leads to the synthesis of well-defined homopolymers and block copolymers. The concept of living polymerizations, that is, polymerization reactions that are lacking the termination step and therefore polymeric chains can be preserved in the active (living) form for prolonged periods of time (under appropriate experimental conditions after consumption of the monomer(s)), was initially introduced and demonstrated through the use of anionic polymerization schemes (Szwarc 1956). In this type of addition, polymerization anions are produced by the use of mainly organometallic compounds that are able to polymerize vinylic-type monomers (styrenic, dienic, (meth)acrylic, etc.) or heteroatom-containing cyclic monomers (oxiranes, lactones, siloxanes, etc.) (Hsieh and Quirk 1996). Therefore, polymerization proceeds by addition of monomer to the carbanions (or oxoanions) present at the end of the growing chains. A wide variety of organometallic initiators have been reported, but those based on organolithium compounds are the most frequently employed for most of the monomers able to polymerize anionically. After completion of the polymerization, the living anions can be used (1) for further polymerization of another monomer (in order to produce block copolymers), (2) for functionalization reactions with reagents of appropriate chemical reactivity and function that would alter the chemistry of the active chain end and will thus incorporate functional groups at this chain end (the end-functional groups can be utilized for coupling with other polymer chains of complementary end functionality or used for other types of functionalization and immobilization reactions), and (3) for coupling reactions with a suitable multifunctional coupling agent that acts as a linking point for several polymer chains, as it may carry a definite number of reactive groups able to react with the active chain end. Alternatively, reaction with a difunctional monomer carrying two polymerizable bonds (e.g., divinylbenzene, DVB) can lead to the production of microgel cross-linked cores acting as branching points. In these cases, star polymers can be produced. It is apparent from the discussion so far that the designed and judicious utilization of these synthetic steps can lead to a variety of polymer structures as will be exemplified through the following examples. Similar possibilities exist for other living polymerization mechanisms.

Livingness of the polymerization and the reactivity of the chain ends are ensured by the use of high vacuum techniques or inert atmosphere setups (Hadjichristidis et al. 2000, Uhrig and Mays 2005). Extensive purification of reagents is needed in order to exclude undesired impurities from the polymerization mixture. By the use of glass sealed reactors, robustness of the polymerization system is maintained and reactions can be performed for long times (in the range of months) without compromising the living nature of the systems. The role of solvents and additives in the kinetics and the living character of several systems, as well as on the characteristics of the polymers produced by the particular methodology, have also been investigated extensively, and therefore, anionic polymerization has been utilized for the synthesis of many interesting polymers (Hsieh and Quirk 1996).

The major drawback for anionic polymerization is its low tolerance to polar chemical functionalities present on the monomers to be polymerized. Functional groups that can undergo proton exchange, for example, hydroxyl, carboxylic, and amine groups, can react with the propagating species and terminate or complicate the polymerization process. Although some of the problems can be resolved by appropriate chemical protection of functional groups (or the use of additives and different more complex initiating systems), applicability of the anionic polymerization methodology to a larger gamut of monomers still remains a challenge.

Nevertheless, a number of very important polymers and block copolymers for biomedical uses can be synthesized by anionic polymerization. For example, poly(ethylene oxide) (PEO), which is a water-soluble, neutral, and biocompatible polymer, is usually prepared by anionic polymerization. PEO (or PEG, polyethylene glycol) is utilized as the corona-forming block in block copolymer–based micellar drug nanocarriers, due to its excellent stealth properties that stem from its physicochemical characteristics. PEO chains show little interaction with plasma proteins and other components of blood, due to the nonionic character and the solvation characteristics (interaction with water molecules) of this particular polymer. Linear PEO chains functionalized by one or two end groups can be utilized as macroinitiators for the synthesis of useful and important block copolymers via other living/controlled polymerization routes. Poly(propylene oxide) (PPO) is another member of the cyclic alkyleneoxide family of monomers that shows greater hydrophobicity, and its water solubility changes rapidly with an increase in molecular weight. Diblock and ABA triblock copolymers of ethylene oxide (EO) and propylene oxide (PO), the well-known Pluronics or poloxamers, are also commercially available in a large variety of molecular weights and hydrophobic/hydrophilic compositions and show reversible self-assembly (micellization) with temperature in water, as a result of changes in the solvation characteristics of the PPO block. Pluronics have been utilized in several studies as nanocarriers of poorly water-soluble drugs due to their physicochemical characteristics, high biocompatibility, nontoxicity, and great availability (Alexandridis and Lindman 2000).

Polysiloxane polymers can also be produced by anionic polymerization in a controlled fashion and are considered biocompatible, due to the chemical inertness of the particular polymer backbone. They can be used as hydrophobic components in amphiphilic block copolymer systems and in a number of different cases as bulk materials for biomedical uses. Poly(ε-caprolactone) (PCL), a member of the polyester family, is a hydrophobic, crystalline, biocompatible, and biodegradable polymer. These characteristics make it a very promising candidate for encapsulation of hydrophobic drugs as in the case of poly(ethylene oxide)-*b*-poly(ε-caprolactone) (PEO-*b*-PCL) micellar nanocarriers. Its synthesis can be accomplished also through more conventional ring-opening polymerization (ROP) schemes. Polydiene polymers, like polyisoprene (PI) and polybutadiene (PB), can be synthesized anionically in a variety of microstructures and in a wide range of molecular weights (up to million daltons) and are hydrophobic. High *cis*-1,4-PI, produced by butyllithium initiation in nonpolar hydrocarbon solvents, has great structural similarities with natural rubber and in some sense can be regarded as biocompatible. The presence of reactive C=C double bonds in the chain structure of PI and PB contributes to the chemical instability of these polymers, for example, toward oxidation, but on the other hand gives also the opportunity of conducting selective functionalization reactions. The introduction of functional groups along the polymer chain can lead to the production of novel materials with potential biomedical interest, although their utilization is not necessarily limited to this particular field. The families of styrenic and (meth)acrylic monomers present large opportunities for imparting hydrophobicity/hydrophilicity to several block copolymers, due to the availability of several side groups of differing chemical identity and polarity, as well as protected functionalities that can lead to polar functionalities after completion of polymerization of the selected monomers. The latter properties are widely utilized in the synthesis of homopolymers and block copolymers of polyelectrolyte nature. The main disadvantage of polymers prepared by anionic polymerization from vinylic monomers is their nonbiodegradability. This property may be of some advantage in particular cases, but in most circumstances, this is a disadvantage, for example, when block copolymers containing these monomers are utilized in nanocarrier systems for water-insoluble drugs. The disadvantage may be overcome by utilization of amphiphilic block copolymers with low enough molecular weights, below the renal excretion limit, which will allow their systemic elimination eventually.

1.2.2 Cationic Polymerization

Living/controlled cationic polymerization was first introduced in the mid-1980s when Higashimura and coworkers reported the first living/controlled cationic polymerization of vinyl ethers (Miyamoto et al. 1984). Until that time, molecular weight and chain structural control of cationic polymerization systems were compromised by the inherent reactivity and instability of polymeric carbocations and the tendency of cationically polymerized monomers to give undesired side reactions, like chain transfer, carbocation rearrangement, and termination reactions (Matyjaszewski 1996). At present, the use of appropriately chosen stabilizing counterions and/or the presence of coordinating Lewis bases, in the course of polymerization reactions, allows in many cases the synthesis of well-defined homopolymers and block copolymers with controllable molecular characteristics. This undoubtedly opened the way to uniform polymers based on isobutylene and vinyl ethers, two classes of important monomers that can be polymerized only through cationic polymerization. It should be borne in mind that cationic polymerization, being an ionic polymerization, is relatively incompatible with some chemical groups on the monomer. Nevertheless, recent developments have shown the successful utilization of carbocationic initiation in the synthesis of several interesting polymers.

Polyisobutylene (PIB) is a hydrophobic polymer, containing no labile C=C bonds in its main chain, and therefore is chemically more stable compared to PI and PB. This particular polymer can serve as the core-forming block in block copolymer–based micellar drug carriers, since it does not show any undesired side effects when present in living organisms. This should be attributed to its pure hydrocarbon nature that dictates its chemical inertness. PIB itself is also a valuable biomaterial for several applications as a bulk material and for the modification of surfaces.

The vinyl ether family of monomers also presents opportunities for the synthesis of several nonpolar and polar homopolymers, including polyelectrolytes, and of course amphiphilic block copolymers, because different side groups, with variable polarity, reactivity, and chemical affinity, can be attached to the main vinyl ether moiety. The specific properties of such side groups can be utilized directly or after deprotection following the main polymerization step.

Another family of monomers prone to cationic polymerization is that of 2-oxazolines (Kobayashi and Uyama 2002). Some monomers of 2-substituted alkyl 2-oxazolines are commercially available, whereas others containing more exotic side groups can be easily synthesized following well-established organic synthesis schemes. Initiators based on alkylhalogenides, tosylates, or triflates can be used giving well-defined homopolymers and copolymers. The polymerization of the 2-substituted alkyl 2-oxazoline monomers proceeds in a controlled manner in most cases, following different mechanisms that depend on the monomer structure and the nature of the initiating system. Hydrophilic and hydrophobic homopolymers, as well as polymers with temperature-sensitive solubility, can be produced depending on the monomer structure. For example, poly(2-methyl-2-oxazoline) and poly(2-ethyl-2-oxazoline) (PEOz) are water soluble; poly(2-isopropyl-2-oxazoline) is thermoresponsive (shows lower critical solution temperature [LCST] within the usually attainable temperature range and close to body temperature), while homopolymers from monomers with more hydrophobic alkyl/aryl chains, like poly(2-*n*-butyl-2-oxazoline) and poly(2-phenyl-2-oxazoline) or polyoxazolines carrying fluorinated side chains, are water insoluble. Polymers obtained from substituted 2-oxazolines can be considered in several cases as pseudo-polyamino acids (pseudo-polypeptides), since their structure is in several cases isomeric to the corresponding poly(amino acid)s (e.g., PEOz is isomeric to polyvaline and poly(2-phenyl-2-oxazoline) is isomeric to poly(phenyl alanine)) and due to the presence of amide bonds in each polymer segment. However, in the case of polyoxazolines, the amide bond is present in the side chain. This gives to substituted poly(2-oxazolines) biocompatibility properties similar to synthetic and natural poly(amino acid)s (Adams and Schubert 2007). Hydrophilic poly(2-oxazolines) are regarded as alternatives to PEO (PEG) chains for stealth function in blood-circulating nanoparticles and bioconjugates (Barz et al. 2011). The chemical structure of poly(2-substituted oxazoline) side groups can be altered effectively by selective postpolymerization reactions resulting in carboxyl, amine, hydroxyl, thiol, alkyne, and azide groups and further functionality expression (Schlaad et al. 2010). The opportunities for the synthesis of block copolymer structures by combining different monomers, which belong to the oxazoline family of monomers only, are significant.

1.2.3 Controlled Free-Radical Polymerization

Traditional free-radical polymerization is characterized by rather ill-defined products in terms of molecular and compositional homogeneity. Despite this fact, it is widely used in the production of industrially important polymers for everyday use. It was not until the mid-1990s when living/controlled polymerization methodologies (living radical polymerization [LRP] or controlled radical polymerization [CRP]) based on radical mechanisms were reported (Matyjaszewski 1998) leading to polymers and copolymers with narrow molecular weight distributions and prescribed molecular characteristics. The basic idea that revolutionized progress in controlled free-radical polymerization schemes is the use of ways to achieve equilibrium between two interchangeable chemical forms of the active/propagating chain ends, that is, a dormant, deactivated form, unable to polymerize the monomer, and an active form that is able to polymerize the monomer. This goal can be accomplished by the use of additional agents, besides the initiator and the monomer, in order to kinetically control the polymerization reaction. In this way, the extent of side reactions, that is, termination and chain transfer, is dramatically decreased. In several cases, these agents are specially synthesized unimolecular radical initiators (usually organic compounds) or mixed initiator-/coinitiator-type systems that can be of organic or organometallic nature (Hawker et al. 2001). Therefore, control of the polymerization reaction is achieved by reversibly capping, or by temporal deactivation, of the propagating radicals with the appropriately designed agents. The dormant form is either capped via formation of a reversible labile covalent bond or stabilized in a lower-reactivity radical form unable to promote monomer polymerization. By judicious choice of reagents and reaction conditions, the propagating radicals spend most of their lifetime in the dormant form. In the short time that they stay in their active form, the propagating radicals can be extended by reaction with the monomer. The low concentration of active radicals in the system decreases the chance for occurrence of termination or chain transfer reactions, which would be catastrophic for the control of the linear structure and the width of the polydispersity of the resulting polymer. This situation resembles a living polymerization system, since side reactions and especially termination are reduced considerably. For these reasons, controlled free-radical polymerization schemes allow the control of the polymerization reaction and give polymers with controlled molecular weights and relatively low polydispersities.

Depending on the detailed scheme that is used in order to transiently deactivate the propagating radical, three major classes of LRP/CRP methodologies can be discriminated: (1) Nitroxide-mediated radical polymerization (NMRP), where the deactivation is achieved by the use of stable nitroxide-type radicals that form a labile covalent bond with the propagating radicals, prone to reversible thermal cleavage (Hawker et al. 2001). (2) Atom transfer radical polymerization (ATRP), where a transition metal complex is used as a mediator/transfer agent for the capping agent at the end of the growing polymer chain (usually a halide atom) (Braunecker and Matyjaszewski 2007). In most cases, Cu(I) complexes are employed, together with an alkyl halide that serves as the initiator. Their reversible catalytic action involves atom transfer (usually of a halogen atom) between the dormant radical and the Cu complex center, giving the active radical and a Cu(II) complex, with the transferred atom as one of the ligands of the metal center. The dormant propagating radicals are actual macromolecular halides that cannot polymerize the monomers present in the reaction medium. Other metals, besides Cu, can also be utilized in different chemical forms and the variety of available initiating species at present allows the implementation of the polymerization reaction in the presence of oxygen or in aqueous media (Braunecker and Matyjaszewski 2007, Freischmann et al. 2010). Trace amounts of the metal catalyst/cocatalyst have been shown to be easily and effectively removed from the final polymers, thus alleviating concerns for using ATRP-produced (co)polymers in nanomedicinal applications. (3) Reversible addition fragmentation transfer polymerization (RAFT), where a reversible chain transfer agent (CTA) is used in order to deactivate the propagating radicals (McCormick and Lowe 2004, Moad et al. 2005). A reversible chain transfer step, involving reaction of the propagating radical with the CTA, transforms the propagating radical to a less reactive species, with a parallel reinitiation step where evolving radicals participate. A chain equilibration step allows for a redistribution of propagating radicals between active and dormant states, giving the required advantages of a controlled polymerization to the system, that is, suppression of termination and chain transfer reactions during the propagation step.

In all cases, polymerization conditions including temperature, chemical structure of the capping/deactivating species, molecular ratios between deactivating species and initiator, chemical structure of initiating and ligating molecules, and polarity of the solvent should be chosen according to the structure of the monomer(s) to be polymerized. It should be borne in mind that in CRP schemes, control over molecular characteristics of the produced polymer chain is relatively lower than in the case of ionic polymerizations. Also, comparatively lower molecular weights can be attained with reasonable control, compared, for example, to anionic polymerization. However, the range of polymerizable monomers is considerably wider, in comparison to anionic and cationic polymerization schemes, including monomers with polar functionalities. This particular feature, together with the rather less stringent experimental requirements in polymerization conditions, explains the high current popularity of the CRP techniques among polymer chemists. Several classes of vinylic monomers can be polymerized by CRPs including the well-known styrenic and (meth)acrylate families, which represent a large library of monomers with diverse chemical structure and polymer properties.

1.2.4 Group Transfer Polymerization

Group transfer polymerization (GTP) has been widely applied for the synthesis of homopolymers and block copolymers based mainly on (meth)acrylic monomers (Webster 2004). Other monomers like acrylonitrile, *N,N*-dimethylacrylamide (DMA), and butyrolactones have been polymerized by GTP. This Michael-type catalyzed addition polymerization uses silyl ketene acetal compounds as initiators. In each addition step, the silane group is transferred to the growing chain end (GTP), that is, from the initiator to the monomer. In its variation of aldol group transfer polymerization (AGTP), an aldehyde is used as the initiator and the silyl group is transferred from the monomer to the initiator. These end functionalities are generally tolerable in biological media. GTP and AGTP have been sometimes regarded as types of anionic polymerization, although they generally give polymers of comparatively lower molecular weights. An advantage of GTP and AGTP is the fact that they can be used for the polymerization of (meth)acrylic monomers at room temperature and under less stringent experimental conditions, compared to anionic polymerization. Some examples relevant to medicinal applications of GTP are related to the synthesis of block copolymers incorporating (2-dimethylamino)ethyl methacrylate (DMAEMA) positively charged blocks, which can be utilized as gene carriers, through the formation of complexes with nucleic acids, in DNA and RNA delivery schemes (Georgiou et al. 2006).

1.2.5 Ring-Opening Metathesis Polymerization

Homopolymers and block copolymers from strained cyclic alkene monomers, of the norbornene family, can be obtained by ring-opening metathesis polymerization (ROMP) (Bielawski and Grubbs 2007). Norbornene-type monomers are characterized by a hydrocarbon, chemically rather inert, main skeleton and may include a variety of polar or protected side-chain functionalities as in the case of (meth)acrylic or vinyl ether-type families of monomers. ROMP is regarded as a transition-metal-mediated coordination addition polymerization and can give well-defined (co)polymers, if the transition metal initiator, coinitiator, and polymerization conditions are chosen properly. Organometallic initiators based on titanium, tungsten, and molybdenum have been utilized successfully so far in well-behaving, controlled polymerization schemes. Polymerization proceeds through the ring-opened carbene form of the monomer complexed with the metal center. With appropriate design of the initiator structure, monomers with polar functionalities can be polymerized in a controlled manner allowing the synthesis of (co)polymers that cannot be synthesized by any other controlled polymerization mechanism. The solvent and the ligands on the metal center play a decisive role in the course of ROMP. Although some concerns may exist regarding the use of ROMP-produced homopolymers and block copolymers in nanomedicinal applications, mainly due to the presence of trace amounts of the metal-based catalyst in the final products, there are a number of reports on the use of amphiphilic block copolymers derived from ROMP as micellar nanocarriers and gene transfer nanovehicles (Smith et al. 2007).

1.2.6 Other Ring-Opening Polymerization Schemes

There are several other living ROP schemes that are applied to particular classes of monomers. One of them is the ROP of lactides and lactones, which is initiated/catalyzed by metal compounds and can be characterized as a coordination–insertion polymerization mechanism (Dechy-Cabaret et al. 2004). As has been discussed earlier, polylactides and polylactones offer the advantage of biocompatibility and biodegradability, and therefore block copolymers containing such types of blocks show significant importance for application in nanomedicine. The cationic ROP of 2-oxazolines has been discussed earlier.

A very important polymer synthesis scheme, for producing polymers and copolymers for medicinal uses, is the ROP of *N*-carboxyanhydrides of β-amino acids. Polyamino acids (or polypeptides) show chemical and structural analogies with biological macromolecules, and they are regarded as biocompatible and biodegradable polymers. Synthetic polyamino acids mimic the chemical structure and physicochemical properties of proteins; for example, they can undergo reversible rodlike α-helix to random coil conformational transitions by changes of temperature and/or pH of the surrounding solution. ROP of *N*-carboxyanhydrides of β-amino acids can be initiated by metal catalysts or by molecules carrying primary amine groups, although the detailed polymerization mechanism is different in each case. Initiation with primary amines gives faster initiation compared to the propagation step and thus leads to polyamino acids of narrow molecular weight distributions. It also gives polyamino acids and polyamino acid–based block copolymers (Deming 1997, 2007, Hadjichristidis et al. 2009, Marsden and Kros 2009, Osada et al. 2009) of several thousand daltons in molecular weight and of narrow molecular weight distribution under appropriate experimental conditions. Amphiphilic or double-hydrophilic block copolypeptides (by metal-catalyst or primary amine initiation) as well as hybrid block copolymers with other non-amino acid synthetic blocks can be synthesized by appropriate choice of available *N*-carboxyanhydride monomers and synthetic protocols, as it will be presented later on (Deming 1997, Hadjichristidis et al. 2009, Osada et al. 2009). Initiation via organic primary amine groups is also preferred over the metal-catalyzed initiation, since (1) the presence of trace amounts of metal catalysts in the final copolymers is avoided and (2) a wider range of end functionalities can be introduced by use of appropriate amines. Some problems of the amine-initiated ROP of *N*-carboxyanhydrides related to the occurrence of side reactions, due to impurities, have been resolved by the use of high vacuum techniques that allow better purification of the monomers and the reagents utilized and provide a clean and isolated environment for conducting the polymerization reaction (Hadjichristidis et al. 2009). An alternative initiation route where protonated amines are utilized as initiators has been also proposed and shown to have great promise in suppressing side reactions and alternative routes of propagation in the course of monomer polymerization (Dimitrov and Schlaad 2003). Silylated amine initiators have also been used giving well-defined polyamino acids with relatively high degrees of polymerization, in comparatively shorter reaction times at room temperature and in quantitative yields (Lu and Cheng 2007).

1.3 BLOCK COPOLYMERS

Block copolymers are synthetic macromolecules that, in the simplest case, are composed of two dissimilar, in chemical nature and consequently in solubility and other physicochemical properties, homopolymer chains connected through their ends by a covalent bond. This is the case of a so-called diblock copolymer. Block copolymers self-organize in nanometer-scale assemblies in solutions (Hamley 1998,

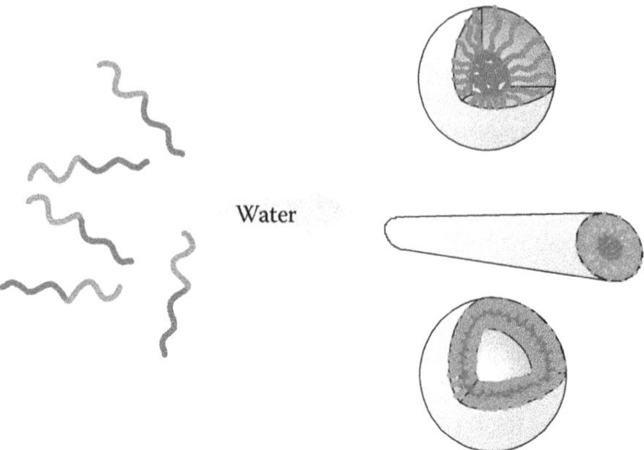

FIGURE 1.1 Self-assembled structures of amphiphilic diblock copolymers in water. Right side from top to bottom: spherical micelles, cylindrical micelles, and vesicles (gray: hydrophobic block, light gray: hydrophilic block).

Hadjichristidis et al. 2003), in analogy to low molecular weight surfactants, a property that attracts interest and makes feasible several nanomedicinal applications of this kind of synthetic macromolecules. More specifically in selective solvents, that is, solvents that dissolve one of the blocks, block copolymers self-assemble in nanosized particles, core-shell micelles, or other aggregates, of variable shape and size, like spherical and cylindrical micelles and vesicles (Zhang and Eisenberg 1996, Gohy 2005, Blanazs et al. 2009) that can be regarded as a category of versatile and functional polymeric nanoparticles (Figure 1.1). The most versatile block copolymers in this regard are amphiphilic block copolymers that are able to self-assemble in aqueous media, forming core-shell micelles and therefore can act as nanocarriers of pharmaceuticals (Gaucher et al. 2005, Harada and Kataoka 2006, Nishiyama and Kataoka 2006, Wagner and Kloeckner 2006, Jeong et al. 2007, Rapaport 2007, Tong and Cheng 2007, Itaka and Kataoka 2009, Wiradharma et al. 2009), since the desired compounds (usually water-insoluble drugs) can be encapsulated in the hydrophobic cores of the micelles.

In another approach, one of the blocks (usually a polyelectrolyte block) of a fully water-soluble block copolymer (i.e., a double-hydrophilic block copolymer) can interact, for example, electrostatically, with a low molecular weight drug or a peptide drug or nucleic acid, carrying complementary charges. In this case, the complexes form the insoluble core and polyion complex micelles are formed carrying the pharmaceutical compound. Shape and size of the nanostructures can be directed through the composition and molecular weight of the copolymer and the individual blocks, as well as the solution conditions (solvent selectivity and composition, concentration, temperature, pH, ionic strength) (Hamley 1998, Hadjichristidis et al. 2003, Gaucher et al. 2005, Harada and Kataoka 2006, Rapaport 2007, Wiradharma et al. 2009). The particular classes of block copolymer–based polymeric nanoparticles will be discussed in detail in Chapters 2 and 3. Here we focus on the synthesis of block copolymers and we make some connection to the desired chemical properties (which in turn determine the physicochemical properties of the macromolecules) that block copolymers should possess in order to facilitate their use in nanomedicine.

An important fact to bear in mind is that the block copolymer–based assemblies/nanoparticles can be controlled through the designed block copolymer synthesis. Furthermore, these nanoparticles should be constructed in a way that will allow them to avoid extravasation to normal tissues and recognition by the reticuloendothelial system cells. In this way, their circulation time after systemic injection increases and passive targeting and drug delivery to inflamed tissues, via the enhanced permeability and retention effect, is enhanced. Therefore, the choice of the block to be incorporated in the corona part is also crucial in order to give stealth properties to the nanoparticle. So far, PEO has been the most frequent choice for that purpose, but several other polymers have been proposed as alternatives (Barz et al. 2011). More interestingly, active targeting to tumors is made possible through modification of the micelle's surface with specific ligands (targeting groups), which can be introduced at the free end of the water-soluble block during the termination step or after polymerization. Development of stimuli-responsive micellar nanocarriers, that is, nanocarriers that respond by changing their structure (or by altering their distribution within the body) to external stimuli, like tumor pH, heat, sound, light, and magnetic fields, can also be achieved by suitable macromolecular engineering and synthesis of functional block copolymers. These types of block copolymer–based nanocarriers have in principle advanced targeting and controlled drug delivery capabilities and should be considered as advanced drug delivery nanosystems.

Besides diblock copolymers, linear block copolymers with a higher number of blocks (three, triblocks; four, tetrablocks; etc.) or having nonlinear architectures (star-block copolymers, graft copolymers, etc.) can also be synthesized. The nature of each block can be varied at will, by choosing the nature of monomer in each case and a suitable polymerization route, as will be exemplified later on through selected examples from the literature. Some basic requirements for a block copolymer to be used in the construction of micellar drug nanocarriers in the form of a nanoparticle in the general sense of the term (and to some extend as a part of a biomedical nanodevice) include the following:

1. Aqueous solubility and in several cases amphiphilicity. Solubility in water in molecular or supramolecular form ensures the application of the formulations in the blood stream, while amphiphilicity, that is, one block being hydrophilic and the other hydrophobic, ensures the formation of compartmentalized self-organized nanostructures containing hydrophobic domains. Since a large number of low molecular weight therapeutic molecules are poorly water soluble, they would be accommodated inside these hydrophobic domains of the self-assembled nanostructure using the appropriate encapsulation route. In some cases, amphiphilicity can be induced in a fully water-soluble block copolymer, that is, both blocks being water soluble (a double-hydrophilic block copolymer), by changes in the aqueous environment, for example, changes in pH, temperature, and ionic strength, or through complexation with other compounds, including low molecular

weight and macromolecular therapeutic compounds, which in turn lowers the solubility of the blocks that respond to these environmental changes or participate in the complexation process.

2. Biocompatibility and if possible biodegradability. At least one, and desirably all, of the blocks should be biocompatible in virtue of their chemical structure, in order to avoid undesired reactions in the human body and rejection by the patient's immune system. Stealth properties are desirable at least for the water-soluble block that actually covers the formed nanoparticles. A low overall molecular weight of the copolymer helps its renal excretion. Biodegradability favors eventual removal of the copolymer from the blood circulation and is also desirable for enhancing/modulating the drug release profile in several cases depending on the structure of the block copolymer–based nanoparticles.
3. Adequate chemical design that allows tunable interactions with the drugs to be encapsulated, so as to enable maximum carrying ability that in turn will permit efficient therapy. Tunable interactions with the environment are also desirable as discussed previously for producing stimuli-responsive polymeric nanoparticles.
4. Targeting chemical groups, usually situated at the free end of the water-soluble block in order to enhance specific interactions with infected cell types. This in turn will allow for maximum availability of the drug in the desired (infected) tumor tissues and will enable effective therapy and drug dosing.

1.3.1 General Synthetic Strategies Leading to Block Copolymers

The synthesis of block copolymers is usually accomplished through living/controlled polymerization synthesis schemes that have been outlined previously. Ideally, living polymers can be produced through these methodologies, under certain experimental conditions depending on the nature of the polymerization mechanism itself. This generally means that during the initiation step, all initiator molecules present in the system produce an equal number of polymer chains. Subsequently, in the propagation step, the polymerization reaction proceeds in the absence (or near absence) of termination or other competitive side reactions, and each chain grows in length at the same rate. When all monomer present is consumed and completely incorporated into polymer chains, the chain ends are still active and able to polymerize a second batch of monomer under similar kinetic conditions and grow further in molecular weight (Morton 1983). If the rate of initiation reaction is larger than the rate for the propagation step, then all polymer chains start growing essentially at the same time, and the result is a polymer sample containing chains of essentially the same chain length, that is, the polymer produced has a narrow molecular weight distribution. Under such conditions, it is easy to control the molecular weight of the polymer within the desired range, through the ratio of the mass of monomer and the moles of initiator utilized, M_n = g monomer/moles of initiator, where M_n is the number average molecular weight of the polymer chain.

The experimental conditions that ensure the livingness of the polymerization reaction mainly concern (1) appropriate choice of the reaction temperature, polymerization medium, nature and reactivity of initiating system in correspondence to the monomer(s) used, etc., aiming at the best kinetic control of the polymerization reaction itself, in order to achieve the best possible results, as far as the molecular characteristics and the resulting polymer uniformity are concerned; and (2) exclusion of impurities that may interfere with the polymerization reaction, for example, substances that can cause termination, chain transfer, deactivation of primary initiator molecules, side reactions that lead to undesired or ill-defined polymerization routes and polymeric products, etc. Therefore, the solvent, monomers, and other reagents used in the course of the reaction should be free from such substances. This in turn calls for the utilization of certain reagent purification protocols before the polymerization. In parallel, exclusion of such disturbing impurities should be ensured during the polymerization reaction, through the use of vacuum or inert atmosphere systems in some cases (but not necessarily so in others). Exclusion of impurities is directly related to the kinetic control of the polymerization reaction, but since special efforts and attention should be paid toward this direction, it sometimes becomes a separate and very important issue that influences the livingness of the polymerization reaction (see, e.g., Hadjichristidis et al. 2000, Uhrig and Mays 2005).

The main advantage of living/controlled polymerizations is that they provide uniform polymers, that is, polymers that have a narrow molecular weight distribution, uniform chemical structure in terms of monomer orientation, stereoregularity, and sequence within the polymer chain; uniform type and number of chemical functionalities along the chain; uniform chemical composition in the case of copolymers; and uniform chain shape and architecture in the case of nonlinear (co)polymers with more complex architectures. In other words, the chemical characteristics of the polymers produced can be controlled to the highest possible degree.

Generally speaking, AB diblock copolymers (with A and B representing different types of blocks/sequences of monomers) can be formed by sequential addition of two different monomers (of types A and B) to the same initiator solution, according to a living polymerization scheme. In this reaction procedure, the first monomer can be polymerized quantitatively giving a living/active polymer chain. Then the second monomer is introduced, and due to the living/controlled nature of the polymerization steps, it is fully polymerized giving the desired diblock copolymer (after termination with a suitable termination agent). Some issues that should be taken into account concern the ability of the active chain end, formed in the first polymerization step, to initiate the polymerization of the second monomer, in an appropriately fast rate in order to give a second block with narrow distribution, comparable to that of the first block. This ensures uniformity of the final diblock copolymer in terms of molecular weight and composition (Hadjichristidis et al. 2003), that is, all macromolecular chains possess nearly the same length (molecular weight uniformity) and content of each monomer (compositional uniformity). Obviously, the average composition of the diblock can be controlled through the ratio of the quantities of the two monomers, whereas the total block copolymer molecular weight and those of individual blocks are controlled through the relative amounts of initiator and monomers used. This is a direct consequence of the living nature of both polymerization steps, that is, absence of termination/deactivation and chain transfer reactions, and a condition that has to be fulfilled in all cases in order to have molecularly uniform block copolymers.

Block copolymers with three chemically different blocks, ABC triblock terpolymers, can be synthesized in an analogous manner if a third monomer is used (Hadjichristidis et al. 2003, 2005). Now three living/controlled polymerization steps need to take place sequentially with appropriate choice of the order of monomer addition, in order to fulfill the kinetic requirements and to achieve a well-defined final triblock copolymer. It becomes apparent that as the number of monomer additions increases, the requirements for monomer purity and the absence of chain termination become increasingly important. Otherwise mixtures of polymer chains are produced and the polymer sample is not uniform. BAB-type block copolymers can be produced by a two sequential living polymerization procedure by the use of a difunctional initiator. In this case, the A block is formed first, giving a difunctional living polymer (a linear polymer that can initiate polymerization of the second monomer from its two chain ends), and the B monomer is polymerized in the second step. Alternatively BA diblock chains can be coupled by a suitable coupling agent through their A active chain ends. More complicated synthetic schemes will be discussed in the following in order to produce block copolymer with more complex (nonlinear) macromolecular architecture.

1.3.2 Examples from Syntheses of Block Copolymers

We now turn to some examples for the synthesis of block copolymers starting from linear and going to nonlinear macromolecules. Linear block copolymers, especially diblocks, are more easily synthesized, and their synthetic protocol may be in some cases scaled to large quantities. On the other hand, nonlinear architectures need more elaborate synthetic schemes and it is not so easy to be prepared in large quantities. However, they show greater interest from the synthesis point of view and also because of their different properties compared to the linear ones. Emphasis is given on the ways to synthesize water-soluble and amphiphilic block copolymers and mostly on block copolymers that show some significance for biomedical application, due to their chemical structure and physicochemical properties.

1.3.2.1 Linear Block Copolymers by Sequential Controlled Monomer Polymerization

Polyisoprene-*b*-poly(ethylene oxide) (PI-*b*-PEO) block copolymers, like other amphiphilic block copolymers containing PEO as the hydrophilic block, have received some attention lately. PI of 1,4-microstructure is a highly hydrophobic polymer with a chemical structure similar to that of squalene, a natural substance of plant origin, and has a low glass transition temperature. In addition, the presence of the C=C double bonds allows for certain chemical modifications of the original structure of the PI blocks, leading to block copolymer with different structure and properties starting from the same precursors. PI-*b*-PEO block copolymers can be prepared in a variety of molecular weights and compositions by anionic polymerization, following the synthetic route outlined in Scheme 1.1. Briefly, isoprene is polymerized first at room temperature in hydrocarbon solvents using s-BuLi as the initiator, followed by polymerization of EO at higher temperature in the presence of phosphazene base (Pispas 2006). Block copolymers with predictable

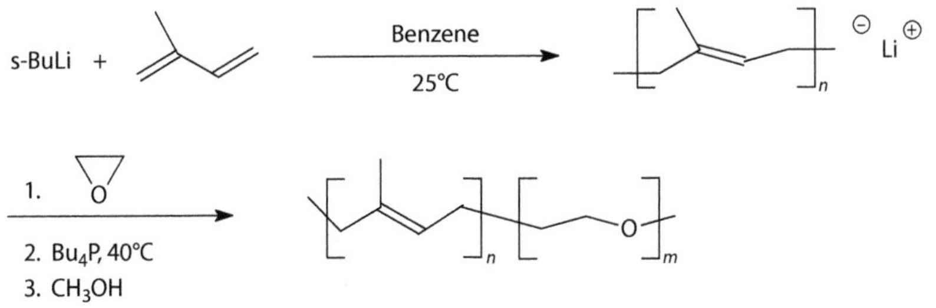

s-BuLi + → (Benzene, 25°C) → Li⊕

1. (ethylene oxide) 2. Bu_4P, 40°C 3. CH_3OH

SCHEME 1.1 Synthesis of amphiphilic PI-*b*-PEO block copolymers by anionic polymerization.

molecular weights and very low molecular weight distributions ($M_w/M_n < 1.1$) can be obtained in this way. Polybutadiene-*b*-poly(ethylene oxide) (PB-*b*-PEO) (Pispas and Hadjichristidis 2003) or PS-*b*-PEO diblock copolymers can be prepared in a similar manner.

Another interesting example is poly(2-vinylpyridine)-*b*-poly(ethylene oxide) (P2VP-*b*-PEO) block copolymers (Fragouli et al. 2002), also synthesized by sequential addition of monomers via anionic polymerization (with 2-vinylpyridine being the first monomer). In these block copolymers, the nonionic PEO block is water soluble at all pH values, while the solubility of the P2VP block changes with pH (protonation at pH > 5 leads to water solubility) (Mahltig et al. 2002). The complexing properties of the P2VP blocks toward metal ions are also of significant importance for the synthesis of hybrid organic–inorganic or organic–biological nanostructures.

Poly(2-vinylpyridine)-*b*-poly((dimethylamino)ethyl methacrylate) (P2VP-*b*-PDMAEMA) diblock copolymers can also be synthesized via living anionic polymerization. In this case, 2-vinylpyridine is polymerized first in a polar solvent and DMAEMA is polymerized in a second step, making use of the higher reactivity of the poly(2-vinyl pyridine)lithium chain end toward the polymerization of methacrylate-type monomers. The polymerization process is well controlled, giving block copolymers of narrow molecular weight distributions and desired composition. Both monomers are basic but have different pK_a values. Therefore, simple changes in solution pH may result in protonation/deprotonation of the two blocks inducing amphiphilicity in the copolymer and promoting different nanostructures in solution, depending also on the composition of the polymer chain. In the low pH regime, both blocks are protonated. At pH 5, P2VP becomes hydrophobic, due to deprotonation, and micelles are formed. At higher pH, PDMAEMA coronas can become deprotonated but remain hydrophilic (Gohy et al. 2001a).

Intrinsically organic–inorganic poly(ferrocenyldimethylsilane-*b*-2-(dimethylamino) ethyl methacrylate) (PFS-*b*-PDMAEMA) amphiphilic diblock copolymers can be obtained through anionic polymerization techniques (Wang et al. 2005). (*tert*-Butyldimethylsilyloxy)-1-propyl lithium, carrying a protected hydroxyl functionality, was used as the initiator for the polymerization of [1]silaferocenophane in THF. After completion of the polymerization, the end hydroxyl group of the PFS block was deprotected in the presence of $[Bu_4N]F$. In the next step, end hydroxyl groups were activated by KH and PFS was utilized as an anionic macroinitiator for the polymerization of DMAEMA in THF at room temperature, resulting in PFS-PDMAEMA

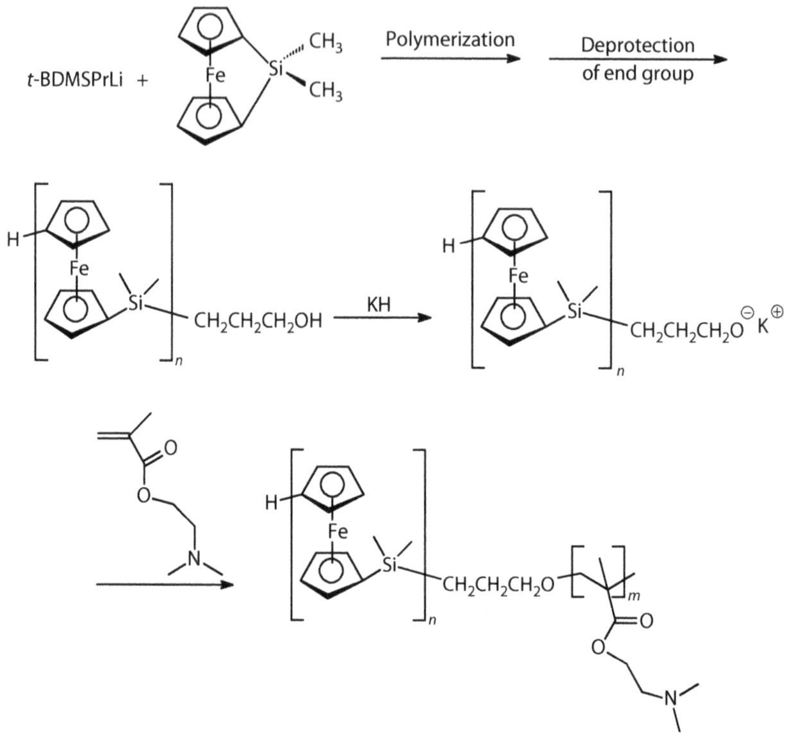

SCHEME 1.2 Synthesis of intrinsically organic–inorganic PFS-*b*-PDMAEMA amphiphilic diblock copolymers by anionic polymerization.

with a long PDMAEMA block (Scheme 1.2). Apart from the pH responsiveness of the PDMAEMA block, these copolymers present the opportunity for the creation of ferrous containing hybrid nanoassemblies by the chemical transformation of the PFS block into iron oxide nanoparticles or nanowires, within the framework of the self-assembled nanoparticles.

Linear block copolymers based on oxazoline monomers, which can be very efficiently synthesized by cationic polymerization routes, have also received considerable attention. The main chemical structure of oxazoline monomers contains the polar amide moiety as in the case of natural amino acids. In particular, 2-ethyloxazoline and 2-phenyloxazoline are isomeric to the amino acids valine and phenylalanine, respectively. In this context, the chemical structures of oxazoline-based polymers and block copolymer show similarities with the structures of polyamino acids and polypeptides. Therefore, the particular block copolymers may be considered as biocompatible and can be utilized in the construction of self-assembled polymeric nanoparticles (micelles, vesicles, etc.). The water solubility and hydrophobic/hydrophilic nature of the copolymers are dictated by the side group substitution of each block and their respective ratios in the copolymer. Poly(2-methyloxazoline) and poly(2-ethyloxazoline) are water soluble and their biocompatibility, due mainly to their neutral chemical nature, makes them very interesting candidates for substituting PEO (PEG) blocks as the outer corona blocks of block copolymer micelles imparting stealth properties to, for example, micellar drug carriers (Adams and Schubert 2007). Poly(2-isopropyloxazoline) and poly(2-butyloxazoline) are considered thermosensitive, showing an LCST ca. 45°C and 25°C, respectively. Poly(2-oxazolines) with higher aliphatic or phenyl side groups are hydrophobic and can be utilized as core-forming blocks in 2-oxazoline block copolymer–based micellar nanocarriers. As a typical example of 2-ozaxoline-based amphiphilic block copolymers, we present the synthesis of copolymers based on 2-butyl-2-oxazoline and 2-ethyl-2-oxazoline (Luxenhofer et al. 2010). In the particular synthetic scheme, 2-ethyl-2-oxazoline monomer was polymerized under inert atmosphere and microwave irradiation conditions, in dry acetonitrile, at room temperature, using methyl trifluoromethylsulfonate as the cationic initiator. Then 2-butyl-2-oxazoline was introduced to the reaction mixture and the polymerization was terminated with piperazine (Scheme 1.3).

Several amphiphilic and thermoresponsive diblock copolymers containing vinyl ether monomers can be prepared depending on the choice of the side group. For example, 2-ethoxyethyl vinyl ether (EOVE) can be used as the first monomer, using the 1-(isobutoxy)ethyl acetate/ethylaluminum sesquichloride/ethyl acetate initiating system, in hexane at 0°C. Next, t-butyldimethylsilyloxyethyl vinyl ether (BMSiVE) can be utilized as the second monomer to produce PEOVE-*b*-PBMSiVE diblock copolymers. Hydrolysis of the PBMSiVE blocks gives poly(2-ethoxyethyl vinyl ether)-*b*-poly(vinyl ethyl alcohol). In a similar way, different block copolymers with polyalcohol blocks were obtained (Aoshima and Hashimoto 2001, Osaka et al. 2007).

(Meth)acrylate monomers with oligoethylene glycol side groups give also polymers that can be used as water-soluble components in the construction of block copolymers (Lutz 2008). The particular polymers have significant physicochemical similarities with PEG (PEO) in terms of water solubility and biocompatibility. Other monomers can be chosen in order to produce the hydrophobic blocks of amphiphilic block copolymers, like methacrylates with aliphatic side chains or phenyl side groups or cyclic lactones (Luzon et al. 2010). These vinyl monomers

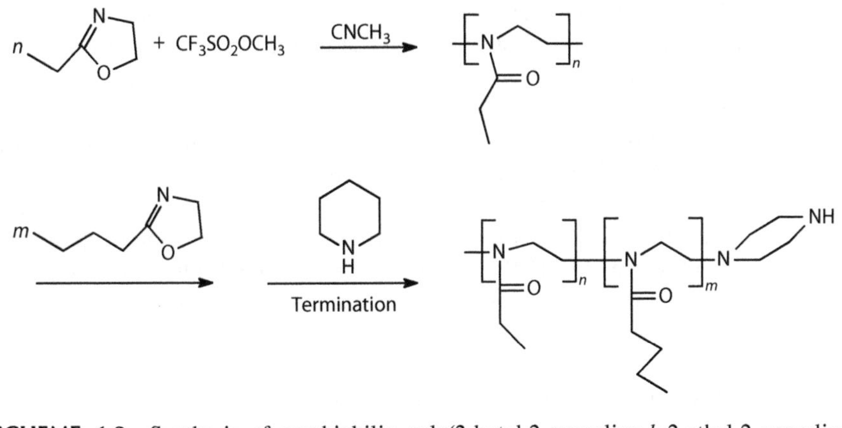

n + $CF_3SO_2OCH_3$ $\xrightarrow{CNCH_3}$ n

m —— —— Termination —— n m NH

SCHEME 1.3 Synthesis of amphiphilic poly(2-butyl-2-oxazoline-*b*-2-ethyl-2-oxazoline) copolymers by cationic polymerization.

can be polymerized by different polymerization mechanisms, but most frequently CRP methodologies are utilized. Double thermoresponsive poly(oligoethylene glycol methacrylate)-*b*-poly(*N*-isopropyl acrylamide) (POEGMA-*b*-PNIPAM) block copolymers have been synthesized by RAFT polymerization. First, oligoethylene glycol methacrylate was polymerized by AIBN in the presence of a functionalized CTA agent in dioxane. Then the POEGMA macro-CTA was used in the presence of AIBN for the polymerization of NIPAM leading to the desired block copolymers (Jochum et al. 2010) (Scheme 1.4). The different thermoresponsive behavior of POEGMA and PNIPAM blocks allows for the construction of a variety of self-assembled nanoparticles in aqueous solutions as a function of temperature. The POEGMA-*b*-PNIPAM diblocks were also functionalized with a biotin end group at the POEGMA terminus imparting targeting properties to the nanoassemblies. Using a similar reaction route, the authors prepared poly(oligoethylene glycol methacrylate)-*b*-poly(pentafluorophenyl methacrylate) (PPFPMA-*b*-POEGM) diblock copolymers, which are typical amphiphilic block copolymers (due to the high hydrophobicity of the perfluorinated PPFPMA block) but also possess a thermosensitive character, due to the POEGMA block.

A diblock copolymer containing sulfonated styrene (SS) monomers and OEGM segments was also synthesized by RAFT (Cortez et al. 2010). In this case, SS was polymerized first using 4-cyanopentanoic acid dithiobenzoate (CPADB) as the CTA and the water-soluble initiator 4,4′-azobis-4-cyanopentanoic acid (ACPA). Polymerization was conducted in a mixture of ethanol and water at 70°C. Then the PSSRAFT macro-CTA was utilized in the preparation of the second POEGMA block. The resulting double-hydrophilic block copolymer had the desired molecular weight and composition and narrow molecular weight distribution.

The polymerization of the new methacrylate monomer 2-(methacryloyloxy) ethyl phosphorylcholine (MPC), via ATRP methodologies, created new synthetic routes for the preparation of novel pH-responsive amphiphilic block copolymers (Salvage et al. 2005, Licciardi et al. 2008). The poly[2-(methacryloyloxy)ethyl phosphorylcholine] (PMPC) blocks are biocompatible, due to their zwitterionic nature and their chemical similarity to the lipidic components (phospholipids) of the cell membranes. Among other uses, they are suggested as alternatives to PEO/PEG corona blocks for imparting stealth properties to the block copolymer micellar nanocarriers. A synthetic pathway toward diblock copolymers containing PMPC blocks is outlined in Scheme 1.5. The specific block copolymers were functionalized with folate groups at the end of the PMPC blocks, which were introduced by the functional ATRP initiator utilized for monomers' polymerization, imparting cell-targeting abilities to the micellar nanocarriers (Licciardi et al. 2008). MPC was polymerized first using an FMOC-protected initiator in a methanol/isopropanol mixture, in the presence of Cu(I)Br catalyst and the 2,2′-bipyridine ligand, at 20°C, under a nitrogen atmosphere. Subsequently 2-(diidopropylamino)ethyl methacrylate (PDPM) monomer was added and the polymerization was allowed to proceed to completion. Deprotection of the end group gave the H_2N-end-functionalized diblock, which was later on conjugated to folic acid.

As it was mentioned in a previous example, controlled free-radical polymerization schemes are used extensively for the synthesis of block copolymers with PNIPAM

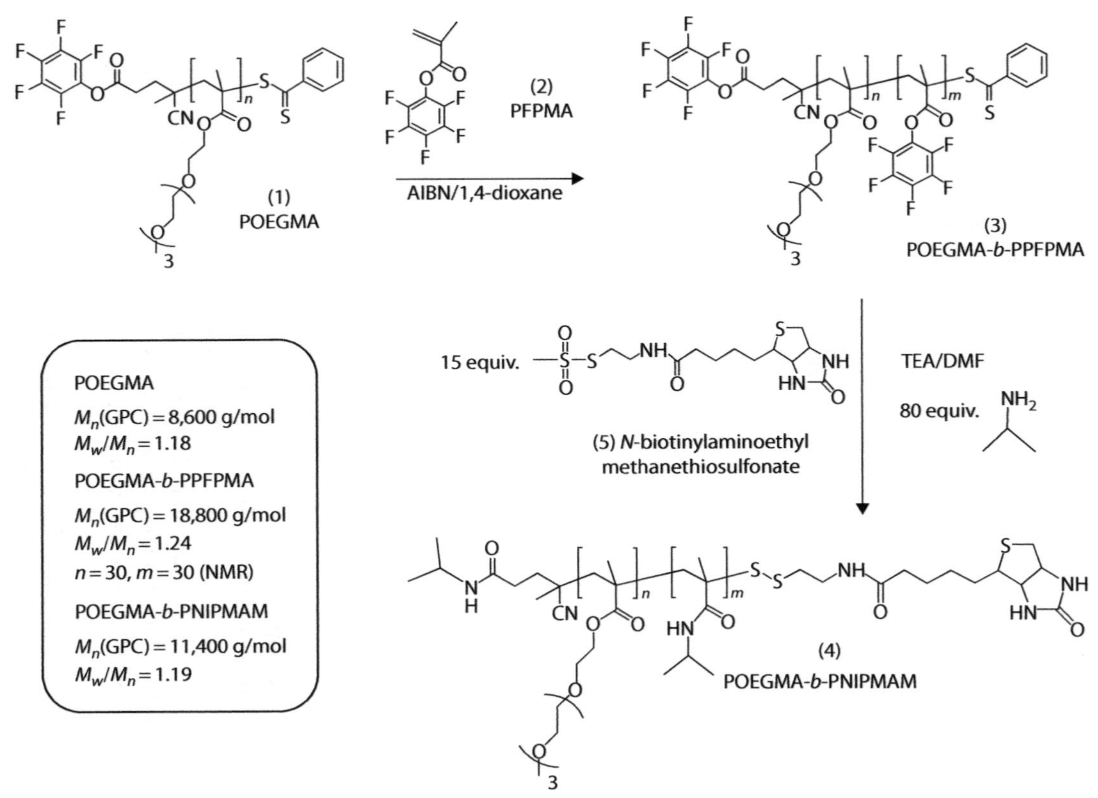

(1) POEGMA

(2) PFPMA

AIBN/1,4-dioxane

(3) POEGMA-*b*-PPFPMA

15 equiv.

(5) *N*-biotinylaminoethyl methanethiosulfonate

TEA/DMF

80 equiv. NH_2

(4) POEGMA-*b*-PNIPMAM

POEGMA

M_n(GPC) = 8,600 g/mol
M_w/M_n = 1.18

POEGMA-*b*-PPFPMA

M_n(GPC) = 18,800 g/mol
M_w/M_n = 1.24
n = 30, m = 30 (NMR)

POEGMA-*b*-PNIPMAM

M_n(GPC) = 11,400 g/mol
M_w/M_n = 1.19

SCHEME 1.4 Synthesis of double thermoresponsive POEGMA-*b*-PNIPAM block copolymers by RAFT polymerization.

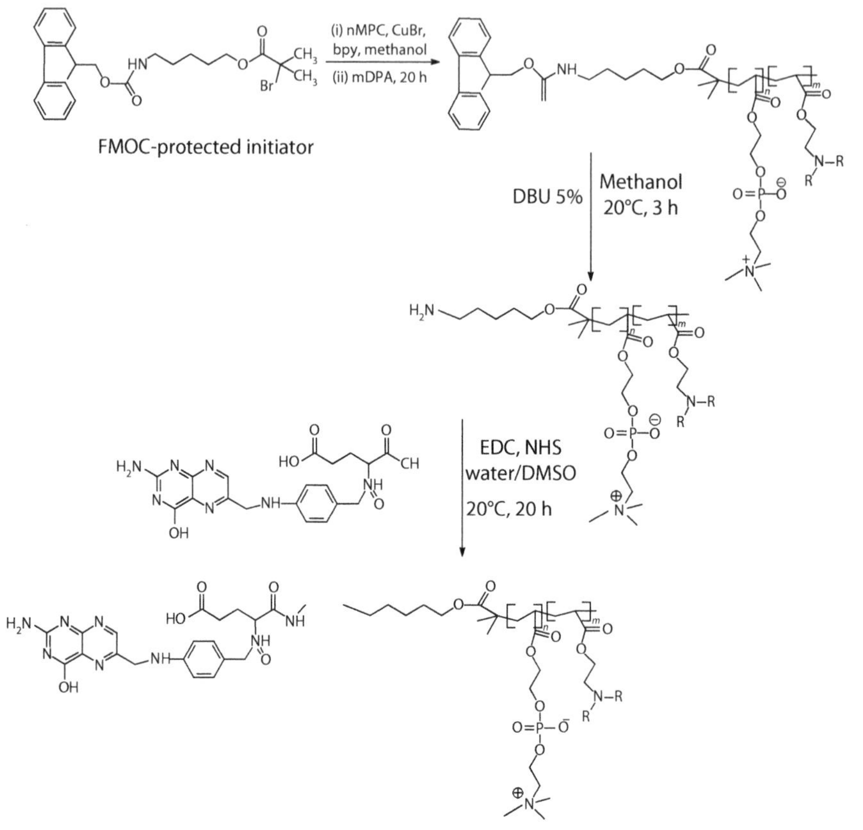

SCHEME 1.5 Synthesis of poly[2-(methacryloyloxy)ethyl phosphorylcholine-*b*-2-(diidopropylamino)ethyl methacrylate] (PMPC-*b*-PDPEM) diblock copolymers with folate end groups.

blocks, which are imparting thermoresponsive properties to the block copolymer and, therefore, allow for the manipulation of self-assembled nanoparticle properties through temperature variations. PNIPAM shows an LCST very close to body temperature. LCST can be manipulated within a certain temperature range by copolymerization with polar or nonpolar comonomers. Azido-terminated block copolymers of DMA and NIPAM were synthesized by RAFT polymerization (De et al. 2008). The azido-functionalized CTA agent 2-dodecylsulfanylthiocarbonylsulfanyl-2-methyl-propionic acid 3-azido propyl ester was utilized for the polymerization of DMA with AIBN as the radical initiator. The N_3-PDMA macro-CTA agent was used in the second step for the polymerization of NIPAM in dioxane (Scheme 1.6). The azide functionality on the PDMA chain end was utilized for the conjugation of folate groups via click chemistry in order to introduce cell-targeting end groups to the copolymer.

In PNIPAM containing amphiphilic block copolymers, incorporation of pH-sensitive monomers, either within the thermosensitive block or the hydrophobic block

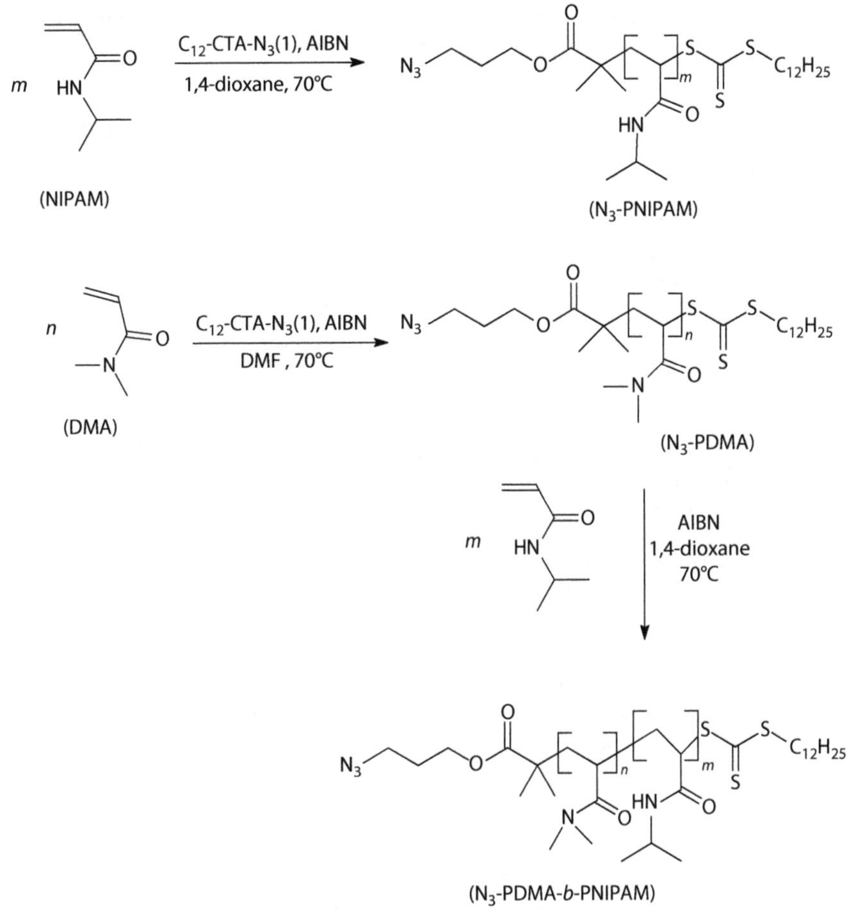

SCHEME 1.6 Synthesis of poly(*N*,*N*-dimethylacrylamide-*b*-*N*-isopropylacrylamide) (PDMA-*b*-PNIPAM) block copolymers.

of the block copolymer, can induce pH sensitivity to the macromolecular chains and consequently to the micellar nanoparticles formed by these copolymers. Along these lines, the synthesis of diblock copolymers where one of the blocks is PNIPAM and the other block is a random copolymer with *tert*-butyl acrylate and acrylic acid segments provides pH responsivity to the block copolymer nanoassemblies in aqueous solutions in conjunction to the thermoresponsive properties imparted by the PNIPAM block (Li et al. 2006a,b) (Scheme 1.7).

Stimuli-responsive diblock copolymers based on PEG of a molecular weight ca. 2000, as the hydrophilic, temperature-sensitive block, and poly(*N*-acryloyl-2,2-dimethyl-1,3-oxazolidine), as the hydrophobic and acid labile block (PEG-*b*-PADMO copolymers), were prepared by RAFT polymerization of ADMO monomer using a macro-PEG-CTA agent (Cui et al. 2011). ADMO polymerizations were conducted in dioxane at 70°C in the presence of AIBN. Block copolymers with

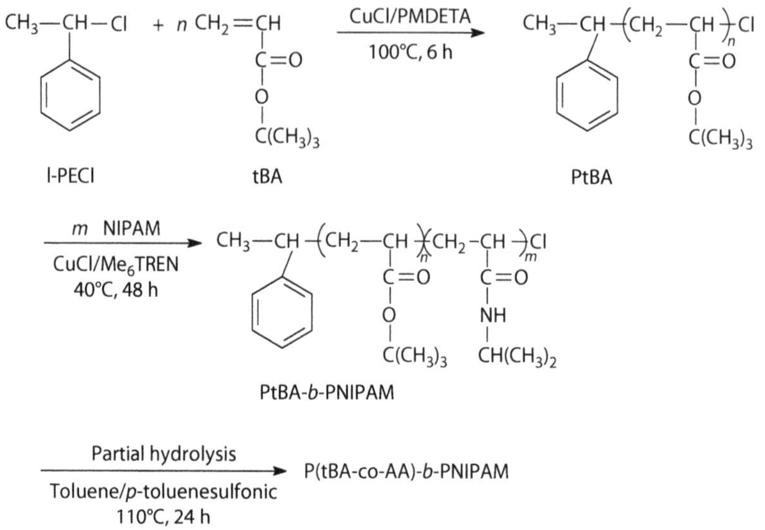

1-PECl + n tBA → (CuCl/PMDETA, 100°C, 6 h) → PtBA

PtBA → (m NIPAM, CuCl/Me$_6$TREN, 40°C, 48 h) → PtBA-*b*-PNIPAM

→ (Partial hydrolysis, Toluene/*p*-toluenesulfonic, 110°C, 24 h) → P(tBA-co-AA)-*b*-PNIPAM

SCHEME 1.7 Synthesis of poly{[(*tert*-butyl acrylate)-co-acrylic acid]-*b*-*N*-isopropylacrylamide} (PtBAcoPAA-*b*-PNIPAM) copolymers.

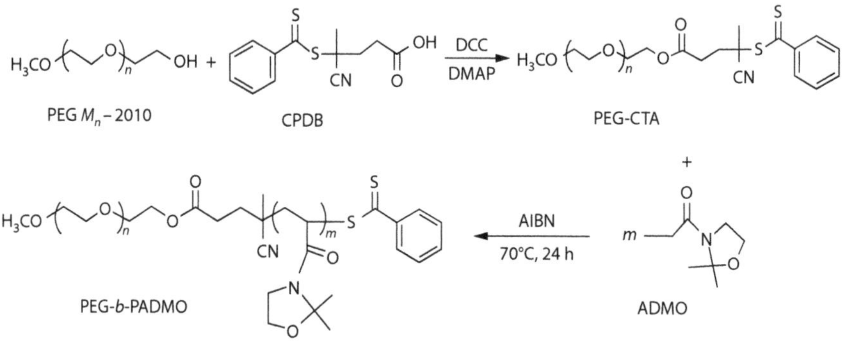

PEG M_n - 2010 + CPDB → (DCC, DMAP) → PEG-CTA

PEG-CTA + m ADMO → (AIBN, 70°C, 24 h) → PEG-*b*-PADMO

SCHEME 1.8 Synthesis of poly[ethylene glycol-*b*-(*N*-acryloyl-2,2-dimethyl-1,3-oxazolidine)] (PEG-*b*-PADMO) copolymers.

controlled characteristics were obtained (Scheme 1.8). These block copolymers form micelles in water with PADMO cores. In acidic media, the PADMO block becomes hydrophilic, due to hydrolysis to poly(2-hydroxyethyl acrylamide) blocks, and micelles are dissociated. This is an important property for drug release from block copolymer micellar nanocarriers.

ATRP has been utilized for the synthesis of block copolymers containing fluorinated monomers. Using the ethyl 2-bromopropionate (EBrP)/CuBr/*N*,*N*,*N'*,*N''*,*N''*-pentamethyl diethylene triamine (PMDETA) initiating system, it was possible to prepare diblock copolymers with a poly(*tert*-butyl acrylate) block and a block containing *n*-butyl acrylate and fluoroacrylate monomers (Peng et al. 2009). Hydrolysis of the poly(*tert*-butyl acrylate) block into poly(acrylic acid) (PAA) resulted in

PAA-*b*-(poly(*n*-butyl acrylate-co-fluoroacrylate)) amphiphilic block copolymers. Due to the presence of the NMR-active ^{19}F nuclei in these block copolymers, their micelles in water were evaluated as magnetic resonance imaging (MRI) contrast agents. In a complementary work, poly(*tert*-butyl acrylate)-*b*-poly(2-[(perfluorononenyl)oxy] ethyl methacrylate) (PtBA-*b*-PFNEMA) block copolymers were prepared via ATRP by the ethyl-2-bromoisobutyrate/CuBr/PMDETA system, starting with the synthesis of the PtBA block. The PtBA-Br macroinitiator was used for the polymerization of FNEMA in dimethylformamide (DMF) in a second step, giving well-defined block copolymers with perfluorinated blocks (Qin et al. 2011).

pH- and salt-responsive diblock copolymers of poly(sodium 2-acrylamido-2-methyl-1-propanesulfonate) and poly(*N*-acryloyl-L-alanine) (PAMPS-*b*-PAAL) were prepared by RAFT polymerization (Kellum et al. 2010). AMPS monomer was polymerized in water under slightly acidic conditions using CPADB (CTP) as the CTA and 4,4′-azobis-(4-cyanopentanoic acid) (V-501) as the water-soluble radical initiator. The monomer to CTA ratio was chosen in order to obtain the desired molecular weights for the PAMPS block at low AMPS monomer conversion (ca. 30%), ensuring negligible loss of chain end functionality during the first polymerization step. Subsequently, the first block was isolated and purified by dialysis against water. The resulting PAMPS macro-CTA was further used for the polymerization of AAL monomer. Again the polymerization reaction was controlled kinetically since macro-CTA/monomer ratios were chosen in order to obtain the desired molecular weights at ca. 60% AAL monomer conversion for increasing the purity of the final copolymer. Molecular weight distributions of the resulting diblocks were lower than 1.3 in all cases, indicating the success of the polymerization protocol. This is also a nice example of the controlled polymerization of acrylate monomers carrying amino acid side groups. Such a chemical feature is expected to increase the biocompatibility of the copolymer.

Biocompatible amphiphilic diblock copolypeptides of the poly(L-glutamic acid)-*b*-poly(L-phenylalanine) type (PGA-*b*-PPA copolymers) have also been synthesized recently by ROP (Kim et al. 2009). In this approach, *n*-butylamine hydrochloride was used as the initiator for the polymerization of the cyclic *N*-carboxyanhydride (NCA) of the γ-glutamic acid benzyl ester in the first step, followed by polymerization of the NCA of phenylalanine. A simple postpolymerization functionalization step including the removal of the benzyl protecting groups in acidic conditions resulted in the desired PGA-*b*-PPA diblock copolypeptides (Scheme 1.9).

Linear triblock copolymers of the ABA type can be synthesized by the use of a difunctional (macro)initiator and sequential addition of monomers. PEO-*b*-PI-*b*-PEO amphiphilic triblock copolymers were synthesized by naphthalene-based initiators in tetrahydrofuran in the absence of any other additives, since carbanions with Na or K as counterions can directly polymerize EO in polar solvents (Batra et al. 1997, Pispas 2006). In this case, the microstructure of PI is that of high 3,4-content with a higher T_g, closer to room temperature.

PEO-*b*-P2VP-*b*-PEO double-hydrophilic and pH-responsive block copolymers were prepared by anionic polymerization by using the difunctional anionic initiator sodium tetraphenyl diisobutane (Karanikolas et al. 2008). 2-Vinyl pyridine was the first monomer to be polymerized at −78°C. Then temperature was raised to

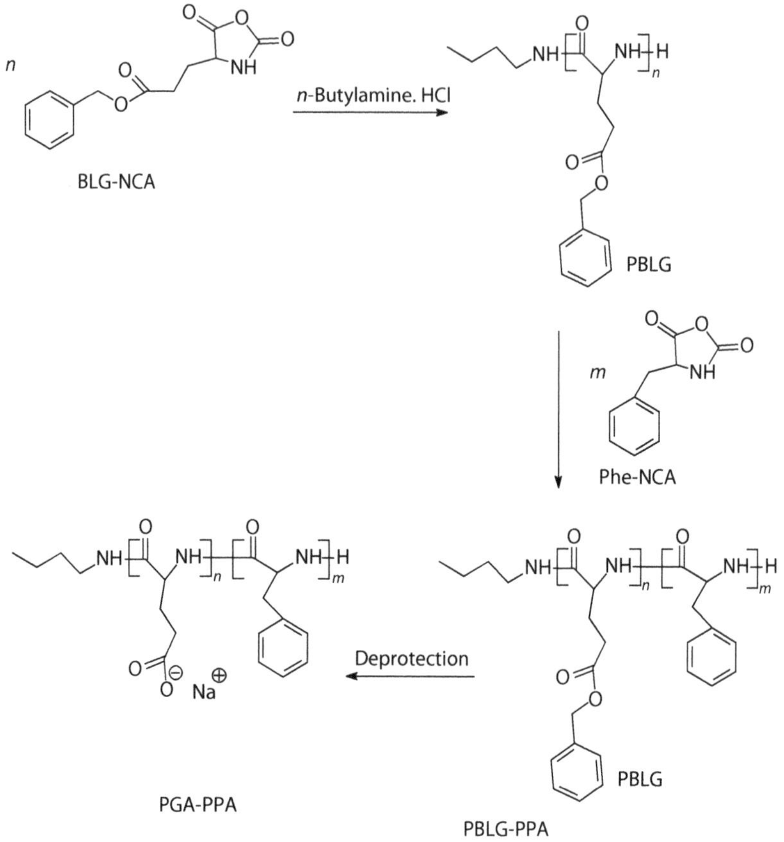

BLG-NCA

n-Butylamine. HCl

PBLG

Phe-NCA

Deprotection

PGA-PPA

PBLG-PPA

SCHEME 1.9 Synthesis of PGA-*b*-PPA block copolymers.

–30°C and EO was introduced to the polymerization mixture. Temperature was then allowed to reach 30°C–35°C slowly and the second polymerization step was completed after 24 h at this temperature. The pH responsiveness of the systems is stemming from the basic character of 2VP segments that can be protonated in aqueous media at pH lower than 5 (becoming hydrophilic), whereas they are deprotonated at larger pH values, becoming hydrophobic and leading to the formation of micelles with P2VP cores and PEO coronas.

A difunctional low molecular weight PCL macroinitiator was utilized for the synthesis of poly(oligoethylene glycol methacrylate)-*b*-poly(ε-caprolactone)-*b*-poly(oligoethylene glycol methacrylate) (POEGM-*b*-PCL-*b*-POEGM) triblock copolymers, possessing a biodegradable hydrophobic inner block and biocompatible hydrophilic outer blocks (Luzon et al. 2010). The hydroxyl end groups of the PCL difunctional initiator were functionalized with 2-bromoisobutyryl bromide. Next OEGM was polymerized with an ATRP scheme involving the use of CuBr and 2,2′-bipyridine (Scheme 1.10). The copolymer was isolated after removing the catalyst by passing through an activated basic alumina column giving colorless copolymers.

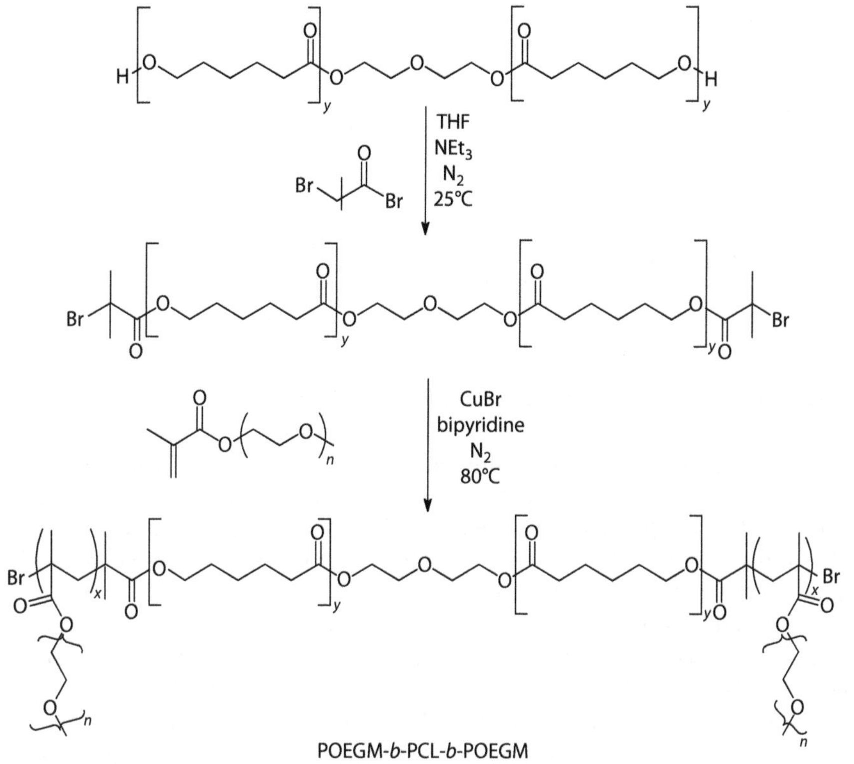

THF
NEt_3
N_2
25°C

CuBr
bipyridine
N_2
80°C

POEGM-*b*-PCL-*b*-POEGM

SCHEME 1.10 Synthesis of POEGM-*b*-PCL-*b*-POEGM triblock copolymers.

Linear triblock terpolymers can be synthesized by the sequential addition of three different monomers. Poly(isoprene-*b*-2-vinylpyridine-*b*-ethylene oxide) (PI-*b*-P2VP-*b*-PEO) triblock terpolymers were synthesized by anionic polymerization using benzyl potassium as the initiator in THF (Koutalas et al. 2004). This particular initiator is able to polymerize all three monomers in polar solvents. Monomers were added to the polymerization mixture according to their reactivity toward the next monomer to be polymerized. Isoprene was polymerized first giving a PI block with high 3,4-microstructure. 2-Vinylpyridine was added as the second monomer. Finally, EO was polymerized third at 40°C for almost 3 days, due to the slower rate of the third polymerization step (Scheme 1.11). In an analogous manner, poly(styrene-2-vinylpyridine-*b*-ethylene oxide) triblock terpolymers have also been synthesized (Gohy et al. 2001b).

Linear terpolymers with polyampholyte and amphiphilic character of the ABC linear architecture have been synthesized by anionic polymerization methodologies. The copolymers comprise a poly(2-vinylpyridine) (P2VP) basic block, a PAA acidic block, and a hydrophobic poly(*n*-butylmethacrylate) (PnBMA) block and have the sequence P2VP-*b*-PAA-*b*-PnBMA (Tsitsilianis et al. 2008). Polymerization started with 2-vinylpyridine as the more reactive monomer of the three in THF at −78°C. Then *tert*-butyl acrylate and *n*-butyl acrylate were introduced to the polymerization

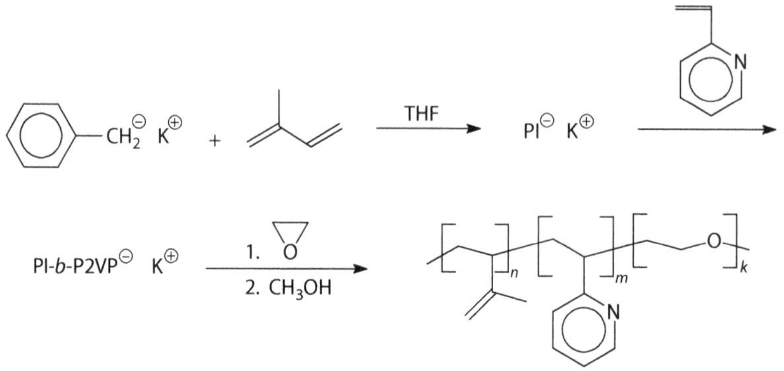

SCHEME 1.11 Synthesis of PI-*b*-P2VP-*b*-PEO triblock terpolymers.

solution producing the second and third block, respectively. After termination of the polymerization, the poly(*tert*-butyl acrylate) block was hydrolyzed to PAA and the triblock polyampholyte could be obtained. The triblock terpolymer showed novel pH-sensitive self-assembly in aqueous media due to the protonation–deprotonation equilibria of P2VP and PAA blocks.

PEG macroinitiators were utilized for the synthesis of poly(ethylene glycol)-*b*-poly(2-(dimethylamino)ethyl methacrylate)-*b*-poly(2-(diethylamino)ethyl methacrylate) (PEG-*b*-PDMAEMA-*b*-PDEAEMA) terpolymers via ATRP (Tang et al. 2003). The hydroxyl end group of the monomethyl-capped PEG was functionalized with 2-bromoisobutyryl bromide in order to incorporate the ATRP initiating site at the end of PEG chains. Then DMAEMA was polymerized in water in the presence of CuBr and 2,2′-bipyridine. DEAEMA in methanol was subsequently added and the polymerization was terminated by exposure to air (Scheme 1.12).

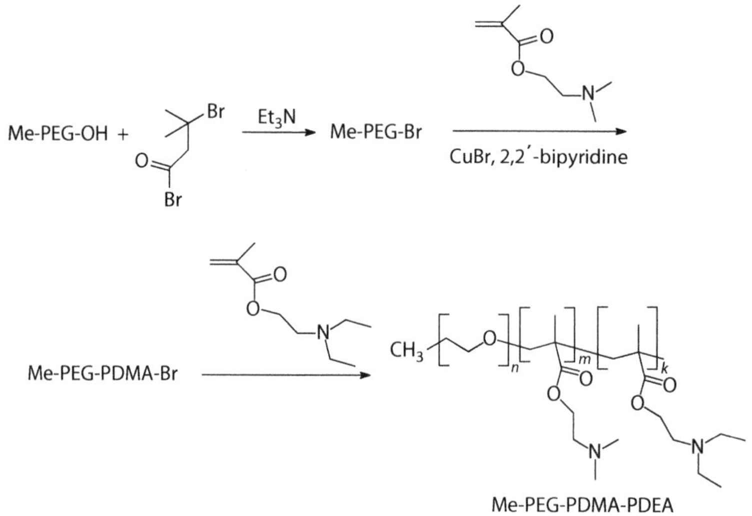

SCHEME 1.12 Synthesis of PEG-*b*-PDMAEMA-*b*-PDEAEMA terpolymers via ATRP.

Poly(ethylene glycol)-*b*-poly(*N*-isopropyl acrylamide)-*b*-poly(sodium styrene sulfonate) (PEG-*b*-PNIPAM-*b*-PSS) triple hydrophilic and thermoresponsive terpolymers were synthesized by ATRP starting from PEG macroinitiators (Kjoniksen et al. 2011). In this work, 2-bromoisobutyryl bromide was also used to functionalize the hydroxyl end group of monomethyl-capped PEG. NIPAM was then polymerized in a water/DMF mixture at 25°C in the presence of $CuCl/CuCl_2/Me_6TREN$. At ~90%, conversion sodium styrene sulfonate monomer was added as a water/DMF solution. The pure triblock terpolymer was isolated after extensive dialysis against water.

RAFT was utilized for the synthesis of hydrophilic–lipophilic–fluorophilic linear ABC triblock terpolymers by sequential addition of monomers. The excellent choice of monomers and of the polymerization scheme allowed the preparation of triblocks with different sequences (ABC, ACB, and BAC). Poly(oligoethylene glycol methyl ether acrylate) was used as the hydrophilic component, poly(benzyl acrylate) as the lipophilic component, and poly(2,2,3,3,4,4,4-heptafluorobutyl acrylate) as the fluorophilic component (Marsat et al. 2011). A newly synthesized CTA 4-(trimethylsilyl)benzyl-4-(trimethylsilyl)butane dithioate was used in conjunction with AIBN in different solvents depending on the nature of the monomer to be polymerized. Typically, the polymerizations of oligoethylene glycol methyl ether acrylate and benzyl acrylate could be conducted in THF, while trifluorotoluene was used for the polymerization of the perfluorinated acrylate monomer. A variety of self-assembled nanostructures, including multicompartment micelles, were observed in aqueous solutions due to the dissimilar solubility characteristics of the three blocks.

Thermoresponsive double-hydrophilic triblock terpolymers of poly(ethylene glycol), poly(*n*-butyl acrylate), and poly(*N*-isopropyl acrylamide) (PEG-*b*-PBA-*b*-PNIPAM copolymers) have also been synthesized by aid of designed macro-PEG CTAs. The particular block copolymers showed interesting temperature-sensitive self-assembly behavior in aqueous solutions. Other (meth)acrylamide monomers including acrylamide, *N*,*N*-diethylacrylamide, and *N*-(2-hydroxypropyl)methacrylamide have been incorporated as the third block under controlled polymerization conditions resulting in well-defined triblock terpolymers (Walther et al. 2008).

A triblock copolymer of the structure poly(ε-caprolactone)-*b*-poly(ethylene oxide)-*b*-poly(2-vinylpyridine) (PCL-*b*-PEO-*b*-P2VP) was prepared by sequential anionic and ROP. P2VP was the first block to be synthesized and PEO the second one. ROP of CL monomer was achieved by metal-catalyzed initiation from the hydroxyl groups of the P2VP-*b*-PEO-OH copolymer. The biotinylated analogue of the triblock was also synthesized, with the biotin groups attached on the P2VP ends of the copolymer aiming at the development of micellar drug nanocarriers with targeting abilities. Both block copolymers showed pH-sensitive micellization in aqueous media (Van Butsele et al. 2009).

Triblock terpolymers of PEG and biodegradable polycarbonate based blocks were synthesized via ROP employing organometallic catalysts (Yang et al. 2010). The use of polycarbonate blocks carrying carboxylic acid functionalities (by the use of 2,2-bis(methylol)propionic acid monomers bearing pendant benzyloxycarbonyl groups and/or ethyloxycarbonyl groups) allowed for the preparation of terpolymers with different solubility characteristics. Depending on the monomer addition method and choice of chemical side group functionality, the synthesis of terpolymers with different segment topology along the chain was possible. The obtained amphiphilic block copolymers self-assembled into micelles in water and were evaluated as nanosized drug carriers for doxorubicin.

Linear pentablock copolymers of the type ABCBA and having the detailed structure poly(methyl methacrylate)-*b*-poly(acrylic acid)-*b*-poly(2-vinylpyridine)-*b*-poly(acrylic acid)-*b*-poly(methylmethacrylate) (PMMA-*b*-PAA-*b*-P2VP-*b*-PAA-*b*-PMMA) were synthesized by anionic polymerization (Tsitsilianis et al. 2008). A difunctional initiator was used together with three sequential monomer additions, according to monomer reactivity, in order to give the desired double-hydrophilic polyampholyte. The copolymer possessed hydrophobic PMMA outer block and pH-sensitive hydrophilic PAA and P2VP inner blocks and showed rich self-assembly aqueous solutions.

The synthesis of a tetrablock amphiphilic ABCD linear quaterpolymer and a pentablock amphiphilic ABCDE quintopolymer by anionic polymerization methodologies has been reported in the literature showing the strength of controlled polymerization methodologies in producing complex and multifunctional amphiphilic block copolymer structures (Ekizoglou and Hadjichristidis 2002).

1.3.2.2 Other Synthetic Schemes for the Synthesis of Linear Block Copolymers

So far, living/controlled polymerization strategies involving the sequential monomer addition method have been discussed. There are other alternative synthetic strategies that allow for the synthesis of a large number of block copolymers (Hadjichristidis et al. 2003). Combination of two different modes of polymerization is utilized when monomers that are polymerized via different polymerization mechanisms should be incorporated as separate blocks in the same block copolymer. In such cases, a transformation from one mechanism to another is necessary. Usually, the first block is synthesized first; it is isolated and is subsequently used as a macroinitiator for the polymerization of the second monomer (Gaucher et al. 2005, Marsden and Kros 2009, Osada et al. 2009). End-group functionality of macroinitiator (i.e., the nature of the end group and the extent of chain end functionalization) plays a decisive role in the success of the block copolymer synthesis. Some relevant examples have been presented incrementally in the previous section, and some more are specifically discussed in the following.

PEO macroinitiators have been widely applied in the synthesis of medically important block copolymers, since this polymer imparts several advantages to the final macromolecule, including stealth properties to the micellar nanocarriers, as has been discussed earlier. A typical example is the use of monohydroxyl end-functionalized PEG for the synthesis of PEO-*b*-PCL block copolymers in the presence of stannous octoate (Shuai et al. 2001) and the use of monoamino end-functionalized PEG for the synthesis of PEO block copolymers with polyamino acids as the second block (Osada et al. 2009). The first example was the synthesis of PEG-*b*-poly(aspartate) copolymers (PEG-PAsp) (Scheme 1.13). Doxorubicin was conjugated to the carboxylic acid groups of the poly(aspartate) (PAsp) block in a second step. PEG-*b*-poly(glutamate) (PEG-PGlu) block copolymers with a more hydrophobic PGlu block have also been synthesized and used for the conjugation or encapsulation of anticancer drugs.

$$CH_3O\!\left[CH_2CH_2O\right]_n\!NH_2 \; + \; m \text{ (}\beta\text{-benzyl-L-aspartate NCA)} \xrightarrow{\text{DMF}}$$

$$CH_3O\!\left[CH_2CH_2O\right]_n\!NH\!\left[\overset{O}{\overset{\|}{C}}\text{—CH(CH}_2\text{COOCH}_2\text{C}_6\text{H}_5\text{)—NH}\right]_m\!H \xrightarrow[\text{Hydrolysis}]{\text{Alkaline}} CH_3O\!\left[CH_2CH_2O\right]_n\!NH\!\left[\overset{O}{\overset{\|}{C}}\text{—CH(CH}_2\text{COO}^{\ominus}\text{Na}^{\oplus}\text{)—NH}\right]_m\!H$$

SCHEME 1.13 Synthesis of poly(ethylene glycol)-*b*-poly(aspartate) (PEG-PAsp) block copolymers by the macroinitiator method.

Polybutadiene-*b*-poly(glutamic acid) (PB-*b*-PGA) diblock copolymers were synthesized by combination of anionic polymerization and ring-opening NCA polymerization (Checot et al. 2003, 2005). Butadiene was polymerized first in hydrocarbon solvents using s-BuLi as initiator, giving a PB chain with high 1,4 microstructure. The living PBLi chains were treated with diphenylethylene and subsequently deactivated with 1-(3-chloropropyl)-2,2,5,5-tetramethyl-1-aza-2,5-disilacyclopentane (CTMADSP). The CTMADSP end groups were hydrolyzed in acidic conditions giving the ω-amino-terminated PB, which was used as the macroinitiator for the ROP of the γ-benzyl-L-glutamate *N*-carboxyanhydride in DMF. Well-defined block copolymers of different composition were obtained. Hydrolysis of the benzyl groups of the glutamate block resulted in the PB-*b*-PGA hybrid block copolymers, which were found to self-assemble in a variety of nanostructures in aqueous media (Scheme 1.14).

Amphiphilic diblock copolymers of poly(2-ethyl-2-oxazoline) and PCL (PEOz-*b*-PCL copolymers) were prepared by using cationic and ring-opening catalyzed polymerization. 2-Ethyl-2-oxazoline was polymerized in acetonitrile using methyl *p*-toluenesulfonate as the cationic initiator. Termination with methanolic KOH resulted in a hydroxyl group on the POz chain end. The hydroxyl terminus was used as the initiating site for the polymerization of ε-caprolactone in chlorobenzene in the presence of a catalytic amount of stannous octoate, resulting in PEOz-*b*-PCL block copolymers (Kim et al. 2002).

Amphiphilic poly(D,L-lactide)-*b*-poly(2-(dimethylamino)ethyl methacrylate (PDLL-*b*-PDMAEMA) diblock copolymers were synthesized by combination of ROP and ATRP (Karanikolopoulos et al. 2010). D,L-lactide was polymerized first using *n*-decanol as the initiator and stannous octoate as the catalyst. The hydroxyl end groups of PDLL were reacted with bromoisobutyryl bromide to be transformed

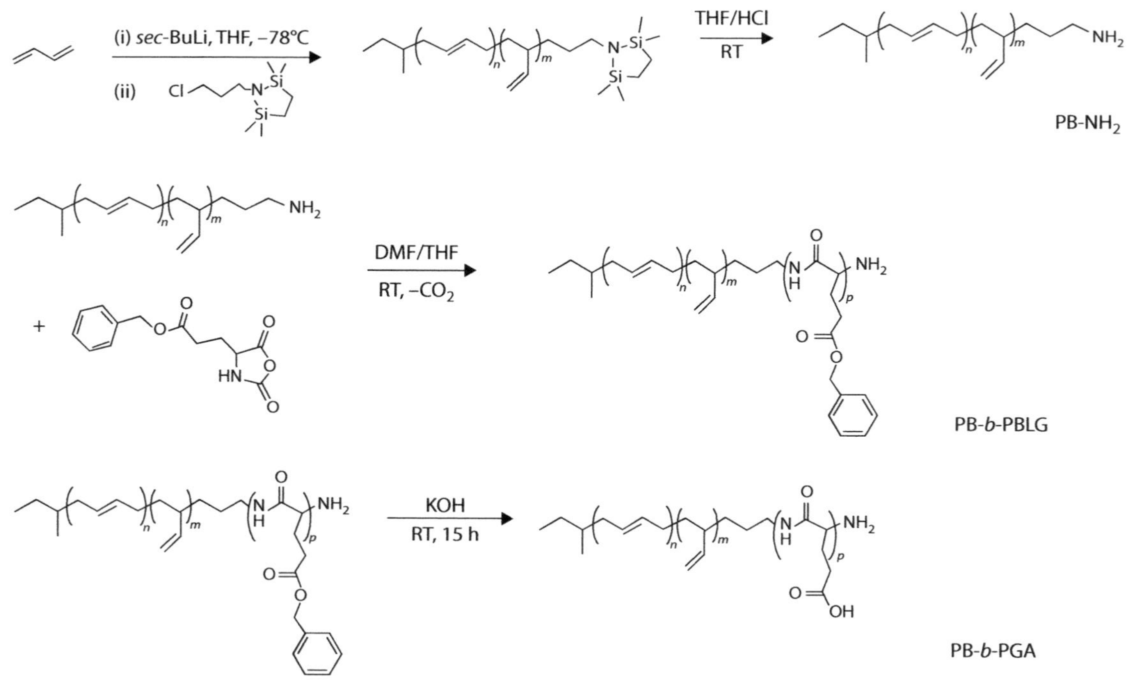

(i) *sec*-BuLi, THF, −78°C

(ii) Cl

THF/HCl

RT

PB-NH_2

+

DMF/THF

RT, $-CO_2$

PB-*b*-PBLG

KOH

RT, 15 h

PB-*b*-PGA

SCHEME 1.14 Synthesis of PB-*b*-PGA diblock copolymers.

to tertiary bromide groups able to act as ATRP initiating groups. Then DMAEMA was polymerized using the CuBr/HMTETA catalyst/ligand system. Size-exclusion chromatography (SEC) analysis of the produced diblocks showed monomodal peaks with molecular weight distributions close or lower than 1.3.

ABA triblock copolymers where A is a poly(L-lactide) block and B a poly(2-ethyl-2-oxazoline) block (PLA-*b*-PEOz-*b*-PLA copolymers) were synthesized by a combination of cationic polymerization of 2-ethyl-2-oxazoline and ROP of L-lactide (Wang and Hsiue 2003). A difunctional macroinitiator was initially synthesized by polymerizing 2-ethyl-2-oxazoline with 1,4-dibromo-2-butene as the initiator, in dry acetonitrile at 100°C. The POz end groups were converted to hydroxyl groups by treating the polymer with methanolic KOH. After drying, the HO-POz-OH polymer was used as the macroinitiator for the polymerization of L-lactide in the presence of stannous octoate in chlorobenzene at 140°C (Scheme 1.15). The final copolymer was isolated with precipitation in ethyl ether. Single peaks were observed in SEC characterization of the materials indicating the absence of homopolymers and the purity of the triblock copolymers.

Amphiphilic poly(ethylene glycol methyl ether)-*b*-poly(γ-methyl-ε-caprolactone)-*b*-poly(2-(dimethylamino)ethyl methacrylate) ABC triblock copolymers (mPEG-*b*-PMCL-*b*-PDMAEMA) were prepared by a combination of ROP and ATRP using monohydroxyl end-functionalized PEG as the macroinitiator (Matter et al. 2011). The γ-methyl-ε-caprolactone block was grown directly from the hydroxyl functionality of PEG using triethylaluminum/pyridine as the catalytic system. After completion of polymerization, the chains were terminated with methanolic HCl. The end groups of the mPEG-*b*-PMCL diblock were converted to ATRP initiators using appropriate functionalization reactions, as described previously. The obtained bromide end-functionalized copolymer, mPEG-*b*-PMCL-Br, was used in the polymerization of DMAEMA in a THF/methanol mixture, in the presence of CuCl/PMDETA. Several terpolymers with different compositions were prepared and some of them were studied in respect to the formation micellar nanoparticles in aqueous solutions.

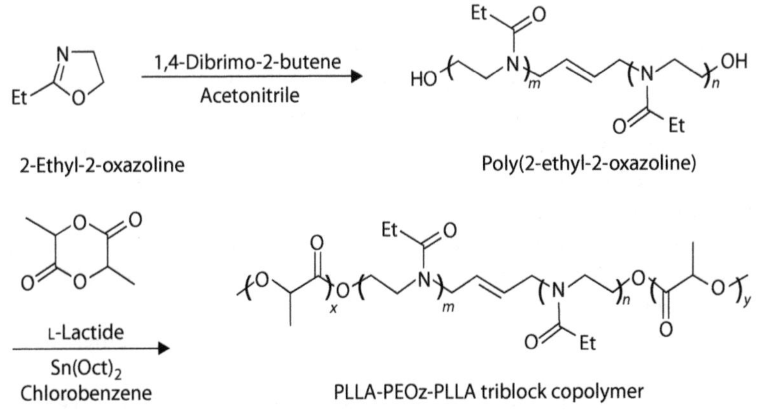

SCHEME 1.15 Synthesis of poly(L-lactide-*b*-2-ethyl-2-oxazoline-*b*-L-lactide) (PLLA-*b*-PEOz-*b*-PLLA) triblock copolymers.

Dual initiators, that is, molecules having two dissimilar initiating sites, able to perform polymerization reactions via two different polymerization mechanisms, can be employed for the synthesis of AB diblock copolymers. In this case, A- and B-type monomers are polymerized by two distinct polymerization mechanisms in the same reactor in a rather sequential manner. It is necessary that the two reactions do not interfere with each other, that is, the conditions required for one of them do not have a deleterious effect on the conditions and the products of the other (Bernaerts et al. 2006).

Using a designed dual initiator, namely, 2-hydroxylethyl 2-bromoisobutyrate, well-defined poly(L-lactide)-*b*-poly(2-(dimethylamino)ethyl methacrylate (PLLA-*b*-PDMAEMA) block copolymers were prepared (Mao et al. 2011). The hydroxyl functionality of the initiator was first utilized for the ROP of LLA in toluene catalyzed by an organometallic catalyst. After work-up, the bromine-terminated PLLA macroinitiator was used for the atom transfer polymerization of DMAEMA in the presence of CuBr/HMTETA in THF at 60°C. Diblock with controlled molecular characteristics and different lengths of the PDMAEMA blocks were successfully synthesized.

Coupling reactions, where preformed homopolymer chains in a living or an activated form are joined together end by end, can also lead to the formation of AB diblock copolymers (Hadjichristidis et al. 2003). An activated form of a homopolymer can be in general a homopolymer carrying an appropriate functional end group that can react effectively and quantitatively with a functional end group of another homopolymer. The route of coupling between living homopolymers with appropriately selected coupling agents is used for the synthesis of BAB triblock copolymers. A BA diblock copolymer of half the molecular weight can be formed by the use of a monofunctional initiator, and then individual BA* living chains (*denotes the position of the active chain end) can be coupled with a suitable coupling/linking agent that allows a permanent covalent bond to be formed between BA chains, on the A side ends. Other more complex nonlinear block copolymer structures can be synthesized by utilization of living polymer chains and sophisticated coupling reactions. Recently, coupling reactions based on click chemistry gain attention in the synthesis of medicinally interesting amphiphilic block copolymers. Click reactions between azides and alkynes (or nitriles) using copper salt–based catalysts lead to the formation of C–C (or C–N) bonds under mild experimental conditions, in the presence of a variety of other functional groups and in high yields. Block copolypeptides were prepared through click coupling of α-azido-PBLG with α-alkyne-poly(trifluoroacetyl-L-lysine) in DMF (Scheme 1.16). The starting homopolypeptides were synthesized by ROP of the respective *N*-carboxyanhydrides at room temperature in DMF using the appropriate ω-amino-containing α-alkyne and α-azido difunctional initiators (Agut et al. 2008).

In a similar reaction route, rod–coil hybrid diblock copolymers containing a rigid PBLG block and a flexible poly[2-(dimethylamino)ethyl methacrylate] block were prepared by click coupling of α-alkyne-poly[2-(dimethylamino)ethyl methacrylate] with α-azido-PBLG or α-alkyne-PBLG with α-azido-poly[2-(dimethylamino)ethyl methacrylate] (Agut et al. 2007).

Polystyrene-*b*-poly(ethylene oxide)-*b*-poly(*tert*-butyl acrylate) (PS-*b*-PEO-*b*-PtBA) triblock terpolymers were synthesized by coupling reaction (Jing et al. 2011). The use of a heterofunctional PEO having a free hydroxyl group and an

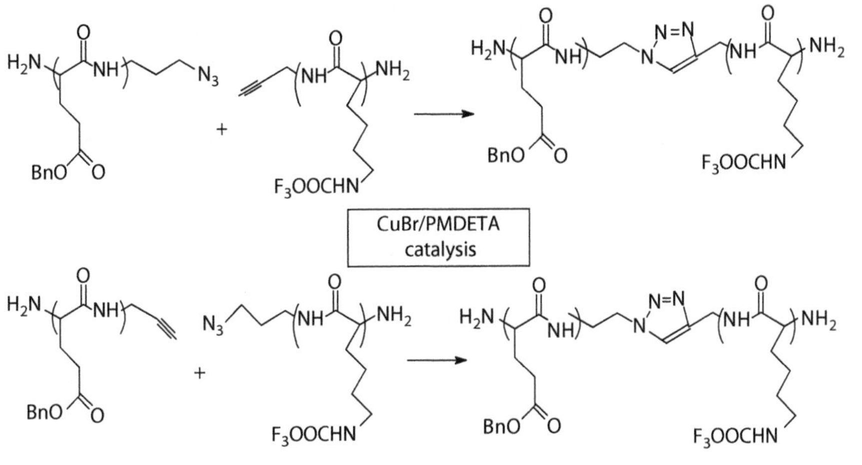

SCHEME 1.16 Synthesis of block copolypeptides through click coupling.

ethoxyethyl-protected hydroxyl at each chain end was the key point of the particular synthesis. It was produced by polymerization of EO with potassium 2-(1-ethoxyethoxy)ethoxide as the initiator. Heterofunctionalized PEO, that is, Br-PEO-alkyne, was then obtained by nucleophilic substitution of the free hydroxyl group into an alkyne group, followed by hydrolysis of the ethoxyethyl group into hydroxyl group and final esterification of the recovered hydroxyl group into 2-bromoisobutyryl group, respectively, as it is shown in Scheme 1.17. Next, azide end-functionalized polystyrene (PS) (N_3-PS) was prepared via ATRP polymerization of styrene and transformation of the Br end group of the resulting PS-Br into an azide by reaction with NaN_3. In the last step of the synthesis, the N_3-PS was coupled to the alkyne functionality of the heterofunctional PEO, while *tert*-butyl methacrylate was polymerized from the –Br end of the PEO in the presence of Cu(0) and PMDETA. This a very illustrative example of the simultaneous use of a macroinitiator and a coupling reaction in order to produce a triblock copolymer with a sequence of blocks that cannot be attained by sequential addition of monomers during several polymerization steps.

Using anionic polymerization and ATRP in combination with click coupling reactions, the synthesis of a poly(2-vinylpyridine)-*b*-poly(ethylene oxide)-*b*-poly(2-(dimethylamino) ethyl methacrylate) (P2VP-*b*-PEO-*b*-PDMAEMA) triblock terpolymer was synthesized. The P2VP-*b*-PEO diblock copolymer was prepared by anionic polymerization and the living P2VP-*b*-PEO-Li was reacted with 2-azidoisobutyryl chloride, to give P2VP-*b*-PEO-N_3. ATRP polymerization of DMAEMA with an alkyne-functionalized initiator gave alkyne end-functionalized PDMAEMA, which was subsequently coupled to P2VP-*b*-PEO-N_3 via a click reaction. In an analogous reaction scheme, alkyne end-functionalized POEGMA chains, containing methacrylate monomers of different oligo-EO length as a side chain, were coupled to P2VP-*b*-PEO-N_3, giving the respective triblock terpolymers (Reinicke and Schmalz 2011).

Other coupling reactions between polymer chains leading to block copolymers have been utilized. A recent example is the synthesis of poly(L-lactide)-*b*-poly(L-lysine)-*b*-poly(ethylene glycol) (PLA-*b*-PLLys-*b*-PEG) triblock copolymers

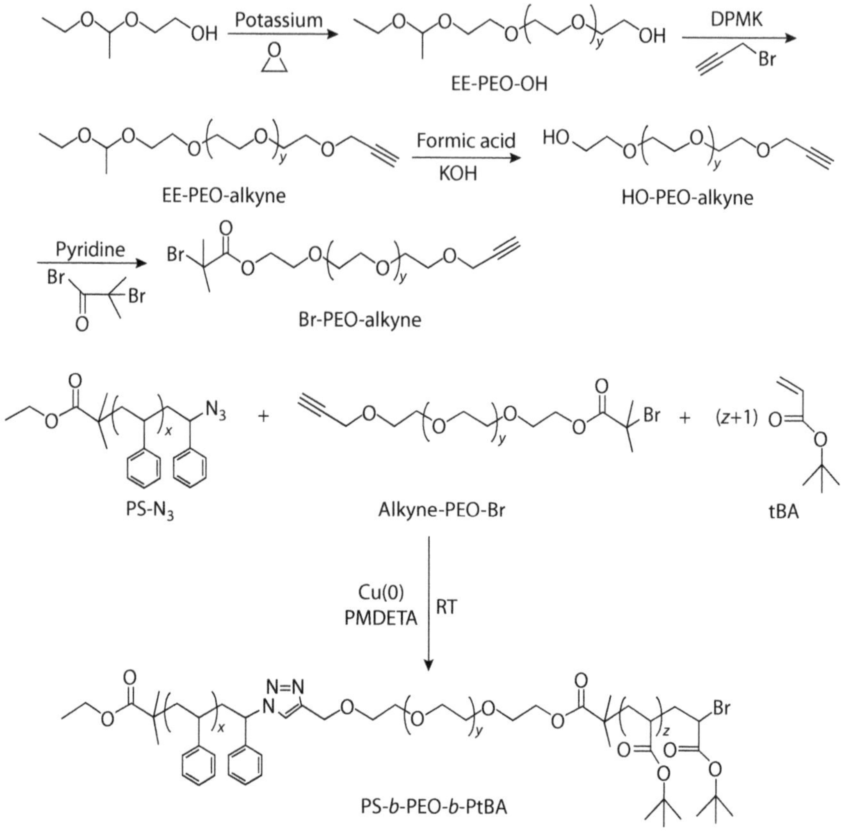

SCHEME 1.17 Synthesis of poly(styrene-*b*-ethylene oxide-*b*-*tert*-butyl acrylate) (PS-*b*-PEO-*b*-PtBA) triblock terpolymers.

(Xiang et al. 2011). A carboxyl-end-functionalized PLA was reacted with excess *N*-hydroxysuccinimide (NHS) in the presence of dicyclohexylcarbodiimide (DCC). After completion of the reaction, $H_2NCH_2CH_2NH_2$ was added to the PLA-NHS end-functionalized homopolymer giving the amino end-functionalized PLA (PLA-NH_2). PLA-NH_2 was then used as a macroinitiator for the polymerization of the Z-protected L-lysine-NCA resulting in the PLA-PZLLys-NH_2 diblock copolymer. In parallel, reaction of MPEG-OH with succinic anhydride gave the carboxyl-end-functionalized MPEG (MPEG-COOH), and in a second reaction with NHS, the MPEG-NHS end-functionalized PEG was obtained. In the last step, a coupling reaction between PLA-PZLLys-NH_2 and MPEG-NHS was performed in dry $CHCl_3$. The desired PLA-PLLys-MPEG triblock was obtained after deprotection of the lysine segments.

1.3.2.3 Nonlinear Block Copolymers

So far discussion has been focused only on linear block copolymers. Changing the topology/architecture of block copolymers to nonlinear can impart significant changes in the behavior of block copolymers and mainly in the way block copolymer molecules organize within self-assembled nanostructures and nanoparticles. Geometrical characteristics, size, and shape of block copolymer micelles can be altered by altering the macromolecular architecture of the block copolymer. Therefore, molecular architecture is one of the ways that allow tuning of the properties of their self-assembled nanoparticles. Examples on this topic will be discussed in the next chapters. Here we focus our discussion on examples concerning the synthesis of block copolymers possessing nonlinear architectures (e.g., star-block copolymers, miktoarm stars, graft copolymer).

Several star-block copolymers, that is, star macromolecules with block copolymers as the arms, have been synthesized by anionic polymerization some decades ago (Bi and Fetters 1976, Nguyen et al. 1986). Stars with well-defined number of arms were prepared by using chlorosilane compounds. Although the importance of these nonlinear macromolecules was tremendous for establishing structure–properties relationships, this methodology gives little room for preparing amphiphilic star molecules directly. An alternative strategy involved the use of difunctional monomers like DVB where the number and distribution of arms are not that well defined. In this case, the possibilities for synthesizing amphiphilic star macromolecules are greater, as it will become evident from the following discussion. Additionally, the development of CRP techniques allowed for the application of similar strategies to a wider number of monomers. Novel linking reactions and a number of reaction schemes, allowing for the synthesis of star and other nonlinear block copolymers, have also been invented in more recent years. Some more recent examples are presented in order to delineate the progress in this field of synthetic polymer chemistry.

Star-block copolymers composed of PS-*b*-poly(*tert*-butylstyrene) arms were obtained by linking PS-*b*-PtBS-Li living diblocks with DVB. In this case, the anionic polymerization of the difunctional monomer DVB initiated by the PS-*b*-PtBS-Li macroinitiators leads to the formation of a cross-linked DVB core, which acts as the linking point for the PS-*b*-PtBS$^{(-)}$ active chains, and a star copolymer with diblock arms is formed. Due to the monomer addition sequence followed, PS blocks are the outer parts of the arms of the particular star-block copolymers. The PS blocks of the stars were selectively sulfonated using SO_3 (Yang and Mays 2002a) giving sulfonated poly(styrene)-*b*-poly(*tert*-butylstyrene) arms. The degree of sulfonation of the star-block copolymers was determined to be in the range 85%–95% by elemental analysis. Neutralization with sodium methoxide in methanol gave the salt form of the amphiphilic star-block polyelectrolytes.

Utilization of ATRP methodologies and a trifunctional initiator molecule, namely, 1,1,1-tris (4-(2-bromoisobutyryloxy)phenyl) ethane (TBriBPE), led to the synthesis of 3-arm star-block copolymers having diblock arms of poly(*n*-butyl acrylate) and polystyrene ($(PBA\text{-}PS)_3$ star copolymers). First, *n*-butyl acrylate was polymerized with TBriBPE in the presence of CuBr/$CuBr_2$/PMEDTA. The trifunctional star $(PBA)_3$ macroinitiator was then used for the polymerization of styrene under similar conditions (Pakula et al. 2011).

A functionalized β-cyclodextrin (heptakis[2,3,6-tri-O-(2-bromo-2-methylpropionyl]-β-cyclodextrin) (21Br-β-CD) was used as the core initiating molecule for the synthesis of star-block copolymers with PBA inner blocks and PS outer blocks. The cyclodextrin initiator had 21 active sites for initiation of the corresponding number of diblock arms via ATRP routes in the presence of CuBr and PMEDTA. Copolymers with narrow molecular weight distributions were obtained. Hydrolysis of the PBA blocks in trifluoroacetic acid resulted in $(PAA\text{-}PS)_{21}$ star-block copolymers with an amphiphilic character and well-defined macromolecular architecture (Pang et al. 2011).

GTP was utilized for the synthesis of star-block copolymers based on positively ionizable hydrophilic DMAEMA and the hydrophobic tetrahydropyranyl methacrylate (THPMA), which carries hydrolyzable side chains. In this case, ethylene glycol dimethacrylate (EGDMA) was chosen as the difunctional monomer, serving as a coupling agent, for producing the core of the star molecules. 1-Methoxy-1-trimethylsiloxy-2-methyl propene was the initiator, while the placement of blocks was varied (Georgiou et al. 2006). After polymerization and isolation, the THPMA units were hydrolyzed in order to produce ionizable hydrophilic methacrylic acid (MAA) units carrying negative charges, thus yielding star-block polyampholytes.

3-Arm star-block copolymers based solely on block polypeptides were synthesized by employing a coupling reaction (Aliferis et al. 2005). The linear diblock copolymer arms were prepared first by polymerization of γ-benzyl-L-glutamate *N*-carboxyanhydride (Glu-NCA) and ε-benzyloxycarbonyl-L-lysine *N*-carboxyanhydride (Lys-NCA) using *n*-hexylamine as the initiator in DMF. The resulting diblock copolypeptides were coupled with the trifunctional agent triphenylmethane-(4,4′,4″-triisocyanate) (MTI) giving the star-block copolymers. The sequence of blocks in respect to the arm center could be reversed based on the sequence of monomer addition in the preparation of the arms. The linking reaction took ca. 4 weeks to reach completion and the final 3-arm star-block copolymer was separated from the linear arms by a salting-out procedure. Detailed molecular characterization of the star copolymers by SEC coupled with light scattering detector and NMR revealed the high homogeneity of the copolymer samples and confirmed their macromolecular architecture.

Different architectures of amphiphilic graft copolymers have been presented in the literature. In the cases where each backbone segment carries a grafted side chain, the macromolecules are denoted as molecular brushes. The "grafting to," "grafting from," and "grafting through" techniques have been used successfully depending on the nature of monomers to be polymerized and of the graft copolymer to be synthesized (Pitsikalis et al. 1998, Hadjichristidis et al. 2003, Zhang and Muller 2005). The "grafting from" technique was utilized in order to prepare cylindrical bottle brushes, that is, molecular brushes with long backbones, having diblock copolymers as side chains. Poly(2-hydroxyethylmethacrylate) (PHEMA) homopolymer was synthesized by ATRP or via anionic polymerization of silyl-protected HEMA and subsequent hydrolysis. The hydroxyl groups of PHEMA were esterified and the introduced α-bromoester side groups were used as initiating sites for the polymerization of *tert*-butyl acrylate and then *n*-butyl acrylate giving brush copolymers with PtBA-*b*-PnBA grafted diblock chains. Acidic hydrolysis of the PtBA blocks resulted in amphiphilic brush copolymers with PAA-*b*-PBA side chains (Borner et al. 2001, Cheng et al. 2001, Zhang et al. 2003). The synthesis of the copolymeric brushes was verified by SEC and NMR measurements.

A novel graft copolymer of the structure poly(ε-caprolactone)-*g*-poly(4-vinyl pyridine) (PCL-*g*-P4VP), containing a biodegradable hydrophobic PCL backbone and potentially hydrophilic P4VP grafted side chains, was synthesized by initially reacting a PCL homopolymer with lithium isopropyl amide at low temperature. This procedure creates active anionic sites along the PCL chain, which can be used in the next step for the polymerization of 4-vinylpyridine. Termination of the living side chains gives the PCL-*g*-P4VP copolymer. Quaternization of the P4VP side chains with a large excess of iodomethane in DMF (Nottelet et al. 2008) resulted in amphiphilic graft copolymers of the type poly(ε-caprolactone)-*g*-quaternized poly(4-vinyl pyridine) (PCL-*g*-QP4VP) with permanently cationic water-soluble side chains. In a similar fashion, the poly(ε-caprolactone)-*g*-quaternized poly(2-(dimethylamino) ethyl methacrylate) (PCL-*g*-PQDMAEMA) graft copolymer has been synthesized, through the polymerization of DMAEMA from the PCL$^{(-)}$ multifunctional macroinitiator and subsequent quaternization of PDMAEMA side chains. The quaternized graft copolymers showed improved water solubility and better defined micellar aggregates than their precursors. The same methodology was utilized from the synthesis of doubly grafted PCL-*g*-(P4VP:PEO) copolymers. In this case, PCL-*g*-P4VP copolymers were again reacted with lithium isopropyl amide in order to create new anionic sites along the backbones. Then end bromoacetylated MeO-PEGs were coupled to the formed anions resulting in graft copolymers having P4VP and PEG side grafted chains (Scheme 1.18).

Neutral amphiphilic block–graft copolymers of the type polystyrene-*b*-poly(hydroxy styrene-*g*-ethylene oxide) (PS-*b*-(PHS-*g*-PEO)) were synthesized by a combination of traditional and metal-free anionic polymerization. First, diblock copolymers of PS and poly(*tert*-butoxystyrene) were synthesized using s-BuLi as the initiator and sequential addition of *tert*-butoxystyrene and styrene. The PS-*b*-PtBOS copolymers were subjected to acid hydrolysis in order to remove the *tert*-butoxy groups, and the PS-*b*-poly(hydroxyl styrene) copolymers were obtained in near quantitative modification. The phenolic groups of the PHOS block were activated with phosphazene base at low temperature and transformed to oxoanions. Then the multifunctional diblock macroinitiator was used for the polymerization of EO giving the desired amphiphilic block–graft copolymers, which were purified by fractional precipitation (Zhao et al. 2009) (Scheme 1.19).

The same approach also led to the synthesis of graft copolymers with a PHOS backbone and poly(ethylene oxide-*b*-propylene oxide) side arms (PHOS-*g*-(PEO-*b*-PPO) copolymers). Activated PHOS multifunctional macroinitiators were used for the sequential polymerization of EO and PO, and the monomer addition sequence determines the topology of the side arms. In this way, graft–block copolymers with PEO inner side parts and PPO outer parts (or vice versa) could be obtained (Zhao et al. 2010a). Amphiphilic graft copolymers with random copolymers of EO and PO as side arms were also prepared in different compositions using similar synthetic routes (Zhao et al. 2010b).

An amphiphilic graft copolymer with PEO as main chain and poly(methyl acrylate) grafted chains was synthesized by using anionic polymerization and ATRP (Li et al. 2006b). The backbone was obtained by anionic copolymerization of an ethyl vinyl ether protected glycidol monomer, that is, 2,3-epoxypropyl-1-ethoxyethyl ether (EPEE), and EO, using diphenylmethyl potassium as the initiator. The obtained poly(ethylene oxide-co-2,3-epoxypropyl-1-ethoxyethyl ether) (poly(EO-co-EPEE))

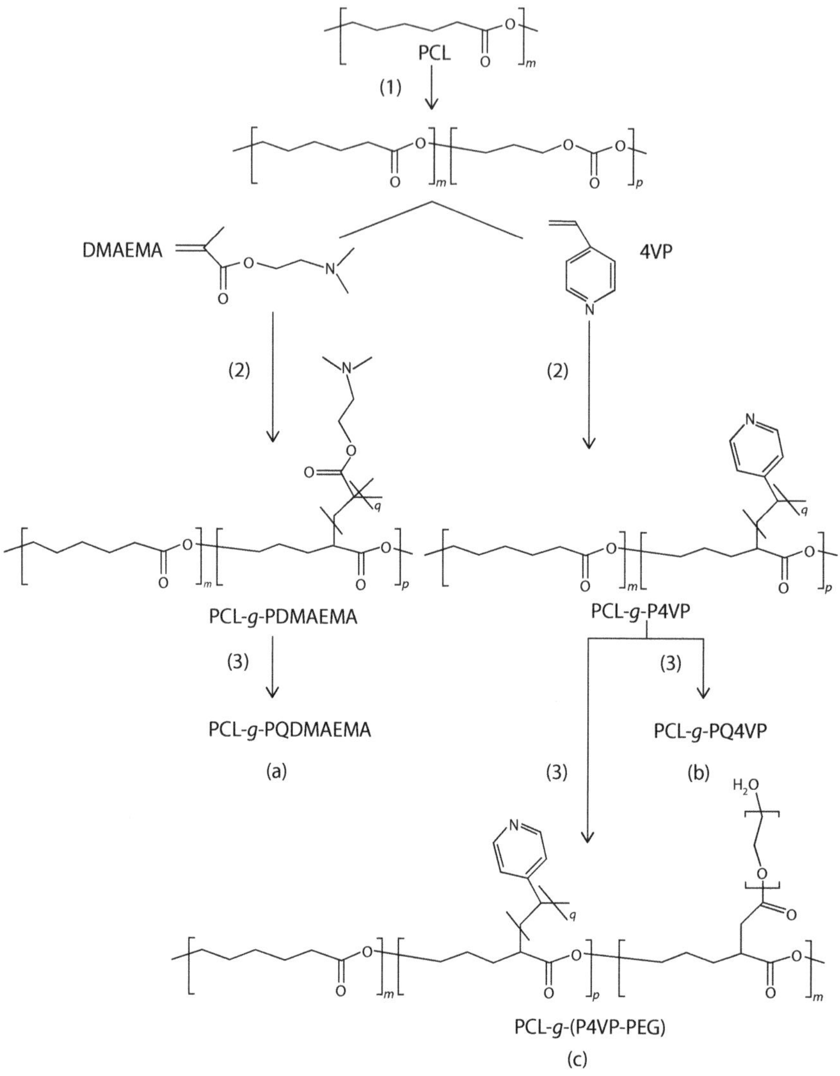

SCHEME 1.18 Synthesis of doubly grafted PCL-*g*-(P4VP-PEO) copolymers.

was treated with acid in order to deprotect the hydroxyl groups in the EPEE segments. Subsequent esterification with 2-bromoisobutyryl bromide produced a multifunctional ATRP macroinitiator that was utilized for the polymerization of methyl acrylate (MA). Detailed characterization of the final graft copolymer by SEC, NMR, MALDI-TOF-MS, and FTIR confirmed the efficient synthesis of the target copolymers.

Several graft copolymers having the architecture of molecular brushes have been reported (Cheng et al. 2001, Zhang and Muller 2005, Xu et al. 2008). Graft copolymers with a gradient grafting density along the backbone have been synthesized

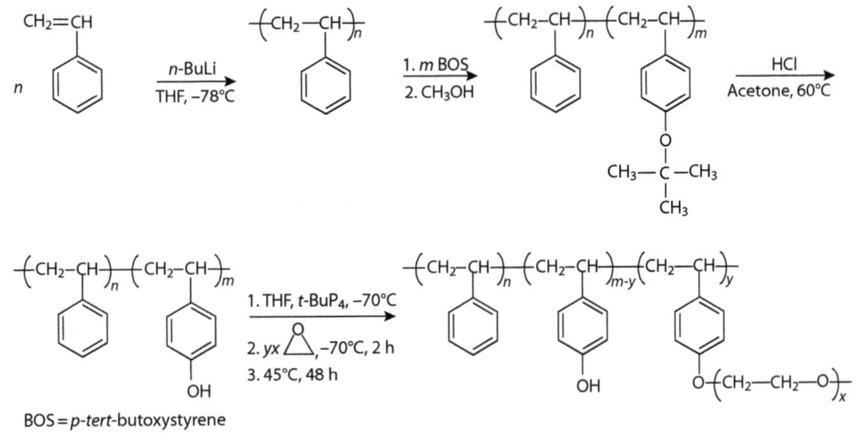

SCHEME 1.19 Synthesis of PS-*b*-(PHS-*g*-PEO) block–graft copolymers.

via ATRP schemes. First, a gradient copolymer was synthesized by ATRP polymerization of methyl methacrylate and (2-trimethylsilyloxy)ethyl methacrylate. After hydrolysis of the silyl protecting group, the resulting hydroxyl groups were esterified and a gradient copolymer of methyl methacrylate and 2-(2-bromopropionyloxy)ethyl methacrylate) was obtained. Side chains were grafted by ATRP "grafting from" polymerization of appropriate monomers (Lord et al. 2004).

The synthesis of well-defined amphiphilic copolymer brushes possessing alternating poly(methyl methacrylate) and poly(*N*-isopropylacrylamide) (poly(PMMA-alt-PNIPAM)) grafts has been also described (Yin et al. 2009). The first synthetic step involves the alternating copolymerization of *N*-[2-(2-bromoisobutyryloxy) ethyl]maleimide (BIBEMI) with 4-vinylbenzyl azide (VBA) and results in the production of poly(BIBEMI-alt-VBA). Poly(BIBEMI-alt-VBA) is a multifunctional macroinitiator bearing bromine and azide moieties arranged in an alternating manner. Therefore, it is possible to initiate ATRP and at the same time participate in click reactions. In this way, polymerization of methyl methacrylate leads to poly(PMMA-alt-VBA) copolymer brushes. Then, amphiphilic poly(PMMA-alt-PNIPAM) copolymer brushes participated in the click reaction with an excess of alkynyl-terminated PNIPAM (alkynyl PNIPAM) affording the graft copolymer with alternating PMMA and PNIPAM grafted side chains.

Amphiphilic miktoarm star copolymers where the chemical structure of the arms of the star is different have been reported in a number of occasions via the use of several polymerization mechanisms, their combinations, and the use of ingenious synthesis routes.

Double-hydrophilic A_2B miktoarm stars were prepared by ATRP methodologies. A Jeffamine carrying a short poly(alkyl oxylene) chain was reacted initially with excess 2-hydroxylethyl acrylate and then with 2-bromoisobutyryl bromide. In this way, the bis-hydroxy-functionalized Jeffamine was transformed to a difunctional ATRP macroinitiator that was used for the polymerization of DMAEMA giving a miktoarm star with one neutral arm and two arms carrying basic groups

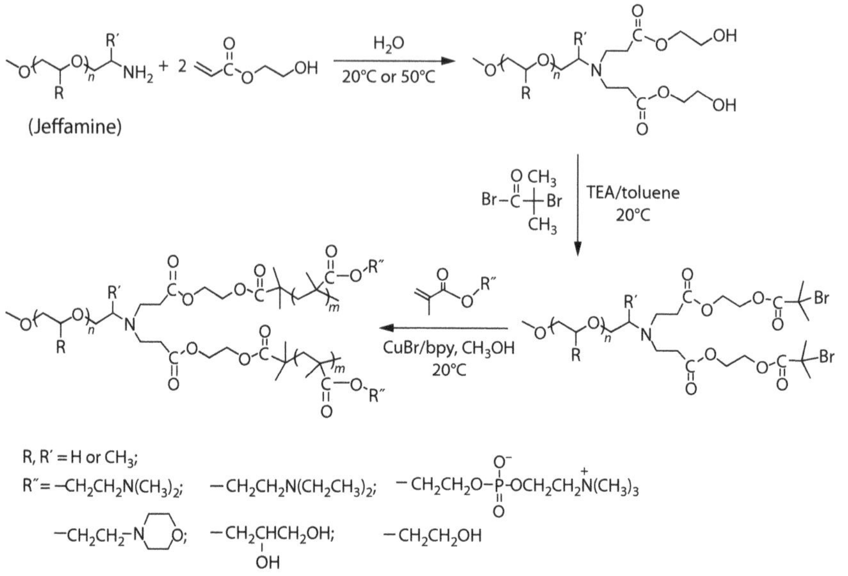

(Jeffamine)

R, R′ = H or CH_3;

R″ = $-CH_2CH_2N(CH_3)_2$; $-CH_2CH_2N(CH_2CH_3)_2$; $-CH_2CH_2O-P(=O)(O^-)-OCH_2CH_2N^+(CH_3)_3$

$-CH_2CH_2$-N(morpholine)O; $-CH_2CH(OH)CH_2OH$; $-CH_2CH_2OH$

SCHEME 1.20 Synthesis of poly(alkyl oxylene)-[2-dimethylamino)ethyl methacrylate]$_2$ (PAO-PDMAEMA$_2$), A_2B-type miktoarm star copolymers.

(Cai and Armes 2005) (Scheme 1.20). In a similar reaction scheme, the difunctional Jeffamine macroinitiator was used for the polymerization of (2-hydroxy)ethyl methacrylate, and the hydroxyl groups of the two PHEMA arms were transformed to hydrophilic chains with acidic groups after reaction with succinic anhydride.

Amphiphilic miktoarm stars with a hydrophobic biodegradable PCL arm and two poly(2-dimethylamino)ethyl methacrylate) arms (or vice versa) have been synthesized by the combination of ROP and ATRP techniques. The key step of the synthesis was the preparation of dual initiators with one or two hydroxyl groups and two or one tertiary bromide moiety, respectively. Hydroxyl groups were utilized for the polymerization of the caprolactone monomer in a first step followed by the polymerization of the methacrylate monomer (Liu et al. 2007). Well-defined copolymers with narrow monomodal molecular weight distributions were obtained.

The synthesis of A_2B type miktoarm stars based on biodegradable caprolactone monomers has been presented. First, ε-caprolactone has been polymerized by 2-methyl-2-(prop-2-yn-1-yl)propane-1,3-diol in the presence of stannous octoate, utilizing the two hydroxyl groups of the initiator. Then the terminal hydroxyl groups of PEL were protected with acetyl groups. 3-Azidopropan-1-ol was reacted with the middle alkyne group of PCL. The resulting hydroxyl group in the middle of PCL chains was used as the initiator for the copolymerization of γ-bromo-ε-caprolactone and ε-caprolactone. The pendant bromide groups were transformed to azides, and finally, amine groups were introduced to one of the arms via cycloaddition of *N,N*-dimethylprop-2-yn-1-amine to the azide groups (Riva et al. 2011) (Scheme 1.21). The synthetic steps were followed by SEC and NMR verifying the success of the synthetic protocol.

AB_2 copolymers, where A and B are polyamino acid chains, were synthesized by combining ROP of *N*-carboxy anhydrides of amino acids and click chemistry. In this case, A is poly(L-lysine) (PLLys) and B is poly(L-glutamic acid) (PGA) (Rao et al. 2008). Two functional primary amine initiators were synthesized for this purpose. *N*-aminoethyl 3,5-bis(propargyloxyl)-benzamide was used for the polymerization of the NCA of L-lysine giving dialkynyl-terminated poly(ε-benzyloxycarbonyl-L-lysine) (DA-PZLLys). 3-Azidopropylamine was used for the polymerization of the NCA of L-glutamic acid giving azide-terminated poly(γ-benzyl-L-glutamate) (N_3-PBLG). DA-PZLLys and N_3-PBLG were coupled together via the copper(I)-catalyzed cycloaddition reaction resulting in the PZLLys-$(PBLG)_2$ miktoarm stars. Hydrolysis of the protecting groups gave the desired PLLys-$(PLG)_2$ polyampholytic miktoarm stars.

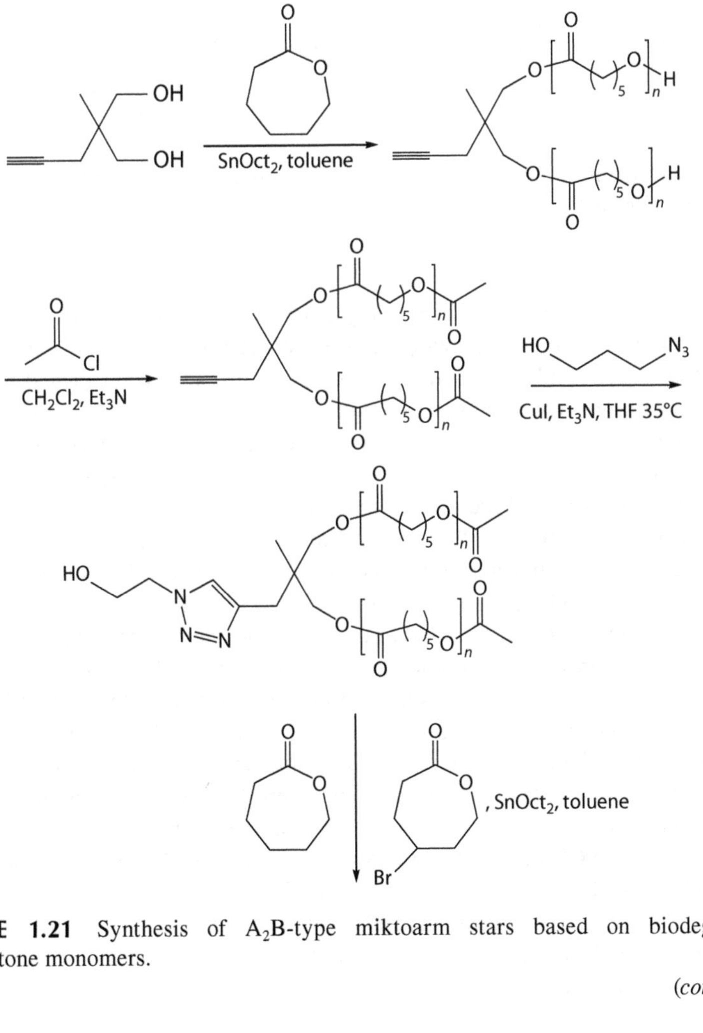

$SnOct_2$, toluene

CH_2Cl_2, Et_3N

CuI, Et_3N, THF 35°C

, $SnOct_2$, toluene

SCHEME 1.21 Synthesis of A_2B-type miktoarm stars based on biodegradable caprolactone monomers.

(continued)

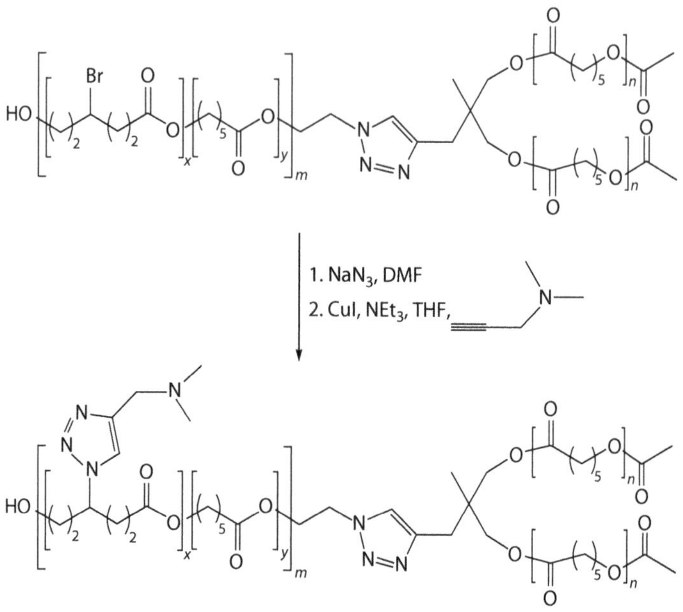

SCHEME 1.21 (continued) Synthesis of A_2B-type miktoarm stars based on biodegradable caprolactone monomers.

In another approach, polystyrene-(poly(glutamic acid))$_2$ (PS-(PGA)$_2$) stars were obtained following a four-step synthesis protocol (Babin et al. 2008). First, PS was prepared by ATRP and the terminal bromide groups were transformed to one or two primary amine groups by reaction with ethylenediamine or tris-(2-aminoethyl) amine. Then the amine groups were used for the ROP of γ-benzyl-L-glutamate *N*-carboxyanhydride. Finally, hydrolysis of the protective functionalities in the polypeptide block afforded the miktoarm copolymers.

Miktoarm stars with two PS arms and one polypeptide arm were synthesized by first preparing a primary amine in-chain-functionalized PS and the subsequent utilization of the middle primary amine for the ROP of the NCA of the appropriate amino acid (Karatzas et al. 2008). Suitable diphenylethylenes were used for the preparation of the in-chain-functionalized PS via anionic polymerization. A similar methodology was also used for the synthesis of A_2B_2 miktoarm stars with A being PS and B a polypeptide (PBLG and poly(ε-*tert*-butyloxycarbonyl-L-lysine) (PBLLys)).

Amphiphilic calixarene-centered A_2B_2 miktoarm stars with two PCL arms and two PEG arms were synthesized using a tetrafunctional calixarene with two hydroxyl functionalities and two alkyne functionalities. The hydroxyl groups were used to prepare a PCL middle functionalized by the two alkyne groups. An azide-functionalized PEG was synthesized in parallel by transforming the hydroxyl end group to azide. The azide-terminated PEGs and the middle-alkyne-functionalized PCL were coupled by click reaction to give the desired miktoarm copolymer (Gou et al. 2010) (Scheme 1.22).

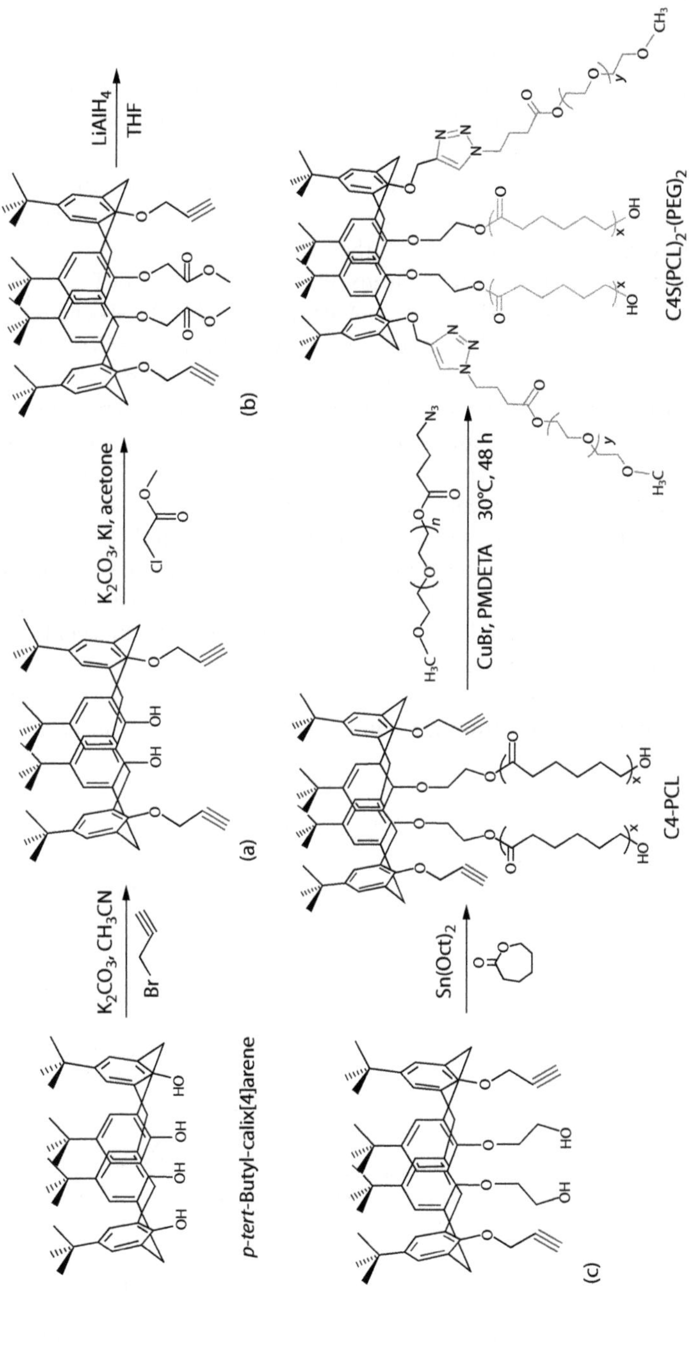

SCHEME 1.22 Synthesis of amphiphilic calixarene-centered A_2B_2 miktoarm stars with two PCL arms and two PEG arms.

In another case, A_2B_2 miktoarm stars were prepared by the combination of anionic polymerization, ATRP, and Glaser coupling in order to obtain a copolymer with two PS or PI arms and PAA arms (Wang et al. 2010).

A AB_4 miktoarm star with one PNIPAM arm and four PDEAEMA arms was synthesized starting from a PNIPAM-based tetrafunctional ATRP macroinitiator. The macroinitiator was synthesized via addition reaction of monoamino-terminated PNIPAM with glycidol, followed by esterification with excess 2-bromoisobutyryl bromide. The polymerization of 2-(diethylamino)ethyl methacrylate proceeded smoothly, and a well-defined double-hydrophilic miktoarm star copolymer, PNIPAM-*b*-$(PDEA)_4$, was finally obtained as evidenced by the molecular characterization results for the copolymer (Ge et al. 2007) (Scheme 1.23).

Miktoarm stars with three different arms were prepared via ATRP schemes combined with click reaction steps (Liu et al. 2009a) (Scheme 1.24). An alkyne-terminated PEG was transformed to Cl-hydroxyl heterofunctional PEG (PEG(-Cl)-OH) via click reaction. The Cl group of PEG(-Cl)-OH was transformed to azide (PEG(-N_3)-OH). Then the hydroxyl group was transformed to a tertiary bromide group through esterification (PEG(-N_3)-Br). The bromide functionality initiated the ATRP of *tert*-butyl methacrylate giving the junction-point azide-functionalized PEG-PtBMA diblock (PEG(-N_3)-*b*-PtBMA). Akynyl end-functionalized PDEAEMA chains were coupled to PEG(-N_3)-*b*-PtBMA via click reaction, and the 3-arm ABC miktoarm star terpolymer was obtained. This novel macromolecule showed pH-responsive and polyampholytic character after hydrolysis of the PtBMA arm to PMAA. Due to the presence of basic PDEAEMA and acidic PMAA chains, different micellar morphologies were obtained in aqueous solutions depending on the pH.

Other pH-responsive miktoarm star terpolymers are those comprising PS, PEO, and poly(2-(dimethylamino)ethyl acrylate) (PDMAEA) arms (Liu et al. 2009b).

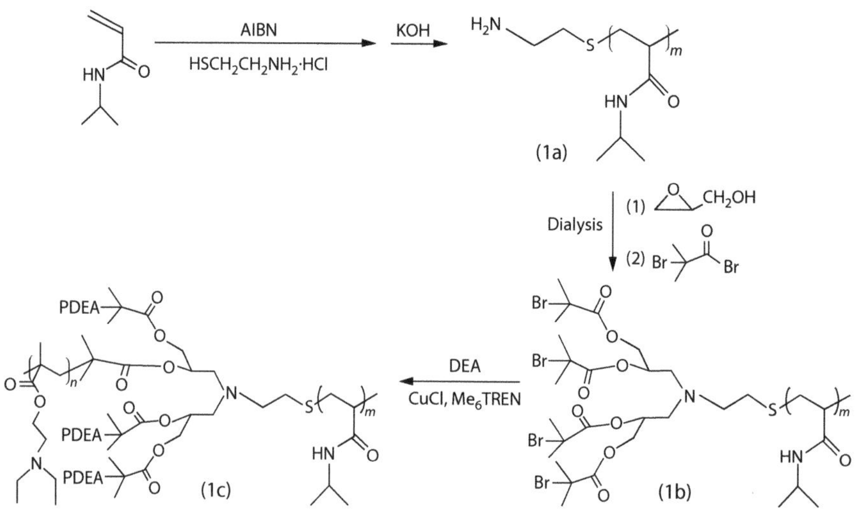

AIBN
$HSCH_2CH_2NH_2{\cdot}HCl$
KOH
(1a)
Dialysis
(1) glycidol (CH_2OH)
(2) 2-bromoisobutyryl bromide
(1b)
DEA
CuCl, Me_6TREN
(1c)

SCHEME 1.23 Synthesis of AB_4 miktoarm star with one PNIPAM arm and four PDEAEMA arms.

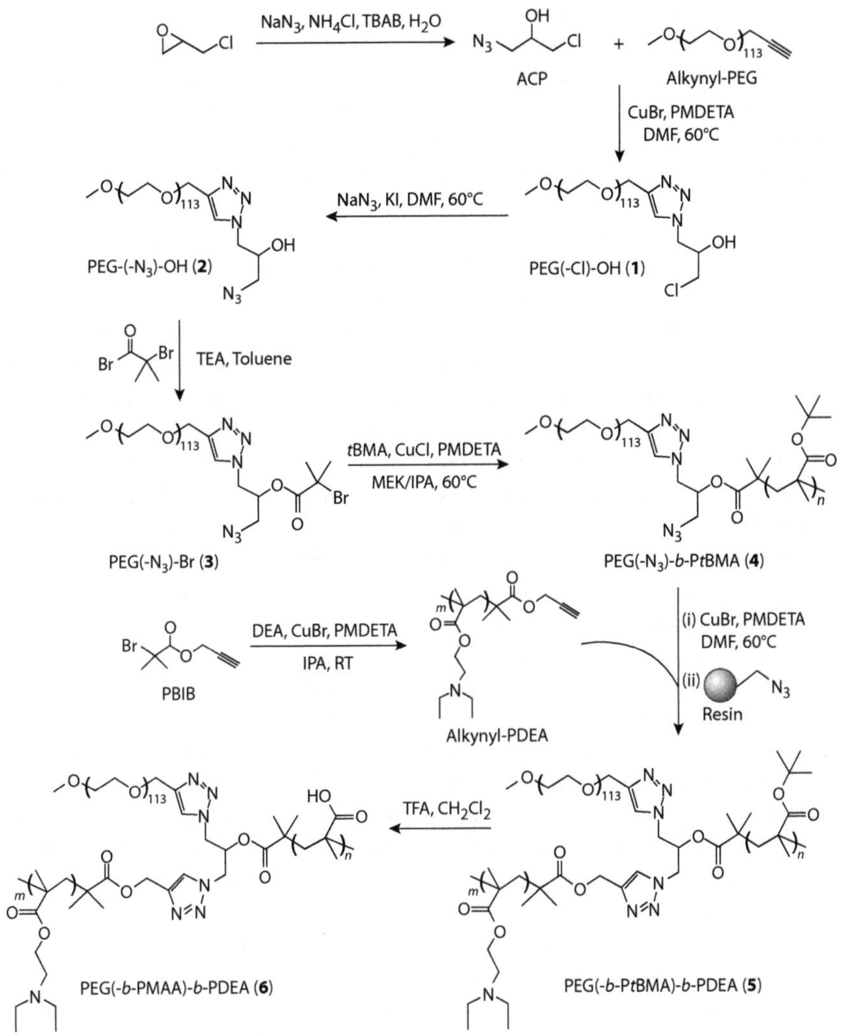

SCHEME 1.24 Synthesis of miktoarm stars with three different arms via ATRP combined with click reactions.

They have been synthesized by the combination of anionic polymerization and RAFT polymerization methodologies. Living polystyryllithium was reacted with 2-methoxymethoxymethyloxirane (MMO) giving a heterofunctional macroinitiator with a free hydroxyl and one protected hydroxyl group. The free hydroxyl group was activated with diphenylmethyl potassium to afford an anionic macroinitiator for the polymerization of EO. After deactivation, the junction-point hydroxyl functionality in the PS-PEO diblock was deprotected under acidic conditions and was used to attach the appropriate CTA functionality to the diblock. The macro-CTA was subsequently utilized for the polymerization of (2-(dimethylamino)ethyl acrylate) yielding the miktoarm terpolymers, bearing basic amino groups on one of the arms. ABC miktoarm star terpolymers with poly(ethylethylene) (PEE), PEO, and poly(γ-methyl-ε-caprolactone) (PMCL) arms were prepared by utilizing the hydroxyl functionality, located at the junction point of a PEE-PEO diblock copolymer, obtained from anionic polymerization, for the controlled ROP of γ-methyl-ε-caprolactone (Saito et al. 2008). These miktoarm stars contain two hydrophobic arms, one of them being biodegradable, and were found to form multicompartment micelles in water.

Thermoresponsive ABC miktoarm star terpolymers were also reported (Zhang et al. 2009). They are composed of PS, PNIPAM, and PCL arms. First, a difunctional PS bearing an alkynyl and a primary hydroxyl moiety at the same chain end (PS-alkynyl-OH) was synthesized by reaction of an azide-terminated PS with an excess of 3,5-bis(propargyloxy)benzyl alcohol (BPBA) under click conditions. The PS-alkynyl-OH macroinitiator was utilized in the subsequent ROP of ε-caprolactone through the hydroxyl end group resulting in the PS(-alkynyl)-*b*-PCL junction-point-functionalized diblock copolymer. The (PS)(PNIPAM)(PCL) amphiphilic ABC miktoarm star terpolymers were then obtained by the click reaction between PS(-alkynyl)-*b*-PCL diblock and an excess of azide-terminated PNIPAM (PNIPAM-N_3). Follow-up of the synthesis steps by SEC, NMR, and FTIR gave evidence of the successful synthesis of the target miktoarm terpolymers. Miktoarm star terpolymers with a polypeptide arm have been also synthesized (Karatzas 2008).

$(PS)_n(P2VP)_n$ heteroarm stars have been synthesized by anionic polymerization, using DVB for linking preformed PS arms and subsequent polymerization of 2VP monomer using the anionic active sites in the DVB core of the formed star (the "in–out" method). These star copolymers show amphiphilic character due to the protonation/deprotonation equilibrium of the P2VP arms at low and high pH, respectively. They show interesting self-assembly behavior, as well as complexation properties toward inorganic salts (Tsitsilianis et al. 2000, Gorodyska et al. 2003) and also toward biomacromolecules giving large opportunities for the creation of hybrid polymeric nanoparticles.

Using the "in–out" method with DVB, a heteroarm star copolymer of the type $(PS)_n$ $(PtBA)_n$ was synthesized. Hydrolysis and then neutralization with NaOH of the hydrophobic heteroarm star copolymer gave the amphiphilic $(\text{polystyrene})_n$ (poly(sodium acrylate)$_n$, $(PS)_n(NaPAA)_n$, star copolymer (Voulgaris and Tsitsilianis 2001). A high degree of hydrolysis was achieved. The solubility and self-assembly behavior of the $(PS)_n(NaPAA)_n$ star were found to depend on the relative length of the PS/PAA arms. In an analogous reaction scheme, ampholytic $(P2VP)_n(PAA)_n$ heteroarm stars with basic P2VP and acidic PAA arms were also synthesized, from $(P2VP)_n(PtBA)_n$ precursors (Hammond et al. 2009). The wide combination of monomer sequences that can be utilized with the "in–out" methodology was also demonstrated through the synthesis of multiarm star-shaped copolymers having the architecture $A_n(B\text{-}b\text{-}C)_n$, where A = PS, B = P2VP, and C = PAA. The amphiphilic/polyampholytic heteroarm star terpolymers obtained after postpolymerization modification of hydrophobic precursors showed interesting behavior in terms of solubility and gelation, under different solution conditions (Stavrouli et al. 2008).

Amphiphilic branched copolymers of the types A_2BA_2 and A_4BA_4 were obtained by ATRP using appropriate macroinitiators (Xu et al. 2006). In particular, amine

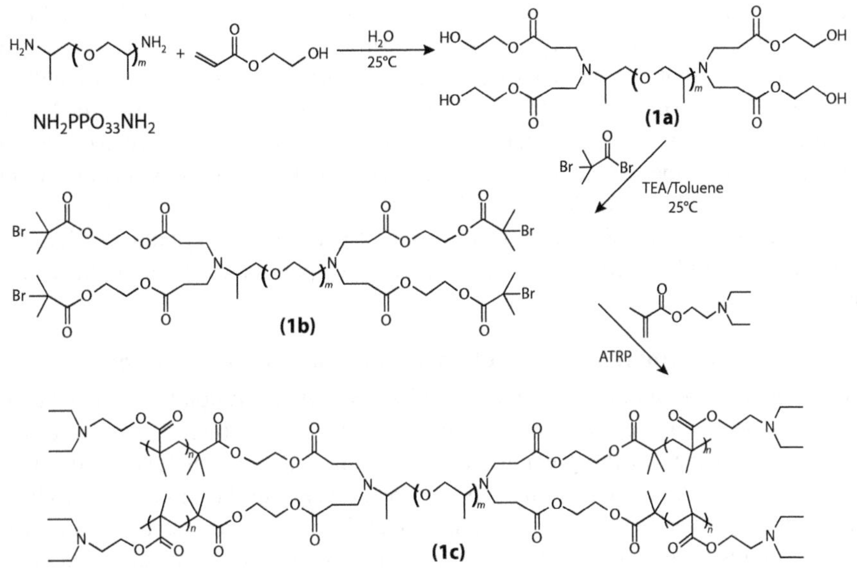

SCHEME 1.25 Synthesis of amphiphilic branched copolymers of the A_2BA_2 type, (1c), by ATRP and the use of macroinitiator (1a, 1b).

difunctional linear PPO was reacted with 2-hydroxyethylacrylate and glycidol to give PPO with two or four hydroxyl groups at each chain end followed by esterification with excess 2-bromoisobutyryl bromide (or 2-bromopropionyl bromide), in order to give PPO chains with two or four ATRP initiating sites at each end. These macroinitiators were used for the polymerization of 2-(diethylamino)ethyl methacrylate and allowed for the formation of the two and four side arms of the copolymers (Scheme 1.25). Molecular characterization results confirmed the structure of the double-hydrophilic branched copolymers. Due to the basic character of the PDEAEMA chains, the particular macromolecules showed pH-responsive character in aqueous solutions.

Amphiphilic and thermoresponsive linear–cyclic diblock copolymers (tadpole-shaped macromolecules), where the linear part is PCL and the cyclic PNIPAM, were obtained via ATRP and ROP (Wan et al. 2011). The reaction scheme employed started from a trifunctional core molecule containing alkynyl, hydroxyl, and bromine moieties. Using ATRP conditions via the bromide functionality, a linear PNIPAM was synthesized. In the next step, bromide end group was transformed to azide and intramolecular click reaction between the alkynyl and azide end groups led to the formation of a macrocyclic poly(*N*-isopropylacrylamide) (c-PNIPAM). The remaining single hydroxyl functionality on the c-PNIPAM chain was used for the subsequent polymerization of ε-caprolactone resulting in the targeted linear–cyclic diblock copolymer, (c-PNIPAM)-*b*-PCL copolymer.

Amphiphilic block copolymers of the linear-dendron architecture were synthesized starting from end-functional linear PEGs and following a stepwise sequence that led to the formation of the dendron block. Dendrons of polylysine and polyester were attached to the linear PEGs. The end groups at the periphery of the dendrons were functionalized by hydrophobic acid–sensitive cyclic acetal groups in order to induce pH sensitivity to the macromolecules and their supramolecular assemblies, through end-group hydrolysis-driven changes in block solubility (Gillies et al. 2004).

In another work, dendritic–linear copolymers composed of poly(benzyl ether) monodendrons of the second or the third generation and linear PNIPAM chains were prepared (Ge et al. 2006). A dendron-based RAFT CTA was synthesized first and it was then used for the RAFT polymerization of *N*-isopropylacrylamide (NIPAM). The presence of PNIPAM blocks imparted thermoresponsive character to the hybrid macromolecules.

1.3.2.4 Postpolymerization Functionalization of Block Copolymers

Block copolymers can also be synthesized by postpolymerization functionalization of precursor copolymers already prepared by one of the controlled polymerization routes outlined in the previous sections. Some examples of functionalization reactions on block copolymers have been already given in the discussion of block copolymer synthesis, for example, simple hydrolysis reactions for removing protecting groups of monomers after polymerization. Some general remarks and basic principles of this particular synthetic methodology are also given here together with additional examples.

In postpolymerization functionalization of block copolymers usually established, organic chemistry reactions are performed selectively on one of the blocks in order to change the chemical nature of its segments. Such a chemical transformation leads to changes in the physicochemical properties of the block copolymer. The particular synthetic strategy is useful when monomers of the appropriate chemical structure and functionality are not available for producing the desired polymer or their controlled polymerization (or block copolymerization in the desired sequence) is not possible for a number of reasons. Successful and useful postpolymerization functionalization reactions should meet two basic criteria (Hadjichristidis et al. 2003): (1) The transformation reaction should not in any way alter the chain-like structure of the precursor block copolymer, for example, via chain scission and inter- or intramolecular cross-linking. In the opposite case, materials with a broadened molecular weight distribution and considerably altered chemical structure are obtained, which do not correspond to the desired chain topology and functionality. Ideally, the chemical modification should lead to block copolymers with the same number of monomeric segments, chain architecture, and narrow molecular weight distribution as the precursor block copolymer. Only changes in the functional groups attached to the monomer segments of the selected block should take place, and these changes should correspond to the desired chemical functionality that should be introduced on the part of the copolymer (i.e., modifications that lead to undesired chemical groups along the chain are not the optimum ones). However, it is not always so easy or simple to transfer reactions known for low molecular weight compounds to macromolecules, mainly because the kinetics of the reactions may change and because certain by-products that are unavoidable but separable in the case of small molecules may become disastrous by-products in the case of macromolecules. (2) The chosen functionalization reactions should be mild and effective, that is, should lead to almost 100% functionalization of the selected block with the desired new functionality. This criterion is not always met, but the properties of the materials may change considerably toward the desired direction, for example, water solubility, even with chemical modifications of less than 100%. In any case, careful and thorough characterization of the modified product should be realized in order to establish the extent and nature of functionalization, since even for 100% modification, the success of the modification should always be verified. Chemical modification of block copolymers usually involves deprotection of protected functionalities already existing on the polymerized monomers, for example, hydrolysis of ester bonds on monomer side chains, as in the case of protected amino acid monomers, which leads to the preparation of water-soluble and pH-sensitive polypeptide blocks (see previous examples), nucleophilic addition to C–C double bonds of the main or side chains or on aromatic rings of side chains and multistep reactions (e.g., epoxidation of C–C double bonds with subsequent nucleophilic substitution). In several cases, a simple protonation/deprotonation equilibrium in aqueous solution can make a block copolymer amphiphilic or alter its solubility and self-assembling characteristics at different pH values (e.g., in the case of amine or carboxylic acid functionalities existing on the segments of the copolymer chain exhibiting basic or acidic properties). Chemical functionalization can take place selectively on one of the blocks or on more than one of them simultaneously or in steps, providing that the reaction conditions employed allow successive modification steps on separate blocks. It is obvious that judicious choice of monomers and existing block sequences, as well as of the overall chemical modification synthetic strategy, allows for great flexibility and variability on the resulting block copolymer structures.

Sulfonation is a reaction that has been frequently utilized for the chemical modification of diblock copolymers prepared by anionic polymerization routes, since it can be used for the modification of PS, PI, and PB blocks. It is also a good example to show the precautions that have to be taken and the different conditions that should be used in order to arrive in a successful chemical modification of a block copolymer, depending on its structure. The conditions for the sulfonation of the phenyl rings of PS are very harsh and can lead to sulfonation of polydiene blocks. For this reason, sulfonation of PS is used mainly for the synthesis of amphiphilic diblock copolymers on precursor diblocks that contain either a hydrogenated polydiene block or a poly(*tert*-butyl styrene) (PtBS) block as the second block (Valint and Bock 1988, Yang and Mays 2002a, Yang et al. 2002b, Muller et al. 2004) (Scheme 1.26). In the second case, the PtBS block segments possess a sterically protected phenyl

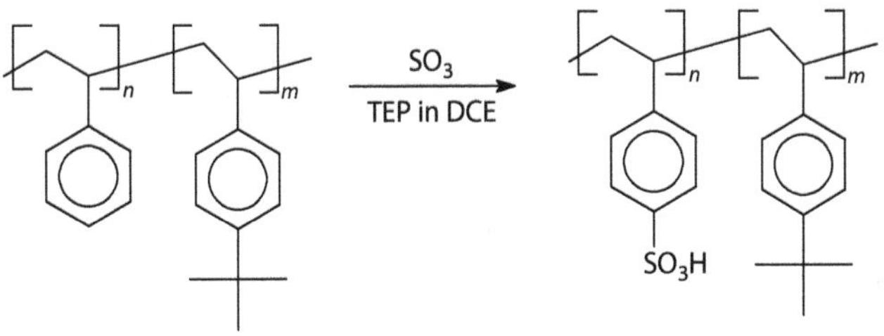

SO_3

TEP in DCE

SO_3H

SCHEME 1.26 Sulfonation of the PS block of poly(styrene-*b*-*tert*-butylstyrene) copolymers.

ring that resists sulfonation. The structure of micelles formed by such amphiphilic block copolymers in aqueous media has been studied in detail (Guenoun et al. 1996, Muller et al. 2004).

Block copolymers of PS and sulfonated PI (PS-*b*-SPI) were successfully synthesized, utilizing anhydrous sulfuric acid in dioxane. Under these reaction conditions, a sulfur trioxide–dioxane complex is prepared, which acts as a mild sulfonation agent for the PI block (Szczubialka et al. 1999), with no sulfonation of the PS block taking place. The amphiphilic PS-*b*-SPI diblock copolymers synthesized were asymmetric with a short PS block and a much longer SPI block, having different degrees of polymerization.

A similar sulfonation procedure was adopted for the synthesis of semi- or fully double-hydrophilic block copolymers of sulfonated polybutadiene (SPB) and PEO (Tauer et al. 2002). The degree of sulfonation in the PB block could be varied from 5% to 80%, as it was demonstrated by elemental analysis of the modified copolymers, giving different degrees of water solubility to the PB block. The sulfonated block copolymers became water soluble if the degree of sulfonation was about 15% or higher. These novel amphiphilic block copolymers were utilized as macromolecular stabilizers in emulsion polymerization.

Quaternization is a common reaction in order to introduce permanent cationic charges to block copolymer carrying blocks with tertiary amine groups. Quaternization of the PDMAEMA block of an asymmetric PFS-*b*-PDMAEMA diblock copolymer with excess CH_3I in THF (Wang et al. 2005) resulted in an amphiphilic block copolymer with a short hydrophobic organometallic block (PFS) and a long hydrophilic, cationic block (QPDMAEMA). The degree of quaternization was quantitative, as indicated by ^1H NMR in D_2O, which selectively solubilizes the QPDMAEMA blocks. Quaternization of PDMAEMA block was found to influence the self-assembly of diblock copolymer and especially the micellar structures in water.

Following a different reaction scheme, β-lactam-functionalized PI-*b*-PEO block copolymers were synthesized recently (Kaditi and Pispas 2010). β-Lactam groups were introduced in the PI block by reaction of the C=C bonds of the isoprene segments with chlorosulfonyl isocyanate (CSI), a widely used reagent for the modification of alkene compounds and subsequent hydrolysis of the chlorosulfonyl group under basic conditions with Na_2SO_3 (Scheme 1.27). Close to 80% functionalization of the double bonds was achieved based on NMR analysis. The solubility

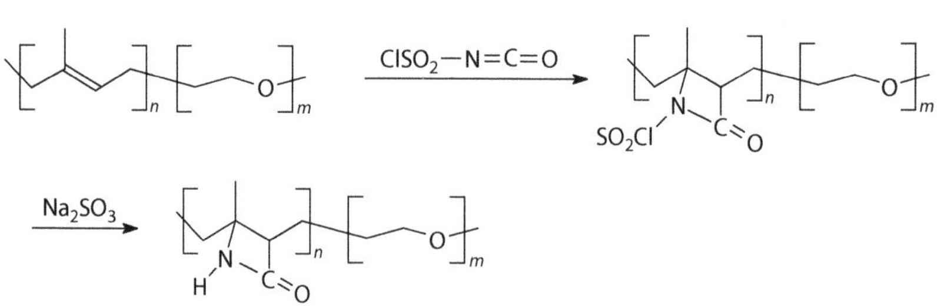

$ClSO_2—N=C=O$

SO_2Cl N C=O

Na_2SO_3

H N C=O

SCHEME 1.27 Synthesis of β-lactam-functionalized PI-*b*-PEO block copolymers.

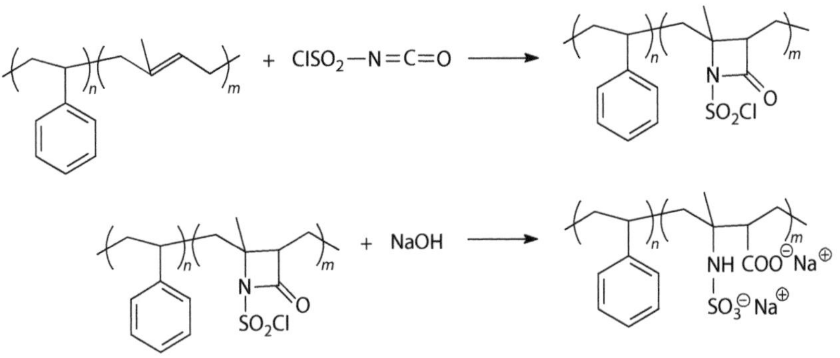

SCHEME 1.28 Synthesis of poly[((sodium sulfamate/carboxylate)isoprene)-*b*-styrene] (SCPI-PS) diblock copolymers by the reaction of poly(isoprene-*b*-styrene) copolymers with CSI and subsequent hydrolysis.

characteristics of the block copolymers in aqueous media were altered, due to the incorporation of the polar β-lactam groups, which are able to participate in hydrogen bond formation. This was demonstrated mainly by the lower aggregation numbers of the micellar nanoparticles obtained for the β-lactam-functionalized PI-*b*-PEO block copolymers compared to their PI-PEO precursors.

Amphiphilic block copolymers carrying a biocompatible poly[(sulfamate/carboxylate)isoprene] (SCPI) block were synthesized from PI-*b*-PS block copolymers also by reaction of the PI block with chlorosulfonylisocyanate, giving a β-lactam ring on each modified isoprene segment. The β-lactam rings were opened and the chlorine atom removed from the chlorosulfonyl groups in one step under basic hydrolysis conditions, giving the desired sulfamate and carboxylate functionalities on the PI block, with a yield ca. 70% (Scheme 1.28) (Uchman et al. 2008). The SCPI block is considered as biocompatible, since it possesses functional groups similar to those of heparin (Pispas 2006), a natural polysaccharide, and it is proposed as an alternative to PEO blocks for imparting stealth properties to micellar nanoparticles.

The same reaction has been utilized for the synthesis of poly[(sulfamate/carboxylate)isoprene]-*b*-poly(ethylene oxide) (SCPI-PEO) double-hydrophilic block copolymers (Pispas 2006), carrying essentially two biocompatible synthetic blocks. It is worth mentioning that the SCPI blocks have an anionic polyelectrolyte character, which allows for complexation with different complementary charged building blocks (like cationic surfactants, polyelectrolytes, and proteins) leading to the formation of polymeric nanoparticles stabilized by the nonionic PEO blocks (Pispas 2007a,b, Gao et al. 2010).

An effective and versatile modification of the PB block in PB-*b*-PEO and PB-*b*-PS diblock copolymers has been reported, involving the use of ω-functional mercaptans (Justynska et al. 2005). A wide variety of functional mercaptans are commercially available and several of them have been shown that can be utilized for modification of the C=C double bonds of the PB-1,2 side chains. In this way, amphiphilic and double-hydrophilic block copolymers containing several different polar functional groups were synthesized and subsequently characterized. The obtained diblock

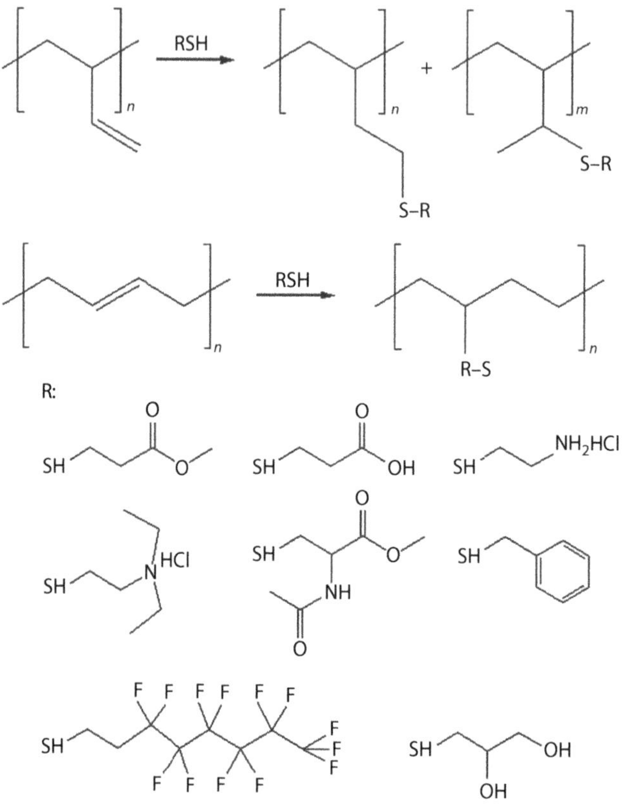

SCHEME 1.29 Versatile modification of the PB block of block copolymers by reaction with ω-functional mercaptans.

copolymers present narrow polydispersity (between 1.05 and 1.15 in most cases) as determined by matrix assisted laser desorption ionization-time of flight mass spectroscopy (MALDI-TOF MS) and SEC. Conversion of double bonds was usually complete, and the degree of functionalization of the diblock copolymer, in respect to the desired functionality, was in the range of 50%–85% (typically 70%–80%), since some cyclization of two neighboring groups leads to some loss of the introduced chemical functionality (Scheme 1.29).

Amphiphilic PB-*b*-PEO copolymers functionalized with (L,L)-cysteine-phenylalanine hydrophobic dipeptides on the PB block were obtained by the free-radical addition reaction of mercaptans, in this case utilizing peptide ω-functionalized mercaptans (Geng et al. 2006) that were synthesized for this purpose. The reaction was shown to be as effective as the ones involving lower molecular weight mercaptan molecules. The hybridization of the synthetic copolymers with peptides leads to a higher level of sophistication and functionalities of the observed self-assembled structures and induces further biocompatibility in the macromolecular system. A variety of self-assembled nanostructures, including spherical micelles, wormlike micelles and vesicles were formed by these hybrid block copolymer (Scheme 1.30).

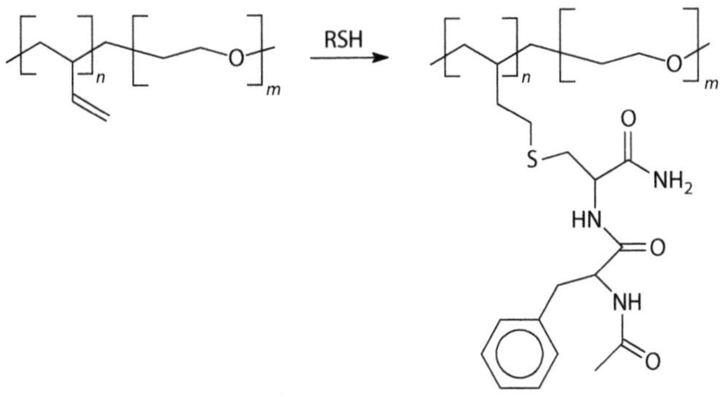

SCHEME 1.30 Functionalization of PB-*b*-PEO copolymers with (L,L)-cysteine-phenylalanine hydrophobic dipeptides on the PB block, via free-radical addition reaction of mercaptans.

A two-step postpolymerization functionalization scheme was followed in order to prepare polyethylene-*b*-poly(methacrylic acid) (PE-*b*-PMAA) copolymers. In the first step, polyethylene-*b*-poly(*tert*-butyl methacrylate) (PE-*b*-PtBMA) diblock copolymers were synthesized by selective hydrogenation of the PB block in polybutadiene-*b*-poly(*tert*-butyl methacrylate) (PB-*b*-PtBMA) copolymers, by using the Wilkinson catalyst system (Pispas et al. 2002). In the second step, the PtBMA block of PE-*b*-PtBMA was hydrolyzed using stoichiometric excess of HCl over the ester groups. The PE block is a semicrystalline hydrophobic material and the PMAA block is hydrophilic. PMAA can be made water soluble by transformation to the salt form, resulting in a series of interesting materials in terms of physical properties.

A dual-responsive poly(acrylic acid)-*b*-poly(*N*,*N*-diethylacrylamide) (PAA-*b*-PDEAAm) diblock copolymer has also been reported. The particular diblock copolymer was synthesized by hydrolysis of the PtBMA block of a PtBMA-*b*-PDEAAm copolymer, prepared by anionic polymerization (Andre et al. 2005). The PAA block imparts pH responsiveness to the system, while the PDEAAm block shows a thermoresponsive character. Therefore, the PAA-*b*-PDEAAm block copolymer formed micelles of different structures in aqueous solutions depending on the pH and the temperature.

Block copolymers containing phosphonic acid groups on one of the blocks were also synthesized using postpolymerization modification. The polystyrene-*b*-poly(vinylphosphonic acid) (PS-*b*-PVPA) diblock copolymers were prepared from the corresponding polystyrene-*b*-poly(diethyl vinyl phosphonate) (PS-*b*-PDEVP) diblock copolymer by acidic hydrolysis of the PDEVP block (Perrin et al. 2009).

Double-hydrophilic poly(*p*-hydroxystyrene-*b*-methacrylic acid) (PHOS-*b*-PMAA) copolymers were synthesized by acidic hydrolysis of the precursor poly(*p*-*tert*-butoxystyrene-*b*-*tert*-butyl methacrylate) diblocks, which were prepared by anionic polymerization (Mountrichas and Pispas 2006). The chemical modification was realized in a single step since similar reaction conditions were possible to lead to deprotection of both blocks, utilizing fivefold molar excess of HCl in dioxane. The presence of acidic phenolic and carboxyl functional groups on the same block copolymer with different pK_a values imparted a pH-responsive character and a complex self-assembling behavior for the hydrophilic block copolymers in aqueous solutions (Scheme 1.31).

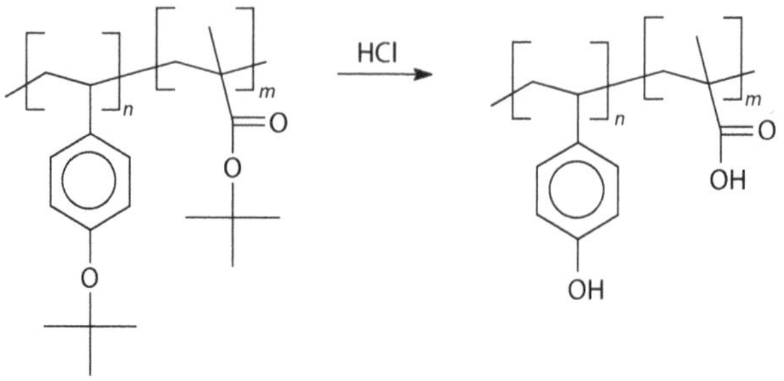

SCHEME 1.31 Synthesis of double-hydrophilic PHOS-*b*-PMAA copolymers via acidic hydrolysis of precursor poly(*p-tert*-butoxystyrene-*b*-*tert*-butyl methacrylate) diblocks in one step.

Amphiphilic glycopeptide block copolymers were synthesized by reacting the L-lysine segments of poly(L-lysine)-*b*-polytetrahydrofuran-*b*-poly(L-lysine) (PLLys-*b*-PTHF-*b*-PLLys) triblock copolymer with D-glyconolactone and lactobionolactone in varying ratios in the presence of diisopropylethylamine (Tian et al. 2008a). Reaction of the same PLLys-*b*-PTHF-*b*-PLLys copolymers with linoleic acid resulted in the linoleic acid modified amphiphilic polypeptide copolymer (Tian et al. 2008b).

It should be added that postpolymerization functionalization reactions are also carried out on amphiphilic block copolymers with the aim of introducing the appropriate end group at the free end of the water-soluble block, either for further conjugation reactions with bio(macro)molecules or for targeting purposes when the attached end groups have targeting abilities against living cells, antibodies, etc. In this manner, after micelle or nanoparticle formation, the targeting groups are present at the periphery of the nanoparticle and they are able to perform their task enhancing the functionality of the nanoparticle (Licciardi et al. 2008). Alternatively, the targeted groups can be introduced in the polymerization reaction, in the initiation step, through the initiator molecule. Since in many cases the targeting functionalities may be deteriorated in the course of the polymerization reaction initiators with protected functionalities or initiators carrying precursors of the required, functionalities are utilized (Yoo and Park 2004, Licciardi et al. 2008, Lam et al. 2009, Zhao et al. 2010c).

1.4 HETEROGENEOUS POLYMERIZATION TECHNIQUES FOR POLYMERIC NANOPARTICLE SYNTHESIS

Emulsion polymerization (or macroemulsion polymerization) is one type of a liquid-phase heterogeneous polymerization process that is traditionally utilized for the preparation of polymeric materials and polymeric particles in the micrometer range, even in industrial scale, through traditional free-radical polymerization schemes. Generally, in heterogeneous liquid-phase polymerization processes, also including microemulsion, miniemulsion, and dispersion polymerization, several chemical and physical events take place that determine the size and nature of the resulting polymeric particles, and these features strongly discriminate the particular synthetic methodologies from the homogeneous polymerization reactions that have been outlined so far (Nomura et al. 2005, Klapper et al. 2008, Ho et al. 2010). Nevertheless, the miniemulsion polymerization process has, in recent years, become a very flexible and effective technique for the synthesis of nanoparticles (with sizes well below the micrometer range) of various structures and morphologies (Schork et al. 2005, Landfester et al. 2010). The key step in this particular polymerization technique is the formation of very small, but stable, droplets of the dispersed phase by application of high shear stress, for example, via ultrasonication. In this way, the initial macrodroplets dispersed in the continuous phase, as of a traditional macroemulsion, are converted to nanodroplets, having a narrow size distribution, and sizes in the range 50–500 nm. The droplet size is also determined by the chemical nature and the concentration of the emulsifier used, which is an amphiphilic surfactant molecule. Usually a costabilizer is utilized in conjunction to the emulsifier. This is a substance that has the lowest solubility in the continuous phase, compared to the other components of the droplet. In the case of an oil-in-water miniemulsion, the costabilizer is a highly hydrophobic substance, while in the case of a water-in-oil emulsion, the costabilizer is a very hydrophilic molecule. Therefore, the costabilizers produce an osmotic pressure to the droplets (osmotic pressure agents). The formed nanodroplets act as nanocontainers/nanoreactors in the sense that the polymerization reaction is confined in their spatial domain (in their interior or at their interface) and no diffusion of reactants (e.g., monomers) is taking place, opposite to the case of a macroemulsion polymerization. The small size of the droplets determines, in some ways, the size of the formed polymeric nanoparticles after completion of the polymerization. Due to the confinement effects on the reaction process, other types of polymerization mechanisms can be used besides traditional free-radical polymerization, and nanoparticles of variable chemical structure can be produced (e.g., polyamide nanoparticles via anionic polymerization, polynorbonene nanoparticles through ROMP, and others). Many excellent recent monographs have been published on heterogeneous liquid-phase polymerizations and in particular on miniemulsion processes (Nomura et al. 2005, Schork et al. 2005, Klapper et al. 2008, Landfester 2009, Min and Matyjaszewski 2009, Oh et al. 2009, Ho et al. 2010, Landfester et al. 2010). Here, we discuss only a few examples on the synthesis of polymeric nanoparticles with structural variability for highlighting the synthetic possibilities offered by such methodologies.

Water-soluble non-cross-linked nanoparticles based on oligoethylene glycol methacrylate monomers were prepared by activators generated by electron transfer (AGET) ATRP inverse miniemulsion polymerization using PEO-Br initiators. The choice of metal complex and surfactant/costabilizer system is of crucial importance. Using the same polymerization technique, water-swellable non-cross-linked nanoparticles based on 2-(hydroxyl)ethyl methacrylate were prepared in the presence of a low molecular weight poly(ethylene-co-butylene)-*b*-PEO diblock copolymer as surfactant (Oh et al. 2007, 2009).

Cross-linked nanoparticles (nanogels) based on oligoethylene glycol methacrylate monomers and containing degradable disulfide linkages were also prepared by AGET ATRP inverse miniemulsion polymerization, using dithiopropionyl PEG dimethacrylate as a water-soluble disulfide-functionalized cross-linker. Cyclohexane was the continuous phase and PEO-Br was the macromolecular initiator utilized. When the ratio of cross-linker to initiator was above 4:1, nanoparticles insoluble

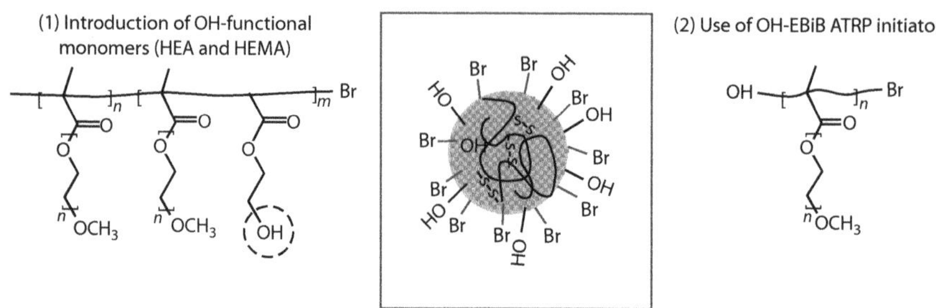

FIGURE 1.2 Synthetic approaches for the preparation of biodegradable POEOMA300-co-PHEMA nanogels, using PEG-co-poly(glycolic acid) dimethacrylate as the hydrolyzable cross-linker. (Reprinted from *J. Am. Chem. Soc.*, 128, Oh, J.K., Tang, C., Gao, H., Tsarevsky, N.V., and Matyjaszewski, K., 5578–5587, Copyright 2006, with permission from Elsevier.)

in organic solvents and water were obtained. In a similar approach, biodegradable POEOMA300-co-PHEMA nanogels were prepared by the use of PEG-co-poly(glycolic acid) dimethacrylate as the hydrolyzable cross-linker (Figure 1.2). The particular nanogels, with a diameter ca. 203 nm, were degraded upon hydrolysis into water-soluble polymers (diameter ca. 9 nm) and could be utilized for drug encapsulation and trafficking since their degradability to smaller components facilitates their excretion from the body via renal filtration (Oh et al. 2006).

Stimuli-responsive nanogels composed of a cross-linked PDEAEMA core and PEG tethered chains were prepared by emulsion polymerization methodology (Oishi and Nagasaki 2007). The use of EGDMA as a cross-linker produces an all methacrylate core. Due to the inherent properties of PDEAEMA, the nanogels showed reversible changes in their size by altering pH, ionic strength, and the temperature of the aqueous solutions. As a result of the basic character of PDEAEMA, nanogel cores swell at low pH and bear positive charges due to protonation of the tertiary amine groups. Ionized PDEAEMA cores at low pH shrink as a result of an increase in solution ionic strength due to charge screening. The thermosensitivity of PDEAEMA cores also results to nanogel shrinkage as solution temperature increases (the PDEAEMA parts become progressively insoluble in water). The tethered PEG chains induce colloidal stability and solubility to the nanogels, even under conditions where PDEAEMA parts become insoluble. The PEG tethers can carry also a carboxylic acid end group as a versatile chemical group for the installation of bio-tags.

Thermosensitive nanogels with sizes in the submicrometer range have been produced by emulsion polymerization using NIPAM as the monomer, methylene bisacrylamide as the cross-linker, ammonium persulfate as the free-radical initiator, and dodecylbenzenesulfonic acid sodium salt as the surfactant (Guan and Yongjun Zhang 2011). These nanoparticles utilize the well-known thermoresponsive character of PNIPAM chains and present volumetric changes with temperature (swelling at low temperatures). Although this is a typical example, PNIPAM nanogels have been prepared in a number of structures using various techniques. In one occasion, core-shell nanogels were synthesized by dispersion polymerization in a two-step process (Li et al. 2004). Initially, the cross-linked PNIPAM cores were produced in water using *N*,*N*′-methylenebis(acrylamide) as the cross-linker and the cationic surfactant dodecyltrimethylammonium chloride as the dispersant. They were used in the next step as nuclei for the dispersion polymerization of 4-vinylpyridine under similar reaction conditions. In this way, a pH-responsive P4VP shell was formed on the surface of the PNIPAM nanogels. Nanosized particles were obtained. Their morphology was evidenced by TEM measurements. The core-shell nanoparticles showed pH- and temperature-dependent sizes in water, as a result of the different responsiveness of the core and shell components to the respective stimuli.

In the cases where a mixture of functionalized and nonfunctionalized monomers is used in a miniemulsion copolymerization scheme, surface-functionalized nanoparticles can be obtained depending on the choice or reaction conditions and monomers. PS and PMMA nanoparticles with surface functionalized with phosphonate groups were prepared by miniemulsion copolymerization of vinyl phosphonic acid with styrene and methyl methacrylate, respectively (Ziegler et al. 2009). The surface concentration of phosphonate groups could be varied by adjusting the ratio of the monomers used and their density through the vinyl phosphonic acid concentration. Surface functionalization was found to be more effective in the case of MMA/VPA pair, something that was correlated to the water solubility of the comonomers, as well to the nucleation mechanism. The size of the nanoparticles was dependent on the amount and chemical nature of the surfactant (ionic vs. nonionic).

Miniemulsion copolymerization also allowed for the synthesis of PS nanoparticles with anionic and cationic surface groups. Copolymerization of styrene with acrylic acid or 2-aminoethyl methacrylate hydrochloride introduced carboxyl and protonated amine functionalities on the surface of the nanoparticles (Musyanovych et al. 2007). The effect of ionic or nonionic surfactant on the size distribution of resulting nanoparticles was investigated. In such surface-functionalized nanoparticles, surface charge can be varied as a function of pH due to protonation/deprotonation equilibria of the surface groups.

Shell-cross-linked fluorinated nanocapsules were obtained by RAFT miniemulsion polymerization (Chen and Luo 2011). First, the amphiphilic macromolecular RAFT CTA poly(methacrylic acid-co-dodecafluoroheptyl acrylate) (MAA-co-DFHA) was prepared by homogeneous RAFT solution polymerization. To the aqueous solution of MAA-co-DFHA, $NaNO_2$ was added as the radical scavenger in the aqueous phase. The oil phase included the DFHA monomer, DVB as the cross-linker, hexadecane, hexylacetate, and AIBN in appropriate quantities. The two phases were mixed and pre-emulsified by magnetic stirring. Miniemulsions were obtained by ultrasonication, and then SDS surfactant was added as an aqueous solution to prevent coagulation of the nanodroplets. After degassing, the free-radical polymerization was initiated at 70°C. Core-shell nanoparticles were obtained with hexadecane and hexylacetate as the core materials and a cross-linked PDFHA shell. Their average diameter was found by TEM to be ca. 114 nm and the average shell thickness ca. 16 nm. The core to shell ratio could be easily controlled by changing the ratio of the components for up to 1:1 ratio. For nanocapsules with a 1:2 or 1:1 ratio, the particles were deformed and an increased amount of cross-linker had to be utilized in order to form well-shaped nanocapsules.

Nanocapsules with aqueous cores and polyurea, polythiourea, or polyurethane (PU) or starch shells were obtained by interfacial polycondensation or via

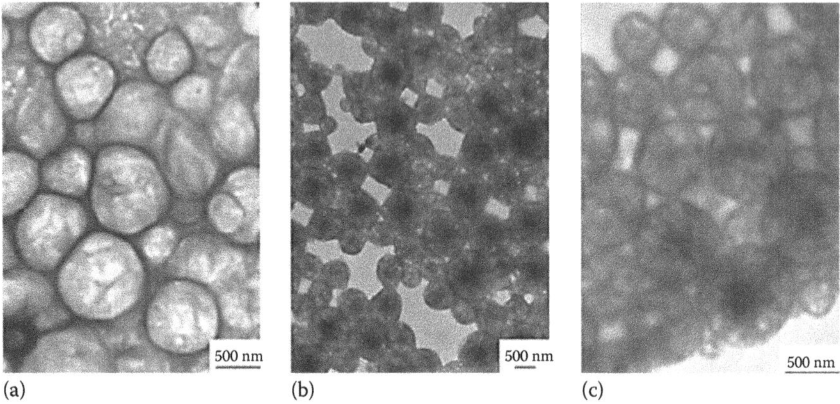

FIGURE 1.3 TEM micrographs of different polymeric nanocapsules based on cross-linked starch. (a) Sample P31 (water as liquid core) and (b, c) sample P24 (formamide as liquid core). (Reprinted with permission from Crespy, D., Stark, M., Hoffmann-Richter, C., Ziener, U., and Landfester, K., *Macromolecules*, 40, 3122–3135, 2007. Copyright 2007, American Chemical Society.)

cross-linking reactions in inverse miniemulsion (Crespy et al. 2007). Cyclohexane was the nonpolar continuous phase and amines or alcohols were used as the dispersed phase. Appropriately chosen hydrophobic diisocyanate or diisothiocyanate monomers, soluble in the continuous phase, were allowed to participate in polycondensation or cross-linking reactions taking place at the interface of the droplets. Variations in the quantities of the reactants facilitated the control of the shell thickness. It was observed that the chemical nature of the monomers and of the continuous phase is important for the formation of the capsules, since it dictates interface formation and properties, as well as location and reactivity of the monomers in respect to the interface. TEM observations led to the conclusion that more spherical nanocapsules are formed when formamide is the liquid core, whereas more elongated nanocapsules were obtained when water was used as the core component (Figure 1.3).

The use of a surface active monomer, namely, polyisobutylene-succinimide pentamine (Lubrizol U), which can act also as the stabilizer, facilitated the synthesis of polyurea nanocapsules with aqueous cores through interfacial miniemulsion polymerization. The presence of amine groups on the macromonomer allowed its covalent incorporation into the PU interfacial layer, leading to the preparation of nanocapsules with less permeable shells (Rosenbauer et al. 2009). Nanocapsules in the size range 250–440 nm were prepared when equal molar ratios of the comonomers (1,6-diaminohexane, Lubrizol U, and toluene-2,4-diisocyanate) were used. Shell thicknesses in the range 15–30 nm were calculated from TEM images and they were found to increase with an increase in the amount of toluene-2,4-diisocyanate in the monomer mixture. Water-soluble fluorescent dyes and magnetite nanoparticles were efficiently encapsulated in these nanocapsules.

Janus nanoparticles, that is, nanoparticles containing two different nanophase-separated polymers as two opposite facets of the same particle, have been prepared by emulsion and miniemulsion polymerizations (Pfau et al. 2002, Kietzke et al. 2007,

Wurm and Kilbinger 2009). One general strategy consists of using the nanoparticles of one polymer as the seeds for the heterogeneous polymerization of the second monomer. Incompatibility between the two chosen polymers is the main factor that drives the formation of Janus morphologies. Other morphologies of composite polymeric nanoparticles can be produced by emulsion polymerization techniques, where phase separation of the polymeric components within the same nanoparticle is observed (Tiarks et al. 2001, Yurko Duda and Flavio Vazquez 2005, Kietzke et al. 2007).

Hairy nanoparticles were prepared by aid of photoemulsion polymerization in a three-step process. PS nanoparticles of narrow size distribution were initially produced by emulsion polymerization. Then a polymerizable photoinitiator was polymerized on the surface of the PS latex forming a thin layer. In this stage, the concentration of the functional monomer was kept low in order to avoid formation of new polymer particles. In a subsequent step, a water-soluble monomer was introduced in the latex suspension and radicals were created by light illumination of the surface layer allowing for the growth of water-soluble chains on the particle's surface. Following analogous strategies, several well-defined spherical polyelectrolyte brushes carrying anionic or cationic groups were synthesized (Ballauff 2007, Xu et al. 2008, Polzer et al. 2011, Wang et al. 2011). Polymeric nanoparticles with grafted glycopolymers were also reported (Pfaff et al. 2011). Polyelectrolyte spherical brushes have been utilized for the immobilization of enzymes acids, as well as in the synthesis of metal nanoparticles/polymer hybrids (Ballauff 2007, Henzler et al. 2010, Polzer et al. 2010, Becker et al. 2011).

Molecularly imprinted nanoparticles (MINs) can be prepared by heterogeneous polymerization methodologies (mini- and microemulsion polymerizations, precipitation polymerization, and others). The general synthetic approach involves the use of the desired template molecule that interacts reversibly with a suitable polymerizable monomer, via (1) reversible covalent bonds, (2) electrostatic interactions, (3) hydrophobic or van der Waals interactions, and (4) covalently attached polymerizable binding groups that can be activated for noncovalent interaction by template cleavage. These interactions are established with complementary functional groups or structural elements of the template. Then the monomer is polymerized in the presence of a cross-linker. In this way, a porous polymeric matrix (the 3D polymeric nanoparticle) is created in which the template molecules are engulfed. Then the template is removed from the matrix polymer/nanoparticle through disruption of the existing interactions between the template and the polymer and is extracted from the matrix (Figure 1.4). Thus, complementary cavities with respect to shape and functional groups remain inside the polymeric nanoparticle. The same template molecule or structurally/functionally analogous molecules can selectively rebind to the polymer into the cavities originally formed by the template molecule (Poma et al. 2010).

The degree of cross-linking and the nature and strength of the template–monomer interactions are some of the factors that influence the preparation protocol to be followed in order to obtain MINs and of course the functionality of the prepared MINs. The high surface area and the high accessibility of the functional cavities in solution promoted the use of MINs as drug delivery systems, enzyme and antibody substitutes, as column material for capillary electrophoresis, as well as in biosensors.

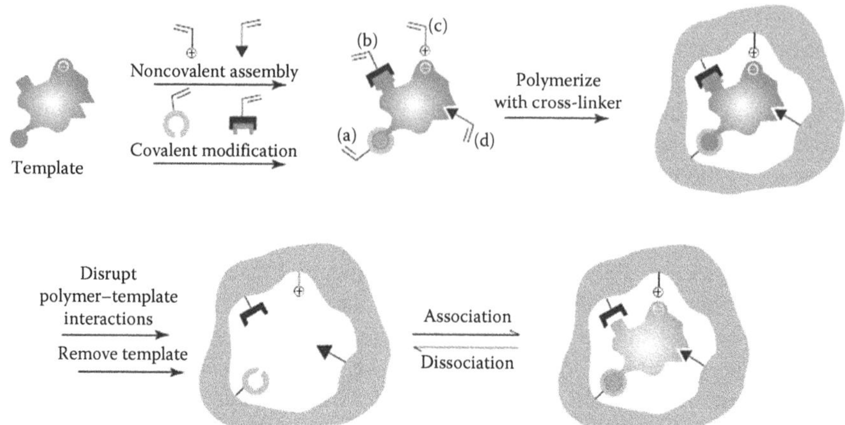

FIGURE 1.4 Scheme depicting the different stages of the molecular imprinting process. (Reprinted from *Trends Biotechnol.*, 28, Poma, A., Turner, A.P.F., and Piletsky, S.A., 629–637, Copyright 2010, with permission from Elsevier.)

In a recent work, inverse microemulsion polymerization was used in order to obtain spherical MINs of ca. 28 nm size imprinted with a small hydrophilic peptide, GFP-9, coupled to fatty acid chains of different lengths (C5, C13, and C15) (Zeng et al. 2010). Polymeric nanoparticles were formed from acrylamide monomer using *N,N′*-ethylene bisacrylamide as the cross-linker and AOT and Brij 30 as surfactants. The aqueous monomer/cross-linker solution was dispersed in hexane and polymerization was initiated at room temperature by ammonium persulfate. The resulting nanoparticles were purified by washing with ethanol and subsequent dialysis, in order to remove unreacted monomers, surfactants, and the peptide template. After nanoparticle purification, it was observed that only nanoparticles imprinted with the peptides coupled to C13 and C15 exhibited specificity and affinity properties. The authors concluded that most probably the C5 chain was too short and the least hydrophobic to correctly act as a surfactant template. For this reason, the C5 chains did not confine the template to the surface of the nanoparticles. Continuing efforts in this field aim at producing novel functional polymeric nanoparticles.

REFERENCES

Adams, N., Schubert, U. S. 2007. *Adv. Drug Deliv. Rev.* 59:1504–1520.

Agut, W., Agnaou, R., Lecommandoux, S., Taton, D. 2008. *Macromol. Rapid Commun.* 29:1147–1155.

Agut, W., Taton, D., Lecommandoux, S. 2007. *Macromolecules* 40:5653–5661.

Alexandridis, P., Lindman, B., Eds. 2000. *Amphiphilic Block Copolymers. Self Assembly and Applications.* Amsterdam, the Netherlands: Elsevier.

Aliferis, T., Iatrou, H., Hadjichristidis, N. 2005. *J. Polym. Sci. Part A: Polym. Chem.* 43:4670–4673.

Andre, X., Zhang, M., Muller, A. 2005. *Macromol. Rapid Commun.* 26:558–563.

Aoshima, S., Hashimoto, K. 2001. *J. Polym. Sci. Part A: Polym. Chem.* 39:746–752.

Babin, J., Taton, D., Brinkmann, M., Lecommandoux, S. 2008. *Macromolecules* 41:1384–1392.

Ballauff, M. 2007. *Prog. Polym. Sci.* 32:1135–1151.
Barz, M., Luxenhofer, R., Zentel, R., Vicent, M. J. 2011. *Polym. Chem.* 2:1900–1916.
Batra, U., Russel, W. B., Pitsikalis, M., Sioula, S., Mays, J. W., Huang, J. S. 1997. *Macromolecules* 30:6120–6126.
Becker, A. L., Welsch, N., Schneider, C., Ballauff, M. 2011. *Biomacromolecules* 12:3936–3944.
Bernaerts, K. V., Willet, N., Van Camp, W., Jerome, R., Du Prez, F. E. 2006. *Macromolecules* 39:3760–3769.
Bi, L.-K., Fetters, L. J. 1976. *Macromolecules* 9:732–742.
Bielawski, C. W., Grubbs, R. H. 2007. *Prog. Polym. Sci.* 32:1–29.
Blanazs, A., Armes, S. P., Ryan, A. J. 2009. *Macromol. Rapid Commun.* 30:267–277.
Borner, H. G., Beers, K., Matyjaszewski, K., Sheiko, S. S., Moller, M. 2001. *Macromolecules* 34:4375–4383.
Braunecker, W. A., Matyjaszewski, K. 2007. *Prog. Polym. Sci.* 32:93–146.
Cai, Y., Armes, S. P. 2005. *Macromolecules* 38:271–279.
Checot, F., Brulet, A., Oberdisse, J., Gnanou, Y., Mondain-Monval, O., Lecommandoux, S. 2005. *Langmuir* 21:4308–4315.
Checot, F., Lecommandoux, S., Klok, H.-A., Gnanou, Y. 2003. *Eur. Phys. J. E* 10:25–32.
Chen, H., Luo, Y. 2011. *Macromol. Chem. Phys.* 212:737–743.
Cheng, G., Boker, A., Zhang, M., Krausch, G., Muller, A. H. E. 2001. *Macromolecules* 34:6883–6888.
Cortez, C., Quinn, J. F., Hao, X., Gudipati, C. S., Stenzel, M. H., Davis, T. P., Caruso, F. 2010. *Langmuir*, 26:9720–9727.
Crespy, D., Stark, M., Hoffmann-Richter, C., Ziener, U., Landfester, K. 2007. *Macromolecules* 40:3122–3135.
Cui, Q., Wu, F., Wang, E. 2011. *Polymer* 52:1755–1765.
De, P., Gondi, S. R., Sumerlin, B. S. 2008. *Biomacromolecules* 9:1064–1070.
Dechy-Cabaret, O., Martin-Vaca, B., Bourissou, D. 2004. *Chem. Rev.* 104:6147–6176.
Deming, T. J. 1997. *Nature* 390:386–389.
Deming, T. J. 2007. *Prog. Polym. Sci.* 32:858–875.
Dimitrov, I., Schlaad, H. 2003. *Chem. Commun.* 2944–2945.
Ekizoglou, N., Hadjichristidis, N. 2002. *J. Polym. Sci. Part A: Polym. Chem.* 40:2166–2170.
Fragouli, P. G., Iatrou, H., Hadjichristidis, N. 2002. *Polymer* 43:7141–7144.
Freischmann, S., Rosen, B. M., Percec, V. 2010. *J. Polym. Sci. Part A: Polym. Chem.* 48:1190–1196.
Gao, G., Yan, Y., Pispas, S., Yao, P. 2010. *Macromol. Biosci.* 10:139–146.
Gaucher, G., Dufresne, M.-H., Sant, V. P., Kang, N., Maysinger, D., Leroux, J.-C. 2005. *J. Control. Release* 109:169–188.
Geng, Y., Discher, D., Justynska, J., Schlaad, H. 2006. *Angew. Chem. Int. Ed.* 45:7578–7581.
Ge, Z., Cai, Y., Yin, J., Zhu, Z., Rao, J., Liu, S. 2007. *Langmuir* 23:1114–1122.
Ge, Z., Luo, S., Liu, S. 2006. *J. Polym. Sci. Part A: Polym. Chem.* 44:1357–1371.
Georgiou, T. K., Phylactou, L. A., Patrickios, C. S. 2006. *Biomacromolecules* 7:3505–3512.
Gillies, E. R., Jonsson, T. B., Frechet, J. M. 2004. *J. Am. Chem. Soc.* 126:11936–11943.
Gohy, J. F., Antoun, S., Jerome, R. 2001a. *Macromolecules* 34:7435–7440.
Gohy, J.-F. 2005. *Adv. Polym. Sci.* 190:65–136.
Gohy, J.-F., Willet, N., Varshney, S., Zhang, J.-X., Jerome, R. 2001b. *Angew. Chem. Int. Ed.* 113:3314–3319.
Gorodyska, G., Minko, S., Tsitsilianis, C., Kiriy, A., Stamm, M. 2003. *Nano Lett.* 3:365–368.
Gou, P.-F., Zhou, W.-P., Shen, Z.-Q. 2010. *J. Polym. Sci. Part A: Polym. Chem.* 48:5643–5651.
Guan, Y., Yongjun Zhang, Y. 2011. *Soft Matter* 7:6375–6383.
Guenoun, P., Davis, H. T., Tirrell, M., Mays, J. W. 1996. *Macromolecules* 29:3965–3969.
Hadjichristidis, N., Iatrou, H., Pispas, S., Pitsikalis, M. 2000. *J. Polym. Sci.: Part A: Polym. Chem.* 38:3211–3234.

Hadjichristidis, N., Iatrou, H., Pitsikalis, M., Pispas, S., Avgeropoulos, A. 2005. *Prog. Polym. Sci.* 30:725–782.
Hadjichristidis, N., Iatrou, H., Pitsikalis, M., Sakellariou, G. 2009. *Chem. Rev.* 109:5528–5578.
Hadjichristidis, N., Pispas, S., Floudas, G. A. 2003. *Block Copolymers: Synthetic Strategies, Physical Properties and Applications.* Hoboken, NJ: John Wiley & Sons, Inc.
Hamley, I. W. 1998. *The Physics of Block Copolymers.* Oxford, U.K.: Oxford University Press.
Hammond, R., Li, C., Tsitsilianis, C., Mezzenga, R. 2009. *Soft Matter* 5:2371–2377.
Harada, A., Kataoka, K. 2006. *Prog. Polym. Sci.* 31:949–982.
Hawker, C. J., Bosman, A. W., Harth, E. 2001. *Chem. Rev.* 101:3661–3688.
Henzler, K., Haupt, B., Lauterbach, K., Wittemann, A., Borisov, O., Ballauff, M. 2010. *J. Am. Chem. Soc.* 132:3159–3163.
Ho, K. M., Li, W. Y., Wong, C. H., Li, P. 2010. *Colloid Polym. Sci.* 288:1503–1523.
Hsieh, H. L., Quirk, R. P. 1996. *Anionic Polymerization: Principles and Practical Applications.* New York: Marcel Dekker, Inc.
Itaka, K., Kataoka, K. 2009. *Eur. J. Pharm. Biopharm.* 71:475–483.
Jeong, J. H., Kim, S. W., Park, T. G. 2007. *Prog. Polym. Sci.* 32:1239–1274.
Jing, R., Wang, G., Zhang, Y., Huang, J. 2011. *Macromolecules* 44:805–810.
Jochum, F. D., Roth, P. J., Kessler, D., Theato, P. 2010. *Biomacromolecules* 11:2432–2439.
Justynska, J., Hordyjewucz, Z., Schlaad, H. 2005. *Polymer* 46:12057–12064.
Kaditi. E., Pispas, S. 2010. *J. Polym. Sci. Part A: Polym. Chem.* 48:24–33.
Karanikolas, A., Tsolakis, P., Bokias, G., Tsitsilianis, C. 2008. *Eur. Phys. J. E* 27:335–343.
Karanikolopoulos, N., Zamurovic, M., Pitsikalis, M., Hadjichristidis, N. 2010. *Biomacromolecules* 11:430–438.
Karatzas, A., Iatrou, H., Hadjichristidis, N., Inoue, K., Sugiyama, K., Hirao, A. 2008. *Biomacromolecules* 9:2072–2080.
Kellum, M. G., Smith, A. E., York, S. K., McCormick, C. L. 2010. *Macromolecules* 43:7033–7040.
Kietzke, T., Neher, D., Kumke, M., Ghazy, O., Ziener, U., Landfester, K. 2007. *Small* 3:1041–1048.
Kim, C., Lee, S. C., Kwon, I. C., Chung, H., Jeong, S. Y. 2002. *Macromolecules* 35:193–200.
Kim, M. S., Dayananda, K., Choi, E. K., Park, H. J., Kim, J. S., Lee, D. S. 2009. *Polymer* 50:2252–2257.
Kjoniksen, A.-L., Zhu, K., Behrens, M. A., Pedersen, J. S., Nystrom, B. 2011. *J. Phys. Chem. B* 115:2125–2139.
Klapper, M., Nenov, S., Haschick, R., Muller, K., Mullern, K. 2008. *Acc. Chem. Res.* 41:1190–1201.
Kobayashi, S., Uyama, H. 2002. *J. Polym. Sci.: Part A: Polym. Chem.* 40:192–209.
Koutalas, G., Pispas, S., Hadjichristidis, N. 2004. *Eur. Phys. J. E* 15:457–464.
Lam, J. K. W., Armes, S. P., Lewis, A. L., Stolnik, S. 2009. *J. Drug Targeting* 17:512–523.
Landfester, K. 2009. *Angew. Chem. Int. Ed.* 48:4488–4507.
Landfester, K., Musyanovych, A., Mailander, V. 2010. *J. Polym. Sci. Part A: Polym. Chem.* 48:493–515.
Li, G., Shi, L., An, Y., Zhang, W., Ma, R. 2006a. *Polymer* 47:4581–4587.
Li, X., Zuo, J., Guo, Y., Yuan, X. 2004. *Macromolecules* 37:10042–10046.
Li, Z., Li, P., Huang, J. 2006b. *Polymer* 47:5791–5798.
Licciardi, M., Crapano, E. F., Giammona, G., Armes, S. P., Tang, Y., Lewis, A. L. 2008. *Macromol. Biosci.* 8:615–626.
Liu, C., Hillmyer, M. A., P. Lodge, T. P. 2009b. *Langmuir* 25:13718–13725.
Liu, H., Li, C., Liu, H., Liu, S. 2009a. *Langmuir* 25:4724–4734.
Liu, H., Xu, H., Jiang, J., Yin, J., Narain, R., Cai, Y., Liu, S.-Y. 2007. *J. Polym. Sci. Part A: Polym. Chem.* 45:1446–1462.
Lord, S. J., Sheiko, S. S., LaRue, I., Lee, H.-I., Matyjaszewski, K. 2004. *Macromolecules* 37:4235–4240.

Lu, H, Cheng, J. 2007. *J. Am. Chem. Soc.* 129:14114–14115.
Lutz, J.-F. 2008. *J. Polym. Sci.: Part A: Polym. Chem.* 46:3459–3470.
Luxenhofer, R., Schulz, A., Roques, C., Li, S., Bronich, T. K., Batrakova, E. V., Jordan, R., Kabanov, A. V. 2010. *Biomaterials* 31:4972–4979.
Luzon, M., Corrales, T., Catalina, F., San Miguel, V., Ballesteros, C., Peinado, C. 2010. *J. Polym. Sci.: Part A: Polym. Chem.* 48:4909–4921.
Mahltig, B, Gohy, J. F., Antoun, S., Jerome, R., Stamm, M. 2002. *Colloid Polym. Sci.* 280:495–502.
Mao, J., Ji, X., Bo, S. 2011. *Macromol. Chem. Phys.* 212:744–752.
Marsat, J. N., Heydenreich, M., Kleinpeter, E., Berlepsch, H., Bottcher, C., Laschewsky, A. 2011. *Macromolecules* 44:2092–2105.
Marsden, H. R., Kros, A. 2009. *Macromol. Biosci.* 9:939–951.
Matter, Y., Enea, R., Casse, O., C. Lee, C. C., Baryza, J., Meier, W. 2011. *Macromol. Chem. Phys.* 212:937–949.
Matyjaszewski, K. Ed. 1996. *Cationic Polymerization: Mechanism, Synthesis and Application.* New York: Marcel Dekker, Inc.
Matyjaszewski, K. Ed. 1998. *Controlled Free Radical Polymerization.* Washington, DC. ACS Symposium Series Vol. 685.
McCormick, C. L., Lowe, A. B. 2004. *Acc. Chem. Res.* 37:312–325.
Min, K., Matyjaszewski, K. 2009. *Cent. Eur. J. Chem.* 7:657–674.
Miyamoto, M., Sawamoto, M., Higashimura, T. 1984. *Macromolecules* 17:265–268.
Moad, G., Rizzardo, E., Thang, S. H. 2005. *Aust. J. Chem.* 58:379–410.
Morton, M. 1983. *Anionic Polymerization: Principles and Practice.* New York: Academic Press.
Mountrichas, G., Pispas, S. 2006. *Macromolecules* 39:4767–4774.
Muller, F., Guenoun, P., Delsanti, M., Deme, B., Auvray, L., Yang, J., Mays, J. W. 2004. *Eur. Phys. J. E* 15:465–472.
Musyanovych, A., Rossmanith, R., Tontsch, C., Landfester, K. 2007. *Langmuir* 23:5367–5376.
Nguyen, A. B., Hadjichristidis, N., Fetters, L. J. 1986. *Macromolecules* 19:768–773.
Nishiyama, N., Kataoka, K. 2006. *Pharmacol. Ther.* 112:630–648.
Nomura, M., Tobita, H., Suzuki, K. 2005. *Adv. Polym. Sci.* 175:1–128.
Nottelet, B., Vert, M., Coudane, J. 2008. *Macromol. Rapid Commun.* 29:743–750.
Oh, J. K., Bencherif, S. A., Matyjaszewski, K. 2009. *Polymer* 50:4407–4423.
Oh, J. K., Dong, H., Zhang, R., Matyjaszewski, K., Schlaad, H. 2007. *J. Polym. Sci. Part A: Polym. Chem.* 45:4764–4775.
Oh, J. K., Tang, C., Gao, H., Tsarevsky, N. V., Matyjaszewski, K. 2006. *J. Am. Chem. Soc.* 128:5578–5587.
Oishi, M, Nagasaki, Y. 2007. *React. Funct. Polym.* 67:1311–1329.
Osada, K., James, C. R., Kataoka, K. 2009. *J. R. Soc. Interface* 6:S325–S339.
Osaka, N., Miyazaki, S., Okabe, S., Endo, H., Sasai, A., Seno, K., Aoshima, S., Shibayama, M. 2007. *J. Chem. Phys.* 127:094905.
Pakula, T., Koynov, K., Boerner, H., Huang, J., Lee, H., Pietrasik, J., Sumerlin, B., Matyjaszewski, K. 2011. *Polymer* 52:2576–2583.
Pang, X., Zhao, L., Akinc, M., Kim, J. K., Lin, Z. 2011. *Macromolecules* 44:3746–3452.
Peng, H., Blakey, I., Dargaville, B., Rasoul, F., Rose, S., Whittaker, A. K. 2009. *Biomacromolecules* 10:374–381.
Perrin, R., Elomaa, M., Jannasch, P. 2009. *Macromolecules* 42:5146–5154.
Pfaff, A., Shinde, V. S., Lu, Y., Wittemann, A., Ballauff, M., Muller, A. H. E. 2011. *Macromol. Biosci.* 11:199–210.
Pfau, A., Sander, R., S. Kirsc, S. 2002. *Langmuir* 18:2880–2887.
Pispas, S. 2006. *J. Polym. Sci. Part A: Polym. Chem.* 44:606–613.
Pispas, S. 2007a. *J. Phys. Chem. B* 111:8351–8359.

Pispas, S. 2007b. *J. Polym. Sci. Part A: Polym. Chem.* 45:509–520.
Pispas, S., Hadjichristidis, N. 2003. *Langmuir* 19:48–54.
Pispas, S., Siakali-Kioulafa, E., Hadjichristidis, N., Mavromoustakos, T. 2002. *Macromol. Chem. Phys.* 203:1317–1327.
Pitsikalis, M., Pispas, S., Hadjichristidis, N., Mays, J. W. 1998. *Adv. Polym. Sci.* 135:1–137.
Polzer, F., Heigl, J., Schneider, C., Ballauff, M., Borisov, O. V. 2011. *Macromolecules* 44:1654–1660.
Polzer, F., Kunz, D. A., Breu, J., Ballauff, M. 2010. *Chem. Mater.* 22:2916–2922.
Poma, A., Turner, A. P. F., Piletsky, S. A. 2010. *Trends Biotechnol.* 28:629–637.
Qin, S., Li, H., Yuan, W., Zhang, Y. 2011. *Polymer* 52:1191–1196.
Rao, J., Zhang, Y., Zhang, J., Liu, S. 2008. *Biomacromolecules* 9:2586–2593.
Rapaport, N. 2007. *Prog. Polym. Sci.* 32:962–990.
Reinicke, S., Schmalz, H. 2011. *Colloid Polym. Sci.* 289:497–512.
Riva, R., Lazzari, W., Billiet, L., Du Prez, F., Jerome, C., Lecomte, P. 2011. *J. Polym. Sci. Part A: Polym. Chem.* 49:1552–1563.
Rosenbauer, E.-M., Landfester, K., Musyanovych, A. 2009. *Langmuir* 25:12084–12091.
Saito, N., Liu, C., Lodge, T. P., Hillmyer, M. A. 2008. *Macromolecules* 41:8815–8822.
Salvage, J. P., Rose, S. F., Phillips, G. J., Hanlon, G. W., Lloyd, A. W., Ma, I. Y., Armes, S. P., Billingham, N. C., Lewis, A. L. 2005. *J. Control. Release* 104:259–270.
Schlaad, H., Diehl, C., Gress, A., Meyer, M., Demirel, A. L., Nur, Y., Bertin, A. 2010. *Macromol. Rapid Commun.* 31: 511–525.
Schork, F. J., Luo, Y., Smulders, W., Russum, J. P., Butti, A., Fontenot, K. 2005. *Adv. Polym. Sci.* 175:129–255.
Shuai, X., He, Y., Na, Y. H., Inoue, Y. 2001. *J. Appl. Polym. Sci.* 80:2600–2608.
Smith, D., Pentzer, E. B., Nguyen, S. T. 2007. *Polym. Rev.* 47:419–459.
Stavrouli, N., Kyriazis, A., Tsitsilianis, C. 2008. *Macromol. Chem. Phys.* 209:2241–2247.
Szczubialka, K., Ishikawa, K., Morishima, Y. 1999. *Langmuir* 15:454–462.
Szwarc, M. 1956. *Nature* 178:1168–1169.
Tang, Y. Q., Liu, S. Y., Armes, S. P., Billingham, N. C. 2003. *Biomacromolecules* 4:1636–1645.
Tauer, K., Zimmermann, A., Schlaad, H. 2002. *Macromol. Chem. Phys.* 203:319–327.
Tian, Z., Wang, M., Zhang, A., Feng, Z. 2008a. *Polymer* 49:446–454.
Tian, Z., Zhang, A., Ye, L., Wang, M., Feng, Z. 2008b. *Biomed. Mater.* 3:0441116.
Tiarks, F., Landfester, K., Antonietti, M. 2001. *Langmuir* 17:908–918.
Tong, R., Cheng, J. 2007. *Polym. Rev.* 47:345–381.
Tsitsilianis, C., Roiter, Y., Katsampas, I., Minko, S. 2008. *Macromolecules* 41:925–934.
Tsitsilianis, C., Stavrouli, N., Bocharova, V., Angelopoulos, S., Kiriy, A., Katsampas, I., Stamm, M. 2008. *Polymer* 49:2996–3006.
Tsitsilianis, C., Vourgaris, D., Stepanek, M., Podhajecka, K., Prochazka, K., Tuzar, Z., Brown, W. 2000. *Langmuir* 16:6868–6876.
Uchman, M., Prochazka, K., Stepanek, M., Mountrichas, G., Pispas, S., Spirkova, M., Walther, A. 2008. *Langmuir* 24:12017–12025.
Uhrig, D., Mays, J. W. 2005. *J. Polym. Sci.: Part A: Polym. Chem.* 43:6179–6222.
Valint, P. L., Bock, J. 1988. *Macromolecules* 21:176–179.
Van Butsele, K., Cajot, S., Van Vlierberghe, S., Dubruel, P., Passirani, C., Benoit, J.-P., Jerome, R., Jerome, C. 2009. *Adv. Funct. Mater.* 19:1416–1425.
Voulgaris, D., Tsitsilianis, C. 2001. *Macromol. Chem. Phys.* 202:3284–3292.
Walther, A., Millard, P.-E., Goldmann, A. S., Lovestead, T. M., Schacher, F., Barner-Kowollik, C., Muller, A. H. E. 2008. *Macromolecules* 41:8608–8619.
Wan, X., Liu, T., Liu, S. 2011. *Biomacromolecules* 12:1146–1154.
Wang, C.-H., Hsiue, G.-H. 2003. *Biomacromolecules* 4:1487–1490.
Wang, G., Hu, B., Huang, J. 2010. *Macromolecules* 43:6939–6942.
Wang, X., Winnik, A., Manners, I. 2005. *Macromolecules* 38:1928–1935.

Wang, X., Wu, S., Li, L., Zhang, R., Zhu, Y., Ballauff, M., Lu, Y., Guo, X. 2011. *Ind. Eng. Chem. Res.* 50:3564–3569.
Wagner, E., Kloeckner, J. 2006. *Adv. Polym. Sci.* 192:135–173.
Webster, O. W. 2004. *Adv. Polym. Sci.* 167:1–34.
Wiradharma, N., Zhang, Y., Venkataraman, S., Hendrick, J. L., Yang, Y. Y. 2009. *Nanotoday* 4:302–317.
Wurm, F., F. M. Kilbinger, A. F. M. 2009. *Angew. Chem. Int. Ed.* 48:8412–8421.
Xiang, L., Shen, L.-J., Long, F., Yang, K., Fan, J.-B., Li, Y.-J., Xiang, J., Zhu, M.-Q. 2011. *Macromol. Chem. Phys.* 212:563–573.
Xu, Y., Bolisetty, S., Drechsler, M., Fang, B., Yuan, J., Ballauff, M., Muller, A. H. E. 2008. *Polymer* 49:3957–3964.
Xu, J., Ge, Z., Zhu, Z., Luo, S., Liu, H., Liu, S. 2006. *Macromolecules* 39:8178–8185.
Yang, C., Tan, J. P. K., Cheng, W., Attia, A. B. E., Ting, C. T. Y., Nelson, A., Hedrick, J. L., Yang, Y.-Y. 2010. *Nanotoday* 5:515–523.
Yang, J., Mays, J. W. 2002a. *Macromolecules* 35:3433–3438.
Yang, J. C., Jablonsky, M. J., Mays, J. W. 2002b. *Polymer* 43:5125–5132.
Yin, J., Ge, Z., Liu, H., Liu, S. 2009. *J. Polym. Sci. Part A: Polym. Chem.* 47:2608–2619.
Yoo, S. H., Park, T. G. 2004. *J. Controll. Release* 96:273–283.
Yurko Duda, Y., Flavio Vazquez, F. 2005. *Langmuir* 21:1096–1102.
Zeng, Z., Hoshino, Y., Rodriguez, A., Yoo, H., Shea, K. J. 2010. *ACS Nano* 4:199–204.
Zhang, L. F., Eisenberg, A. 1996. *J. Am. Chem. Soc.* 118:3168–3181.
Zhang, M., Breiner, T., Mori, H., Muller, A. H. E. 2003. *Polymer* 44:1449–1458.
Zhang, M., Muller, A. H. E. 2005. *J. Polym. Sci. Part A: Polym. Chem.* 43:3461–3481.
Zhang, Y., Liu, H., Dong, H., Li, C., Liu, S. 2009. *J. Polym. Sci. Part A: Polym. Chem.* 47:1636–1650.
Zhao, H., Duong, H. H. P., Yung, L. Y. L. 2010c. *Macromol. Rapid Commun.* 31:1163–1169.
Zhao, J., Mountrichas, G., Zhang, G., Pispas, S. 2009. *Macromolecules* 42:8661–8868.
Zhao, J., Mountrichas, G., Zhang, G., Pispas, S. 2010a. *Macromolecules* 43:1771–1777.
Zhao, J., Zhang, G., Pispas, S. 2010b. *J. Polym. Sci. Part A: Polym. Chem.* 48:2320–2328.
Ziegler, A., Landfester, K., Musyanovych, A. 2009. *Colloid Polym. Sci.* 287:1261–1271.

2 Polymeric Nanoparticles from Pure Block Copolymers

2.1 GENERAL ASPECTS OF COPOLYMER SELF-ASSEMBLY

2.1.1 Complex Phase Behavior of Low-Molecular-Weight Amphiphilic Molecules

Surfactants are naturally occurring or synthetic organic compounds with at least one *lyophilic* group and one *lyophobic* group in the molecules. Due to their amphiphilic nature, these compounds have a high affinity toward surfaces and interfaces. In fact, the term *surfactant* is an abbreviation for *surface active agent.* If the solvent in which the surfactant is to be used is water or an aqueous solution, as in biologic applications, the lyophilic and lyophobic groups of the molecules are termed *hydrophilic* and *hydrophobic*, respectively. In aqueous media, their dual preference for water presents a problem. On one hand, the hydrophobic groups (typically hydrocarbon tails) do not mix well with water and exhibit a strong tendency to minimize the contacts with water. The formation of a pure surfactant phase is also unfavorable, on the other hand, since the hydrophilic groups (typically charged, polar, or hydrogen bonded to water) prefer the contact with water instead of being immersed in a nonpolar medium provided by the hydrophobic portions of the molecules. Two phenomena result from these opposing forces: A small part of the molecules may migrate to and arrange as a monolayer at the air–water or solid–water interface (the process is referred to as *adsorption*), whereas the majority find an alternative way of limiting the unfavorable contacts by aggregating in the bulk solution with the hydrophilic groups oriented toward the aqueous phase. The latter process, referred to as *self-assembly, micellization, micelle formation, or self-association*, can be viewed as an alternative to adsorption. It is a very important phenomenon since the behavior of particular self-assembled structures composed of many surfactant molecules is qualitatively different from that of the unassociated free surfactant molecules called *unimers.*

The surfactant molecules associate into a rich variety of self-assembled structures. At low surfactant concentration, these are discrete and well separated, whereas upon increasing concentration, different liquid-crystalline phases occur. The self-assembled structures vary in shape from spherical to cylindrical to lamellar depending on the conditions: concentration, temperature, pressure, and presence of additives. However, the main shape-determining factor is the molecular parameters of the surfactant molecule described by the so-called *surfactant packing parameter, S*. The concept of a surfactant packing parameter, defined by Equation 2.1, takes into account the molecular geometry in terms of hydrophobic volume, chain length, and head-group area (Israelachvili et al. 1976). It is widely invoked in the literature to explain and predict the optimal aggregate structure in solution:

$$S = \frac{\nu}{a_0 l} \tag{2.1}$$

Here, ν is the volume of the hydrophobic tails, a_0 is the effective head-group area, and l is the length of the fully extended hydrocarbon tail. If N surfactant molecules packed in a sphere with a nonpolar core of radius r are considered, for the total volume and area of the core, one may write, respectively,

$$\frac{4\pi r^3}{3} = N\nu \quad \text{and} \quad 4\pi r^2 = Na_0$$

If no hole is allowed in the center, the distance from the center to the interface, that is, r, cannot exceed l. Thus, the relation for packing of surfactant molecules into spheres can be easily obtained:

$$\frac{\nu}{a_0 l} = \frac{1}{3} \tag{2.2}$$

Analogically for a cylinder and a bilayer, Equations 2.3 and 2.4, respectively, can be derived:

$$\frac{\nu}{a_0 l} = \frac{1}{2} \tag{2.3}$$

$$\frac{\nu}{a_0 l} = 1 \tag{2.4}$$

In other words, the value of the surfactant packing parameter relates the geometry of the molecule to the mean curvature and the shape of the self-assembled structures, respectively. Logically, S gives a value that must not be exceeded for certain geometries. The relationship between the value of S and the optimal aggregate structure is graphically presented in Figure 2.1. Thus, small values of S (e.g., below 1/3), which corresponds to molecules with a relatively large polar head and a small hydrocarbon tail, imply highly curved aggregates such as small spherical micelles. The geometry of molecules with S between 1/3 and 1/2 can be approximated to a truncated cone; such surfactants are expected to form aggregates of a cylindrical or rodlike shape. Upon a further increase of S to a value around unity, corresponding to the cylindrical geometry of the molecule, bilayer or sheetlike structures are formed. Surfactants with $S>1$ form reverse aggregates and phases. By convention,

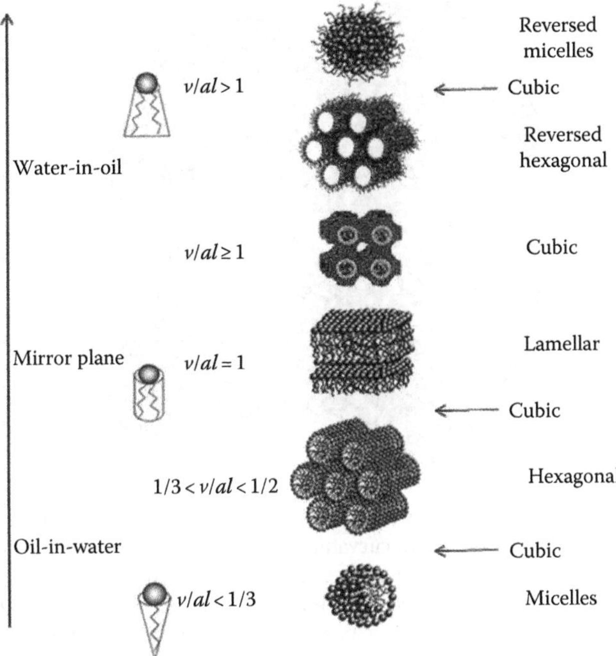

FIGURE 2.1 Critical packing parameter of surfactant molecules and preferred aggregate morphology for geometric packing reasons. (Holmberg, K., Jonsson, B., Kronberg, B., and Lindman, B.: *Surfactants and Polymers in Aqueous Solution*. ISBN: 0-471-49883-1. 2002. Copyright Wiley-VCH Verlag GmbH & Co. KGaA. Reproduced with permission.)

the curvature of the aggregates is positive, and the aggregate is *normal* when the nonpolar domain is in the interior. If the polar part is inside the aggregate, the latter is termed *reverse* and the curvature is negative.

The concept of spontaneous curvature is another recent approach to predict the shape of the structures that are formed upon self-assembly. If the surfactant aggregates are considered to be built up of surfactant films, then the resulting structures depend on the curvature of the films. The curvature, which is the inverse of the radius, decreases, and, accordingly, the radius of the *normal* structures increases with increasing hydrophobic volume and/or decreasing head-group area. Both the normal and reverse micelles are highly curved, and the curvatures are, respectively, positive and negative, whereas the planar films (e.g., a lamellar phase) have zero curvature and infinite radii. There are surfactants that form the so-called *bicontinuous* structures. These are characterized by a complex *saddle-shape* geometry with curvatures of opposite signs, but the mean curvature is zero. Such a structure is shown in Figure 2.2.

If the concentration of the surfactant is sufficiently high, the discrete aggregates can arrange into particular arrays, thus forming condensed liquid-crystalline phases. For instance, the spherical micelles are organized to create cubic ordering, whereas the rodlike micelles assemble into a hexagonal array. An idealized

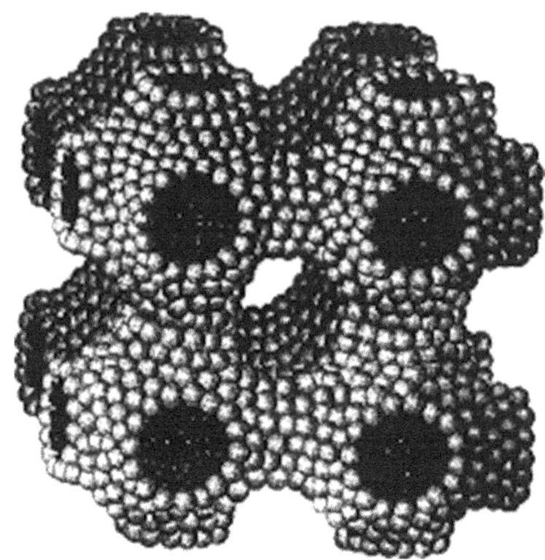

FIGURE 2.2 A bicontinuous structure with the surfactant molecules aggregated into interconnected films characterized by two curvatures of opposite sign resulting in small (zero for a minimal surface structure) mean curvature. (Holmberg, K., Jonsson, B., Kronberg, B., and Lindman, B.: *Surfactants and Polymers in Aqueous Solution*. ISBN: 0-471-49883-1. 2002. Copyright Wiley-VCH Verlag GmbH & Co. KGaA. Reproduced with permission.)

sequence of self-assembled structures and liquid-crystalline phases as a function of water content is presented in Figure 2.3. This figure also illustrates the dependence of the structures on composition, that is, the surfactant packing parameter. The symmetry of the curvature of the aggregates around the lamellar phase is noteworthy as well. With the exception of the lamellar phase, each structure has a reverse counterpart in which the polar and nonpolar parts change their roles and positions in the self-assembled structures. The phases indicated in Figure 2.3 are the most commonly occurring ones. More exotic and less important phases have been observed as well.

2.1.2 Driving Forces, Critical Concentrations, Thermodynamics, and Kinetics of Self-Assembly of Low-Molecular-Weight Surfactants

In aqueous solution at low concentration, the amphiphilic molecules are freely suspended as unimers. The water molecules arrange around the hydrophobic portion to form a quasicrystalline surface. Those that are involved in the formation of the "cage" or solvation shell have substantially reduced mobility. Above a certain concentration, referred to as critical micellization concentration (CMC) or, more generally, critical aggregation concentration (CAC), to minimize the unfavorable interactions, the amphiphilic molecules spontaneously associate, thus driving the hydrophobic parts "out" of the aqueous solution and sequestering them within the interior of the aggregate. As a result, the water molecules ordered around the hydrophobic portions are released, which generates a favorable entropy gain. Figure 2.4 illustrates the hydrophobic effect. When released into the bulk, the water

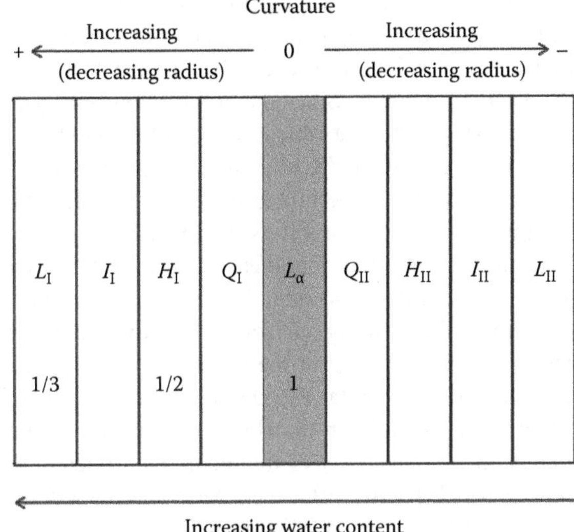

FIGURE 2.3 Sequence of liquid-crystalline phases as a function of water content and surfactant packing parameter. *L*, *I*, *H*, and *Q* are the notations for the micellar, cubic micellar, hexagonal, and bicontinuous cubic phases, respectively. The subscripts "I" and "II" denote normal and reverse phases, respectively. The mirror plane (symmetry of curvature around the lamellar phase, L_α) is presented in gray.

molecules can reform all possible hydrogen bonds, thereby generating a negative but very small enthalpic contribution. The contribution of entropy to the free energy change (Equation 2.5), however, is overwhelming. Therefore, the self-assembly in aqueous solutions is typically an entropy-driven process:

$$\Delta G^0 = \Delta H^0 - T\Delta S^0 \tag{2.5}$$

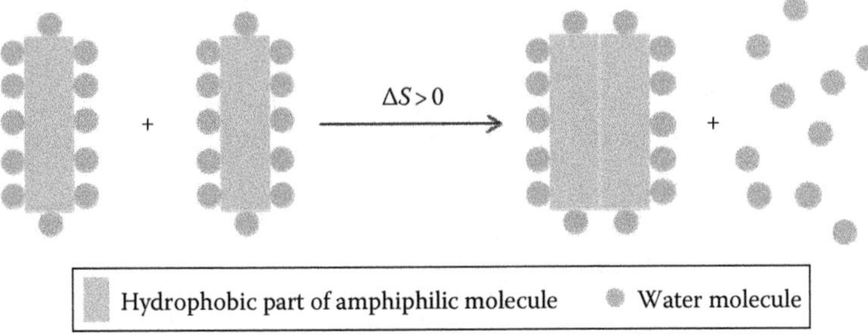

FIGURE 2.4 Simplified schematic representation of the hydrophobic effect and molecular self-assembly. Entropy gain originates from the release of the water molecules from the quasicrystalline surfaces (cages) formed around the hydrophobic parts by hydrogen bonding into bulk solution.

The CAC, defined as the surfactant concentration above which aggregates are spontaneously formed, represents a fundamental parameter describing the self-assembly of amphiphilic molecules in solution. The CAC is unique for each surfactant and indicates the onset of self-association. Since the first-formed aggregates are generally spherical micelles, the concentration at which the latter begin to form is also referred to as CMC. In the following, CAC and CMC are used as synonyms. The CMC is a physical characteristic that can be ascertained in many ways. Basically, each physical property that changes abruptly upon self-assembly and shows discontinuity when plotted against concentration can serve for determining the CMC. Figure 2.5 shows a schematic representation of the concentration dependence of a certain physical property for solutions of a micelle-forming surfactant. Ideally, the different techniques used for CMC determination should give the same results. In practice, however, it is not always the case, especially as far as polymers are concerned. Their inherent polydispersity may broaden the unimer–micelle transition, so that the exact value of CMC largely depends on the sensitivity of the technique to detect the formation or the presence of aggregates. Typically, after the CMC is surpassed, the concentration of micelles increases in a linear fashion (Hiemenz 1986).

The Gibbs energy of micellization can be derived using several approaches. The most common methods are based on a phase-separation model and a mass action model. In the former model, the micellization can be considered to have some features in common with the formation of a separate liquid phase. For ideal phases, the state of equilibrium in such a complex system can be defined via the chemical potential:

$$\mu_i = \mu_i^0 + RT \ln X_i \tag{2.6}$$

Here, μ_i is the chemical potential of a defined component in the system, μ_i^0 is the standard chemical potential for an isolated molecule of type i, R is the gas constant,

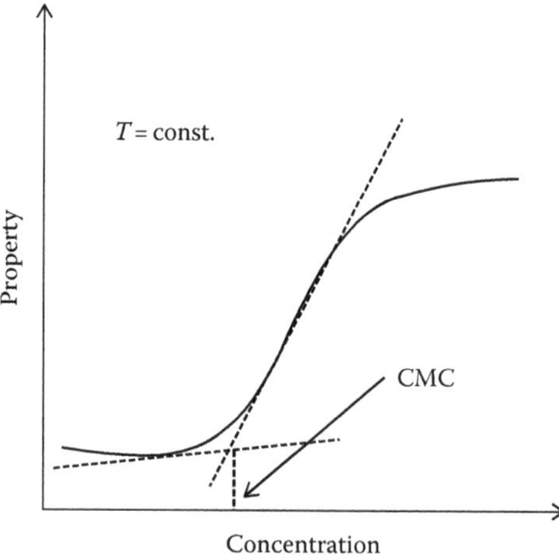

FIGURE 2.5 Concentration dependence of a certain physicochemical property of a surfactant in solution and determination of the CMC.

T is the absolute temperature, and X_i is the mole fraction. Accordingly, the chemical potential of the dissolved surfactant at low concentrations can be written as

$$\mu_{sur}(\text{solvent}) = \mu_{sur}^{0} + RT \ln X_{sur} \tag{2.7}$$

At concentrations equal to the CMC, the chemical potentials of the surfactant in the micelles and dissolved surfactant are equal, that is, $\mu_{sur}(\text{micelle}) = \mu_{sur}(\text{solvent})$, which leads to

$$\mu_{sur}(\text{micelle}) = \mu_{sur}^{0} + RT \ln \text{CMC} \tag{2.8}$$

The free energy of micellization directly measures the free energy difference per mole between surfactant molecules in micelles and in water:

$$\Delta G_{m} = \mu_{sur}(\text{micelle}) - \mu_{sur}^{0} = RT \ln \text{CMC} \tag{2.9}$$

Equation 2.9 can be used to calculate the standard free energy of micellization for nonionic surfactants. For ionic surfactants, however, the change of dissociation of charges from the head groups affects the results.

Applying the mass action model, which assumes a dissociation–association equilibrium between the unimers and micelles, requires determining an equilibrium constant. The equilibrium for a nonionic surfactant, that is, in the absence of charge effects, can be simply presented by Equation 2.10 and the equilibrium constant of micellization by Equation 2.11:

$$nS \overset{K_m}{\rightleftarrows} Sn \tag{2.10}$$

$$K_m = \frac{[Sn]}{[S]^n} \tag{2.11}$$

At equilibrium,

$$\Delta G^0 = -RT \ln K_m = -RT \ln[Sn] + nRT \ln[S] \tag{2.12}$$

which can be rewritten to give the free energy for transferring single surfactant molecule from an aqueous solution to a micelle, ΔG_m^0:

$$\frac{\Delta G^0}{n} = \Delta G_m^0 = -\frac{RT}{n} \ln[Sn] + RT \ln[S] \tag{2.13}$$

When n is large enough, for example, $n > 50$, the first term on the right-hand side of Equation 2.13 approaches zero and can be neglected, which results in

$$\Delta G_{\mathrm{m}}^{0} = RT\ln[S] = RT\ln \mathrm{CMC} \tag{2.14}$$

Equation 2.14 is identical to Equation 2.9, which is derived by applying the phase-separation model.

In the case of ionic surfactants, the micelles attract a substantial portion of counterions that form an attached layer around the micelles. For many ionic surfactants, the following equation has been derived:

$$\Delta G_{\mathrm{m}}^{0} = 1.8\,RT\ln \mathrm{CMC} \tag{2.15}$$

Comparing this with Equation 2.14 immediately shows that for surfactants with similar hydrophobic residues and CMCs, $\Delta G_{\mathrm{m}}^{0}$ of the ionic surfactant is 1.8 times more negative than that of the nonionic one. This is also illustrated in Figure 2.6 for a series of hypothetical surfactant couples. On the other hand, for similar $\Delta G_{\mathrm{m}}^{0}$ values, the CMCs of the ionic surfactants are invariably higher.

The kinetics of micellization of surfactants in dilute aqueous solutions has been studied extensively mainly in the 1970s and 1980s. It is a general consensus that micelles are formed in a stepwise process. The elementary step is the equilibrium between a unimer and a micellar aggregate:

$$A_{n-1} + A \underset{K_{n,n-1}}{\overset{K_{n-1,n}}{\leftrightarrow}} A_n \tag{2.16}$$

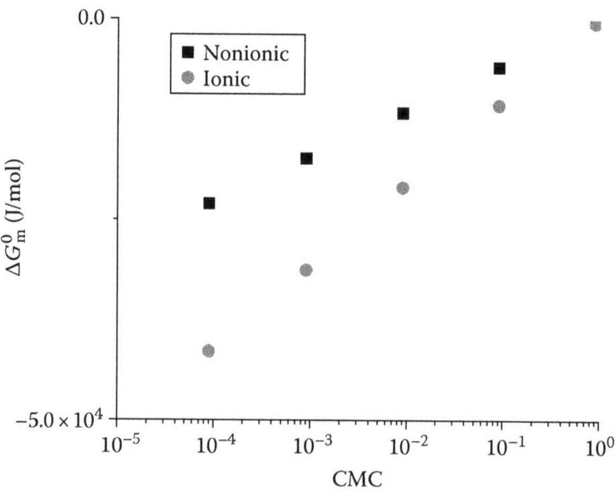

FIGURE 2.6 CMC variations of $\Delta G_{\mathrm{m}}^{0}$ for hypothetical couples of ionic–nonionic surfactants with similar hydrophobic residues and CMCs.

The rate constant of association, $k_{n-1,n}$, is diffusion controlled and slightly dependent on the surfactant type and micellar size, whereas the dissociation rate constant, $k_{n,n-1}$, strongly depends on the hydrophobic residue, for example, alkyl chain length and micellar size. For nonionic surfactants, the theory (Aniansson and Wall 1974, 1975, Kahlwelt and Teubner 1980) predicts two relaxation processes: a fast process arising from a rapid rearrangement of the size distribution and a slower one that arises from the formation of new micelles. Accordingly, from fast kinetic measurements, two relaxation times characterizing the molecular processes in micellar solutions have been determined. The time constant of the fast process, τ_1, measures the rate at which surfactant molecules exchange between micelles. The theory predicts a linear increase of τ_1^{-1} with the total surfactant concentration according to the following equation, whereas the experimentally found values of τ_1 are of the order of μ_s:

$$\tau_1^{-1} = -k_{n,n-1}(n-1) + \frac{nk_{n,n-1}a}{[A]} \tag{2.17}$$

where

$a = [A] + n\,[A_n]$ is the total surfactant concentration
$[A]$ is the unimer concentration at equilibrium, that is, CMC
n is the aggregation number
$[A_n]$ is the micelle concentration

The slow process is characterized by the time constant τ_2, which measures the rate at which micelles form and disintegrate. Typically, τ_2 is longer than a millisecond and also dependent on concentration, though the experimentally found concentration dependence (considerably) disagrees with theoretical prediction (Hermann and Kahlewelt 1980, Kahlewelt and Teubner 1980):

$$\tau_2^{-1} = M\left(\frac{b_n m^2}{d}\right) X^{\nu} \frac{\left[1 + (p/m^2)X^{-1}\right]}{\left[1 + (\sigma^2/p)X\right]} \tag{2.18}$$

Here, M denotes a proportionality constant, b_n the mean dissociation rate constant, d the effective width of the dissociation rate determining region, p the mean aggregation number, σ the width of micelle size distribution, and m^2 is defined as $(p^2+\sigma^2)$, whereas ν is the ratio between the aggregation number of the nucleus, n, and the mean aggregation number, p. X is the ratio between the number of unimers incorporated into micelles and that of free unimers, that is, $(N - N_\gamma)\,N_\gamma$, where N denotes the total number density of surfactant molecules and N_γ that of free surfactant molecules. At equilibrium N_γ corresponds to the number density of unimers at the CMC.

As seen from Equation 2.18, the slow relaxation process is critically dependent on the micellar size distribution. Therefore, as noted elsewhere (Holmberg et al. 2002), kinetic measurements can be used to determine the standard deviation of the distribution.

It is noteworthy that both relaxation processes slow down with increasing surfactant alkyl chain length. Kinetic studies of surfactant micellar solutions should lead to a better understanding of their properties.

2.1.3 Generalities, Common Features, and Differences in the Self-Assembly of Amphiphilic Copolymers

2.1.3.1 Generalities

Amphiphilic block, graft, starlike, and telechelic copolymers are composed of at least one solvophilic moiety and at least one solvophobic moiety. In the following, we will consider mainly aqueous solutions, for most of the biomedical applications of polymeric and hybrid particles (Chapter 5) are in water-based media. Accordingly, *hydrophilic* and *hydrophobic* instead of *solvophilic* and *solvophobic*, respectively, will be used.

In a selective solvent, the amphiphilic copolymers self-assemble in a much similar fashion as surfactants and lipids. The geometry of the self-assembled structures is influenced by the difference in the solubility of the constituent moieties and by the constraint imposed by the chemical linkage between them. The canonical sequence of the self-assembled structures, that is,

> *spherical micelles – cylindrical or wormlike micelles – planar or disklike micelles and vesicles – reverse aggregates,*

is generally dictated by the proportions of the constituent moieties described by the surfactant packing parameter, *S*, defined earlier. The latter determines the curvature of the hydrophilic–hydrophobic interface. The Gaussian curvature, *K*, and the mean curvature, *H*, defined by the two principal radii of the interfacial curvature via Equations 2.19 and 2.20, respectively, have been introduced to describe the particle 3D geometry (Figure 2.7). The two curvatures are linked together via *S* (Equation 2.21) as shown elsewhere (Hyde 1990). A distinctive feature of the copolymers is that the parameters determining *S* are sensitive to the properties of the surrounding media. Thus, changes in the environmental conditions can cause changes in any of ν, a_0, or l and, consequently, in the surfactant packing parameter and interfacial curvature, which may lead to gradual or sharp and sudden changes in the dimensions

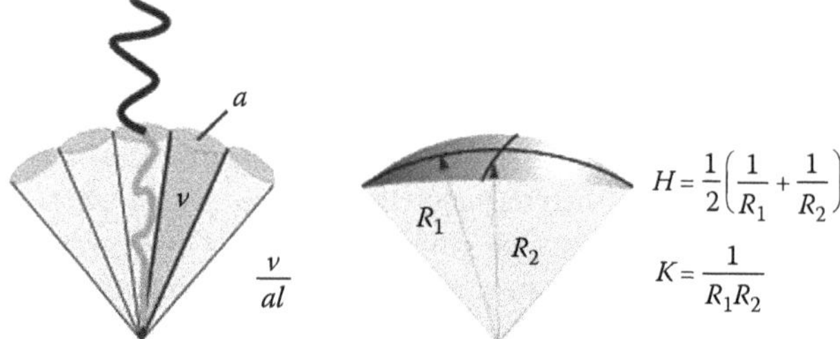

FIGURE 2.7 Description of the amphiphile shape in terms of surfactant packing parameter and its relations to the interfacial mean curvature, *H*, and Gaussian curvature, *K*. (Antonietti, M. and Forster, S.: Vesicles and liposomes: A self-assembly principle beyond lipids. *Adv. Mater.* 2003. 15. 1323–1333. Copyright Wiley-VCH Verlag GmbH & Co. KGaA. Reproduced with permission.)

of the self-assembled structures or even their complete reorganization and transition to other morphology:

$$K = \frac{1}{R_1 R_2} \tag{2.19}$$

$$H = \frac{1}{2}\left(\frac{1}{R_1} + \frac{1}{R_2}\right) \tag{2.20}$$

$$S = 1 - Hl + \frac{Kl^2}{3} \tag{2.21}$$

For polymers, however, it is more convenient to characterize the preferred geometry by the hydrophilic fraction, f, which is synthetically more accessible than S. An excellent correlation between S and f has been found by Discher et al. (2007) on the basis of numerous results from both experiment and simulation:

$$f \approx e^{-S/\beta} \tag{2.22}$$

where $\beta = 0.66$. The type of the structure results from the inherent molecular curvature, which is governed by the surfactant packing parameter or the corresponding hydrophilic fraction. The relations between the geometry of the self-assembled structures, molecular curvature, S, and f are presented in Figure 2.8.

The polymer chain entropy and entropy loss during self-assembly can influence significantly the resultant structures at the equilibrium. Therefore, the pure geometric considerations are frequently not adequate to describe the self-assembly of amphiphilic macromolecules. The obtained morphology results also from the minimization of the free energy involving the interfacial hydrophilic–hydrophobic energy and the loss of entropy of the polymer chains. Thus, for instance, for stiff polymer chains, characterized by low conformation entropy, the minimization of the interfacial area is dominant due to the low entropy loss during the self-association and, therefore, the composition range of stability of, for example, planar structures formed by copolymers with shape-persistent hydrophobic block (such as rigid rods or blocks developing H-bonding or Coulombic or π–π interactions) will be extended beyond the limits set by S or f (see more in Section 2.2.3). In addition, factors such as the inherent polydispersity of the polymers, metastability of the structures, and slow kinetics associated with the large molar masses can also contribute to the extending of the composition range of stability of certain morphology and to coexistence of different structures.

2.1.3.2 Critical Concentrations and Temperatures

Like surfactants, amphiphilic copolymers form micelles/aggregates above certain critical concentration via the so-called closed association mechanism. Below CMC,

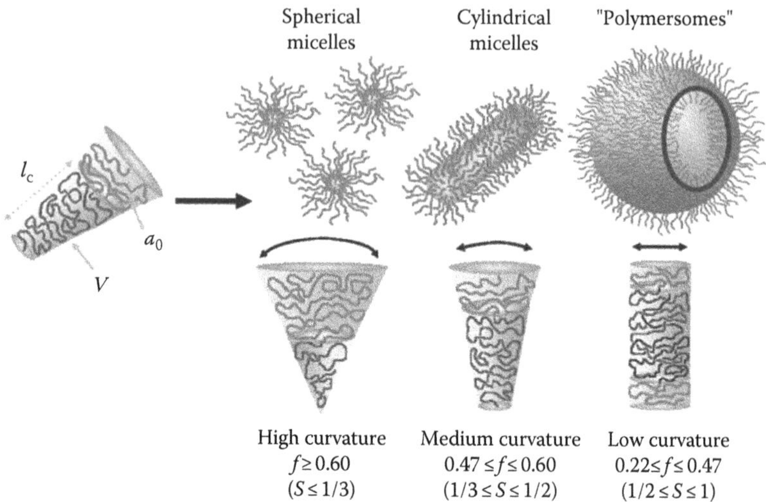

FIGURE 2.8 Interrelations between the self-assembled structures formed by amphiphilic copolymers in a selective solvent (water) and their solvophilic (hydrophilic) fractions, *f*. The values of the corresponding surfactant packing parameter, *S*, are given in parentheses. (Blanazs, A., Armes, S.P., and Ryan, A.J.: Self-assembled block copolymer aggregates: From micelles to vesicles and their biological applications. *Macromol. Rapid Commun.* 2009a. 30. 267–77. Copyright Wiley-VCH Verlag GmbH & Co. KGaA. Reproduced with permission.)

only unimers exist, whereas above CMC, the multimolecular aggregates are in equilibrium with the unimers. In water, the hydrophobic block or moieties are densely packed (but not necessarily free of water) in the interior of the aggregates, whereas the hydrophilic ones are highly swollen and exposed toward the outer parts of the structures, thus forming a shell or corona around the hydrophobic domains (Figure 2.8). Typically, the CMCs are located in the low micromolar region and even lower, so that sometimes it is not possible to be determined even by highly sensitive techniques. These extremely low CMCs indicate that some copolymers cannot be dissolved as unimers; they exist only as aggregates since only traces are required to self-associate. This could be advantageous for many applications, since high dilution is typically problematic for the classical surfactants. Micelle formation can also be induced at a fixed concentration by manipulation of the solvent quality via changing environmental parameters such as temperature, pH, and ionic strength. Thus, copolymers that contain moieties showing an increased hydrophobicity at elevated temperatures self-associate above certain temperature as referred to as critical micellization temperature (CMT). The most extensively studied copolymers that exhibit temperature-induced micellization are the commercially available poly(ethylene oxide)–poly(propylene oxide)–poly(ethylene oxide) (PEO-PPO-PEO) block copolymers as well as copolymers based on poly(*N*-isopropylacrylamide) (PNIPAM). Strictly speaking, the term *micelle* refers to the equilibrium structures; these are stable in a significant range of environmental conditions and are characterized with constant aggregation number, size, and shape. However, the structures that are formed at temperatures below T_g of the core-forming block are definitely not equilibrium structures, and, therefore, they should be called *micelle-like aggregates*. Nevertheless, the term *micelle* has been extensively (but erroneously) used in the literature to describe also *micelle-like aggregates*. Therefore, in the following, we will use *micelle* and *aggregate* interchangeably; the term *particle* will be used when more general situation is described.

Basically, the experimental techniques for CMC determination of low molecular weight surfactants are also applicable for copolymer micellar systems. The precise location of the CMCs, however, can be hindered due to the very low CMC values, so that some techniques might not be sensitive enough to detect aggregates. Furthermore, due to the slow kinetics of the polymer chains, equilibrium structures are not attainable at all or attained after a long time period. The composition polydispersity, which could be appreciable even for copolymers of low molar mass distribution, in addition broadens the unimer-to-micelle transition, so that no sharp CMC (or CMT) could be observed, which also complicates the measurements.

2.1.3.3 Aggregate Evolution and Micellization Kinetics

The evolution of the copolymer micelles has been typically considered to be analogous to that of the surfactant micelles (Bromberg 2001). Accordingly, two relaxation processes are identified: a fast process related with the exchange of unimers between micelles and a slow one assigned to micelle formation and dissociation. Despite the similarities, the kinetics and mechanisms of copolymer self-assembly are more complicated than those of the conventional surfactants. Kinetic and mechanistic studies are relatively scarce, and information has been obtained by fast reaction techniques such as fluorescence nonradiative energy transfer (Liu 1995, Wang et al. 1995, Rager et al. 1997), temperature-jump measurements (Hecht and Hoffmann 1995, Honda et al. 1996), time-resolved light scattering (Honda et al. 1994, Michels et al. 1997), and pulsed-field gradient NMR (Fleischer 1993). Despite the increasing number of reports during the last decade (Gohy 2005, Zana 2005, Zhu et al. 2005, Lund et al. 2006, 2009, Rao et al. 2008), a complete understanding of the micellization kinetics of copolymers is still lacking. The component exchange in polymeric micelles is expected to be extremely slow due to the low CMCs, which practically means that no free chains are present in the solution. Furthermore, most of the kinetic studies involve systems of amphiphilic block copolymers with glassy micellar cores. In such systems, the slow dynamics has been associated with the vitrified nature of the core-forming block, and even for relatively low molar mass copolymers, no exchange of unimers has been observed (Riess and Hurtrez 1996, Weber 1996). However, Creutz et al. estimated the exchange rate constants for copolymers composed of poly(sodium methacrylate) as the water-soluble block and poly((dimethylamino)ethyl methacrylate) ($T_g = -6°C$) as the hydrophobic block on the order of 10^{-3} s^{-1}, which, compared to those typical for the conventional surfactants, that is, 10^{6}–10^{8} s^{-1}, makes 10^{-9}–10^{-11} times slower exchange of chains than in surfactant micelles (Creutz et al. 1997). It is remarkable that for micelles with liquid cores consisting of low T_g polymers such as polyisoprene or polybutadiene (PB), the structures initially formed are locked in or exhibit immeasurably slow rates of intermicellar chain exchange (Yu et al. 1997, Schillen et al. 1998, Won et al. 2003, Jain and Bates 2004). It appears that it is not the T_g

nor even the coil dimensions but the high amphiphilicity resulting in strong segregation and incompatibility of the core from the solvent that governs the micellar dynamics (Schillen et al. 1998, Rager et al. 1999, Won et al. 2003, Jain and Bates 2004). Indeed, the closer inspection of the systems for which micelle dynamics is limited to an experimentally favorable and measurable range of times (seconds to hours) reveals that in nearly most of the cases, the copolymer/solvent pairs offer marginal amphiphilicity.

Some theoretical approaches (Halperin and Alexander 1989, Nyrkova and Semenov 2005), developed to predict the micelle growth mechanism for diblock copolymers in a selective solvent, favor a mechanism, in which micelles grow by a stepwise insertion of unimers into micelles (unimer exchange mechanism). Other models (Lessner et al. 1981, Kahlweit 1982) favor the fission/fusion mechanism, according to which the growth is mainly due to fragmentation and recombination of micelles. A joint analytical model combining these two models has been proposed (Dormidontova 1999, Li and Dormidontova 2010), according to which the micellization is a three-stage process involving (1) rapid coupling of unimers into small aggregates (submicelles), (2) fusion of submicelles resulting in a broad (bimodal) aggregation number distribution, and (3) slow adjustment of the aggregation number distribution to the equilibrium one by unimer exchange and fission/fusion of non-equilibrium in size micelles. According to the simulation results, the contribution of the micelle fusion is substantial, which is somewhat contradictory to the mechanism for surfactant micelles (Aniansson et al. 1976).

2.1.3.4 Computer Simulation and Modeling, Scaling and Mean-Field Theories, and Mathematical Approaches to Copolymer Self-Assembly

Various theories, models, and mathematical approaches have been developed to derive relations between the macromolecular parameters such as degrees of polymerization of the constituent blocks, total degree of polymerization (DP), the Flory–Huggins parameters of the interaction between the blocks and between the blocks and the solvent, and the structural parameters of the aggregates (CMC, aggregation number, radius of the core, corona thickness, hydrodynamic radius). In some of the approaches, the minimization of the total Gibbs free energy governed by the balance of contributions from the chain stretching in the core, the interfacial energy, and the repulsion between the coronal chains, with respect to aggregate parameters, is used to derive relations between the copolymer and micellar characteristics. Other approaches are based on the scaling concept of Alexander–de Gennes and on the mean-field theories (Leibler et al. 1983, Noolandi and Hong 1983, Nagarajan and Ganesh 1989, Hurter et al. 1993). A concise overview on the thermodynamic background and the different approaches is presented by Riess (2003) and Borisov et al. (2011) and earlier by Tuzar and Kratochvil (1993), Hamley (1998), and Linse (2000).

Linse has considered two limiting cases examined for a monodispersed *AB* diblock copolymer in a selective for block *A* solvent (Linse 2000): *hairy micelles* for which $N_A >> N_B$ and, accordingly, $L > R_c$ and *crew-cut micelles* with $N_B >> N_A$

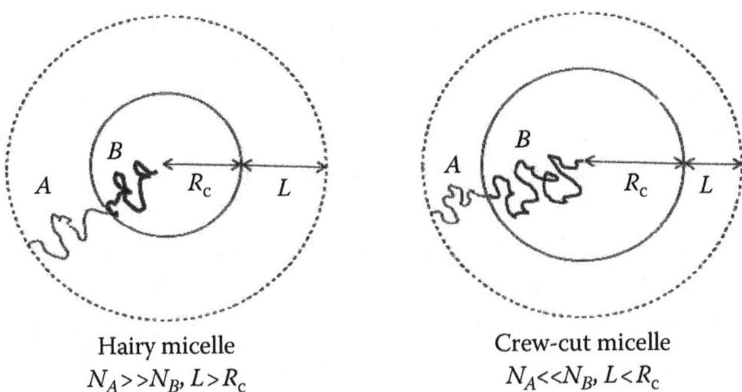

FIGURE 2.9 Schematic representation of hairy and crew-cut micelles formed by an AB diblock copolymer in a selective for block A solvent. N_A, N_B, R_c, and L correspond to the degrees of polymerization of blocks A and B, core radius, and corona thickness, respectively.

and $R_c > L$, where N_A, N_B, R_c, and L denote the degrees of polymerization of blocks A and B, core radius, and corona thickness, respectively (Figure 2.9). For the hairy micelles, based on the star polymer theory of Daoud and Cotton (1982), scaling relations for the aggregation number, N_{agg}, and L have been derived:

$$N_{agg} \sim N_B^{4/5} \tag{2.23}$$

$$L \sim N_{agg}^{1/5} N_A^{3/5} \tag{2.24}$$

A similar result for N_{agg} has been obtained by Zhulina and Birshtein (1986) and Halperin (1989). The scaling for R_c (scales as $N_B^{3/5}$) and L ($\sim N_A^{3/5} N_B^{6/25}$) is in agreement with the relations of Daoud and Cotton. For the total micellar radius, R, Halperin (1987) has obtained the following relation:

$$R \sim N_A^{3/5} N_B^{4/25} \tag{2.25}$$

For deriving scaling relations for the crew-cut micelles (Equations 2.26 and 2.27), the theories for the polymer brush of Alexander and de Gennes have been used assuming micelles with an aggregation number N_{agg} and uniformly stretched chains in the core with radius R_c:

$$R_c \sim \gamma^{1/3} N_B^{2/3} a \tag{2.26}$$

$$N_{agg} \sim \gamma N_B \tag{2.27}$$

Here, γ is the interfacial tension between the blocks A and B, and a is the segment length.

The scaling models, however, are not able to predict finite chain effects and polymer–solvent interactions and to assess numerical values of the micellar characteristics, since they are restricted to long polymer chains in good solvents and only predict the trends of variation of the micellar parameters with the molecular ones. More detailed mean-field calculations and computer simulations complement the scaling models.

With respect to the semianalytical mean-field models, Noolandi and Hong (1983) and Leibler et al. (1983) derived the micellar characteristics by minimizing the Gibbs energy of, respectively, isolated micelles and the whole micellar system. By considering the molar volumes of the solvent (S) and blocks A and B and knowing the interfacial tension, γ_{BS}, and the interaction parameter, χ_{AS}, Nagarajan and Ganesh (1989) derived scaling equations for PEO-PPO block copolymers in water (Equations 2.28 through 2.30) from which the strong influence of the coronal block on the micellar characteristics is seen:

$$R_c \sim N_A^{-0.17} N_B^{0.73} \tag{2.28}$$

$$L \sim N_A^{0.74} N_B^{0.06} \tag{2.29}$$

$$N_{agg} \sim N_A^{-0.51} N_B^{1.19} \tag{2.30}$$

The lattice self-consistent mean-field theory has been used to calculate the effects of the copolymer architecture on the association behavior of nonionic and charged copolymers (Linse 2000, Monzen et al. 2000, Shusharina et al. 2000).

Computer simulations have the advantage to proceed without pre-assumptions of micelle geometry and chain conformation and make it possible to vary intermolecular forces. Various aspects of the self-assembly and basic characteristics of micelles of copolymers of different, for example, AB, ABA, BAB, and starlike A_2B_2, architectures have been established and compared (Linse 2000, Kim and Jo 2001, Kim et al. 2002). Chain-length dependence (Milchev et al. 2001), dynamics of chain exchange (Haliloglu and Mattice 1996), formation of surface micelles (Milchev and Binder 1999), solubilization of low molecular weight substances (Hurter et al. 1995, Chen and Shew 2001), chain density profiles, and thickness of the interfacial region (Linse 2000) have been considered by computer simulations as well.

Theoretical and computational research regarding the self-assembly and morphology of amphiphilic copolymers of more complex architecture in dilute solution has been carried out as well. On the basis of the scaling theory, Borisov et al. predicted that the aggregate morphology of graft copolymers with hydrophobic backbone and hydrophilic graft chains varies from pearl-necklace micelles for dense grafted copolymers to spherical and cylindrical micelles in the case of spare grafting (Borisov and Zhulina 2005b). Using a Brownian dynamics simulation approach, Kim et al. investigated various star-block, Π-shaped, and triblock copolymers and found qualitative agreement with the predictions formulated by the mean-field theory, in particular, that the CMC increases in the order of star-block < triblock < regular Π-shaped copolymers and that the micellar formation of the latter is more difficult than that of the others (Kim et al. 2003). The microstructures formed by graft copolymers in solvents that are poor for the backbone and good for the graft chains and vice versa have been investigated using a 2D real-space algorithm of the self-consistent field theory (Zhang et al. 2007b). The 2D calculations have been shown able to catch the effects of architecture parameters on the self-assembly behavior of graft copolymers, the order of morphological transitions, the structure of aggregates, and the density distributions of the free end and inner blocks of the backbone.

Statistical thermodynamic theories, complemented by a numerical self-consistent approach, which describe the self-assembly in dilute solution of amphiphilic ionic/hydrophobic diblock copolymers with both strongly and weakly dissociating ionic blocks, have been recently overviewed by Borisov et al. (2011). The responsive behavior of nanoaggregates formed by copolymers with pH-sensitive polyelectrolyte blocks has been analyzed as well. As the ionization of the polyelectrolyte block introduced long-ranged repulsive interactions in the corona of a micelle, the models of the micellar corona have to be modified. Analytical expressions for R_c, L, N_{agg}, CMC as a function of the degrees of polymerization of the ionic and hydrophobic blocks, the degree of ionization of the coronal block, Flory–Huggins parameters, interfacial tension, etc., have been derived for starlike and crew-cut micelles with polyelectrolyte corona. For example, on the basis of the scaling theory, for starlike micelles, it is found that the corona thickness is controlled solely by the length and degree of ionization of the coronal block; the aggregation number exhibits a strong decrease, whereas CMC increases with increasing length and ionization of the latter. For the crew-cut micelles, the aggregation number and core radius are given by the same expressions as for the starlike micelles. The corona thickness, however, is independent of the aggregation number, that is, independent of area per polyelectrolyte block at the core–corona interface. Complementary expressions have been derived on the basis of mean-field theory and self-consistent field modeling.

In salt-free solutions, the transition in morphology, for example, from lamella to cylinder and from cylinder to sphere, occurs upon an increase in the degree of ionization of the coronal block and/or an increase in N_A/decrease in N_B, with N_A and N_B being the degrees of polymerization of the polyelectrolyte and hydrophobic blocks, respectively. In salt-dominated solutions, the progressive increase in the salt concentration leads to sphere-to-cylinder and to cylinder-to-lamella transitions. Spherical micelles are the stable aggregate morphology in a wide range of salt concentration for copolymers with N_A and N_B ranging from $N_A \gg N_B$ to $N_A \leq N_B$. In contrast, cylindrical micelles and vesicles are dominant only in a narrow range of high salt concentration for strongly asymmetric copolymers with $N_A \leq N_B$. For each morphology, the increase in salt concentration is predicted to produce a progressive decrease in the corona thickness and a simultaneous increase in the core radius resulting in a smooth decrease of the L/R_c ratio, interrupted by jumps at the transition points. The authors emphasize that it is the reduction in elastic stretching of the core blocks B that drives the change in micelle morphology (Borisov et al. 2011).

The canonical sequence of morphological transformations might be inverted to *lamella–cylinder–sphere* in micelles with weakly dissociating, pH-sensitive coronae upon increasing salt concentration (Borisov and Zhulina 2005a). This typically occurs at low salt concentrations when the coronal ionization is strongly coupled to the conformations of the soluble block A. To specify binodals, separating the regions of thermodynamic stability of the aggregates of different morphologies, Borisov et al. considered the coronal free energy of a crew-cut aggregate with a weakly curved core, that is, $R_c \gg L$ (Borisov et al. 2011). From the asymptotic expressions of the binodals, it follows that an increase in the salt concentration triggers the lamella–cylinder–sphere transitions as a result of enhanced ionization of the coronal blocks, driven by the substitution of the hydrogen ions by the salt ions, thereby decreasing local pH inside the corona. On aggregate, the spherical shape is stabilized as a result of a decreased ionic contribution to the coronal free energy.

In salt-dominated solutions, the addition of salt leads to an enhanced screening of electrostatic interactions in the corona, and block copolymers with a pH-sensitive block exhibit the conventional morphological sequence. Diagrams in DP of the hydrophobic block versus salt concentration coordinates that localize the stability regions of the three main morphologies at fixed DP of the pH-sensitive block and degree of ionization have been constructed (Borisov et al. 2011). The comparison of the diagrams indicates that a certain morphology can be tuned by variation in both salt concentration and solution pH. A theoretical study (Victorov et al. 2010) predicts the existence of more complex morphologies such as branched cylinders in the vicinity of the sphere–cylinder binodal curve. Branched structures and networks have been also predicted elsewhere (Netz 1999).

Finally, we remark the universality of the self-assembly mechanism. On the basis of experimental results for micellization of various diblock, triblock, graft, and star heteroarm amphiphilic block copolymers as well as low molecular weight ionic and nonionic surfactants, a scaling relation (Equation 2.31) relating the aggregation number and polymerization degrees of the constituent blocks has been derived (Forster et al. 1996):

$$N_{agg} = N_{agg,0}\, N_A^{-0.8}\, N_B^2 \tag{2.31}$$

When plotted as $N_{agg}N_A^{0.8}/N_{agg,0}$ versus N_B, all data fall into a common master curve spanning the range of three orders of magnitude in block length. Importantly, $N_{agg,0}$ is related to the surfactant packing parameter and is known for many block copolymer and surfactant systems (Forster et al. 1996), which allow tuning of the aggregation number by variations in the molar masses of the constituent blocks.

The theoretical results and predictions from the scaling theories as well as the computer simulations and models are frequently compared with the available experimental data. Good agreement but also contradictions between the theory and experiment have been found. Undoubtedly, both contribute to our understanding of the phenomenon of self-assembly.

2.1.3.5 Summary and Conclusions

In summary, similar to the low molecular weight surfactants, the amphiphilic block, graft, and telechelic copolymers self-assemble in selective solvents (particularly in water) into aggregates of diverse structures and morphologies above a certain CMC. Their self-associative characteristics are directly related to the macromolecular structure and segmental incompatibility. However, the copolymers offer much wider variations in the molecular composition and architecture than the conventional surfactants. Furthermore, as remarked earlier, the proportions of the constituent moieties, described by the packing parameter, are sensitive to the properties of the surrounding media, so that by adding external variables such as temperature, pressure, pH, and presence of salt or other additives, the resulting interactions, self-association, properties, and morphologies become sufficiently complex, versatile, and, importantly, tunable. Spherical and cylindrical micelles as well as vesicles are the most commonly observed morphologies also for the copolymers. However, complex self-assembled structures, such as toroids, helices, multicore particles, disks, tubules, and multicompartment micelles that the latter can form, have rarely been observed for the classical surfactants. The phenomena of coexisting morphologies, which are believed to derive from a number of inherent for synthetic copolymer characteristics such as dispersity in chain lengths, are also worth mentioning as a distinctive feature. The dispersity can contribute to the thermodynamic stabilization of certain structures, for example, polymer vesicles, by segregation of a subpopulation of chains to the inner or outer leaflet of the membrane (Luo and Eisenberg 2001a). The composition dispersity may also result in a broad transition from unimers to micelles; that is why, even for copolymers with a narrow molar mass distribution, less sharp CMCs (or CMTs) compared to these of the surfactants have been observed.

The amphiphilic copolymers offer potential advantages over the classical surfactants in their extremely low CACs and longer lifetime indicating greater stability. Furthermore, the structures that they form are generally more robust and larger. Particles with dimensions spanning the range from tens of nanometers to (hundreds of) microns can be obtained depending not only on the molecular parameters but also on the fabrication procedure. The moieties of different solubility and compatibility are separated in different microdomains sizing of typically 5–100 nm depending on the chain length.

The self-assembled structures are characterized by certain CMC and/or CMT; specific morphology; particle molar mass and aggregation number; dimensional parameters such as radius of gyration, hydrodynamic radius, core radius or the corresponding membrane thickness for the polymersomes and disklike micelles, and radius for cylindrical micelles; and thickness of the corona. Noteworthy to mention is the unique structural *core–corona* composition, consisting of a core of certain morphology (spherical, cylindrical, and lamellar) in which instead of a surfactant's relatively low alkyl chains, long entangled polymer chains are present, engulfed by a corona built up of hydrophilic chains. For the conventional surfactant micelles, we can speak only of a shell, that is, a very thin layer of hydrophilic heads on the surface of the particles. In contrast, the corona in the copolymer self-assembled structures is a distinctive entity characterized by certain parameters and specific properties. The corona, depending on the macromolecular characteristics, can be thick and largely contributing to the overall particle dimensions as in the hairy micelles or relatively thin as in the crew-cut and cylindrical micelles and polymersomes. If the moieties building the core and the corona are highly incompatible, the core–corona interface is sharp and distinct as documented by small-angle neutron scattering (SANS) and small-angle x-ray scattering (SAXS), whereas for more compatible blocks, a partial mixing in the boundary region can be displayed. Similar to the core, the corona is also able to accommodate active substances of appropriate nature and to serve as a carrier. By carefully designing of its composition, intelligent properties can be conferred to the whole aggregate so that desired structural transitions and rearrangements take place upon applying an external stimulus.

The numerous repeating hydrophobic units in each chain result in an energy penalty to transfer a single copolymer macromolecule from an aggregate to solution. An exponential dependence of the activation barrier on chain length has been shown to precipitously decline the exchange kinetics with increasing molecular weight (Choi et al. 2010). This arrested molecular exchange is a unique, polymer-related aspect of copolymer micellar kinetics. The slow or practically the absence of chain exchange leads to kinetically trapped in a nonequilibrium state structures that do not interact each other. Such a kinetic stability of the copolymer aggregates is attractive as far as their application as drug carriers in vivo is concerned. Furthermore, the propensity for formation of kinetically frozen nonequilibrium structures implies sensitivity on the processing route and, consequently, provides possibilities for controlling over the resulting structures and for obtaining a wide diversity of structures from a limited set of building blocks (Hayward and Pochan 2010).

Some copolymers, especially those of large enough hydrophobic blocks, in a nonaggregated state, form structures that are known as *unimolecular micelles*. These consist of only one macromolecule, whose hydrophobic moiety is compactly coiled in the interior and surrounded by the chain(s) of the hydrophilic block(s). Unimolecular micelles have not been reported for the conventional low molecular weight surfactants.

2.2 SELF-ASSEMBLED POLYMERIC AGGREGATES: FROM MICELLES TO VESICLES AND MORE COMPLEX STRUCTURES

2.2.1 Spherical Micelles

2.2.1.1 General Features

The spherical morphology is the simplest and most frequently observed aggregate morphology. Spherical micelles have been identified in many studies dealing with self-assembly of copolymers in solutions. These are typically prepared from copolymers of the common linear diblock chain architecture and highest hydrophilic fraction, $f \geq 0.60$ (Figure 2.8). However, copolymers of more complicated architecture—multiblock, nonlinear, starlike, as well as dendritic and graft copolymers—have also been found to form spherical core–corona micelles. The molar masses of the spherical micelle-forming copolymers span an interval of orders of magnitude—from oligomers such as hydrophobically modified short chains to copolymers with molar masses reaching millions of Da. Well-defined micelles with narrow size distribution have been successfully obtained by precisely synthesized copolymers exhibiting the lowest possible molecular weight distribution accessible by the contemporary synthetic polymer chemistry but also by copolymers displaying considerably broader dispersity indexes. The constituent moieties in various combinations, for example, synthetic–synthetic, synthetic–natural, and natural–natural, can be chemically rather distinct, thus allowing the tendency of segregation and microphase separation. Spherical micelles in aqueous solution are generally produced by one of the following procedures:

1. Direct rehydration of a solid copolymer sample. The micellar dispersion is then left to anneal by standing or stirring. Depending on nature of the constituent copolymer moieties, micellization can be induced by applying external stimuli (e.g., change in temperature or pH). The stimuli can also induce phase and morphological transitions (see later in this chapter).
2. In the second technique, the copolymer is molecularly dissolved in a good for the constituent moieties solvent to which water, which is precipitant for the hydrophobic moiety, is gradually added. The common solvent is removed by extensive dialysis against water or via other appropriate methods.

In both techniques, an equilibrium situation is not necessarily reached, especially when the hydrophobic moiety has a high glass transition temperature. In that case, kinetically "frozen" micelles, which may have a spherical shape and core–corona structure but exhibit large dispersity in size, composition, and aggregation number, are formed. Depending on the hydrophilic block/moiety, the copolymers can be classified in three categories: (1) nonionic, mainly based on PEO; (2) anionic, copolymers with poly((meth)acrylic acid); and (3) cationic, copolymers containing cationic or cationizable units/groups. The dimensions of the spherical micelles span a range from just a few nm—typical for hydrophobically modified oligomers—to structures reaching hundreds of nm in diameter observable for micelles, prepared from copolymers of considerably high molar masses. Depending on the nature of the hydrophobic block/moiety, the cores either may be completely free of water and strongly segregated with a well-pronounced, sharp boundary with the hydrophilic corona or may contain appreciable amounts of water and a diffusive barrier permitting penetration of monomer units from the corona-forming block in the core, and vice versa. Hairy micelles, for which the corona is normally larger than the core, represent the more common spherical morphology than the crew-cut micelles. Both structures are depicted in Figure 2.9. Key parameters such as CMC, size, aggregation number, and morphology (hairy or crew cut) are largely dependent on the relative block lengths, molar masses, chain architecture, and properties of the constituent moieties.

Most of the current understanding on copolymer self-assembly in aqueous solution is based on the knowledge accumulated on the PEO-PPO-PEO block copolymers (chemical structures of the monomer units are shown in Figure 2.10). Their micellization has been extensively studied and reviewed in a number of articles (Wanka et al. 1994, Almgren et al. 1995, Chu and Zhou 1996, Hamley 1998, Booth and Attwood 2000, Booth et al. 2000). These copolymers mainly in the ABA-type chain architecture are commercially available in a range of molecular weights and PEO contents varying from 10 to 80 wt.%. Copolymers of reverse, that is, BAB, and X-shaped chain architectures are also available, however, in more limited molecular weight and composition ranges. In many aspects, these copolymers behave like

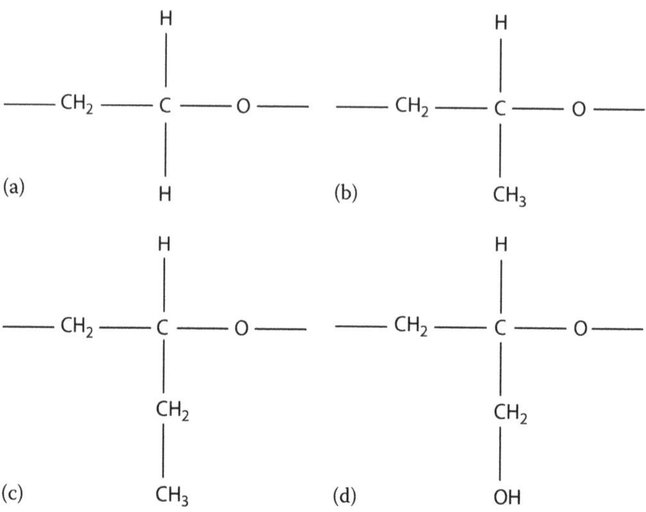

FIGURE 2.10 Monomer units of (a) PEO, (b) PPO, (c) PBO, and (d) PG.

classic low molecular weight hydrocarbon nonionic surfactants. Their aggregation behavior, however, is much more temperature dependent: The CMC, for example, is shifted orders of magnitude within a very small increase in temperature. In general, the PEO and the PPO blocks are mutually miscible and water soluble at low, for example, below 10°C, temperatures. Upon increasing temperature, the PPO blocks become more hydrophobic giving rise to segregation in micellar cores dominated by PPO. The experimental data for copolymers of PEO contents of ≥40–50 wt.% obtained from numerous studies from different research groups are compatible with a picture of spherical micelles with dehydrated PPO blocks in the core and fully hydrated PEO chains in the corona. Refined models, based on numerical calculations, predict water penetration in the micellar core as well as moderate miscibility of PPO and PEO (Linse 1993, Malmsten et al. 1993). Later on, this has been experimentally observed by Goldmint et al. who, using SANS, concluded that the PPO blocks make up only 40%–60% of the micellar core (Goldmint et al. 1999).

Wanka et al. have shown that the energy for transfer of one propylene oxide (PO) unit from the aqueous medium to the micellar interior is about 4–5 times lower than the corresponding energy for transfer of a CH_2 group of the classic surfactants indicating that the PO unit is 4–5 times less hydrophobic than CH_2 (Wanka et al. 1994). In the same paper, it has been concluded that the block copolymer macromolecule must contain at least about 10 PO units in order to be able to form micelles in aqueous solution. Analogous to classic surfactants, the magnitude of the CMC is mainly determined by the length of the PPO block (Wanka et al. 1994). PEO-PPO-PEO copolymers in general form relatively small in size micelles 8–15 nm in radius depending on the composition. The dimensions of the micelles are independent of the copolymer concentration. Interestingly, upon increasing temperature, the aggregation number has been found to strongly increase, whereas the dimensions of the micelles show only slight (if any) temperature dependence. This somewhat contradictory behavior has been attributed to compensation effects associated with gradual dehydration of both PPO and PEO blocks leading to increasing aggregation number due to increasing hydrophobicity and, hence, the size of the micelles, on one hand, and decreasing the dimensions of the individual macromolecules building the micelles, on the other. The core radii are typically in the 3.5–6.0 nm range depending on copolymer composition and temperature. For a series of copolymers with similar size of PPO block and increasing degrees of polymerization of the flanking PEO blocks, the core radius, R_c, has been found to scale with the reduced temperature as (Mortensen and Brown 1993)

$$R_c \sim (T - \mathrm{CMT})^{0.2} \quad (2.32)$$

At elevated temperatures, the diameter of the core becomes comparable to the length of fully stretched PPO block, which is considered as the driving force for sphere to prolate ellipsoid and rod morphological transitions (Mortensen and Pedersen 1993). The sphere-to-rod transition temperature depends on the PEO block size (Mortensen 2001b): the large PEO blocks stabilize the spherical morphology, and transitions to rod might not be observed.

At low concentrations, but above the CMC, the copolymers of reverse chain architecture, PPO-PEO-PPO, also form spherical micelles. Typically, they self-associate at higher concentrations and temperatures (higher values of CMC and CMT) than those of the corresponding copolymers of ABA chain architecture. Since the middle block is hydrophilic, it adopts a looping conformation so that *flowerlike* spherical micelles are formed (Figure 2.11). At concentrations approaching the overlap

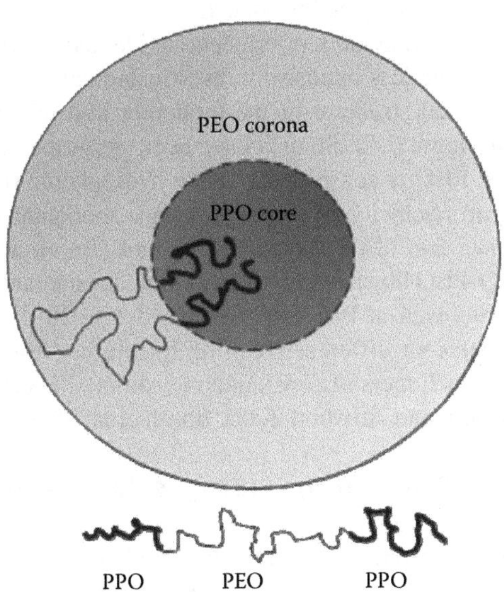

FIGURE 2.11 Schematic presentation of a flowerlike micelle formed in dilute aqueous solution by PPO-PEO-PPO block copolymers.

$$\begin{array}{l} (EO)_m-(PO)_n \\ (EO)_m-(PO)_n \end{array} \!\!> N-CH_2-CH_2-N <\!\! \begin{array}{l} (PO)_n-(EO)_m \\ (PO)_n-(EO)_m \end{array}$$

FIGURE 2.12 Chemical structure of X-shaped PEO-PPO copolymers. EO and PO stand for ethylene oxide and PO monomer units, respectively.

concentration C*, the PPO blocks from one macromolecule may reside in the cores of two adjacent micelles, which results in formation of branched or network-like clusters in which the individual micelles are interconnected by the PEO chains. The clusters coexist with the individual flowerlike micelles, but above a certain critical concentration, all micelles become part of the network.

The chemical structure of X-shaped PEO-PPO copolymers is presented in Figure 2.12. In fact, they are composed of four identical diblock copolymer chains that are chemically connected to a central ethylenediamine moiety. These copolymers are also commercially available, however, in more limited assortments of molecular weights and composition. Similarly to the linear block copolymers, they self-associate into (spherical) micelles above some specific CMCs and CMTs, which are typically higher than those of the corresponding linear PEO-PPO-PEO copolymers. The micelles are characterized with less compact structure (Nivaggioli et al. 1995, Perreur et al. 2001), which is due to the presence of a diamine junction in the middle of the macromolecule, thus introducing extra free volume and hindrance in forming of a compact structure. The hindrance encountered by the X-shaped copolymers is reflected in smaller micellar dimensions (about 11–12 nm in the 40°C–50°C temperature range and smaller at lower temperatures) and lower aggregation number (Nivaggioli et al. 1995). In addition, the tertiary diamine moiety may be protonated, and thus water into the core may be introduced.

Diblock and triblock copolymers of ethylene oxide and butylene oxide are also commercially available but less intensively investigated compared to the PEO-PPO copolymers. The chemical structure of the monomer unit of poly(butylene oxide) (PBO) is depicted in Figure 2.10. Studies have been performed also on laboratory-prepared copolymers. PBO is considerably more hydrophobic than PPO; based on CMC values, the hydrophobicity of the oxybutylene monomer unit has been calculated to be six times that of the oxypropylene unit (Booth and Attwood 2000). Accordingly, the PEO-PBO block copolymers exhibit considerably lower CMCs and ability to self-associate even at low, for example, below 20°C, temperatures. The copolymers are available in different chain architectures—diblock, triblock, and reverse triblock. For copolymers of constant composition, the CMC has been found to vary as follows (Booth and Attwood 2000, Booth et al. 2000):

$$(EO)_m-(BO)_n \ll (BO)_{n/2}-(EO)_m-(BO)_{n/2} \le (EO)_{m/2}-(BO)_n-(EO)_{m/2}$$

Morphological transitions from spherical to elongated wormlike micelles have been observed for these copolymers; however, they are shifted to lower temperatures and typically take place at about 40°C–50°C, even at lower (below 20°C) temperatures for copolymers of higher PBO content. Similarly to the PEO-PPO block copolymers, the spherical micelles, which these copolymers form, are relatively small in size (radii in the 5–10 nm range) and aggregation number (30–120 macromolecules per micelle). In contrast to the former, the latter increases strongly with increasing temperature due to dehydration of both PBO and PEO blocks (see preceding text). In spite of the high hydrophobicity of PBO, the cores of the micelles still contain appreciable amounts of water: degrees of hydration of 28% at 25°C and 26% at 35°C have been determined (Fairclough et al. 2006).

Spherical micelles have been formed by copolymers that can be considered as analogues of the commercially available PEO-PPO-PEO copolymers in which the flanking PEO blocks are substituted with blocks of linear polyglycidol (PG) (Rangelov et al. 2008). The latter is structurally similar to PEO and differs in that each monomer unit bears a hydroxymethylene group (Figure 2.10). In particular, following a slight collapse in the 25°C–40°C range, the micelles of the copolymers of the highest PG contents, $(G)_{63}(PO)_{68}(G)_{63}$ and $(G)_{135}(PO)_{68}(G)_{135}$, where G stands for glycidol, did not change their dimensions (11–14 nm in hydrodynamic radii) in the temperature interval 40°C–70°C. Those of $(G)_{40}(PO)_{68}(G)_{40}$ were constant (hydrodynamic radii 13–14 nm) in the temperature range 25°C–50°C and upon further heating undergo transition to anisotropic, elongated particles as implied by the R_g/R_h ratio well above 2. The aggregation numbers of $(G)_{63}(PO)_{68}(G)_{63}$ and $(G)_{135}(PO)_{68}(G)_{135}$ were below or slightly above 100. The temperature variations of N_{agg} are presented in Figure 2.13, where, in contrast to PEO-PPO and PEO-PBO copolymers for which a gradual increase of N_{agg} is observed (see preceding text), maxima and breaks are seen. This indicates that the magnitude of N_{agg} and its variations with temperature are governed by the superposition of two opposite processes: the increase in hydrophobicity of the PPO block, on one hand, and the enhancement of the solubility of the PG moieties, on the other (Rangelov et al. 2008). The SANS experiments revealed a gradual increase by 7%–10% with increasing temperature from 25°C to 60°C of the radii of gyration of the PPO cores of the micelles, which are in the range 5.9–7.2 nm.

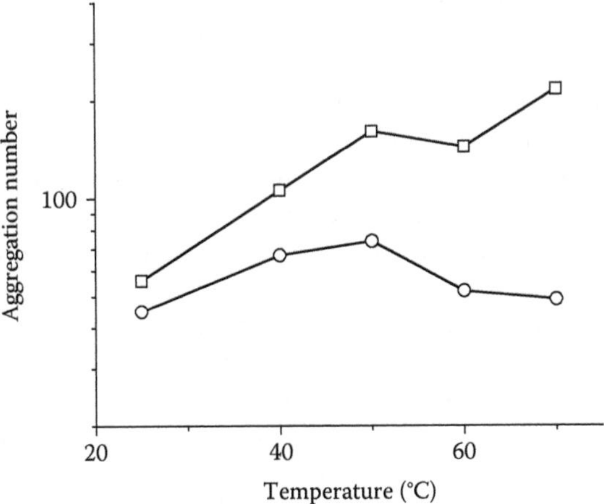

FIGURE 2.13 Temperature dependence of the aggregation numbers of micelles formed in aqueous solution by $(G)_{63}(PO)_{68}(G)_{63}$ (squares) and $(G)_{135}(PO)_{68}(G)_{135}$ (circles).

The results for corona thickness, calculated as difference between the hydrodynamic radii from dynamic light scattering (DLS) and the radii of the equivalent spheres from SANS, indicate that in the temperature interval, 40°C–60°C are either stable or slightly increase in the 4.2–5.8 nm range.

2.2.1.2 Chain Architecture and Nature of Constituent Blocks of Spherical Micelle-Forming Copolymers

2.2.1.2.1 Diblock Copolymers

The literature on the spherical polymeric micelles is vast. It is not possible to encompass all publications; therefore, we focus readers' attention on the most recent papers—those published in the last 2 years. The aim is to demonstrate the abundance of copolymers of different compositions and chain architectures that are able to form spherical micelles of the familiar core–corona structure as well as to give information about important characterization parameters of the micelles, their properties, and implications for possible applications. We start this section with the simplest chain architecture—diblock. The hydrophilic blocks typically used can be truly hydrophilic, that is, they are always hydrophilic, or can be sensitive to external stimuli such as temperature and pH and are able to undergo (reversible) transition from soluble to insoluble state. Such polymers are widely used to convey intelligent properties to the micelles. Hydrophilic polymer chains covalently connected to a strongly hydrophobic residue (not necessarily of polymeric nature) as well as just the opposite—hydrophobic polymer chains conjugated to a hydrophilic residue—have also been considered as copolymers of diblock chain architecture. Good examples for the latter type are the biodegradable copolymers based on synthetic polypeptides—poly(L-leucine) or poly(γ-benzyl-L-glutamate)—conjugated to β-cyclodextrin, which is a cyclic oligosaccharide composed of seven glucopyranose units (Wu et al. 2010, Zhang et al. 2010c). The molar masses of the copolymers described in these articles were below 10,000. They exhibited low CMCs: ~3 mg/L and lower. A solvent displacement method was used to prepare spherical micelles of size below 200 nm, which was found to decrease with increasing proportion of the hydrophobic part.

The PEGylated lipids are representatives of the former type of diblock copolymers, that is, a phospholipid- or lipid-mimetic residue conjugated to a hydrophilic PEG chain. The structure and dynamics of micelles of self-assembled 1,2-distearoyl-sn-glycero-3-phosphatidylethanolamine-*N*-[methoxy(polyethylene glycol) 2000] (DSPE-PEG$_{2000}$) in aqueous media have been studied elsewhere (Vukovic et al. 2011). The micelles have a hydrophobic core, an ionic interface, and a semipolar PEG layer. The atomistic molecular dynamic simulations revealed that the observed assemblies have different aggregation numbers and dimensions in saline solution and in pure water due to the effects of screening of their charged PO_4^- groups (Figure 2.14). The local density and conformations of the PEG chains in the corona also depended on N_{agg}. For the micelles with low aggregation number (N_{agg} = 10), the simulation results showed that ~30% of the core was always fully exposed to water, whereas the micelles with N_{agg} = 90 had a more homogeneous PEG corona, but ca. 10% of the core was still fully exposed to water. The hydrophobic cores of the low aggregation number micelles had a relatively sharp boundary, and the PEG chains

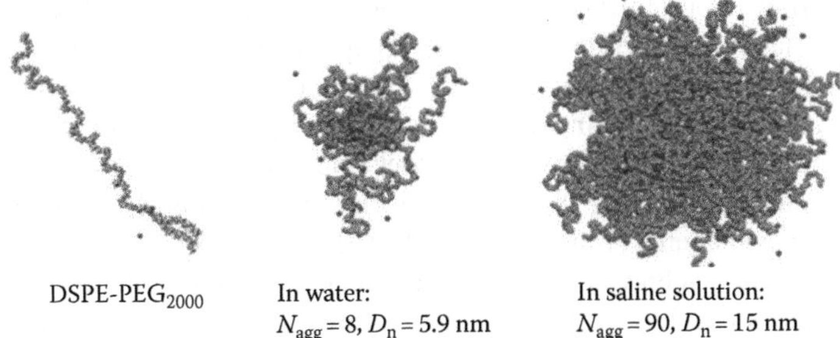

FIGURE 2.14 Snapshots of equilibrated micelles of self-assembled 1,2-distearoyl-sn-glycero-3-phosphatidylethanolamine-*N*-[methoxy(polyethylene glycol) 2000] (DSPE-PEG_{2000}) with different aggregation numbers and sizes in water and saline solution. (Reprinted with permission from Vukovic, L., Khatib, F.A., Drake, S.P., Madriaga, A., Brandenburg, K.S., Kral, P., and Onyuksel, H., *J. Am. Chem. Soc.*, 133, 13481–13488, 2011. Copyright 2011, American Chemical Society.)

in the corona adopted *mushroom* conformation in contrast to the high aggregation number micelles in which they had more *brushlike* conformation. The average thickness of the corona was 3.8 nm, which was in very good agreement with the value of 3.5 nm experimentally found earlier (Johnsson et al. 2001). The aggregation numbers and hydrodynamic radii of the DSPE-PEG_{2000} micelles found in the two studies were also very close.

In contrast to DSPE-PEG_{2000} and other commercially available PEG-derivatized phospholipids, the PEGylated lipids described by Rangelov et al. are nonionic (Rangelov et al. 2002b, p. 7074–7081). These authors have synthesized a number of nonionic PEG-lipids via anionic polymerization of ethylene oxide using a lipid-mimetic alcohol—1,3-didodecyloxy-propane-2-ol—as a starting material. The resulting copolymers are entirely nonionic and differ in the degrees of polymerization of the PEG moieties—from 30 to 92, corresponding to molar masses in the 1300–4000 range. The copolymers spontaneously self-associate in water forming small core–corona micelles. The latter were parameterized by DLS and static light scattering (SLS): particles of hydrodynamic radii 50–70 Å and weight average molar masses up to 185.8×10^3 g/mol, corresponding to N_{agg} in the 24–79 range, were formed. The aggregation numbers were found to increase with decreasing hydrophilicity of the PEG-lipids (i.e., with decreasing DP of the PEG chain). At elevated temperatures, more compact and denser micelles were formed. The rod–coil polymer, described elsewhere (Yang et al. 2011b), also belongs to this group of amphiphilic macromolecules. The hydrophobic residue has a rigid, rodlike structure composed of three biphenyls and one styrene unit. It is linked to a short PEO chain (DP = 34), thus forming an amphiphilic rod–coil macromolecule. Lamellar structures are reported in the dry, crystalline state, whereas in aqueous solution, the macromolecules self-associate into spherical micelles that can be used as nanoreactors for Suzuki coupling reaction at room temperature.

As repeatedly noted throughout this chapter, PEO is the hydrophilic block of choice to create in combination with various in nature hydrophobic moieties self-assembling amphiphilic copolymers. From these, the PEO-*b*-polydiene diblock copolymers have long been used for material applications and objects of intensive investigations. The preparation of these copolymers by anionic polymerization, however, has hampered their wide-scale availability. A recent paper reports on the preparation of PEO-*b*-polyisoprene block copolymers via reversible addition–fragmentation chain transfer (RAFT) polymerization of isoprene using PEO macrochain extension agents (Bartels et al. 2010). The assembly of these copolymers in aqueous solution achieved by solvent displacement resulted in uniform micelles of varying sizes (18–154 nm in diameter) following the general trends observed for other copolymers, that is, longer block length PEO relative to polyisoprene gave smaller particles, and shorter overall polymer chain length afforded smaller particle assemblies.

Other strongly hydrophobic polymers employed in combination with PEO to form amphiphilic block copolymers are polyesters such as poly(ε-caprolactone) (PCL), polylactide (PLA), poly(lactic acid-co-glycolic acid), and their derivatives. The latter are nontoxic, biocompatible, biodegradable, and bioadsorbable in vivo; therefore, most of the numerous studies deal with possible biomedical applications of such copolymers, in particular investigations of the properties of the self-assembled structures—size, morphology, solubilization capacity, pharmacokinetics, biodegradation, etc.—as well as investigations in biomedical fields of controlled drug delivery, tissue engineering, etc. For example, Richter and coauthors in a recent paper study polymeric micelles as a drug delivery system for sagopilone (a poorly water-soluble anticancer drug) with respect to a passive tumor targeting (Richter et al. 2010). Two series of copolymers—PEO-PCL and PEO-PLA at different PEO-to-polyester ratios—were investigated to identify suitable copolymers. Micelles were prepared via film formation and sonication and thoroughly investigated with regard to their size and morphology, thermal properties, and solubilization capacity. Three of the copolymers with compositions PEO_{2000}-PLA_{2200}, PEO_{2000}-PCL_{2600}, and PEO_{5000}-PCL_{5000} (the numbers in subscript corresponding to molar mass of the blocks), notably all in PEO-to-polyester ratios close to unity, were identified as the most suitable sagopilone delivery systems. They formed well-defined, monodisperse micelles with dimensions below 100 nm (Figure 2.15) and exhibited high solubilization capacity.

Modifications of either hydrophobic or hydrophilic blocks have been envisaged as well to improve particular properties of the micellar systems. Falamarzian and coauthors developed PEO-PCL containing stearyl substituents on the PCL block to increase the solubilization capacity of the micelles with regard to amphotericin B (Falamarzian and Lavasanifar 2010). Encapsulated in predominantly spherical nanosized micelles, amphotericin B showed reduced hemolytic activity against rat red blood cells. In another paper, octreotide was conjugated to the distal end of the PEO block of a PEO_{6000}-PLA_{5000} diblock copolymer to bind to somatostatine receptors overexpressed on tumor cells aiming at enhancing intracellular drug delivery and improving therapeutic efficacy (Zhang et al. 2011a). PEO-*b*-poly(L-lactide-co-β-malic acid) has been galactosylated, as galactose is considered to be a promising ligand that is able to specifically recognize and bind asialoglycoprotein receptor, to obtain target delivery of doxorubicin to HepG2 cells (Suo et al. 2010).

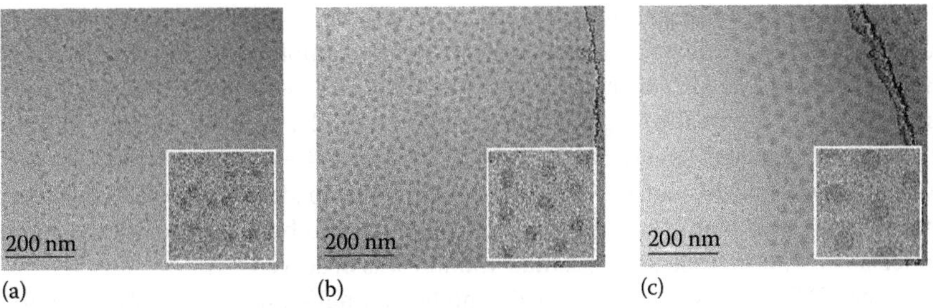

FIGURE 2.15 Cryo-TEM images of polymeric micelles composed of (a) PEO_{2000}-PLA_{2200}, (b) PEO_{2000}-PCL_{2600}, and (c) PEO_{5000}-PCL_{5000} at a copolymer concentration of 20 g/L. The inset shows a small region (100 nm^2) of the image at a larger scale. (Reprinted from *Int. J. Pharm.*, 389, Richter, A., Olbrich, C., Krause, M., and Kissel, T., 244–253, Copyright 2010, with permission from Elsevier.)

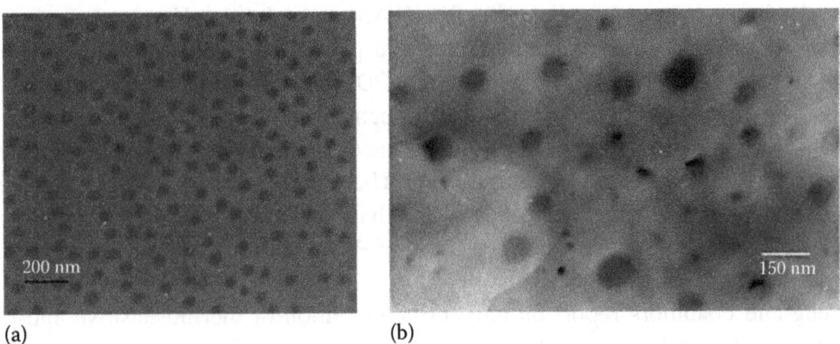

FIGURE 2.16 TEM images of octreotide-conjugated PEO_{6000}-b-PLA_{5000} micelles (a) and galactosylated PEO-*b*-poly(L-lactide-co-β-malic acid) micelles (b). (Reprinted with permission from Zhang, Y. et al., *Pharm. Res.*, 28, 1167, 2011c; Suo, A. et al., *Int. J. Nanomed.*, 5, 1029, 2010. With permission from Pharmaceutical/Dove Medical Press.)

Transmission electron microscopy (TEM) images of nanosized objects are presented in Figure 2.16. Polymeric micelles composed of drug conjugated to a diblock copolymer have been developed as well. Jin and coworkers have shown the synthesis and preparation of spherical micelles from PEG-poly(D,L-lactic acid-co-glycolic acid) to which doxorubicin is chemically attached (Jin et al. 2010). In addition, the micelles were decorated with a targeting agent, so that the accumulation of the targeted micelles in tumor tissue depended on dual effect of passive and active targeting.

A number of copolymers based on PNIPAM have recently been developed. A novel photosensitive polymer comprising PNIPAM as predominantly hydrophilic moiety conjugated with a relatively short but very hydrophobic coumarin residue has been prepared (Jiang et al. 2010b). Spherical micelles with diameters of 30–50 nm were formed in aqueous solution. Upon irradiation at 365 nm, the photodimerization of the coumarin end groups resulted in switching from micellar morphology into hollow spheres, that is, vesicles, with diameters of 200–350 nm. The photoswitching was reversible as the micellar morphology was recovered upon irradiation at 254 nm.

Via ring-opening polymerization of tetramethylene carbonate with hydroxyl-terminated PNIPAM as a macroinitiator, thermosensitive amphiphilic block copolymers have been synthesized (Lee and Chen 2011). CMCs ranging in the 1.11–22.9 mg/L interval depending on the length of the constituent blocks were determined. A core–corona structure and spherical morphology of the self-assembled structures were evident from ^{1}H NMR and TEM. The micelles can be loaded with active substances. Interestingly, the blank and drug-loaded micelles exhibited no difference in the dimensions—an average size less than 130 nm was observed.

When PNIPAM is combined with another stimuli-responsive polymer into a single polymer chain, *schizophrenic* aggregation may be observed. Such dually responsive copolymers have been synthesized by Smith and coworkers (2010). The copolymers consist of PNIPAM and poly(*N,N*-diethylaminoethyl methacrylate) (PDEAEMA), which is pH sensitive. Their self-assembly can be controlled by the block length, temperature, and pH. Spherical micelles sizing 21–25 nm in radii with a PDEAEMA core and corona of PNIPAM were formed at temperatures below lower critical solution temperature (LCST) of PNIPAM and solution pH greater than p*K*(a) of PDEAEMA. At elevated temperatures and $pH<7.5$, micelles of practically the same size but with PNIPAM in the core and PDEAEMA chains in the corona were formed. *Schizophrenic* micelles have also been reported by Zhang and coauthors (2011b). In contrast to the previously mentioned copolymers, the copolymer these authors have prepared is composed of two pH-sensitive blocks—poly(acrylic acid) (PAA) and poly(*N,N*-dimethylaminoethyl methacrylate) (PDMAEMA). In acidic pH ($pH=3.4$), the core of the micelles was built of PAA, whereas at basic pH ($pH=9.2$), PDMAEMA chains were in the core.

Peng and coauthors report on another combination of thermosensitive and pH-sensitive poly(ethylene glycol methacrylate) and poly(2-diisopropylaminoethyl methacrylate), respectively, into a single chain of diblock architecture (Peng et al. 2010). The copolymers were capable of self-assembly in aqueous solution to form pH-sensitive nanoparticles, which were spherical in shape with diameter reaching 132 nm. CMCs at pH 7.4 in the interval $4.5–8.9\times10^{-3}$ wt.% and critical aggregation pH (5.8–6.6) were determined. The micelles were able to solubilize water-insoluble photosensitizer, meso-tetra(hydroxyphenyl) chlorine. The loaded nanoparticles have potential in photodynamic therapy.

Spherical micelles with high aggregation number have been formed by doubly stimuli-responsive diblock copolymers comprising PEG with $DP=45$ and poly(*N*-acryloyl-2,2-dimethyl-1,3-oxazolidine) (PADMO) with $DP=18–47$ (Cui et al. 2011b). The latter is hydrophobic but labile at acidic conditions so that PADMO can be easily converted into the hydrophilic poly(2-hydroxyethyl acrylamide) (see Figure 2.54). According to these authors, the thermosensitive transitions were attributed to the densely packed PEG chains in the corona resulting in strong clustering attractive interactions and insufficient hydration of the PEG chains (Figure 2.17).

Jiang and coauthors describe the synthesis and properties of the self-assembled structures of diblock copolymers based on PEG and hydrophobic polyacrylates bearing pendant six-member cyclic ketal groups (Jiang et al. 2011). The copolymers were prepared by atom transfer radical polymerization (ATRP) and varied in type of the ketal group and PEG-to-polyacrylate ratios. Two populations of spherical micelles

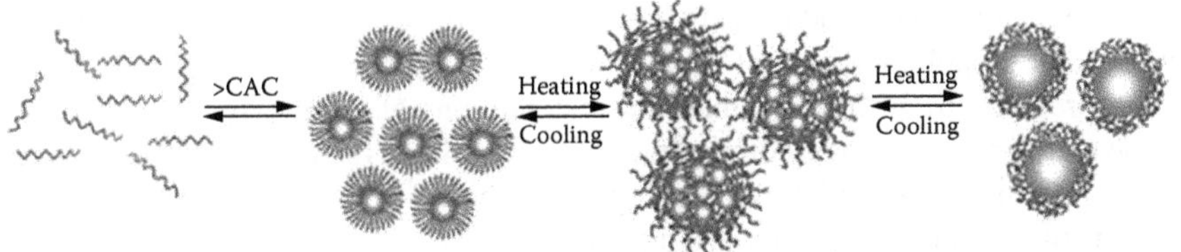

FIGURE 2.17 Structural transitions of spherical micelles formed in water by PEG-PADMO diblock copolymers. (Reprinted with permission from Cui, Q., Wu, F., and Wang, E., *J. Phys. Chem. B*, 115, 5913–5922, 2011b. Copyright 2011, American Chemical Society.)

of presumably core–corona structure were typically observed—one with particle diameters ranging from 20 to 70 nm and the other within the 110–410 nm range. At acidic conditions (pH = 3.5), the cyclic ketal groups were cleaved, which resulted in transformation from amphiphilic to doubly hydrophilic copolymers and disruption of the aggregates. The release behavior can be controlled by a combination of pH and temperature.

Starting from a bifunctional initiator, firstly employing ATRP of a new acetal-protected glycerol monomethacrylate monomer (*cis*-1,3-benzylidene glycerol methacrylate) and, in the second step, organo-based catalyzed polymerization of L- or D-lactide, Wolf and coauthors prepared well-defined diblock copolymers—molecular weights 7,000–30,000 g/mol and polydispersity index (PDI) 1.12–1.17 (Wolf et al. 2010). Acidic hydrolysis of the acetal protecting groups transformed the poly(*cis*-1,3-benzylidene glycerol methacrylate) blocks into water-soluble poly(1,3-dihydroxypropyl methacrylate), which exclusively contains primary hydroxyl groups. The resulting amphiphilic block copolymers self-assembled into spherical aggregates with dimensions of about 20 nm.

By selective hydrolysis of the hemiketal groups of a novel PCL-poly(methoxymethyl acrylate) copolymers, Li and coworkers synthesized PCL-PAA diblock copolymer (Li et al. 2011a). The size and size distribution of the micelles were found to vary with pH and ionic strength of the aqueous solutions.

Destruction of self-assembled structures (spherical aggregates with an average size of 160 nm) formed in water by a photoresponsive poly(dialkoxycyanostilbene methacrylate)-b-PEO diblock copolymer upon UV irradiation is described elsewhere (Menon et al. 2011). The stilbene chromophore undergoes transformation from a linear *trans*-isomer to a bent *cis*-isomer, which would hinder self-aggregation on one hand, and, on the other hand, the *cis*-isomer exhibits a higher dipole moment, which contributes to the reduction of the hydrophobic character of the copolymer and loss of the amphiphilic properties.

An electric-field-triggered sphere-to-cylinder transition is presented elsewhere (Lee and Park 2010). The system consists of ionic-*b*-neutral copolymers, namely, poly(styrene sulfonate-*b*-methylbutylene) in dilute solutions. Upon the application of small electric field (ca. 30 V/cm), the spherical micelles initially formed transformed into interconnected spheres that were eventually converted into aggregates with cylindrical shape. The process was reversible as the spherical micelles were recovered when the electric field was switched off.

Nonionic and composed of truly hydrophilic and truly hydrophobic blocks is the amphiphilic copolymer poly(vinyl alcohol)-*b*-polystyrene (Jeon and Youk 2010). It was prepared by hydrolysis under basic conditions of a poly(vinyl acetate)-*b*-PS copolymer synthesized by a combination of Co-mediated radical polymerization and RAFT. The micelles, which the amphiphilic copolymer formed, were spherical in shape with an average diameter around 44 nm.

By a combination of polymerization techniques—ring-opening polymerization and xanthate-mediated RAFT—PCL-*b*-poly(N-vinyl pyrrolidone) (PNVP) block copolymers have been successfully prepared (Mishra et al. 2011). In water, the copolymer PCL(63)-*b*-PNVP(90) (the figures refer to DPs of the constituent blocks) formed spherical micelles with ~34 nm diameter. A regular spherical shape and diameters of less than 50 nm have been reported for the micelles prepared from PLA-*b*-poly(2-methacryloyloxyethyl phosphorylcholine) (Liu et al. 2011a). The excellent cytocompatibility of the micelles has been attributed to the zwitterionic phosphorylcholine groups.

Wan and coworkers studied the effects of chain topology of thermoresponsive block copolymers on the self-assembly, thermal phase transition, and controlled release (Wan et al. 2011). They prepared *tadpole*-shaped, linear-cyclic diblock copolymers comprising linear PCL as a hydrophobic block and cyclic PNIPAM. Linear analogues were also synthesized for comparison. Spherical micelles consisting of PCL cores and solvated coronae of cyclic or linear PNIPAM segments were formed at lower, for example, 20°C, temperatures. The thermoinduced collapse of the coronae was observed at temperatures that were lower for the micelles of the tadpole copolymer compared to its linear analogue. The micelles of the linear-cyclic copolymers also exhibited lower toxicity.

A hydrophilic polymer currently under investigation for its use in polymer–drug conjugates is poly(2-hydroxypropyl methacrylate) (see Figure 2.18 for the structural formula of the monomer). A recent review summarizes its broad pharmaceutical and biomedical applications (Talelli et al. 2010). In the field of polymer micelles, being hydrophilic, poly(2-hydroxypropyl methacrylate) serves either as a micellar stealth corona or as a micellar core if hydrophobically modified.

An important family of thermosensitive and biocompatible polymers is the poly(*N*-vinyl lactams). In particular, the seven-membered ring analogue, namely, poly(*N*-vinylcaprolactam) (PVCL), has LCST 34°C–37°C—close to the physiological temperature. Its aqueous solution properties have been modified via copolymerization with other vinyl or acrylic monomers. A recent paper reports

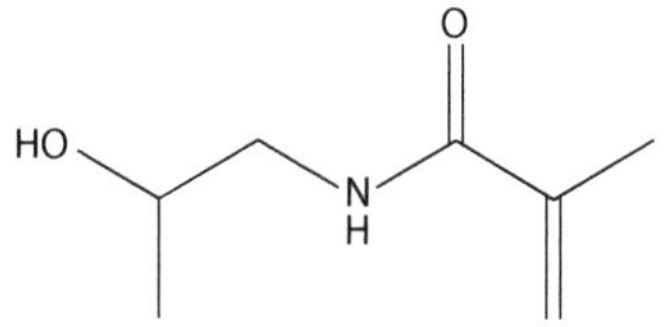

FIGURE 2.18 Structural formula of 2-hydroxypropyl methacrylate.

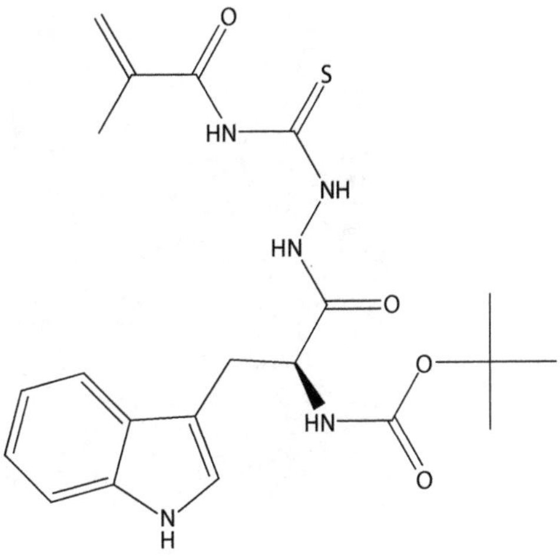

N-t-Boc-tryptophanamido-N′-methacryl thiourea

FIGURE 2.19 Structural formula of *N-t*-Boc-tryptophanamido-*N*′-methacryl thiourea.

on the preparation of amphiphilic copolymers based on PVCL as a hydrophilic and thermoresponsive block and poly(*N-t*-Boc-tryptophanamido-*N*′-methacryl thioureas) (Zhang et al. 2010d). The latter moieties (see Figure 2.19 for the structural formula) not only provide the hydrophobic character but are also known to strongly bind with DNA (Love et al. 1995, Bochkarev et al. 1997). The copolymer self-associated in water into spherical micelles with diameters around 30–40 nm from TEM.

A novel polymer, poly(*N*-vinylpiperidone), has been prepared by RAFT polymerization using xanthate as a chain-transfer agent (Ieong et al. 2011). The polymer is considered as the missing vinyl lactam polymer since very few reports on the polymerization of *N*-vinylpiperidone (Figure 2.20) and studies of the properties of the resulting polymers are available in the scientific literature in contrast to the polymers of the five- and seven-membered ring analogues—poly(*N*-vinylpyridone) and PVCL, respectively. The novel polymer exhibited LCST properties—cloud points (CPs) in the 68°C–87°C range depending on the molecular weight—noncytotoxicity and high biocompatibility. The chain extension with vinyl acetate afforded well-defined amphiphilic diblock copolymers, which self-associated into relatively small (d<24 nm) spherical micelles (Figure 2.20). The latter also exhibited thermal sensitivity.

Via ring-opening polymerization of 2-ethoxy-2-oxo-1,3,2-dioxaphospholane (EP) and 2-isopropoxy-2-oxo-1,3,2-dioxaphospholane (IPP) initiated by propargyl alcohol, amphiphilic copolyphosphates (Figure 2.21) have been synthesized (Zhai et al. 2011). Driven by hydrophobic interactions, the copolymers, prepared in varying EP/IPP ratios, self-associated into spherical micelles in which the core was built of poly(2-isopropoxy-2-oxo-1,3,2-dioxapholane) (PIPP), whereas the corona was composed of the more hydrophilic poly(2-ethoxy-2-oxo-1,3,2-dioxaphospholane) (PEP) chains. Particle size distributions and a representative TEM image are shown in Figure 2.21. Diameters of 89, 150, and 189 nm, depending on the composition,

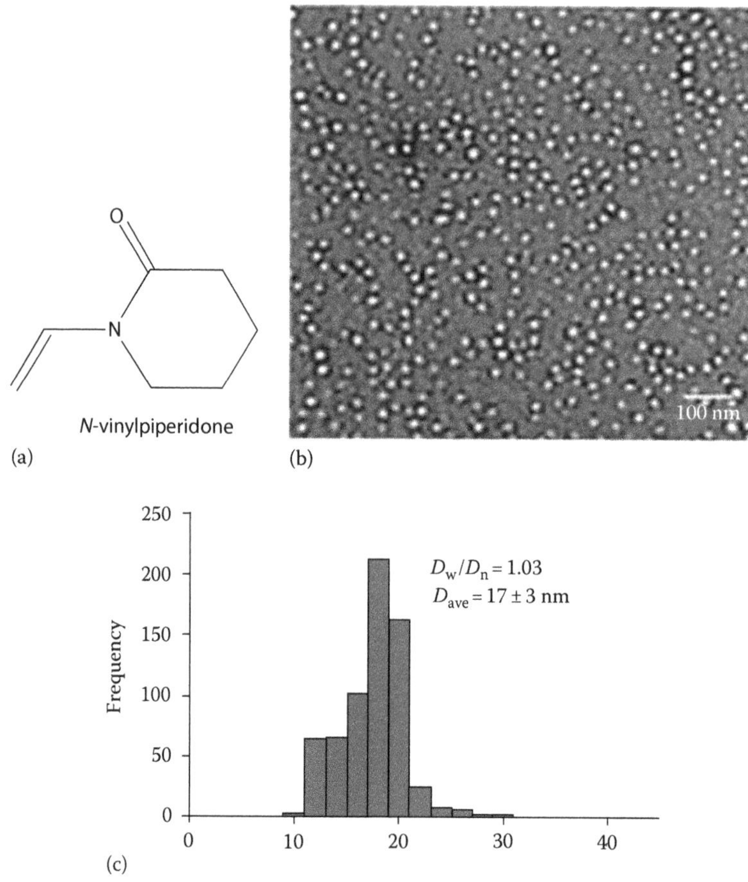

FIGURE 2.20 Structural formula of *N*-vinylpiperidone (a), representative TEM image (b), and particle size distribution from DLS (c) of spherical micelles obtained from diblock copolymer poly(*N*-vinylpiperidone)-*b*-poly(vinyl acetate) with DPs of 62 and 21, respectively. (Reprinted with permission from Ieong, N.S., Redhead, M., Bosquillon, C., Alexander, C., Kelland, M., and O'Reilly, R.K., *Macromolecules*, 44, 886–893, 2011. Copyright 2011, American Chemical Society.)

were measured by DLS. The reports on the polymeric micelles fully composed of polyphosphates, in particular, for biomedical applications, are seldom. In Chapter 5, more information about their biocompatibility, biodegradability, cell internalization, and intracellular delivery of drugs is presented.

2.2.1.2.2 Triblock and Multiblock Copolymers

There is a large body of works particularly with respect to ABA triblock copolymers that form spherical micelles with either nonionic cores and coronae (see preceding text for the PEO-PPO-PEO copolymers) or nonionic and entirely hydrophobic cores and charged coronae. The unique structure of the copolymers studied elsewhere (Caba et al. 2010), however, allows for the formation of micelles with a charged block

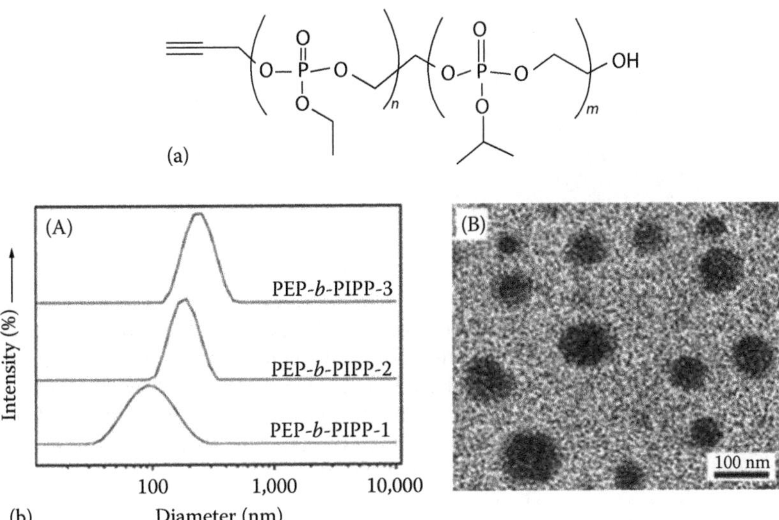

FIGURE 2.21 (a) Structural formula of poly(2-ethoxy-2-oxo-1,3,2-dioxaphospholane)-*b*-poly(2-isopropoxy-2-oxo-1,3,2-dioxapholane), PEP-b-PIP. (b) Particle size distribution (A) and a representative TEM micrograph (B) of PEP-b-PIPP micelles in water. 1, 2, and 3 in (A) denote EP/IPP ratios 1.06, 0.37, and 0.24, respectively; total M_w of 1, 2, and 3 corresponds to 13,660, 24,000, and 25,060, respectively. (Zhai, X., Huang, W., Liu, J., Pang, Y., Zhu, X., Zhou, Y., and Yan, D.: *Macromol. Biosci.* 2011. 11. 1603–1610. Copyright Wiley-VCH Verlag GmbH & Co. KGaA. Reproduced with permission.)

in the core under conditions at which the middle block is not fully neutralized. Such copolymers consist of a central polyurethane block bearing pendant carboxylic acid groups and flanking PEO blocks. At low pH, protonation of the carboxylates in the core leads to micelle formation driven by both hydrophobic and hydrogen-bonding interactions. The authors used SANS to elucidate the fine structure of the micellar cores of a series of PEO-polyurethane-PEO copolymers that were relatively close in composition, molar mass, and mass fraction of PEO (within the 0.75–0.90 range). Despite the similarities, no single model could be used to describe the micelles, which was attributed to the variation in N_{agg} (from 9 to 32). Thus, at high N_{agg}, the core can be represented as a solid core of polyurethane, whereas for lower aggregation number micelles, the core is better modeled as semidilute polymer chains emanating from a central point (starlike polymer model). The two limiting cases are depicted in Figure 2.22. The micelles are generally small in size—R_h within the 6.2–9.0 nm range.

Polyanhydrides are characterized by anhydride bonds connecting the repeat units of the backbone. They are considered to be biocompatible as they degrade into nontoxic diacid products that are metabolized and safely eliminated from the body. Poly(octadecanoic anhydride) has been prepared by melt polycondensation of octadecanoic dicarboxylic acid. The resulting polymers varying in molecular weight were reacted with methoxy poly(ethylene glycol) yielding ABA triblock copolymers with M_n ranging in the 7,600–15,300 interval (Xing et al. 2011). In aqueous

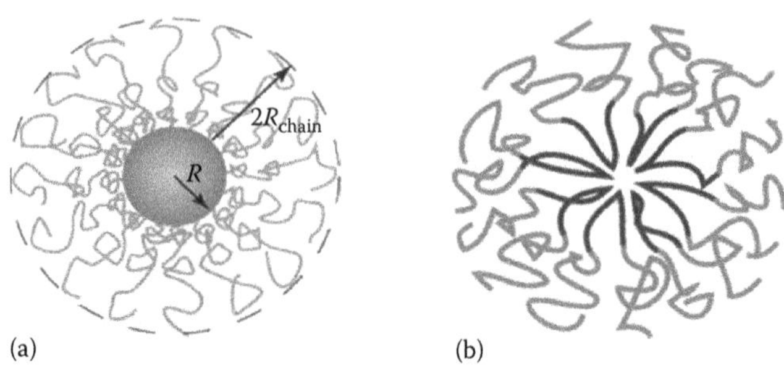

FIGURE 2.22 Schematic representations of the triblock model consisting of a homogeneous core with a corona of noninteracting Gaussian chains (a) and star polymer model (b) from micelles formed by PEO-polyurethane-PEO copolymers with higher (a) and lower (b) aggregation numbers. (Reprinted from *J. Colloid Interface Sci.*, 344, Caba, B.L., Zhang, Q., Carroll, M.R.J., Woodward, R.C., St Pierre, T.G., Gilbert, E.P., Riffle, J.S., and Davies, R., 81–89, Copyright 2010, with permission from Elsevier.)

solution, the copolymers were able to encapsulate paclitaxel with encapsulation efficiency >75%. The resulting nanoparticles were with dimensions 170–200 nm in diameter.

Kyeremateng and coworkers report on ABA triblock copolymers end capped with perfluoroalkyl segments (Kyeremateng et al. 2010). Solketal methacrylate was polymerized via ATRP using PPO as a macroinitiator. The terminal bromine units of the resulting ABA-type copolymers were substituted with N_3 and then coupled with perfluorinated segments, nonadecafluoro-1-decyl hex-5-ynoate, in particular. By acid hydrolysis of the hydrophobic poly(solketal methacrylate), moieties were converted into hydrophilic poly(glycerol monomethacrylate) (PGMA), thus making *triphilic* copolymers. The copolymers were of a fixed DP of PPO and differed in the DP, respectively, lengths of the PGMA blocks, which markedly influences their aqueous solution properties. Schematic representations of the structures formed by the two copolymers at different concentrations and temperatures are shown in Figure 2.23. In aqueous solution, F_9-$PGMA_{42}$-PPO_{27}-$PGMA_{42}$-F_9, that is, the copolymer with longer PGMA blocks, formed micelles and aggregates of micelles, which disintegrated into single *flowerlike* micelles at high temperature. The PGMA blocks, in looping conformation, formed the corona, while the PPO block and the F_9 segments formed the core. The immiscibility between the F_9 segments and the PPO blocks within the core resulted in a compartmentalized core where they formed the inner and the outer cores, respectively (Figure 2.23a). In contrast, single spherical micelles with PPO core and F_9-terminated PGMA coronal chains were formed by the copolymer with shorter PGMA blocks, F_9-$PGMA_{24}$-PPO_{27}-$PGMA_{24}$-F_9, in aqueous solution above CMC (Figure 2.23b). The lack of the F_9 segments within the core of micelles is due to the inability of their short rigid PGMA blocks to loop.

As repeatedly noted throughout the book, formation of *flowerlike* micelles can be expected from copolymers of BAB chain architecture, that is, with water-insoluble flanking blocks, or from multiblock copolymers containing BAB chain sequence, provided that the middle water-soluble A block is long enough to be able to adopt a looping conformation. BAB triblock copolymers with truly hydrophobic

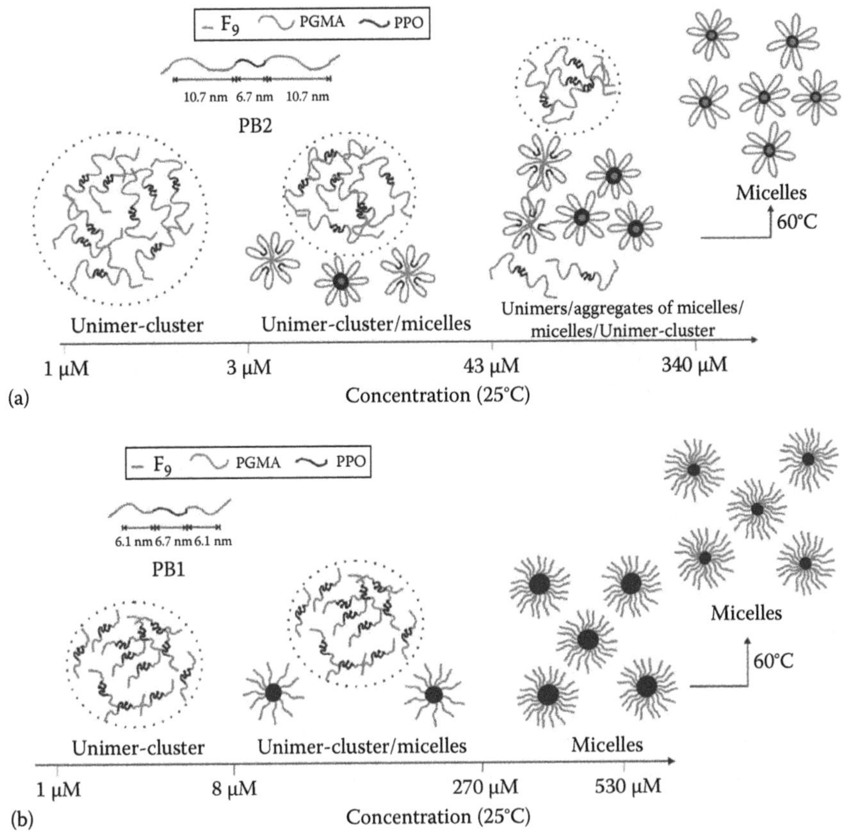

FIGURE 2.23 Schematic representation of micelle structures obtained as a result of increasing polymer concentration and temperature of aqueous solutions of (a) F_9-$PGMA_{42}$-PPO_{27}-$PGMA_{42}$-F_9 (PB2) and (b) F_9-$PGMA_{24}$-PPO_{27}-$PGMA_{24}$-F_9 (PB1). The length scales shown below the single chains denote the contour lengths of the respective blocks. In PB1, the short length of the PGMA blocks prevents it from looping; therefore, the F_9 segments cannot be sequestered from the aqueous environment. (Reprinted with permission from Kyeremateng, S.O., Henze, Th., Busse, K., Kressler, J., *Macromolecules*, 43, 2502–2511, 2010. Copyright 2010, American Chemical Society.)

flanking blocks such as polystyrene (PS), poly(methyl methacrylate) (PMMA), and PCL have been recently prepared (Das et al. 2011, Wang et al. 2011c). Linear poly(butyl acrylate)-PEO-poly(butyl acrylate) and its cyclized analogue have been found to self-associate in aqueous solution into *flowerlike* micelles (Honda et al. 2010). The chemical structures of the two copolymers as well as the corresponding micelles are presented in Figure 2.24. Noteworthy, CMCs, determined by viscometry, as well as hydrodynamic dimensions (D_h of about 20 nm) and size distribution of the micelles were not affected by the chain topology. Despite no apparent differences in the micellar structures and parameters were observed, the region of thermal stability was dramatically expanded for the micelles formed by the cyclized copolymer (Figure 2.24). This has been attributed to bridging between

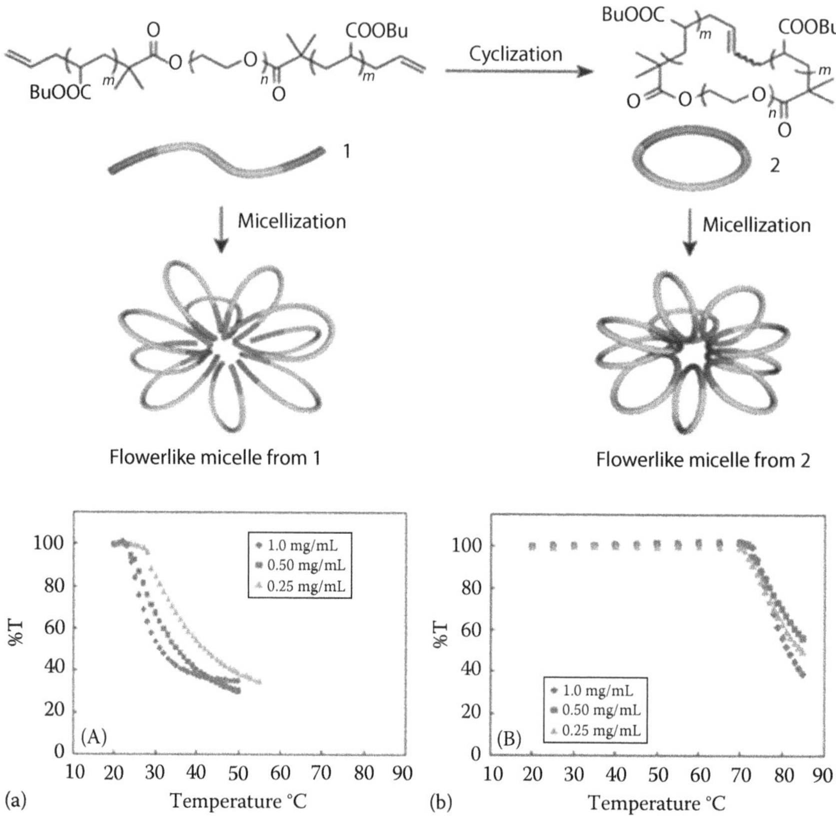

FIGURE 2.24 (a) Chemical structures of linear poly(butyl acrylate)-PEO-poly(butyl acrylate) and cyclic poly(butyl acrylate)-PEO amphiphilic block copolymers and schematic structures of *flowerlike* micelles they form. (b) Temperature variations of turbidity of micellar solutions of poly(butyl acrylate)-PEO-poly(butyl acrylate) (A) and its cyclic analogue (B). (Reprinted with permission from Honda, S., Yamamoto, T., and Tezuka, Y., *J. Am. Chem. Soc.*, 132, 10251–10253, 2010. Copyright 2010, American Chemical Society.)

the micelles of the linear copolymer resulting in formation of large physically cross-linked structures. Bridging is obviously not possible for the micelles of the cyclized copolymer.

A new synthetic methodology leading to sticker building blocks based on gradient hydrophobic/ionic sequences is presented elsewhere (Borisova et al. 2011). The copolymers were of BAB chain architecture in which the *A* block is PAA, whereas the *B* blocks were gradient poly(styrene-grad-acrylic acid). The presence of acrylic acid monomer units in the terminal blocks enables by variation of pH to tune the strength of the net cohesive interactions between these blocks. Copolymers of total M_n 12,500 and 17,000 and fraction of acrylic acid 72%–77% were found to form molecular solutions at high pH. Association of the terminal styrene-rich blocks was induced by a pH decrease resulting, in dilute solution, in formation of spherical flowerlike micelles with a dense styrene-rich core. The size of the micelles was found to increase with decreasing pH and/or increasing ionic strength of the solutions but did not exceed 12 nm in radii. Macroscopic gelation was observed at concentrations exceeding the overlap threshold. It is noteworthy that both association–dissociation and solgel transition were fully reversible. *Flowerlike* micelles were also formed by double hydrophilic poly(methacrylic acid)-PEO-poly(methacrylic acid) triblock copolymers at low pH (Liu et al. 2011c). The size of the micelles (48–310 nm) was strongly influenced by the solution pH and concentration.

Studies on the self-assembly of amphiphilic ternary (linear) block copolymers are less common. Such copolymers with three incompatible blocks or, most typically, with one hydrophilic and two mutually incompatible hydrophobic blocks may form aggregates with internally compartmentalized hydrophobic microdomains (see Section 2.4.1). The linear architecture in particular offers the possibility to modify the self-assembly behavior both by variation of the nature and relative size of the constituent blocks and by changing block sequence, that is, ABC, ACB, and CAB, within the macromolecules. Marsat and coauthors have recently reported on a series of triphilic, that is, A (hydrophilic), B (lipophilic), and C (fluorophilic) linear terpolymers in which the block sequences were systematically varied (Marsat et al. 2011). Oligo(ethylene glycol) monomethyl ether acrylate, benzyl acrylate, and 1*H*,1*H*-perfluorobutyl acrylate were copolymerized by RAFT in varying sequences to produce ABC, ACB, and BAC terpolymers. DLS revealed formation of spherical micellar aggregates with hydrodynamic diameters ranging 70–200 nm. Local phase separation of the cores and multiple morphologies were observed by cryo-TEM. Selective solubilization of hydrocarbon and fluorocarbon substances by the multicompartment particles has been demonstrated.

Presumably onion-like are the CO-releasing micelles prepared from ABC triblock copolymers composed of a hydrophilic PEG block, a poly[Ru(CO)$_3$Cl(ornithinate acrylamide)] block, which is capable of releasing CO, and hydrophobic poly(*n*-butylacrylamide) (Hasegawa et al. 2010): the most hydrophobic block—poly(*n*-butylacrylamide)—drove the micellization of the copolymers and formed the inner domain of the hydrophobic cores that was surrounded by the less hydrophobic moieties of poly(Ru(CO)$_3$Cl(ornithinate acrylamide), whereas the corona was composed by the PEG chains. The micelles were spherical as evidenced by TEM with diameter of about 30 nm and aggregation numbers 63–88 macromolecules per particle. More information about their CO-release properties is given in Chapter 5. Quite similar is the situation with the PEG-poly(5-methyl-5-propargyloxycarbonyl-1,3-dioxan-2-one)-PLA triblock copolymers that are designed as oxygen carriers (Li et al. 2011c). These copolymers formed spherical particles with diameters ranging 32–94 nm depending on the copolymer composition. PLA formed the inner domain of the hydrophobic core, whereas the poly(5-methyl-5-propargyloxycarbonyl-1,3-dioxan-2-one) moieties were situated between the PLA core and the PEG corona so that the propargyl groups to which azided hemoglobin was attached were on the surface of the micelles (see more in Chapter 5).

Sun and coauthors prepared a number of biocompatible and biodegradable ABC and ABCBA copolymers of PEO, PCL, and PLA (Sun et al. 2010a). In solid state, the copolymers formed microphase-separated crystalline materials. Starlike and flowerlike micelles with spherical morphology were formed in aqueous solution.

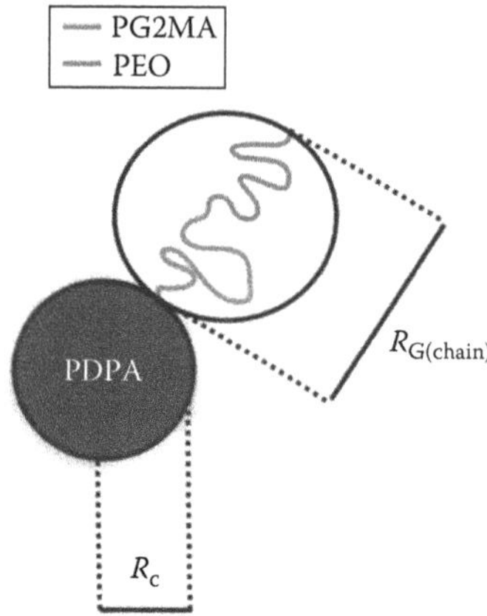

FIGURE 2.25 Schematic representation of a spherical micelle formed by PEO-PGMA-poly[2-(diisopropylamino)ethyl methacrylate] triblock copolymer in aqueous solution. (Giacomelli, F.C., Stepanek, P., Giacomelli, C., Schmidt, V., Jaeger, E., Jaeger, A., and Ulbrich, K., *Soft Matter*, 7, 9316–9325, 2011. Reproduced by permission of The Royal Society of Chemistry.)

They demonstrated a biphasic drug-release profile implying multicompartment arrangement of the core.

In contrast to the previously mentioned ABC copolymers, two of the blocks of the triblock copolymer PEO-PGMA-poly[2-(diisopropylamino)ethyl methacrylate] described elsewhere are hydrophilic (Giacomelli et al. 2011). The cores of the micelles that these copolymers formed in aqueous solution were composed of poly[2-(diisopropylamino)ethyl methacrylate] to which hydrophilic chains consisting of PEO and PGMA were attached (Figure 2.25). The core-forming block is able to undergo a sharp hydrophilic/hydrophobic pH-induced transition. The PEO chains are expected to reduce the protein adsorption, whereas the PGMA segments can be used as a cross-linking platform to create nanoparticles with covalently stabilized shell.

Unlikely, the hydrophobic block of PLA is located between the two hydrophilic blocks of the recently prepared PEO-PLA-poly(ethyl ethylene phosphate) triblock copolymers (Wu et al. 2011a). Poly(ethyl ethylene phosphate) is a hydrophilic polyphosphoester, which receives considerable attention due to its biodegradability and biocompatibility. The copolymers were of fixed PEO-PLA composition and increasing DP of poly(ethyl ethylene phosphate). They formed spherical aggregates with diameters of 97–116 nm.

There are a number of micelle-forming block copolymers with blocks in which comonomer units, different from those building the block, are introduced, that is, (A–co–B)–C, A–(B–co–C), A–(B–co–C)–A, and A–(B–co–C)–D chain architectures. The introduction of monomer units that can be isolated or form short segments renders new functions, properties, functionality, or sensitivity of the copolymers that the parent copolymers do not possess. Such copolymers are those described elsewhere (Cai et al. 2011). These authors prepared by anionic polymerization a series of PEO-*b*-poly(EO-stat-BO)-*b*-polyisoprene with M_n ranging from 9.0 to 14.4 kg/mol and isoprene content 14–35 wt.%. At 25°C, they self-associated into near spherical aggregates, which transformed to vesicles at elevated temperature. The thermal sensitivity here is rendered by the BO units. Even a faster transformation compared to related PEO-*b*-PNIPAM-*b*-polyisoprene systems was reported. In another paper, poly(methyl methacrylate-co-methacrylic acid)-*b*-poly[poly(ethylene glycol) methyl ether methacrylate] copolymers have been synthesized (Yang et al. 2011a). The copolymers self-associated into core–corona micelles with coronae of poly[poly(ethylene glycol) methyl ether methacrylate], whereas poly(methyl methacrylate-co-methacrylic acid) formed the core, which was sensitive to the pH changes of the environment. The micelles can swell or even dissociate as the pH increases from the strong acid stomach conditions to the neutral intestine conditions and provide the entrapped contents a fast release rate in the intestinal tract as schematically depicted in Figure 2.26.

Similar is the approach of Gan and coworkers, who by copolymerizing NIPAM with acrylic acid significantly altered the LCST of triblock copolymers PEO-*b*-poly(NIPAM-co-acrylic acid)-*b*-PMMA (Gan et al. 2011). The copolymers self-assembled into thermoresponsive micelles with spherical shape and average diameters ~120 nm, which exhibited good biocompatibility and ability to release drugs (folic acid) in a controllable manner.

Convertine and coworkers have developed a new class of diblock copolymers that facilitate siRNA delivery (Convertine et al. 2010). The diblock copolymers consisted of a cationic PDMAEMA block to mediate siRNA binding and a second pH-responsive and membrane-disruptive block composed of a mixture of positively charged DMAEMA residues, negatively charged propylacrylic acid residues, and hydrophobic butyl methacrylate residues in specific ratios (Figure 2.27). Propylacrylic acid and DMAEMA ensured the transition from an inactive conformation at physiological pH to a more hydrophobic membrane-interactive conformation in response to the endosomal pH drop, whereas the addition of hydrophobic butyl methacrylate

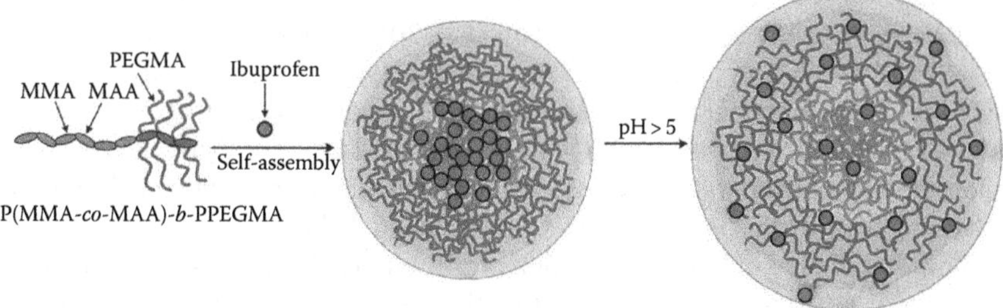

FIGURE 2.26 Schematic representation of micellization of poly(methyl methacrylate-co-methacrylic acid)-*b*-poly[poly(ethylene glycol) methyl ether methacrylate] copolymers and pH-dependent drug release. (Reprinted with permission from Yang, Y.Q., Zheng, L.Sh., Guo, X.D., Qian, Y., and Zhang, L.J., *Biomacromolecules*, 12, 116–122, 2011a. Copyright 2011, American Chemical Society.)

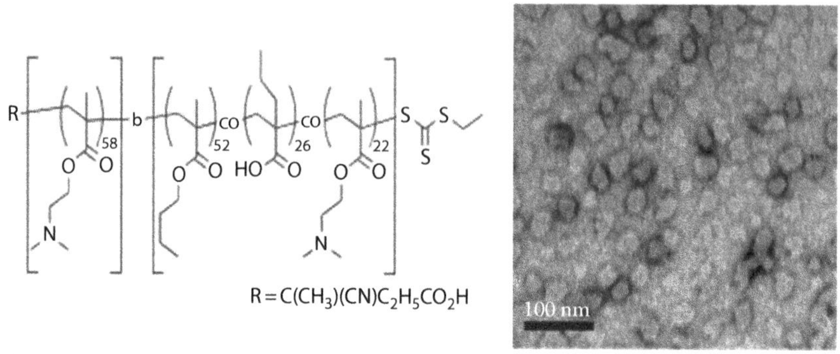

(a) (b)

FIGURE 2.27 (a) Chemical structure of the "diblock" copolymers poly(DMAEMA)-*b*-poly[(butyl methacrylate)-*co*-(DMAEMA)-*co*-(propylacrylic acid)]. (b) TEM image of micelles formed from an aqueous solution at c = 0.5 mg/mL in PBS. (Reprinted with permission from Convertine, A.J., Diab, C., Prieve, M., Paschal, A., Hoffman, A.S., Jonhson, P.H., and Stayton, P.S., *Biomacromolecules*, 11, 2904–2911, 2010. Copyright 2010, American Chemical Society.)

residues increased the hydrophobicity of the copolymer, increasing the pK_a of the propylacrylic acid carboxylate residues, thus raising the pH at which protonation occurs. The optimal incorporation of hydrophobic butyl methacrylate residues was found to be between 40 and 50 mol.%. The copolymers spontaneously formed spherical micelles in the size range of 40 nm, which appeared somewhat smaller (Figure 2.27) due to the collapse of the PDMAEMA corona under the anhydrous conditions of the TEM experiments. pH had a strong effect on the CMC values: Due to increased ionization of DMAEMA units from both corona and core resulting in enhancement of the overall hydrophilicity, CMC changed drastically from ~5 μg/mL at physiological conditions (pH 7.4) to ~100 μg/mL at conditions found within the endosomal compartments (pH 4.7).

A coumarin-functionalized block copolymer PEO-*b*-poly(*n*-butyl methacrylate-co-4-methyl-[7-(methacryloyl)oxyethyloxy]-coumarin) has been synthesized via ATRP using PEO-Br as macroinitiator (Jin et al. 2011). Coumarin and its derivatives are well-known linking groups in photoreactive polymers. Its role in the hydrophobic block of the copolymer is to chemically bind the anticancer drug 5-fluorouracil under UV irradiation at wavelength >310 nm (Figure 2.28). The drug-attached copolymer self-assembled into spherical aggregates with diameter of ~90 nm. The release of the drug was photo triggered by UV irradiation at 254 nm.

The effect of the gradient composition profile on the pH responsivity and micelle formation of a series of copolymers of methyl methacrylate and methacrylic acid has been studied (Zhao et al. 2011). All copolymers have the same average chain lengths (DP = 200) and a methyl methacrylate fraction of 0.5 but differ in their end-to-end composition profiles: uniform, linear gradient, triblock with linear gradient midblock, and diblock. Spherical micelles at high pH have been found for linear gradient, triblock, and diblock copolymers. Methyl methacrylate units were found in the hydrophilic corona of the micelles formed by the linear gradient

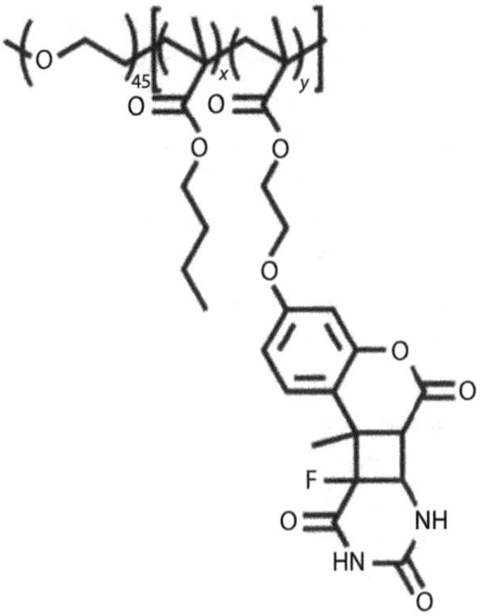

FIGURE 2.28 Chemical structure of 5-fluorouracil-attached PEO-*b*-poly(*n*-butyl methacrylate-*co*-4-methyl-[7-(methacryloyl)oxyethyloxy]-coumarin). (Reprinted with permission from Jin, Q., Mitschang, F., and Agarwal, S., *Biomacromolecules*, 12, 3684–3691, 2011. Copyright 2011, American Chemical Society.)

copolymer, whereas for the micelles of the diblock and triblock copolymers, they entirely resided in the cores.

By selective partial hydrolysis of amphiphilic copoly(2-oxazoline)s, copolymers with varying ratios of poly(2-methyl-2-oxazoline), poly(2-phenyl-2-oxazoline), and poly(ethylenimine) have been obtained (Lambermont-Thijs et al. 2011). The copolymers exhibited both temperature and pH responsiveness: At acidic conditions, due to the protonation of the ethylenimine units, the copolymers formed micelles at both ambient and elevated temperatures. At basic conditions, however, spherical micelles were evidenced only at elevated temperatures. Bloksma and coauthors investigated the self-assembly of chiral copolymers (Bloksma et al. 2012). Copolymers with gradient and block chain architectures have been prepared by copolymerizing hydrophilic 2-ethyl-2-oxazoline with hydrophobic chiral R-2-butyl-4-ethyl-2-oxazoline or racemic RS-2-butyl-4-ethyl-2-oxazoline. Spherical micelles were observed for the gradient enantiopure copolymers. The self-assembled structures of the block copolymers exhibited richer structural polymorphism: Spherical micelles, cylindrical micelles, sheets, and vesicles were observed depending on the hydrophobic-to-hydrophilic ratio within the copolymer chain. When the enantiopure block was replaced by the corresponding racemic block, only spherical micelles were observed.

Multiblock copolymers able to self-associate into spherical core–corona micelles are occasionally mentioned in the previous text. Indeed, despite their numerous advantages and fascinating structures such as unimolecular and multimolecular *flowerlike* micelles, relatively little, compared to di- and triblock copolymers, has been reported with regard to their synthesis and self-assembly. In order to expand the utility of current polymeric micellar systems, Green and coauthors have developed multiblock copolymers containing alternating blocks of PAA and PS (Green et al. 2011).

The copolymers were prepared by copper-catalyzed azide–alkyne cycloaddition of heterotelechelic (α-alkyne, ω-azide) poly(*tert*-butyl acrylate)-*b*-PS diblock copolymers followed by cleavage of the *tert*-butyl moieties. The resulting (PS-b-PAA)$_n$ contained up to nine diblock repeats. The copolymers were self-assembled by applying the solvent exchange method. A very low CMC (~2×10^{-4} mg/mL) was found for the copolymer with $n=9$ and $M_n=73.3$ kg/mol. The latter copolymer assembled into particles with fairly broad size distribution (PDI = 0.35) and an average hydrodynamic diameter of 11 nm. The shorter copolymer, $n=4$ and $M_n=43.8$ kg/mol, formed better defined in terms of size distribution (PDI = 0.16) but larger (diameter ~35 nm) particles. Based on TEM analysis, inclusion of PAA chains and production of multicompartments within the assembled structures were suggested. Considerably larger particles (198–257 nm) but invariably spherical in shape have been obtained from biotin and PLA-modified PEO-PPO-PEO block copolymers (Li et al. 2010b). Biotin was firstly conjugated to three differing in composition PEO-PPO-PEO copolymers with PEO contents of 50% and 70%. The resulting species bearing a hydroxyl group were used to initiate ring-opening polymerization of *l*-lactide in the presence of stannous octoate, thus producing tetrablock copolymers.

2.2.1.2.3 Nonlinear Copolymers

A variety of copolymers with sophisticated nonlinear chain architecture have become available in the recent years thanks to the significant progress in the controlled polymerization techniques, design of multifunctional initiators, and control in the coupling reactions. These are miktoarm, starlike, H-shaped, "palm-tree" and dumbbell-structured, and dendritic copolymers. As the constituent moieties can possess different solubilities in water, the amphiphilic character typically results in formation of self-assembled structures. The nonlinear copolymers typically exhibit fundamentally different self-assembly behaviors compared to the linear block copolymers (Ramzi et al. 1997, Pispas et al. 2000, Sotiriou et al. 2002, Yun et al. 2003), which provokes research interest in the preparation of a variety of nonlinear copolymers with varying chain topology, composition, arm numbers, etc. In the following, recent examples of self-assembly in aqueous solution of copolymers of nonlinear chain architecture are presented.

A_2B-type miktoarm copolymers, where A and B denote PEG and PCL, respectively, have been constructed on a core with orthogonal functionalities that facilitate the performance of "click" chemistry followed by ring-opening polymerization (Soliman et al. 2010). The copolymers self-assembled into spherical micelles into which a hydrophobic drug (nimodipine) can be loaded with high encapsulation efficiency. The same approach has been employed to construct ABC-type miktoarm copolymers, where A and B are as mentioned earlier and C is PS (Khanna et al. 2010). The core size of the spherical micelles formed in aqueous solution and hydrodynamic diameter were found to be inversely proportional to the length of the PEG arm.

Yang and coworkers designed Y-shaped miktoarm copolymers composed of PEG and two hydrophobic poly(solketal acrylate) chains (Yang et al. 2010). The copolymers were of fixed DP of PEG of 44 and increasing from 14 to 35 DP of the poly(solketal acrylate) chains. The micelles that were formed in water were with diameters in the 100–200 nm range and exhibited thermosensitivity: Sharp transitions in the transmittance of the micellar dispersions were observed in the temperature range 30°C–60°C depending on the copolymer composition. Compared to the linear analogue, the transition was shifted to lower temperatures. The sizes of the micelles were found to decrease with increasing temperature, which was attributed to abatement of hydrogen-bonding effect between ketal groups of poly(solketal acrylate) segments and water molecules, which increased hydrophobic interactions supposedly leading to formation of more compact and condensed core and an overall decrease in aggregate dimensions. Also thermosensitive is the ABC miktoarm copolymer with arms of PEG (45), PCL (49), and poly(ethyl ethylene phosphate) (242 or 81), where the numbers in parentheses denote the DP of the arms (Yuan et al. 2011). The thermosensitivity was rendered by poly(ethyl ethylene phosphate), which is known to undergo a tunable temperature collapse in aqueous solution. At lower temperatures (~20°C) and above the CMC, PEG and poly(ethyl ethylene phosphate) formed the corona of the spherical self-assembled structures with diameter around 34 nm. At elevated temperatures (55°C, which is above the CP of the copolymer), the ^{1}H NMR data in D_2O indicated strong dehydration of the poly(ethyl ethylene phosphate) arm, and the dispersion turned into strongly opalescent. Rearrangement into short rodlike particles due to the collapse of poly(ethyl ethylene phosphate) chains resulting in changes in the hydrophobic/hydrophilic volume ratios was observed at prolonged heating. The chemical pathway for the synthesis of the copolymers, schematic illustration of the morphological transition, as well as DLS data and TEM images of the self-assembled structures are presented in Figure 2.29.

An interesting topology has been described elsewhere (Lonsdale and Monteiro 2011). These authors demonstrate the synthesis of miktoarm block copolymers of AB, A_2B, and AB_2 types where A and B denote poly(*tert*-butyl acrylate) and cyclic PS, respectively, which after removing of the *tert*-butyl groups afforded amphiphilic miktoarm structures. Spherical core–corona micelles were formed by the copolymers of longer (e.g., 46 units) PAA chains. The AB_2 miktoarm copolymer packed more densely into the core compared to its linear analogue, thus forcing the PAA chain to adopt much extended and stretched conformation.

Palm-tree miktoarm copolymers consisting of one PEG chain attached to a hyperbranched polyglycerol core and several PCL chains as shown in Figure 2.30 have been found to self-associate into micelles of spherical shape (Zhang et al. 2010a). The potential of the latter as drug carriers was investigated: Higher loading capacity, entrapment efficiency, and more sustained drug release compared to those of micelles formed by linear PEG-PCL diblock copolymers were exhibited.

Starlike copolymers with arms composed of block copolymer chains, thus forming star-block copolymers (Figure 2.31), have been reported by a number of authors. Typically, the hydrophobic moieties are located in the very interior of the macromolecule, attached at one end to the branching point (core), whereas the other end is connected to a hydrophilic block. Nguyen-Van and coauthors described the synthesis of starlike copolymers prepared by ring-opening polymerization of ε-caprolactone using pentaerythritol as initiator followed by chain extension by polymerization of cyclic ethyl ethylene phosphate (Nguyen-Van et al. 2011). The resulting self-assembled structures were with spherical shape and diameters of 150 nm and

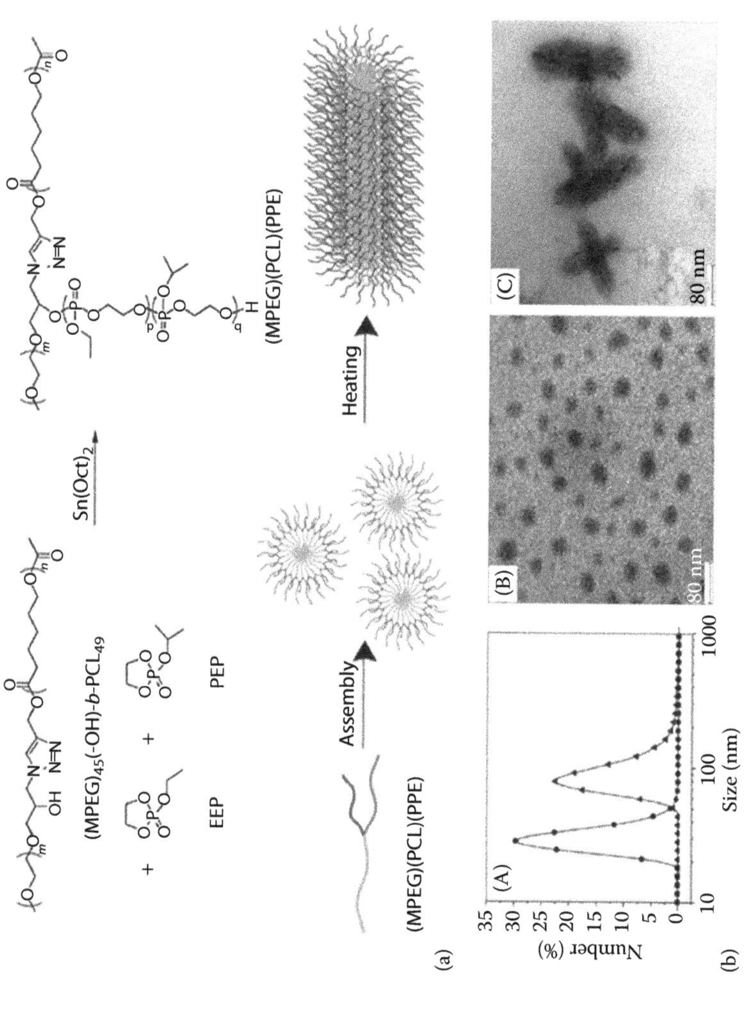

FIGURE 2.29 (a) Chemical pathway for the synthesis of the ABC miktoarm copolymers and schematic illustration of the morphological transition. (b) Size distribution from DLS (A) and TEM images of the self-assembled structures (B and C) formed in aqueous solution at 25°C (circles in A and B) and 50°C (triangles in A and C). (Reprinted from *Colloids Surf. B Biointerfaces*, 85, Yuan, Y.-Y and Wang, J., 81–85, Copyright 2011, with permission from Elsevier.)

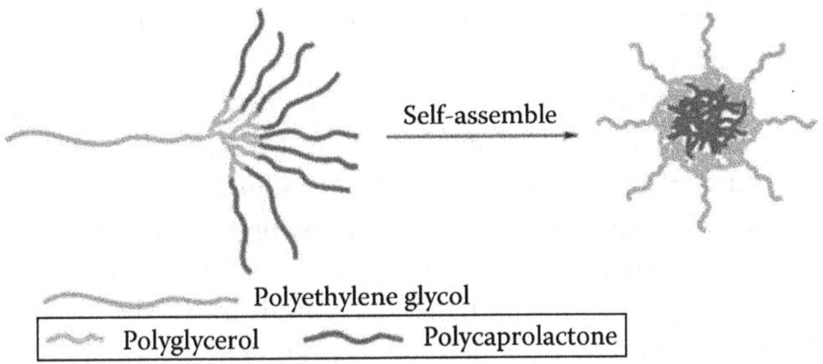

FIGURE 2.30 Self-assembly of a *palm-tree* miktoarm copolymer, composed of PEG, core of hyperbranched polyglycerol and PCL chains. The micelles were obtained by adding water to a THF solution followed by removal of THF by dialysis. (Reprinted with permission from Zhang, X., Cheng, J., Wang, Q., Zhong, Zh., and Zhuo, R., *Macromolecules*, 43, 6671–6677, 2010a. Copyright 2010, American Chemical Society.)

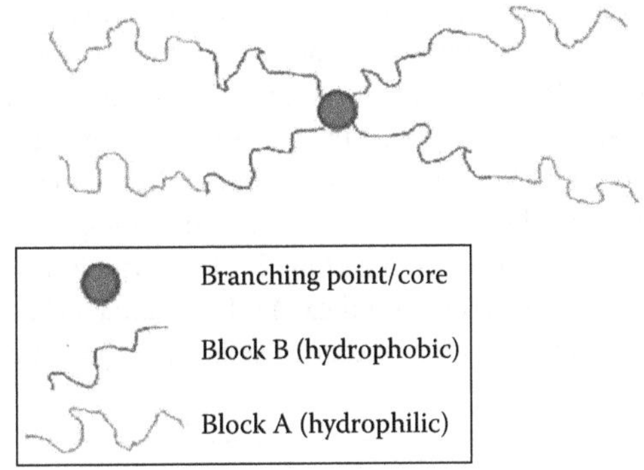

FIGURE 2.31 Schematic presentation of a four-arm star-block copolymer.

showed potential for treatment of human breast cancer. Six-arm star-block copolymers are developed for encapsulation of poorly water-soluble drugs and enhancement of their cellular uptake (Tu et al. 2011). They were composed also of PCL as a hydrophobic moiety, whereas the hydrophilic one was zwitterionic—poly(2-methacryloyloxyethyl phosphorylcholine), in particular. The copolymers exhibited relatively low CMC. The size of micelles ranged from 80 to 170 nm and increased after loading with paclitaxel.

The arms of the star-block copolymers may consist of more than two different in nature block entities. Such four-arm block copolymers have been described elsewhere (Zhu et al. 2011). They contained a hydrophobic PCL segment (block A), a hydrophilic poly[oligo(ethylene oxide) methacrylate] segment (block B), and a thermoresponsive poly[di(ethylene glycol) methyl ether methacrylate] (block C). By reversing the order of polymerizations, two series of copolymers with different block sequences, that is, $(ABC)_4$ and $(ACB)_4$, were prepared. The copolymers of both series formed micelles in water at room temperature. The thermal responses of the micelles, however, strongly depended on the sequence distribution of the blocks along the arms: The copolymers with the thermoresponsive segment on the periphery, that is, $(ABC)_4$, underwent reversible solgel transitions at temperatures between 22°C and 37°C.

The arms of the pentaarmed star-block copolymers were composed of biodegradable PCL and PLA, which formed the cores of the micelles, and PDMAEMA, which provided double (pH and thermal) sensitivity of the corona (Li et al. 2010a). The stimuli sensitivity was examined by following the optical transmittance of micellar dispersions of one particular copolymer ($M_n \sim 45{,}000$) as a function of temperature or pH. The optical transmittance decreased drastically in the 43°C–53°C range with LCST of around 48°C, which is consistent with the LCST of PDMAEMA blocks. The transition as a function of pH at ambient temperature and $c = 10$ mg/mL was also sharp with an inflection point at pH 7. The variation of the hydrodynamic radii as a function of pH is presented in Figure 2.32. As seen, the micelles gradually shrank due to protonation/deprotonation of the tertiary amine groups in the PDMAEMA segments.

The topology of A_8B_4 star-block copolymers described by Nabid and coworkers (Nabid et al. 2011) is schematically presented in Figure 2.33. They have been prepared by a macromolecular coupling reaction between carboxyl-terminated PEG and fourarm star-shaped PCL macromers with eight hydroxyl end groups. The latter were obtained by reacting the starlike PCL with 3-hydroxy-2-(hydroxymethyl)-2-methylpropanoic acid. In water, spherical shaped micelles with sizes ranging from 30 to 50 nm were observed by TEM. Quite similar is the topology of the dendrimer–star copolymers with Y-shaped arms (Zhang et al. 2011) in which the two PEG segments were substituted by poly[2-(2-methoxyethoxy) ethyl methacrylate-co-oligo(ethylene glycol) methacrylate]. As the latter is thermosensitive, the spherical micelles that

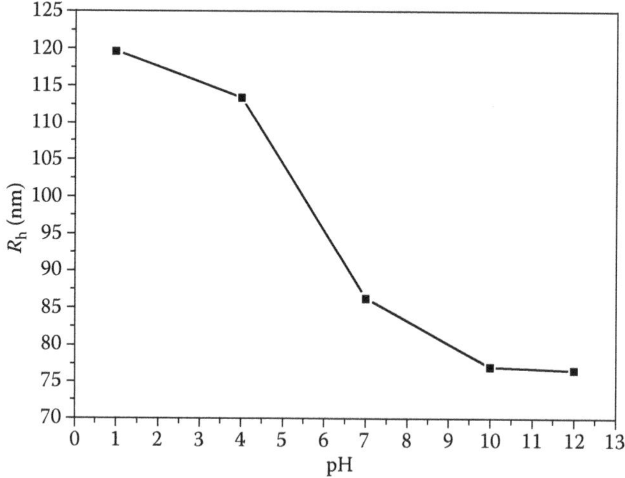

FIGURE 2.32 pH dependence of hydrodynamic radius (R_h) of micelles formed in aqueous solution by a pentaarm star-block copolymer with total M_n of about 45,000 and arms composed of PCL-*b*-PLLA-*b*-PDMAEMA. Measurements were done at ambient temperature and $c = 2$ mg/mL. (Reprinted from *Polymer*, 51, Li, J., Ren, J., Cao, Y., and Yuan, W. 1301–1310, Copyright 2010, with permission from Elsevier.)

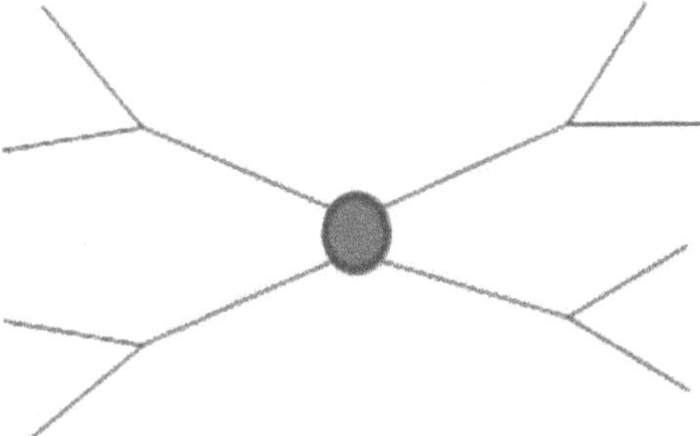

FIGURE 2.33 Schematic presentation of A_8B_4 star-block copolymer, where A and B denote PEG and PCL, respectively.

were formed in water also exhibited tunable thermosensitive properties observed by transmittance, DLS, and TEM.

Kowalczuk and coauthors prepared pH-sensitive amphiphilic starlike copolymers with a large number of arms (Kowalczuk et al. 2011). The copolymers were prepared by ATRP of *tert*-butyl (meth)acrylate using 2-bromoesterified hyperbranched poly(arylene oxindole) as a macroinitiator. The average number of the arms, experimentally determined, was 27.4—a value very close to the number of bromine initiating groups (28), which indicated that the 2-bromoesterified poly(arylene oxindole) core was an efficient macroinitiator. The DPs of the arms varied from 16 to 67. Starlike amphiphilic structures were obtained by acidic hydrolysis. They yielded aggregates in water with dimensions reaching 130 nm and aggregation numbers of up to 280 stars per particle. The dimensions and aggregation numbers were found to decrease with increasing pH as shown in Figure 2.34.

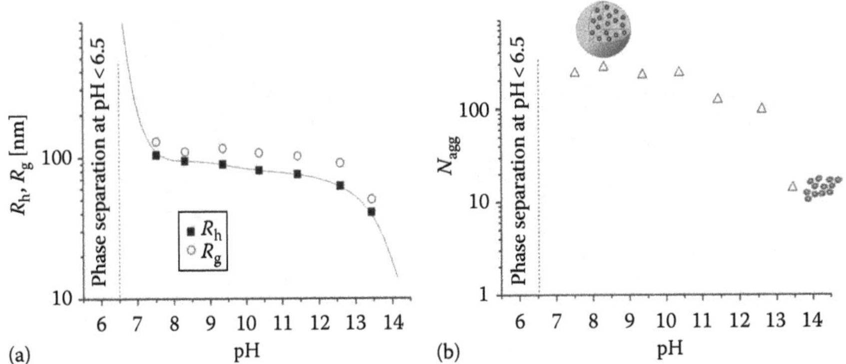

FIGURE 2.34 Variations of (a) the radii of gyration, R_g, the hydrodynamic radii, R_h, and (b) the aggregation number, N_{agg}, of particles formed by a starlike copolymer composed of poly(arylene oxindole) core and ~27 arms of PAA of DP of 37 and total M_n=94,250 g/mol in water with changing pH. (Kowalczuk, A., Trzebicka, B., Rangelov, S., Smet, M., Dworak, A.: *J. Polym. Sci. A—Polym. Chem.* 2011. 49. 5074–5086. Copyright Wiley-VCH Verlag GmbH & Co. KGaA. Reproduced with permission.)

A second-generation poly(benzyl ether) dendron with oligo(ethylene glycol) chains at the periphery has been reported to self-associate in aqueous solution into spherical micelles with ~50 nm in diameter (Xu et al. 2010). Dendritic block copolymers with hydrophobic core and hydrophilic periphery (shell) have been prepared by a multistep process based on anionic ring-opening polymerization (Libera et al. 2011). Amphiphilic stars with four or six arms of poly(*tert*-butyl glycidyl ether)-*b*-polyglycidol were used in the first step. The synthesis of dendritic copolymers with four arms is schematically presented in Figure 2.35. The final copolymers, denoted 3-D_i, represented a third generation of dendritic copolymers and varied in composition, number of hydroxyl groups (from 432 to 1170), and total M_w (90,000–226,000 g/mol). In good solvent (*N,N*-dimethylformamide (DMF)),

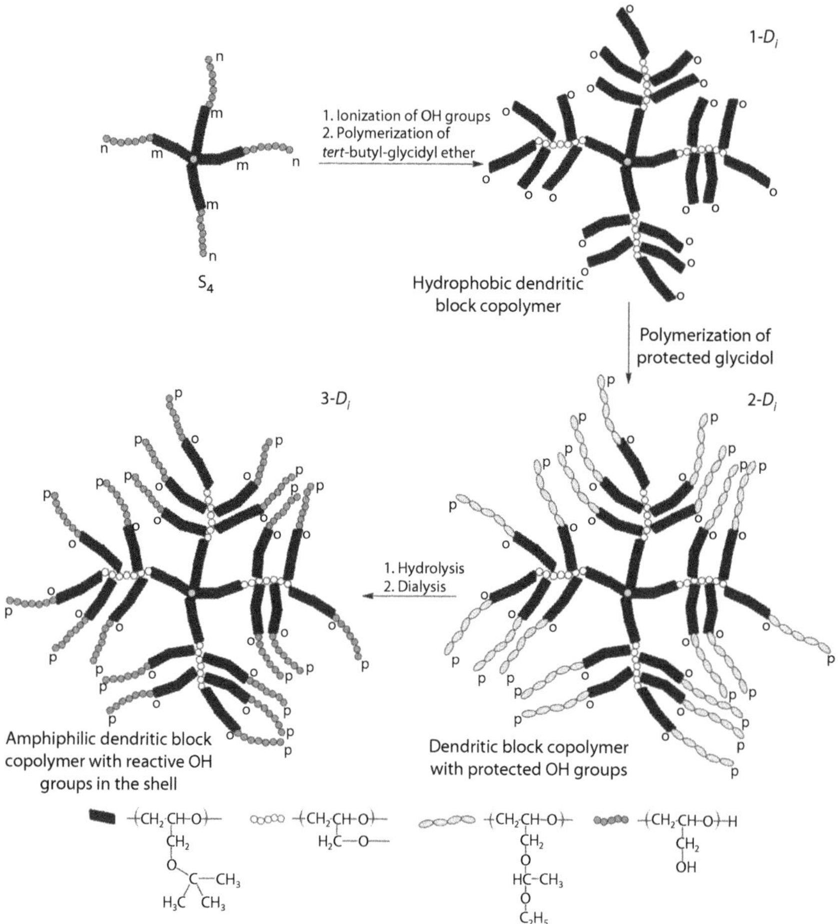

FIGURE 2.35 Schematic representation of the synthesis of dendritic copolymers with star-like macroinitiator with four arms. (Reprinted from *Polymer*, 52, Libera, M., Walach, W., Trzebicka, B., Rangelov, S., and Dworak, A., 3526–3536, Copyright 2011, with permission from Elsevier.)

their radii were in the 6.4–8.0 nm range, which reflected the compact structure. The authors found that the behavior of the dendritic copolymers in dilute aqueous solution was mainly dependent on the number and density of OH groups in the periphery, rather than on the mass ratio of the hydrophobic (*tert*-butyl glycidyl ether) to hydrophilic (glycidol) components. The copolymer with the most numerous OH groups at the periphery did not self-associate neither at 25°C nor at 55°C. That, with the less numerous OH groups, formed at 25°C huge particles of molar mass hundreds of millions corresponding to N_{agg} of almost 3000. The copolymers of intermediate number of OH groups were in aggregated state at 25°C, but the particles consisted only of a few macromolecules. The situation was dramatically different at 55°C at which all light scattering parameters drastically changed indicating secondary aggregation and formation of considerably larger in molar mass and aggregation number particles.

Multiarm copolyphosphates, the hydrophobic core of which is composed of hyperbranched polyphosphate, whereas the hydrophilic moieties consist of linear polyphosphate arms, have been prepared (Liu et al. 2011b). These hyperbranched structures exhibited amphiphilicity and in aqueous solution formed spherical nanoparticles with tunable size about 70–100 nm via adjusting the molar mass of the polyphosphate arms. The hydrophobic core of the self-assembled structures was redox responsive, and in reductive environment, the micellar structures could be destructed, thus triggering drug-release behavior. In contrast to the hyperbranched copolyphosphates, the cores of the branched structures of Wu and coworkers (2011b) and Zhang and coworkers (2011d) are hydrophilic. The former authors describe hyperbranched poly(amine ester), which was chain extended by polymerization of ε-caprolactone. By oil-in-water emulsion technique, nanoparticles were prepared and loaded with camptothecin. β-Cyclodextrin-functionalized hyperbranched polyglycerol has been prepared by the latter authors. Self-assembled structures with size ranging from 200 to 300 nm, high loading capacity, and encapsulation efficiency of paclitaxel were formed in water.

A large group of spherical micelle-forming copolymers of nonlinear topology is the *graft* amphiphilic copolymers. These can be composed of a hydrophilic backbone to which hydrophobic polymeric chains are grafted or just opposite—a hydrophobic backbone with hydrophilic grafts. Some mixed topologies such as block–graft, backbone composed of a copolymeric chain to which homopolymer chains are grafted, and homopolymer backbone with grafted polymeric chains composed of different monomer units organized or not in blocks are also possible. PCL and PEG have been the most frequently employed to design such graft copolymers as hydrophobic and hydrophilic building segments, respectively, but other polymers have also been used.

Sun and coworkers have thoroughly determined the CMC, surface tension at the CMC, as well as dimensions and morphology of spherical micelles formed by a comblike surfactant PMMA-graft-methoxy PEG (Sun et al. 2011). Light-responsive 2-diazo-1,2-naphthoquinone groups, which are hydrophobic, have been incorporated into the structure of comblike PEG (Chen et al. 2011a). Spherical micelles with an average size of about 135 nm were formed in water, which could be disrupted upon exposure to light because the hydrophobic 2-diazo-1,2-naphthoquinone groups were converted into hydrophilic 3-indenecarboxylic acid. Also disassemblable are the micelles prepared from poly(amide amine)-graft-PEG (Sun et al. 2010b). The micelles were spherical of less than 50 nm in diameter and able to load doxorubicin in their cores. The disassembling mechanism involved cleavage of the disulfide linkages throughout the backbone under reductive conditions.

The backbone of the series of copolymers described elsewhere is a cationic fluorine-containing ABC copolymer comprising poly(hexafluorobutyl methacrylate), poly(methacryloxyethyl trimethylammonium chloride), and PS (Xiong et al. 2010). PEG chains were grafted to provide sufficient amphiphilicity and steric stability of the self-assembled structures. The spherical micelles, formed in water, were shown to possess excellent dispersive stability and low toxicity. The potential of the micelles as nonviral gene vectors was investigated (see more in Chapter 5).

Heterograft copolymers have been synthesized by ring-opening copolymerization of ε-caprolactone and cyclic phosphoester functionalized PEG (phosPEG) and PCL as schematically shown in Figure 2.36 (Zhu et al. 2012). The main strategy for the preparation of these copolymers involved the "grafting through" method consisting of polymerization of macromonomers modified with terminal functionality. The authors varied the DP ratios of the macromonomers while keeping constant the DP of PCL in the backbone, which was reflected in variations in the CMC values (in the 0.69–1.09 mg/L interval) and dimensions of the self-assembled structures. The latter

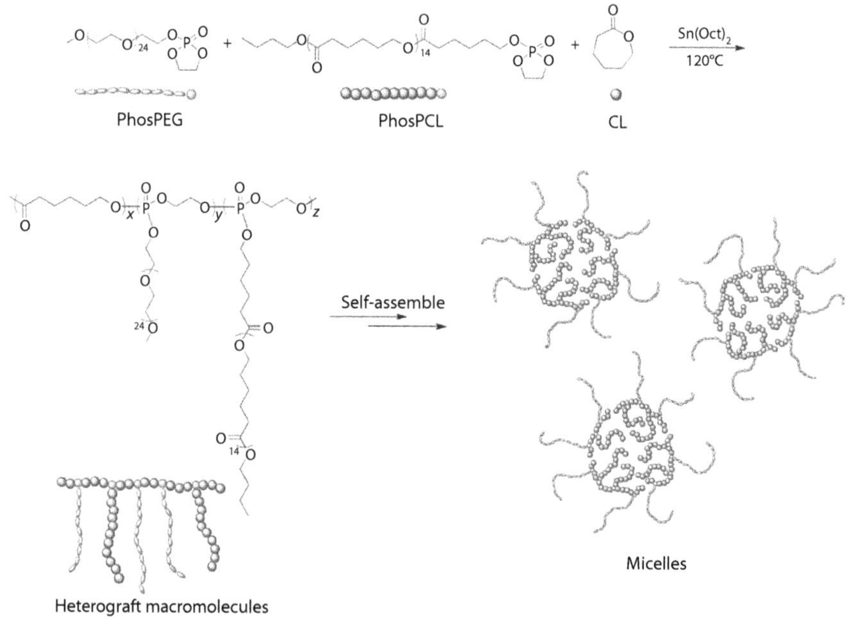

FIGURE 2.36 "Grafting through" method for preparation of heterograft amphiphilic copolymers consisting of cyclic phosPEG, cyclic phosphoester functionalized PCL (phosPCL), and linear PCL and their self-assembly into spherical micelles in water. (Zhu, W., Sun, Sh., Xu, N., Gou, P., and Shen, Zh.: *J. Appl. Polym. Sci.* 2012. 123. 365–374. Copyright Wiley-VCH Verlag GmbH & Co. KGaA. Reproduced with permission.)

are schematically presented in Figure 2.36 and were found to increase in size from diameters of 35.4 to 53.5 nm with increasing the total PCL content.

Using multifunctional poly(*tert*-butyl methacrylate)-co-poly(2-hydroxy-3-azido-propyl methacrylate) bearing hydroxyl and azide groups from junction points and combining a "grafting from" strategy by ring-opening polymerization of EP and "grafting to" strategy by *click reaction* with α-propargyl-ω-acetyl-PCL, heterograft copolymers with PCL and poly(ethyl ethylene phosphate) segments have been prepared (Yuan et al. 2010). These *centipede-like* brush copolymers were amphiphilic and formed in water and above the CMC of ~10^{-3} mg/mL spherical micelles with diameters of 50–90 nm.

A graft copolymer with side chains of PNIPAM, attached to poly(2-oxepane-1,5-dione)-co-PCL, has been reported by He and coworkers (2010). The copolymer was characterized with lower CMC compared to its linear analogue and smaller in size spherical core–corona micelles. Furthermore, besides the thermoresponsiveness due to the PNIPAM chains, pH sensitivity, attributed to the acid-cleavable property of the hydrazone bond, has been shown.

Poly(β-amino ester)s are a class of cationic polymers containing amino groups that have been originally developed for gene delivery. They are hydrophilic and show good pH sensitivity with short responsive times and sharp transitions. Generally, linear poly(β-amino ester)s can be prepared by Michael addition polymerization of diacrylate with secondary bisamine. Novel approaches to synthesize amphiphilic graft copolymers with hydrophilic PEG-based poly(β-amino ester) backbone and hydrophobic side chains of PCL (Li et al. 2011b) or octadecyl groups (Chen et al. 2011b) have been recently reported. By a DCC coupling method carboxyl-terminated PCL with DP=22 was grafted to a PEG-based poly(β-amino ester) chain with an average grafting degree of ~20%. The copolymer self-associated into spherical objects at concentrations above 3 mg/L with dimensions that were slightly sensitive to pH variations unlike the ζ-potential (Figure 2.37). Practically the same (~72.5 nm) were the dimensions of the micelles prepared from the poly(β-amino ester) grafted with octadecyl groups (Chen et al. 2011b). These copolymers, however, exhibited higher CMC values—in the 25–45 mg/L range depending on the degree of grafting, which was controlled by the feeding ratios and reaction time. Stable micelles were obtained at ~40% content of octadecyl chains.

Block–graft is the topology of the copolymers of Lee and coworkers (Lee and Wu 2011). Hydroxyl-terminated PNIPAM was used to initiate the ring-opening polymerization of α-chloro-ε-caprolactone, thus making diblock (PNIPAM-*b*-PCL) copolymers with pendant chlorides. The latter were substituted by sodium azide, and following *click reaction* with various kinds of terminal alkynes, the PCL was modified. The copolymers self-associated at very low CMCs (~2–10 mg/L) into spherical micelles of core–corona structure and dimensions around 75–145 nm. The solutions exhibited also LCST behavior with reversible changes in optical properties.

2.2.1.2.4 Hybrid Copolymers

Among the hybrid copolymers that self-associate into spherical micelles, those prepared from linear or branched polysaccharides are the most numerous due to availability, biocompatibility, and bioinertness as well as immunoneutrality of

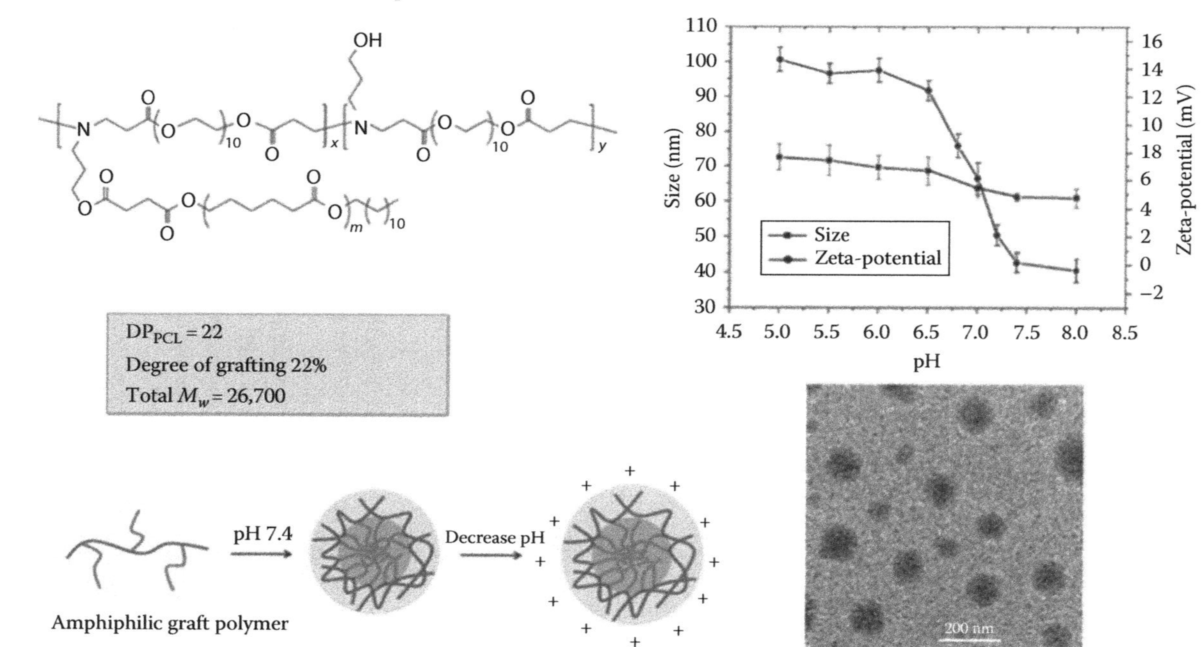

FIGURE 2.37 Structural formula and micellar properties of a pH-sensitive PCL-grafted PEG-based poly(β-amino ester). (Reprinted from *Polymer*, 52, Li, Y., Heo, H. J., Gao, G.H., Kang, S.W., Huynh, C.T., Kim, M.S., Lee, J.W., Lee, J. H., and Lee, D.S., 3304–3310, Copyright 2011, with permission from Elsevier.)

these natural products. They are typically modified with hydrophobic chains such as PCL, PLA, PLGA (poly(D,L-lactic-co-glycolic acid)), or other hydrophobic moieties. Trimethylated chitosan, a linear polysaccharide composed of randomly distributed β-(1–4)-linked D-glucosamine (deacetylated unit) and *N*-acetyl-D-glucosamine (acetylated unit), of varying molecular weights grafted with PCL has been synthesized (Zhang et al. 2011a). The authors studied the effect of chitosan molecular weight, PCL grafting density, and trimethylation degree on the organosolubility, self-assembly, and cytotoxicity of the copolymers. Spherical cationic micelles with average diameter 25–55 nm and uniform morphology were formed in pH 7.4. By reducing the trimethylation degree and/or increasing PCL grafting levels, the cytotoxicity of the cationic micelles could be suppressed. Also grafted by PCL are the copolymers based on oligoagarose (Bhaw-Luximon et al. 2011). These graft copolymers exhibited amphiphilic behavior and formed in water small (10–20 nm) spherical micelles into the cores of which ketoprofen was loaded.

Dextran, a branched polysaccharide, the straight chain of which consists of α-1,6-glycosidic linkages between glucose units, while branches begin from α-1,3-linkages, has been frequently used as a suitable modification platform to create, for example, drug carrier systems. In a recent study, a block copolymer composed of dextran and PLGA has been prepared by reacting aminated dextran of molecular weight ~5200 and *N*-hydroxysuccinimide-terminated PLGA ($M_w = 4900$) (Jeong et al. 2011). Spherical aggregates of dimensions about 80 nm were prepared in water by dialysis. They were able to incorporate doxorubicin with contents up to 12.3%. The incorporation resulted in a considerable increase of particle dimensions—twice for the highest doxorubicin content. Dextran of lower molecular weight (1500 Da) has been used to prepare novel amphiphilic copolymers (Wang et al. 2010b, 2011b). In the former paper, 1,2-dipalmitoyl-*sn*-glycero-3-phosphoethanolamine (DPPE) was reacted with activated dextran to prepare hydrophobically modified dextran-DPPE. The latter authors firstly prepared dextran-PLA to which DPPE was attached, thus forming dextran-PLA-DPPE copolymer. The composition of the copolymers could be controlled by adjusting the ratios of the lactic acid and/or DPPE to activated dextran. Owing to good biocompatibility of all components, the copolymers showed promises as nanocontainers for hydrophobic drugs. Both copolymers self-associated in water into spherical, as evidenced by TEM, particles above certain CMC. The latter was approximately an order of magnitude higher (6.42×10^{-2} vs. 2.45×10^{-3} mg/mL) for dextran-DPPE. This copolymer also formed smaller in size (30–60 nm) micelles compared to those of the more hydrophobic dextran-PLA-DPPE (~150 nm). By ATRP of (meth)acrylate monomers, diblock and triblock (ABA-type) glycopolymers consisting of poly(butyl acrylate) and poly(2-{[(*d*-glucosamine-2-*N*-yl) carbonyl}oxy} ethyl methacrylate) have been prepared (Leon et al. 2011). DLS and TEM, performed to study their self-assembly in aqueous solution, indicated coexistence of micelles and polymeric vesicles. The self-assembled structures contained glucose moieties in their coronae, which were valuable for recognition and binding reactions.

Polypeptides in combination with synthetic blocks or moieties constitute an interesting class of *hybrid copolymers* able to self-associate particularly in spherical micelles. The secondary interactions along the polypeptide chains precisely controlled by the specific peptide sequences allow formation of well-defined nano- and microstructures. In addition, the synthetic polymer moieties provide control over the biologic activity of the polypeptide and may also induce self-assembly. Coupling (Lutz et al. 2006), "grafting from" (Hentschel et al. 2007), and macromonomer (Ayers et al. 2003) approaches have been the most popular synthetic approaches for preparation of polymer–polypeptide conjugates. In the coupling approach, the peptide segments are coupled to one or multiple reactive sites of a polymer block/moiety. The "grafting from" approach uses polypeptide to initiate polymerization of a desired monomer, whereas the macromonomer approach involves polymerization of polypeptide segments bearing a polymerizable group. There exists also an inverse conjugation strategy, where peptide is sequentially elongated on a synthetic polymer chain (Eckhardt et al. 2005).

Via ring-opening polymerization of benzyl glutamate *N*-carboxyanhydride in the presence of PEG as a macroinitiator, a series of diblock copolymers of PEG and poly(γ-benzyl-L-glutamate) (PBLG) were prepared (Thambi et al. 2011). The copolymers were of fixed PEG length and varying block length of the PBLG moieties and in aqueous solution self-assembled into spherical micelles sizing ca. 20–100 nm in diameter depending on the PBLG block length. Importantly, the copolymers bear a disulfide bond, which is cleavable under acidic conditions resulting in fast decomposition of the micelles. Hybrid amphiphilic copolymers based on poly(L-alanine) with PEG as a hydrophilic block are described elsewhere (Jiao et al. 2011). The copolymers were conjugated with folate to produce folate receptor-targeted drug carriers. CMCs in the micromolar range, spherical shape, and nanoscale dimensions (55–75 nm in diameter) of the micelles were exhibited. The copolymers described in the study of Qiao and coworkers are also based on poly(L-alanine) (Qiao et al. 2010). The hydrophilic PEG blocks, however, were substituted by PNIPAM, which is thermosensitive and exhibits LCST properties. Spherical core–corona micellar structure and reversible phase transition within a very narrow temperature range have been reported. Both thermal and pH sensitive are the copolymers described elsewhere (Lee et al. 2010a). They have been prepared by condensation polymerization of trans-4-hydroxy-L-proline initiated by hydroxyl-terminated PNIPAM in the presence of $Sn(Oct)_2$ as a catalyst. TEM showed spherical morphology and core–corona structure also proven by ^1H NMR. *Schizophrenic* self-assembly has been reported for PPO-poly(L-lysine) diblock copolymers (Naik et al. 2011). These materials exhibited both pH and temperature sensitivities resulting in formation of various structures in which the constituent moieties exchanged their locations (Figure 2.38).

Park and coauthors have found significantly different temperature sensitivities as well as different nanoassemblies in water of two copolymers of the same composition but with ABC and ACB sequences, where A, B, and C denote PEG, poly(L-alanine), and poly(D,L-alanine), respectively (Park et al. 2011). The location of poly(L-alanine), which has an α-helical secondary structure, in the copolymer chains largely influenced the aqueous solution properties of the two copolymers. Schematic representation of the self-assembled structures and the observed thermotropic transitions are shown in Figure 2.39. The spherical micelles, sizing 10–30 nm from DLS,

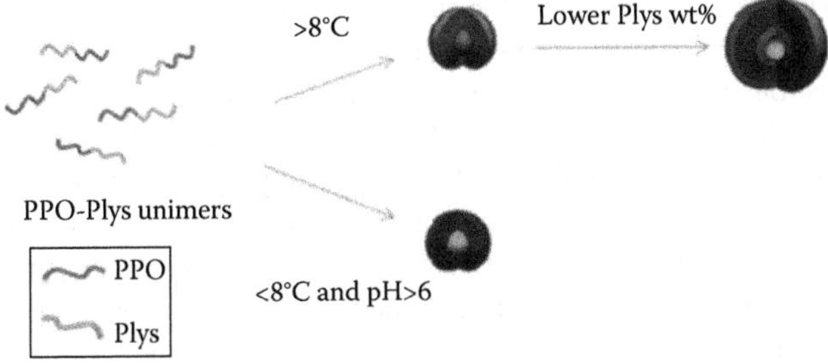

FIGURE 2.38 *Schizophrenic* self-assembly of PPO-*b*-poly(L-lysine) copolymers. (Reprinted with permission from Naik, S., Ray, J.G., and Savin, D., *Langmuir*, 27, 7231–7240, 2011. Copyright 2011, American Chemical Society.)

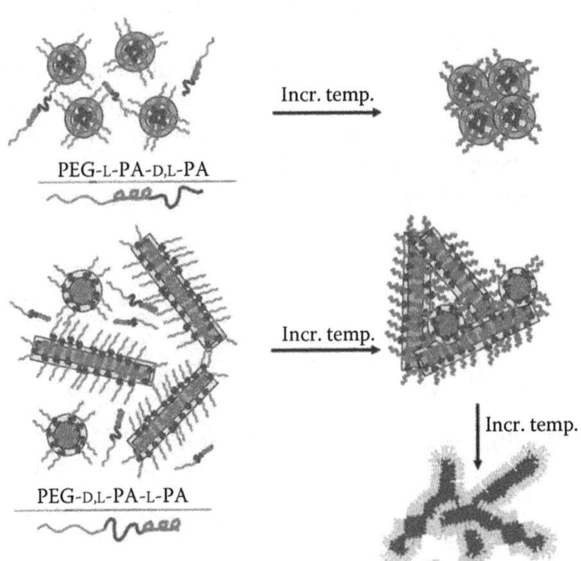

FIGURE 2.39 Schematic presentation of the conformational changes of PEG-poly(D,L-alanine)-poly(L-alanine) and PEG-poly(L-alanine)-poly(D,L-alanine) in water as a function of temperature. Thin gray, black, and thick gray lines indicate the hydrophilic flexible PEG, hydrophobic poly(D,L-alanine), and poly(L-alanine), respectively. As the temperature increases, the PEG (thick dark gray) is partially dehydrated (thick light gray) and slight conformational changes of polyalanine, leading to the formation of a gel. Poly(L-alanine) of PEG-poly(D,L-alanine)-poly(L-alanine) further undergoes conformational change from *a*-helix to random coil and further dehydration of PEG (light gray) to form a squeezed gel. (Park, S.H., Choi, B.G., Moon, H. J., Cho, S.-H., and Jeong, B., *Soft Matter*, 7, 6515–6521, 2011. Reproduced by permission of The Royal Society of Chemistry.)

were compartmentalized, whereas the nonspherical particles, also compartmentalized, were considerably larger: ~150 nm and above.

Stimuli-responsive micelles with an α-helical core and a thermoresponsive corona have been obtained in water from a thermoresponsive graft polypeptide (Ding et al. 2011). The copolymer, poly(L-glutamate)-graft-poly[2-(2-methoxyethoxy)ethyl methacrylate], was prepared by ring-opening polymerization of the corresponding *N*-carboxyanhydride followed by ATRP. The copolymer adopted α-helical conformations both in aqueous solution at 25°C and 60°C and in solid state, which was preserved also in the micellar state. The phase transition temperature was controlled simply by varying the concentration of NaCl in the solution.

A heterograft copolymer based on polyaspartylhydrazide (see the formula in Figure 2.40) has been prepared by Liccardi and coworkers (Liccardi et al. 2010). Polyaspartylhydrazide is a water-soluble, nontoxic, and biocompatible polymer that finds a number of biomedical applications. The authors chose palmitic acid and PEG 2000 to, respectively, graft the polymer in order to provide hydrophobic interactions and sterically stabilize the self-assembled particles. The latter as well as polyaspartylhydrazide molecular weight and degrees of derivatizations of PEG 2000 and palmitic acid are schematically presented in Figure 2.40. The particles were of spherical morphology and an average diameter of about 30 nm.

The synthesis and characterization of a well-defined thermosensitive block conjugate consisting of fluorescently labeled pentapeptide, glycine–arginine–lysine–phenylalanine–glycine–dansyl, and poly[di(ethylene glycol) monomethyl ether methacrylate] have been recently reported (Trzcinska et al. 2012). Here, the pentapeptide is hydrophilic, whereas the thermosensitive properties were conveyed by

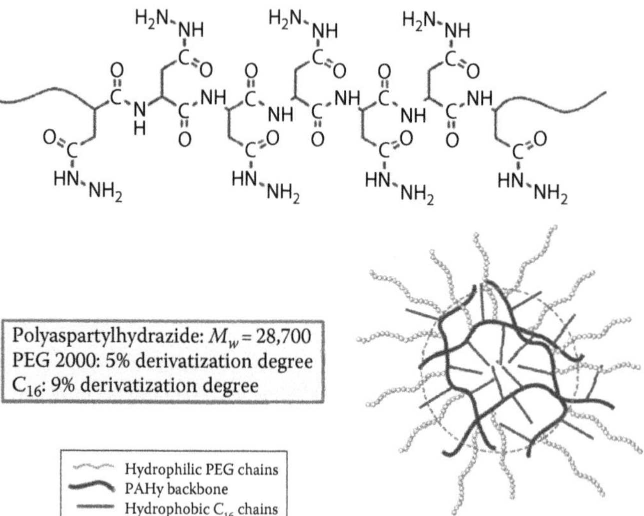

FIGURE 2.40 Structural formula of polyaspartylhydrazide and schematic representation of a micelle formed by a heterograft copolymer based on polyaspartylhydrazide. (Reprinted from *Int. J. Pharm.*, 396 (1–2), Liccardi, M., Cavallaro, G., Di Stefano, M., Pitarresi, G., Fiorica, C., and Giammona, G., 219–228, Copyright 2010, with permission from Elsevier.)

poly[di(ethylene glycol) monomethyl ether methacrylate], which is an LCST polymer showing a transition temperature of 26°C. The pentapeptide was obtained by the standard solid-phase peptide synthesis and was modified by 2-bromopropionic acid to initiate ATRP of di(ethylene glycol) monomethyl ether methacrylate (Figure 2.41). In aqueous solution, the block conjugate underwent a sharp phase transition, which was shifted by approximately 4°C to higher temperatures compared to homo poly[di(ethylene glycol) monomethyl ether methacrylate] due to the presence of the hydrophilic pentapeptide moiety. The mode of heating—abrupt or gradual—was found to strongly influence the size of the particles formed above the transition temperatures (Figure 2.41). The aggregates consisted of a core formed by partially dehydrated, collapsed poly[di(ethylene glycol) monomethyl ether methacrylate] chains surrounded by a thin corona formed by the attached peptide chains. The light scattering parameters were determined at 45°C (Figure 2.41). The particle aggregation number corresponded to approximately 1620 macromolecules per particle. A value of 57.5 nm was obtained for the hydrodynamic radius, which gave an R_g/R_h value of 0.89. The latter is above the theoretical value for hard spheres (0.778) and close to what is typically observed for micelles with large and bulky cores and relatively thin hydrophilic coronae, crew-cut micelles. Noteworthy, even in an aggregated state, the peptides were fully accessible to enzymatic digestion.

Soy protein isolate, which has recently attracted great attention, has been used to prepare thermally responsive graft copolymers with PNIPAM (Li et al. 2011d). Both soy protein isolate and the graft copolymer formed in water huge aggregates (100–700 nm in radius and molar masses reaching 10×10^6 Da). Their dimensions and morphology were dependent on temperature, pH, ionic strength, and concentration.

Generally, studies on nucleic acid/synthetic polymer conjugates, their self-assembly, and possible and real applications are relatively scarce. Decoration of the self-assembled structures with DNA or RNA chains would afford novel biofunctions such as targeted drug/gene delivery, biodiagnostics, and gene silencing that cannot be provided by the traditional polymeric micelles. Besides the various lipophiles such as cholesterol, fatty acids, and bile acids, siRNA has been chemically conjugated with PEG via a reducible disulfide linkage to form *doubly hydrophilic* conjugates that have been complexed with cationic species like poly(ethyleneimine), fusogenic peptides, solid lipid nanoparticles, and gold nanoparticles (Lee et al. 2007, 2008, Kim et al. 2008a,b). Preparation of *amphiphilic* hybrid diblock conjugates based on siRNA and their micellization in water have been reported elsewhere (Lee and Park 2011). The copolymer was synthesized by conjugation of PLGA, modified by 3-(2-pyridyldithio) propionyl hydrazide, to thiol-modified siRNA as shown in Figure 2.42. In aqueous solution, the conjugate formed core–corona micelles with PLGA in the core and negatively charged siRNA in the corona. Well-defined spherical aggregates were observed by atomic force microscopy (AFM) with an average diameter from DLS of 22 nm and zeta-potential of −16.6 mV (Figure 2.42).

In the last decade, linear and graft polymer–DNA conjugates have been prepared (Jeong and Park 2001, Mori and Maeda 2002, 2004, Li et al. 2004, Oishi et al. 2005, Alemdaroglu and Herrmann 2007, Ding et al. 2007, Fluegel and Maskos 2007, Teixeira et al. 2007). A recent paper reports on the synthesis of DNA-functionalized

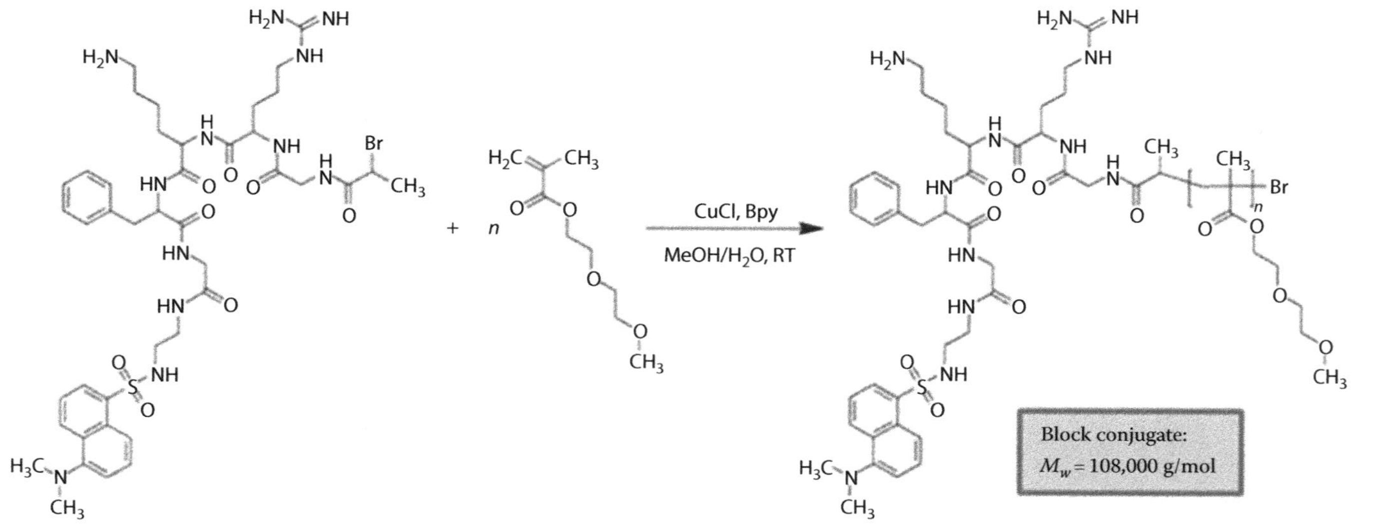

CuCl, Bpy

MeOH/H_2O, RT

Block conjugate:
M_w = 108,000 g/mol

(a)

FIGURE 2.41 (a) Synthesis of thermosensitive block conjugate consisting of fluorescently labeled pentapeptide, glycine–arginine–lysine–phenylalanine–glycine–dansyl (GRKFG-Dns), and poly[di(ethylene glycol) monomethyl ether methacrylate] (P(DEGMA-ME)) via ATRP.

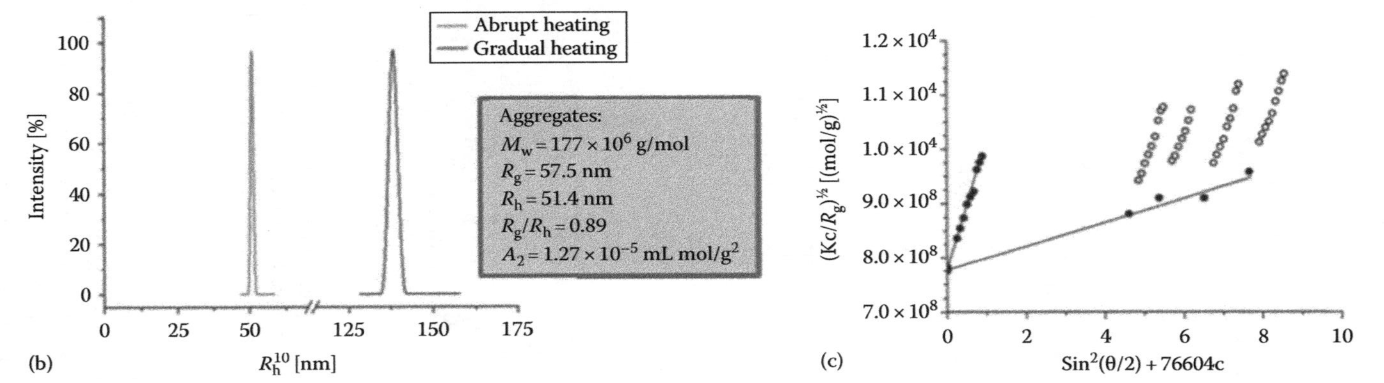

FIGURE 2.41 (continued) (b) Aggregate size distribution measured at 45°C upon abrupt (light gray) and gradual (dark gray) heating in aqueous solution with $c = 0.1$ g/L). (c) Berry plot of P(DEGMA-ME)-GRKFG-Dns mesoglobules in water at 45°C. Concentration interval, $c = 0.1 \div 0.047$ g/L, with abrupt heating of the initial dispersion and in situ dilution to the desired concentration. Block conjugate molar mass and static and DLS parameters of the aggregates are presented in the blue boxes. (Trzcinska, R., Szweda, D., Rangelov, S., Suder, P., Silberring, J., Dworak, A., and Trzebicka, B.: *J. Polym. Sci. A—Polym. Chem.* 2012. 50(15). 3104–3115. Copyright Wiley-VCH Verlag GmbH & Co. KGaA. Reproduced with permission.)

DCC/NHS
PDP-Hydrazide (PDPH)
in MC

siRNA-SH

(a)

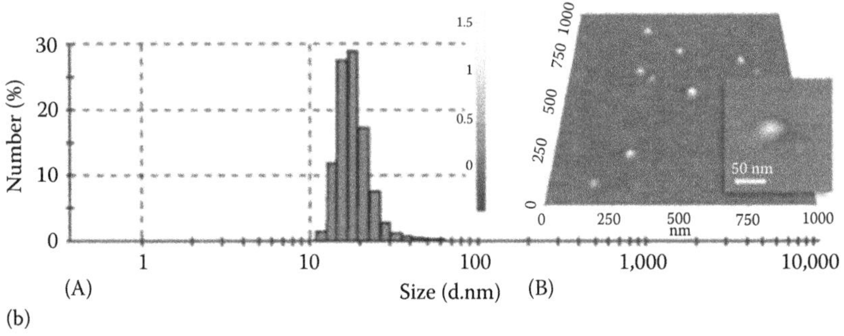

(b)

FIGURE 2.42 (a) Synthetic route for preparation of PLGA-siRNA conjugate. (b) Size distribution (A) and AFM images (B) of micelles formed by the conjugate. (Reprinted from *J. Control. Release*, 152, Lee, Y. and Park, T. G., 152–158, Copyright 2011, with permission from Elsevier.)

diblock and miktoarm starlike copolymers by ATRP and *click* chemistry (Pan et al. 2011). The copolymers consisted of PNIPAM and single-stranded DNA (ssDNA). They were of comparable M_n—29,200 and 26,000 Da for the diblock and miktoarm copolymer, respectively. Their aqueous solution did not show clouding, but at temperatures above the LCST of PNIPAM, small particles of presumably core–corona structure as schematically depicted in Figure 2.43 were registered.

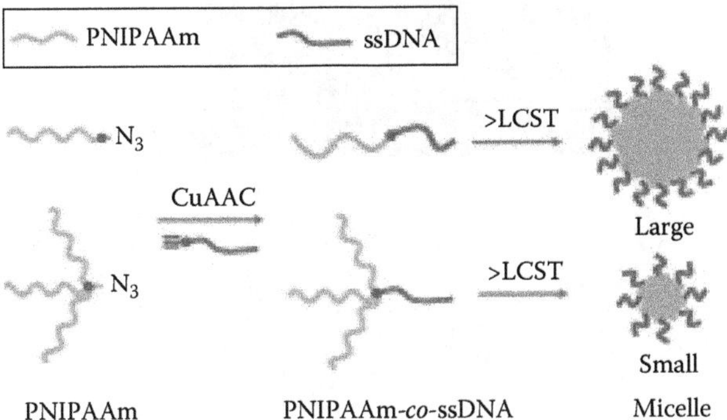

FIGURE 2.43 Schematic presentation of the self-assembly of linear diblock and miktoarm starlike bioconjugate copolymers composed of PNIPAM and ssDNA. (Reprinted from *Polymer*, 52, Pan, P., Fujita, M., Ooi, W.-Y., Sudesh, K., Takarada, T., Goto, A., and Maeda, M., 895–900, Copyright 2011, with permission from Elsevier.)

The miktoarm copolymer formed smaller in size ($R_h = 15.2$ vs. 24.3 nm) and aggregation number (67 vs. 137) micelles that appeared rather compact judging from the R_g/R_h ratio (0.71–0.73).

2.2.2 Wormlike Aggregates

2.2.2.1 General Features

The wormlike micelles are the second of the three basic morphologies—sphere, cylinder, and lamella. They are often referred to as threadlike, rodlike, polymerlike, or cylindrical micelles and have a mean curvature between those of spherical micelles and vesicles. The range of S allowable for this morphology is between 1/3 and 1/2, corresponding to hydrophilic weight fraction in the 0.47–0.60 range (see Figure 2.8). This range is two-to-three-fold narrower compared to those of either spherical micelles or vesicle bilayers, which explains why aggregates with wormlike morphology are relatively less frequently observed.

The wormlike micelles are classic 1D self-assembled structures for which the length is (at least) several times greater than their cross-sectional diameter. These micelles exhibit rich structural polymorphism: They can be relatively short, that is, of finite length, or very long, reaching the micrometer length range; they can be also rigid (with high persistent length) or very thin and flexible. Micelles that are relatively short and rigid are usually termed *cylinders* or *rods*, whereas those that are long and thin are called *threadlike* micelles. In the following, we will use these terms as synonyms, and wormlike micelles will be used when a more general situation is described. The wormlike micelles are typically characterized by the contour length, L, and persistence length, l_p. In analogy with linear polymers, the former is defined as the maximum end-to-end distance, whereas the latter as the average projection of the end-to-end vector on the tangent of the chain contour at a chain end in the limit of infinite chain length. l_p is a measure of the rigidity with the higher the persistence length, the more rigid and less flexible the wormlike structure. Another parameter is

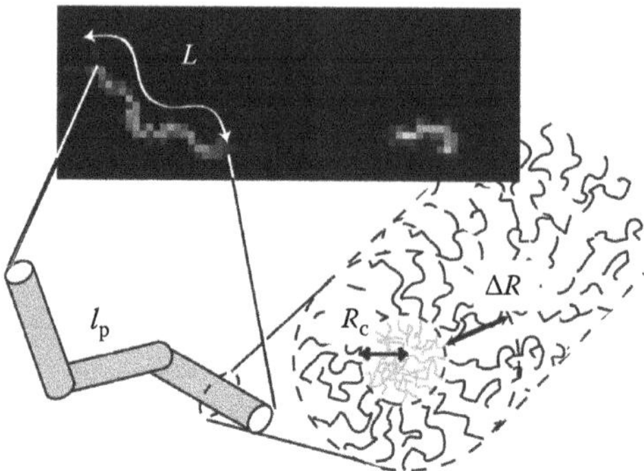

FIGURE 2.44 Fluorescence microscopy image of wormlike micelles and schematic presentation of the contour length, L, persistence length, l_p, core radius, R_c, and corona thickness, ΔR. (Reprinted with permission from Lonetti, B., Tsigkri, A., Lang, P.R., Stellbrink, J., Willner, L., Kohlbrecher, J., and Lettinga, M.P., *Macromolecules*, 44 (9), 3583–3593, 2011. Copyright 2011, American Chemical Society.)

the cross section, which is given as the corona thickness, ΔR, and the core radius, R_c. These characteristics are schematically presented in Figure 2.44.

Ideally, the wormlike micelles could be infinitely long and linear structures, which result from the optimization of the aggregate energy through creating uniform curvature everywhere. As stated elsewhere (Dan and Safran 2006), the system entropy introduces a degree of randomness through bending of the cylindrical micelles and appearance of topological defects in the form of end caps and/or branching. These authors have shown that cylindrical systems characterized by a relatively high interfacial curvature favor end-cap defects resulting in shortening of the length and formation of linear structures, whereas the systems with relatively high spontaneous curvature favor branching and formation of extended network structures. These two structures exhibit significantly different solution properties. In general, compared to the spherical micelles and polymer vesicles, the wormlike aggregates are considerably less stable to perturbations, and phase and morphology transitions have been frequently documented. Their unique elongated shape has been exploited in numerous applications. The susceptibility to transitions sometimes facilitates specific applications and generally motivates deeper insight. Dispersions of wormlike micelles typically behave as non-Newtonian liquids and exhibit elasticity (i.e., storage modulus, $G' >$ loss modulus, G'') with G' and G'' being typically larger than those of, for example, spherical micelles. The wormlike micelles appear surprisingly useful in their ability to reptate through the bloodstream and exhibit a long circulation time. Larger solubilization capacity, implying potentially larger payload per micelle than that of the spherical micelles, has been displayed as well.

2.2.2.2 Copolymer Chain Architecture, Nature of the Constituent Blocks, and Properties of the Resulting Worms

A variety of amphiphilic polymers have been reported to form at appropriate conditions wormlike aggregates or to undergo morphological, for example, sphere to cylinder or lamella to cylinder, transitions upon variation of certain external stimuli or addition of a third component to the binary polymer/water solution. It will be shown in the following that these are di- or triblock copolymers of linear chain architecture as well as starlike block copolymers and miktoarm and graft copolymers. The nature of the constituent moieties is also rather diverse: Synthetic polymers such as PEO, PS, or polyethylene; naturally occurring dextran and chitosan; as well as oligopeptides have been used in combinations to create amphiphilic copolymers that are able to form wormlike aggregates in aqueous solution.

Since their introduction as commercial products in the 1950s, the morphological and phase transitions of one of the most extensively studied polymeric surfactants—PEO-PPO-PEO triblock copolymers—with increasing concentration and/or temperature have been of considerable academic and industrial interest. The rich structural polymorphism and phase behavior are mainly dependent on the copolymer molecular weight and composition as shown in the previous and next sections. The copolymers that, according to *S*, are expected to form wormlike aggregates are those with moderate PEO contents, that is, 30–50 wt.%. Most typically, these copolymers form spherical micelles at ambient temperatures and low (but invariably above CMC) concentrations. Upon heating, a morphological transition to elongated wormlike aggregates is observed. Sphere-to-rod transition has been observed also for copolymers of higher PEO content, for example, 70%–80%; however, it only appears at high temperatures—transitions to cylinders have been reported to take place at ~90°C–95°C (Mortensen and Brown 1993, Mortensen and Talmon 1995) and lower, that is, 80°C and 70°C for copolymers with 50% and 40%, respectively, PEO content (Mortensen 2001a, Guo et al. 2006). These copolymers in particular have been investigated by SANS as a major technique, and the results are best fitted to very long and relatively thin structures. For example, for the best fit of the results at 70°C, the length and core diameter of the wormlike micelles that $EO_{19}PO_{43}EO_{19}$ forms are fixed to 4000 and 70 Å, respectively (Guo et al. 2006). Slightly thinner (50 Å) and considerably shorter (500 Å) but still very long with a length-to-diameter ratio (aspect ratio) of 10 are those of $EO_{25}PO_{40}EO_{25}$ at 80°C (Mortensen 2001a). Generally, the sphere-to-cylinder transition of PEO-PPO-PEO copolymers is observed at temperatures very close to CPs. For copolymers of lower PEO contents, the CPs are accordingly at lower temperatures. The comparative analysis of SANS data based on the spherical and prolate ellipsoidal structure has shown that the $(EO)_{13}(PO)_{30}(EO)_{13}$ (40 wt.% PEO content and CP = 58°C) micelles can be best described by a prolate ellipsoidal structure, the aspect ratios of which increase with increasing temperature, which, together with viscosity and DLS results, suggests anisotropic growth of the micelles to wormlike structures at temperatures near the CP (Ganguly et al. 2009a). Similar behavior has been described for $EO_{20}PO_{70}EO_{20}$ (Ganguly et al. 2009b). These authors report on time-dependent sphere-to-rod micellar growth on approaching CP, the rate of which increases with increasing temperature. Time-dependent sphere-to-rod transition has also been observed for $EO_{20}PO_{70}EO_{20}$ at room temperature as described elsewhere (Denkova et al. 2009). In this paper, the transition is induced by a solvent jump initiated by adding a structure-making salt (KCl) and ethanol. Figure 2.45 shows the evolution of the aggregate structure with time followed by cryo-TEM. The salts in principle act as a dehydrating agent for the hydrophilic

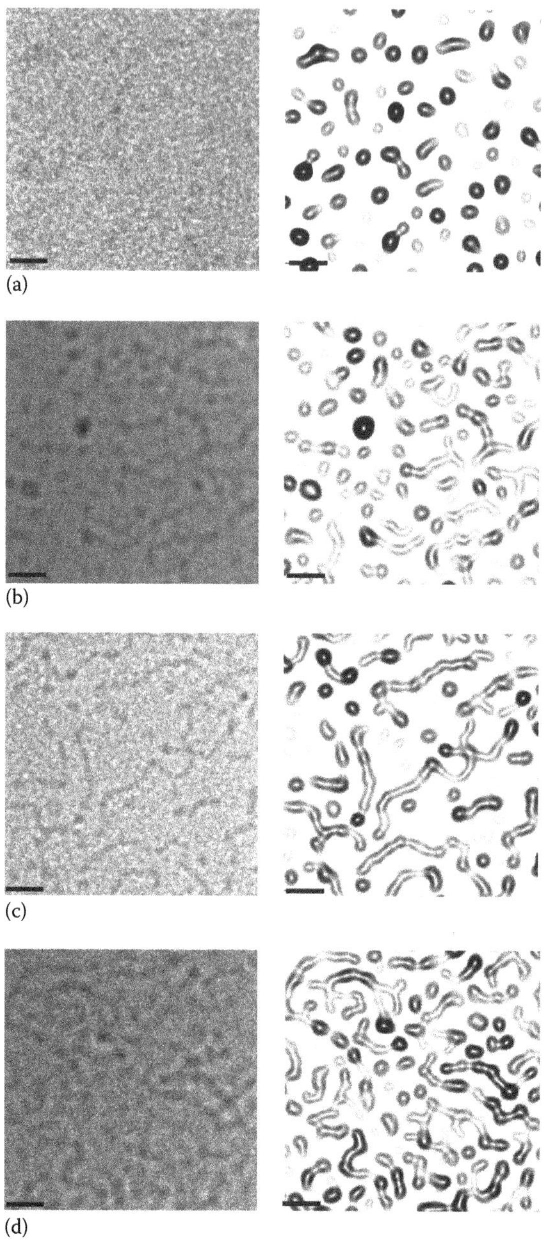

FIGURE 2.45 Time-dependent sphere-to-rod transition of $EO_{20}PO_{70}EO_{20}$ micelles. Samples for cryo-TEM were taken at room temperature from dispersions containing 40 g/L $EO_{20}PO_{70}EO_{20}$, 2.0 M KCl, and 8 vol.% ethanol at 5 min (a), 1 h (b), 2 h (c), and 4.5 h (d). Contrast-enhanced copies are shown on the right. The bar corresponds to 50 nm. (Reprinted with permission from Denkova, A.G., Mendes, E., and Coppens, M.-O., *J. Phys. Chem. B*, 113, 989–996, 2009. Copyright 2009, American Chemical Society.)

PEO corona and have an effect of increasing temperature. That is why wormlike PEO-PPO-PEO micelles can be formed also at room temperature in the presence of small quantities of water structure-making salts as shown for $(EO)_{13}(PO)_{30}(EO)_{13}$ and NaCl (Ganguly et al. 2009a). Heating–cooling cycles to and below the CP of $EO_{20}PO_{70}EO_{20}$ in the presence of NaCl have been reported to produce kinetically stable wormlike micelles at room temperature (Ganguly et al. 2009b). Castalletto and Hamley use a capillary flow cell to align underflow wormlike micelles prepared from 4 wt.% aqueous solution of $EO_{19}PO_{43}EO_{19}$ in the presence of 2M NaCl at temperatures close to the CP, which for this salt concentration is 43°C (Castalletto and Hamley 2006). The flow alignment has been studied by SAXS, small-angle light scattering, and rheology. Surprisingly, at moderate flow rates, the results suggest a decrease in persistence length, that is, adopting a less extended conformation under flow, which corresponds to coexistence of isotropic and ellipsoidal patterns. At higher flow rates, isotropic scattering is observed, which can be related either to the flow-induced breakup of the micelles or to the faster flow rate compared to the micellar recombination dynamics.

The solubilization of hydrophobic organic substances in the hydrophobic micellar core of Pluronic copolymers is able to change the interfacial curvature between the core and corona regions, which, in turn, leads to phase transition. Some recent examples of solubilization of 1-phenylethanol in micelles of $EO_{37}PO_{56}EO_{37}$ (Guo et al. 2006) and pentaerythritol tetraacrylate in micelles of $EO_{20}PO_{70}EO_{20}$ (Petrov et al. 2008) nicely illustrate this peculiarity. The latter authors demonstrate the possibility to "lock" the structure of the wormlike micelles by UV cross-linking of the solubilized in the micellar cores pentaerythritol tetraacrylate. The effects of hydrophobic modification of $EO_{20}PO_{70}EO_{20}$ via esterification with lauric and oleic acids have been evaluated by performing rheological studies of aqueous solutions of the initial and final products (de Rodrigues and Nascimento 2010). Enhancement of the hydrophobic character of PPO by increasing temperature and reduction of the PEO segments by addition of salts together with the hydrophobic modification favored formation of long, wormlike micelles resulting in significant alternation of the rheological behavior compared to that of the initial $EO_{20}PO_{70}EO_{20}$ micelles. Increasing concentration can also be a contributing factor to the structural polymorphism of the PEO-PPO-PEO copolymers although there is no clear distinction between the lyotropic and thermotropic phase behaviors. Phase diagrams displaying regions of rodlike micelles and hexagonal phases can be found in the literature. Recent simulation results show that $EO_{37}PO_{56}EO_{37}$ can form, depending on concentration, several microstructures including wormlike micelles among the others (Dong et al. 2010).

The rodlike morphology observed by the copolymer abbreviated LGP134 has been recently documented by SANS (Rangelov et al. 2008). The copolymer belongs to a family of copolymers that are considered as structural analogues of PEO-PPO-PEO copolymers in which the PEO blocks are substituted by blocks of linear PG. These copolymers exhibit rich structural polymorphism, examples of which are presented throughout this chapter. In particular, LGP134 is based on PPO of molar mass of 4000, and its hydrophilic (i.e., PG) content is 40%. At room temperature, it has been found to form rodlike aggregates with cross-sectional radius of 5.3 nm as evidenced by SANS. Interestingly, the copolymers of higher PG contents are slightly anisotropic (morphology that can be associated with prolate ellipsoids) at room temperature and upon heating become more spherical—just opposite to what has been observed for the PEO-PPO-PEO copolymers. These anomalous, from the view point of the typical behavior of PEO-PPO-PEO copolymers, thermotropic transitions have been attributed to the simultaneous increase in both hydrophobicity of PPO and hydrophilicity of PG causing counteracting effects also reflected in other particle parameters and system behavior (Rangelov et al. 2008).

Another family of commercial nonionic polymeric surfactants based on alkylene oxides is the PEO-PBO copolymers. They are available in AB, ABA, and BAB chain architectures. Predominantly, these copolymers form spherical micelles (see Section 2.2.1); however, there is evidence from various techniques (mainly SANS and light scattering) that some of them at appropriate conditions form elongated self-assembled structures (cylinders or wormlike micelles). As both constituent blocks are thermosensitive and exhibit LCST properties, the transition from spheres to cylinders is typically brought upon increasing temperature. Since PBO is more hydrophobic than PPO and displays LCST that is generally lower than that of PPO, the transition to cylinders is shifted to moderate temperatures compared to PEO-PPO-PEO copolymers. A detailed SANS study on micellar structures formed by a diblock copolymer $(EO)_{18}(BO)_9$ in dilute aqueous solution has been recently presented elsewhere (Norman et al. 2005). The copolymer is of PEO content of 55% and was found to form spherical micelles at low concentrations and low temperatures. On increasing temperature, the observation of a sharp increase in the aggregation number together with the upturn in scattering intensity, $I(q)$, at low scattering vector, q, and the change in the shape of the pair distance distribution function, $p(r)$, suggests formation of elongated structures. Thus, the aggregation numbers of the particles at concentrations 0.2 and 1.0 wt.% changed from about 30–60 in the 10°C–40°C range to >300 at 50°C. Figure 2.46 shows the SANS intensity profile from a solution of 0.2 wt.% of $EO_{18}BO_9$ at 50°C. As noted, the scattered intensity scales as $q^{-5/3}$, which is intermediate of the scaling law for rigid rods (q^{-1}) and Gaussian coils (q^{-2}), and accordingly the experimental results are best fitted to a model of a semiflexible chain. Furthermore, the $p(r)$ function, shown in the inset of Figure 2.46, is no longer symmetric, which also suggests that the micelles are not spherical but instead a morphological transition has taken place. The fitted parameters for $c = 0.2$ wt.% given by the model are radius of 49 ± 1 Å and length of 950 ± 10 Å. Interestingly, the authors believe that the formation of wormlike species is faster in the more dilute systems. This suggestion is based on the results obtained for the 1.0 wt.% solution in the 50°C–80°C range, which are fitted to prolate ellipsoids with slowly increasing long axis with increasing temperature (Table 2.1). Wormlike micelles at this concentration were formed at temperature as high as 90°C. Other data for the same copolymer but of different batches show satisfactory agreement with the results presented earlier (see Table 2.1). The slight differences in the parameters can be attributed to different concentrations and temperatures as well as to batch-to-batch variability, but, as a whole, the data are consistent. It seems probable from these works that there is a gradual transition from polydisperse spherical micelles to prolate ellipsoids and finally to wormlike micelles through mixed

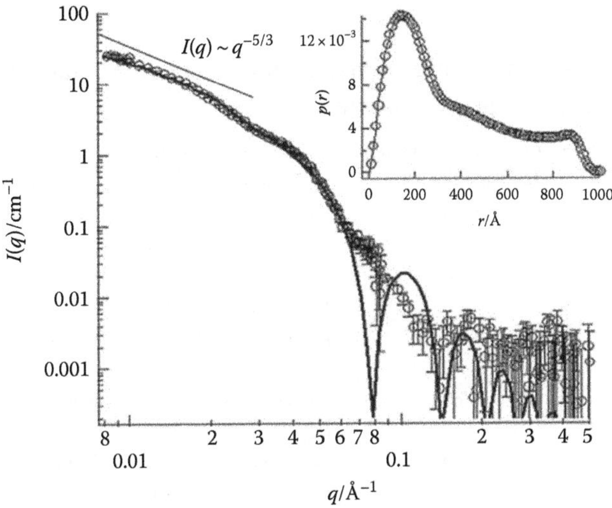

FIGURE 2.46 SANS intensity profile from an aqueous solution of 0.2 wt.% of $EO_{18}BO_9$ at 50°C. The solid line indicates the fit to a semiflexible chain with excluded volume. The inset shows the pair distance distribution function. (Reprinted from *J. Colloid Interface Sci.*, 288, Norman, A.I., Ho, D.L., Karim, A., and Amis, E.J., 155–156, Copyright 2005, with permission from Elsevier.)

systems. It is the general consensus that the decreasing hydration of the corona with temperature leads to the transition from spherical micelles to wormlike micelles. In a recent paper, this hypothesis has been examined, and a particular emphasis has been put on the hydration of the core (Fairclough et al. 2006): Hydration of the order of 23% of the wormlike micelles at 65°C has been found, which is consistent with predictions from PEO-PPO-PEO systems. The effect of methylation of the terminal hydroxyl groups is to lower the temperature of the sphere-to-worm transition from 40°C–50°C for the hydroxyl-terminated sample to 30°C–40°C for the methylated one (Kelarakis et al. 2001).

It has been demonstrated by a number of techniques that the micelles that copolymers of compositions $EO_{17}BO_{12}$, $EO_{13}BO_{10}$, and $EO_{11}BO_8$ form undergo a temperature-induced transition from spherical to elongated micelles and wormlike micelles upon a further increase in temperature (Chaibundit et al. 2002, 2005, Zhou et al. 2007a). The three copolymers are of very close PEO contents (ca. 45%), and, accordingly, the transitions registered as a sudden and sharp increase in certain parameter (see, e.g., Figure 2.47) take place at lower temperatures compared to that of $EO_{18}BO_9$ characterized by 55% of PEO content. The onset temperatures of the sphere-to-cylinder transitions determined as a deviation from the low temperature baseline (Figure 2.47) are approximately 15°C, 20°C, and 30°C for $EO_{17}BO_{12}$, $EO_{13}BO_{10}$, and $EO_{11}BO_8$, respectively. The potential use of wormlike micelles formed by these copolymers to achieve improved drug loading capacity has been explored as well (Zhou et al. 2007a). Also related to the wormlike geometry are the high values of the solubilization capacity of the micellar core of a diblock copolymer of ethylene

TABLE 2.1
Specific Morphology and Parameters of Elongated Structures Formed by $EO_{18}BO_{9-10}$ in Aqueous Solution

Copolymer Composition	Concentration (wt.%), Temperature (°C)	Morphology (Å), Parameters (Å)[a]	References
$EO_{18}BO_9$	0.2, 50	Wormlike $R=49$, $b=145$, $L=950$	Norman et al. (2005)
$EO_{18}BO_9$	1.0, 50	Prolate ellipsoids $R_{short}=130$, $R_{long}=160$	Norman et al. (2005)
$EO_{18}BO_9$	1.0, 65	Prolate ellipsoids $R_{short}=135$, $R_{long}=180$	Norman et al. (2005)
$EO_{18}BO_9$	1.0, 80	Prolate ellipsoids $R_{short}=135$, $R_{long}=200$	Norman et al. (2005)
$EO_{18}BO_9$	1.0, 90	Wormlike $R=42$, $b=100$, $L=1500$	Norman et al. (2005)
$EO_{18}BO_{10}$	2.0, 50	Prolate ellipsoids $R_{short}=37$, $R_{long}=136$	Soni et al. (2002)
$EO_{18}BO_{10}$	5.0, 65	Wormlike $R=30$, $b=123$, $L=1850$	Fairclough et al. (2006)
$EO_{18}BO_{10}$	1.0, 60	Wormlike $R=37$, $b=511$, $L=1500$	Hamley et al. (2001)

[a] R and L are the radius and total length of the wormlike micelles; R_{short} and R_{long} correspond to short and long axis of the prolate ellipsoids; b is the Kuhn segment, which is characteristics of the chain stiffness—the longer the Kuhn segment, the stiffer the chain.

oxide and styrene oxide (SO) with composition $EO_{17}SO_8$ found for furosemide, griseofulvin, and nabumetone (Crothers et al. 2005, 2008).

A sphere-to-rod transition at elevated temperatures has also been reported for triblock copolymer of composition $EO_{16}BO_{10}EO_{16}$ (Yu et al. 1998). The phase boundaries of the detailed phase diagram of this copolymer have been determined by rheometry, SAXS, light microscopy, and tube-inversion method (Figure 2.48). As seen, solution of rodlike micelles is formed at elevated temperatures (>60°C–65°C) and concentration below 40% with 5% being the lowest investigated concentration. No information, however, for the parameters characterizing the rodlike micelles is available.

Another relatively large group of copolymers that are capable of forming wormlike micelles is that with constituent blocks (moieties) of PCL typically combined with PEO as a hydrophilic component. As noted in the previous section, PCL is biocompatible, hydrolytically biodegradable, and already being applied in drug delivery. PCL is hydrophobic, and the combination with PEO or other hydrophilic polymers imparts an amphiphilic character of the resulting copolymers. Besides all those properties, it is semicrystalline; the crystallization of certain moieties generally favors lamellar phases but can be confined within spherical and cylindrical

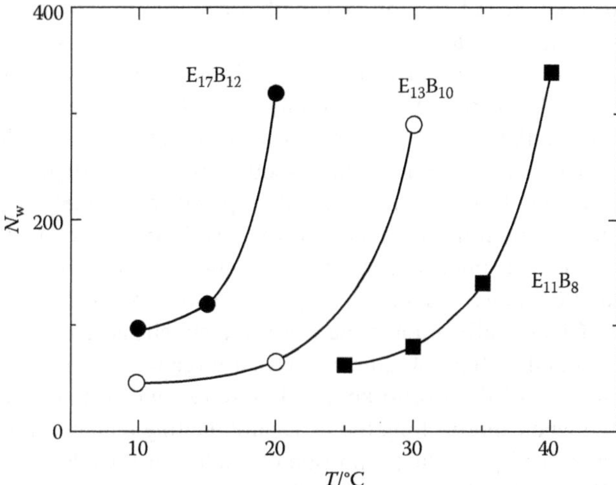

FIGURE 2.47 Temperature dependence of the micellar association number for copolymers $EO_{17}BO_{12}$, $EO_{13}BO_{10}$, and $EO_{11}BO_8$ in aqueous solution. (Reprinted from *Int. J. Pharm.*, 354, Zhou, Zh., Chaibundit, Ch., D'Emanuele, A., Lennon, K., Attwood, D., and Booth, C., 82–87, Copyright 2007, with permission from Elsevier.)

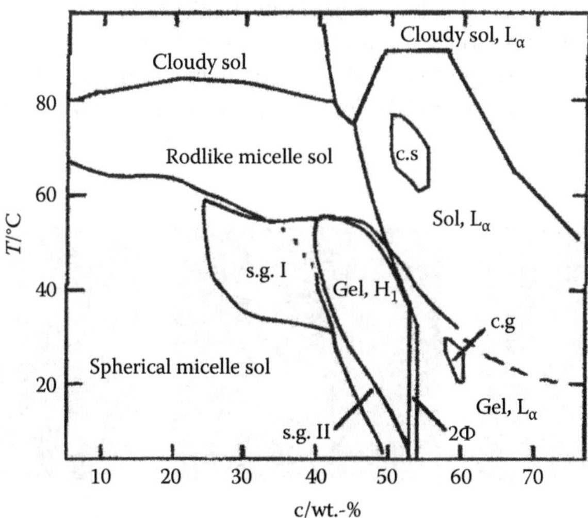

FIGURE 2.48 Phase diagram for aqueous solutions of $EO_{16}BO_{10}EO_{16}$. H_1 = hexagonal phase, L_α = lamellar phase, 2Φ = two phases, s.g.I = soft gel I, s.g.II = soft gel II, c.g. = light cloudy gel, and c.s. = light cloudy sol. (Reprinted with permission from Yu, G.-E., Li, H., Fairclough, J.P. A., Ryan, A.J., McKeown, N., Ali-Adib, Z., Price, C., and Booth, C., *Langmuir*, 14, 5782–5789, 1998. Copyright 1998, American Chemical Society.)

domains. The copolymers based on PCL with propensity to form elongated structures are of varying chain architectures with diblock being the simplest one. In a recent paper, Rajagopal et al. investigate the self-assembly in water of a large set of semicrystalline PEO-PCL diblock copolymers (Rajagopal et al. 2010). Those of them with f_{PEO} in the 0.36–0.46 range form wormlike micelles within a narrow subphase deep in the vesicle region. The crystallization is seemingly suppressed as only a small fraction of the wormlike micelles appears completely rigid. They are observable rotating and translating as rigid rods by fluorescence movie snapshots, which are presented in Figure 2.49. The percentage of these micelles increases with increasing PCL molar mass and is typically below 15% at 25°C. What clearly indicates that PCL crystallization in the wormlike micelle core is the driving force for rigidification is the effect of temperature: At large undercooling, for example, −20°C, the majority of the wormlike micelles were rigid, whereas their fraction nearly vanished when approaching the PCL bulk melting temperature (60°C). The worms are typically >1 μm. The proportion of the longer micelles increases as the size of the PCL block is increased; the diameters were found to scale as $\sim M^{0.67}$. Nanometers in cross section and microns in length are the wormlike micelles prepared from another PEO-PCL block copolymer (Cai et al. 2007). They are flexible with a fluid, rather than glassy or crystalline, core and upon sonication can generate stable spherical micelles. The wormlike micelles have advantages over the spherical ones in terms of solubilization capacity and anticancer activity (see more in Chapter 5).

A series of diblock copolymers in which the hydrophilic PEO is substituted by dextran has been prepared (Zhang et al. 2010b). Dextran is attractive for biomedical applications because of its biocompatibility, biodegradability, hydrophilicity, and ability to interact with biomacromolecules. Within the series, the mass fraction of dextran, f_{DEX}, varied from 0.16 to 0.45. The copolymer with $f_{DEX}=0.40$ was found to form in aqueous solution wormlike micelles in coexistence with a fraction of small spherical aggregates.

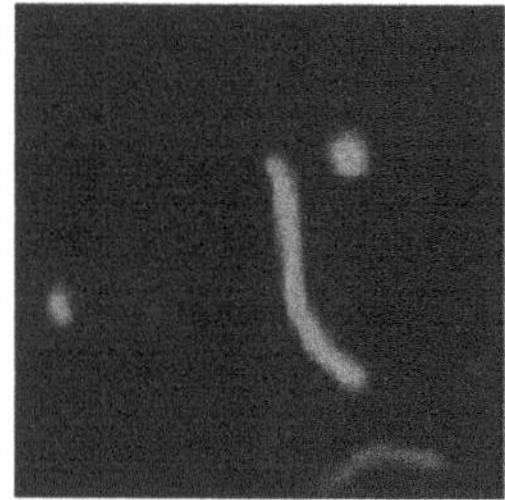
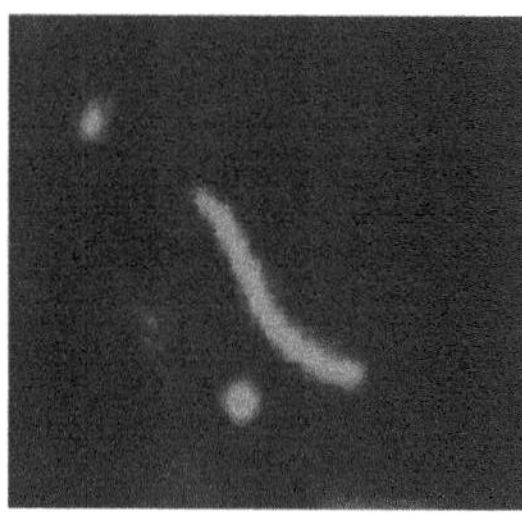
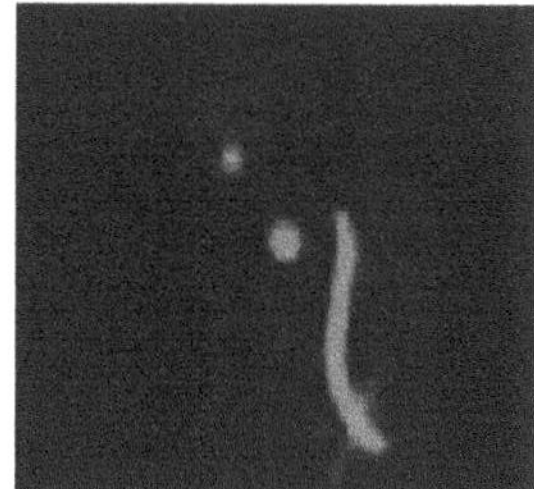

FIGURE 2.49 Fluorescence microscopy snapshots representing a wormlike micelle obtained from a PEO-PCL diblock copolymer with $f_{PEO}=0.41$. Snapshots were taken at 4 s intervals. About 10% of the micelles exhibit rigid-body motion. (Reprinted with permission from Rajagopal, K., Mahmud, A., Christian, D.A., Pajerowski, J.D., Brown, A.E.X., Loverde, S.M., and Discher, D.E., *Macromolecules*, 43, 9736–9746, 2010. Copyright 2010, American Chemical Society.)

Series of block copolymers with middle water-soluble blocks of PEO and outer PCL blocks of linear, that is, BAB, and nonlinear (starlike with 4 arms) chain architectures have been reported to form spherical *flowerlike* micelles (Lu et al. 2007, Zhang et al. 2007a). At higher concentrations, morphology transitions to necklace-like and wormlike aggregates have been observed by AFM. The authors propose that wormlike micelles were firstly formed and their close packing caused gelation of the aqueous solutions.

Yang and coworkers describe the synthesis and self-assembly of copolymers of unconventional dendron–linear–dendron chain architecture consisting of flanking dendron-like PCL and linear PEO in the middle (Yang et al. 2009). When prepared by the nanoprecipitation method, these copolymers formed spherical *flowerlike* micelles, the size and morphology of which were largely independent of the composition (PEO content from 12.6 to 41.5 wt.%) and dendritic topology (three generations). However, when dialysis was used to prepare doxorubicin-loaded particles, predominantly wormlike micelles were observed, also for the linear PCL-PEO-PCL analogue of the highest PEO content.

Also of nonlinear topology is the miktoarm star terpolymer described elsewhere (Saito et al. 2010). The chemical structure of the copolymer is illustrated in Figure 2.50a. In dilute neutral aqueous solution, elongated multicompartment micelles with alternating dark and light stripes were observed by cryo-TEM (Figure 2.50b). The latter were attributed to poly(γ-methyl-ε-caprolactone) and poly(ethyl ethylene) (PEE) domains, respectively, which have different electron densities. The length of worms ranged between 0.1 and 1.0 μm, whereas the average width of the structures was about 36 ± 10 nm, which exceeded two times the fully stretched PEE block, thus contributing to the suggestion for segmented, rather than cylindrically symmetric, morphology (Figure 2.50c). Spherical-like end caps possibly shortening the structures are clearly seen. When subjected to chemical degradation at high pH, a remarkable transition from segmented wormlike micelles

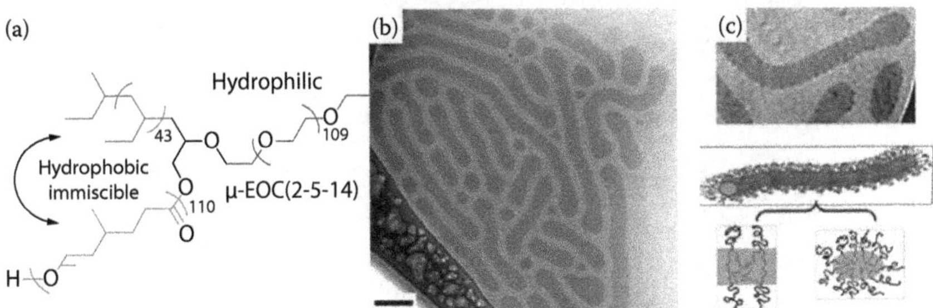

FIGURE 2.50 (a) Chemical structure of miktoarm star terpolymer, μ-EOC(2-5-14). E, O, and C denote PEE, PEO, and poly(γ-methyl-ε-caprolactone) chains, respectively, whereas the numbers refer to the molar masses of the corresponding arms in kDa. (b) Cryo-TEM image of a 0.5 wt.% dispersion in neutral water. Scale bar corresponds to 100 nm. (c) Schematic presentation of the structure of the segmented wormlike micelles. (Reprinted with permission from Saito, N., Liu, Ch., Lodge, T.P., and Hillmyer, M.A., *ACS Nano*, 4(4), 1907–1912, 2010. Copyright 2010, American Chemical Society.)

to raspberrylike vesicles or disks due to cleavage of poly(γ-methyl-ε-caprolactone) chains was revealed by cryo-TEM.

Wormlike aggregates prepared from copolymers based on another hydrophobic polyester—PLA—should be mentioned as well (Du et al. 2009). PLA is also very attractive and widely used for biomedical applications. In spite of relatively high T_g and degree of crystallinity of the PLA block, the wormlike aggregates prepared from copolymer with PEO appeared rather flexible (Figure 2.51). This notion was also corroborated by DLS, the results from which, in particular, the slow relaxation mode at long relaxation times, may be associated with disentanglement of the wormlike micelles. Figure 2.51 shows branched and circular wormlike micelles with fairly uniform diameters of ca. 11 nm. Wormlike micelles prepared from a PLA-PEO diblock copolymer, however, with a longer PLA block, to which different cancer-targeting moieties are attached, are described elsewhere (Lee et al. 2010b). More information about their ability to recognize different breast cancer cells is available in Chapter 5.

The possibility of a rational control through synthetic manipulation of the molecular characteristics on the aggregate morphology has been illustrated using a set of PEO-based copolymers (Won et al. 2002). The materials chosen for hydrophobic moieties were PBD and its saturated analogue PEE. The copolymers were of relatively low molecular weight (<13.1 kg/mol); those that were found to self-associate in cylindrical morphology were within the 0.39–0.53f_{PEO} range. Wormlike micelles with length that exceeded 5 μm and a cross-sectional diameter of 15 nm were observed by cryo-TEM for the copolymer with $f_{PEO}=0.45$. In contrast to the micellar and vesicular dispersions, the aqueous dispersion of the wormlike micelles from the same copolymer behaved as a non-Newtonian liquid and displayed considerably

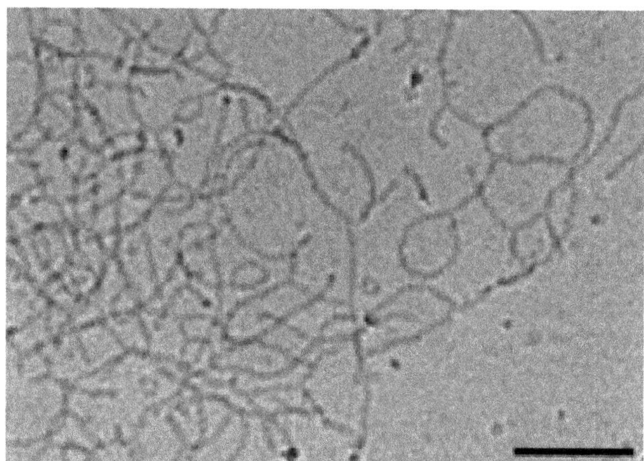

FIGURE 2.51 Cryo-TEM image of wormlike micelles obtained from PLA_{56}-PEG_{44} in aqueous solution. The content of hydrophilic component and scale bar correspond to 0.33 and 200 nm, respectively. (Reprinted with permission from Du, B., Mei, A., Yin, K., Zhang, Q., Xu, J., and Fan, Zh., *Macromolecules*, 42, 8477–8484, 2009. Copyright 2009, American Chemical Society.)

higher zero shear viscosity and storage and loss moduli. Coexisting morphologies, for example, spherical and cylindrical micelles or cylindrical micelles and bilayered vesicles, were observed also within f_{PEO} ranges that are typical for certain morphology, thus implying nonergodicity in the self-assembly behavior (Jain et al. 2004, Hayward and Pochan 2010).

One of the most attractive features of the polymeric amphiphiles is the possibility to control the polymer–solvent interactions, and consequently their self-assembly, by several factors such as solvent quality, temperature, pH, and ionic strength. The polyelectrolytes are sensitive to the last two factors, and when covalently bound to a hydrophobic polymer, morphological changes can be induced. The morphological changes can be understood in terms of, for example, affecting the packing factor of the amphiphilic copolymer through the area per hydrophilic chain caused by small variations in pH and/or ionic strength. Such morphological changes have been recently observed for copolymers of different chain architectures. Fernyhough et al. investigated a PBD_{24}-$PMAA_{10}$ copolymer of simple diblock chain architecture, where the subscripts denote the degrees of polymerization of the PBD and poly(methacrylic acid) (PMAA) blocks; the total number-average molecular weight was 2000 g/mol, and the weight and volume fractions were calculated to 0.40 and 0.33, respectively (Fernyhough et al. 2009). The phase sequence as a function of pH is presented in the phase diagram shown in Figure 2.52a. The transitions were not sharp and the coexisting morphology regions are indicated. The authors have found an inversely proportional relation between the packing parameter and the degree of ionization of PMAA chains, which are presented in Figure 2.52b as a function of pH. As seen from Figure 2.52a and b, cylindrical aggregates, a TEM image of which is shown in Figure 2.52c, are formed in a relatively narrow pH interval.

Of considerably higher total molecular weight, different composition and chain architecture is the copolymer described elsewhere (Strandman et al. 2007). The copolymer is of starlike architecture consisting of four block arms of composition $PMMA_{73}$-PAA_{143} with outer polyelectrolyte blocks. In saline solution (ionic strength 80 mM) at low pH (4.5), wormlike aggregates are formed according to DLS and cryo-TEM studies. The wormlike aggregates disintegrate upon increasing pH via a mechanism of formation of a pearl-necklace structure in agreement with the results of coarse-grained computer simulations a snapshot of which is presented in Figure 2.53. The changes in the morphology were also attributed to the higher degree of ionization of the polyelectrolyte blocks in addition to swelling of the corona owing to the higher osmotic pressure by trapped counterions.

In recent molecular thermodynamic modeling for solutions of diblock copolymers composed of a hydrophobic block and a weak polyelectrolyte block, phase diagrams have been constructed using the self-consistent field theory in the strong-segregation approximation (Victorov et al. 2010). The phase diagrams are presented as regions of stable aggregate morphologies in the pH–saline planes. The morphology stability maps demonstrate structural response of the aggregates to changing pH and salinity of the environment. Cylindrical aggregates with spherical end caps and branched cylinders are formed in relatively narrow regions. In the regions where the branched cylindrical micelles are about to transform into lamellae, the model predicts that the

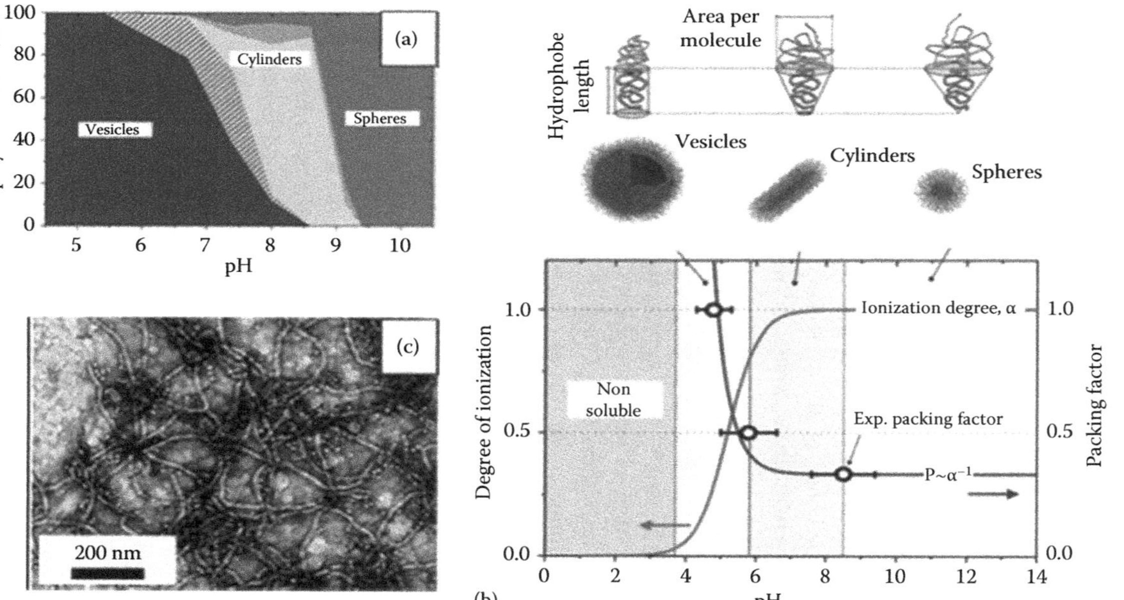

FIGURE 2.52 (a) Phase diagram (morphology vs. pH) of PBD_{24}-$PMAA_{10}$ in dilute aqueous solution. Coexistence regions are denoted with lines. (b) Packing factor and degree of ionization as a function of pH. (c) TEM micrograph of negatively stained wormlike structures taken from 0.1 wt.% dispersion of PBD_{24}-$PMAA_{10}$ at pH 8. (Fernyhough, C., Ryan, A.J., and Battaglia, G., *Soft Matter*, 5, 1674–1682, 2009. Reproduced by permission of The Royal Society of Chemistry.)

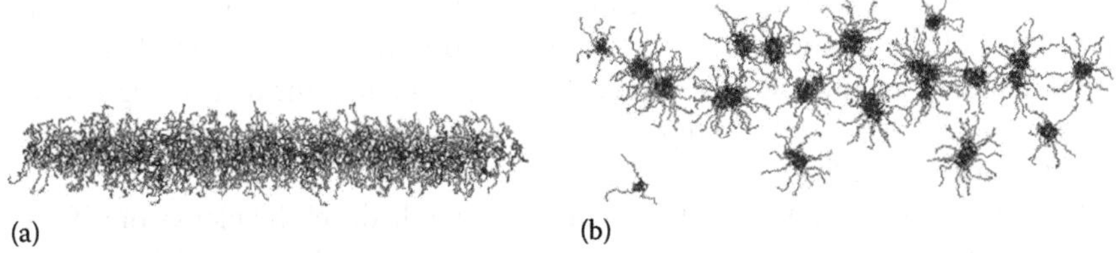

FIGURE 2.53 Snapshots of the simulated self-assembled structures at low pH (4.5, noncharged) (a) and at high pH (12.5, charged) (b) obtained by starlike block copolymer ($PMMA_{73}$-$PAA_{143})_4$. (Reprinted from *Polymer*, 48(24), Strandman, S., Zarembo, A., Darinskii, A. A., Loflund, B., Butcher, S. J., Tenhu, H., 7008–7016, Copyright 2007, with permission from Elsevier.)

branches are the most stable structures. They further proliferate, whereas the cylindrical parts disappear, and the system would possibly produce highly connected spongelike aggregates (Victorov et al. 2010).

Another strategy for preparation of pH-sensitive polymers is the introduction of acid-labile linkages into the copolymer backbone or between the backbone and the side chains. Such copolymers are, for example, those based on PEO and PADMO (see Figure 2.54 for the chemical structure of the oxazolidine moiety) (Cui et al. 2011a). The self-assembly of these copolymers is controlled by the PADMO content, and the morphology of the particles varies from spherical to cylindrical to vesicular with increasing the PADMO block molar mass at constant length of the PEO block of 2000. Wormlike micelles have been registered in coexistence with spherical ones for the copolymer of f_{PEO} of about 0.31 calculated from the molecular weight data. Under acid conditions, the hydrophobic pendant oxazolidine groups are converted to water-soluble β-hydroxyl amine derivatives (Figure 2.54) leading to formation of a double hydrophilic diblock copolymer and disruption of the micelles.

In another study, the latter block has been combined with both a thermoresponsive polymer (PNIPAM) and a relatively short but with rigid molecular structure polymer based on alkynyl functionalized polyfluorene, thus forming a series of rod–coil–coil triblock copolymers (Lin et al. 2009a). The copolymers displayed reversible micelle

FIGURE 2.54 Chemical structures of the amphiphilic diblock copolymer PEO-*b*-PADMO and the resulting from the acid hydrolysis double hydrophilic PEO-*b*-PHEAM. Here, PADMO and PHEAM denote poly(*N*-acryloyl-2,2-dimethyl-1,3-oxazolidine) and poly(2-hydroxyethyl acrylamide), respectively. Wormlike micelles in aqueous solution have been observed for PEO-*b*-PADMO at $n=45$ and $m=47$. (Reprinted from *Polymer*, 52(8), Cui, Q., Wu, F., and Wang, E., 1755–1765, Copyright 2011, with permission from Elsevier.)

morphologies in the heating–cooling cycles. In particular, the strong π–π electron interactions of the fluorene moieties resulted in different micelle morphologies of worms, bundles, and hollow tubes. Schematic illustration of the morphological variations upon heating as well as a representative TEM image of wormlike micelles are presented in Figure 2.55.

Bis-hydrophilic block terpolymers with two outer hydrophilic blocks of different chemistry and an inner hydrophobic block have been synthesized (Walther et al. 2008). To previously form diblock copolymers of PEO and poly(*n*-butyl acrylate) as a hydrophobic block of constant lengths and ratio, various (meth)acrylamide derivatives have been polymerized. Depending on the hydrophilic-to-hydrophobic balance, a variety of self-assembled structures including wormlike micelles have been obtained. The aggregates are of dynamic character and can undergo fusion and fission processes induced by temperature and time.

A variety of *hybrid* block and graft copolymers have been reported to form wormlike micelles. Many of them exhibit structural polymorphism, and we refer to them in other sections of the book. The biohybrid copolymers combine advantageous features of synthetic polymers and biologically relevant polypeptides, sugars, carbohydrates, and polysaccharides. The biohybrid copolymers can self-associate into variety of morphologies. Their self-assembly can be controlled not only by changing copolymer composition but also by utilizing some typical properties for biologically relevant polymers such as, for example, hydrogen-bonding and secondary structure interactions in polypeptides. Wang et al. demonstrate how by changing the length of poly(L-leucine) (a well-known polypeptide exhibiting α helix-forming properties when long enough) in a series of block copolymers with PEO both the hydrophobicity and the secondary structure of poly(L-leucine) can be varied, which in turn affect the aggregation behavior (Wang et al. 2011a). The copolymer with a poly(L-leucine) block that adopts an α-helical conformation has been shown to form preferentially wormlike micelles. In another study, two tetrapeptides—tetraphenylalanine and tetravaline—have been conjugated to PEO chains of increasing molar masses, thus making two series of biohybrid copolymers (Tzokova et al. 2009). The authors used different approaches, for example, dialysis of a THF solution against water or direct aqueous rehydration, to achieve self-assembly. In all cases, strongly elongated structures—nanotubes, fibers, wormlike micelles, as well as structures that are reminiscent to those formed by the polypeptides alone—were obtained, which could be related to the propensity of the tetrapeptide moieties to form β-sheets. It is noteworthy that f_{PEO} of the copolymer that was found to form wormlike micelles is 0.71 (!)—much higher than the typical range for wormlike micelles obtained from, for example, coil–coil copolymers, suggesting that the packing of the tetrapeptide moiety significantly influences the overall self-assembly behavior.

The synthesis of dendron-like biohybrids has been recently reported in detail (Peng et al. 2009). The copolymers are of different topologies—asymmetric, that is, dendron–linear, and symmetric, that is, dendron–linear–dendron (Figure 2.56). The linear segments were based on PEO, whereas the dendron-like moieties were based on highly branched poly(γ-benzyl-L-glutamate). Wormlike micelles have been

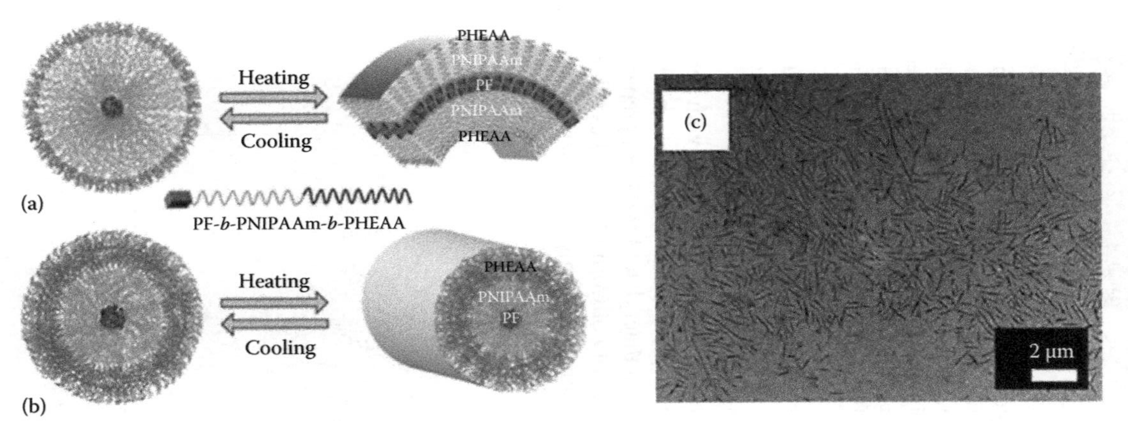

FIGURE 2.55 Schematic illustration of morphological variations from spherical micelles to (a) hollow tubes and (b) wormlike micelles and vice versa upon heating–cooling cycles for copolymers with compositions PF_7-*b*-$PNIPAM_{120}$-*b*-$PHEAA_{30}$ (a) and PF_7-*b*-$PNIPAM_{75}$-*b*-$PHEAA_{75}$ (b). TEM micrograph of wormlike micelles taken from aqueous dispersion of PF_7-*b*-$PNIPAM_{120}$-*b*-$PHEAA_{30}$ at 40°C (c). Here PF and PHEAA denote poly[2,7-(9,9-dihexylfluorene)] and poly(*N*-hydroxyethyl acrylamide), respectively. (Lin, S.-T., Fuchise, K., Chen, Y., Sakai, R., Satoh, T., Kakuchi, T., Chen, W.-C., *Soft Matter*, 5(19), 3761–3770, 2009. Reproduced by permission of The Royal Society of Chemistry.)

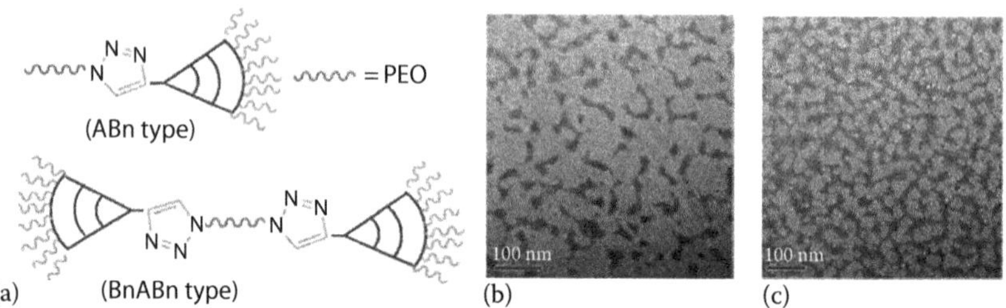

FIGURE 2.56 (a) Schematic representation of the topology of asymmetric and symmetric dendron-like/linear biohybrids. TEM images of self-assembled nanostructures (wormlike micelles) obtained from the linear, D0-PBLG$_{80}$-*b*-PEO (b), and dendron-like, D1-PBLG$_{41}$-*b*-PEO (c), analogues. Both copolymers have similar f_{PEO}. PBLG denotes poly(γ-benzyl-L-glutamate). (Reprinted with permission from Peng, S.-M., Chen, Y., Hua, C., and Dong, C.-M., *Macromolecules*, 42(1), 104–113, 2009. Copyright 2009, American Chemical Society.)

observed in aqueous solution for copolymers with f_{PEO} = 12–22 wt.% for both the linear and dendron-like analogues (Figure 2.56) implying that the dendron-like topology has no apparent effect on the aggregate morphology.

Well-defined diblock copolymers composed of blocks with pendant carbohydrate residues and hydrophobic noncarbohydrate blocks based on PDEAEMA, poly(*n*-butyl acrylate), and poly(*n*-butyl methacrylate) have been prepared (Cameron et al. 2008). A phase consisting of a majority of wormlike micelles (Figure 2.57) is formed in aqueous solution by one such amphiphilic glycopolymer with hydrophilic volume fraction of 0.35 as revealed by TEM.

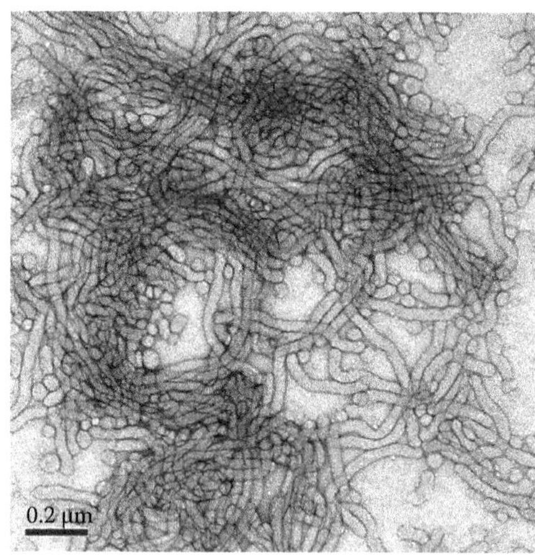

FIGURE 2.57 Wormlike morphology observed by TEM from aqueous solution of p(GalEMA$_{25}$-*b*-BA$_{100}$). Here, GalEMA and BA denote 2-(*b*-D-galactosyloxy)ethyl methacrylate and *n*-butyl acrylate, respectively. (Cameron, N.R., Spain, S.G., Kingham, J.A., Weck, S., Albertin, L., Barker, C.A., Battaglia, G., Smart, T., and Blanazs, A., *Faraday Discuss.*, 139, 359–368, 2008. Reproduced by permission of The Royal Society of Chemistry.)

Formation of wormlike micelles has also been reported for one of the most popular polysaccharides—chitosan that has been chemically modified by grafting with poly(sodium 4-styrene sulfonate) (Jiang et al. 2010). As expected, not only morphology but the ion-exchange properties in general can be effectively controlled by adjusting the poly(sodium 4-styrene sulfonate) graft content.

We will close this section with copolymers combining biomimetic, that is, glycopolymer and polypeptide, blocks into one chain and their self-assembly behavior in aqueous solution (Dong and Chaikof 2005). The copolymers are of BAB chain architecture and are composed of a hydrophilic lactose-bearing central block flanked by hydrophobic poly(L-alanine) end blocks (Figure 2.58). The morphology of the aggregates was somewhat concentration dependent and changed from sphere to lamellae and then to wormlike micelles. Figure 2.58 shows inner-connected spheres, which look like "worm." The aggregates are surfaced by lactose shell, which makes them useful as artificial polyvalent ligands in studies of carbohydrate–protein recognition and for the design of site-specific drug delivery systems.

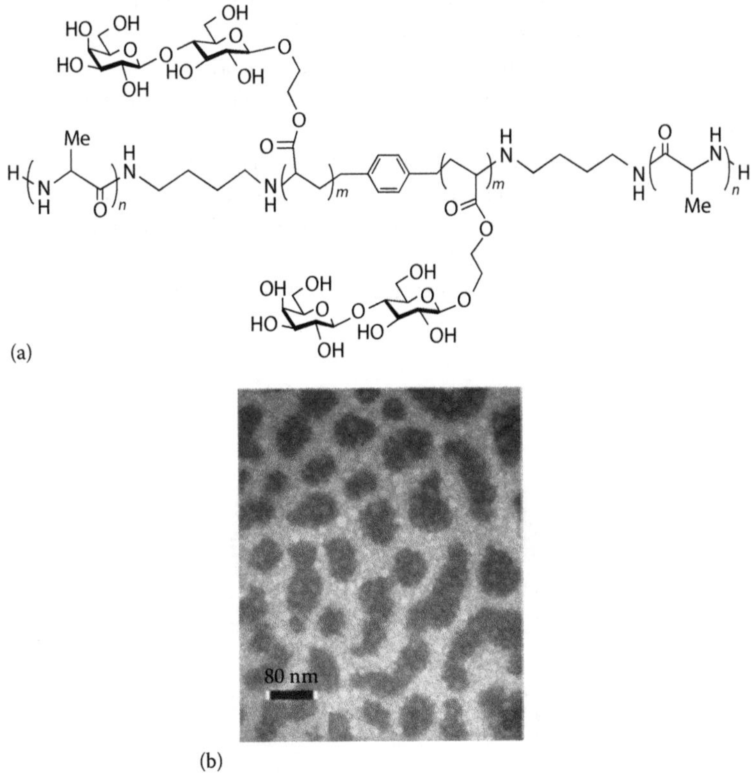

(a)

(b)

FIGURE 2.58 (a) Chemical structure of a biomimetic glycopolymer–polypeptide triblock copolymer poly(L-alanine)-*b*-poly(2-acryloyl-oxythyllactoside)-*b*-poly(L-alanine) and (b) TEM micrograph of aggregates from the triblock copolymer with composition 22-52-22 formed in water at c=5 mg/mL and pH 6.69. (With kind permission from Springer Science+Business Media: *Colloid Polym. Sci.*, 283(12), 2005, 1366–1370, Dong, C.-M. and Chaikof, E.L.)

2.2.2.3 Toroidal Micelles

In the absence of topological defects (end caps and branch junction points), the cylindrical micelles should be infinitely long, linear structures. A degree of randomness can be introduced through bending of the cylindrical micelles. Closing up their ends would result in formation of ring-shaped structures free of topological defects called toroids (Figure 2.59); however, this would be accompanied by entropy loss. The latter could be compensated by formation of relatively small toroids, thus resulting in increasing the number of the particles, or by formation of regions with different local curvatures resulting in more complicated toroidal structures (supertoroids), thus increasing the number of possible configurations.

Most of the toroidal micelles reported are typically formed by amphiphilic block copolymers with AB, ABA, and ABC chain architectures in organic solvent/water mixtures or upon the addition of a third component. Their formation has been proposed to arise by different pathways: through elimination of high-energy end caps of cylindrical micelles, via perforation of disklike micelles (Figure 2.60) (Cui et al. 2009), or via more complicated transformations through different morphologies. The latter has been observed for poly(4-vinylpyridine)-*b*-PS-*b*-poly(4-vinylpyridine) in a dioxane/water mixture from which the dioxane is removed by dialysis: The high stirring rates and annealing times have been found to lead to increasing numbers of toroids formed through a cylinder–sphere–vesicle–toroid transformation (Zhu et al. 2004, Yu and Jiang 2009).

The regions in which ring-shaped structures are formed are very narrow. As a consequence, the observations of toroidal structures have been rather scarce. Typically, the toroidal micelles have a wide size distribution and coexist with cylindrical micelles and/or cylindrical–toroidal networks (see Figures 2.51 and 2.61). It is hoped that uniformity in size and morphology might be reached in the near future.

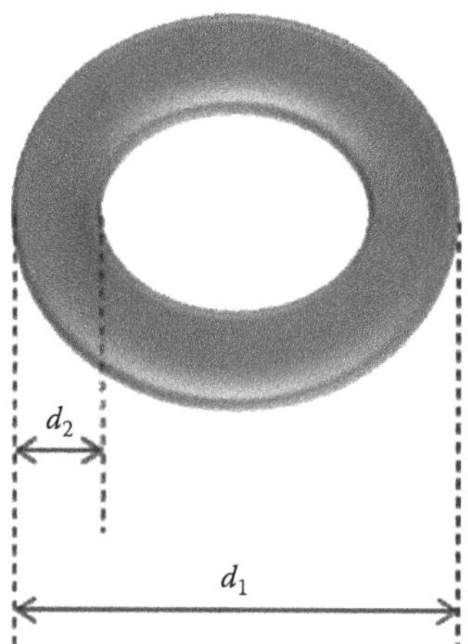

FIGURE 2.59 Schematic presentations of a toroid with the typical parameters characterizing its dimensions—overall diameter, d_1, and cross-sectional diameter, d_2.

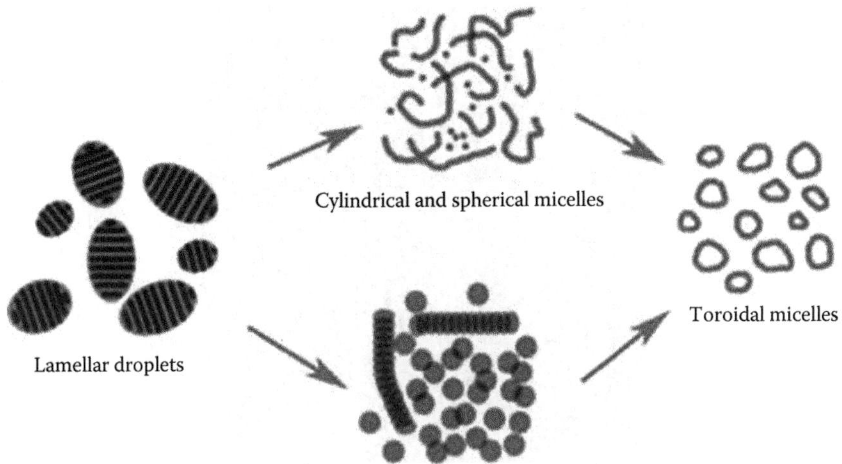

FIGURE 2.60 Proposed routes to toroidal assembly from PAA-*b*-poly(methyl acrylate)-*b*-PS in mixed THF/water solution via interaction with organic diamines. (Cui, H., Chen, Zh., Wooley, K.L., and Pochan, D., *Soft Matter*, 5, 1269–1278, 2009. Reproduced by permission of The Royal Society of Chemistry.)

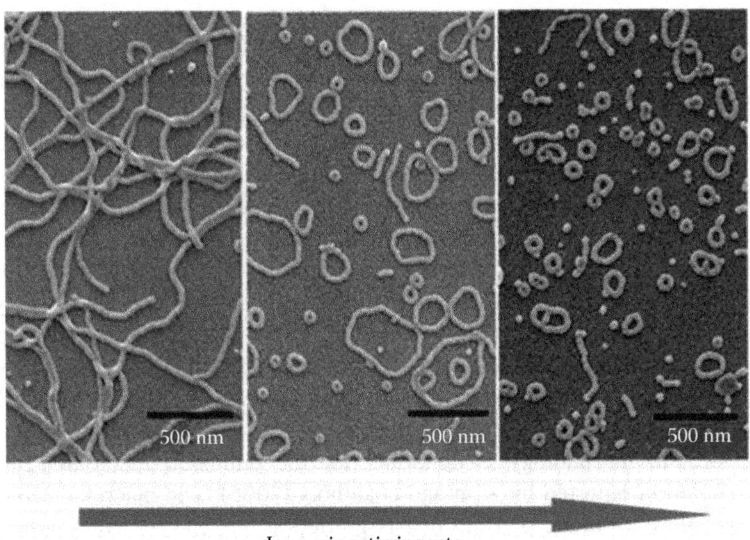

FIGURE 2.61 Effect of shear flow (stirring rate) on the morphology of micelles formed by poly(4-vinylpyridine)-*b*-PS-*b*-poly(4-vinylpyridine) block copolymer in aqueous solution. Sizable toroids are formed at moderate stirring rates, whereas the higher stirring rates produce smaller in size and larger in number toroids. Notable features are the apparently wide size distribution and coexisting morphology. (Reprinted with permission from Yu, H. and Jiang, W., *Macromolecules*, 42, 3399–3404, 2009. Copyright 2009, American Chemical Society.)

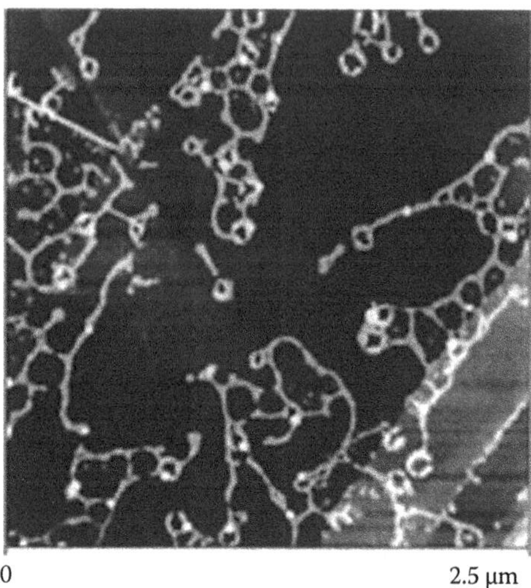

FIGURE 2.62 AFM image (height profile) of connected toroids formed in aqueous solution by PEE(144)-PSSH(136) at salt concentration of 0.05 mol/L. (Reprinted with permission from Forster, S., Hermsdorf, N., Leube, W., Schnablegger, H., Regenbrecht, M., and Akari, S., *J. Phys. Chem. B*, 103, 6657–6668, 1999. Copyright 1999, American Chemical Society.)

Fractal toroid–micronetworks have been reported to form from polyelectrolyte block copolymers based on poly(ethyl ethylene-*b*-styrenesulfonic acid) (Forster et al. 1999). The combination of a soft hydrophobic block and a highly charged polyelectrolyte block helps self-assembly in aqueous solution. Fusion of polyelectrolyte micelles upon increasing copolymer concentration and salt content was observed to produce formation of toroidal networks (Figure 2.62) in which the toroids are of diameters ranging from 80 to 110 nm, whereas the cross-sectional diameter of the rims is about 27 nm. Toroidal nanostructures, induced by regulation of electrostatic and hydrophobic interactions in aqueous solution systems of an ABC terpolymer, have been reported elsewhere (Tsitsilianis et al. 2008). The terpolymer is composed of poly(2-vinylpyridine)-PAA-poly(*n*-butyl methacrylate), and the diversity of structural organizations, including toroids, has been achieved by tuning both pH and copolymer concentration.

The toroidal morphology is dominant for the dumbbell macromolecules in aqueous solution reported by Kim and coauthors (2006). The macromolecules are composed of an aromatic scaffold that is grafted at one end by hydrophilic polyether dendrons and hydrophobic branches containing alkyl chains at the other (Figure 2.63). Notably, only the dumbbell macromolecules with dodecyl chains form toroidal structures; those with hexyl and tetradecyl chains form spheres and short cylinders and long wormlike micelles, respectively. According to the authors, this behavior originates from side-by-side connections of discrete bundles through combination of strong hydrophobic interactions and anisotropic aggregation of rod segments. Comparisons with the length of a fully extended macromolecule suggest a bilayer packing within cylindrical domains.

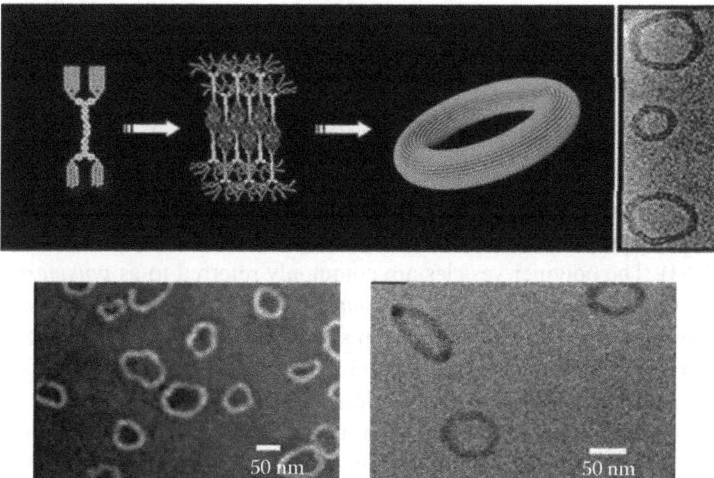

FIGURE 2.63 Schematic representation and TEM and cryo-TEM images of toroidal structures from amphiphilic dumbbell macromolecules. (Reprinted with permission from Kim, J.-K., Lee, E., Huang, Zh., and Lee, M., *J. Am. Chem. Soc.*, 128, 14022–14023, 2006. Copyright 2006, American Chemical Society.)

2.2.3 Polymer Vesicles (Polymersomes)

The field of polymer vesicles is an emerging area of research that is able to attract the attention of an increasing number of scientists. The recent advances in polymer chemistry and physics, the complex hierarchical structure allowing design and control on the size and interactions with environment, resemblance of the natural cellular bilayer membranes, tailorable membrane properties, versatile potential applications spanning from (bio)electronics and catalysis to medical therapy among the others make the polymer vesicles attractive and promising objects of investigations. Compared to polymer micelles and lipid vesicles (liposomes), the publication activity in this field is considerably less intensive with a total number of publications since the first mention of polymer vesicles in 1995 (Zhang and Eisenberg 1995) of about 2500. Over the last decade, the publication productivity has been found to sustainably grow from just several to a few hundred papers published annually. Undoubtedly, future development relying on exploiting unique property–performance relations is to be witnessed in the years to come.

There are surprisingly high number of reviews covering different aspects of polymer vesicle preparation, properties, and applications (Discher and Eisenberg 2002, Opsteen et al. 2004, Ahmed et al. 2006, Discher et al. 2007, Levine et al. 2008, Christian et al. 2009, Li and Keller 2009, LoPresti et al. 2009, Meng et al. 2009, Onaca et al. 2009, Zhang et al. 2009a, Malinova et al. 2010, Massignani et al. 2010, Le Meins et al. 2011). In this section, we try to systematically summarize the existing knowledge on polymer vesicles focusing on their specific and distinctive properties as well as factors that govern their structure, size, and morphology. Promising approaches to modulate the basic polymersome properties and performance by, for instance, selecting copolymers with appropriate chain architecture, molar mass, ratio, and nature of the constituent entities or applying diverse preparation methods are envisaged as well.

2.2.3.1 Definition, Structure, Morphology, and Dimensions

Polymer vesicles are hollow particles with a membrane built from amphiphilic block copolymers. The membrane is typically composed of entangled chains and encloses an aqueous compartment (core), thus separating the latter from the outside medium (Figure 2.64). The polymer vesicles are commonly referred to as *polymersomes*, in analogy to the lipid-based vesicles, *liposomes*. In the following, *polymersomes* and *polymer vesicles* will be used as synonyms. In contrast to liposomes, the building elements of the polymersomes are amphiphilic copolymers, which are of (considerably) higher molar mass than those building liposomes, which imparts distinctly different properties such as superior stability and toughness, reduced permeability, restricted chain mobility within the membrane, and better resistance to dissolution. These properties are largely controlled by the copolymer molar mass, which, in addition, dictates the membrane thickness. The latter ranges typically between several and less than 50 nm, thus contrasting the liposome bilayer membranes having thicknesses of 3–5 nm. The higher thickness makes the polymersomes less susceptible to defects and fluctuations compared to liposomes (Discher et al. 2002). Another very strong contrast with the liposomes is that the concept of *bilayer*, that is, an arrangement of two monolayers of lipid molecules with their hydrophobic tails pointed toward the interior of the membrane, is hardly applicable for polymersomes as precisely stated elsewhere (Le Meins et al. 2011) since entanglement and interdigitation can occur between the hydrophobic blocks.

The polymersome membrane has a sandwich structure consisting of a hydrophobic layer with thickness *d* in the middle and two hydrophilic layers (Figure 2.65). The influence of copolymer molar mass on the total membrane thickness is schematically

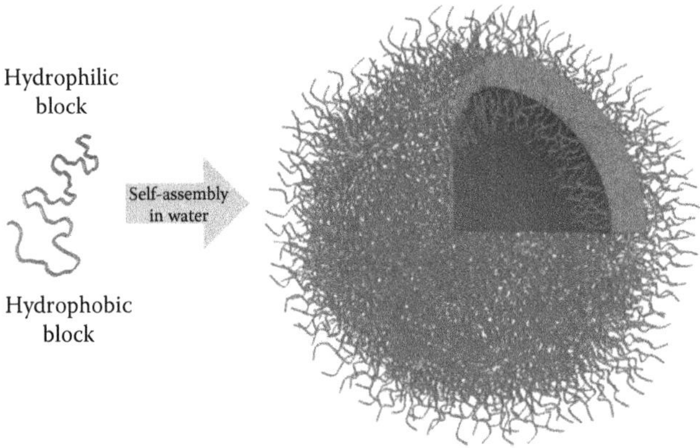

FIGURE 2.64 Schematic representation of a polymer vesicle. (LoPresti, C., Lomas, H., Massignani, M., Smart, T., and Battaglia, G., *Mater. Chem.*, 19, 3576–3590, 2009. Reproduced by permission of The Royal Society of Chemistry.)

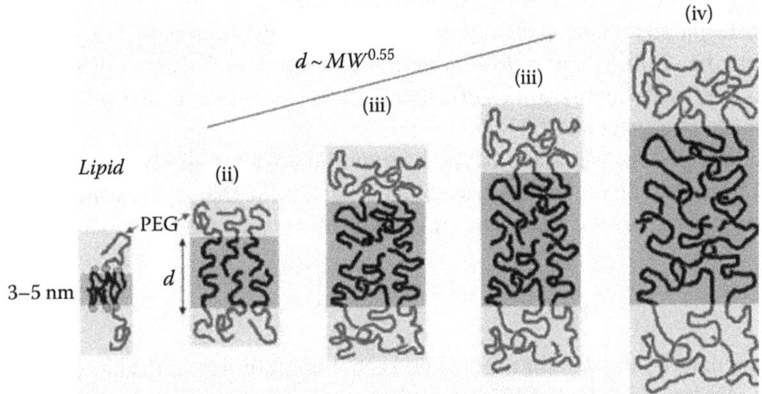

FIGURE 2.65 Schematic scaling of polymersome membrane thickness with copolymer molecular weight. (Ahmed, F., Photos, P.J., and Discher, D.E., *Drug Dev. Res.* 2006. 67. 4–14. Copyright Wiley-VCH Verlag GmbH & Co. KGaA. Reproduced with permission.)

presented in Figure 2.65. Based on the experimental data for a series of different copolymers (Aranda-Espinoza et al. 2001, Bermudez et al. 2002, Battaglia and Ryan 2005), a scaling equation using molecular dynamics simulation (Srinivas et al. 2004) has been developed (Equation 2.33). The equation relates the thickness d and the molar mass of the hydrophobic block, $Mw_{\mathrm{hydrophobic}}$:

$$d \sim Mw_{\mathrm{hydrophobic}}^{b} \tag{2.33}$$

Scaling exponents, b, for polymersomes based on PEE-PEO (Srinivas et al. 2004) and PB-PEO (Bermudez et al. 2002) have been found to equal to 0.55. In good agreement with them, (b=0.5) is data obtained for a series of PS-PAA polymersomes (Chen et al. 2009), whereas a value of 0.66 has been determined by Battaglia et al. for polymer vesicles of PEO and PBO copolymers (Battaglia and Ryan 2005), however, for d<7 nm. It seems that a scaling exponent of 0.5, which corresponds to a nonperturbed state of the chains, is universal for coil–coil block copolymer-based vesicles. Below certain system-specific critical molar mass of the hydrophobic block or layer thickness, higher values of the exponent, reflecting a more extended chain conformation, have been observed; besides the previously mentioned PBO-PEO polymersomes for which b=0.66, the same value of the scaling exponent has been determined for PEE-PEO polymersomes, noteworthy, for d<7 nm and below a critical DP of the PEE block (Srinivas et al. 2004). The scaling exponent close to 0.83 found by numerical simulation for membrane thickness below 7 nm (Srinivas et al. 2004) is also in conformity with the existence of a critical thickness or molar mass. Obviously, at low membrane thickness (corresponding to low molar mass of the hydrophobic block), the chains are stretched and thus are characterized by an exponent corresponding to a strong-segregation state, that is, 2/3. Upon increasing membrane thickness (molar mass of the hydrophobic block), the strong segregation and stretching are opposed by interdigitation and melting, which results in a gradual reduction of the scaling exponent to a value of 1/2 that is compatible with a nonconstraint state. One exception from the rule is the finding of Leson et al. who have found a scaling exponent of unity, corresponding to fully stretched chains, for a series of poly(2-vinylpyridine)-PEO block copolymers of a relatively limited molar mass range (Leson et al. 2007b).

Interestingly, the thickness of the hydrophilic layers, t, has been found to scale almost linearly with the molar mass (Smart et al. 2008), that is, the scaling exponent $b=1$, indicating that the chains are in brush conformation (de Gennes 1980a,b):

$$t \sim Mw_{\text{hydrophilic}}^{b} \tag{2.34}$$

The boundary conditions for the scaling exponent according to the block copolymer theory (Matsen and Bates 1995) and its positions for the chains in the hydrophilic and hydrophobic layers are shown in Figure 2.66. $b=1/2$ and $b=1$ are the two limits (boundary conditions) corresponding to random coil and fully stretched (brush) conformations, respectively. The chains that build the hydrophobic layer are at intermediate conditions, $1/2<b<1$, typical for bulk strongly segregating systems (Forster et al. 1999), whereas those building the hydrophilic layers are almost fully stretched. The latter is expected to contribute to the enhancement of the blood circulation time of the polymersomes in vivo. Furthermore, the area per macromolecule on the boundary between the hydrophilic and hydrophobic layers has been found by Massignani et al. to follow a power law with molar mass with an exponent of 1/3 (Masignani et al. 2010). If there is a slight mismatch in the effective interfacial area per hydrophilic moiety compared to the hydrophobic one, the membrane can spontaneously close into a vesicle rather than forming a planar lamellar structure (Hamley 2005).

In principle, due to the low mobility and slow kinetics of the large, compared to lipid molecules, polymer chains, the polymer vesicles represent metastable or kinetically frozen, near-equilibrium-trapped structures. It has been possible to observe intermediate structures of morphological transformations (Yu et al. 1999), which shed light on the mechanism of the transitions. Formation of vesicles under equilibrium conditions has been seldom observed. Luo et al. have shown formation of

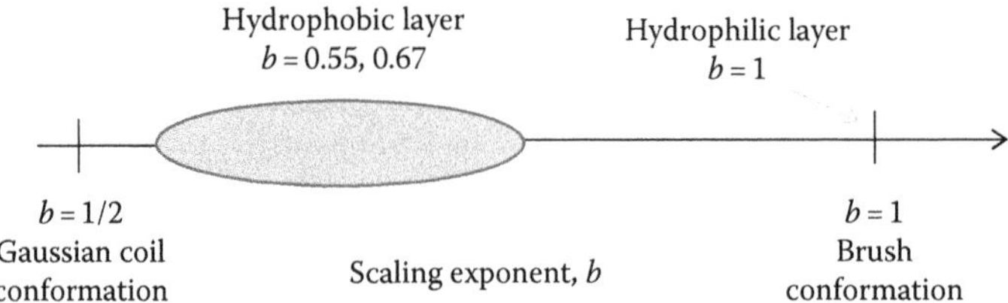

FIGURE 2.66 Experimental values of the scaling exponent, b, found for copolymer chains building hydrophobic and hydrophilic layers of polymersome membranes. The theoretical boundary conditions at $b=1/2$ and $b=1$ correspond to random Gaussian coil conformation and brush conformation, respectively.

equilibrium polymersomes based on PS-PAA in tetrahydrofuran/dioxane/water mixtures under thermodynamic control (Luo and Eisenberg 2001b): the size of the vesicles can be changed reversibly upon increasing/decreasing water content, which indicates that the vesicles are equilibrium structures.

The general vision of ideal polymer vesicles is for unilamellar spherical objects that are uniform in size and shape. Such an idealized vesicle is depicted in Figure 2.64, however, seldom observed in practice: Small uniform and unilamellar vesicles from PS-PAA are shown in Figure 2.67a (Zhang and Eisenberg 1996). Being in their great majority nonequilibrium structures, the polymersomes typically exhibit dispersity in size, deviating from sphericity shapes, and coexistence of

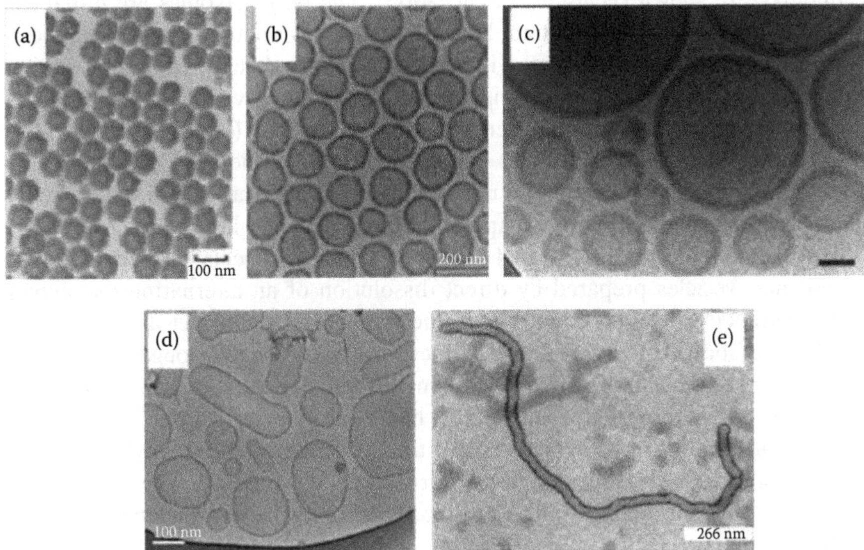

FIGURE 2.67 Cryo-TEM images of unilamellar polymer vesicles of various morphologies, size, and size distribution. (a) Small uniform vesicles from PS-PAA. (Reprinted with permission from Zhang, L. and Eisenberg, A., *Macromolecules*, 29, 8805–8818, 1996. Copyright 1996, American Chemical Society.) Scale bar 100 nm. (b) Polymersomes prepared from poly(2-vinylpyridine)-PEO. (Hauschild, S., Anja Rumplecker, U.L., Borchert, U., Rank, A., Schubert, R., and Forster, S.: *Small*. 2005. 1. 1177–1180. Copyright Wiley-VCH Verlag GmbH & Co. KGaA. Reproduced with permission.) Scale bar 200 nm. (c) Polymersomes prepared from PEO-*b*-poly((4″-acryloxybutyl)2,5-di(4′-butyloxybenzoyloxy)benzoate) upon the addition of water to a dioxane solution. (Yang, J., Levy, D., Deng, W., Keller, P., and Li, M.-H., *Chem. Commun*., 4345, 2005. Reproduced by permission of The Royal Society of Chemistry.) Scale bar 50 nm. (d) Polymer vesicles prepared by direct dissolution of an alternating copolymer of *N*-*n*-tetradecylmaleimide and vinyl gluconamide. (Reprinted with permission from Fenimore, S. G., Abezgauz, L., Danino, D., Ho, Ch.-Ch., and Co, C.C., *Macromolecules*, 42, 2707–2707, 2009. Copyright 2009, American Chemical Society.) Scale bar 100 nm. (e) Tubular vesicles from poly(2-methyloxazoline-*b*-dimethylsiloxane-*b*-2-methyloxazoline) prepared in aqueous media. (Grumelard, J., Taubert, A., and Meier, W., *Chem. Commun*., 13, 1462–1463, 2004. Reproduced by permission of The Royal Society of Chemistry.) Scale bar 266 nm.

different morphologies. There are a number of factors that have a critical impact on the polymersome morphology. Besides the nature of the copolymers and their constituent blocks, for example, T_g, the presence of ionizable groups or groups capable of hydrogen-bonding, donor–acceptor, or electrostatic interactions; the experimental conditions such as temperature, concentration, and pH; the presence of additives (salts, acids, bases, cosolvents, surfactants); as well as the method of preparation influence even stronger the polymorphism and size dispersity of the polymer vesicles. The strong influence of the experimental conditions and methods of preparation is related to the fact the polymersomes are not equilibrium structures. Figures 2.67 and 2.68 represent different polymersome morphologies. In Figure 2.67b, polymer vesicles from poly(2-vinylpyridine)-PEO prepared using color printing technology are shown (Hauschild et al. 2005). The polymersomes are unilamellar and of narrow size distribution, however, as clearly seen, not perfectly spherical. The polymer vesicles depicted in Figure 2.67c have been prepared by adding water to a dioxane solution of an amphiphilic copolymer in which the hydrophobic block is a nematic side-on liquid-crystalline polymer poly((4′-acryloxybutyl)2,5-di(4′-butyloxybenzoyloxy)benzoate) and the hydrophilic one is PEO (Yang et al. 2005). The polymersomes, though unilamellar and spherical, are of undoubtedly much broader size distribution compared to those depicted in Figure 2.67a and b. Coexistence of nearly spherical and elongated vesicular objects has been reported for polymer vesicles prepared by direct dissolution of an alternating copolymer of *N-n*-tetradecylmaleimide and vinyl gluconamide (Fenimore et al. 2009). The vesicles are shown in Figure 2.67d. Tubes, that is, strongly elongated vesicles, are another possible morphology. Unilamellar tubular vesicles are depicted in Figure 2.67e. The object in Figure 2.67e has been prepared by self-assembly of amphiphilic triblock copolymer, poly(2-methyloxazoline-*b*-dimethylsiloxane-*b*-2-methyloxazoline), in aqueous media (Grumelard et al. 2004). The diameter of this presumably soft and flexible tube is ca. 10 of nm, whereas the length is in the millimeter range.

Multilamellar polymersomes represent another frequently identified by TEM vesicular morphology. The presence of additional membranes increases the hydrophobic volume of the multilamellar polymersomes, which implies an enhancement of the solubilization capacity toward hydrophobic substances and sustained release of hydrophilic substances from the aqueous compartments. Representative micrographs of multilamellar polymersomes are shown in Figure 2.68. The bilamellar polymer vesicles (Figure 2.68a) represent the simplest multilamellar morphology (Won et al. 2002). The two hydrophobic membranes—the outer and the inner—are concentric. When the membranes are more than two and more or less concentric (Figure 2.68b), the polymersomes are termed *onion-like* (Shen and Eisenberg 2000a). Interestingly, the concentric walls are separated with uniform in thickness water-filled spaces. Solid *onions* in which there is no spacing between the vesicle walls have been prepared as well (Shen and Eisenberg 2000a). Typically, the multilamellar vesicles coexist with polymersomes of different morphologies as shown in Figure 2.68c in which giant unilamellar vesicles are observed together with multilamellar polymersomes (Kickelbick et al. 2003). These authors have investigated

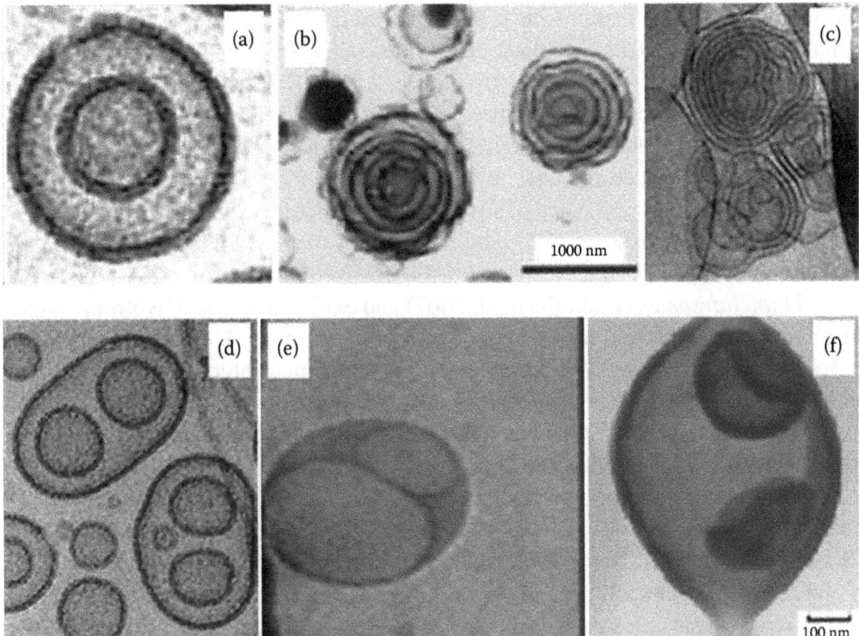

FIGURE 2.68 Cryo-TEM images of multilamellar and multivesicular polymersomes of various morphologies, size, and size distribution. (a) Bilamellar polymersome. (Reproduced with permission from Won, Y.-Y., Brannan, A.K., Davis, H.T., and Bates, F.S., *J. Phys. Chem. B*, 106, 3354–3364, 2002. Copyright 2002, American Chemical Society.) (b) Onion-like vesicles from PS-*block*-poly(4-vinylpyridine decyliodide). (Shen, H. and Eisenberg, A.: *Angew. Chem., Int. Ed.* 2000a. 39. 3310–3312. Copyright Wiley-VCH Verlag GmbH & Co. KGaA. Reproduced with permission.) (c) Polydisperse multilamellar vesicles in coexistence with big unilamellar vesicles from a dilute (0.12 wt.%, corresponding to 3.3xCAC) solution of a short-chain poly(dimethylsiloxane)-*b*-PEO diblock copolymer in water. (Reproduced with permission from Kickelbick, G., Bauer, J., Husing, N., Andersson, M., and Palmqvist, A., *Langmuir*, 19, 3198–3201, 2003. Copyright 2003, American Chemical Society.) (d) Vesicles formed by PEO-PEE at 1% in water. (Reproduced with permission from Won, Y.-Y., Brannan, A.K., Davis, H.T., and Bates, F.S., *J. Phys. Chem. B*, 106, 3354–3364, 2002. Copyright 2002, American Chemical Society.) (e) Entrapped vesicles from an aqueous dispersion of PG-PPO-PG triblock copolymer. (Reproduced with permission from Rangelov, S., Halacheva, S., Garamus, V.M., and Almgren, M. *Macromolecules*, 41, 8885–8894, 2008. Copyright 2008, American Chemical Society.) (f) Entrapped vesicles from PS-PEO in water. (Reproduced with permission from Yu, K. and Eisenberg, A., *Macromolecules*, 31, 3509–3518, 1998. Copyright 1998, American Chemical Society.)

two short-chain poly(dimethylsiloxane)-*b*-PEO diblock copolymers in water and reached to the conclusions that increasing the length of the poly(dimethylsiloxane) block or copolymer concentration leads to an increased tendency for formation of multilamellar vesicles. The structures in Figure 2.68d through f are compound vesicles that consist of two (or more) small vesicles enclosed by a larger one. These multivesicular polymersomes have been prepared by block copolymers of rather different nature: PEO-PEE diblock copolymer in Figure 2.68d (Won et al. 2002), PG-PPO-PG triblock copolymer in Figure 2.68e (Rangelov et al. 2008), and PS-PEO diblock copolymer in Figure 2.68f (Yu and Eisenberg 1998).

Besides the vesicular morphologies represented in Figures 2.67 and 2.68, more complex lamellar structures can be formed. Figure 2.69 shows examples of dispersed multilamellar aggregates of such complex structures. The aggregate depicted in Figure 2.69a displays parallel going membranes that are folded rather than concentric as in the onion-like vesicles (cf. Figure 2.68b). Such disperse objects are termed *lamellarsomes* (Battaglia et al. 2007) and exhibit features of lyotropic lamellar phase normally found at high copolymer concentration. The irregular in shape vesicles shown in Figure 2.69b have been prepared from a triblock copolymer based on PEO containing flanking short blocks of lipid-mimetic units. The particles consist of two different parts: One has a regular vesicular structure, whereas the other represents folded structures with numerous interlamellar connections (Rangelov et al. 2004). Large compound vesicles have been obtained as a result of fusion of small unilamellar vesicles (Zhang et al. 1997). The vesicles were prepared from PS-*b*-PAA in water/DMF mixture, whereas fusion has been induced by the addition of NaCl. The bilayer shell around the periphery is clearly seen in the images, so, in that aspect, the particles are vesicles with internal structure. They further develop with time to form larger structures of hexagonally packed hollow hoops. The object shown in Figure 2.69c is an intermediate structure representing a tube-walled vesicle with hollow regions in the wall running parallel to the surface (Yu et al. 1999). The morphological transition has been induced by addition of water to vesicular dispersions of PS-*b*-PEO in a tetrahydrofuran/water mixture. A spontaneous formation of large pores in the membrane has been observed (Haluska et al. 2002, Chen et al. 2007). In Figure 2.69d, a TEM image of perforated vesicles with highly folded membranes is shown.

Polymersomes cover a wide range of sizes varying from a few nanometers to microns. A notable feature is that the actual size is largely dependent on the preparation procedure (see more in Section 2.2.3.3) and, to date, it is indeed unclear what exactly controls the size. Reasonable explanations of the experimental findings have been given using curvature considerations (Fernyhough et al. 2009) and crystal growth argumentations (Battaglia and Ryan 2006a). The drop of the average vesicle diameter of PB-PMAA vesicles upon increasing hydrophilic-to-hydrophobic ratio in the polymersome-forming region has been ascribed to the former considerations. In other words, the more hydrophilic the copolymer, the more curved structures, that is, of lower diameters, are formed. Using crystal growth argumentations, it has been shown that the dimensions of polymersomes are inversely proportional to the concentration gradient, that is, the higher the concentration, the smaller the dimensions. Both considerations are schematically presented in Figure 2.70, but it should be always kept in mind that the polymersomes can be subjected to post-formation control (see Section 2.2.3.3).

Assuming a mechanism for vesicle formation via slowly closing up disklike micelles (Uneyama 2007), Le Meins et al. have reached an expression (Equation 2.35) relating the minimal radius, R_{min}, with the bending modulus, κ, and line

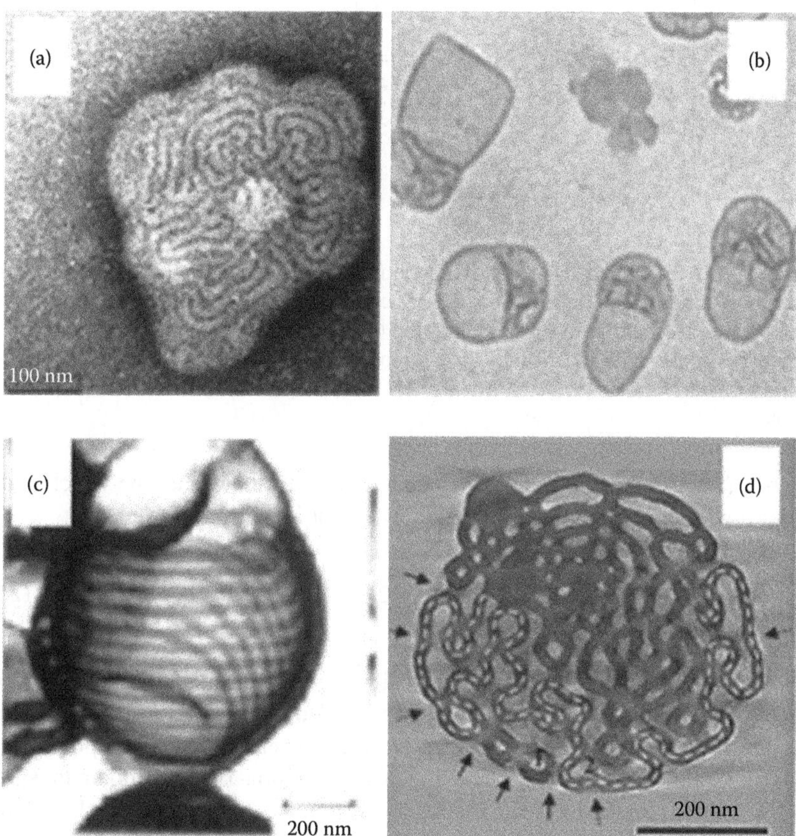

FIGURE 2.69 Micrographs of polymer vesicles of complex morphologies. (a) TEM image of a multilamellar aggregate from PEO-PBO. (Battaglia, G., Tomas, S., and Ryan, A.J., Lamellarsomes: Metastable polymeric multilamellar aggregates, *Soft Matter.*, 3, 470–475, 2007. Reproduced by permission of The Royal Society of Chemistry.) (b) Cryo-TEM micrograph of vesicles taken from water dispersions of $(DDGG)_7(EO)_{45}(DDGG)_7$. (Reprinted with permission from Rangelov, S., Almgren, M., Edwards, K., and Tsvetanov, Ch., Formation of normal and reverse bilayer structures by self-assembly of nonionic block copolymers bearing lipid-mimetic units, *J. Phys. Chem. B*, 108, 7542–7552, 2004. Copyright 2004, American Chemical Society.). (c) Tube-walled vesicle from PS-*b*-PEO. (Reprinted with permission from Yu, K., Bartels, C., and Eisenberg, A., Trapping of intermediate structures of the morphological transition of vesicles to inverted hexagonally packed rods in dilute solutions of PS-b-PEO, *Langmuir*, 15, 7157–7167, 1999. Copyright 1999, American Chemical Society). (d) TEM image of perforated vesicles prepared from PEO-poly(3-(trimethoxysilyl)propyl methacrylate) diblock copolymer. (Reprinted with permission from Chen, Y., Du, J., Xiong, M., Guo, H., Jinnai, H., and Kaneko, T., Perforated block copolymer vesicles with a highly folded membrane, *Macromolecules*, 40, 4389–4392, 2007. Copyright 2007, American Chemical Society.)

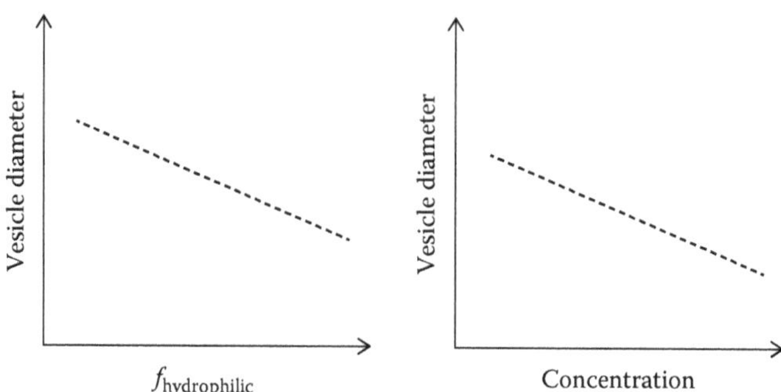

FIGURE 2.70 Simplified schematic representation of the curvature considerations (a) and crystal growth argumentations (b) for the effects of the copolymer chemical composition expressed as a hydrophilic-to-total weight ratio ($f_{hydrophilic}$) and solution concentration on the vesicle dimensions. With increasing $f_{hydrophilic}$, more curved vesicles, that is, of lower diameter, are formed. Similar to crystal growth, vesicles of larger diameters are formed in dilute solution.

tension, γ (Le Meins et al. 2011). The equation clearly shows that the vesicle radius decreases with decreasing bending modulus and/or increasing line tension:

$$R_{min} = 2\frac{\kappa}{\gamma} \tag{2.35}$$

Since the bending modulus is known to increase with molar mass, the copolymers of higher molar masses could be expected to form vesicles easier. This has been observed experimentally by Shen et al. for a series PS-PAA copolymers of similar hydrophilic ratio (Shen and Eisenberg 2000b). Besides molar mass, the polydispersity of the hydrophilic blocks, in particular, can affect the polymersome size and size distribution. This occurs via segregation of the chains with preferential accommodation of the shorter ones in the inner layer, while the outer layer is enriched with the longer ones (Luo and Eisenberg 2001a). Such an arrangement could simultaneously lower the curvature energy and increase the membrane curvature (Terreau et al. 2003, 2004). By blending copolymers with a similar PS block and different in length PAA blocks, thus varying polydispersity of the hydrophilic chains, these authors nicely demonstrate that the polymer vesicles tend to decrease their dimensions with increasing polydispersity. An interesting and somewhat counterintuitive observation is that the vesicle size distribution is found to decrease as well. A particular observation reported elsewhere (Ma and Eisenberg 2009) is that the membrane thickness decreases with decreasing vesicle size below a certain critical value of the radius of PS-PAA polymersomes. Supposedly, this relation is linked to repulsive interactions between the PAA chains in the internal layer and may be applicable only for polymersomes with polyelectrolyte hydrophilic coronae.

In general, the vesicles in the range 100–1000 nm have been extensively investigated. The most appropriate size for biomedical applications is 100–400 nm.

2.2.3.2 Copolymer Chain Architecture and Constituent Blocks

As shown in the beginning of Chapter 2, amphiphilic block copolymers comprising hydrophilic and hydrophobic moieties (blocks) are able to self-associate in aqueous solution into variety of structures such as spherical and wormlike micelles as well as vesicular or lamellar structures. The morphology (aggregate geometry) of the self-assembled structures is dictated by the hydrophilic block-to-total molecular weight ratio, f. The most appropriate ratio for formation of lamellar structures is $f = 35 \pm 10\%$ as Discher and Eisenberg have recently concluded based on the data collected for coil–coil block copolymers (Discher and Eisenberg 2002). However, this rule is not strictly observed, and the ratio has been found to vary with chain chemistry and architecture, molecular weight, and solvent conditions, thus broadening the range of f in which formation of lamellar structures is anticipated. Examples from the recent literature given in the succeeding text demonstrate the deviating behavior of vesicle-forming copolymers of completely different chemical nature and chain architecture. These are typically copolymers with hydrophobic shape-persistent blocks or moieties able to interact electrostatically (Coulombic interactions), via π–π or H-bonding. PEG-based diblock copolymers with a hydrophobic block consisting of a nematic liquid-crystalline polymer have been found to form polymer vesicles at hydrophilic/hydrophobic ratios ranging from 40/60 to 18/82 (Yang et al. 2006). A nonionic triblock copolymer $EO_5PO_{70}EO_5$ and a PEG-poly(*N*-acryloylpiperidine) diblock copolymer, both of ca. 10 wt.% content of hydrophilic component, are also reported to form vesicles in aqueous solution (Schillen et al. 1999, Jo et al. 2008, Li et al. 2008), whereas grafted with hydrophilic chains, polyphosphazenes require considerably higher (40%–50%) hydrophilic contents to form vesicular structures (Zhang et al. 2009b). Vesicles are the dominating morphology of dextran-block-PCL when the mass fraction of dextran equals to 0.16 (Zhang et al. 2010b).

The copolymers yielding polymersomes typically have average molar masses ranging from 2,000 to 20,000 Da. The chain architecture and chemical composition can be either very simple or rather complex. These are factors of key importance to impart the corresponding polymer vesicles their peculiar properties. The linear diblock AB architecture, where A and B denote the hydrophilic and hydrophobic blocks, respectively, is the simplest chain architecture, and the majority of polymersome-forming copolymers are of such a chain architecture. Copolymers of more complex linear chain architectures such as ABA, BAB, ABC, ABCA, and ABABA (here C can be either hydrophilic or hydrophobic block) have also been reported to yield polymer vesicles. Depending on the macromolecular architecture, the chains can adopt different conformations in the membrane. The latter for various architectures are schematically presented in Figure 2.71 in a very comprehensive manner (LoPresti et al. 2009). For AB diblock copolymers, there is only one possible conformation with the B block situated in the hydrophobic layer and the A block protruding out from the wall on both sides. However, the membrane formed by a hypothetical diblock copolymer displayed in Figure 2.71 is interdigitated, in which the B blocks are laterally oriented and robust entanglement within the hydrophobic layer takes place. The other possible arrangement (not presented) involves an *end-to-end* orientation of B blocks

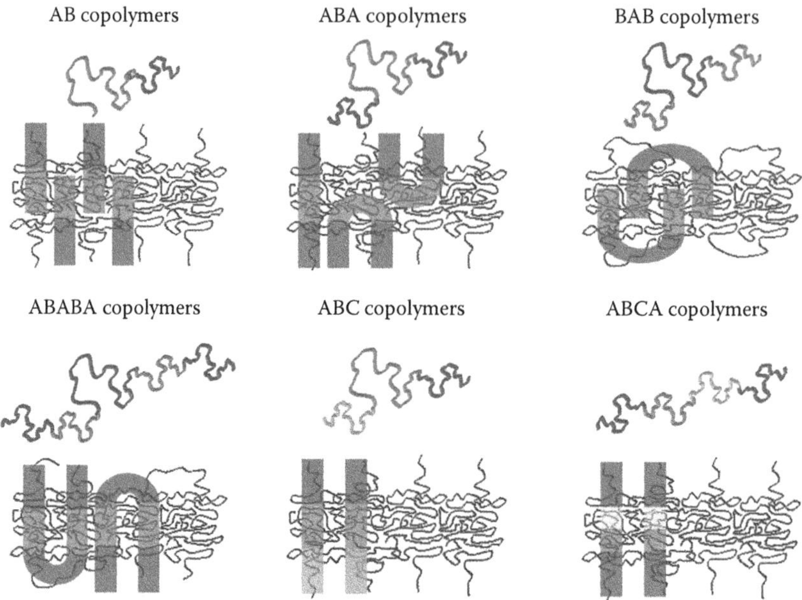

FIGURE 2.71 Schematic representation of possible conformations of block copolymer chains in polymersome membranes. (LoPresti, C., Lomas, H., Massignani, M., Smart, T., and Battaglia, G., *J. Mater. Chem.*, 19, 3576–3590, 2009. Reproduced by with permission of The Royal Society of Chemistry.)

from different chains. Supposedly, in such an arrangement, the B blocks are less entangled, and the hydrophobic membrane is clearly segregated in two monolayers similarly to the liposomal bilayer membranes. Two are the possible chain conformations of ABA-type triblock copolymers—*trans* and U shape (Figure 2.71). Packing arguments may predict that the amount of the chains in U-shape conformation is lower in the inner monolayer (Schillen et al. 1999). An increased probability of finding A block end segments on the same side of the membrane, that is, U-shape conformation, with increasing hydrophobic domain size (proportional to the length or molar mass of the B block in binary systems) has been suggested as well (Noolandi et al. 1996; Svensson et al. 1999). The hydrophilic A block must adopt a looping conformation so that the two flanking B blocks of copolymers of BAB-type chain architecture assemble into the same membrane (Figure 2.71). Alternatively, provided that the A block is long enough, the B blocks of the same chain may anchor in two different adjacent membranes, thus causing physical cross-linking. Obviously, similar conformations may be anticipated for pentablock copolymers (Figure 2.71). The introduction of a third or even fourth block, thus making ABC, ABCA, ABCD, ABCDA, etc., architectures, provides additional dimensions to control the polymersome properties and performance. Possible arrangements and chain conformations for such architectures are presented in Figure 2.71.

As noted earlier, polymer vesicles have been mostly produced from block copolymers of AB and ABA chain architectures. In the following, by giving examples from the recent literature, we illustrate the ability of copolymers with chain architectures different from the latter two to self-associate into polymersomes and how the membrane properties can be engineered by employing such copolymers. The examples are not only restricted to linear chain copolymers but also extended to star-like, grafted, and branched copolymers as well as copolymers containing dendritic moieties, ω-terminated, and end-capped (co)polymers.

Xu et al. have prepared hydrophilic homopolymers of *N*,*N*-dimethylacrylamide and *N*-(2-hydroxypropyl)methacrylamide as well as statistical copolymers with *N*-acryloxysuccinimide of various molar masses containing between 6 and 23 wt.% hydrophobic end groups based on pyrene, cholesterol, or octadecane (Xu et al. 2011). The presence of two spatially close rigid rings at the ω-terminus has been shown to be crucial in vesicle formation (tubular vesicles and spherical polymersomes are reported) since the homopolymer with two octadecyl end groups yields polymeric micelles. In a broader aspect, these functional polymers can be considered as AB diblock copolymers in which the ω-end groups play the role of the hydrophobic block.

The BAB chain architecture, that is, with flanking hydrophobic blocks, is not much preferable for self-assembly. The latter is hindered because of entropy penalty due to the looping conformation that the middle hydrophilic block has to adopt. As noted earlier, physical cross-linking resulting from anchoring of the B blocks in different membranes may be observed as well. Nevertheless, BAB copolymers have been prepared and reported to yield polymer vesicles. Aggregation of polymersomes prepared from PBO-PEO-PBO as well as other AB and ABA copolymers induced by addition of homopolymer (PEO) has been investigated by Smart et al. (2010). The polymersomes were spontaneously formed in water, and narrow size distributions were ensured by passing dispersions through extruder. Any hindrance or peculiarities related with the specificities of the BAB architecture were reported. The hydrophobic blocks of the BAB copolymer described by Rangelov et al. (2004) are composed of repeating lipid-mimetic units, DDGG. These moieties, although relatively short, are much more hydrophobic than the PBO blocks of the former BAB copolymer. Therefore, to obtain polymer vesicles, the initial aqueous dispersions were subjected to freeze–thaw cycles followed by extrusion through 200 nm pore-size filters (see more in Section 2.2.3.3). Besides unilamellar polymersomes, fractions of multilamellar vesicles as well as vesicles with folded internal structures shown in Figure 2.69b have been identified (Rangelov et al. 2004).

The BAB sequence is also present in pentablock architectures of the types BABAB and ABABA. A BABAB copolymer comprising poly(methylphenylsilane) and PEO as hydrophobic and hydrophilic segments, respectively, has been prepared and reported to form a variety of well-defined morphologies in water-based solvent systems including vesicles (Holder et al. 1998, Sommerdijk et al. 2000). A parallel arrangement of the rigid poly(methylphenylsilane) blocks in the membrane interior and looping conformation of the two PEO blocks have been proposed. Although the ABABA chain architecture cannot be considered as exotic and a number of copolymers having such a chain sequence have been described in the literature, to the best of authors' knowledge, vesicle formation by ABABA copolymers has not been reported.

The introduction of a third constituent moiety provides an extra level of control and diversifies the polymersome properties. At a favorable for vesicle formation hydrophilic/hydrophobic ratio, ABC terpolymers may self-assemble into asymmetric membranes with chemically different internal and external hydrophilic layers. Such a segregation of the two hydrophilic blocks has been observed by Liu et al. for PAA-PS-poly(4-vinylpyridine) triblock copolymer in DMF/THF/water mixtures (Lui and Eisenberg 2003). The segregation is based on the difference in repulsive interactions within the PAA or poly(4-vinylpyridine) corona under different pH conditions. Thus, at lower pH, vesicles with poly(4-vinylpyridine) chains on the outside and PAA chains on the inside are formed, whereas just the opposite arrangement was observed at higher pH. Due to the presence of organic solvents, the hydrophobic layer of the membranes is swollen, and the vesicles are under dynamic conditions, which facilitates the reversible inversion of vesicles with poly(4-vinylpyridine) on the outside to vesicles with PAA on the outside upon increasing/decreasing pH. *Chimaeric* polymersomes, as the monstrous female creature composed of parts of multiple animals Χίμαιρα in Greek mythology, have been prepared from a series of biamphiphilic ABC copolymers of composition PEO-PCL-poly(2-(diethylamino)ethyl methacrylate), in which only the molar mass of poly(2-(diethylamino)ethyl methacrylate) was varied (Liu et al. 2010). Polymer vesicles of average sizes 130–175 nm were formed by the film rehydration method. The close to neutral zeta-potential as well as results from NMR suggested preferential location of the longer nonionic PEO chains on the outside of the vesicles, while the shorter chains of poly(2-(diethylamino)ethyl methacrylate) formed the internal hydrophilic layer. Obviously, this arrangement was favorable to ensure high protein loading most probably due to effective electrostatic and/or hydrogen-bonding interactions between the exogenous proteins and poly(2-(diethylamino)ethyl methacrylate) chains. Also segregated are the PEO and PAA chains, primarily forming the inner and outer interfaces, respectively, of polymersomes prepared from PEO-PCL-PAA (Wittemann et al. 2007). In another study, Blanazs et al. demonstrate how the chain expression at the polymersome surface can be tailored (Blanazs et al. 2009b). The authors use polymersomes prepared from a series of amphiphilic ABC copolymers of PEO-poly(2-(diisopropylamino)ethyl methacrylate)-poly(2-(dimethylamino)ethyl methacrylate). By varying the relative hydrophilic volume fraction via systematic variation of pH, the asymmetric polymersome membranes express either cationic polymer chains at the outer hydrophilic layer or nonionic chains to the interior and vice versa as shown in Figure 2.72. Triblock terpolymers of PEO-PAA-PNIPAM, in which the molar masses and ratios between the constituent blocks are varied, have been prepared by Xu et al. (2009). At room temperature, the copolymers are soluble in water as unimers, but when raising the temperature to 37°C, which is above the LCST of PNIPAM, the latter turns hydrophobic and nanosized (about 220 nm) vesicles are formed. What is interesting here is that the hydrophobic block at these conditions is the flanking rather than the middle block and, strictly speaking, the chain architecture is ACB. The role of the PAA moieties is that they allow cross-linking at the vesicle interface using cystamine via carbodiimide chemistry.

In contrast to the aforementioned ABC copolymers in which one or two of the hydrophilic blocks are charged, both water-soluble blocks of the copolymers

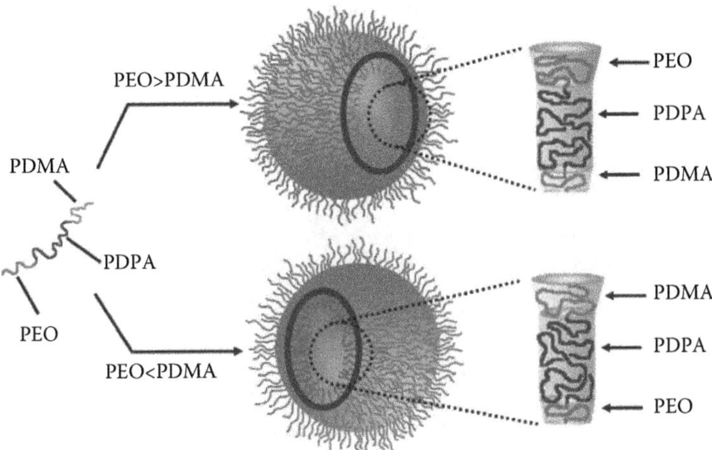

FIGURE 2.72 Tailoring of chain expression at the polymersome surface via varying the relative volume fractions of the two hydrophilic blocks. PEO, PDPA, and PDMA correspond to poly(ethylene oxide), poly(2-(diisopropylamino)ethyl methacrylate), and poly(2-(dimethylamino)ethyl methacrylate), respectively. At higher pH, the PDMA block is collapsed; the hydrophilic volume of the PEO block is higher than that of PDMA, and PEO chains are oriented toward the outer side of the vesicles (above). At lower pH, PDMA chains are protonated and of relatively higher volume fraction compared to the volume fraction of the PEO chains; accordingly, they are expressed at the polymersome surface. (Blanazs, A., Massignani, M., Battaglia, G., Armes, S.P., and Ryan, A.J.: *Adv. Func. Mater.* 2009b. 19. 2906–2914. Copyright Wiley-VCH Verlag GmbH & Co. KGaA. Reproduced with permission.)

described by Stoenescu et al. are nonionic (Stoenescu et al. 2002, 2004). These authors have synthesized PEO-poly(dimethylsiloxane)-poly(2-methyloxazoline) triblock copolymers in which the DP of the middle (hydrophobic) block was kept constant, whereas the lengths of the flanking hydrophilic blocks were varied. The copolymers have been found to self-associate into vesicles with asymmetric membranes in which, in accordance with geometric considerations, the less voluminous hydrophilic chains are segregated toward the inner side of the membrane, whereas the longer, bulkier, and of higher volume chains of the other block are oriented toward the outside of the vesicles. This is demonstrated in Figure 2.73, but it should be kept in mind that PEO and poly(2-methyloxazoline) are not compatible, which, undoubtedly, contributes to the segregation and formation of asymmetric membranes.

Breaking the membrane symmetry by using ABC biamphiphilic copolymers in which upon vesicle formation the hydrophilic block segregates to the outside surface of the membrane potentially dictates the spontaneous curvature and, hence, the vesicle size. Brannan et al. have shown that the bilayer symmetry can be broken using architecturally asymmetric block copolymers of the type ABCA (Brannan and Bates 2004). In these tetrablock copolymers, the two middle blocks, B and C, are hydrophobic—PS and PB, respectively. The flanking hydrophilic blocks are PEO of equal length, but within a series of four ABCA copolymers, its content was varied from 32 to 66 wt.%. The effect of symmetric tetrablock architecture has been manifested

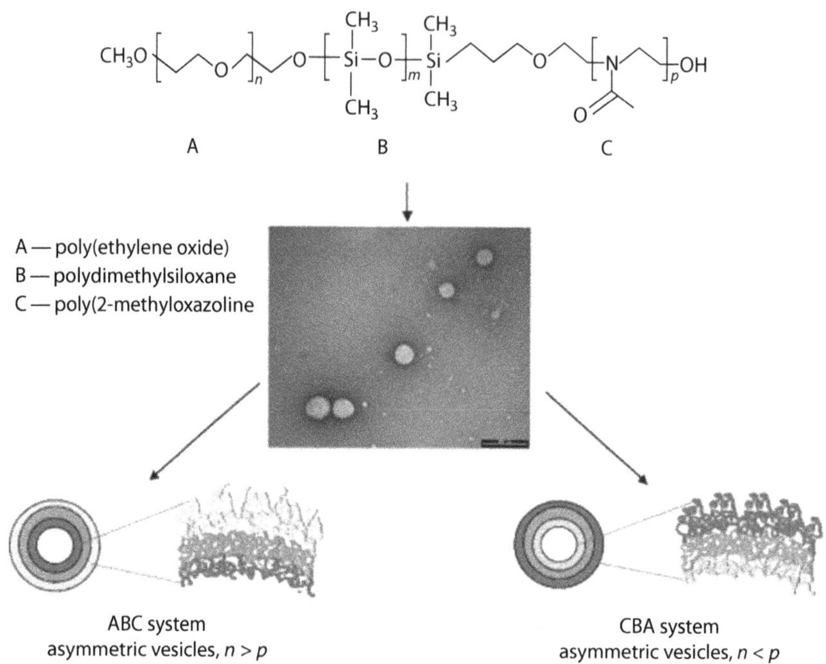

FIGURE 2.73 Vesicles with asymmetric membranes formed in water by ABC copolymers. The chains of the hydrophilic blocks A and C are segregated with bulkier and of larger volume chains oriented toward the outside of the vesicles. Scale bar 500 nm. (Stoenescu, R., Graff, A., and Meier, W.: *Macromol. Biosci.* 4. 2004. 930. Copyright Wiley-VCH Verlag GmbH & Co. KGaA. Reproduced with permission.)

in segregation of the hydrophobic layer of the membrane as shown in Figure 2.74. For PEO wt. fraction < 0.50, the hydrophobic layer is split in two parallel going sublayers of PS and PB. The copolymers of higher PEO contents displayed mixed morphologies, mostly wormlike micelles and a minority of vesicles with thinner membranes. For the latter, in-plane periodicity suggesting either bicontinuous or hexagonally arranged state of segregation of PS and PB was observed (Figure 2.74).

The chain topology of the copolymers described later is also linear; however, the monomer units are not connected to form blocks. These are alternating, regular, and random copolymers as well as end-capped polymers. The examples demonstrate that the ability for vesicle formation is not the exclusive realm of the amphiphilic block copolymers. By free-radical polymerization, Wu et al. have prepared alternating copolymers of alkylmaleates (hydrophobic) and polyhydroxy vinyl ethers (hydrophilic) that have appropriate amphiphilicity to self-associate in polymersomes (Wu et al. 2008). The polymers are capable of forming ultrasmall vesicles in water with thin flexible shells. Phosphorylcholine-substituted polyolefin vesicles have been produced from copolymers obtained by ring-opening metathesis polymerization (Kratz et al. 2009). The polymers contain a phosphorylcholine residue at every eighth backbone carbon atom. A distinctive feature of these polymersomes is their narrow distribution. 2-(Diethylamino)ethyl methacrylate and 3-(trimethoxysilyl)

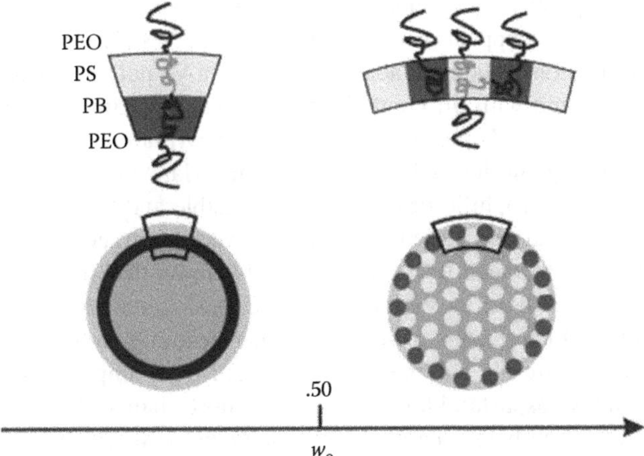

FIGURE 2.74 Structural models of vesicles prepared from asymmetric PEO-PS-PB-PEO tetrablock copolymers in which PS and PB that form the hydrophobic layer of the membranes are segregated. w_o is the PEO weight fraction. (Reprinted with permission from Brannan, A.K. and Bates, F.S., *Macromolecules*, 37, 8816–8819, 2004. Copyright 2004, American Chemical Society.)

propyl methacrylate have been copolymerized to obtain a random copolymer that is coupled to PEG to form diblock copolymer of poly(ethylene glycol)-*b*-poly((2-(diethylamino)ethyl methacrylate-*s*-(3-(trimethoxysilyl)propyl methacrylate) (Du and Armes 2005). In aqueous/THF mixtures, the copolymer spontaneously forms 200–400 nm in size vesicles. The trimethoxysilyl groups can be hydrolyzed and subsequently reacted to produce siloxane cross-links. In addition, the swelling of the membranes and, hence, the permeability of the vesicles can be controlled by changing the ionization state via pH variations. Random copolymers of acrylic acid and distearylglyceryl acrylate have been obtained from partial transesterification of poly(*N*-acryloxysuccinimide) with distearin followed by thorough hydrolysis to obtain acrylic acid monomer units (Chiu et al. 2008). The copolymer is of considerable high molar mass (ca. 300 kg/mol), thus contrasting the typical molar mass of polymersome-forming copolymers. End capped with β-cyclodextrin, PS has been found to form vesicular structures in water (Felici et al. 2008). The cyclodextrins on the surface of the polymersomes form inclusion complexes with hydrophobic compounds and have been shown to participate in recognition reactions with specific membrane receptors.

Copolymers of nonlinear, that is, comblike, dendritic, grafted, and starlike copolymers, have also been found to form vesicles. Linear–dendritic is the architecture of the copolymers reported by Del Barrios et al. (2010). The hydrophilic block is PEG, whereas the hydrophobic moieties are different generations of azobenzene-containing dendrons based on 2,2-*bis*(hydroxymethyl) propionic acid. A generation-dependent aggregation behavior was observed for the series: Formation of polymersomes with diameters in the range 300–800 nm has been reported for the fourth generation of dendritic moiety, corresponding to hydrophilic/hydrophobic ratio equal to 20/80. The aggregates are photoresponsive as revealed by the appearance of wrinkles in the vesicle membrane resulting from UV illumination. Biodegradable and pH-sensitive polymersomes have been prepared from grafted with different chains copolymers (Kim and Lee 2010). On the biodegradable and pH-sensitive backbone of poly(β-amino ester) obtained by reacting equimolar quantities of 1,4-butanediol diacrylate and 3-amino 1-propanol, biocompatible and hydrophilic PEG and biodegradable and hydrophobic poly(D,L-lactide) have been grafted. Vesicular structures were observed for copolymers of hydrophilic contents between 10% and 15%, notably lower than the typical 35 ± 10%. The ionization of poly(β-amino ester) loosens the packing state of the biodegradable poly(D,L-lactide), which renders pH-dependent permeability on hydrophilic solutes. Also biodegradable are the polymersomes obtained from amphiphilic copolymers poly(2-hydroxyethyl aspartamide) grafted with short chains of oligo(lactic acid) (Lee et al. 2006). The backbone of these copolymers is hydrophilic, whereas the grafted chains are hydrophobic just opposite to the series of amphiphilic polyphosphazenes grafted with hydrophilic PEG chains (Zheng et al. 2009). To enhance the rigidity of the highly flexible backbone, hydrophobic ethyl-*p*-aminobenzoate side groups have been grafted as well (Figure 2.75a). Polymersomes (Figure 2.75b) are formed at the typical hydrophilic weight fraction, which can be conveniently adjusted by tuning the PEG to ethyl-*p*-aminobenzoate ratio.

Hybrid, that is, comblike–dendritic is the structure of the amphiphilic copolymers of Tian et al. (2006). The copolymers consist of poly(γ-*n*-dodecyl-L-glutamate) in α-helical conformation, which imparts a rodlike character of the hydrophobic block, and a polyester dendron with a numerous hydroxyl end groups (Figure 2.76). In contrast to the typically fluctuating, flexible linear copolymer macromolecules, the geometric shape of the macromolecule depicted in Figure 2.76 is somewhat restricted. The high shape persistence yielded unusually low size dispersity of the resulting vesicles. Comblike PEG has been prepared by ATRP of α-methylacryloyl-ω-hydroxyl-PEO

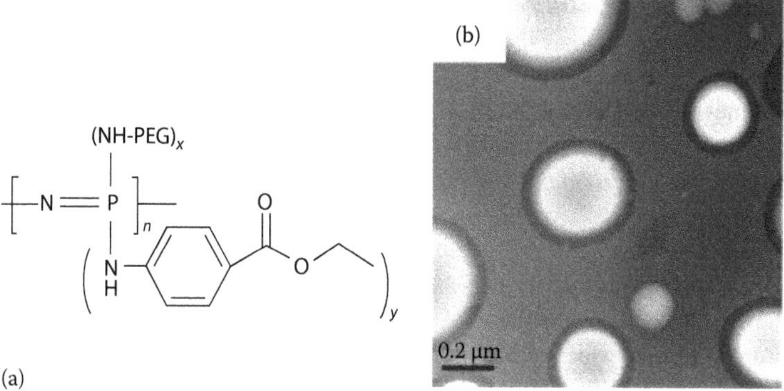

FIGURE 2.75 Chemical structure (a) and (b) TEM image of polymersomes formed by a copolymer of composition (mole fractions) $x=0.31$ and $y=1.69$, PEG molar mass 1100, and $f_{PEG}=0.49$. (Reprinted from *Polymer*, 50, Zheng, C., Qiu, L., and Zhu, K., 1173–1177, Copyright 2009, with permission from Elsevier.)

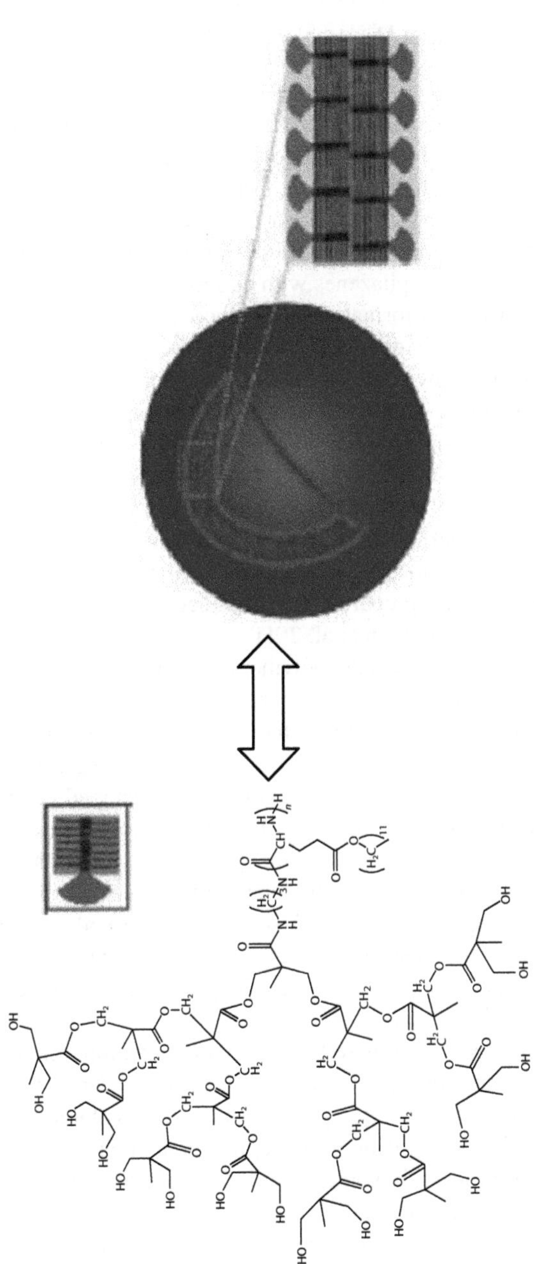

FIGURE 2.76 Chemical structure of comb–dendritic copolymers with poly(γ-*n*-dodecyl-L-glutamate) as a hydrophobic comblike block and a polyester dendron as a hydrophilic moiety. The inset represents the persistent shape of the macromolecule. (Tian, L., Nguyen, Ph., and Hammond, P.T., *Chem. Commun.*, 3489–3491, 2006. Reproduced by permission of The Royal Society of Chemistry.)

(Li et al. 2006b). In water, the polymer spontaneously self-associates into vesicles with diameters ranging from 200 to 500 nm and quite thick (about 50 nm) walls. The comblike polymer has been further reacted with cholesteryl chloroformate, thus producing cholesterol-grafted derivatives (Li et al. 2007). Notably, very small amounts of grafted cholesterol are able to improve stability of the vesicles in the presence of hydrophobic drugs.

3-Miktoarm star copolymers of the type AB_2, where A and B correspond to PEG and poly(L-lactic acid), respectively, have been shown able to form polymersomes (Yin et al. 2009). The range of PEG volume fractions in which vesicle formation has been observed is much broader than that of the linear diblock counterparts. The same is the number of the arms of the amphiphiles of Jun and coworkers (2010). The authors have grafted cyclotriphosphazenes with equimolar amounts of PEG and oligopeptide in *cis*-nongeminal conformation to produce tripodal starlike copolymers. As the grafted components are in equimolar amounts, the self-assembly of the tripodal amphiphiles can be tuned by controlling the hydrophobicity of the oligopeptide. Three PEG chains of molar mass of about 1100 have been attached by esterification to citric acid, and the resulting product has been reacted at its hydroxyl terminus with different amounts of D,L-lactide to obtain 4-miktoarm amphiphilic copolymers of the type A_3B, where A and B correspond to PEG and poly(D,L-lactic acid), respectively (Jain and Kumar 2010a). Formation of polymersomes has been observed at PEG contents 10%–30%. Drug loading, protein adsorption, degradation, and release kinetics of the polymer vesicles prepared from these copolymers have been further studied (Jain and Kumar 2010b, Ayen et al. 2011).

pH-responsive and size-controllable polymersomes have been prepared from carboxyl-terminated commercial hyperbranched polyesters (Shi et al. 2008). Polymersomes with diameters about 200 nm have been formed at mildly acidic conditions. Their dimensions progressively increase with decreasing pH, and, remarkably, the process is reversible. Zhou and coauthors have developed giant (exceeding 100 μm) vesicles generated from an amphiphilic starlike copolymer consisting of a core of poly(3-ethyl-3-oxetanemethanol) and many PEO arms (Zhou and Yan 2004). The polymer vesicles are formed at specific temperatures depending on copolymer composition (Zhou et al. 2007b). Figure 2.77 shows the chemical structure of these hyperbranched polyesters and possible models of membrane packing. The vesicles have been shown to undergo sequential morphology changes at temperatures near the LCST.

Briefly, polymersomes can be prepared from any kind of amphiphilic macromolecules provided that the appropriate balance and conditions for vesicle formation are found. Most of polymersomes are typically intended for biomedical applications. Therefore, the major prerequisite for the constituent moieties is to be biocompatible, nonharmful, or, at least, biologically inert. In addition, abilities to control the release of the encapsulated substances, to degrade to nonharmful products, and to destabilize the structures at desired time or place are often required, which, to a large extend, directs the efforts of the researchers involved to develop appropriate copolymers. Undoubtedly, the main factors contributing to all these properties are the nature and chemical composition of the block copolymers. The constituting blocks may be soft and flexible, rigid, amorphous or crystalline, glassy and

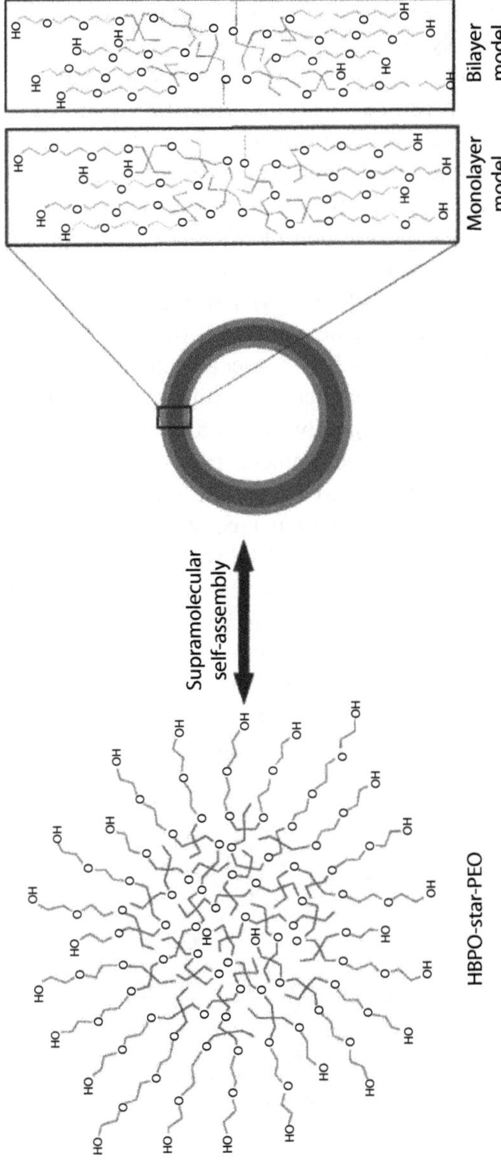

FIGURE 2.77 Chemical structure of the hyperbranched polyester-star-PEG (HBPO-star-PEG) and monolayer and bilayer molecular packing models. HBPO and PEG moieties are denoted with blue and red, respectively. (Reprinted with permission from Meng, F., Zhong, Zh., and Feijen, J. 2009. *Biomacromolecules*., 10, 197–209, 2009. Copyright 2009, American Chemical Society.)

vitreous, liquid crystalline, charged or nonionic, and synthetic or naturally occurring, in conformations of random coil or helix, and all combinations between the blocks to form macromolecular amphiphiles, for example, soft–soft and soft–rigid, are possible. The constituent blocks of the polymersome-forming macromolecular amphiphiles are, in generally, the same as those of the micelle-forming ones. Coil–coil block copolymers such as PEO-PB have been the most extensively studied polymersome-forming copolymers, and most of the current knowledge on polymer vesicles is based on such copolymers. If one or more blocks are replaced by a polymer block of different nature, for example, rodlike, glassy, crystalline, liquid crystalline, and helix, the shape anisotropy and additional order introduced thereby strongly influence the self-assembly and properties of the polymersomes. The change of the mechanical properties, particularly the membrane viscosity, of giant polymersomes upon the substitution of the low T_g hydrophobic block with a glassy liquid-crystalline one has been nicely demonstrated by Mabrouk et al. (2009). Figure 2.78 reveals the gel-like behavior of giant polymersomes prepared from poly(ethylene oxide)-poly(4′-methoxyphenyl 4-(6′-acryloyloxy)hexyloxy)benzoate), the hydrophobic block of which is a smectic liquid-crystalline polymer of high T_g. In particular, the vesicle released after micropipette aspiration shows a persistent deformation since the tongue, which is a molded replica of the micropipette, does not change its length and diameter over a period exceeding 10 min, in contrast to the polymersomes prepared from the referent PEO-PB copolymer, which undergo a rapid (within less than 40 ms) recovering of their spherical shape. Although the method of preparation, as will be shown in the next section, is an important factor to control the size of the vesicles, owing to their high persistence length, the copolymers containing rigid block(s) typically form giant polymersomes with thick membranes. Such giant vesicles have been prepared from PS-poly(L-isocyanoalanine-(2-thiophen-3-ylethyl)amide) using electroformation (Vriezema et al. 2003, 2004). The thiophene units are attached laterally to the helical backbone and stack on top of each other due to hydrogen bonding (Figure 2.79), which

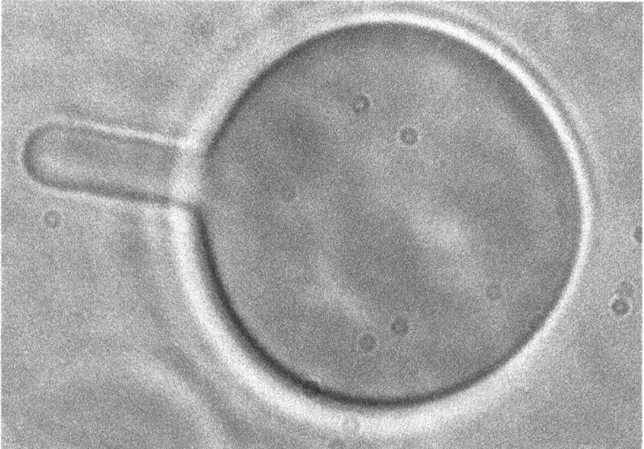

FIGURE 2.78 Giant vesicle (tens of μm) prepared from poly(ethylene oxide)-poly(4′-methoxyphenyl 4-(6′-acryloyloxy)hexyloxy)benzoate) copolymer in water showing persistent deformation after micropipette aspiration. (Mabrouk, E., Cuvelier, D., Pontani, L.-L., Xu, B., Levy, D., Keller, P., Brochard-Wyart, F., Nassoy, P., and Li, M.-H., *Soft Matter*, 5, 1870–1878, 2009. Reproduced by permission of The Royal Society of Chemistry.)

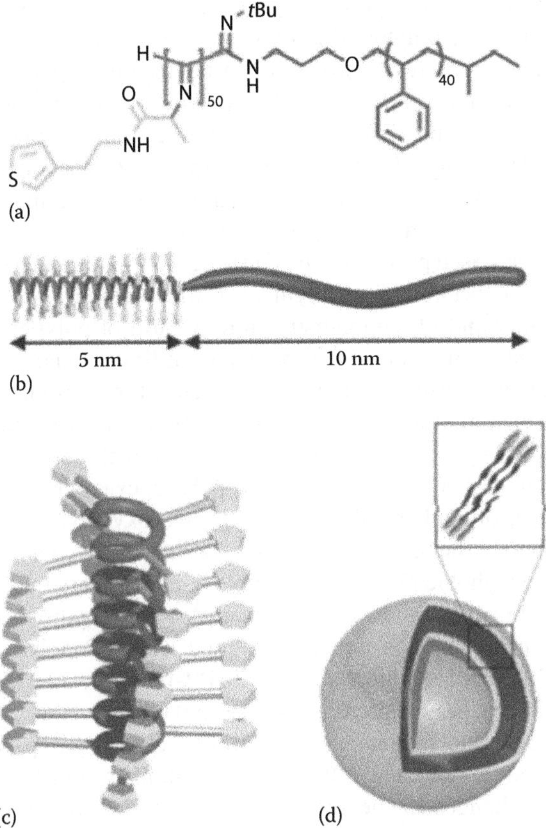

FIGURE 2.79 Chemical structure (a) and schematic presentations of the PS-poly(L-isocyanoalanine-(2-thiophen-3-ylethyl)amide) diblock copolymer (b), poly(L-isocyano-alanine-(2-thiophen-3-ylethyl)amide) block showing stacks of thiophene groups (c), and a vesicle formed in water (d). (Vriezema, D.M., Hoogboom, J., Velonia, K., Takazawa, K., Christianen, P.C.M., Maan, J.C., Rowan, A.E., and Nolte, R.J.M.: *Angew. Chem. Int. Ed.* 2003. 42. 772–776. Copyright Wiley-VCH Verlag GmbH & Co. KGaA. Reproduced with permission.)

additionally increases the rigidity and persistence length of the block. Micron-size polymer vesicles are formed in water with membrane thickness 30 ± 10 nm corresponding to ca. twice the length of a single copolymer macromolecule. The most probable membrane structure is depicted in Figure 2.79d. Despite their rigid rod block copolymer nature, these giant aggregates have a fluidic membrane and are capable of encapsulating solutes (Vriezema et al. 2003, 2004).

In principle, the secondary structures that polypeptides are able to form (α-helix, β-sheet) are rigid. Therefore, the typical structures of the polypeptide-containing polymersome-forming block copolymers are of the type coil–rod or rod–rod. The polymer vesicles prepared from polypeptide-containing copolymers are termed *peptosomes*. Not only does the ability to form secondary structures but also a number of other properties make the polypeptides attractive and interesting for building blocks of vesicle-forming amphiphilic copolymers. These are their biocompatibility, functionalities, and stimuli responsiveness. Most of the polypeptides studied are able to undergo either reversible or irreversible transitions upon variations of pH, ionic strength, and/or temperature. The transitions produce rearrangement of the vesicle membrane related to changes in the size and morphology, appearance of channels in the membranes, destabilization of the vesicular structure, etc., that may be used to prepare intelligent systems, materials, and nanoreactors.

The polypeptide-based polymersome-forming block copolymers may be divided in two groups: (1) hybrid block copolymers in which at least one of the blocks is a synthetic polymer and (2) copolymers entirely built of polypeptides. Although the number of studies on peptosomes is somewhat limited and recent reviews on polymer vesicles and capsules only partially focus on peptosomes (Li and Keller 2009, van Dongen et al. 2009), the list of polypeptides used as building blocks of vesicle-forming copolymers is quite long—poly(L-lysine), poly(L-leucine), poly(L-glutamic acid), poly(γ-benzyl-L-glutamate), poly(isocyanoalanyl-alanine methyl ester), oligovaline, oligophenylalanine, poly(2-hydroxyethyl aspartamide), polysarcosine, and many others. Synthetic polymers such as PB, polyisoprene, and PS constitute the hydrophobic block in hybrid block copolymers in which the hydrophilic moieties are polypeptides. Polymersomes with a hydrophilic polypeptide corona have been prepared in aqueous solution from PB-poly(L-glutamic acid) (Checot et al. 2002, 2005, Kukula et al. 2002), PB-poly(L-lysine) (Sigel et al. 2007, Gebhardt et al. 2008), polyisoprene-poly(L-lysine) (Checot et al. 2007), as well as PS with an oligopeptide sequence as a hydrophilic head group (see Dirks et al. 2005). For the first three systems, the variations in pH or temperature have been shown to induce changes in the secondary structure of the polypeptide hydrophilic layers of the membranes (charged coils, α-helix, β-sheet). Sanson et al. have reported on highly stable, nonpermeable to water, stimuli-responsive, and size-tunable polymersomes prepared from poly(trimethylene carbonate)-poly(L-glutamic acid) (Sanson et al. 2010b). In addition, the poly(trimethylene carbonate) block that forms the hydrophobic layer of the membrane can be rapidly degraded by enzymatic hydrolysis. Vesicle budding and fission have been observed at temperatures above the poly(trimethylene carbonate) melting temperature (34°C–35°C), whereas fusion has occurred upon decreasing temperature (Sanson et al. 2010a). Double hydrophilic, but sensitive to both pH and temperature, are the copolymers poly(2-(dimethylamino)ethyl methacrylate)-poly(glutamic acid) (Agut et al. 2010). At room temperature and close to the isoelectric point, direct or inverse electrostatic polymersomes are formed. At basic conditions, poly(2-(dimethylamino)ethyl methacrylate) is uncharged but exhibits LCST around 40°C above which formation of polymersomes has been observed. No stimuli responsiveness has been reported for diblock copolymers of poly(L-lactic acid)-polysarcosine (Makino et al. 2007). Various self-assembled structures, including vesicles, have been obtained by varying the hydrophilic/hydrophobic balance. Deviating from the linear architecture are the copolymers suggested by Lee et al. (2006). These authors have prepared comblike copolymers by grafting lactic acid on hydrophilic poly(2-hydroxyethyl aspartamide) leading to a hydrophilic polypeptide backbone to which hydrophobic side chains of synthetic polymer are grafted. Just the opposite is the structure of the amphiphilic polyelectrolyte formed by grafting PB with cystein (Hordyjewicz-Baran et al. 2007). The resulting random copolymer assembled into pH-responsive vesicles in which the hydrophobic backbone is parallel to the membrane.

Polymersomes formed by hybrid copolymers of "reverse" structure, that is, a polypeptide as a hydrophobic block and a synthetic polymer as a hydrophilic one, have been occasionally reported. Kimura et al. (1999) have conjugated the naturally occurring, hydrophobic 15-mer peptide antibiotic, gramicidin A, to PEG, thus obtaining an amphiphilic hybrid polypeptide. In aqueous solution, unilamellar vesicles are formed with peptide fragments in an antiparallel double-helix conformation building the hydrophobic core of the membrane.

Copolymers in which hydrophobic oligopeptides, in particular, oligovaline and oligophenylalanine, as side chains of polymethacrylate backbone coupled to a hydrophilic PEO chain have been shown to self-assemble to form vesicle using a solvent exchange (Adams et al. 2008). Double hydrophilic and, consequently, soluble at room temperature is the copolymer prepared by Luo et al. (2010). Above the LCST of the block bearing α-aspartic acid derivatives as side chains, formation of vesicles has been observed. Note that the chain architecture of the former and the latter copolymers is not linear.

Peptosomes can also be formed from copolymers containing two distinct peptide blocks. The vesicle size and structure can be controlled not only by adjustment of the macromolecular characteristics such as hydrophobic content or overall copolymer chain length but primarily by the ordered conformations of the polymer segments as shown by Bellomo et al. for poly(*N*-2-(2-(2-methoxyethoxy) ethoxy)acetyl-L-lysine)-poly(L-leucine) (Bellomo et al. 2004). The pH responsiveness, achieved by replacing 70% of the L-leucine in the hydrophobic domains with L-lysine, of the hydrophilic block was demonstrated via destabilization of the vesicles resulting from *a*-helix-to-coil transition in L-leucine-rich domains upon reduction of pH. Double responsive peptosomes have been formed from a series of ABA-type copolypeptides poly(L-lysine)-poly(γ-benzyl-L-glutamate)-poly(L-lysine) (Iatrou et al. 2007). Upon a pH and temperature increase from, respectively, 7.4 to 11.7 and 25°C to 37°C, the flanking poly(L-lysine) blocks undergo transitions *charged coil-to-α-helix-to*-β-sheet, whereas the middle block of poly(γ-benzyl-L-glutamate) retains its α-helix conformation (Figure 2.80). While the transitions of the secondary structure of the block forming the hydrophilic layers of the membrane induce some size variations, the release process seems to have been barely connected with these size variations. As noted elsewhere, the effects of secondary structure on polymersome assemblies should be pronounced when the stimuli-responsive polypeptide constitutes the hydrophobic layer of the membrane (Li and Keller 2009). Helix-to-coil transitions or other kinds of destabilization of the hydrophobic layer resulting from the response of the constituent moieties to an external stimulus may lead to formation of pores in the membrane, thus triggering of a release process, or even complete dissociation of the structures.

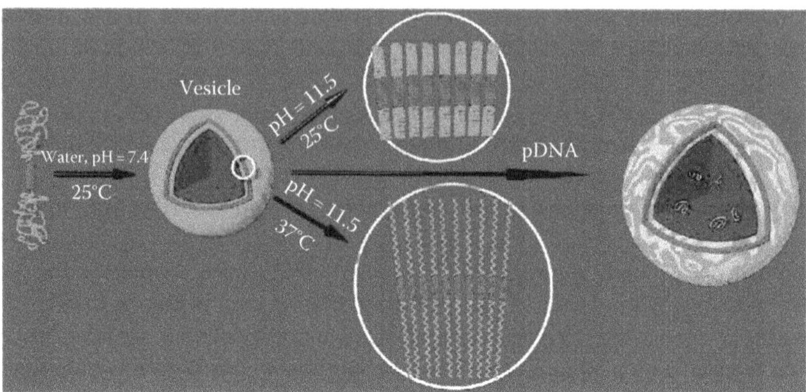

FIGURE 2.80 Formation of peptosomes from double responsive ABA-type copolypeptides, poly(L-lysine) (light gray), and poly(γ-benzyl-L-glutamate) (dark gray). The transitions *charged coil-to-α-helix-to-*β-sheet of the poly(L-lysine) block upon increasing pH and temperature are schematically presented. Simultaneous condensation in the poly(L-lysine) layer and encapsulation inside the vesicles of plasmid DNA are shown as well. (Reprinted with permission from Iatrou, H., Frielinghaus, H., Hanski, S., Ferderigos, N., Ruokolaninen, J., Ikkala, O., Richter, D., Mays, J., and Hadjichristidis, N., *Biomacromolecules*, 8, 2173–2181, 2007. Copyright 2007, American Chemical Society.)

Schizophrenic vesicles, that is, vesicles in which the constituent blocks build either the hydrophilic or the hydrophobic layer of the membrane depending on pH, have been described by Rodriguez-Hernandez et al. who used a zwitterionic block copolypeptide, poly(L-glutamic acid)-poly(L-lysine) (Rodriguez-Hernandez and Lecommandoux 2005). At pH in the range 5–9, the copolypeptide is soluble. At acidic conditions ($pH<4$), poly(L-glutamic acid) is in a helical conformation and forms the hydrophobic layer of the membrane, just opposite to the situation at high pH ($pH>10$) at which poly(L-lysine) adopts a rodlike α-helical conformation and swops the place of poly(L-glutamic acid). The transitions and structure of the vesicles at low and high pH are depicted in Figure 2.81. The vesicular assemblies composed of relatively short-chain copolypeptides, poly(L-lysine)-poly(L-leucine) or poly(L-glutamic acid)-poly(L-leucine), exhibit membrane fluidity most probably resulting from disrupting the sheet formation, which poly(L-leucine) tends to form, by charge–charge repulsion in between the hydrophilic block (Holowka et al. 2005). Besides stimuli responsiveness, fine adjustment of vesicle dimensions by applying extrusion techniques and solute encapsulation is also possible. A series of rod–rod diblock copolypeptides based on poly(γ-benzyl-L-glutamate) of varying degrees of polymerization, which is the structure-forming hydrophobic block, have been developed (Marsden et al. 2010). The hydrophilic blocks are sequence specific and are based on glycine, lysine, isoleucine, alanine, leucine, and glutamic acid. The application of a water addition/solvent evaporation method to produce polymersomes with controllable sizes has been demonstrated.

Other naturally occurring polymers that have been used as a constituent block of polymersome-forming copolymers are the polysaccharides. The latter introduce

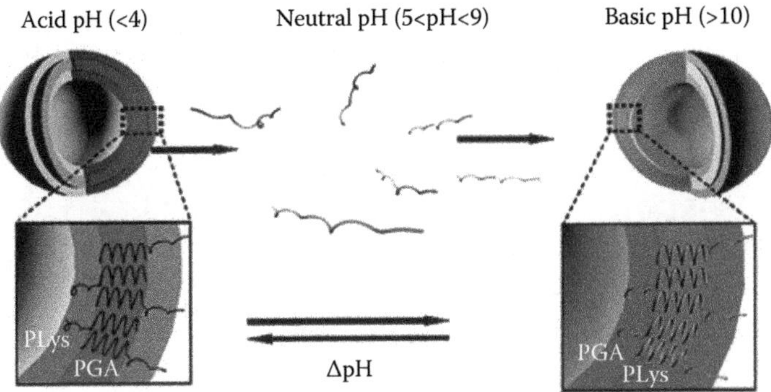

FIGURE 2.81 Schematic representation of the structure of the vesicles formed by a zwitterionic block copolypeptide, poly(L-glutamic acid)-poly(L-lysine), at different pH. PGA and PLys denote poly(L-glutamic acid) and poly(L-lysine), respectively. (Reprinted with permission from Rodriguez-Hernandez, J. and Lecommandoux, S., *J. Am. Chem. Soc.*, 127, 2026–2027, 2005. Copyright 2005, American Chemical Society.)

biologic functionality, bioactivity, biodegradability, and biocompatibility into or onto the polymersomes. It should also be noted that the carbohydrates not only participate as building elements of the cell walls and serve as an energy source for metabolism but also play vital roles in the cell–cell recognition or signal transmission—properties and functions that might be extremely useful for biomedical applications. Compared to the abundance of polypeptides and synthetic polymers, the number of carbohydrate-based polymers used for vesicle formation is quite limited. The few examples comprising polymers with pendant sugar moieties (glycopolymers) and sugar-terminated linear or starlike polymers have been recently reviewed (van Dongen et al. 2009). These are, for example, β-cyclodextrin-terminated PS (Felici et al. 2008) and polyether imide (Guo et al. 2008). These polymeric amphiphiles are able to form polymersomes in which the β-cyclodextrin, being sufficiently polar, is presented on the polymersome surface. β-Cyclodextrin has a hydrophobic interior and is able to act as a host in host–guest inclusion reactions, which has been employed by these authors to create polymersomes with a viable surface for immobilization of adamantyl-conjugated PEG and enzymes. Rather stable polymersomes with a lactose surface have been obtained from starlike block copolymers of PCL and poly(lactobionamidoethyl methacrylate) (Zhou et al. 2008). The structural formula is shown in Figure 2.82. Compared to the linear analogues, these star-block copolymers exhibit much lower CAC, which is reflected in high stability.

Sugar-functionalized polymersomes have also been prepared from diblock copolymers in which one of the block is a glycopolymer such as poly(2-glucosyloxyethyl methacrylate) and poly(2-(β-D-glucopyranosyloxy) ethyl acrylate), whereas the other represents a synthetic polymer—poly(diethyleneglycol methacrylate) and PS, respectively (Li et al. 2001, Pasparakis and Alexander 2008).

A different approach to carbohydrate-containing polymer vesicles is to use copolymers in which all saccharidic residues are linked within distinct polysaccharide

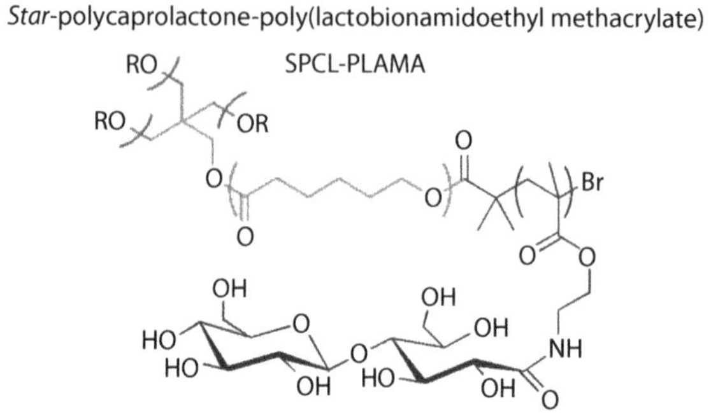

Star-polycaprolactone-poly(lactobionamidoethyl methacrylate)

SPCL-PLAMA

FIGURE 2.82 Structural formula of a starlike block copolymer of PCL and poly(lactobionamidoethyl methacrylate). (Zhou, W., Dai, X.H., and Dong, C.M.: *Macromol. Biosci.* 2008. 8. 268–278. Copyright Wiley-VCH Verlag GmbH & Co. KGaA. Reproduced with permission.)

chains. Such a chain can be connected to another chain of a (synthetic) polymer, thus forming a linear diblock copolymer. Alternatively, a synthetic monomer can be polymerized from reactive functions carried by a polysaccharide backbone, or, reverse to that, polysaccharide chains can be grafted onto a preformed synthetic polymer backbone. The advantage of these structures is that they fully, or at least to a large extent, preserve the properties of the constituent chains. In spite that the inherent immiscibility between the synthetic blocks and polysaccharides and rigidity of the latter are expected to favor the microphase separation and formation of vesicles, in particular, so far very few studies have reported on polymersome formation from polysaccharide-based diblock copolymers. One of them is previously mentioned in the context of deviating from the typical for vesicle formation hydrophilic content study of Zhang et al. (2010b). These authors have prepared a series of well-defined amphiphilic diblock copolymers of dextran and PCL. The morphology of the self-assembled structures formed in water changed from spherical micelles at the highest dextran content to polymersomes at the lowest. Also based on dextran are the copolymers of Houga et al. (2009). The hydrophobic block is PS, which in DMSO forms the hydrophobic layer of the membrane. The spontaneously formed polymersomes can be "frozen" by transferring to water via dialysis. Thermoresponsive polymersomes of about 300 nm in diameter have been prepared from a double hydrophilic diblock copolymer maltoheptaose-poly(*N*-isopropylacylamide) at 45°C (Otsuka et al. 2010). Figure 2.83 shows the chemical structure and proposed architecture of the giant polymersomes formed by graft copolymers (Gao et al. 2008). Chitooligosaccharide—a hydrophilic polymer based on the naturally occurring chitin—has been grafted with hydrophobic chains consisting of PCL. To the best of our knowledge, no other of this kind polysaccharide-based copolymers have been reported to form polymersomes.

Interesting combinations *polysaccharide–polypeptide* have been reported in several studies on self-assembly in solution of dextran-poly(γ-benzyl-L-glutamate) and hyaluronan-poly(γ-benzyl-L-glutamate) (Schatz et al. 2009, Upadhyay et al. 2009). Strong interactions between the rigid and hydrophobic α-helical polypeptide blocks

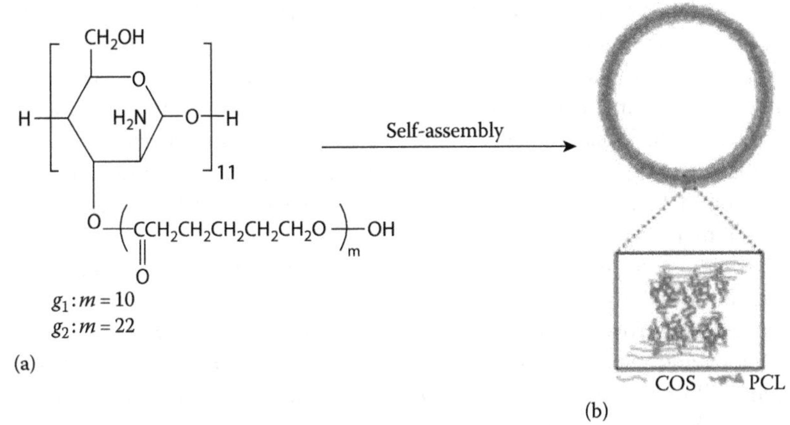

(a) $g_1: m = 10$, $g_2: m = 22$ — Self-assembly — (b) COS PCL

FIGURE 2.83 Chemical structure and proposed architecture of the vesicles formed by chitooligosaccharide-*graft*-PCL copolymers. COS and PCL denote chitooligosaccharide and poly(ε-caprolactone), respectively. (Gao, K.J., Li, G., Lu, X., Wu, Y.G., Xu, B.Q., and Fuhrhop, J.H., *Chem. Commun.*, 12, 1449–1451, 2008. Reproduced by permission of The Royal Society of Chemistry.)

favor formation of flat and interdigitated membranes with a thickness in the range 9–20 nm depending on the DP of the poly(γ-benzyl-L-glutamate) block, whereas solubility and fluidity have been imparted by the hydrophilic polysaccharides. Polymersomes of dimensions of around 100 and 400 nm in diameter for the dextran- and hyaluronan-containing copolymers, respectively, have been produced by the solvent displacement method. Biomedical applications of the hyaluronan-based polymersomes related to loading and delivery of anticancer drugs are presented in Chapter 5.

Finally, to finish with the bio-inspired amphiphilic polymers, we should refer to the study of Teixeira Jr. et al. who in 2007 for the first time prepared a hybrid nucleo-copolymer by covalently linking PB of molar mass 2000 to an oligonucleotide sequence (Teixeira et al. 2007). The latter has never been used as the hydrophilic segment to build amphiphilic block copolymers. The nucleotide sequence, cytidine$_{12}$, comprises 12 bases that are enough to provide stability to formation of double helix upon hybridization with complementary nucleotide strands. In water, the nucleo-copolymer has been shown to self-associate into vesicles of ca. 80 nm in diameter as well as larger (micron size) multivesicular structures.

2.2.3.3 Methods for Preparation

The methods for preparation of polymersomes are basically the same as those used for preparation of liposomes and polymer micelles. However, compared to liposomes, the formation of polymer vesicles is generally more difficult and slower due to the larger molar mass of the block copolymer chains. Polymer vesicles can be prepared via direct dissolution of the amphiphilic copolymers in aqueous solution, as referred to as *solvent-free method*, or via displacement of an organic solvent by water—the so-called *phase-inversion method.* None of them yields exclusively vesicles; on the contrary, a variety of supramolecular structures (spherical, rod- or disklike micelles, vesicles, lamellae, tubes) can be formed applying any of the earlier techniques. Whether polymersomes will be formed or not depends, as discussed earlier, on a number of factors such as content of hydrophilic component, nature of the constituent blocks, molar mass and dispersity, presence of additives, and temperature. Typically, the input of energy in the systems via vigorous stirring or vortexing, freeze–thaw cycling, sonication, extrusion, and applying of an electric current leads to formation of homogeneous in size vesicular dispersions and decrease of dimensions and lamellarity.

The *solvent-free preparation method* can be divided in two techniques: bulk swelling and thin-film rehydration. Both are applicable for spontaneous formation of polymersomes from amphiphilic copolymers of typically low T_g of the hydrophobic block and low total molar mass. For polymeric amphiphiles of higher molar mass, an additional energy input is needed to facilitate and speed up the vesicle formation. The bulk swelling involves direct hydration of the polymer, which can be with the consistency of powder, solid, or viscous liquid. The thin film is formed by casting from a dilute polymer solution in an appropriate solvent. The latter is removed by rotary evaporation or under a gentle stream of nitrogen leaving a thin polymer film on the surface (typically glass), which is then hydrated by the addition of an aqueous buffer or pure water. The swelling of the film or the polymer bulk phase is caused by water permeation through pores or defects being driven by hydration forces. The driving force for polymersome formation upon hydration is the concentration gradient between the copolymer front, diffusing in water, and the water front, diffusing in the copolymer, which decreases linearly with time, and, as a result, the lamellar structures may not have sufficient time to unbind (Battaglia and Ryan 2006a). Upon applying of an additional energy input, the concentration gradient is kept constant, and the mutual diffusion is enhanced, which results in complete unbinding of the lamellar structures. Furthermore, the stiffness of the membrane, which, in turn, is governed by the copolymer nature and molecular weight, dictates the membrane undulations and unbinding. Thus, more flexible membranes are expected to produce earlier unbinding and formation of multilamellar vesicles, while unbinding occurs at lower concentrations as the membrane flexibility is reduced forming unilamellar vesicles. In the same paper, the authors have shown that the polymersome-forming block copolymers swell in water following complex kinetics according to two qualitatively different regimes: subdiffusional regime of a mutual diffusion as a result of molecular-level arrangement of the amphiphilic membranes and Fickian diffusion of the membranes after having reached their equilibrium morphology (Battaglia and Ryan 2006a,c). The same authors in another paper (Battaglia and Ryan 2006b) studied in detail the phase diagrams and morphology formed as a function of molecular weight and concentration of two polymersome-forming PEO-PBO block copolymers. The evolution of the phase sequences for the two copolymers is presented in Figure 2.84 from which is seen that the smaller copolymer forms dispersed vesicles at higher concentrations, whereas the copolymer of larger molecular weight initially forms vesicular clusters that eventually break up into dispersed vesicles. Furthermore, the larger copolymer was found to form exclusively unilamellar vesicles in contrast to the smaller copolymer, which gave rise to multilamellar vesicles.

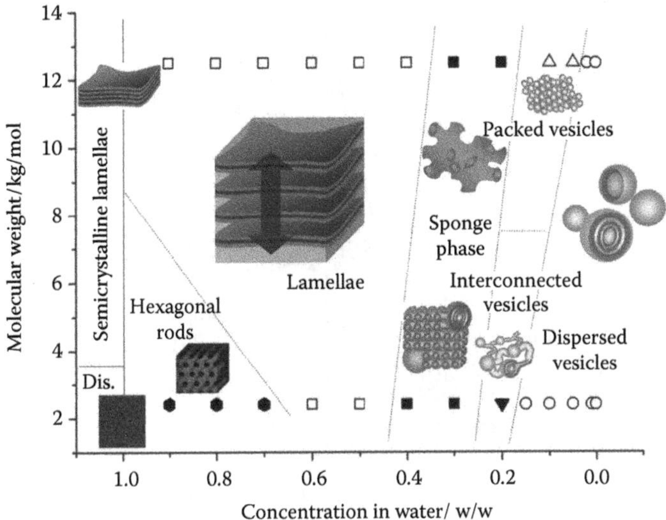

FIGURE 2.84 Phase diagram of polymersome-forming PEO-PBO block copolymers as a function of their concentration in water and molecular weight. (Reprinted with permission from Battaglia, G. and Ryan, A., *Macromolecules*., 39, 798–805, 2006. Copyright 2006, American Chemical Society.)

The *phase-inversion method* is applicable for glassy and crystalline copolymers and involves dissolution of the copolymer in a suitable solvent that is good for all constituent blocks followed by mixing with an aqueous solution/phase. The displacement of the organic solvent by water triggers the copolymer self-assembly. Noteworthy, this technique is also applicable for preparation of micelles; moreover, morphological transitions *spherical micelles–cylindrical micelles–vesicles* have been induced by increasing water content (Shen and Eisenberg 1999). The presence of an organic solvent decreases T_g and fluidizes the membranes, which results in a decreased stability. Typical for the method is the broad size distribution of the polymersomes, which may be overcome by repeated extrusion. The most essential drawback, however, is that the residual solvent can induce biologic toxicity, thereby restricting the biomedical applications of the polymer vesicles prepared via the phase-inversion technique. Ultrafiltration or extensive dialysis against water should be undertaken to exclude residual solvent.

Other methods and post-formation treatments have been employed to change some characteristics of the polymersomes such as size, size distribution, and lamellarity aiming at improving both their performance and reproducibility of the techniques applied. The post-formation treatments are typically based on the use of mechanical forces, which disrupt the large multilamellar vesicles into small membrane patches that eventually close to form smaller unilamellar vesicles than the initial ones. These comprise freeze–thawing (dispersions are subjected to repeating cycles of submerging in liquid nitrogen followed by fast heating to >50°C), sonication (treatment with ultrasound in an ultrasonic bath for desired time period), extrusion (dispersions are passed through membranes with controlled pore size),

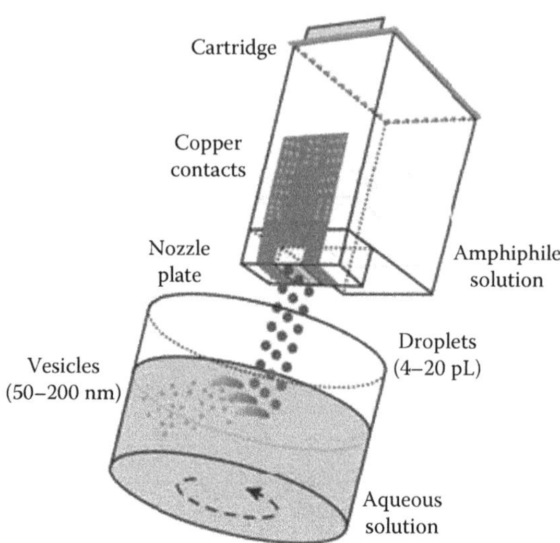

FIGURE 2.85 Schematic presentation of the preparation of polymersomes using inkjet printing technology. (Hauschild, S., Lipprandt, U., Rumplecker, A., Borchert, U., Rank, A., Schubert, R., and Forster, S.: *Small.* 2005. 1. 1177–1180. Copyright Wiley-VCH Verlag GmbH & Co. KGaA. Reproduced with permission.)

and electroformation (mechanical stress is introduced by inducing periodic motions and by increasing interlayer repulsion through electroviscous/electrostatic effects). Single-step formation of monodispersed polymersomes so far has been achieved by phase inversion using inkjet printing (Hauschild et al. 2005) and microfluidic devices (Shum et al. 2008). The former authors have utilized the inkjet printing technology to prepare and load small unilamellar vesicles based on poly(2-vinylpyridine)-PEO diblock copolymers. The method is schematically presented in Figure 2.85: The cartridge is filled with a copolymer solution in ethanol, which is "printed" into a stirred aqueous solution where the self-assembly into polymersomes takes place. The modern inkjet printers produce with high reproducibility droplets with volumes in the picoliter range, which allows vesicle formation of high precision.

The microfluidic device for simultaneous preparation and loading of giant (tens of μm) polymersomes from PEO-poly(lactic acid) diblock copolymers using water-in-oil-in-water (W/O/W) double emulsion templates is shown schematically in Figure 2.86a (Shum et al 2008). The geometry requires the outer phase (consisting of poly(vinyl alcohol) in water) to be immiscible with the middle phase (diblock copolymer in toluene/chloroform mixture), which is in turn immiscible with the inner phase (aqueous solution containing encapsulates). The inner drops form at the tip of the small injection tube in a co-flow geometry, while the middle oil stream, containing the inner drops, breaks up into drops in the collection tube. The double emulsion drop is composed of an aqueous core surrounded by a layer of diblock copolymer in organic solvent. Schematic of the acorn-like structure of a double emulsion drop is presented in Figure 2.86b. Polymersomes are formed by evaporation of the organic solvent.

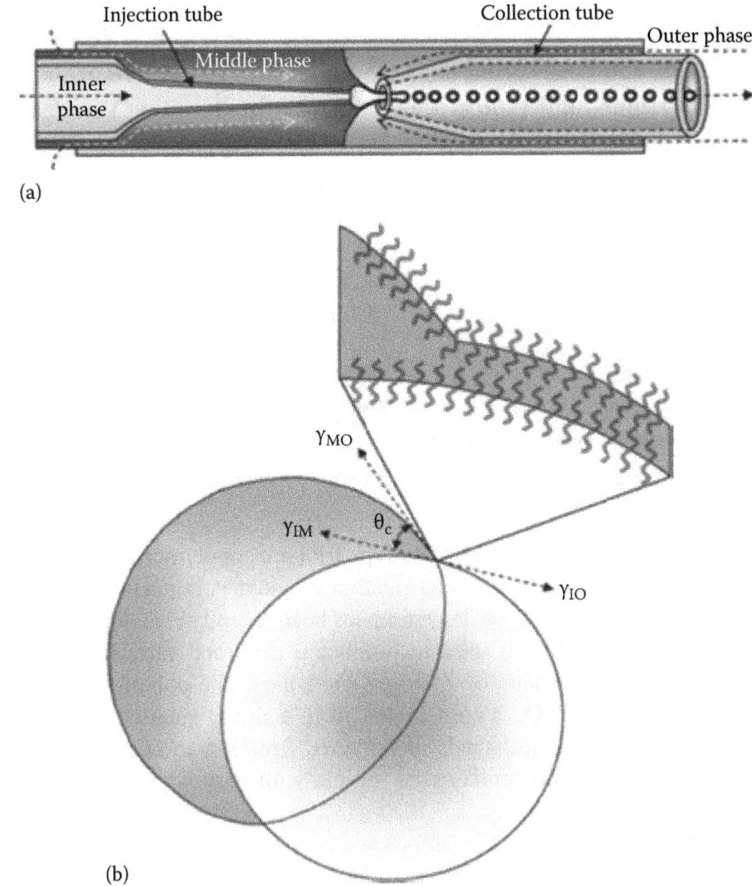

FIGURE 2.86 Schematic presentations of (a) the microfluidic device generating double emulsions and (b) proposed structure of a double emulsion drop. The organic phase in (b) is depicted with blue. γ_{IO}, γ_{IM}, and γ_{MO} in (b) are the surface tensions of the inner–outer, the inner–middle, and the middle–outer interfaces, respectively. (Reproduced with permission from Shum, H.Ch., Kim, J.-W., and Weitz, D.A., *J. Am. Chem. Soc.*, 130, 9543–9549, 2008. Copyright 2008, American Chemical Society.)

The W/O/W double emulsion technique is frequently used for preparation of vesicles from low molecular weight surfactants and lipids. It is also applicable for preparation of polymer vesicles (Pautot et al. 2003) as well as for polymersomes with asymmetric membranes where the internal and external leaflets have different compositions as suggested elsewhere (Massignani et al. 2010). Another method to produce polymersomes involves participation of a surfactant, which facilitates dispersing the polymer in water and either may be subsequently removed or together with the polymer may form vesicular complexes (Kelarakis et al. 2008). The sudden change of a certain external parameter may also trigger vesicle formation as recently shown elsewhere (Blanazs et al. 2009a). These authors reported on the generation of

TABLE 2.2
Methods and Post-Formation Techniques for Preparation of Polymersomes and Typical Size Ranges of the Resulting Vesicles

Method/Technique	Size Range
Rehydration	<100 nm to >1 μm
Phase inversion	Hundreds of nm–tens of μm
Sonication	<100 nm
Extrusion	100, 200, 400—depending on the pore size
W/O/W	<100 nm to >1 μm
Inkjet	50–100 nm
Electroformation	>1 μm
Microfluidic device	Tens of μm
Photolithography/dewetting/rehydration	>1 μm

asymmetric polymersomes from an ABC-type triblock copolymer by a sharp switch of pH. An interesting approach for fabrication of giant polymersomes combining photolithography with bulk phase dewetting has been recently reported (Howse et al. 2009). UV photolithography is used to produce a patterned surface consisting of hydrophilic squares surrounded by hydrophobic bands. The polymersome-forming block copolymer, PEO-PBO, is spin casted from a dilute solution and following dewetting forms well-defined islands of polymer, from which upon hydration and detachment from the surface unilamellar vesicle with narrow size distribution are spontaneously formed.

In brief, there exist a variety of methods and post-formation techniques for preparation of polymersomes. Using these methods and techniques, polymersomes with sizes ranging from nanometers to tens of micrometers with relatively high control of size distribution, lamellarity, and morphology can be formed. Each of them is associated with specific/typical dimensions of the polymer vesicles as shown in Table 2.2. It should be always kept in mind that, as discussed earlier, different factors such as absolute and relative block lengths, nature of the constituent blocks, polydispersity, especially that of the hydrophilic block, presence of additives, pH, and temperature may also influence the dimensions, lamellarity, and morphology of the polymersomes.

2.2.3.4 Polymersome Physical and Mechanical Properties

The polymersomes are typically compared to the conventional phospholipid vesicles (liposomes). Undoubtedly, they have advantages over the latter in terms of enhanced chemical, thermal, and mechanical stability; low permeability; high robustness; etc., and, in that aspect, they are closer to the viral capsids. Moreover, the molar masses of the proteins building the capsids and block copolymers constituting the polymersomes are within the same range. In general aspects, the properties of the polymersomes directly reflect the chemical composition, the block length ratio, and the molar mass of the amphiphilic copolymers from which they are prepared. Therefore, the careful selection of the copolymer type and molar mass is important as it imparts polymersomes with a broad range of tunable properties. As shown earlier, the membrane thickness can be controlled by the molar mass of the hydrophobic block of the copolymer (see Figure 2.65). Furthermore, properties such as membrane bending rigidity and electromechanical stability have been found to scale with the squared thickness, that is, $\sim d^2$ (Goetz et al. 1999, Discher et al. 2007), but chain interactions within the membranes could be reflected in these dependences as well. Related to the molar mass are also other properties such as lateral mobility of polymer chains within the membrane, permeability, stability, membrane viscosity and fluidity, and toughness. Due to the higher molar mass of the polymers as compared to the lipids, the polymersome membranes are generally thicker, more stable, less permeable and less mobile, tougher, and stronger. Le Meins et al. have demonstrated this by collecting a number of mechanical and physical properties of liposomes and polymersomes, which are presented in Table 2.3 (Le Meins et al. 2011). The scaling exponents of the properties with the membrane thickness and molar mass are shown as well. The effects of molar mass are nicely demonstrated in Figure 2.87 (Ahmed et al 2006), in which variations of some typical physical properties are presented as a function of molar mass of the constituting amphiphiles. The changes of some of the mechanical and physical properties are particularly pronounced when the chains become entangled, that is, at $M_w > M_e$, where M_e is the entanglement molar mass. Evidence for, for example, copolymer diffusivities, D, yielding numerical comparisons can be found elsewhere (Lodge and Dalvi 1995, Lee et al. 2002): According to these studies, D

TABLE 2.3
Physical and Mechanical Properties of Liposomes and Polymersomes and Scaling with Membrane Thickness, *d*, and Copolymer Molar Mass, M_w

Property	Liposomes	Polymersomes	Scaling
Bending modulus (kT)	11–30 (Rawicz et al. 2000)	40–460 (Bermudez et al. 2004; Dimova et al. 2002)	$\sim d^2$ (Bermudez et al. 2004)
Stretching modulus (mN/m)	250 ± 2 (Rawicz et al. 2000)	80–100 (Bermudez et al. 2002)	$\sim d^0$ (Bermudez et al. 2002)
Lysis strain (%)	5 (Bermudez et al. 2002)	20–50 (Bermudez et al. 2002)	$\sim d^{0.6}$ for $M_w < M_e$ (Bermudez et al. 2002)
Membrane surface shear viscosity (mN/ms)	10^{-5} (Evans et al. 2003)	15×10^{-3} (Dimova et al. 2002)	—
Water permeability (μm/s)	15–150 (Lorenceau et al. 2005)	0.7–10 (Lorenceau et al. 2005)	$\sim d^{-1}$ (Battaglia et al. 2006c)
Lateral diffusion coefficient (μm²/s)	3.8 (Lee et al. 2002)	0.12–0.0024 (Lee et al. 2002)	$\sim M_w^{-1}$ (Rouse scaling for $M_w < M_e$) $\sim M_w^{2-3}$ (bulk reptation for $M_w > M_e$)

Source: Adapted from Le Meins, J.-F. et al., *Eur. Phys. J. E*, 34, 14, 2011. With permission.

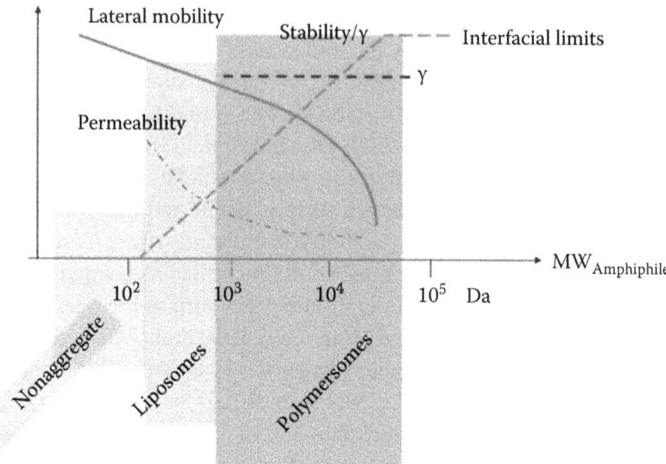

FIGURE 2.87 Variations of some typical physical properties with the molar mass of the constituting amphiphiles. Here, elasticity, γ, appears independent of molar mass. (Ahmed, F., Photos, P.J., and Discher, D.E.: *Drug Dev. Res.* 2006. 67. 4–14. Copyright Wiley-VCH Verlag GmbH & Co. KGaA. Reproduced with permission.)

scales with molar mass as $D \sim M_w^{-a}$, where $a = 1$, thus being consistent with the Rouse model or 2–3 (bulk reptation model) for molar masses, respectively, lower and higher than M_e. The entanglements, as shown by Lee et al., are clearly responsible for the reduced mobility of the chains, which, in turn, contributes to the enhancement of the resistance to detergents (Lee et al. 2002). The polymersome membranes exhibit far higher viscosity (Dimova et al. 2002, Dalhaimer et al. 2003) and lower lateral diffusivity (Lee et al. 2002) than those of liposomes. Accordingly, a higher-energy barrier for pore formation and a longer lifetime of the pores once formed have been documented (Nardin et al. 2000, Bermudez et al. 2003). Furthermore, the water permeability has been found considerably lower, reaching even an order of magnitude, than that typically observed for liposomes (Table 2.3). Although the data available in the literature for water permeability are somewhat limited to just a few copolymers, the absolute values have been found to vary with the nature of the copolymers used and their molar masses. Thus, for poly(2-methyloxazoline)-poly(dimethylsiloxane)-poly(2-methyloxazoline) of molar mass 10,700 Da, a value of 0.8 μm/s has been determined (Kumar et al. 2007). Higher values, that is, 2.5 and 7 μm/s, respectively, for PEO-PEE of molar mass 3900 Da (Discher et al. 1999) and poly(butyl acrylate)-PAA of molar mass 5500 Da (Lorenceau et al. 2005) have been measured, which are still considerably lower than those of the liposomes.

Drastic alterations of the properties can also occur upon substitution of a copolymer block (moiety) of low T_g with a glassy or crystalline one or (which is equivalent) upon temperature-induced phase transitions. The reduced mobility of the entangled and glassy polymers is expected also to slow down lateral or in-plane arrangements that are related to morphological changes of the polymersomes. Switch in polarity either via changing of an external factor, for example, temperature or pH, or via substitution of a more polar block with a less polar one can have similar effects: variations of the water permeability through the membranes of different types of polymer vesicles have been attributed to different thicknesses and polarities of the hydrophobic layer of the membranes (Leson et al. 2007a). The thickness and polarity profiles of the membranes can be affected by small organic molecules. The addition of ethanol to polyisoprene-PEO vesicles (Bauer et al. 2006) and glycerol to PEO-polydimethylsiloxane-PEO vesicles (Yan et al. 2007) has been found to substantially alter the permeability. Whereas the effect of glycerol is associated with a simultaneous increase of polarity and flexibility, the role of dioxane is to decrease the glass transition temperature of PS, that is, to plasticize the walls of polymersomes based on a PS-PAA block copolymer (Choucair et al. 2005). By varying the amounts of dioxane added, the extent of loading and release of doxorubicin can be controlled independently. Cosolvents can also be used to swell and fluidize the membrane core helping to minimize the cost of "flip-flop," that is, the transfer of polymer chains from one monolayer to another (Discher et al. 2002).

Modern computational methodologies have also provided insights into the effects of block lengths on different physical properties. In particular, the molar mass dependences of properties such as the area elastic modulus, membrane thickness, and lateral chain mobility have been studied by using the coarse-grained model (Srinivas et al. 2004). The simulation data agree well with the experimental results reported (Bermudez et al. 2002). They have been used to calculate the absolute value of the elastic modulus, which, in good agreement with the experiment, has been found independent from molar mass (Figure 2.87).

The distinctive properties highlighted earlier are strongly influenced by both the copolymer nature and molar mass. In addition, by varying external parameters and the type and amount of additives, a broad range of polymersome properties can be controlled. Biologically relevant properties such as adhesion, fusion and fission, responsiveness, stealth abilities, surface chemistry, targetability, loading, and release are discussed in Chapter 5.

2.2.4 Other Morphologies

Spherical and cylindrical (wormlike) micelles as well as polymersomes are the most common structures formed by amphiphilic copolymers. However, the research in the past decade has focused not only on these basic morphologies and manipulating their properties but also on creating other morphologies. Recent approaches to controlling the self-assembly of amphiphilic block copolymers with a view to obtaining novel aggregate morphologies have been overviewed by Holder and Sommerdijik (2011).

2.2.4.1 Multicompartment Micelles

The cores of the multicompartment micelles are internally segregated into different (types of) domains. Such morphology could be advantageous for simultaneous delivery of pharmaceutical and bioactive agents of different types and solubilities that are otherwise incompatible. Presumably, the simplest copolymer architecture to form such multicompartment self-assembled structures is the ABC-type architecture. However, aggregates with segregated cores can be formed also from AB,

BAB, and ABA copolymers. Rangelov et al. have described spherical core–corona particles formed in aqueous solution from AB and BAB block copolymers of ethylene oxide and a lipid-mimetic monomer 1,3-didodecyloxy-2-glycidyl-glycerol (DDGG) (Rangelov et al. 2002a). The cores of the particles appeared too large considering the molar masses and volumes of the component groups of the hydrophobic poly(DDGG) moieties, which implied that there was space in the cores that was not occupied by the hydrophobic anchors (Figure 2.88). A picture of coronae built of PEO in a brush or looping conformation and cores of poly(DDGG) in which PEO/water domains are randomly distributed was derived from the data. The existence of aqueous compartments within the hydrophobic cores was unambiguously proved by the successful entrapment of a hydrophilic dye, 5(6)-carboxyfluorescein: The dye leaked out slowly with time from the aqueous compartment, and upon a surfactant addition, the aggregates were destroyed and all 5(6)-carboxyfluorescein released (Figure 2.88).

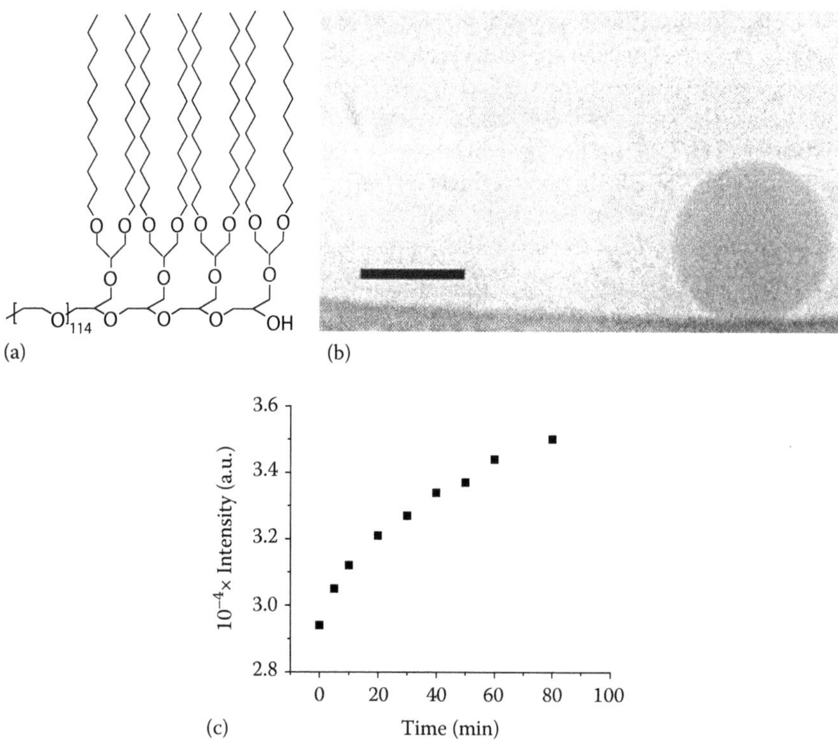

FIGURE 2.88 (a) Chemical structure of a copolymer bearing a short block of lipid-mimetic units of composition $(EO)_{114}(DDGG)_4$. (b) Cryo-TEM micrograph of a nanoparticle taken from dispersions of $(DDGG)_{3.6}(EO)_{454}(DDGG)_{3.6}$. (Reproduced with permission from Rangelov, S., Almgren, M., Tsvetanov, Ch., and Edwards, K., *Macromolecules*, 35, 4770–4778, 2002a. Copyright 2002, American Chemical Society.) (c) Leakage of 5(6)-carboxyfluorescein from the PEO/water domains randomly distributed in the hydrophobic cores of nanoparticles formed by $(EO)_{114}(DDGG)_4$ in water. Bar in (b) corresponds to 100 nm.

"Negative" to this arrangement is the structure of the large compound particles formed in water by PG analogues of the PEO-PPO-PEO copolymers (Rangelov et al. 2007, Halacheva et al. 2007, 2010). The results indicate that the PG-PPO-PG (LGP) copolymers of molar masses of the PPO block of 2000 and lower form particles that are considerably larger in size and aggregation number than those of the corresponding PEO-PPO-PEO copolymers. Hydrophobic interactions via PPO and interactions via hydrogen bonding between the numerous hydroxyl groups of the PG chains are equally involved in the formation of the large compound particles. The latter are well-separated and spherical objects with interior consisting of discontinuous hydrophobic PPO domains of a slightly prolate spherical shape randomly distributed in a strongly hydrogen-bonded continuous medium consisting of PG and water (Figure 2.89). Individual copolymer chains were found to coexist with the PPO domains in the interior. The former are not involved in the formation of the PPO domains but mediate interdomain interactions via hydrogen bonding. The number of PPO domains per large compound particle was found to vary from 3 to 69 depending on the copolymer composition and temperature (Halacheva et al. 2007).

Branched or network-like structures due to interchain association of the terminal blocks can be formed by copolymers with a middle solvent-affinitive block. At dilute conditions, discrete and well-separated structures of nanometer dimensions could be formed and, in that aspect, can be considered as multicompartment micelles with numerous hydrophobic domains interconnected by hydrophilic chains. Such a model for chain arrangement has been proposed by Dimitrov et al. for triblock copolymers poly(ethoxyethyl glycidyl ether)-PPO-poly(ethoxyethyl glycidyl ether) (PEEGE-PPO-PEEGE (Dimitrov et al. 2004). Interestingly, both blocks exhibit LCST properties, but the more hydrophobic one, PEEGE, was found to dictate the cloud points, CMC values, and LCST properties of the copolymers. The authors propose a model for formation of physically cross-linked structures (Figure 2.90a) at low, but invariably above the LCST of PEEGE, temperatures. Upon increasing temperature, due to gradual dehydration of the chains, they are converted into core-like structures,

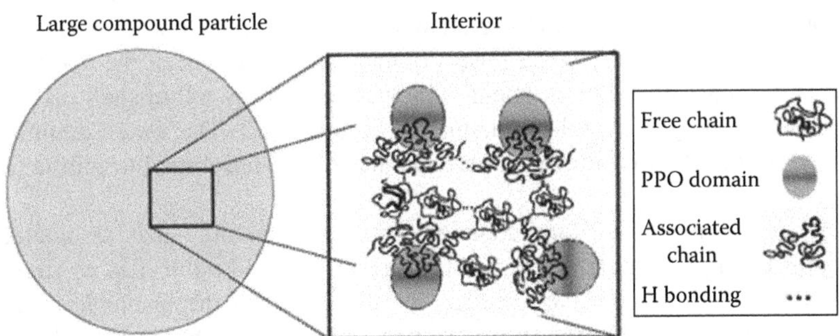

FIGURE 2.89 Schematic representation of the interior of a large compound particle formed by the LGP copolymers in aqueous solution. (Reprinted with permission from Halacheva, S., Rangelov, S., and Garamus, V., *Macromolecules*, 40, 8015–8021, 2007. Copyright 2007, American Chemical Society.)

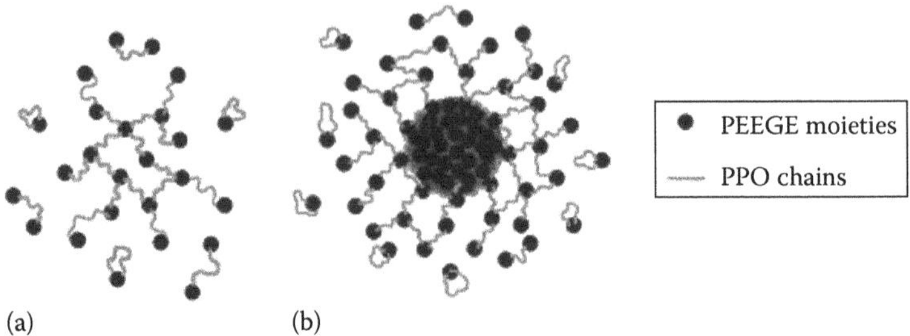

FIGURE 2.90 Schematic representation of possible states of chain arrangement of PEEGE-PPO-PEEGE copolymers in water at low (a) and elevated (b) temperatures. (Reprinted with permission from Dimitrov, Ph., Rangelov, S., Dworak, A., and Tsvetanov, Ch., *Macromolecules*, 37, 1000–1008, 2004. Copyright 2004, American Chemical Society.)

whereas unimers of lower EEGE content that coexist in the solution are accommodated in the outer parts of the particles, thus forming multicompartment coronae of branched or network-like structure (Figure 2.90b).

Another type of structure for multicompartment micelles is that of a core–shell–corona morphology sometimes referred to as layered or onion-like micelles. Most commonly, it is formed by linear ABC block copolymers for which A is hydrophilic, whereas B and C are hydrophobic blocks with poor thermodynamic compatibility (Stahler et al. 1999, Kubowicz et al. 2005, Lutz and Laschewsky 2005). The high Flory–Huggins interaction parameter, χ, is a prerequisite to drive the phase separation between blocks B and C within the hydrophobic core. Linear tetrablock copolymers of the type ABCA have been reported to form lamellar structures—vesicles (Brannan et al. 2004) and plates or disks (Gomez et al. 2005) with laterally segregated membranes as shown in the previous section (Figure 2.74). Due to topological constraint, the layered structure is the preferred morphology for aggregates that a linear symmetric pentablock copolymer ABCBA forms (Thunemann et al. 2006). Two-compartment particles have been prepared from a symmetric pentablock copolymer for which A, B, and C correspond to highly immiscible each other PEO, poly(γ-benzyl-L-glutamate), and poly(perfluoro ether), respectively. The concentric or coaxial inner domains are composed of the perfluorated block as the innermost compartment, which is covered by a thin shell (ca. 2 nm) of β-sheets of poly(γ-benzyl-L-glutamate) (Figure 2.91). The water solubility is provided by the PEO corona. The particles have been developed to mimic transport proteins.

ABC terpolymers of more complex chain architecture such as miktoarm, miktoarm block, and graft are also able to form onion-like micelles as shown in Figure 2.92. In the onion-like micelles, the domains that the insoluble blocks B and C form are concentric. Other arrangements of the hydrophobic cores, however, are also possible. Zhu and Jiang (2005), for example, report on the self-assembly of linear PS-poly(2-vinylpyridine)-PEO triblock copolymer into giant segmented wormlike micelles composed of disks with constant thickness of ca. 65 nm and different diameters connected through threads (Figure 2.93). These giant particles

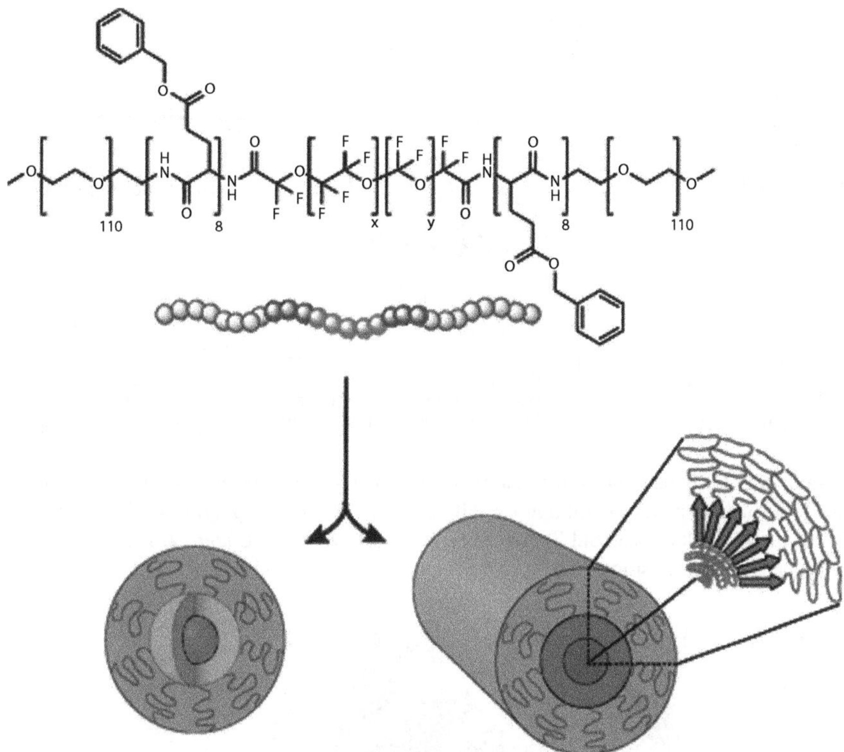

FIGURE 2.91 Spherical and cylindrical morphologies of two-compartment particles with concentric (for the spherical particles) and coaxial (for the cylindrical ones) inner compartments obtained from a linear symmetric ABCBA pentablock copolymer. A, B, and C correspond to PEO, PBLG, and poly(perfluoro ether), respectively. (Reprinted with permission from Thunemann, A.F., Kubowicz, S., von Berlepsch, H., and Mohwald, H., *Langmuir*, 22, 2506–2510, 2006. Copyright 2006, American Chemical Society.)

are of a typical shuttle-like structure, that is, larger diameter in the middle that gradually decreases toward the two ends. The individual disks are layered as well as the interconnecting threads with PS chains in the interior surrounded by a shell of poly(2-vinylpyridine), whereas the hydrophilic PEO chains form the corona (Figure 2.93). The authors performed a kinetic study and suggested a possible mechanism for formation involving initial formation of spherical onion-like micelles that join together to form a shuttle-like contour followed by rearrangement of the spheres to disks (Zhu and Jiang 2005).

The use of hydrocarbon and fluorocarbon blocks as hydrophobic immiscible components is a widespread approach to create multicompartment micelles. Linear triblock copolymers poly(4-vinylbenzyl-*N*-methylmorpholinium chloride)-*b*-PS-*b*-poly(4-vinylbenzyl pentafluorophenol ether) have been reported to form in aqueous solution multicompartment micelles consisting of a raspberrylike hydrophobic core over which nanometer-scale fluorocarbon domains are dispersed that is stabilized by a hydrophilic corona built of poly(4-vinylbenzyl-*N*-methylmorpholinium

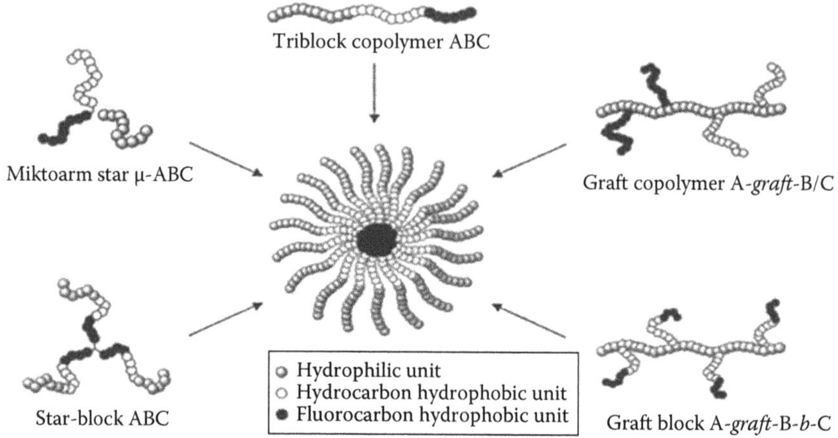

FIGURE 2.92 Schematic representation of a core–shell–corona onion-like spherical micelle formed in aqueous solution by self-assembly of ABC terpolymers of various chain architectures. (Holder, S.J. and Sommerdijk, N.A.J.M., *Polymer Chem.*, 2, 1018–1028, 2011. Reproduced by permission of The Royal Society of Chemistry.)

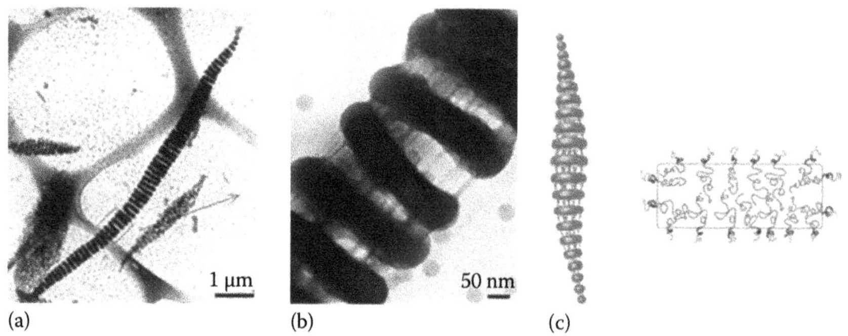

FIGURE 2.93 (a) Representative TEM image of a giant segmented wormlike micelle. (b) Enlarged boxed section in (a) viewing individual disks and interconnecting threads. (c) Schematic drawing a giant segmented wormlike micelle (left) and proposed chain organization within the disks (right). PS chains (in the interior) are surrounded by a shell of poly(2-vinylpyridine). The hydrophilic PEO chains form the corona. (Reprinted with permission from Zhu, J. and Jiang, W., *Macromolecules*, 38, 9315–9323, 2005. Copyright 2005, American Chemical Society.)

chloride) chains (Kubowicz et al. 2005, von Berlepsch et al. 2009). Figure 2.94 shows the chemical composition of the copolymer used, sketch, and cryo-electron tomography (cryo-ET) of raspberrylike multicompartment micelles.

Multicompartment particles of complex morphology can be prepared by copolymers of nonlinear chain architecture as well. Li and coworkers used miktoarm terpolymers comprising again a fluoropolymer, in particular poly(perfluoropropylene oxide), as a superhydrophobic moiety, which is immiscible with the other hydrophobic moiety—PEE—whereas the hydrophilic component is PEO (Li et al. 2006a). A variety of

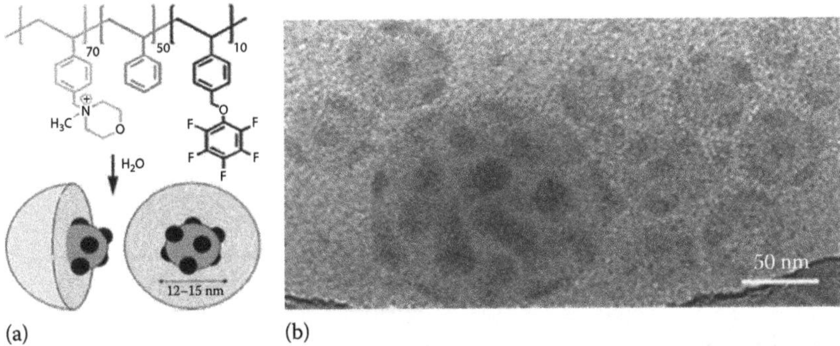

FIGURE 2.94 (a) Chemical composition of poly(4-vinylbenzyl-N-methylmorpholinium chloride)-*b*-PS-*b*-poly(4-vinylbenzyl pentafluorophenol ether) triblock copolymer and sketch of the multicompartment micelles it forms in aqueous solution. (Kubowicz, S., Baussard, J.F., Lutz, J.F., Thunemann, A.F., von Berlepsch, H., and Laschewsky, A.: *Angew. Chem. Int. Ed.* 2005. 44. 5262–5265. Copyright Wiley-VCH Verlag GmbH & Co. KGaA. Reproduced with permission.) (b) Cryo-ET of the particles. (von Berlepsch, H., Bottcher, Ch., Skrabania, K., and Laschewsky, A., *Chem. Commun.*, 17, 2290–2292, 2009. Reproduced by permission of The Royal Society of Chemistry.)

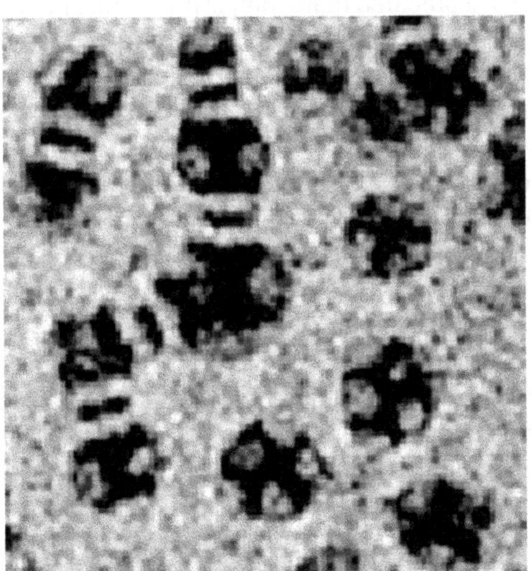

FIGURE 2.95 Morphologies of multicompartment micelles observed in aqueous dispersions of miktoarm terpolymers, μ-PEE-PEO-poly(perfluoropropylene oxide). (Reproduced with permission from Li, Zh., Hillmyer, M.A., and Lodge, T.P., *Langmuir*, 22, 9409–9417, 2006a. Copyright 2006, American Chemical Society.)

aggregate morphologies including segmented wormlike micelles as well as raspberrylike micelles have been observed some of them depicted in Figure 2.95. A small number of structural elements, however, can be invoked to relate the aggregate structure to copolymer composition. Upon the addition of THF, which is a selective solvent for both PEO and PEE moieties, the structures evolve from multicompartment

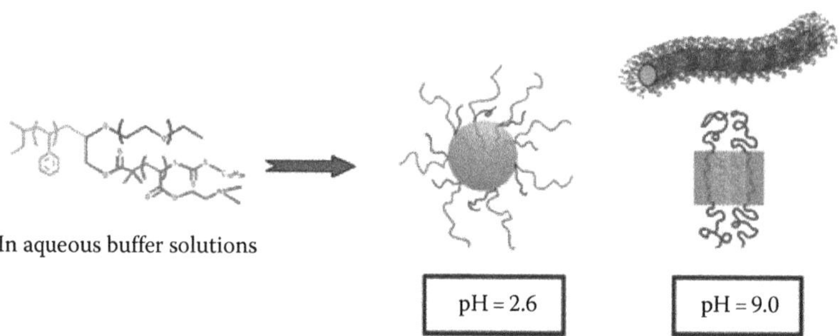

FIGURE 2.96 Evolution of spherical micelles with mixed corona prepared in aqueous solution by ABC miktoarm copolymer, μ-PS-PEO-poly(2-(dimethylamino)ethyl acrylate), to wormlike micelles with segmented core with increasing pH. (Reprinted with permission from Liu, Ch., Hillmyer, M.A., and Lodge, T.P., *Langmuir*, 25, 13718–13725, 2009. Copyright 2009, American Chemical Society.)

micelles to core–shell–corona micelles to micelles with mixed corona composed of PEO and PEE (Liu et al. 2008). By manipulating the ionic character of an ABC miktoarm copolymer, μ-PS-PEO-poly(2-(dimethylamino)ethyl acrylate), through pH variations, Liu et al. have prepared multicompartment micelles (Liu et al. 2009). The authors demonstrate the evolution of predominantly spherical micelles with mixed corona to wormlike micelles with alternating domains of PS and PDMAEA with increasing pH (Figure 2.96).

A recent paper reports on simulations of solution state self-assembly of miktoarm *ABC* terpolymers composed of a solvophilic *A* arm and two solvophobic *B* and *C* arms (Kong et al. 2009). A morphological phase diagram is constructed (Figure 2.97), which reveals, not unexpectedly, that the overall micellar morphology is largely controlled by the volume fraction of the solvophilic arm, f_A, whereas the internal compartmentalization depends on the ratio between volume fractions of the two solvophobic arms, f_B and f_C. Generally, with decreasing f_A, the overall morphology changes from spheres to cylinders to lamellar structures, which is the same as that observed for the amphiphilic diblock copolymers. More precisely, as depicted in Figure 2.97, a morphological sequence *hamburger micelles* → segmented bilayer sheets → segmented semivesicles → laterally structured vesicles → layer structures with decreasing f_A when $f_B \approx f_C$ is predicted. The sequence is different, for example, *raspberry micelles* → multicompartment worms → multicompartment onions, for $f_B < f_C$ upon f_A decreasing. The morphological transition within the solvophobic interior of the particles is consistent with confined self-assembly of *BC* diblock copolymers. For terpolymers with equal or nearly equal length of *B* and *C* arms, several previously unknown structures, including vesicles with novel lateral structures (helices or stacked donuts), segmented semivesicles, and elliptic or triangular bilayer sheets, are discovered.

An interesting approach to multicompartment micelles has been suggested elsewhere (Stahler et al. 1999). These particles have been prepared by aqueous radical polymerization of acrylamide (water-soluble) and polymerizable hydrocarbon and fluorocarbon surfactants (surfmers), which are in a micellar state during the polymerization, rather than by self-assembly of previously synthesized copolymers.

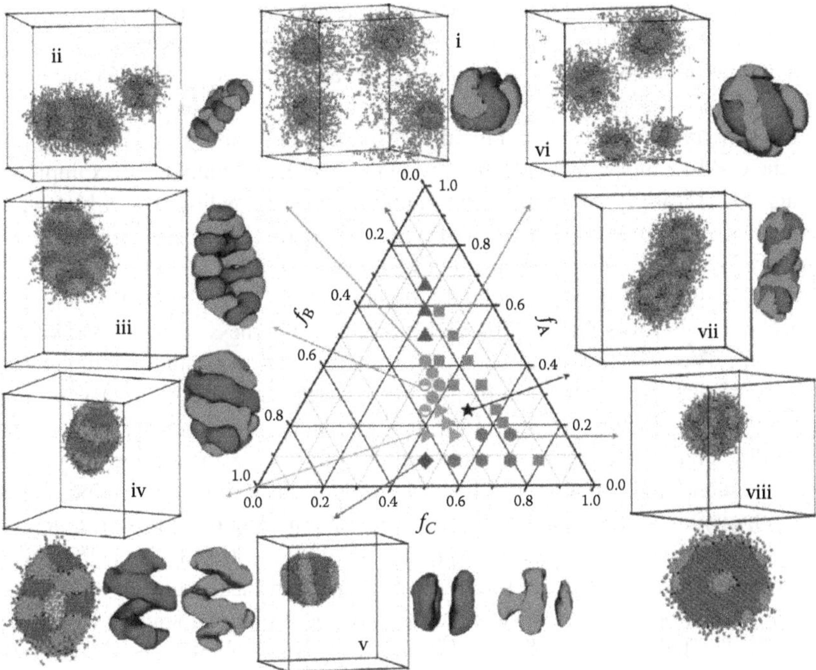

FIGURE 2.97 Morphological phase diagram of *ABC* miktoarm terpolymers in a selective for the *A* arm solvent. f_A, f_B, and f_C denote the volume fractions of the *A*, *B*, and *C* arms, respectively. Representative snapshots of the micellar morphologies are shown in the boxes, whereas the cross-sectional view or views of *B* and/or *C* domains are shown separately on the right or below the corresponding snapshot. The latter are labeled as follows: (i) hamburger micelles, (ii) segmented bilayer sheets, (iii) segmented semivesicles, (iv) laterally structured vesicles, (v) layer structures, (vi) raspberry micelles, (vii) multicompartment worms, and (viii) multicompartment onions. (Reprinted with permission from Kong, W., Li, B., Jin, Q., Shi, A.-Ch., *J. Am. Chem. Soc.*, 131, 8503–8512, 2009. Copyright 2009, American Chemical Society.)

2.2.4.2 Disklike Micelles

Disklike micelles and spheroids (oblate or prolate) have been relatively rarely observed. According to the surfactant packing parameter model of Israelachvili for simple hydrocarbon amphiphiles, they are not stable morphologies because the peripheral regions are highly curved, while the central regions are too thick leading to high-energy packing. As stated elsewhere, "the formation of oblate spheroidal and disklike morphologies should be due to additional factors other than straightforward considerations of the packing parameter of the hydrophobic chains" (Holder et al. 2011). Indeed, disklike micelles are formed from mixed (polymer–polymer or polymer–surfactant) systems (see Chapter 3), in mixed solvents, or upon the addition of organic or inorganic substances that are able to alter not only the quality of the solvent but also the volume and interfacial energy of the constituent moieties. Pure disklike micelles formed by amphiphilic block copolymers in water have been rarely target objects for investigations. Most frequently, they are mentioned as transient intermediate structures registered during shape transformation of aggregates upon the action of external stimuli. However, the demands of numerous applications require focusing of research not only on manipulating the aggregate properties via chemistry of the amphiphilic copolymers but also on controlling of the overall shape of the aggregates including those beyond the most common morphologies. Such morphology is the disklike morphology. In the following, we give a number of examples to illustrate the versatility of amphiphilic copolymers and conditions at which disklike micelles are formed as well as the driving forces, mechanism of formation, and parameters involved in directing the process of disk formation.

SANS data from a 5 wt.% aqueous solution of $EO_{19}PO_{43}EO_{19}$ at 80°C can be best fitted to large thin disks with radius of 1000 Å and thickness of about 40 Å (Guo et al. 2006). The morphological changes result from the well-known temperature sensitivity of this important class of polymeric surfactants. Upon increasing temperature, the hydrophobicity of both PPO and PEO increases, which leads to packing of more PPO chains in the core and shrinking of PEO in the corona, which, in turn, reduces the interfacial curvature and interfacial energy. The simultaneous increase in hydrophobicity and decrease in interfacial energy are the origin of the morphological transformations particularly leading to disk formation at 80°C for $EO_{19}PO_{43}EO_{19}$. Similar transformations, that is, from spherical to cylindrical and then to disklike morphology, can be achieved through addition of increasing amounts of hydrophobic alcohol (1-phenylethanol) to $EO_{37}PO_{56}EO_{37}$ aqueous solution at constant temperature of 23°C (Guo et al. 2006). The mechanism behind this transformation is the interfacial curvature change between the core and the corona regions with temperature or added alcohol. Thus, the lowered solubility of the PEO chains and the shrinking in their volume tend to flatten the micelle interface curvature between the core and corona giving rise to formation of particles of cylindrical and disklike morphologies that reduce the surface energy. In fact, the morphological transformations of the PEO-PPO-PEO copolymer micelles have been well documented long time ago. In particular, the change of the micellar structure from spherical to oblate or prolate ellipsoids takes place exclusively at elevated temperatures and, according to Mortensen and Pedersen, leads to decreasing intermicellar interactions (Mortensen and Pedersen 1993). Recent simulation studies, however, predict formation of oblate ellipsoids at ambient temperatures for $EO_{17}PO_{60}EO_{17}$ in semiconcentrated (e.g., 16–20 vol.%) aqueous solutions (Yang et al. 2008). These structures derive from micellar clusters that are formed as a result of coalescence of spherical micelles. Note that $EO_{17}PO_{60}EO_{17}$ has relatively short PEO chain that would favor such coalescence. The PEO chains among different micelles twist together, and the micelles rearrange in one direction, thus forming oblate ellipsoids.

Formation of disklike micelles has been reported from ABC-type amphiphilic copolymers with an ionic hydrophilic block in the presence of organic counterions (diamino derivatives) that interact with the polyelectrolyte corona and a cosolvent (THF) acting as a plasticizer of the core (Li et al. 2005). The micellar assemblies have been formed through self-organization of PAA-*b*-poly(methyl acrylate)-*b*-PS copolymers. The authors use organic diamines as counterions to change the conformation of the polyelectrolyte chains in the corona and, consequently, the volume of the latter. Since the counterions bind to the polyelectrolyte chains, they can decrease the

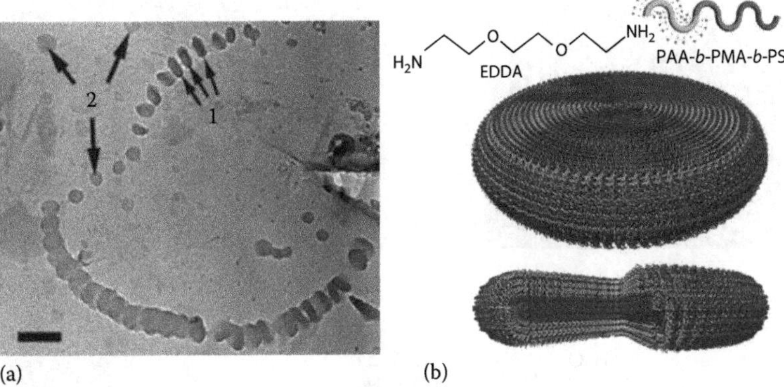

FIGURE 2.98 (a) Cryo-TEM image showing disks parallel (arrows 1) and perpendicular (arrows 2) to electron beam axis. Disks have been formed from PAA-*b*-poly(methyl acrylate)-*b*-PS copolymer in aqueous solution in the presence of ethylenedioxy-bis-ethylenediamine (EDDA) and THF. The bar corresponds to 200 nm. (b) Idealized cartoon of disklike micelles with cross section showing a segregated core, consisting of PS and poly(methyl acrylate) and hydrophilic corona of PAA with closely associated EDDA counterions. (Reprinted with permission from Li, Zh., Chen, Zh., Cui, H., Hales, K., Qi, K., Wooley, K.L., and Pochan, D., *Langmuir*, 21, 7533–7539, 2005. Copyright 2005, American Chemical Society.)

effective corona volume due to screening electrostatic repulsions and collapsing the chains. The decrease is compensated by an effective increase of the corona volume as the counterions remain in the corona regions. The resultant micellar morphology (spherical, cylindrical, or disklike) has been found to largely depend on the prevailing effect (overall decrease or increase in corona volume), which, by proper selection of the counterion and its quantity, allows a reversible access to desired regions of the block copolymer phase diagrams. Particularly, for disk formation, multivalent and small diamine is required (Li et al. 2005). Figure 2.98 shows a cryo-TEM image of disklike aggregates as well as a cartoon of idealized disklike morphology.

Core–corona micelles of disklike morphology have been prepared in aqueous solution from AB- and ABA-type block copolymers whose hydrophilic blocks are also ionic: PAA is used as the hydrophilic A blocks, whereas the hydrophobic block, which is identical for the two copolymers, consists of a side-chain liquid-crystalline polymer, poly(*n*-octadecyl acrylate), hereinafter PODA (Akiba et al. 2010). In aqueous solution, both copolymers form disks that slightly differ in thickness (32 vs. 36 nm for the triblock and diblock, respectively) as determined by synchrotron SAXS, which is attributed to the different chain architectures. The core thickness of the micelles of the triblock copolymer is limited by the length of the extended PODA block that adopts either U-type or bridging conformation. Such conformations are not possible for the diblock copolymer, and its macromolecules are arranged to form a bilayer the thickness of which is governed by twice the length of the extended PODA chain. Upon heating to 40°C, which is above the melting temperature of PODA, the micelles of the diblock copolymer undergo a disk-to-sphere transformation, whereas those of the triblock copolymer maintain their disklike morphology.

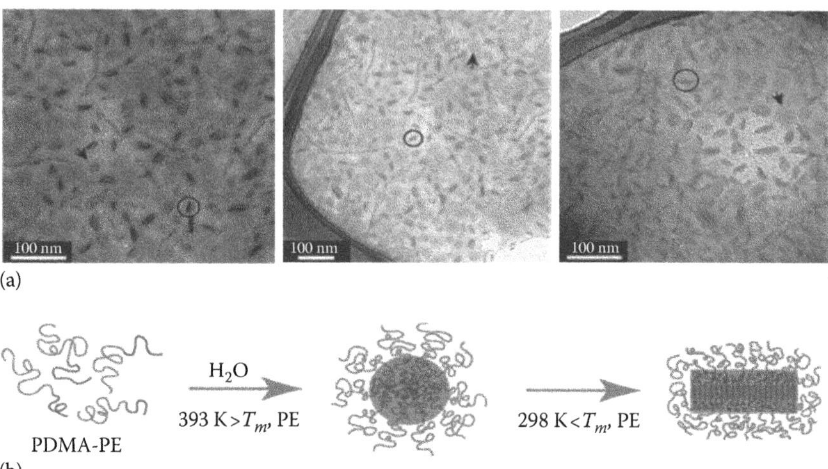

FIGURE 2.99 (a) Representative cryo-TEM images from 0.5 wt% dispersion of a poly(*N*,*N*-dimethylacrylamide)-polyethylene diblock copolymer in water. The disklike micelles are observed both edge on and face on. A small portion of long, threadlike aggregates is observable as well. (b) Hypothesized mechanism toward the formation of disklike micelles. The block copolymer self-assembled into spherical micelles at 120°C, above the melting of polyethylene. When cooling down to 25°C, the crystallizable polyethylene drove their morphology into disk-like micelles. (Reprinted with permission from Yin, L. and Hillmyer, M.A., *Macromolecules*., 44, 3021–3028, 2011. Copyright 2011, American Chemical Society.)

The different temperature responsiveness is attributed to the conformational constraint of the PODA block imposed by the ABA-type chain architecture. No information, however, about the size of the disklike micelles is provided.

Disklike micelles have been obtained by direct dispersion of a poly(*N,N*-dimethylacrylamide)–polyethylene diblock copolymer into water above the melting transition of polyethylene followed by cooling to 25°C (Yin and Hillmyer 2011). The disklike morphology, visualized by cryo-TEM (Figure 2.99a), was ascribed to the crystallization of polyethylene, driving the formation of disklike structures in a stepwise "micellization–crystallization" protocol. Interestingly, the copolymer has a relatively short hydrocarbon block and a long water-soluble block corresponding to weight fraction of 0.75 (!). The preparation protocol, schematically presented in Figure 2.99b, involves heating of the dispersion in a pressure vessel to 120°C, which is above the melting transition of the polyethylene block, holding at this temperature for 5–7 days, and cooling down to 25°C at a rate of 1°C–2°C/min. As a result "frozen" disklike micelles with a lateral core radius of 14.3 ± 2.4 nm and thickness of 5.5 ± 0.8 nm were prepared.

Also driven by the internal crystallization of the insoluble block is the formation of disklike micelles described elsewhere (Wang et al. 2010a). The micelles are formed in aqueous solution from a coil–brush diblock copolymer consisting of linear polyethylene and poly[oligo(ethylene glycol) methyl ether methacrylate] (POEGMA). The former crystallizes in the interior of the micelles, whereas the swollen POEGMA brushlike chains form the outer stabilizing corona. The same driving force, that is,

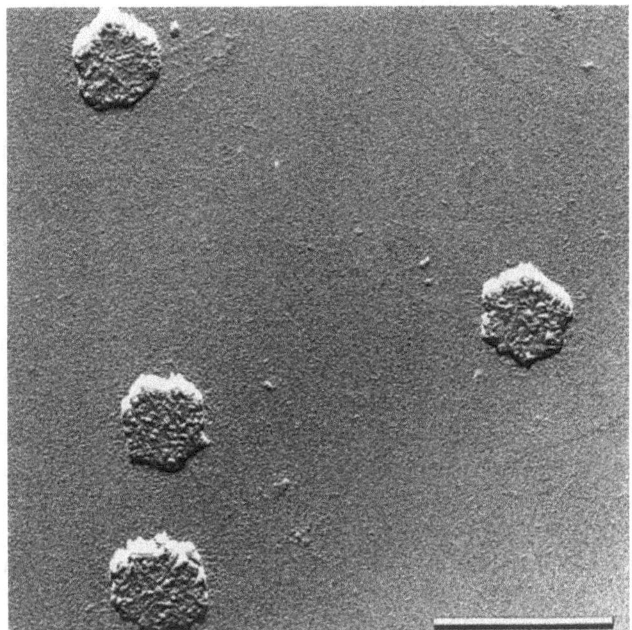

FIGURE 2.100 TEM micrograph of individual, hexagonally shaped platelets obtained from poly(sodium glutamate-*b*-leucine) by air-drying of a dilute aqueous suspension. The bar corresponds to 500 nm. (Reprinted from *J. Coll. Interf. Sci.*, 217, Constancis, A., Meyrueix, R., Bryson, N., Huille, S., Grosselin, J.-M., Gulik-Krziwicki, Th., and Soula, G., 357–368, Copyright 1999, with permission from Elsevier.)

crystallization, is behind the formation of platelets from a copolymer whose constituent blocks are of completely different nature from that of the previously mentioned ones (Constancis et al. 1999). These authors describe the formation of discrete platelets from a polypeptidic diblock copolymer of poly(leucine-block-glutamate). The platelets are hexagonally shaped (Figure 2.100) with molar mass reaching 7×10^8 g/mol and number-average diameter and thickness of about 190 and 45 nm, respectively. The core of the platelets is composed of crystalline, helical leucine segments, whereas the ionized polyglutamate forms polyelectrolyte brushes that extend from the two interfaces into the aqueous phase. The particles exhibit ability to absorb proteins directly from the solution (see more in Chapter 5). Entirely in line with these experimental results are the Brownian molecular dynamics simulations revealing that the disk morphology is the preferred aggregate morphology for diblock copolymers with a rigid core-forming block (Lin et al. 2009b). Other examples of particles displaying flat, platelet, or disklike structures with crystalline or semicrystalline cores formed in nonaqueous solution can be found in the literature. However, the book is restricted to aqueous media, and nonaqueous solutions are not considered.

Fundamentally distinct behavior and driving force from those of the disklike micelles formed as a result of either crystallization of the core-forming block or addition of organic counterions to ionic copolymers are those exhibited by block copolymers with (super) strong segregation between the blocks. The super strong-segregation regime (SSSR) of phase behavior has been identified in the 1990s (Nyrkova et al. 1993, Semenov et al. 1995). According to the SSSR, the interactions between blocks A and B become so strong that the interfacial energy overwhelms the conformational entropy (in the bulk) or the coronal crowding (in solution), which favors the formation of flat interfaces. This regime has been predicted to be experimentally accessible only using ionic polymers (Semenov et al. 1995). However, unprecedented formation of flat disk morphology by a nonionic, flexible coil ABC-type block copolymer has been recently reported (Lodge et al. 2004). The copolymer consists of PEO, PS, and a fluoropolymer obtained via grafting perfluorohexyl iodide groups to a 1,2-PB backbone, thus, in aqueous solution, creating internally segregated micelles with two interfaces (fluoropolymer/PS and PS/PEO + water), thereby significantly augmenting the role of interfacial tension. A representative cryo-TEM image is shown in Figure 2.101a. The round objects with uniform contrast correspond to side-on oriented disklike micelles, whereas the objects denoted with the long arrow and in the inset have been identified as disklike micelles viewed edge on. In the inset, the variation of the contrast indicates a segregation of the hydrophobic interior of the micelles into a darker core built of the fluoropolymer moieties and slightly lighter shell of PS. The characteristic halo of PEO corona, which is highly solvated and, therefore, not well visible, is denoted with arrowheads. The structure is schematically drawn in Figure 2.101b. It is noteworthy that the parent copolymer, PEO-PS-PB, from which the fluorinated copolymer has been prepared, is reported to form conventional spherical core–corona micelles with mixed PS-PB cores.

In brief, the commonly encountered mechanisms of formation of disklike micelles are (1) extremely strong interfacial tension typically experienced by fluor-containing copolymers, (2) change of the interfacial curvature upon manipulation of the volumes of the core and corona regions, (3) presence of rodlike corona- and/

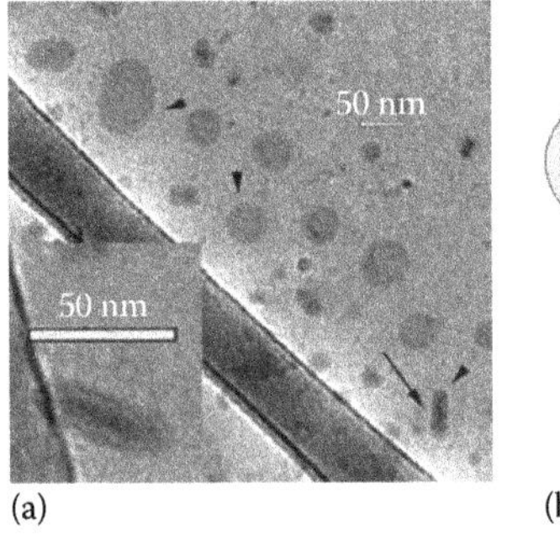

(a)

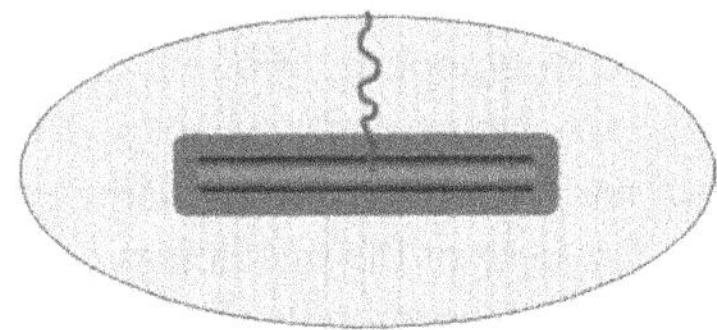

(b)

FIGURE 2.101 (a) Cryo-TEM images of PEO-PS-fluoroPB micelles with disklike morphology formed in water. The long arrow denotes a micelles viewed edge on. A magnified image of an edge-on viewed micelle is shown in the inset. The arrowheads indicate the extent of the PEO corona. (b) Schematic drawing of a core–shell–corona disklike micelle obtained from the same ABC copolymer in aqueous solution. (Reprinted with permission from Lodge, T.P., Hillmyer, M.A., and Zhou, Zh., *Macromolecules*, 37, 6680–6682, 2004. Copyright 2004, American Chemical Society.)

or core-forming block leading to parallel arrangement of the chains in the corona or core, and (4) presence of crystallizable core-forming block developing closely packed, flat structures.

2.2.4.3 Bicontinuous Micelles

Particulate and bulk bicontinuous cubic phases have stimulated significant research interest because of their potential for applications in controlled release and drug delivery as well as for biochip and biosensor development. The normal and reverse bicontinuous phases are situated in both sides of the lamellar phase (see Figure 2.3). The geometry of the bicontinuous cubic phase liquid crystal is presented by a 3D lattice built from contorted bilayer that separates two networks of water channels. The latter are identical and continuous but nonintersecting (Scriven 1976) and have a diameter of about 5 nm in the fully hydrated state when the phase is formed by low molecular weight compounds and larger if polymers are involved. The lipophilic part of the liquid crystal consists of infinite bilayer periodically curved in three dimensions (Larsson 1989, Lindblom and Rilfors 1989), which forms the so-called infinite periodic minimal surface (Patton and Carey 1979, Patton et al. 1985). Three types of infinite periodic minimal surfaces, namely, primitive (P), diamond (D), and gyroid (G) surfaces, corresponding to cubic phases C_P, C_D, and C_G are known. They are associated with the space groups Im3m, Pn3m, and Ia3d, respectively. Nodal surfaces as approximations of minimal surfaces (von Schnering and Nesper 1991) are presented in Figure 2.102. It has been proposed that the dynamic structure of the cubic phases, for example, thermal oscillations, is better described by the nodal surfaces (Larsson 1997).

The bicontinuous cubic phase liquid crystal is solid like. Macroscopically, the cubic phase is transparent; optically isotropic; that is, nonbirefringent; soft; and highly viscous. It is able to swell and shrink upon the addition of active substances. Essential advantages are the very large surface area that has been estimated to be around 400 m^2/g for one of the most extensively studied lipid with nonlamellar propensity—glyceryl monooleate in excess water (Lawrence 1994) and high bilayer

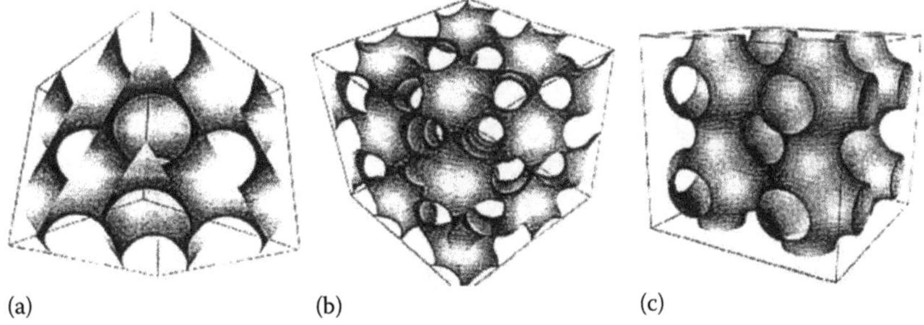

FIGURE 2.102 Mathematically generated minimal surfaces that represent the three bicontinuous cubic morphologies associated with hydrated lipids. (a) Pn3m (diamond), (b) Ia3d (gyroid), and (c) Im3m (primitive). (Reproduced with permission from Benedicto, A.D. and O'Brien, D.F., Bicontinuous cubic morphologies in block copolymers and amphiphile/water systems: Mathematical description through the minimal surfaces, *Macromolecules*, 30, 3395–3402, 1997. Copyright 1997, American Chemical Society.)

content per unit volume. Due to the presence of lipid bilayers and water channels, both water-soluble and oil-soluble substances as well as molecules with amphiphilic character can be solubilized in the cubic phase. A small but important part of research has been directed toward developing methods for dispersing of cubic phases and preparation of discrete aggregates with bicontinuous morphology. Typically, aqueous dispersions of cubic phases are produced by high-energy fragmentation of bulk cubic gel. However, colloidal stabilization is required since the most frequently used amphiphiles with propensity to form bicontinuous phases do not form stable dispersions. A variety of dispersing agents have been used to prepare aqueous dispersions of cubic phases. The reader can refer to Chapter 3 for discrete particles with internal morphology prepared from polymer–amphiphile mixtures.

Bicontinuous phases in binary copolymer–water systems have been largely elusive. Even for PEO-PPO-PEO copolymers, despite their commercial availability in wide compositional ranges, the numerous practical applications, and considerably more academic studies, bicontinuous phases have been only sporadically and occasionally observed. Zhang and Khan, for example, have identified a cubic phase that may possess a bicontinuous-type structure judging by its relative position—an extremely narrow region between the hexagonal and lamellar phases in the phase diagram of $EO_{13}PO_{30}EO_{13}$-water binary systems (Zhang and Khan 1995). Alexandridis and coauthors were the first who unambiguously identified both normal and reverse bicontinuous phases for $EO_{19}PO_{43}EO_{19}$ (Alexandridis et al. 1998). The normal structure occurred at about 62–64 wt.% copolymer content, whereas the reverse one was observed at higher (67–84 wt.%) concentration in the presence of xylene. Both phases were located between the hexagonal (normal or reverse) and lamellar phases. They were properly characterized by SAXS. The diffraction patterns of samples from the normal and reverse bicontinuous regions showed striking similarities and were associated with the gyroid minimal surface. A spongelike (L_3) phase, which is a bicontinuous yet disordered phase (Figure 2.103), has been found for a PEO-PPO-PEO triblock copolymer that corresponds in composition to $EO_5PO_{30}EO_5$ in binary systems (Hecht et al. 1995). In the concentration–temperature phase diagram,

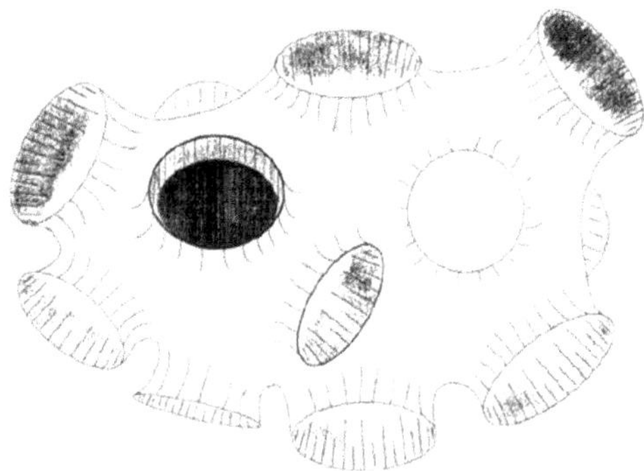

FIGURE 2.103 Schematic drawing of the bicontinuous L_3 (spongelike) phase. (Reprinted with permission from Hecht. E., Mortensen, K., and Hoffmann, H., *Macromolecules*, 28, 5465–5476, 1995. Copyright 1995, American Chemical Society.)

the L_3 phase is located just above the lamellar phase, that is, at slightly higher temperatures, in the concentration range spanning from about 6 to 50 wt.%. To the best of our knowledge, this is the first time that L_3 phases have been found and systematically investigated for PEO-PPO-PEO copolymers. The authors carried out a number of measurements applying DLS and SANS among the others from which, for instance, the interlamellar distance (below 20 nm), the cooperative diffusion coefficients, and the corresponding to hydrodynamic correlation length of about 350 nm were determined. It is noteworthy that copolymers of 20 or 40 wt.% of hydrophilic content only have been found to form cubic or spongelike bicontinuous phases. For a weakly selective solvent and mild segregation between the blocks—conditions that correlate well with concentrated PEO-PPO-PEO aqueous solutions at ambient temperatures—has the possibility for forming discrete aggregates with a bicontinuous morphology been proposed based on self-consistent field simulations (Fraaije and Sevink 2003). The simulations, performed for a diblock copolymer $A_{N-M}B_M$ with $N=20$ and A being solvophobic, resulted in a range of morphologies with internal structure depending on the size ratio $f=M/N$ (Figure 2.104).

Very few experimental observations for formation of discrete particles with internal bicontinuous morphology have been reported. Two of them are in water–organic solvent mixtures and will only be mentioned in the following. These are PS-PAA in a water–DMF mixture reported elsewhere (Yu et al. 1996) and the triblock copolymers

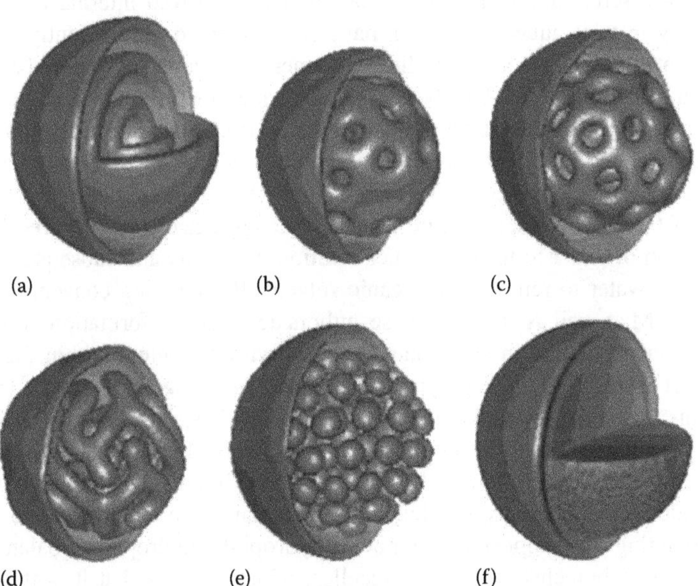

FIGURE 2.104 Morphologies of discrete aggregates generated by self-consistent field simulations for an $A_{N-M}B_M$ diblock copolymer in a weakly selective solvent at size ratios $f=M/N$ equal to 0.35 (a), 0.30 (b), 0.25 (c), 0.20 (d), 0.15 (e), and 0.10 (f). The block A is slightly more solvophobic and B is slightly more solvophilic. (Reprinted with permission from Fraaije, J. and Sevink, G.J.A., *Macromolecules*, 36, 7891–7893, 2003. Copyright 2003, American Chemical Society.)

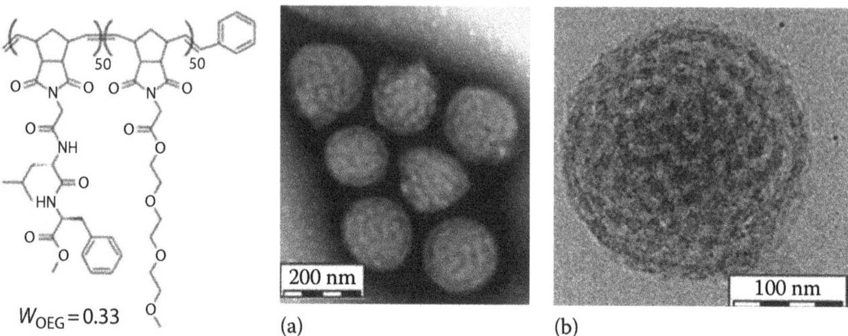

FIGURE 2.105 Left panel: Chemical formula of polynorbornene-based double-comb diblock copolymer with tripeptide and oligo(ethylene glycol) side chains; W_{OEG} denotes the weight fraction of oligo(ethylene glycol) side chains. Right panel: Conventional TEM (a) and cryo-TEM (b) images showing the internal bicontinuous morphology of discrete nanoparticles formed in aqueous solution. (Parry, A.L., Bomans, P.H.H., Holder, S.J., Sommerdijik, N., and Biagini, S.C.G.: *Angew. Chem. Int. Ed.* 2008. 47. 8859–8862. Wiley-VCH Verlag GmbH & Co. KGaA. Reproduced with permission.)

PAA-PMA-PS mentioned in Section 2.4.2 in water–THF mixtures when complexed with 2,2′-(ethylenedioxy)diethylamine (Hales et al. 2008).

The first discrete aggregates with unambiguously proved internal bicontinuous morphology in pure aqueous solution have been obtained only recently. In one of the two contributions, the amphiphilic copolymer used can be considered as a copolymer with polynorbornene backbone to which tripeptide (glycine–leucine–phenylalanine) and oligo(ethylene glycol)methyl ether side chains are grafted (Parry et al. 2008). The tripeptide and oligo(ethylene glycol) moieties are segregated, so that a copolymer of diblock and comblike architecture with oligo(ethylene glycol) weight fraction of 0.33 is obtained (Figure 2.105). The aggregates are prepared by dropwise addition of water to the copolymer solution in DMSO and subsequent dialysis against pure water to remove the organic solvent. By applying conventional TEM and cryo-TEM as well as cryo-ET, these authors demonstrate formation of internally structured and spherical in shape nanoparticles with diameters within the 50–450 nm range (Figure 2.105). The polynorbornene backbone together with the tripeptide side chains forms the hydrophobic regions as ^{1}H NMR spectra reveal. They form a network of branched wormlike domains with diameters of about 20 nm that separate water channels of slightly lower (15 nm) diameters. Interestingly, changing the tripeptide side chains from glycine–leucine–phenylalanine to leucine–valine–leucine, that is, changing the composition but not the hydrophilic/hydrophobic balance, leads to formation of branched wormlike micelles, which implies that it is the specific amino acid sequence rather than the precise value of oligo(ethylene glycol) weight fraction or the packing parameter that determines the formation of discrete particles with internal bicontinuous structure.

Another approach to obtaining discrete particles of internal morphology is proposed by essentially the same authors (McKenzie et al. 2010). They used a diblock copolymer of PEO and poly(octadecyl methacrylate) (PODMA) (Figure 2.106) from

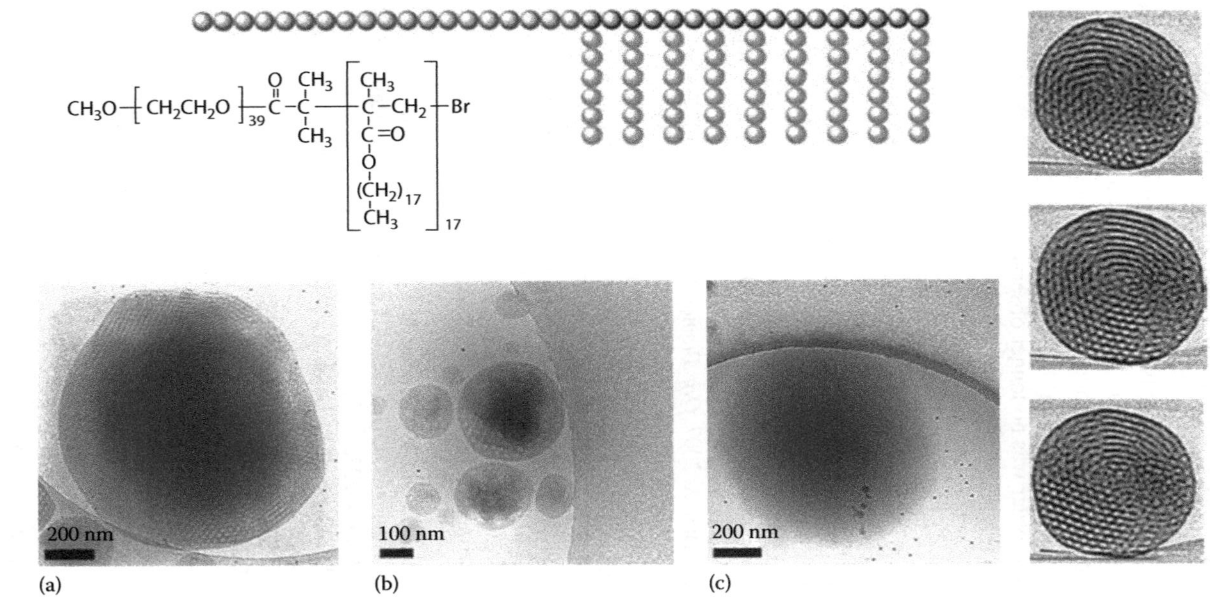

FIGURE 2.106 Upper panel: Chemical formula and schematic structure of $(PEO)_{39}$-*b*-$(PODMA)_{17}$ used. Lower panel: Cryo-TEM images taken from 5 wt.% aqueous dispersion of $(PEO)_{39}$-*b*-$(PODMA)_{17}$ vitrified at 4 (a), 22 (b), and 45°C (c). Right panel: Gallery of *z* slices from cryo-ET showing the internal organization of the particles (c=5 wt%, vitrification at 4°C. (Reprinted with permission from McKenzie, B.E., Nudelman, F., Bomans, P.H.H., Holder, S.J., and Sommerdijik, N., *J. Am. Chem. Soc.*, 132, 10256–10259, 2010. Copyright 2010, American Chemical Society.)

which an aggregate dispersion was prepared via solvent exchange method. The size of the aggregates showed strong concentration dependence: diameters of about 350 nm at 5 wt.% and ~275 nm at 1 wt.% were reported. Cryo-TEM and cryo-ET revealed an ordered internal microphase-separated structure of the samples vitrified at 4°C, which is consistent with a spongelike bicontinuous network structure of intertwined water-filled and carbon-rich channels both of approximately 13 nm in diameter (Figure 2.106). The majority of the structural components were bicontinuous, but lamellar organization was partially observed as Figure 2.106 clearly shows. At higher temperatures, in particular, at 22°C, which is the melting temperature of the PODMA moieties, the internal structure became apparently less ordered (Figure 2.106), whereas order was no longer observed at 45°C.

REFERENCES

Adams, D. J., Atkins, D., Cooper, A. I., Furzeland, S., Trewin, A., Young, I. 2008. *Biomacromolecules* 9:2997–3003.

Agut, W., Brulet, A., Schatz, C., Taton, D., Lecommandoux, S. 2010. *Langmuir* 26:10546–10554.

Ahmed, F., Photos, P. J., Discher, D. E. 2006. *Drug Dev. Res.* 67:4–14.

Akiba, I., Akino, Y., Masunaga, H., Sakurai, K. 2010. *IOP Conf. Series: Mater. Sci. Eng. Synchrotron Rad. Polym. Sci.* 14:012009.

Alemdaroglu, F. E., Herrmann, A. 2007. *Org. Biomol. Chem.* 5:1311–1320.

Alexandridis, P., Olsson, U., Lindman, B. 1998. *Langmuir* 14:2627–2638.

Almgren, M., Brown, W., Hvidt, S. 1995. *Colloid Polym. Sci.* 273:2–15.

Aniansson, E. A. G., Wall, S. N. 1974. *J. Phys. Chem.* 78:1024–1030.

Aniansson, E. A. G., Wall, S. N. 1975. *J. Phys. Chem.* 79: 857–858.

Aniansson, E. A. G., Wall, S. N., Almgren, M., Hoffmann, H., Kielmann, I., Ulbricht, W., Zana, R., Lang, J., Tondre, C. 1976. *J. Phys. Chem.* 80:905–922.

Antonietti, M., Forster, S. 2003. *Adv. Mater.* 15:1323–1333.

Aranda-Espinoza, H., Bermudez, H., Bates, F. S., Discher, D. E. 2001. *Phys. Rev. Lett.* 87:208–301.

Ayen, W. Y., Chintankumar, B., Jain, J. P., Kumar, N. 2011. *Polym. Adv. Technol.* 22:158–165.

Ayers, L., Vos, M. R. J., Adams, P. J. H. M., Shklyarevskiy, I. O., van Hest, J. C. M. 2003. *Macromolecules* 36:5967–5973.

Bartels, J. W., Cauet, S. I., Billings, P. L., Lin, L. Y., Zhu, J., Fidge, Ch., Pochan, D. J., Wooley, K. L. 2010. *Macromolecules* 43:7128–7138.

Battaglia, G., Ryan, A. J. 2005. *J. Am. Chem. Soc.* 127:8757–8764.

Battaglia, G., Ryan, A. J. 2006a. *J. Phys. Chem. B* 110:10272–10279.

Battaglia, G., Ryan, A. J. 2006b. *Macromolecules* 39:798–805.

Battaglia, G., Ryan, A. J., Tomas, S. 2006. *Langmuir* 22:4910–4913.

Battaglia, G., Tomas, S., Ryan, A. J. 2007. *Soft Matter* 3:470–475.

Bauer, A., Kopschuetz, C., Stolzenburg, M. Foerster, S. Mayer, Ch. 2006. *J. Membr. Sci.* 284:1–4.

Bellomo, E. G., Wyrsta, M. D., Pakstis, L., Pochan, D. J., Deming, T. J. 2004. *Nature Mater.* 3:244–248.

Benedicto, A. D., O'Brien, D. F. 1997. *Macromolecules* 30:3395–3402.

Bermudez, H., Aranda-Espinoza, H., Hammer, D. A., Discher, D. E. 2003. *Europhys. Lett.* 64:550–557.

Bermudez, H., Brannan, A. K., Hammer, D. A., Bates, F. S., Discher, D. E. 2002. *Macromolecules* 35:8203–8208.

Bermudez, H., Hammer, D. A., Discher, D. E. 2004. *Langmuir* 20:540–543.
Bhaw-Luximon, A., Meeram, L. M., Jugdawa, Y., Helbert, W., Jhurry, D. 2011. *Polymer Chem.* 2(1):77–79.
Blanazs, A., Armes, S. P., Ryan, A. J. 2009a. *Macromol. Rapid Commun.* 30:267–277.
Blanazs, A., Massignani, M., Battaglia, G., Armes, S. P., Ryan, A. J. 2009b. *Adv. Func. Mater.* 19:2906–2914.
Bloksma, M. M., Hoeppener, S., D'Haese, C., Kempe, K., Mansfeld, U., Paulus, R. M., Gohy, J.-F., Schubert, U. S., Hoogenboom, R. 2012. *Soft Matter* 8:165–172.
Bochkarev, A., Pfuetzner, R. A., Edward, A. M., Frappier, L. 1997. *Nature* 385:179.
Booth, C., Attwood, D. 2000. *Macromol. Rapid Commun.* 21:501–527.
Booth, C., Yu, G.-E., Nace, V. M. 2000. Block copolymers of ethylene oxide and 1, 2-butylene oxide. In: P. Alexandridis, B. Lindman, Eds. *Amphiphilic Block Copolymers: Self Assembly and Applications*. Amsterdam, the Netherlands: Elsevier, pp. 57–86.
Borisov, O. V., Zhulina, E. B. 2005a. *Langmuir* 21:3229–3231.
Borisov, O. V., Zhulina, E. B. 2005b. *Macromolecules* 38:2506–2514.
Borisov, O. V., Zhulina, E. B., Leermakers, F. A. M., Mueller, A. H. E. 2011. *Adv. Polym. Sci.* 241:57–129.
Borisova, O., Billon, L., Zaremski, M., Grassi, B., Bakaeva, Z., Lapp, A., Stepanek, P., Borisov, O. 2011. *Soft Matter* 7:10824–10833.
Brannan, A. K., Bates, F. S. 2004. *Macromolecules* 37:8816–8819.
Bromberg, L. 2001. Hydrophobically modified polyelectrolytes and polyelectrolyte block-copolymers. In: H.S. Nalwa. Ed. *Handbook of Surfaces and Interfaces of Materials*, vol. 4, New York: Academic Press, Chapter 7, pp. 369–404.
Caba, B. L., Zhang, Q., Carroll, M. R. J., Woodward, R. C., St Pierre, T. G., Gilbert, E. P., Riffle, J. S., Davies, R. 2010. *J. Colloid Interface Sci.* 344:81–89.
Cai, S. S., Vijayan, K., Cheng, D., Lima, E. M., Discher, D. E. 2007. *Pharm. Res.* 24:2099–2109.
Cai, Y., Aubrecht, K. B., Grubbs, R. B. 2011. *J. Am. Chem. Soc.* 133:1058–1065.
Cameron, N. R., Spain, S. G., Kingham, J. A., Weck, S., Albertin, L., Barker, C. A., Battaglia, G., Smart, T., Blanazs, A. 2008. *Faraday Discuss.* 139:359–368.
Castalletto, V., Hamley, I. W. 2006. *Polymers Adv. Technol.* 17(3):137–144.
Chaibundit, C., Ricardo, N. M. P. S., Crothers, M., Booth, C. 2002. *Langmuir* 18(11):4277–4283.
Chaibundit, C., Sumanatrakool, P., Chinchew, S., Kanatharana, P., Tattershall, C. E., Booth, C., Yuan, X.-F. 2005. *J. Colloid Interface Sci.* 283:544–554.
Checot, F., Brulet, A., Oberdisse, J., Gnanou, Y., Mondain-Monval, O., Lecommandoux, S. 2005. *Langmuir* 21:4308–4315.
Checot, F., Lecommandoux, S., Gnanou, Y., Klok, H.-A. 2002. *Angew. Chem. Int. Ed.* 41:1339–1343.
Checot, F., Rodriguez-Hernandez, J., Gnanou, Y., Lecommandoux, S. 2007. *Biomol. Eng.* 24:81–85.
Chen, Ch., Liu, G., Shi, Y., Zhu, C., Pang, Sh., Liu, X., Ji, J. 2011a. *Macromol. Rapid Commun.* 32:1077–1081.
Chen, J., Xing, M. M. Q., Zhong, W. 2011b. *Polymer* 52:933–941.
Chen, Q., Schonherr, H., Vansco, G. J. 2009. *Soft Matter* 5:4944–4950.
Chen, Y., Du, J., Xiong, M., Guo, H., Jinnai, H., Kaneko, T. 2007. *Macromolecules* 40:4389–4392.
Chen, Y., Shew, C. Y. 2001. *ACS Polym. Prepr. (Div. Polym. Sci.)* 42:626–627.
Chiu, H.-C., Lin, Y.-W., Huang, Y.-F., Chuang, C.-K., Chern, C.-S. 2008. *Angew. Chem. Int. Ed.* 47:1875–1888.
Choi, S.-H., Lodge, T. P., Bates, F. S. 2010. *Phys. Rev. Lett.* 104:047802.
Choucair, A., Soo, P. L., Eisenberg, A. 2005. *Langmuir* 21:9308–9313.
Christian, D. A., Cai, Sh., Bowen, D. M., Kim, Y., Pajerowski, J. D., Discher, D. E. 2009. *Eur. J. Pharm. Biopharm.* 71:463–474.

Chu, B., Zhou, Z. 1996. Physical chemistry of polyoxyethylene block copolymer surfactants. In: *Nonionic Surfactants: Polyoxyalkylene Block Copolymers*. V. M. Nace, Ed. *Surfactant Science Series* 60, New York: Marcel Dekker, pp. 67–143.

Constancis, A., Meyrueix, R., Bryson, N., Huille, S., Grosselin, J.-M., Gulik-Krziwicki, Th., Soula, G. 1999. *J. Colloid Interface Sci.* 217:357–368.

Convertine, A. J., Diab, C., Prieve, M., Paschal, A., Hoffman, A. S., Jonhson, P. H., Stayton, P. S. 2010. *Biomacromolecules* 11:2904–2911.

Creutz, S., van Stam, J., Antoun, S., De Schryver, F. C., Jerome, R. 1997. *Macromolecules* 30:4078–4083.

Crothers, M., Ricardo, N. M. P. S., Heatley, F., Keith Noxin, S., Attwood, D., Booth, C. 2008. *Int. J. Pharm.* 358:303–306.

Crothers, M., Zhou, Zh., Ricardo, N. M. P. S., Yang, Zh., Taboada, P., Chaibundit, Ch., Attwood, D., Booth, C. 2005. *Int. J. Pharm.* 293:91–100.

Cui, H., Chen, Zh., Wooley, K. L., Pochan, D. 2009. *Soft Matter* 5:1269–1278.

Cui, Q., Wu, F., Wang, E. 2011a. *Polymer* 52(8):1755–1765.

Cui, Q., Wu, F., Wang, E. 2011b. *J. Phys. Chem. B* 115:5913–5922.

Dalhaimer, P., Bates, F. S., Aranda-Espinoza, H., Discher, D. 2003. *C. R. Acad. Sci. (Paris)* 4:251–258.

Dan, N., Safran, S. A. 2006. *Adv. Colloid Interface Sci.* 123–126:323–331.

Daoud, M., Cotton, J. P. 1982. *J. Phys. (Fr)* 43:531–538.

Das, P. J., Barak, A., Kawakami, Y., Kannan, Th. 2011. *J. Polym. Sci. A—Polym. Chem.* 49:1376–1386.

Del Barrios, J., Oriol. L., Sanchez, C., Luis Serrano, J., Di Cicco, A., Keller, P., Li, M.-H. 2010. *J. Am. Chem. Soc.* 132:3762–3769.

Denkova, A. G., Mendes, E., Coppens, M.-O. 2009. *J. Phys. Chem. B* 113:989–996.

Dimitrov, Ph., Rangelov, S., Dworak, A., Tsvetanov, Ch. 2004. *Macromolecules* 37:1000–1008.

Dimova, R., Seifert, U., Pouligny, B., Forster, S., Dobereiner, H. G. 2002. *Eur. Phys. J. E* 7:241–251.

Ding, J., Xiao, Ch., Tang, Zh., Zhuang, X., Chen, X. 2011. *Macromol. Biosci.* 11(2):192–198.

Ding, K., Alemdaroglu, F. E., Borsch, M., Berger, P., Herrmann, A. 2007. *Angew. Chem. Int. Ed.* 46:1172–1175.

Dirks, A. J. T., van Berkel, S. S., Hatzakis, N. S., Opsteen, J. A., van Delft, F. L., Cornelissen, J. J. L. M., Rowan, A. E., van Hest, J. C. M., Rutjes, F. P. J. T., Nolte, R. J. M. 2005. *Chem. Commun.* 33:4172–4175.

Discher, B. M., Won, Y. Y., Ege, D. S., Lee, J. C. M., Bates, F. S., Discher, D. E., Hammer, D. A. 1999. *Science* 284:1143–1146.

Discher, D. E., Eisenberg, A. 2002. *Science* 297:967–973.

Discher, D. E., Ortiz, V., Srinivas, G., Klein, M. L., Kim, Y., Christian, D., Cai, Sh., Photos, P., Ahmed, F. 2007. *Prog. Polym. Sci.* 32:838–857.

Dong, C.-M., Chaikof, E. L. 2005. *Colloid Polym. Sci.* 283(12):1366–1370.

Dong, S., Zhang, H., Cui, X., Sui, Z., Xu, J., Wang, H. 2010. *J. Theoret. Comput. Chem.* 9(4):767–783.

van Dongen, S. F. M., de Hoog, H.-P. M., Peters, R. J. R. W., Nallani, M., Nolte, R. J. M., van Hest, J. C. M. 2009. *Chem. Rev.* 109:6212–6274.

Dormidontova, E. E. 1999. *Macromolecules* 32:7630–7644.

Du, B., Mei, A., Yin, K., Zhang, Q., Xu, J., Fan, Zh. 2009. *Macromolecules* 42:8477–8484.

Du, J. Z., Armes, S. P. 2005. *J. Am. Chem. Soc.* 127:12800–12801.

Eckhardt, D., Groenewolt, M., Krause, E., Börner, H. G. 2005 *Chem. Commun.* 2814–2816.

Evans, E., Heinrich, V., Ludwig, F., Rawiczy, W. 2003. *Biophys. J.* 85:2342–2351.

Fairclough, J. P. A., Norman, A. I., Shaw, B., Mark Nace, V., Heenan, R. H. 2006. *Polym. Int.* 55:793–797.

Falamarzian, A., Lavasanifar, A. 2010. *Macromol. Biosci.* 10:648–656.

Felici, M., Marza-Perez, M., Hatzakis, N. S., Nolte, R. J. M., Feiters, M. C. 2008. *Chem. Eur. J.* 14:9914–9920.

Fenimore, S. G., Abezgauz, L., Danino, D., Ho, Ch.-Ch., Co, C. C. 2009. *Macromolecules* 42:2702–2707.

Fernyhough, C., Ryan, A. J., Battaglia, G. 2009. *Soft Matter* 5:1674–1682.

Fleischer, G. 1993. *J. Phys. Chem.* 97:517–521.

Fluegel, S., Maskos, M. 2007. *Biomacromolecules* 8:700–702.

Forster, S., Hermsdorf, N., Leube, W., Schnablegger, H., Regenbrecht, M., Akari, S. 1999. *J. Phys. Chem. B* 103:6657–6668.

Forster, S., Zisenis, M., Wenz, E., Antonietti, M. 1996. *J. Chem. Phys.* 104:9956–9970.

Fraaije, J., Sevink, G. J. A. 2003. *Macromolecules* 36:7891–7893.

Gan, Y.-Ch., Yuan, J.-F., Liu, X.-J., Wang, P., Gao, Q.-Y. 2011. *J. Bioactive Comp. Polym.* 26:173–190.

Ganguly, R., Choudhury, N., Aswal, V. K., Hassan, P. A. 2009a. *J. Phys. Chem. B* 113(3):668–675.

Ganguly, R., Kumbhakar, M., Aswal, V. K. 2009b. *J. Phys. Chem. B* 113(28):9441–9446.

Gao, K. J., Li, G., Lu, X., Wu, Y. G., Xu, B. Q., Fuhrhop, J. H. 2008. *Chem. Commun.* 12:1449–1451.

Gebhardt, K. E., Ahn, S., Venkatachalam, G., Savin, D. A. 2008. *J. Colloid Interface Sci.* 317:70–76.

de Gennes, P.-G. 1980a. *Macromolecules* 13:1069–1075.

de Gennes, P.-G. 1980b. *Scaling Concepts in Polymer Physics*. Ithaca, NY: Cornell University Press.

Giacomelli, F. C., Stepanek, P., Giacomelli, C., Schmidt, V., Jaeger, E., Jaeger, A., Ulbrich, K. 2011. *Soft Matter* 7:9316–9325.

Goetz, R., Gompper, G., Lipowski, R. 1999. *Phys. Rev. Lett.* 82:221–224.

Gohy, J. F. 2005. *Adv. Polym. Sci.* 190:65–136.

Goldmint, I., Yu, G.-E., Booth, C., Smith, K. A., Hatton, T. A. 1999. *Langmuir* 15:1651–1656.

Gomez, E. D., Rappl, T. J., Agarwal, V., Bose, A., Schmutz, M., Marques, C. M., Balsra, N. P. 2005. *Macromolecules* 38:3567–3570.

Green, A. C., Zhu, J., Pochan, D., Jia, X., Kiick, K. 2011. *Macromolecules* 44:1942–1951.

Grumelard, J., Taubert, A., Meier, W. 2004. *Chem. Commun.* 13:1462–1463.

Guo, L., Colby, R. H., Thiyagarajan, P. 2006. *Physica B* 385–386:685–687.

Guo, M. Y., Jiang, M., Zhang, G. Z. 2008. *Langmuir* 24:10583–10586.

Halacheva, S., Rangelov, S., Garamus, V. 2007. *Macromolecules* 40:8015–8021.

Halacheva, S., Rangelov, S., Tsvetanov, Ch. Garamus, V. M. 2010. *Macromolecules* 43:772–781.

Hales, K., Chen, Z. Y., Wooley, K. I., Pochan, D. J. 2008. *Nano Lett.* 8:2023–2026.

Haliloglu, T., Mattice, W. L. 1996. Monte Carlo simulations of self-assembly in macromolecular systems. In: Webber, S. E., Munk, P., Tuzar, Z. (Eds.). *Solvents and Self-Organization of Polymers. NATO ASI Series, Series E: Applied Sciences*, vol. 327. Dordrecht, the Netherlands: Kluwer Academic Publisher, pp. 167–196.

Halperin, A. 1987. *Macromolecules* 20:2043–2049.

Halperin, A. 1989. *Macromolecules* 22:2403–2412.

Halperin, A., Alexander, S. 1989. *Macromolecules* 22:2403–2412.

Haluska, C. K., Gozdz, W. T., Dobereiner, H.-G., Forster, S., Gompper, G. 2002. *Phys. Rev. Lett.* 89:238302.

Hamley, I. W. 1998. *The Physics of Block Copolymers*, vol 4. Oxford, U.K.: Oxford Science Publications, Chapters 3 and 4, pp. 131–265.

Hamley, I. W. 2005. *Soft Matter* 1:36–43.

Hamley, I. W., Pedersen, J. S., Booth, C., Mark Nace, V. 2001. *Langmuir* 17:6386–6388.

Hasegawa, U., van der Viles, A. J., Simeoni, E., Wandrey, Ch., Hubbell, J. A. 2010. *J. Am. Chem. Soc.* 132:18273–18280.

Hauschild, S., Lipprandt, U., Rumplecker, A., Borchert, U., Rank, A., Schubert, R., Forster, S. 2005. *Small* 1:1177–1180.

Hayward, R. C., Pochan, D. J. 2010. *Macromolecules* 43:3577–3584.

He, Y., Zhang, Y., Xiao, Y., Lang, M. 2010. *Colloids Surf. B Biointerfaces* 80:145–154.

Hecht, E., Hoffmann, H., 1995. *Colloids Surf. A Physicochem. Eng. Aspects* 96:181–197.

Hecht, E., Mortensen, K., Hoffmann, H. 1995. *Macromolecules* 28:5465–5476.

Hentschel, J., Ten Cate, M. G. J., Börner, H. G. 2007. *Macromolecules* 40:9224–9232.

Hermann, C.-U., Kahlewelt, M. 1980. *J. Phys. Chem.* 84:1536–1540.

Hiemenz, P. C. 1986. *Principles of Colloid and Surface Chemistry*, 2nd edn. New York: Marcel Dekker.

Holder, S. J., Hiorns, R. C., Sommerdijk, N. A. J. M., Williams, S. J., Jones, R. G., Nolte, R. J. M. 1998. *Chem. Commun.* 14:1445.

Holder, S. J., Sommerdijk, N. A. J. M. 2011. *Polym. Chem.* 2:1018–1028.

Holmberg, K., Jonsson, B., Kronberg, B., Lindman, B. 2002. *Surfactants and Polymers in Aqueous Solution*. Hoboken, NJ: John Wiley & Sons, Ltd.

Holowka, E. P., Pochan, D. J., Deming, T. J. 2005. *J. Am. Chem. Soc.* 127:12423–12428.

Honda, C., Abe, Y., Nose, T. 1996. *Macromolecules* 29:6778–6785.

Honda, C., Hasegawa, Y., Hirunuma, R., Nose, T. 1994. *Macromolecules* 27:7660–7668.

Honda, S., Yamamoto, T., Tezuka, Y. 2010. *J. Am. Chem. Soc.* 132:10251–10253.

Hordyjewicz-Baran, Z., You, L. C., Smarsly, B., Sigel, R., Schlaad, H. 2007. *Macromolecules* 40:3901–3903.

Houga, C., Giermanska, J., Lecommandoux, S., Borsali, R., Taton, D., Gnanou, Y., Le Meins, J.-F. 2009. *Biomacromolecules* 10:32–40.

Howse, J. R., Jones, R. A. L., Battaglia, G., Ducker, R. E., Leggett, G. J. Tyan, A. J. 2009. *Nature Mater.* 8:507–511.

Hurter, P. N., Alexandridis, P. Hatton, T. A. 1995. Solubilization in amphiphilic copolymer solutions. In: S. D. Christian, F. F. Scamehorn, Eds. *Solubilization in Surfactant Aggregates*. New York: Marcel Dekker, pp. 192–235.

Hurter, P. N., Scheutjens, J. M. H. M., Hatton, T. A. 1993. *Macromolecules* 26:5592–5601.

Hyde, S. T. 1990. *J. Phys. (Paris)* 51(C7):209–228.

Iatrou, H., Frielinghaus, H., Hanski, S., Ferderigos, N., Ruokolaninen, J., Ikkala, O., Richter, D., Mays, J., Hadjichristidis, N. 2007. *Biomacromolecules* 8:2173–2181.

Ieong, N. S., Redhead, M., Bosquillon, C., Alexander, C., Kelland, M., O'Reilly, R. K. 2011. *Macromolecules* 44:886–893.

Israelachvili, J. N., Mitchell, D. J., Ninham, B. W. 1976. *J. Chem. Soc. Faraday Trans. 2 Mol. Chem. Phys.* 72:1525–1568.

Jain, J. P., Kumar, N. 2010a. *Biomacromolecules* 11:1027–1035.

Jain, J. P., Kumar, N. 2010b. *Eur. J. Pharm. Sci.* 40:456–465.

Jain, S., Bates, F. S. 2004. *Macromolecules* 37:1511–1523.

Jaing, J., Hua, D., Jiang, J., Tang, J., Zhu, X. 2010. *Carbohyd. Polym.* 81:358–364.

Jeon, H. J., Youk, J. H. 2010. *Macromolecules* 43:2184–2189.

Jeong, J. H., Park, T. G. 2001. *Bioconjug. Chem.* 12:917–923.

Jeong, Y.-L., Kim, D. H., Chung, Ch.-W., Yoo, J.-J., Choi, K. H., Kim, C. H., Ha, S. H., Kang, D. H. 2011. *Int. J. Nanomed.* 6:1415–1427.

Jiang, J., Shu, Q., Chen, X., Yang, Y., Yi, Ch., Song, X., Liu, X., Chen, M. 2010. *Langmuir* 26:14247–14254.

Jiang, S., Yao, Y., Nie, Y., Yang, J., Yang, J. 2011. *J. Colloid Interface Sci.* 364:264–271.

Jiao, Zh., Wang, X., Chen, Zh., 2011. *Drug Deliver* 18:478–484.

Jin, Ch., Qian, N., Zhao, W., Yang, W., Bai, L., Wu, H., Wang, M., Song, W., Dou, K. 2010. *Biomacromolecules* 11:2422–2431.

Jin, Q., Mitschang, F., Agarwal, S. 2011. *Biomacromolecules* 12:3684–3691.
Jo, Y. S., Van der Vlies, A. J., Gantz, J., Antonijevic, S., Demurtas, D., Velluto, D., Hubbell, J. A. 2008. *Macromolecules* 41:1140–1150.
Johnsson, M., Hansson, P., Edwards, K. 2001. *J. Phys. Chem.* 105:8420–8430.
Jun, Y. J., Park, M. K., Jadhav, V. B., Song, J. H., Chae, S. W., Lee, H. J., Park, K. S., Jeong, B., Choy, J. H., Sohn, Y. S. 2010. *J. Control. Release* 142:132–137.
Kahlweit, M. 1982. *J. Colloid Interface Sci.* 90:92–99.
Kahlwelt, M., Teubner, M. 1980. *Adv. Colloid Interface Sci.* 13:1–64.
Kelarakis, A., Castelletto, V., Krysmann, M. J., Havredaki, V., Viras, K., Hamley. I. W. 2008. *Langmuir* 24:3767–3772.
Kelarakis, A., Mai, Sh.-M., Havredaki, V., Mark Nace, V., Booth, C. 2001. *Phys. Chem. Chem. Phys.* 3:4037–4043.
Khanna, K., Varshney, S., Kakkar, A. 2010. *Macromolecules* 43:5688–5698.
Kickelbick, G., Bauer, J., Husing, N., Andersson, M., Palmqvist, A. 2003. *Langmuir* 19:3198–3201.
Kim, H. R., Kim, I. K., Bae, K. H., Lee, S. H., Lee, Y., Park, T. G. 2008b. *Mol. Pharm.* 5:622–631.
Kim, J.-K., Lee, E., Huang, Zh., Lee, M. 2006. *J. Am. Chem. Soc.* 128:14022–14023.
Kim, K. H., Kim, S. H., Huh, J., Jo, W. H. 2002. *ACS Polym. Prepr. (Div. Polym. Chem.)* 43:438–439.
Kim, K. H., Kim, S. H., Huh, J., Jo, W. H. 2003. *J. Chem. Phys.* 119:5705–5711.
Kim, M. S., Lee, D. S. 2010. *Chem. Commun.* 46:4481–4483.
Kim, S. H., Jeong, J. H., Lee, S. H., Kim, S. W. Park, T G. 2008a. *J. Control. Release* 129:107–116.
Kim, S. H., Jo, W. H. 2001. *Macromolecules* 34:7210–7218.
Kimura, S., Kim, D. H., Sugiyama, J., Imanishi, Y. 1999. *Langmuir* 15:4461–4463.
Kong, W., Li, B., Jin, Q., Shi, A.-Ch. 2009. *J. Am. Chem. Soc.* 131:8503–8512.
Kowalczuk, A., Trzebicka, B., Rangelov, S., Smet, M., Dworak, A. 2011. *J. Polym. Sci. A—Polym. Chem.* 49:5074–5086.
Kratz, K., Breitenkamp, K., Hule, R., Pochan, D., Emrick, T. 2009. *Macromolecules* 42:3227–3229.
Kubowicz, S., Baussard, J. F., Lutz, J. F., Thunemann, A. F., von Berlepsch, H., Laschewsky, A. 2005. *Angew. Chem. Int. Ed.* 44:5262–5265.
Kukula, H., Schlaad, H., Antonietti, M., Forster, S. 2002. *J. Am. Chem. Soc.* 124:1658–1663.
Kumar, M., Grzelakowski, M., Zilles, J., Clark, M., Meier, W. 2007. *Proc. Natl. Acad. Sci. USA* 104:20719–20724.
Kyeremateng, S. O., Henze, Th., Busse, K., Kressler, J. 2010. *Macromolecules* 43:2502–2511.
Lambermont-Thijs, H. M. L., Heuts, J. P. A., Hoeppener, S., Hoogenboom, R., Schubert, U. S. 2011. *Polym. Chem.* 2:313–322.
Larsson, K. 1989. *J. Phys. Chem.* 93:7304–7314.
Larsson, K. 1997. *Chem. Phys. Lipids* 88:15–20.
Lawrence, J. M. 1994. *Chem. Soc. Rev.* 23:417–424.
Lee, A. H., Oh, K. T., Baik, H. J., Lee, B. R., Oh, Y. T., Lee, D. H., Lee, E. S. 2010b. *Bull. Korean Chem. Soc.* 31(8):2265–2271.
Lee, H. J., Yang, S. R., An, E. J., Kim, J.-D. 2006. *Macromolecules* 39:4938–4940.
Lee, J. C. M., Santore, M., Bates, F. S., Discher, D. E. 2002. *Macromolecules* 35:323–326.
Lee, R.-Sh., Chen, W.-H. 2011. *Polym. Int.* 60:255–263.
Lee, R.-Sh., Chen, W.-H., Huang, Y.-T. 2010a. *Polymer* 51:5942–5951.
Lee, R.-Sh., Wu, K.-P. 2011. *J. Polym. Sci. A—Polym. Chem.* 49:3163–3173.
Lee, S. H., Bae, K. H., Kim, S. H., Lee, K. R., Park, T. G. 2008. *Int. J. Pharm.* 364:94–101.
Lee, S. H., Kim, S. H., Park, T. G. 2007. *Biochem. Biophys. Res. Commun.* 357:511–516.

Lee, S. J., Park, M. J. 2010. *Langmuir* 26:17827–17830.
Lee, Y., Park, T. G. 2011. *J. Control. Release* 152:152–158.
Leibler, L., Orland, H., Wheeler, J. C. 1983. *J. Chem. Phys.* 79:3550–3557.
Le Meins, J.-F., Sandre, O., Lecommandoux, S. 2011. *Eur. Phys. J. E* 34:14.
Leon, O., Munoz-Bonilla, A., Bordege, V., Sanchez-Chavez, M., Fernandez-Garcia, M. 2011. *J. Polym. Sci. A—Polym. Chem.* 49:2627–2635.
Leson, A., Filiz, V., Forster, S., Mayer, C. 2007a. *Chem. Phys. Lett.* 444:268–272.
Leson, A., Hauschild, S., Rank, A., Neub, A., Schubert, R., Forster, S., Mayer, C. 2007b. *Small* 3:1074–1083.
Lessner, E., Teubner, M., Kahlweit, M. 1981. *J. Phys. Chem.* 85:1529–1536.
Levine, D. H., Ghoroghchian, P. P., Freudenberg, J., Zhang, G., Therien, M. J., Greene, M. I., Hammer, D. A., Murali, R. 2008. *Methods* 46:25–32.
Li, C., Gu, Ch., Zhang, Y., Lang, M. 2011a. *Polymer Bull.* 68:69–83.
Li, F., Ketelaar, T., Cohen Stuart, M. A., Sudholter, E. J. R., Leermakers, F. A. M. 2008. *Langmuir* 24:76–82.
Li, H.-P., Ma, B.-G., Zhou, S.-M., Zhang, L.-M., Yi, J.-Zh. 2011d. *Colloid Polym. Sci.* 288(14–15):1419–1426.
Li, J., Ren, J., Cao, Y., Yuan, W. 2010a. *Polymer* 51:1301–1310.
Li, M.-H., Keller, P. 2009. *Soft Matter* 5:927–937.
Li, T., Jing, X., Huang, Y. 2011c. *Polym. Adv. Techn.* 22:12266–12271.
Li, X., Ji, J., Shen, J. 2006b. *Macromol. Rapid Commun.* 27:214–218.
Li, X., Ji, J., Wang, X., Wang, Y., Shen, J. 2007. *Macromol. Rapid Commun.* 28:660–665.
Li, Y., Heo, H. J., Gao, G. H., Kang, S. W., Huynh, C T., Kim, M. S., Lee, J. W., Lee, J. H., Lee, D. S. 2011b. *Polymer* 52:3304–3310.
Li, Z., Dormidontova, E. E. 2010. *Macromolecules* 43:3521–3531.
Li, Z., Zhang, Y., Fullhart, P., Mirkin, C. A. 2004. *Nano Lett.* 4:1055–1058.
Li, Z. C., Shen, Y., Liang, Y. Z., Li, F. M. 2001. *Chin. J. Polym. Sci.* 19:297–302.
Li, Z. L., Xiong, X. Y., Li, Y. P., Gong, Y. Ch., Gui, X. X., Ou-Yang, X., Lin, H. Sh., Zhu, L. J., Xie, J. L. 2010b. *J. Appl. Polym. Sci.* 115:1573–1580.
Li, Zh., Chen, Zh., Cui, H., Hales, K., Qi, K., Wooley, K. L., Pochan, D. 2005. *Langmuir* 21: 7533–7539.
Li, Zh., Hillmyer, M. A., Lodge, T. P. 2006a. *Langmuir* 22:9409–9417.
Libera, M., Walach, W., Trzebicka, B., Rangelov, S., Dworak, A. 2011. *Polymer* 52:3526–3536.
Liccardi, M., Cavallaro, G., Di Stefano, M., Pitarresi, G., Fiorica, C., Giammona, G. 2010. *Int. J. Pharmaceut.* 396(1–2):219–228.
Lin, S.-T., Fuchise, K., Chen, Y., Sakai, R., Satoh, T., Kakuchi, T., Chen, W.-C. 2009a. *Soft Matter* 5 (19):3761–3770.
Lin, Sh., He, X., Li, Y., Lin, J., Nose, T. 2009b. *J. Phys. Chem. B* 113:13926–13934.
Lindblom, G., Rilfors, L. 1989. *BBA-Rev. Biomembranes* 988:221–256.
Linse, P. 1993. *J. Phys. Chem.* 97:13896–13902.
Linse, P. 2000. Modeling of self-assembly of block copolymers in selective solvent. In: P. Alexandridis, B. Lindman, Eds. *Amphiphilic Block Copolymer Self-Assembly and Applications*. Amsterdam, the Netherlands: Elsevier, pp. 13–40.
Liu, Ch., Hillmyer, M. A., Lodge, T. P. 2008. *Langmuir* 24:12001–12009.
Liu, Ch., Hillmyer, M. A., Lodge, T. P. 2009. *Langmuir* 25:13718–13725.
Liu, F., Eisenberg, A. 2003. *J. Am. Chem. Soc.* 125:15059–15064.
Liu, G. 1995. *Can. J. Chem.* 73:1995–2003.
Liu, G., Lv, P., Chen, Ch., Hu, X., Ji, J. 2011a. *Macromol. Chem. Phys.* 212:643–651.
Liu, G., Ma, Sh., Li, Sh., Ru, Ch., Meng, F., Liu, H., Zhong, Zh. 2010. *Biomaterials* 31:7575–7585.
Liu, J., Pang, H., Zhu, Zh., Zhu, X., Zhou, Y., Yan, D. 2011b. *Biomacromolecules* 12:2407–2415.
Liu, R., Tao, Y., Zhu, Y. Chen, M., Yang, Ch., Liu, X. 2011c. *Polym. Int.* 60:327–332.

Lodge, T. P., Dalvi, M. C. 1995. *Phys. Rev. Lett.* 75:657–660.
Lodge, T. P., Hillmyer, M. A., Zhou, Zh. 2004. *Macromolecules* 37:6680–6682.
Lonetti, B., Tsigkri, A., Lang, P. R., Stellbrink, J., Willner, L., Kohlbrecher, J., Lettinga, M. P. 2011. *Macromolecules* 44(9):3583–3593.
Lonsdale, D. E., Monteiro, M. J. 2011. *J. Polym. Sci. A—Polym. Chem.* 49:4603–4612.
LoPresti, C., Lomas, H., Massignani, M., Smart, T., Battaglia, G. 2009. *J. Mater. Chem.* 19:3576–3590.
Lorenceau, E., Utada, A. S., Link, D. R., Cristobal, G., Joanicot, M., Weitz, D. A. 2005. *Langmuir* 21:9183–9186.
Love, J. J., Li, X., Case, D. A., Giese, K., Grosschedl, R., Wright, P. E. 1995. *Nature* 376:791.
Lu, Ch., Liu, L., Guo, Sh., Zhang, Y., Li, Z., Gu, J. 2007. *Eur. Polym. J.* 43:1857–1865.
Lund, R., Willner, L., Monkenbusch, M., Panine, P., Narayanan, T., Colmenero, J., Richter, D. 2009. *Phys. Rev. Lett.* 102:188301.
Lund, R., Willner, L., Stellbrink, J., Lindner, P., Richter, D. 2006. *Phys. Rev. Lett.* 96:68302.
Luo, L., Eisenberg, A. 2001a. *J. Am. Chem. Soc.* 123:1012–1013.
Luo, L., Eisenberg, A. 2001b. *Langmuir* 17:6804–6811.
Luo, Ch., Liu, Y., Li, Zh. 2010. *Macromolecules* 43:8101–8108.
Lutz, J.-F., Börner, H. G., Weichenhan, K. 2006. *Macromolecules* 39:6376–6383.
Lutz, J. F., Laschewsky, A. 2005. *Macromol. Chem. Phys.* 206:813–817.
Ma, L., Eisenberg, A. 2009. *Langmuir* 25:13730–13736.
Mabrouk, E., Cuvelier, D., Pontani, L.-L., Xu, B., Levy, D., Keller, P., Brochard-Wyart, F., Nassoy, P., Li, M.-H. 2009. *Soft Matter* 5:1870–1878.
Makino, A., Yamahara, R., Ozeki, E., Kimura, Sh. 2007. *Chem. Lett.* 36:1220–1222.
Malinova, V., Belegrinou, S., De Bruyn Obtoter, D., Meier, W. 2010. *Adv. Polym. Sci.* 224:113–165.
Malmsten, M., Linse, P., Zhang, K. W. 1993. *Macromolecules* 26:2905–2910.
Marsat, J.-N., Heydenreich, M., Kleinpeter, E., Berlepsch, H. V., Boettcher, Ch., Laschewsky, A. 2011. *Macromolecules* 44:2092–2105.
Marsden, H. R., Gabrielli, L., Kros, A. 2010. *Polym. Chem.* 1:1512–1518.
Massignani, M., Lomas, H., Battaglia, G. 2010. *Adv. Polym. Sci.* 229:115–154.
Matsen, M. W., Bates, F. S. 1995. *Macromolecules* 28:8884–8886.
McKenzie, B. E., Nudelman, F., Bomans, P. H. H., Holder, S. J., Sommerdijik, N. 2010. *J. Am. Chem. Soc.* 132:10256–10259.
Meng, F., Zhong, Zh., Feijen, *J. Biomacromol.* 10:197–209.
Menon, S., Thekkayil, R., Varghese, Sh., Das, S. 2011. *J. Polym. Sci. A—Polym. Chem.* 49:5063–5073.
Michels, B., Waton, G., Zana, R. 1997. *Langmuir* 13:3111–3118.
Milchev, A., Bhattacharya, A., Binder, K. 2001. *Macromolecules* 34:1881–1893.
Milchev, A., Binder, K. 1999. *Langmuir* 15:3232–3241.
Mishra, A. K., Patel, V. K., Vishwakarma, N. K., Biswas, Ch., S. Raula, M., Misra, A., Mandal, T. K., Raj, B. 2011. *Macromolecules* 44:2465–2473.
Monzen, M., Kawakatsu, T., Doi, M., Hasegawa, R. 2000. *Comput. Theoret. Polym. Sci.* 10:275–280.
Mori, T., Maeda, M. 2002. *Polym. J.* 34:624–628.
Mori, T., Maeda, M. 2004. *Langmuir* 20:313–319.
Mortensen, K. 2001a. *Polym. Adv. Technol.* 12:2–22.
Mortensen, K. 2001b. *Colloids Surf. A Physicochem. Eng. Aspects* 183–185: 277–292.
Mortensen, K., Brown, W. 1993. *Macromolecules* 26:4128–4135.
Mortensen, K., Pedersen, J. S. 1993. *Macromolecules* 26:805–812.
Mortensen, K., Talmon, Y. 1995. *Micromolecules* 28:8829–8834.
Nabid, M. R., Rezaei, S. J. T., Sedghi, R., Niknejad, H., Entezami, A. A., Oskooie, H. A., Heravi, M. 2011. *Polymer* 52:2799–2809.

Nagarajan, R., Ganesh, K. 1989. *Macromolecules* 22:4312–4325.
Naik, S., Ray, J. G., Savin, D. 2011. *Langmuir* 27:7231–7240.
Nardin, C., Hirt, T., Leukel, J., Meier, W. 2000. *Langmuir* 16:1035–1041.
Netz, R. R. 1999. *Europhys. Lett.* 46:391–398.
Nguyen-Van, C., Hsieh, M.-F., Chen, Y.-Ts., Liau, I. 2011. *J. Biomater. Sci. Polym. Ed.* 22:1409–1426.
Nivaggioli, T., Tsao, B., Alexandridis, P., Hatton, T. A. 1995. *Langmuir* 11:119–126.
Noolandi, J., Hong, K. M. 1983. *Macromolecules* 16:1443–1448.
Noolandi, J., Shi, A.-C., Linse, P. 1996. *Macromolecules* 29:5907–5919.
Norman, A. I., Ho, D. L., Karim, A., Amis, E. J. 2005. *J. Colloid Interface Sci.* 288:155–165.
Nyrkova, I. A., Khokhlov, A. R., Doi, M. 1993. *Macromolecules* 26:3601–3610.
Nyrkova, I. A., Semenov, A. N. 2005. *Macromol. Theory Simul.* 14:569–585.
Oishi, M., Nagasaki, Y., Itaka, K., Nishiyama, N., Kataoka, K. 2005. *J. Am. Chem. Soc.* 127:1624–1625.
Onaca, O., Enea, R., Hughes, D. W., Meier, W. 2009. *Macromol. Biosci.* 9:129–139.
Opsteen, J. A., Cornelissen, J. J. L. M., van Hest, J. C. M. 2004. *Pure Appl. Chem.* 67:4–14.
Otsuka, I., Fuchise, K., Halila, S., Fort, S., Aissou, K., Pignot-Paintrand, I., Chen, Y., Narumi, A., Kakuchi, T., Borsali, R. 2010. *Langmuir* 26:2325–2332.
Pan, P., Fujita, M., Ooi, W.-Y., Sudesh, K., Takarada, T., Goto, A., Maeda, M. 2011. *Polymer* 52:895–900.
Park, S. H., Choi, B. G., Moon, H. J., Cho, S.-H., Jeong, B. 2011. *Soft Matter* 7:6515–6521.
Parry, A. L., Bomans, P. H. H., Holder, S. J., Sommerdijik, N., Biagini, S. C. G. 2008. *Angew. Chem. Int. Ed.* 47:8859–8862.
Pasparakis, G., Alexander, C. 2008. *Angew. Chem. Int. Ed.* 47:4847–4850.
Patton, J. S., Carey, M. C. 1979. *Science* 204:145–148.
Patton, J. S., Vetter, R. D., Hamosh, B., Borgstrom, B., Lindstron, M., Carey, M. C. 1985. *Food Microstruct.* 4:29–42.
Pautot, S., Frisken, B. J., Weitz, D. A. 2003. *Langmuir* 19:2870–2879.
Peng, S.-M., Chen, Y., Hua, C., Dong, C.-M. 2009. *Macromolecules* 42(1):104–113.
Peng, Ch.-L., Yang, L.-Y., Luo, Ts.-Y., Lai, P.-Sh., Yang, Sh.-J., Lin, W.-J., Shieh, M.-J. 2010. *Nanotechnology* 21:155103.
Perreur, C., Habas, J., Peyrelasse, J., Francois, J. 2001. *Phys. Rev. E* 63:031505.
Petrov, P., Yuan, J., Yoncheva, K., Mueller, A. H. E., Tsvetanov, Ch. 2008. *J. Phys. Chem. B* 112(30):8879–8883.
Pispas, S., Hadjichristidis, N., Potemkin, I., Khokhlov, A. 2000. *Macromolecules* 33:1741.
Qiao, P., Niu, Q., Wang, Zh., Cao, D. 2010. *Chem. Eng. J.* 159:257–263.
Rager, T., Meyer, W. H., Wegner, C., 1999. *Macromol. Chem. Phys.* 200:1672–1680.
Rager, T., Meyer, W. H., Wegner, C., Winnik, M. A. 1997. *Macromolecules* 30:4911–4919.
Rajagopal, K., Mahmud, A., Christian, D. A., Pajerowski, J. D., Brown, A. E. X., Loverde, S. M., Discher, D. E. 2010. *Macromolecules* 43:9736–9746.
Ramzi, A., Prager, M., Richter, D., Efstratiadis, V., Hadjichristidis, N., Young, R. N., Allgaier, J. B. 1997. *Macromolecules* 30:717.
Rangelov, S., Almgren, M., Edwards, K., Tsvetanov, Ch. 2004. *J. Phys. Chem. B* 108:7542–7552.
Rangelov, S., Almgren, M., Halacheva, S., Tsvetanov, Ch. 2007. *J. Phys. Chem. C* 111:13185–13191.
Rangelov, S., Almgren, M., Tsvetanov, Ch., Edwards, K., 2002a. *Macromolecules* 35:4770–4778.
Rangelov, S., Almgren, M., Tsvetanov, Ch., Edwards, K., 2002b. *Macromolecules* 35:7074–7081.
Rangelov, S., Halacheva, S., Garamus, V. M., Almgren, M. 2008. *Macromolecules* 41:8885–8894.
Rao, J., Zhang, J., Xu, J., Liu, S. 2008. *J. Colloid Interface Sci.* 328:196–202.

Rawicz, W., Olbrich, K. C., McIntosh, T., Needham, D., Evams, E. 2000. *Biophys. J.* 79:328–340.
Richter, A., Olbrich, C., Krause, M., Kissel, T. 2010. *Int. J. Pharm.* 389:244–253.
Riess, G. 2003. *Progr. Polym. Sci.* 28:1107–1170.
Riess, G., Hurtrez, G. 1996. Block copolymers: synthesis, colloidal properties and application possibilities of micellar systems. In: Webber, S. E., Munk, P., Tuzar, Z. (Eds.) *Solvents and Self-Organization of Polymers. NATO ASI Series, Series E: Applied Sciences*, vol. 327. Dordrecht, the Netherlands: Kluwer Academic Publisher, pp. 33–51.
de Rodrigues Jr., A., Nascimento, R. S. V. 2010. *J. Appl. Polym. Sci.* 116(5):3047–3055.
Rodriguez-Hernandez, J., Lecommandoux, S. 2005. *J. Am. Chem. Soc.* 127:2026–2027.
Saito, N., Liu, Ch., Lodge, T. P., Hillmyer, M. A. 2010. *ACS Nano* 4(4):1907–1912.
Sanson, C., Le Meins, J.-F., Schatz, C., Soum, A., Lecommandoux, S. 2010a. *Soft Matter* 6:1722–1730.
Sanson, C., Schatz, C., Le Meins, J.-F., Brulet, A., Soum, A., Lecommandoux, S. 2010b. *Langmuir* 26:2751–2760.
Schatz, C., Louguet, S., Le Meins, J.-F., Lecommandoux, S. 2009. *Angew. Chem. Int. Ed.* 48:2572–2575.
Schillen, K., Bryskhe, K., Mel'nikova, Y. S. 1999. *Macromolecules* 32:6885–6888.
Schillen, K., Yekta, A., Ni, S., Winnik, M. A. 1998. *Macromolecules* 31:210–212.
von Schnering, H. G., Nesper, R. 1991. *Z. Phys. B Condensed Matter* 83:407–412.
Scriven, L. E. 1976. *Nature* 263:123–125.
Semenov, A. N., Nyrkova, I. A., Khokhlov, A. R. 1995. *Macromolecules* 28:7491–7500.
Shen, H., Eisenberg, A. 1999. *J. Phys. Chem. B* 103:9473–9487.
Shen, H., Eisenberg, A. 2000a. *Angew. Chem. Int. Ed.* 39:3310–3312.
Shen, H., Eisenberg, A. 2000b. *Macromolecules* 33:2561–2572.
Shi, Z. Q., Zhou, Y. F., Yan, D. Y. 2008. *Macromol. Rapid Commun.* 29:412–418.
Shum, H. Ch., Kim, J.-W., Weitz, D. A. 2008. *J. Am. Chem. Soc.* 130:9543–9549.
Shusharina, N. P., Liinse, P., Khokhlov, A. R. 2000. *Macromolecules* 33:8488–8496.
Sigel, R., Losik, M., Schlaad, H. 2007. *Langmuir* 23:7196–7199.
Smart, T., Lomas, H., Massignani, M., Flores-Merino, M. V., Ruiz Perez, L., Battaglia, G. 2008. *Nano Today* 3:38–46.
Smart, T., Ryan, A. J., Howse, J. R., Battaglia, J. 2010. *Langmuir* 26:7425–7430.
Smith, A. E., Xu, X., Kirkland-York, S. E., Savin, D. A., McCormick, Ch. L. 2010. *Macromolecules* 43:1210–1217.
Soliman, G. M., Sharma, R., Choi, A. O., Varshney, S. K., Winnik, F. M., Kakkar, A. K., Maysinger, D. 2010. *Biomaterials* 31:8382–8392.
Sommerdijk, N. A. J. M., Holder, S. J., Hiorns, R. C., Jones, R. G., Nolte, R. J. M. 2000. *Macromolecules* 33:8289–8294.
Soni, S. S., Sastry, N. V., Patra, A. K., Joshi, J. V., Goyal, P. S. 2002. *J. Phys. Chem. B* 106:13069–13077.
Sotiriou, K., Nannou, A., Velis, G., Pispas, S. 2002. *Macromolecules* 35:4106.
Srinivas, G., Discher, D. E., Klein, M. L. 2004. *Nat. Mater.* 3:638–644.
Stahler, K., Selb, J., Candau, F. 1999. *Langmuir* 15:7565–7576.
Stoenescu, R., Graff, A., Meier, W. 2004. *Macromol. Biosci.* 4:930–935.
Stoenescu, R., Meier, W. 2002. *Chem. Commun.* 24:3016–3017.
Strandman, S., Zarembo, A., Darinskii, A. A., Loflund, B., Butcher, S. J., Tenhu, H. 2007. *Polymer* 48(24):7008–7016.
Sun, L., Du. Zh., Wang, W., Liu, Y. 2011. *J. Surf. Deterg.* 14:161–166.
Sun, L., Shen, L.-J., Zhu, M.-Q., Dong, Ch.-M., Wei, Y. 2010a. *J. Polym. Sci. A—Polym. Chem.* 48:4583–4593.
Sun, Y., Yan, X., Yuan, T., Liang, J., Fan, Y., Gu, Zh., Zhang, X. 2010b. *Biomaterials* 31:7124–7131.

Suo, A., Qian, J., Yao, Y., Zhang, W. 2010. *Int. J. Nanomed.* 5:1029–1038.
Svensson, M., Alexandridis, P., Linse, P. 1999. *Macromolecules* 32:5435–5443.
Talelli, M., Rijcken, C. J. F., van Nostrum, C. F., Storm, G., Hennink, W. E. 2010. *Adv. Drug Deliver Rev.* 62:231–239.
Teixeira Jr., F., Rigler, P., Vebert-Nardin, C. 2007. *Chem. Commun.* 11:1130–1132.
Terreau, O., Bartels, C., Eisenberg, A. 2004. *Langmuir* 20:637–645.
Terreau, O., Luo, L., Eisenberg, A. 2003. *Langmuir* 19:5601–5607.
Thambi, Th., Deepagan, V. G., Yoo, Ch. K., Park, J. H. 2011. *Polymer* 52:4753–4759.
Thunemann, A. F., Kubowicz, S., von Berlepsch, H., Mohwald, H. 2006. *Langmuir* 22:2506–2510.
Tian, L., Nguyen, Ph., Hammond, P. T. 2006. *Chem. Commun.* 33:3489–3491.
Trzcinska, R., Szweda, D., Rangelov, S., Suder, P., Silberring, J., Dworak, A., Trzebicka, B. 2012. *J. Polym. Sci. A—Polym. Chem.* 50(15):3104–3115.
Tsitsilianis, C., Roiter, Y., Katsampas, I., Minko, S. 2008. *Macromolecules* 41:925–934.
Tu, S., Chen, Y.-W., Qiu, Y.-B., Zhu, K., Luo, X.-L. 2011. *Macromol. Biosci.* 11:1416–1425.
Tuzar, Z., Kratochvil, P. 1993. Micelles of block and graft copolymers in solution. In: E. Matijevic, Ed. *Surface and Colloid Science*, vol. 15. New York: Plenum Press, Chapter 1, pp. 1–83.
Tzokova, N., Fernyhough, C. M., Butler, M. F., Armes, S. P., Ryan, A. J., Topham, P. D., Adams, D. J. 2009. *Langmuir* 25(18):11082–11089.
Uneyama, T. 2007. *J. Chem. Phys.* 126:114902.
Upadhyay, K. K., Le Meins, J.-F., Misra, A., Voisin, P., Bouchaud, V., Ibarboure, E., Schatz, C., Lecommandoux, S. 2009. *Biomacromolecules* 10:2802–2808.
Victorov, A. I., Plotnikov, N. V., Hong, P.-D. 2010. *J. Phys. Chem. B* 114:8846–8860.
Vriezema, D. M., Hoogboom, J., Velonia, K., Takazawa, K., Christianen, P. C. M., Maan, J. C., Rowan, A. E., Nolte, R. J. M. 2003. *Angew. Chem. Int. Ed.* 42:772–776.
Vriezema, D. M., Kros, A., de Gelder, R., Cornelissen, J. J. L. M., Rowan, A. E., Nolte, R. J. M. 2004. *Macromolecules* 37:4736–4739.
Vukovic, L., Khatib, F. A., Drake, S. P., Madriaga, A., Brandenburg, K. S., Kral, P., Onyuksel, H. 2011. *J. Am. Chem. Soc.* 133:13481–13488.
Walther, A., Millard, P.-E., Goldmann, A. S., Lovestead, T. M., Schacher, F., Barner-Kowollik, C., Mueller, A. H. E. 2008. *Macromolecules* 41(22):8608–8619.
Wan, X., Liu, T., Liu, Sh. 2011. *Biomacromolecules* 12:1146–1154.
Wang, H., Han, S., Sun, J., Fan, T., Tian, C., Wu, Y. 2011b. *Carbohyd. Polym.* 83(3):1408–1413.
Wang, R., Chen, G.-T., Du, F.-S., Li, Z.-C. 2011a. *Colloids Surf. B Biointerfaces* 85(1):56–62.
Wang, T., Xu, Q., Wu, Y., Zeng, A., Li, M., Gao, H. 2010b. *Carbohyd. Polym.* 80(1):303–307.
Wang, Y., Kausch, C. M., Chun, M., Quirk, R. P., Mattice, W. L. 1995. *Macromolecules* 28:904–911.
Wang, Y., Wang, Ch., Fu, Sh., Liu, Q., Dou, D., Lv, H., Fan, M., Guo, G., Luo, F., Qian, Zh. 2011c. *Int. J. Pharm.* 407:184–189.
Wang, W., Liu, R., Li, Zh., Meng, Ch., Wu, Q., Zhu, F. 2010a. *Macromol. Chem. Phys.* 211:1452–1459.
Wanka, G., Hoffmann, H., Ulbricht, W. 1994. *Macromolecules* 27:4145–4159.
Weber, S. E. 1996. Use of fluorescence methods to characterize the interior of polymer micelles. In: Weber, S. E., Munk, P., Tuzar, Z. (Eds.) *Solvents and Self-Organization of Polymers. NATO ASI Series, Series E: Applied Sciences*, vol. 327. Dordrecht, the Netherlands: Kluwer Academic Publisher, pp. 457–78.
Wittemann, A., Azzam, T., Eisenberg, A. 2007. *Langmuir* 23:2224–2230.
Wolf, F. K., Hofmann, A. M., Frey, H. 2010. *Macromolecules* 43:3314–3324.
Won, Y.-Y., Brannan, A. K., Davis, H. T., Bates, F. S. 2002. *J. Phys. Chem. B* 106:3354–3364.
Won, Y.-Y., Ted Davis, H., Bates, F. S. 2003. *Macromolecules* 36:953–955.
Wu, D., Abezgauz, L., Danino, D., Ho, C.-C., Co, C.-C. 2008. *Soft Matter* 4:1066–1071.

Wu, Q., Wang, Ch., Zhang, D., Song, X., Verpoort, F., Zhang, G. 2011b. *Reactive Func. Polym.* 9:980–4.
Wu, Q. H., Liang, F., Wei, T. Z., Song, X. M., Liu, D. L., Zhang, G. L. 2010. *J. Polym. Res.* 17:183–190.
Wu, Y., Wang, T., Li, M., Fan, T., Gao, H., Wu, X. 2011a. *J. Polym. Res.* 18:1147–1158.
Xing, J., Deng, L., Xie, Ch., Xiao, L., Zhai, Y., Jin, F., Li, Y., Dong, A. 2011. *Polym. Adv. Technol.* 22:669–674.
Xiong, Sh., Li, L., Jiang, J., Tong, L., Wu, Sh., Xu, Z., Chu, P. 2010. *Biomaterials* 31:2673–2685.
Xu, H., Meng, F., Zhong, Z. 2009. *J. Mater. Chem.* 19:4183–4190.
Xu, L., Shao, L., Chen, L., Hu, M., Bi, Y. 2010. *Chem. Lett.* 39:1177–1179.
Xu, J., Tao, L., Boyer, C., Lowe, A. B., Davis, T. P. 2011. *Macromolecules* 44:299–312.
Yan, Y., Hoffmann, H., Leson, A., Mayer, C. 2007. *J. Phys. Chem. B* 111:6161–6166.
Yang, Ch., Zhong, K.-L., Wang, Q., Chen, T., Jin, L. Y. 2011b. *Fiber. Polym.* 12:983–988.
Yang, J., Levy, D., Deng, W., Keller, P., Li, M.-H. 2005. *Chem. Commun.* 34:4345–4347.
Yang, J., Pinol, R., Gubellini, F., Levy, D., Albouy, P.-A., Keller, P., Li, M.-H. 2006. *Langmuir* 22: 7907–7911.
Yang, J., Zhang, D., Jiang, Sh., Yang, J., Nie, J. 2010. *J. Colloid Interface Sci.* 352(2):405–414.
Yang, Sh., Zhang, X., Yuan, Sh. 2008. *J. Mol. Model.* 14:607–620.
Yang, Y., Hua, Ch., Dong, Ch.-M. 2009. *Biomacromolecules* 10:2310–2318.
Yang, Y. Q., Zheng, L. Sh., Guo, X. D., Qian, Y., Zhang, L. J. 2011a. *Biomacromolecules* 12:116–122.
Yin, H., Kang, S.-W., Bae, Y. H. 2009. *Macromolecules* 42:7456–7464.
Yin, L., Hillmyer, M. A. 2011. *Macromolecules* 44:3021–3028.
Yu, K., Bartels, C., Eisenberg, A. 1999. *Langmuir* 15:7157–7167.
Yu, K., Eisenberg, A. 1998. *Macromolecules* 31:3509–3518.
Yu, K., Zhang, L. F., Eisenberg, A. 1996. *Langmuir* 12:5980–5984.
Yu, G.-E., Li, H., Fairclough, J. P. A., Ryan, A. J., McKeown, N., Ali-Adib, Z., Price, C., Booth, C. 1998. *Langmuir* 14:5782–5789.
Yu, H., Jiang, W. 2009. *Macromolecules* 42:3399–3404.
Yu, Y., Zhang, L., Eisenberg, A. 1997. *Langmuir* 13:2578–2581.
Yuan, Y.-Y., Du, Q., Wang, Y.-C., Wang, J. 2010. *Macromolecules* 43:1739–1746.
Yuan, Y.-Y., Wang, J. 2011. *Colloids Surf. B Biointerfaces* 85:81–85.
Yuan, W., Zhang, J., Wei, J., Yuan, H., Ren, J. 2011. *J. Polym. Sci. A—Polym. Chem.* 49:4071–4080.
Yun, J., Faust, R., Szilagyi, L. S., Keki, S., Zsuga, M. 2003. *Macromolecules* 36:1717.
Zana, R. 2005. *Dynamics of Surfactant Self-Assemblies: Micelles, Microemulsions, Vesicles, and Lyotropic Phases*. Boca Raton, FL: CRC Press.
Zhai, X., Huang, W., Liu, J., Pang, Y., Zhu, X., Zhou, Y., Yan, D. 2011. *Macromol. Biosci.* 11:1603–1610.
Zhang, C., Qiu, L., Zhu, K. 2009b. *Polymer* 50:1173–1177.
Zhang, G., Liang, F., Song, X., Liu, D., Li, M., Wu, Q. 2010d. *Carbohyd. Polym.* 80:885–890.
Zhang, H., Cai, G., Tang, G., Wang, L., Jiang, H. 2011a. *J. Biomed. Mater. Res. B Appl. Biomater.* 98B(2):290–299.
Zhang, J., Li, S., Li, X. 2009a. *Recent Pat. Nanotechnol.* 3:225–231.
Zhang, K., Khan, A. 1995. *Macromolecules* 28:3807–3812.
Zhang, L., Bartels, C., Yu, Y., Shen, H., Eisenberg, A. 1997. *Phys. Rev. Lett.* 79:5034–5037.
Zhang, L., Eisenberg, A. 1995. *Science* 268:1728–1731.
Zhang, L., Eisenberg, A. 1996. *Macromolecules* 29:8805–8815.
Zhang, L., Liang, Y., Meng, L. 2010c. *Polymer Adv. Technol.* 21:720–725.
Zhang, L., Lin, J., Lin, Sh. 2007b. *J. Phys. Chem. B* 111:9209–9217.
Zhang, X., Cheng, J., Wang, Q., Zhong, Zh., Zhuo, R. 2010a. *Macromolecules* 43:6671–6677.
Zhang, X., Ma, J., Yang, Sh., Xu, J. 2011b. *Micro Nano Lett.* 6:830–831.

Zhang, X., Zhang, X., Wu, Zh., Gao, X., Cheng, C., Wang, Zh., Li, Ch. 2011d. *Acta Biomater.* 7:585–592.
Zhang, Y., Guo, Sh., Lu, Ch., Liu, L., Li, Z., Gu, J. 2007a. *J. Polym. Sci. A—Polym. Chem.* 45:605–613.
Zhang, Y., Wang, X., Wang, J., Zhang, X., Zhang, Q. 2011c. *Pharm. Res.* 28:1167–1178.
Zhang, Y. L., Dou, X. W., Jin, T. 2010b. *Exp. Polym. Lett.* 4:599–610.
Zhao, Y., Luo, Y.-W., Li, B.-G., Zhu, Sh. 2011. *Langmuir* 27:11306–11315.
Zheng, C., Qiu, L., Zhu, K. 2009. *Polymer* 50:1173–1177.
Zhou, W., Dai, X. H., Dong, C. M. 2008. *Macromol. Biosci.* 8:268–278.
Zhou, Y., Yan, D. 2004. *Angew. Chem. Int. Ed.* 43:4896–4899.
Zhou, Y., Yan, D., Dong, W., Tian, Y. 2007b. *J. Phys. Chem. B* 111:1262–1270.
Zhou, Zh., Chaibundit, Ch., D'Emanuele, A., Lennon, K., Attwood, D., Booth, C. 2007a. *Int. J. Pharm.* 354:82–87.
Zhu, J., Jiang, W. 2005. *Macromolecules* 38:9315–9323.
Zhu, J., Liao, Y., Jiang, W. 2004. *Langmuir* 20:3809–3812.
Zhu, W., Nese, A., Matyjaszewski, K. 2011. *J. Polym. Sci. A—Polym. Chem.* 49:1942–1952.
Zhu, W., Sun, Sh., Xu, N., Gou, P., Shen, Zh. 2012. *J. Appl. Polym. Sci.* 123:365–374.
Zhu, Z., Armes, S. P., Liu, S. 2005. *Macromolecules* 38:9803–9812.
Zhulina, E. B., Birshtein, T. M. 1986. *Vysokomol. Soed. Ser. A* 28:773–778.

3 Organic–Organic Hybrid Nanoassemblies

3.1 INTRODUCTORY NOTES

Amphiphilic copolymers of various chain architectures and composition were shown in Chapter 2 to form a variety of self-assembled aggregate structures in dilute aqueous solution. Spherical and wormlike micelles as well as polymer vesicles were the most frequently observed self-assembled structures, but other more or less exotic morphologies such as toroids, disks, multicompartment, and bicontinuous micelles have been documented as well. Most of the research in the last decade was focused on the control of the chemistry of the constituent moieties whereby manipulation of the properties of the self-assembled structures can be achieved. Many examples demonstrating systematic variations of, for instance, morphology through synthetic manipulations of a series of self-associating copolymers can be found in the scientific literature. Surprisingly, however, considerably less attention has been given to another approach to influence the self-assembly, in general, and properties of the self-assembled structures. This approach is the *co-assembly*. It is a straightforward and feasible approach providing an experimentally simple way to modify self-assembled structures and allowing expanding the utility of the latter. The method consists of blending of a given amphiphilic copolymer with other substances, not necessarily of polymeric nature or even not necessarily of amphiphilic character (Figure 3.1), thus providing access to composite, hybrid self-assembled structures. To diversify the approach, mixing of two or more homopolymers that do not self-associate, taken separately or adding of a substance (surfactant, homopolymer, block copolymer, protein, etc.), to preformed self-assembled structures has been considered. Different types of forces like hydrophobic interactions, electrostatic interactions, hydrogen bonding, donor–acceptor interactions, metal–ligand coordination bonds, etc., have been found to facilitate the formation of mixed structures or to contribute to the introduction of new functionality and properties. Whatever the type of interactions, in all cases, the systems rearrange accordingly, thus optimizing their self-organization to approach a thermodynamic equilibrium state. The introduction of additional entities incorporated even in small amounts in the mixed self-assembled structures is an excellent approach to tune the aggregate morphology, to significantly alter the aggregate characteristics, and to produce new functionality and properties. In this chapter, many examples of formation of such hybrid structures are given. Properties originating from their mixed nature will be shown to expand the range of possible applications and utility. In particular, hybrid structures resulting from co-assembly of an amphiphilic copolymer and a surfactant as well as copolymer–lipid, copolymer–homopolymer, and copolymer–copolymer interactions will be considered (Figure 3.1). Recent examples from the scientific

	Self-associating amphiphilic copolymer (self-assembled structures)
Type of co-assembling substance	Surfactants; Lipids; (Co)polymers; Proteins; Oligo- and polynucleotides
Type of interactions or forces	Hydrophobic interactions, electrostatic interactions, hydrogen bonding, donor–acceptor interactions, metal–ligand coordination bonds, screening of electrostatic charges, Coulombic attraction, adsorption, bridging, depletion, recognition, host–guest interactions, stereocomplexation
Type of functionality or property produced	Alteration of aggregate properties, modulation of functional properties, morphology transitions, tuning aggregate shape and curvature, synergistical improvement of performance, charging of nonionic structures, reduction of dimensions, increasing the volume of core or corona domains, imparting polyelectrolyte characteristics, penetration of water in the hydrophobic core, vesicle formation, imparting steric, protective and "stealth" properties, promoting release, formation of channels/pores in lipid membranes, triggering destabilization of structures, disintegration of existing structures, creation of new structures, energy dissipation, nanoscale lubrication, induction of emission

FIGURE 3.1 Co-assembly of amphiphilic copolymers with various species indicated, types of interactions involved, and functionality/properties of the hybrid structures produced.

literature of construction of mixed structures from inherently nonassociating species, formation of nanoreactors by incorporation of proteins as well as hybrid structures formed upon interactions of polymers with oligo- and polynucleotides such as DNA, will be shown. Particular focus will be placed upon the control of the overall shape of the mixed structures, their external and internal morphology, the introduction of desired functions, and alteration of the properties. Adding salts to solutions of polyelectrolytes, self-assembled structures involving polymer–drug conjugates, and lipoplexes are not considered in this chapter.

3.2 POLYMER–SURFACTANT HYBRID STRUCTURES

Compared to the numerous studies on self-assembly of amphiphilic copolymers in selective solvents, the approach of mixing macromolecular amphiphiles with low molecular weight surface-active agents to modulate the properties of the resulting hybrid structures has received considerably less attention despite of its feasibility and effectiveness. In principle, a variety of dispersed particles can be found in copolymer–surfactant aqueous solutions depending on the concentration and surfactant-to-copolymer ratio: from unimers of surfactant and copolymer to clusters of adsorbed surfactant molecules on the hydrophobic moieties of the macromolecules and hybrid aggregates in a wide range of compositions and various morphologies. In an early work, Almgren and coauthors have shown that micelle formation of poly(ethylene oxide)-poly(propylene oxide)-poly(ethylene oxide) (PEO-PPO-PEO) block copolymers was suppressed by the addition of surfactant, sodium dodecyl sulfate (SDS) in particular (Almgren et al. 1991). Several distinct steps of interactions between a nonionic block copolymer ($EO_{97}PO_{69}EO_{97}$) and an anionic surfactant (SDS) have been revealed: SDS-induced micellization, formation of SDS-copolymer aggregates, collapse of the aggregates, and complete inhibition of the micellar growth (Hecht and Hoffmann 1994, 1995, Hecht et al. 1995, Ghoreishi et al. 1999, Li et al. 2001).

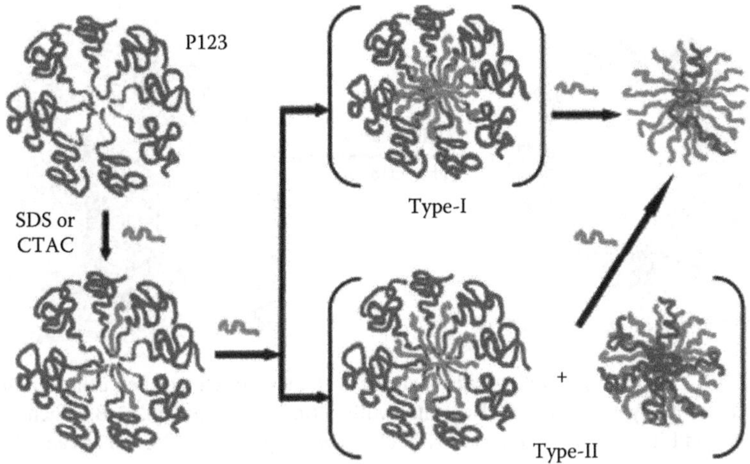

FIGURE 3.2 Schematic presentation of possible aggregation mechanism of block copolymer assemblies and an ionic surfactant. P123 is $EO_{20}PO_{70}EO_{20}$. (Reprinted with permission from Kumbhakar, M., *J. Phys. Chem. B*, 111, 14250–14255. Copyright 2007, American Chemical Society.)

On the basis of the same system, an association model that assumes all possible configurations has been developed (Hecht et al. 1995). According to the model, the mixed micelles progressively break down upon increasing SDS concentration due to enhancement of the repulsive interactions between the charged groups included in the composite structure. The model seems to be applicable also for related mixtures where destabilization of micelles is the dominant effect. The transition from copolymer-rich to surfactant-rich aggregates is schematically shown in Figure 3.2. There is ambiguity concerning the region of intermediate surfactant-to-copolymer ratios (Hecht et al. 1995, Li et al. 2000, Jansson et al. 2004, 2005, Ganguly et al. 2006). The transition could proceed either via gradual incorporation of surfactant molecules into the copolymer micelles, releasing copolymer unimers (type-I, Figure 3.2) or via simultaneous buildup of surfactant-rich micelles and destruction of copolymer-rich micelles (type-II, Figure 3.2). Particularly for $EO_{20}PO_{70}EO_{20}$ and $EO_{100}PO_{70}EO_{100}$ with SDS and hexadecyltrimethylammonium chloride (CTAC), it has been demonstrated by steady-state fluorescence that the interaction mechanism is mainly type-II, but for higher copolymer concentrations, interactions via type-I also operate (Kumbhakar 2007). The latter has been additionally supported by SANS (Ganguly et al. 2006).

Hybrid polymer–surfactant nanostructures have been successfully employed as structure-directing agents for preparation of various inorganic materials such as mesoporous and hollow silica spheres and metal, metal oxide, and metal hydroxide films and particles. To illustrate the applicability, some examples are given in the succeeding text (more information can be found in Chapter 4). Mesoporous silica of bicontinuous *Ia3d* structure has been prepared via an acid-catalyzed silica sol-gel process using templates of mixed surfactants composed of $EO_{20}PO_{70}EO_{20}$ and SDS (Chen et al. 2005). More complicated, that is, composed of three components, is the

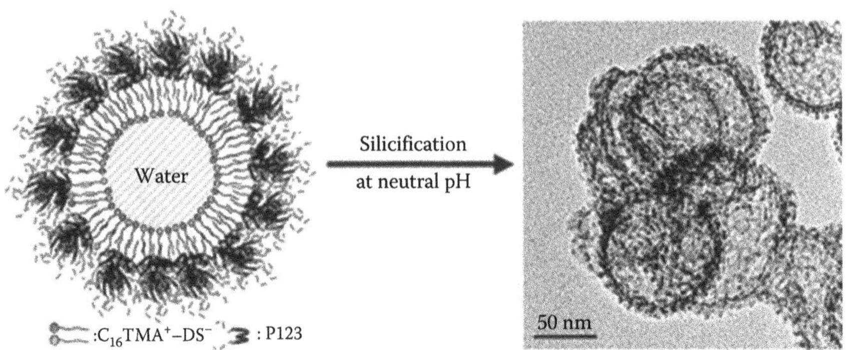

FIGURE 3.3 Synthesis of hollow silica spheres with mesostructured shell using *catanionic*-block copolymer hybrid particles. (Reprinted with permission from Yeh, Y.-B., Chen, B.-Ch., Lin, H.-P., and Tang. Ch.-Y., *Langmuir*, 22(1), 6–9. Copyright 2006, American Chemical Society.)

system used for the preparation of hollow silica spheres with mesostructured shells (Yeh et al. 2006). These authors used ternary surfactant mixtures composed of a *catanionic*, that is, hexadecyltrimethylammonium bromide (CTAB)-SDS, surfactant and $EO_{20}PO_{70}EO_{20}$, which is nonionic. The structure of the hybrid particles is vesicular and consists of a closed bilayer composed of CTA^+-DS^- on which $EO_{20}PO_{70}EO_{20}$ micelles are adsorbed (Figure 3.3). 2D hexagonal mesoporous $Ni(OH)_2$ films have been electrodeposited from dilute surfactant solutions composed of cylindrical SDS micelles to which the chains of macromolecular cosurfactants such as PEO-PPO-PEO copolymers are bound (Tan et al. 2005).

Rodriguez et al. have found complete mixing of $EO_5PO_{68}EO_5$ (a hydrophobic member of the group of commercially available PEO-PPO-PEO copolymers) and sodium bis(2-ethylhexyl)sulfosuccinate (AOT), which is an anionic and also strongly hydrophobic surfactant (Rodriguez et al. 2007, 2010). Both amphiphiles exhibit propensity to form vesicles. When mixed, large hybrid vesicles and two-phase regions were reported to form at both low and high AOT contents, whereas at intermediate AOT contents, the solutions were single phase, transparent, and isotropic. This bilayer–micelle–bilayer transition was observed by a number of experimental techniques. In the isotropic region, evidence was provided by SANS that the mixed aggregates had the shape of small spheres or short ellipsoids. The inner core of the aggregates was occupied by the PPO chains apparently in an extended conformation. This conformation together with the repulsion introduced by the AOT groups led to a decrease of the effective packing parameter which favored the formation of quasi-spherical aggregates. The synergistic effect of mixing was exhibited in a faster adsorption on the air–solution interface and improved solubilization capacity. Upon dilution, the morphology changed to disklike (Rodriguez et al. 2010).

The interactions in water between one of the hydrophilic representatives of the PEO-PPO-PEO copolymers ($EO_{106}PO_{70}EO_{106}$) and cationic vesicles of di-*n*-octadecyldimethylammonium bromide (DODAB) as a function of concentration of the two components and the preparation protocol have been studied by differential scanning calorimetry (DSC) (Feitosa and Winnik 2010). Depending on the ratio between the components, the vesicle/polymer mixtures underwent an irreversible transition into mixed micelles and/or PEO-PPO-PEO-bearing vesicles, formed in particular through polymer adsorption or intercalation.

Unlike the most common pattern of destabilization of polymer micelles upon surfactant addition, a unique case of vesicle formation is reported elsewhere (Kelarakis et al. 2008). The introduction of ionic single-tailed surfactants to a relatively short nonionic copolymer of composition $EO_{18}BO_{10}$ led to a dramatic increase of the hydrodynamic radii from about 7 nm of the initial copolymer micelles to ~100 nm. These large hybrid structures were stable up to molar ratios corresponding to four to five surfactant molecules per single copolymer chain. Studies with SDS, CTAB as well as sodium decyl sulfate indicated that the size and the interval of stability were insensitive to both the length of the surfactant tail and charge. These structures developed in size with time, following a multiequilibrium status between several particles with various sizes and copolymer contents. Laser confocal scanning microscopy revealed that they were vesicles (Figure 3.4). According to the authors, the incorporation of charged surfactant tails to the micellar core can promote water penetration thus giving rise to vesicle formation (Kelarakis et al. 2008).

Other mixed structures assembled through interactions between synthetic block copolymers and low molecular weight surfactants with particular emphasis on the vesicular structures in aqueous solution and in organic solvents have recently been summarized elsewhere (Pispas 2011b). Due to the greater structural variability obtained by synthetic chemistry in both the surfactant and block copolymer structures,

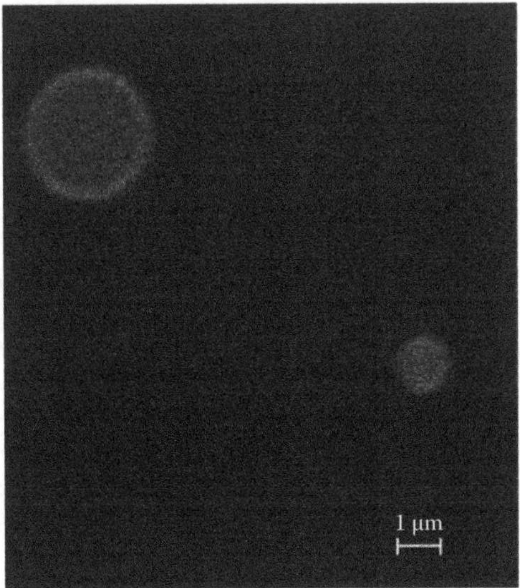

FIGURE 3.4 Laser confocal scanning micrograph of hybrid vesicular structures composed of $EO_{18}BO_{10}$-SDS. Total concentration 2.5 wt.% in aqueous solution, $EO_{18}BO_{10}$-SDS ratio 0.5. (Reprinted with permission from Kelarakis, A., Castelletto, V., Krysmann, M.J., Havredaki, V., Viras, K., and Hamley, I.W., *Langmuir*, 24, 3767–3772. Copyright 2008, American Chemical Society.)

the mixed vesicles offer additional possibilities for structural and functionality control, induction of stimuli responsiveness, as well as tuning and controlling the nature of forces acting in these systems. Based on some examples from the literature to which we also refer, the author discussed future directions and perspectives for further advancement in the field. Improved and innovative organic and polymer synthesis protocols are requirements to achieve control of the vesicles' size and mechanical properties and permeability of the membranes, placement of chemical functionality at the surface and the core of the bilayer, as well as stimuli-triggered and selective responsiveness. Elaboration of the structure and properties through purely physicochemical means and pathways has been envisaged as well (Pispas 2011a,b).

Vesicle formation resulting from imparting amphiphilicity after complexation of negatively charged poly(sodium methacrylate-*b*-ethylene oxide) with cetylpyridinium bromide has been reported by Kabanov et al. (1998). Steric repulsion and hydrophilicity of the PEO chains drove the closure of the lamellar structures in which a hydrophilic fluorescent dye could be efficiently encapsulated. The same approach, that is, complexation between a negatively charged block copolymer and a single-tail cationic surfactant, has recently been employed to form photoresponsive mixed vesicles (Wang et al. 2010). The photoresponsiveness was imparted by an azobenzene group incorporated in the surfactant tail of a cationic surfactant AzoC10. Upon complexation with poly(acrylic acid-*b*-ethylene oxide), vesicles were formed (Figure 3.5), which were simultaneously loaded with hydrophilic (in the vesicles' pool) and hydrophobic (in the vesicles' bilayer) dyes. The trans-to-cis isomerization of the azobenzene group upon UV irradiation destabilized the vesicles and released the encapsulated dyes.

The effects of an oligomeric surfactant of low hydrophobicity (a propylene oxide [PO] oligomer with ethylene oxide [EO] units at each end) on the mechanical properties of unilamellar polymersome membranes formed by poly(ethylene oxide-*b*-ethyl

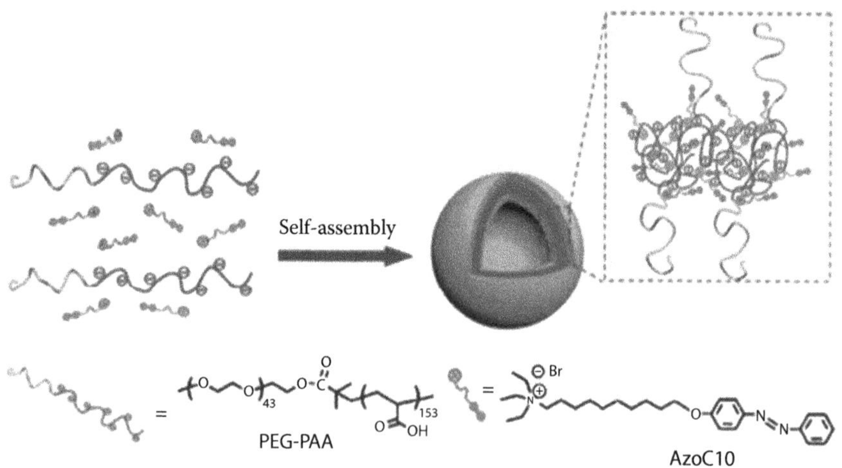

FIGURE 3.5 Schematic illustration of vesicle formation in PEG-PAA/AzoC10 complexes. (Reprinted with permission from Wang, Y., Han, P., Xu, H., Wang, Z., Zhang, X., and Kabanov, A.V., *Langmuir*, 26:709–715. Copyright 2010, American Chemical Society.)

ethylene) have been studied elsewhere (Santore et al. 2002). Reduction of the area expansion modulus, enhancement of water permeability, and tendency to lysis were among the effects, which were proportional to the surfactant concentration. The authors suggested that the locus of surfactant incorporation was at the corona-to-core interface of the bilayer.

The long hydrophobic tail of CTAB produced disruption of poly(*N*-isopropyl acrylamide)-*b*-poly(ethylene oxide) (PNIPAM-*b*-PEO) polymersomes into mixed copolymer/surfactant aggregates, whereas any observable changes in the structural properties were produced with dodecyltrimethylammonium bromide (DTMAB) (Zhao et al. 2009). At higher DTMAB concentrations, mixed vesicles of larger than initial size were formed at higher temperatures.

The interactions between cationic vesicles formed by didodecyldimethylammonium bromide (DDAB) and nonionic poly(isoprene-*b*-ethylene oxide) block copolymers has been studied by Pispas and Sarantopoulou (2007). Surfactant vesicles decorated with block copolymer chains that coexisted with mixed aggregates resulted from these interactions. The size of the vesicles was found to decrease with increasing copolymer concentration. The mixed vesicular structures were temperature sensitive, reminiscent of the behavior of the pure DDAB vesicles. The molecular characteristics of the block copolymer were also found to influence the structure of the mixed vesicles. Via functionalization of the polyisoprene block, double hydrophilic poly[(sulfamate/carboxylate)isoprene-*b*-ethylene oxide] block copolymers have been obtained. Electrostatic complexation between the anionic poly[(sulfamate/carboxylate)isoprene] block and the cationic heads of DDAB was the primary mode of interaction (Pispas 2011a). This complexation imposed additional constrains on the bilayer organization leading to an initial decrease in the vesicle size and full disintegration of the vesicles and formation of complex coacervate core micelles (C3Ms) (see succeeding text) at higher copolymer contents. The mixed vesicular structures were stable to changes in the ionic strength of the solution and exhibited thermoresponsive properties.

Consistent with the commonly observed pattern is the behavior of another PEO-based copolymer having poly(styrene oxide) as a hydrophobic block (Taboada et al. 2005): When SDS was added to a micellar solution of $SO_{20}EO_{67}$, a strong decrease in both light scattering intensity and hydrodynamic radii was observed, which was ascribed to enhanced electrostatic repulsion of the charged SDS-$SO_{20}EO_{67}$ hybrid micelles. Similar is the situation with a series of PB-PEO copolymers in the presence of either anionic (SDS) or cationic (DTMAB) surfactants for which Pispas and Hadjichristidis found that SDS is more powerful in inducing changes than DTMAB (Pispas and Hadjichristidis 2003). The changes, however, were associated with the molecular characteristics of the block copolymers, as some large structures were detected only for the copolymers with shorter chains. These large structures, presumably vesicle-like, were found to coexist with smaller micelles (smaller than the original block copolymer micelles). The rearrangement to vesicular structures was favored as the short hydrophobic chains were not able to fill the intramicellar space increase upon the incorporation of the surfactant molecules.

Strong associations between nonionic surfactants and uncharged amphiphilic block copolymers have been reported elsewhere (Zheng and Davis 2000).

The interactions between nonionic surfactant and nonionic block copolymer have initially been considered as weak and, therefore, uninteresting; however, these authors demonstrated spectacular morphology transitions. They applied cryogenic transmission electron microscopy (cryo-TEM) to mixed micelles of the nonionic surfactant $C_{12}EO_5$ and two block copolymers with diblock and triblock chain architectures. The diblock copolymer—poly(ethylene oxide)-polybutadiene (EO_{126}-B_{45})—formed alone in aqueous solution spherical micelles with clearly visible core–corona structure and total diameter of 47 nm. When $C_{12}EO_5$ was added, the corona disappeared, corresponding to the saturation of surfactant molecules binding to the copolymer micelles. Further addition produced mixed micelles with smaller dimensions and a broad size distribution. At the highest surfactant contents studied, the size distribution narrowed and all particles were about 5 nm in diameter. The morphological transitions are presented in Figure 3.6. Figure 3.7 shows the morphological transitions of the structures obtained from the triblock copolymer—poly(ethylene oxide)-poly(ethyl ethylene)-poly(ethylene oxide) (EO_{21}-EE_{35}-EO_{21})—which alone

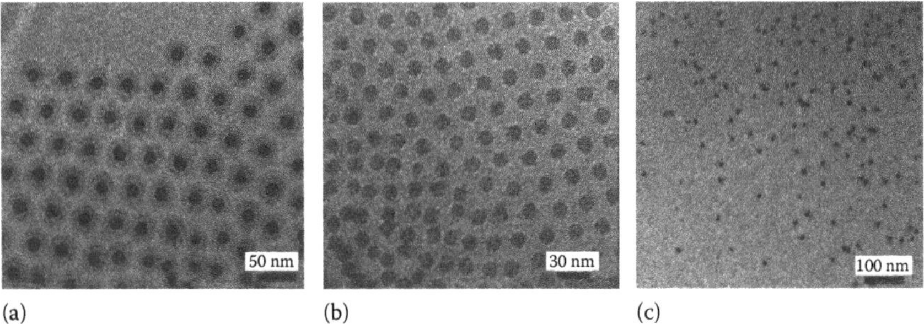

FIGURE 3.6 Cryo-TEM micrographs of the diblock copolymer EO_{126}-B_{45} mixed with $C_{12}EO_5$. (a) Neat EO_{126}-B_{45} micelles. Mixed micelles formed at 1:1 (b) and 20:1 (c) $C_{12}EO_5$/EO_{126}-B_{45} ratios. Total concentration is 1 wt.%. (Reprinted with permission from Zheng, Y. and Davis, H.T., *Langmuir*, 16, 6453–6459. Copyright 2000, American Chemical Society.)

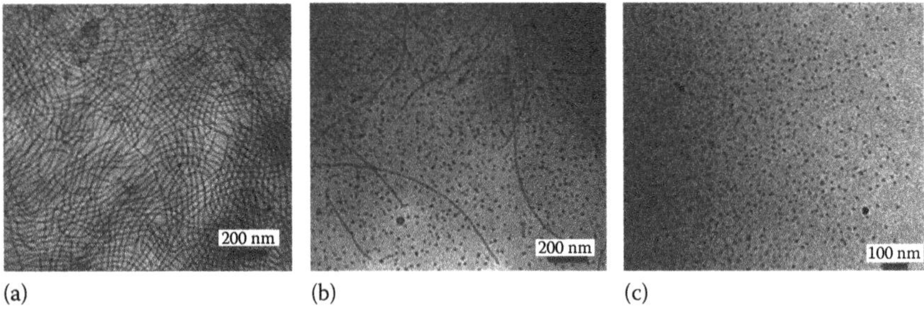

FIGURE 3.7 Cryo-TEM micrographs of the triblock copolymer EO_{21}-EE_{35}-EO_{21} mixed with $C_{12}EO_5$. Neat EO_{21}-EE_{35}-EO_{21} micelles. (a) Mixed micelles formed at 5:1 (b) and 10:1 (c) $C_{12}EO_5$/EO_{21}-EE_{35}-EO_{21} ratios. Total concentration is 1 wt.%. (Reprinted with permission from Zheng, Y. and Davis, H.T., *Langmuir*, 16, 6453–6459. Copyright 2000, American Chemical Society.)

forms long cylindrical micelles. The latter gradually disappeared, whereas more and more spherical micelles were generated and eventually no cylindrical micelles were observed (Zheng and Davis 2000).

Studies on copolymer–surfactant interactions and hybrid particle formation have been extended to other systems having hydrophobic blocks of higher T_g. Polymer–surfactant structures with molar fractions approximately 1–5 have been reported for PEO-PS block copolymers with cationic and anionic surfactants (Bronstein et al. 2000a,b, 2001). The addition of SDS was shown by ^{1}H nuclear magnetic resonance (NMR) to result in loosening of the micellar cores and an increase of the mobility of the PS chains, that is, SDS acted as a plasticizer of the core. At the same time, the alkyl chains of the surfactant partially decreased their mobility as judged from the broadening of the signals related to methylene protons. Data from other experimental techniques, for example, ultracentrifugation and light scattering, proved that comicellization occurred and hybrid micelles were formed. Three populations of hybrid aggregates corresponding to micelles, micellar clusters, and supermicellar aggregates were detected (Bronstein et al. 2000a). An increase in the SDS loading was found to initially result in an increase in block copolymer/surfactant micelle size and weight at the SDS concentration of 0.8×10^{-3} mol/L and in a slight decrease of both parameters at the critical micellization concentration (CMC) and higher concentrations. This decrease was caused by incorporation of SDS molecules in block copolymer micelles followed by charging the PS core and repulsion between similar charges. The effects of the cationic surfactant cetylpyridinium chloride were analogical to those of SDS. In both cases, ion exchange of surfactant counterions in the hybrid PEO-PS/surfactant micelles allowed saturating the micellar structures with appropriate metal salts or complexes which were subsequently reduced to eventually form metal nanoparticles located in the micelles (Bronstein et al. 2000a,b).

It is the general consensus that the morphological changes are promoted by altering the balance between the core chains stretching, interfacial tension between the core and the solvent, and intercorona interactions. Typically, the hydrocarbon tails of the surfactant partition in the core of the aggregates. Zhu et al. considered that it had the dominating role in the morphological transitions of poly(4-vinylpyridine)$_{43}$-*b*-polystyrene$_{366}$-*b*-poly(4-vinylpyridine)$_{43}$ in *N,N*-dimethylformamide (DMF)/water mixtures in the presence of a low molecular weight amphiphile, pentadecylphenol (Zhu et al. 2005). The magnitude of interactions in the corona region largely depends on the nature of the corona-forming block and surfactant type. Thus, hydrogen bonding between the pentadecylphenol head group and poly(4-vinylpyridine) also contributed to the morphological transitions but to a lesser extent; in particular, it weakened the interactions among the corona-forming block and hydrophobic chains of the surfactant inserted into the core of the aggregates. In almost identical complexes composed of an asymmetric diblock copolymer polystyrene-*b*-poly(4-vinylpyridine), mixed with 3-pentadecylphenol, the sufficient mobility of the systems has been reported to enable reorganization to nanostrand networks (Figure 3.8) in response to increased surface pressure (Lu and Bazuin 2005). For polystyrene-poly(acrylic acid) (PS-PAA) in dioxane/water mixtures, the sulfate and sodium ions of SDS screened the electrostatic charges generated by the partially charged PAA chains (Burke and Eisenberg 2001).

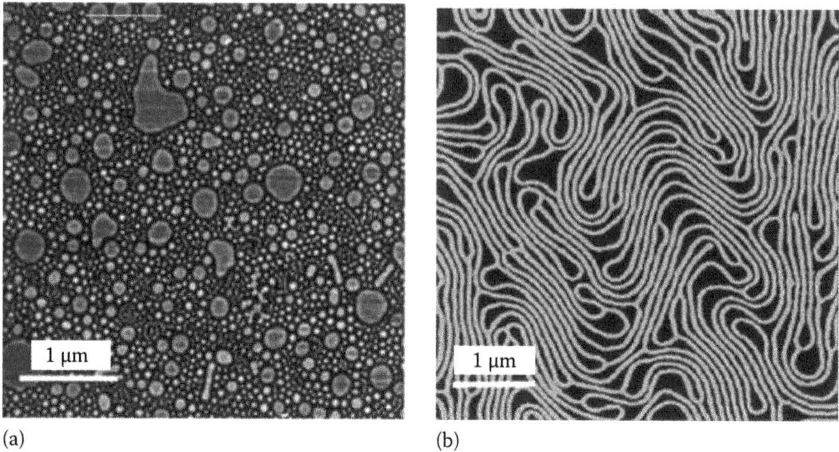

FIGURE 3.8 Formation of nanostrand networks from supramolecular complexes composed of a diblock copolymer polystyrene-*b*-poly(4-vinylpyridine) and 3-pentadecylphenol at the air/water interface (a) before and (b) after compression of the solution immediately after spreading. (Reprinted with permission from Lu, Q. and Bazuin, C.G., *Nano Lett.*, 5(7), 1309–1314. Copyright 2005, American Chemical Society.)

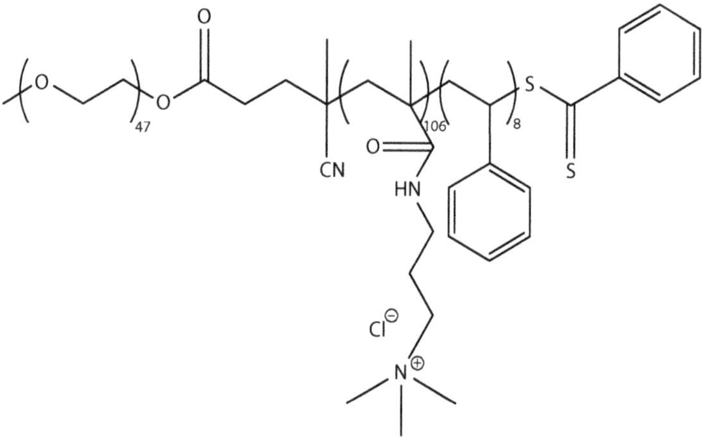

FIGURE 3.9 Chemical structure and composition of PS-*b*-PMAPTAC-*b*-PEO. (Reprinted from Liu, J. et al., *Can. J. Chem.*, 88, 208, 2010. With permission.)

An interesting case of very strong interactions in the corona region is presented elsewhere (Liu et al. 2010). The chemical structure and composition of the ABC copolymer described in this paper are given in Figure 3.9. The self-assembly of this copolymer was considerably hindered due to the short PS block, and accordingly, the copolymer was characterized by sufficiently high CMC (~1.0 g/L). Above the CMC, onion-like micelles composed of a PS core, a cationic shell of poly(3-(methacryloylamino)propyltrimethylammonium chloride) (PMAPTAC), and a PEO corona were formed. Upon the introduction, however, of small quantities of SDS, the CMC dropped down by approximately 3 (!) orders of magnitude due to partial neutralization of the charges resulting in overall hydrophobization of the system. In the absence of SDS, PMAPTAC block adopted an extended-chain conformation due to the repulsion between the positive charges. With the addition of SDS, the hydrodynamic dimensions gradually decreased from about 150 to 125 nm in diameter, indicating reduction of the repulsive interactions and shrinking of the segments in the shell region. Morphology transitions were not detected. Macrophase separation was observed at ca. 60% of apparent degree of neutralization (ADN).

Hybrid wormlike micelles based on CTAB have been obtained upon complexation with oppositely charged polyelectrolytes. Random copolymers of styrene and sodium styrene sulfonate (Nakamura and Shikata 2003, 2004, 2007, Nakamura et al. 2003, 2005) as well as of methyl methacrylate and sodium styrene sulfonate (Oikonomou et al. 2011) have been used in a slight CTAB excess to form wormlike micelles exhibiting viscoelastic properties. The decisive importance of the delicate balance of electrostatic and hydrophobic interactions was demonstrated as the phase separation was restricted in a mixing region around charge stoichiometry, when hydrophobic units were introduced into the polyelectrolyte chain. Specific stoichiometric cation–π electron interactions between the surfactant and both charged and uncharged phenyl rings are decisive for the formation of the polymer-induced hybrid wormlike micelles (Nakamura et al. 2003). Formation of viscoelastic solutions has been observed also for CTAB-based wormlike micelles composed of copolymers containing hydrophobic nonaromatic units such as methyl methacrylate (Oikonomou et al. 2011), indicating that the aforementioned behavior is more general and not restricted to copolymers containing phenyl rings. The latter authors demonstrated that the methyl methacrylate molar content was a crucial factor controlling the phase separation behavior and the rheology of these systems. Zwitterionic amphiphilic polyacrylamides have been shown to induce a strong rheological synergy in solutions of cationic wormlike micelles (Rodriguez et al. 2011). The polyacrylamides, prepared by copolymerization of acrylamide and sulfobetaine methacrylate monomer, were of high molar mass ($1.5-1.8\times10^6$ Da) and rather broad molecular weight dispersity (>4.0). In principle, hydrophobic associations between the polymer backbone and the hydrophobic tail of the surfactant (Goddard and Ananthapadmanabhan 1993), hydrophobic interactions between polymer pendant side chains and the surfactant tail (Panmai et al. 2002, Piculell et al. 2003, Li and Kwak 2004, Gouveia et al. 2009), and attractive electrostatic forces between opposite charges (Rojas et al. 2010) are cooperative interactions that may lead to synergetic rheological effects. In this particular case, the results from the study indicated that the zwitterionic micellar aggregates of the copolymer were formed on the surfactant wormlike micelles; they became points of interaction between wormlike micelles, thus strengthening physical entanglements between the micelles (Rodriguez et al. 2011).

A recent review presents a near-exhaustive overview of the literature on the co-assembly of neutral–ionic copolymers with oppositely charged species including synthetic and natural (co)polymers, multivalent ions, metallic nanoparticles, surfactants, and polyelectrolyte block copolymer micelles (Voets et al. 2009). The resulting co-assembled particles are of core–corona micellar or vesicular structure. The former are stabilized by their corona consisting of neutral water-soluble units that surround

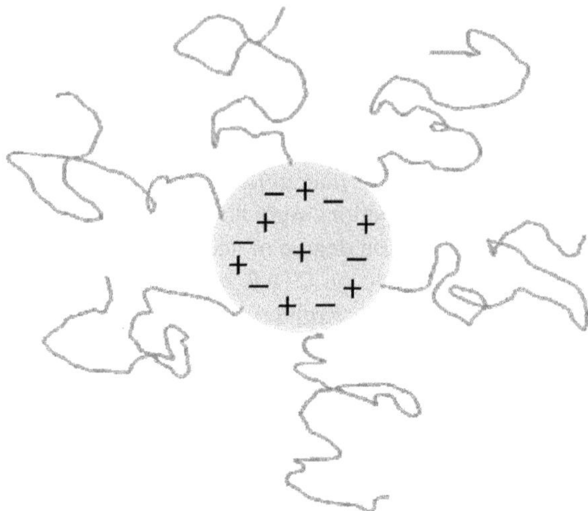

FIGURE 3.10 Schematic representation of a spherical multichain core–corona structure consisting of an ionic surfactant and an oppositely charged ionic–neutral block copolymer. The neutral (nonionic) water-soluble blocks of the copolymer reside in the corona. The charged moieties of the surfactant and block copolymer constitute the core.

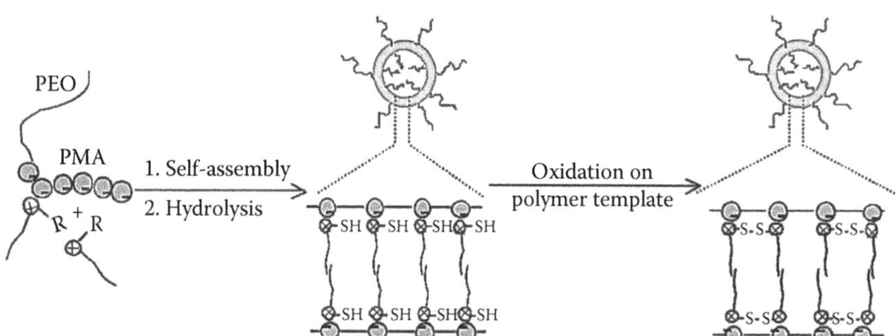

FIGURE 3.11 Hybrid vesicles formed in aqueous solution by co-assembly of an anionic–nonionic diblock copolymer and a reactive single-tail cationic surfactant. (Reprinted with permission from Bronich, T.K., Ouyang, M., Kabanov, V.A., Eisenberg, A., Szoka, Jr., F.C., and Kabanov, A.V., *J. Am. Chem. Soc.*, 124(40), 11872–11873. Copyright 2002, American Chemical Society.)

the water-insoluble core consisting of complexed oppositely charged units. Schematic representation of a spherical multichain core–corona structure consisting of an ionic surfactant and an oppositely charged ionic–neutral block copolymer is shown in Figure 3.10. Vesicular structures formed by co-assembly of a single-tail reactive surfactant and an anionic–nonionic diblock copolymer are presented in Figure 3.11. Small nanosized vesicles were initially formed and further stabilized by dimerization of the surfactant monomers thus forming double-tail surfactant, which builds the bilayer membrane of the vesicles (Bronich et al. 2002). Four different terms are in use to describe these structures: block ionomer complexes (BIC) (Kabanov et al. 1996), polyion complex (PIC) micelles (Harada and Kataoka 1995), C3Ms (van der Burgh et al. 2004), and (inter)polyelectrolyte complexes ((I)PEC) (Gohy et al. 2000, Pergushov et al. 2003). Electrostatic interactions which can be split into Coulombic attraction and entropy gain through counterion release (Kriz et al. 2001, de Vries et al. 2006, Hofs et al. 2006) are the driving force of the co-assembly. Hydrophobic interactions also play a significant role in complexation (Voets et al. 2007). The solution behavior of these particles strongly depends on the molecular characteristics of both the copolymer and the surfactant (Bronich et al. 2000). Rodlike particles are preferentially formed when stiff copolymers are used. A large number of examples of comicellization of ionic–neutral block, graft or random copolymers with oppositely charged surfactants are summarized in the review of Voets et al. (2009).

A very large group of hybrid particles are those prepared from ionic surfactants and oppositely charged polyelectrolytes. The co-association is driven by both electrostatic and hydrophobic interactions frequently leading to well-defined multichain aggregates in which the core, formed by the hydrophobic surfactant tails, is coated by a partially neutralized shell. It is noteworthy that the precipitation threshold does not always correspond to full neutralization, which implies that forces different from the electrostatic ones also play an important role. Polyelectrolyte–surfactant systems have been recently reviewed by Langevin (2009). The author summarized the current research on this topic and made some important conclusions about the influence of factors such as molecular weight, degree of branching, charge density, backbone rigidity, and concentration of the polymer, as well as polar head, chain length and concentration of the surfactant, and the role of the added salt. The observations were that at low polymer/surfactant ratios, when the polymer was moderately rigid with linear chain architecture and the surfactant tail was short, the complexes formed were dense and monodispersed. They were spherical with flexible polymers (low persistent length) and rodlike with rigid polymers (high persistent length). The complexes are not equilibrium systems, as indicated by sensitivity of their dimensions to the preparation methods applied.

3.3 POLYMER–LIPID HYBRID STRUCTURES

Lipids are essential biomolecules for the structure and function of living matter. Fats; waxes; sterols; fat-soluble vitamins; mono-, di-, and triglycerides; phospholipids; glycolipids; and others (see Figure 3.12 for the chemical structures of some common lipids) constitute the broad group of these naturally occurring molecules. One special type—amphiphilic lipids—is the predominant building block of biological membranes. Sterols (mainly cholesterol) and sphingomyelin, which are non-glyceride lipid components, are also found in biological membranes. Similarly to surfactants, the amphiphilic lipids are composed of a hydrophilic head and a hydrophobic moiety consisting of one or two hydrocarbon tails all linked to a glycerol skeleton. Typical bilayer-forming lipids consist of two hydrocarbon chains and a charged (ionic, zwitterionic) or nonionic head group. Such lipid molecules have a packing parameter around unity and, according to the geometrical packing concept (see Chapter 2), self-assemble into bilayers.

The bilayers can bend and self-close, thus forming vesicular aggregates called liposomes. The liposomes, as described by Bangham in the 1960s (Bangham et al. 1965),

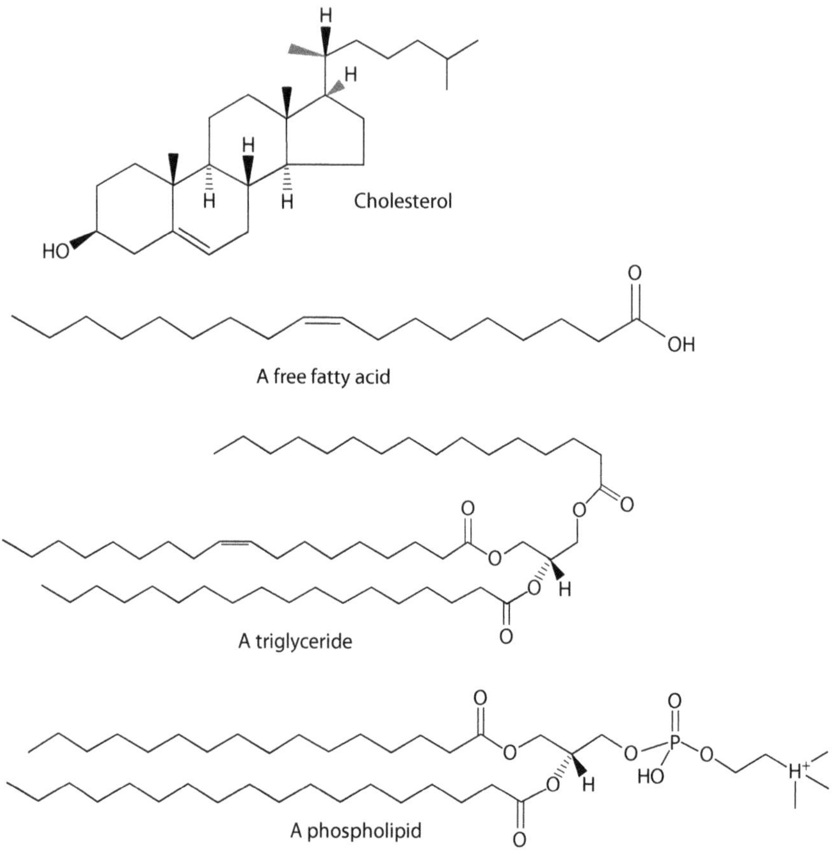

Cholesterol

A free fatty acid

A triglyceride

A phospholipid

FIGURE 3.12 Chemical structures of some common lipids.

are typically spherical self-closed structures composed of curved bilayers, enclosing part of the surrounding media into their interior (Figure 3.13). In that aspect, they closely mimic the biological cells and subcellular compartments. The liposome size ranges from about 20 nm to several micrometers. They may be composed of one or several membranes each with thickness of about 3–5 nm. Depending on the size and lamellarity, the liposomes may be classified into the following types:

- *Unilamellar vesicles* that can be subdivided into *small unilamellar vesicles* (SUVs) with diameters in the 20–100 nm range, *large unilamellar vesicles* (LUVs) with d between 100 nm and 1 μm, and *giant unilamellar vesicles* (GUVs) with $d > 1$ μm.
- *Oligolamellar vesicles* (OLVs), which represent liposomes with dimensions ranging from 100 nm to 1 μm, containing up to 8–10 concentric bilayers and *multilamellar vesicles* (MLVs) with $d > 0.5$ μm and many concentric bilayers.
- *Multivesicular vesicles* (MVVs). These are composed of many SUVs enclosed by a common membrane; their dimensions are typically above 1 μm.

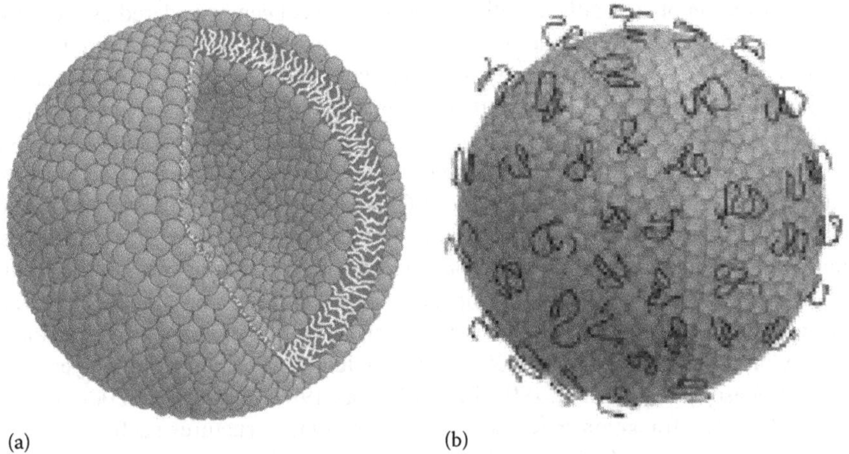

FIGURE 3.13 Schematic representation of conventional unilamellar (a) and sterically stabilized (b) liposomes. (The illustrations were kindly provided and reproduced with permission from G. Karlsson, University of Uppsala, Sweden.)

The amphiphilic character of the liposomes, with their hydrophobic bilayer and hydrophilic interior, enables solubilization and encapsulation of both hydrophobic and hydrophilic substances. The liposomes typically exhibit large solubilization capacity; their relatively easy preparation and possibilities to control the physicochemical properties have made them attractive drug delivery systems. The disadvantages of the use of liposomes in drug delivery are mainly associated with their low colloidal and biological stability. Despite the lipid composition that closely resembles those of cell membranes, the liposomes are rapidly removed from the circulation after their intravenous administration: in vivo, rapid enzymatic degradation is followed by opsonization, which triggers uptake by the mononuclear phagocytic system (Papahadjopoulos et al. 1991). A strategy that has been used to improve the performance and prolong the blood circulation time of liposomes is surface modification by which the membrane is protected, stability increased, and premature release of the encapsulated cargo reduced. This is typically achieved by using polymer-derivatized lipids. These are composed of a lipid residue covalently connected to a hydrophilic polymer chain—usually poly(ethylene glycol) (PEG). The lipid residue is anchored in the bilayer membrane, whereas the PEG chains create a repulsive barrier around liposomes as shown in Figure 3.13. A repulsive force is experienced due to unfavorable entropy associated with compressing (loss of conformational freedom) when two polymer-coated surfaces approach each other (Israelachvili 1992). In addition, an osmotic repulsive force, resulting from the difference in chemical potential between the water in the bulk and in the interaction region, is also induced (de Gennes 1987). The interactions with blood components are also reduced. The mechanism behind has the same origin as the colloidal stability. By preventing the markers for reaching the liposomal surface, the opsonization (i.e., adsorption of marker macromolecules) and taking up the marked species by the macrophages are largely eliminated, which results in prolonged circulation time (Woodle and Lasic 1992, Woodle 1995).

The invention of sterically stabilized liposomes has been considered as the most successful breakthrough for the use of liposomes in pharmaceutical formulations. The PEG-lipids, which are nowadays commercially available in PEG molar masses ranging from 350 to 5000 Da and diverse lipid anchors, are the most commonly used stabilizers. Since then, a large number of liposome-based formulations have been investigated. Some of them have reached the final stage, that is, clinical trials, and have been commercialized. Usually PEG-lipids of PEG molar mass of 2000 and contents of 5–8 mol% are used for formulation of sterically stabilized liposomes. Supposedly, by increasing PEG-lipid contents, the density and thickness of the repulsive barrier can be increased and the liposome circulation time prolonged. However, at a certain critical content, which largely depends on the PEG molar mass and less strongly on lipid composition, the PEG-lipids induce a transition from bilayers to a micellar phase (Hristova et al. 1995, Edwards et al. 1997, Belsito et al. 2000, Marsh et al. 2003). The transition may occur via intermediate structures such as open or perforated liposomes, which are useless as far as encapsulation and delivery of active substances are concerned. Depending on the surface density and molar mass of the PEG chains grafted on the bilayer, several regimes of PEG chain conformation can be identified (Alexander 1977, de Gennes 1980), in which the controlling factor is the distance between the PEG chains in the lipid bilayer (D), defined as $D^2 = A/M$, where A is the area of the lipid molecule and M is molar fraction of the polymer, relative to their Flory dimension (R_F). The latter is given by $R_F = aN^{3/5}$, where N is the degree of polymerization (DP) and a is the size of the monomer unit. The regimes are defined as (1) interdigitated mushrooms, when $D > 2R_F$; (2) mushrooms, when $D < 2R_F$; and (3) brushes, when $D < R_F$. At low grafting density, each polymer chain is in its random coil conformation. When the grafting density is high, the chains overlap and interact laterally to form a uniform continuous brush. In this regime, there is energy stored in the brush which increases with increasing PEG-lipid concentration and PEG molar mass. This energy is expressed as a lateral tension between the brushes. If the polymer is grafted on a solid surface, it is unlikely that the lateral tension would affect its structure since the solids are held together by strong chemical forces. In the lipid bilayer, however, which is characterized by considerably weaker cohesive forces, the strong lateral interactions between the polymer chains may create curvatures, area expansions, and mechanical breakdown. According to Hristova and Needham, there are two possible mechanisms by which the lateral pressure between the polymer chains could be relaxed (Hristova and Needham 1995). One mechanism, observed for highly charged gel-phase bilayers, involves transition to interdigitated bilayers, whereas the other involves curving the grafting surface. The highest curvature, corresponding to the biggest pressure relaxation, will be obtained if the lipids and PEG-lipids pack into mixed cylindrical or spherical micelles.

Even in small amounts added to a vesicle population, the PEG-lipids may give rise to rapid leakage of entrapped water-soluble substances. The membrane permeability passes through a maximum at PEG-lipid contents corresponding to the transition of the PEG chains from the mushroom to the brush regime (Kenworthy et al. 1995, Nikolova and Jones 1996). The increase in permeability has been attributed to the spontaneous formation of defects (holes) in the bilayer (Kaschiev and Exerowa 1983) that form more easily in the transition region between the two conformation regimes.

Furthermore, the macromolecules of the PEG-lipids with their large hydrophilic heads and relatively small lipophilic portions can be approximated to cone and exhibit propensity to form, in aqueous solution, small spherical micelles (Johnsson et al. 2001). Such macromolecules show preference to accommodate at highly curved surfaces such as the rims of the transient holes in the bilayers thus stabilizing them and contributing to the enhancement of the leakage from the liposomes. They also possess stronger solubilization power toward lipid bilayers.

It has been suggested that copolymers containing short sequences of lipid-mimetic units would provide stronger anchoring in the liposomal bilayer (Rangelov et al. 2001). With the larger hydrocarbon volume and hence more cylindrical shape, these copolymers exhibit propensity to self-associate in aqueous solution in less-curved aggregates—compartmentalized particles and polymer vesicles (see Chapter 2). They have been reported to destabilize the bilayer structures upon the incorporation of considerably larger quantities compared to the conventional PEG-lipids. Particularly, for a series of diblock copolymers with close DPs of the PEG chain and varying number of lipid-mimetic units—$DDP(EO)_{92}$, $(DDGG)_2(EO)_{115}$, and $(DDGG)_4(EO)_{114}$, where DDP and DDGG denote the lipid-mimetic anchors, respectively, 1,3-didodecyloxy-propane-2-ol and 1,3-didodecyloxy-2-glycidyl glycerol (Figure 3.14)—it has been shown that the increasing number of lipid-mimetic anchors allows incorporation of more copolymers without damaging the membrane integrity (Rangelov et al. 2003). Figure 3.15 shows cryo-TEM images of egg phosphatidylcholine (EPC) liposomes containing different amounts of the aforementioned copolymers. At low copolymer contents (Figure 3.15b and c), the liposomes were still intact and looked better separated than those without copolymer (Figure 3.15a), which indicated the stabilizing effect of the copolymers. Upon increasing copolymer content, open liposomes and fragments, indicating destabilization of membrane integrity, appeared. It is noteworthy that they appeared at different copolymer contents depending on the number of lipid-mimetic anchors per copolymer chain (Figure 3.15d through f). This finding is better demonstrated in Figure 3.16 in which the copolymer content at which the first indications for membrane destabilization appeared is plotted as a function of the number of the lipid anchors. The results showed that the introduction of more than one lipid anchor per copolymer chain is beneficial as far as increasing quantities of copolymer in the membrane without affecting its integrity is concerned.

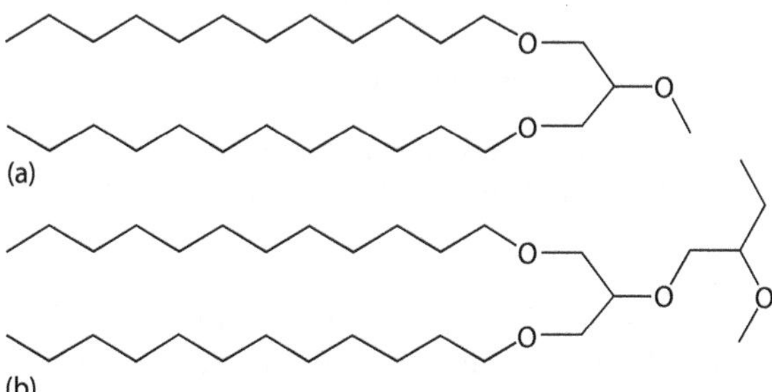

FIGURE 3.14 Structural formulas of the lipid-mimetic units: 1,3-didodecyloxy-propane-2-ol (DDP) (a) and 1,3-didodecyloxy-2-glycidyl glycerol (DDGG) (b).

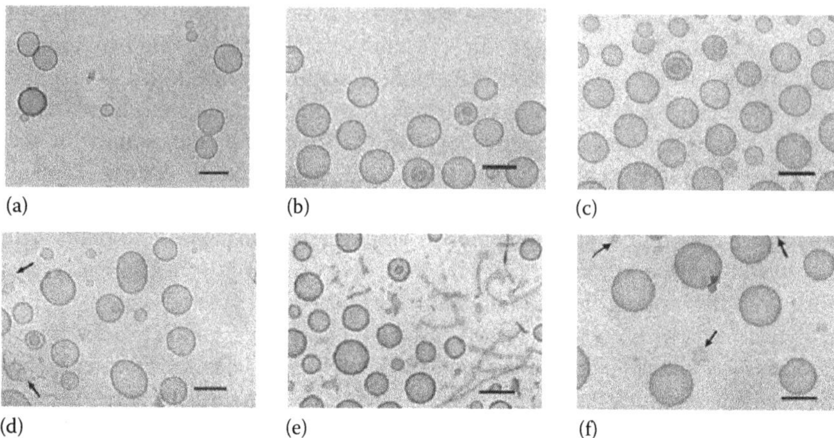

FIGURE 3.15 Cryo-TEM micrographs of pure EPC liposomes (a) and EPC liposomes containing 2.1 mol% of $DDP(EO)_{92}$ (b), 10.0 mol% of $(DDGG)_2(EO)_{115}$ (c), 7.5 mol% of $DDP(EO)_{92}$ (d), 12.1 mol% of $(DDGG)_2(EO)_{115}$ (e), and 14.2 mol% of $(DDGG)_4(EO)_{114}$ (f). [EPC] = 1.0 mM, bars = 100 nm. (Reprinted with permission from Rangelov, S., Edwards, K., Almgren, M., and Karlsson, G., *Langmuir*, 19, 172–181. Copyright 2003, American Chemical Society.)

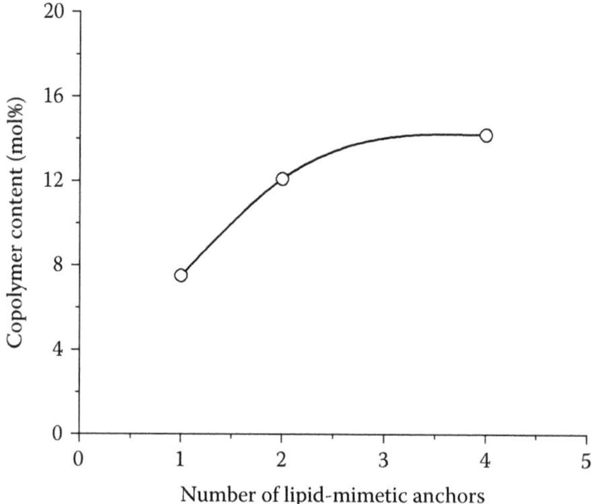

FIGURE 3.16 Copolymer contents in EPC liposomes at which the first fragment, indicative for membrane destabilization, appeared as a function of the number of lipid-mimetic anchors per copolymer chain.

By incorporation of more copolymer, a protective PEG layer of high density and thickness can be formed, which is further expected to have a beneficial effect on the liposomal longevity.

The effects of these copolymers on the morphological properties and membrane integrity of lipid bilayers based on different phospholipids have been investigated as

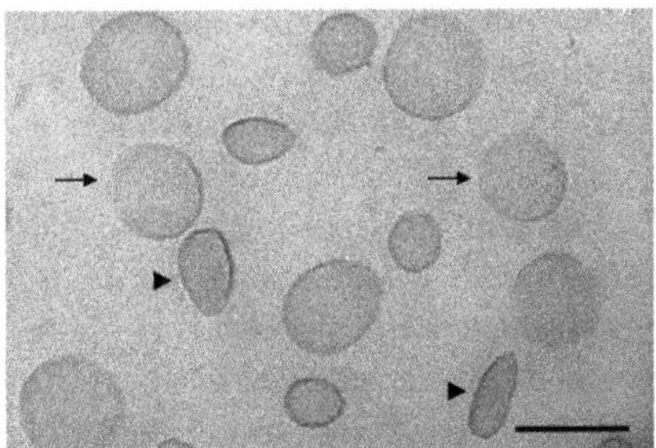

FIGURE 3.17 Cryo-TEM image of a sample based on DSPC stabilized by 7.5 mol% of $(DDGG)_4(EO)_{114}$. Arrows and arrowheads show oblong flattened liposomes observed face on and edge on, respectively. Bar = 200 nm. (Momekova, D. et al., *J. Disper. Sci. Technol.*, 29(8), 1106, 2008. Reprinted with permission from Taylor & Francis Group.)

well (Momekova et al. 2007, 2008, 2010). "Flat," nonspherical liposomes (Figure 3.17) were observed when $(DDGG)_2(EO)_{115}$ or $(DDGG)_4(EO)_{114}$ were incorporated in gel-phase lipid bilayers based on distearoylphosphatidylcholine (DSPC) at copolymer contents of 7.5 mol% and higher (Momekova et al. 2008). It has been speculated that these are kinetically frozen structures with non-even distribution of the stabilizing copolymers in the liposomal membrane resulting in curving in areas of elevated copolymer contents. Demixing of lipid-rich and polymer-rich membrane domains within the same vesicle membrane has been experimentally observed later on for giant lipo-polymersomes composed of 1-palmitoyl-2-oleoyl-*sn*-glycero-3-phosphatidylcholine and the commercially available copolymer polybutadiene-*b*-poly(ethylene oxide) (PBD-*b*-PEO) (Nam et al. 2011). The nonspherical shape did not compromise neither the loading ability of these liposomes nor the membrane permeability (Momekova et al. 2008). The latter was even found to decrease upon the incorporation of the tested copolymers.

The copolymers bearing more than one lipid-mimetic anchors per chain provide effective steric stabilization also of liposomes based on dipalmitoyl phosphatidylcholine (DPPC). Well-separated unilamellar liposomes were the dominant species even at concentrations as high as 10 mol% (Momekova et al. 2010). The liposomes stabilized by these copolymers are typically larger in size than the plain liposomes. This is especially pronounced for the higher copolymer contents. As noted earlier, the less conical and more cylindrical macromolecules are expected to fit better to less-curved surfaces. In addition, the larger hydrophobic part takes a larger area in the lipid membrane, thus ensuring more space in the surface above the lipid bilayer to be taken by the hydrophilic PEG chain, without interaction, or with much weaker interaction with the neighboring PEG chains. The combination of these two factors, that is, the preference to less-curved surfaces and the weak lateral interactions between the PEG chains, may be the reason why the liposomes, stabilized by the former two copolymers, are larger in size (Momekova et al. 2010). It has been shown that the copolymers condition the formation of long-circulating pH-sensitive liposomes based on dioleoylphosphatidylethanolamine (DOPE) (Momekova et al. 2007). DOPE is a non-lamellar and highly fusogenic lipid. A very important advantage of the copolymers was that they facilitated the formation of liposomes and that their incorporation in the membrane did not significantly interfere with the pH sensitivity of the formulation and even in some cases the acid-triggered calcein leakage was optimized.

Structurally related, acid-labile PEG-conjugated vinyl ether lipids have been used to stabilize DOPE as unilamellar liposomes (Shin et al. 2003). Acid-catalyzed hydrolysis of the vinyl ether bonds destabilized these liposomes by removal of the PEG layer, thereby promoting release of the contents on the time scale of hours at pH<5. pH-sensitive liposomes which generate fusion ability under weakly acidic conditions have been developed by surface modification of liposomes with polyglycidol derivatives having carboxyl groups (Kono et al. 1997, Sakaguchi et al. 2008, Yuba et al. 2010). The liposomes were prepared from equimolar mixtures of EPC and DOPE and contained 30 wt.% of the polymers. The latter were hydrophobically modified and, as seen from Figure 3.18, bear an anchoring chain. The liposomes were prepared by vortexing or extrusion. Data for their dimensions and zeta-potential are presented in Figure 3.18.

Formation of hybrid lipid–polymer vesicles with a homogeneous membrane structure has been reported elsewhere (Ruysschaert et al. 2005). Vesicles were obtained from

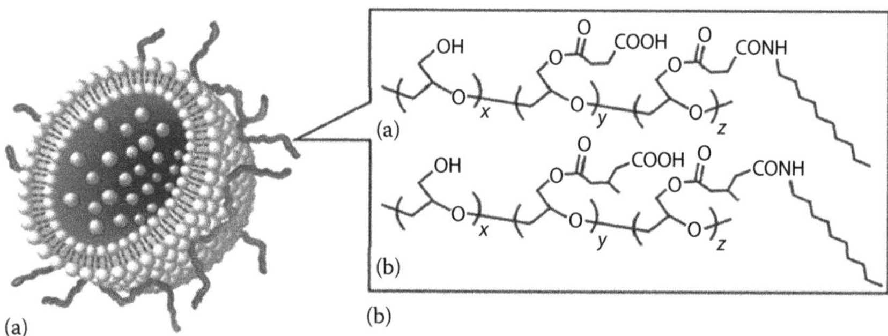

Liposome	Method	Zeta-potential (mV)	Particle size (nm)
Unmodified	Vortex	−0.51 (0.73)	1560 (817)
Unmodified	Extrusion	−0.92 (0.42)	116 (26)
SucPG	Vortex	−11 (2.4)	431 (112)
SucPG	Extrusion	−12 (3.5)	105 (18)
MGluPG	Vortex	−11 (3.5)	523 (78)
MGluPG	Extrusion	−11 (2.4)	110 (23)

FIGURE 3.18 Cartoon of a pH-sensitive polymer-modified liposome loaded with active substance and chemical structures of succinylated polyglycidol (SuPG) (a) and 3-methylglutarylated polyglycidol (MGluPG) (b). Particle size and zeta-potential data are presented in the table. Numbers in parenthesis give the standard deviations. (Reprinted from *Biomaterials*, 31, Yuba, E., Kojima, C., Harada, A., Tana, Watarai, Sh., and Kono, K., 943–51, Copyright 2010, with permission from Elsevier.)

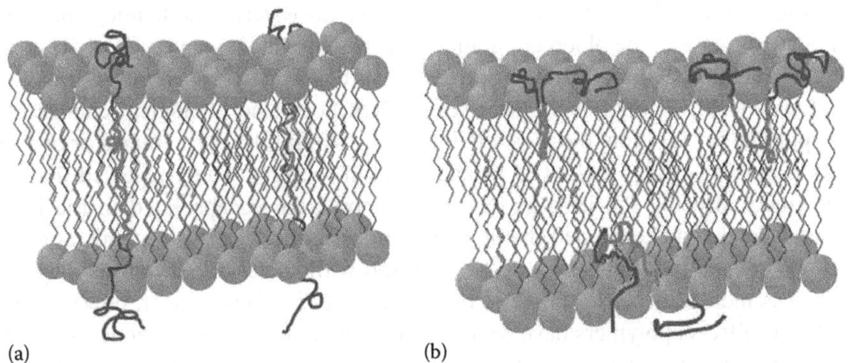

FIGURE 3.19 Schematic of possible modes of Pluronic interactions with lipid bilayers: full (a) and partial (b) insertion of PPO in the lipid bilayer. (Reprinted with permission from Firestone, M.A., Wolf, A.C., and Seifert, S., *Biomacromolecules*, 4, 1539–1549. Copyright 2003, American Chemical Society.)

an amphiphilic triblock copolymer poly(2-methyloxazoline)-poly(dimethylsiloxane)-poly(2-methyloxazoline) (PMOXA-PDMS-PMOXA) (total molecular weight of 9000 g/mol, i.e., a PDMS middle block of 5400 g/mol and two hydrophilic PMOXA blocks each of 1800 g/mol) and lipids such as egg phosphatidylethanolamine, EPC, and dipalmitoylphosphatidylcholine by using different preparation techniques, which gave different dimensions of the particles but invariably a homogeneous distribution of the copolymer and lipid in the membrane. Unimer exchanges on a time scale of minutes were observed.

The effect of PEO-PPO-PEO triblock copolymers on their mode of interaction with model membranes based on dimyristoyl-*sn*-glycero-3-phosphocholine (DMPC) has been evaluated (Firestone et al. 2003). Full insertion (Figure 3.19) was observed when the PPO block was sufficiently long, that is, its chain length commensurate with the acyl chain dimensions of the lipid bilayer. Poor integration of the polymer is observed, allowing the PEO chains to adopt a more flat configuration (i.e., laterally directed orientation), when the PPO block length is less than the bilayer leaflet. The two possible modes of interactions are schematically presented in Figure 3.19. At reduced temperatures, all PEO-PPO-PEO copolymers studied, independently of molar mass of PPO, were expulsed from the bilayer, which hinted a possible approach to controlled release. The dependence of structure on temperature, however, indicates that even with optimized PPO chain length (to promote bilayer insertion), the PEO-PPO-PEO triblock copolymers are generally more weakly held in the lipid bilayer than the PEG-lipids and may therefore not be ideal for use in the steric stabilization of vesicles (Firestone et al. 2003).

Being relatively inexpensive, readily available, and compositionally and architecturally rich, the PEO-PPO-EPO copolymers have been considered as substitutes of the PEG-lipids in preparation of sterically stabilized liposomes for drug delivery. In a series of papers, the interactions of PEO-PPO-PEO copolymers with phosphatidylcholine liposomes have been investigated in order to probe the degree of steric stabilization and the mode of incorporation (Kostarelos et al. 1995, 1998a,b, 1999).

The results obtained indicate that it is possible to induce steric stabilization and even a limited increase in the blood circulation time has been documented (Woodle et al. 1992). Structural studies, however, revealed that inclusion of low amounts (2 mol%) of PEO-PPO-PEO copolymers gave rise to significant morphological changes of the liposome preparation, the most obvious being small bilayer fragments or bilayer disks (Johnsson et al. 1999). The leakage data in the same study revealed that all copolymers affected the leakage of carboxyfluorescein encapsulated in the liposomes. The authors concluded that it is doubtful that effective steric stabilization of phosphatidylcholine liposomes can be achieved without serious structural changes as well as increased permeability of the liposomes. In another paper, however, the PEO-PPO-PEO copolymers have been shown to increase the spontaneous curvature of their mixtures with DOPE—a lipid with propensity to form an inverted hexagonal phase at acidic pH (Bergstrand and Edwards 2004). The inclusion of these copolymers stabilized the lamellar phase, and liposomes that were stable and non-leaky at low pH were produced. In contrast to the case with phosphatidylcholine liposomes, no signs of bilayer disks, fragments, and even open liposomes were observed at copolymer concentrations corresponding to 10 mol%. The difference is likely to be connected to differences in molecular properties and phase propensity of the lipids.

The penetration of nonionic copolymers such as PEO-PPO-PEO into the hydrophobic membranes is primarily driven by hydrophobic interactions. The hydrophilic/hydrophobic balance and the overall hydrophobicity as well as parameters such as CMC and critical micellization temperature (CMT) have been found to drastically affect the polymer incorporation into the lipid bilayer and, more generally, the type of mixed aggregates formed. For example, the interactions of $(EO)_2(PO)_{30}(EO)_2$ with DPPC liposomes were stronger than those of the more hydrophilic $(EO)_{100}(PO)_{65}(EO)_{100}$, which led to a complete disruption of the lipid membrane and formation of crew-cut aggregates as shown in Figure 3.20 (Chieng and Chen 2009). Entire solubilization of the lipid membrane and formation of mixed micelles have

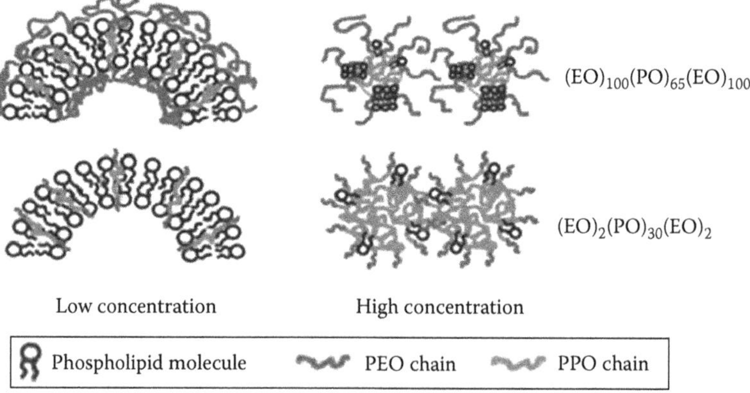

FIGURE 3.20 Schematic presentation of mixed lipid–polymer structures depending on the copolymer composition and concentration. (Reprinted with permission from Chieng, Y.Y. and Chen, S.B., *J. Phys. Chem. B*, 113, 14934–14942. Copyright 2009, American Chemical Society.)

typically been observed when the CMC of a specific PEO-PPO-PEO copolymer was reached at a given temperature (Wu and Lee 2009, Wu et al. 2009). A recent example for association between $(EO)_{100}(PO)_{65}(EO)_{100}$ and liposome reported elsewhere suggested that the formed types of lipid/polymer complexes depended not only on the temperature and relative amounts of both components but also on the thermal pretreatment of the sample system with respect to T_m of the lipid and CMT of the copolymer (Feitosa and Winnik 2010). By isothermal titration calorimetry, Wu et al. have shown that $(EO)_{132}(PO)_{56}(EO)_{132}$ macromolecules are incorporated into fluid-phase DMPC liposomes above their main transition temperature (T_m=24°C), whereas at temperatures below T_m, the gel-phase membranes of DMPC prevent the incorporation (Wu and Lee 2009). Starting from mixed complexes, decreasing the temperature below the T_m of DMPC, the spherically shaped lipid/polymer vesicles change to disk-shaped morphology found by phase separation between the polymer and the lipid gel phase.

Cheng and Tsourkas have recently reported on preparation of polymersomes by co-assembly of PBD-*b*-PEO (PBD and PEO molar masses of 2500 and 1300, respectively) and 1-palmitoyl-2-oleoyl-*sn*-glycero-3-phosphocholine (POPC) (Cheng and Tsourkas 2008). Nanometer-sized vesicles (averaged hydrodynamic diameter of 125 nm) were prepared by hydration of a dried thin film of PBD-*b*-PEO containing 15% POPC followed by freeze-thawing and extrusion through 100 nm pore-size filters. PBD moieties in the interior of the membrane were cross-linked by free-radical polymerization, and afterward, the POPC molecules were extracted from the membrane, thus generating pores in the membrane (Figure 3.21). The resulting porous polymersomes were intended for magnetic resonance imaging applications. The porous structure of the membrane accelerated the water-exchange rate between the water-bound contrast agent (chelated Gd) and the surrounding bulk water, which is crucial for the performance of the contrast agent.

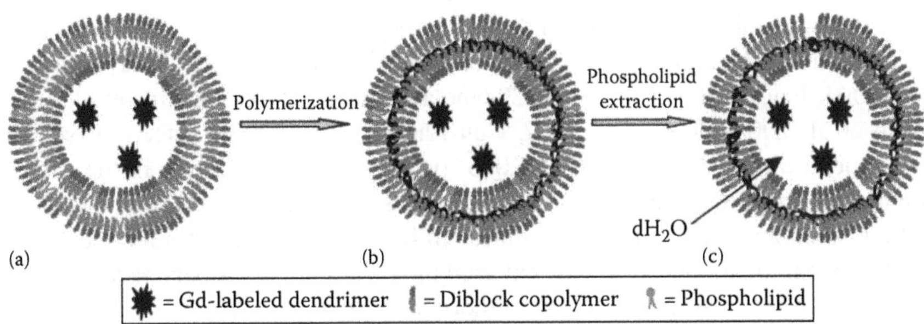

FIGURE 3.21 Schematic diagram illustrating the approach used to synthesize paramagnetic porous polymersomes. (a) Vesicles consisting of the diblock copolymer PBD-*b*-PEO and the phospholipid POPC at a molar ratio of 85:15 were prepared. Second-generation dendrimer conjugated Gd chelates were encapsulated within the aqueous interior during vesicle formation. (b) Cross-linking of PBD-*b*-PEO within the vesicle bilayer was induced by free-radical polymerization. (c) Pores were formed in the polymersome bilayer by extracting the POPC with the surfactant Triton X-100. (Reprinted with permission from Cheng, Zh. and Tsourkas, A., *Langmuir*, 24, 8169–8173. Copyright 2008, American Chemical Society.)

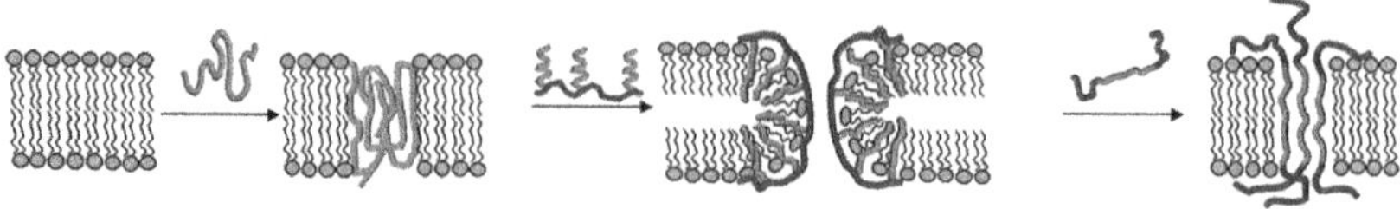

FIGURE 3.22 Pore formation by full insertion of (from left to right) a homopolymer (polyelectrolyte), an amphiphilic graft copolymer, and an amphiphilic triblock copolymer. (Schulz, M., Olubummo, A., and Binder, M., W.H., *Soft Matter*, 8, 4849–4864, 2012. Reproduced by permission of The Royal Society of Chemistry.)

Formation of pores and channels in lipid membranes can be induced simply by insertion of synthetic polymers, that is, without any additional posttreatment such as cross-linking and extraction. Several reviews summarize the important factors for formation of channels and pores (Binder 2008, Tribet and Vial 2008). Figure 3.22 presents pore formation by insertion of (co)polymers of different chain architecture. Recent studies report on nonspecific permeability of lipid bilayers induced by incorporation of hydrophobically modified PAAs (Tribet et al. 2007). Nanometer-sized (1–5 nm) polymer channels were formed depending on polymer concentration with the ability of transmembrane exchange of small molecules such as albumin and dextran. Formation of small pores permeable for low molecular weight solutes upon the adsorption of PAA on liposomes composed of egg lecithin in slightly acidic milieu has been reported (Berkovich et al. 2012). According to these authors, the dipole potential is an important driving force for the insertion of polyacids into biological membranes. Molecular dynamic simulations and complementary experimental methods have shown that a novel polycation, poly(allyl-*N*,*N*-dimethyl-*N*-hexylammonium chloride), can associate and insert into zwitterionic POPC membranes (Kepczynski et al. 2012). The lamellar structure was not influenced; however, formation of hydrophilic pores was promoted, which increased considerably the membrane permeability. The effects of polymer architecture and molecular weight on the generation of pores and permeability of lipid membranes have been studied for a large variety of di-, tri- and star-block copolymers (Cox et al. 1999, Francis et al. 2002, Firestone et al. 2003, Peleshanko et al. 2004, Cheyne and Moffitt 2005, Demina et al. 2005, Logan et al. 2005, Frey et al. 2007, Gau-Racine et al. 2007). Effects of sealing (Firestone et al. 2003, Yasuda et al. 2005) or membrane destruction in the case of cationic polymers (Gabriel et al. 2008, Waschinski et al. 2008) have been observed.

An elegant approach is to apply controlled breakup of the membrane. Such triggered permeabilization achieved by external light irradiation has been recently demonstrated (Tribet et al. 2010). These authors found that azobenzene-modified PAA copolymers embedded in dioleoylphosphatidylcholine GUVs can induce membrane perturbation by *cis–trans* photoconversion.

Mixed aggregates can be obtained upon solubilization of commonly used lipids by polymers. In particular, to a dimyristoylphosphatidylcholine dispersion, undergoing freeze-thawing and extrusion through a 400 nm pore-size filter, a solution of styrene–maleic acid copolymer (approximate molecular weight of 9500 kDa and styrene-to-maleic acid ratio of 3:1) was added to obtain small disks—around 9 nm in diameter by dynamic light scattering (DLS) (Orwick et al. 2012). The system studied is considered as a detergent-free alternative to membrane scaffold protein–stabilized nanodisks and has potential application in biophysical studies with membrane proteins and as delivery vehicles for hydrophobic compounds.

The interactions between charged polymers and lipid membranes are driven by electrostatic forces, thereby forming a polymer layer around the liposome surface. Compared to nonionic copolymers, the deep incorporation of charged polymers into the hydrophobic membrane interior is restricted. Instead, the adsorption of the polymer molecules can have strong effects on the membrane organization such as segregation of lipid molecules, migration between the two monolayers (flip-flop), and membrane disruption (Hong et al. 2006, Menger et al. 2006, 2009, Whitten et al. 2009). The partial neutralization of the liposomal surface leads to reduction of interliposomal repulsion, which may eventually result in interliposomal polymer migration (Menger et al. 2009) and membrane fusion (Menger et al. 2011). The polymer can adopt different conformations on the liposomal surface and accordingly either induce or prevent membrane fusion. This dual effect is presented in Figure 3.23 for two cationic polymers—poly(*N*-ethyl-4-vinylpyridinium bromide) and polylysine. As seen the thin polylysine layer formed on the surface decreased the electrostatic repulsion between the negatively charged vesicles, which ensured liposomal contact, and consequently, the polymer-decorated liposomes became leaky to water, which, in turn, promoted exchange of adjacent membrane parts (Figure 3.23a and b). As opposed to this, poly(*N*-ethyl-4-vinylpyridinium bromide) macromolecules adopted a looping conformation thus forming a physical barrier that prevented fusion between vesicles due to the increased thickness of the polymer shell compared to polylysine (Figure 3.23c and d).

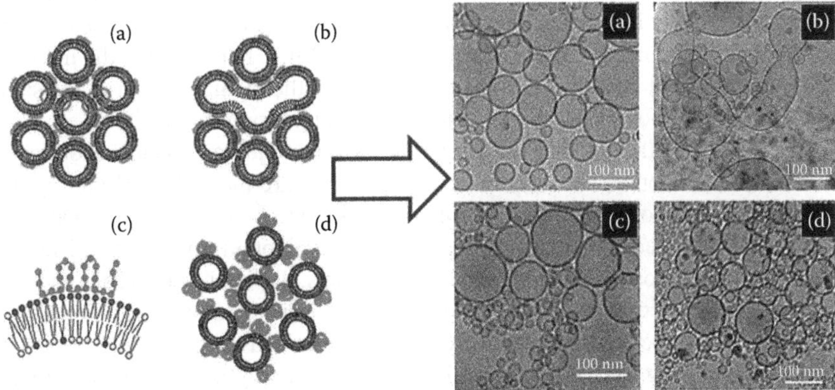

FIGURE 3.23 Schematic illustration (left) and corresponding cryo-TEM images (right) of the polylysine/liposome complexes before (a) and after (b) fusion. Schematic illustration (left) and corresponding cryo-TEM images (right) of looping in a poly(*N*-ethyl-4-vinylpyridinium bromide)/liposome complex, showing lateral phase separation (domain formation) induced by poly(*N*-ethyl-4-vinylpyridinium bromide) (c) and formation of a physical barrier preventing liposomal fusion (d). (Reprinted with permission from Menger, F.M., Yaroslavov, A.A., Sybachin, A.V., Kesselman, E., Schmidt, J., Talmon, Y., and Rizvi, S.A.A., *J. Am. Chem. Soc.*, 133, 2881–2883. Copyright 2011, American Chemical Society.)

Rinaudo et al. report on an effective stabilization of lipid vesicles against pH, osmotic, and salt shocks by electrostatic interactions with polyelectrolytes (Rinaudo et al. 2012). Chitosan and hyaluronan of different molecular weights were used in the study from which it was concluded that the former was adsorbed in a flat conformation, while the latter formed loops and trains. The maximum amount of polymer adsorbed on the external membrane surface corresponded to a low degree of coverage and was interpreted within the frame of a "patch-like" model. Generally, at a low degree of surface coverage, vesicle aggregation is observed. In the excess of polyelectrolyte, a homogeneous polymer layer is formed, which prevents the vesicles from aggregation. Systems based on polyelectrolytes and polyelectrolyte complexes are sensitive to the presence of salts. The effect of salt concentration has been studied elsewhere (Menger et al. 2009, Ngo and Cosa 2010, Sikor et al. 2010). High sensitivity of complexes of polyethyleneimine (PEI) and DMPC against NaCl addition has been observed (Sikor et al. 2010): At low NaCl concentration, aggregation of the complexes was observed, whereas at high salt concentrations, vesicle aggregation was prevented, concomitant with drastic changes in the DMPC chain-melting transition of the hydrocarbon tails, suggesting a strong penetration ability of PEI molecules into the lipid bilayer at high salt concentrations. Ngo and Cosa have investigated the interaction between the negatively charged fluorescent polyelectrolyte poly[5-methoxy-2-(3-sulfopropoxy)-1,4 phenylene vinylene] and DOPC in the presence of mono- and dications (Ngo and Cosa 2010). The addition of Ca^{2+} ions showed a significant increase in lipid/polymer interactions leading to polymer uptake by the DOPC liposomes. The systems exhibited potential as light-harvesting devices.

Undoubtedly, other components of the membrane also influence the vesicle's structure and properties. Yaroslavov et al., for example, studied the adsorption of the synthetic polycation, poly(*N*-ethyl-4-vinylpyridinium bromide), on the surface of three-component liposomes, composed of anionic cardiolipin, electroneutral egg lecithin, and nonionic cholesterol (Yaroslavov et al. 2012). In particular, the incorporation of cholesterol increased the microviscosity of the membrane but at the same time caused formation of defects and made the complexation irreversible.

An interesting class of nanoparticles that have emerged as potent nanocarriers alternative to liposomes and polymeric nanoparticles are the *lipid–polymer hybrid nanoparticles*. In some aspects, they can be considered as "inverted" analogues of polymer-stabilized liposomes, though with their lipophilic interior and hydrophilic shell, they closely resemble emulsion droplets. Typically they are composed of a hydrophobic (co)polymer enveloped by lipid layers. The hydrophobic interior (the polymer) can be loaded with therapeutic or other active agents. Inhalable dry-powder form of drug-loaded nanoparticles composed of poly(lactic-co-glycolic acid) (PLGA), egg lecithin, and levofloxacin as polymer, lipid, and drug models, respectively, has been reported (Wang et al. 2012). In another study, the nanoparticles, designed for anticancer drugs to bypass multidrug resistance in human breast cancer, consist of a core composed of dimethyldidodecylammonium bromide-modified PLGA surrounded by DPPC shell (Li et al. 2012). Very close to the emulsion droplets are the nanocapsules described elsewhere (Sanchez-Moreno et al. 2012). They are of core–shell structure consisting of an olive-oil-filled core and a shell composed of different phospholipids (phosphatidylserine, lecithin) and biocompatible macromolecules such as PEO-PPO-PEO or chitosan. The latter generates an enriched carboxylic surface, which can be linked to specific antibodies, which, in turn, are expected to facilitate the uptake of the nanocapsules by cancer cells.

The lipid molecules considered so far in this section are diglycerides, that is, double-chained lipids, which typically display the propensity to form a lamellar mesophase in binary lipid–water systems and liposomes when dispersed in large excess of water. Depending on the temperature, water content, and mainly, lipid molecular structure (single or double chained, length of fatty acyl chains, degree of saturation), a variety of lyotropic liquid crystals have been observed, the most spectacular examples of which are perhaps the bicontinuous cubic phases. The bicontinuous cubic-phase liquid crystal represents a 3D lattice built from a contorted bilayer that separates two water spaces. The reader can refer to Figures 2.2 and 2.106. The water domains are identical, continuous but nonintersecting (Scriven 1976). The lipophilic part of the liquid crystal consists of an infinite bilayer periodically curved in three dimensions (Larsson 1989, Lindblom and Rilfors 1989), the midplane of which forms the so-called infinite periodic minimal surface (Patton and Carey 1979, Patton et al. 1985). The bicontinuous cubic liquid-crystalline phase is transparent, optically clear, that is, non-birefringent, soft, and highly viscous, and has a unique structure at the nanometer scale. Depending on the lipid composition and type of the mesophase, the diameter of water channels may vary from 4 to 20 nm, whereas the bilayer thickness is 3–5 nm depending on the length of the acyl chain. Essential features are the very large surface area that has been estimated to be around 400 m^2/g (Lawrence 1994) and the high bilayer content per unit volume. Due to the presence of lipid bilayers and water channels, both water-soluble and oil-soluble substances as well as molecules with amphiphilic character can be solubilized in the cubic phase. The substances tend to be stable in the matrix and the unique structure provides control over their release.

A small but important part of the research has been directed toward development of methods for dispersing cubic phases as well as other liquid-crystalline nonlamellar phases to form nanostructured aqueous dispersions. When such a phase is dispersed in water, it is assumed that the highly ordered internal organization and the properties of the liquid-crystalline phase are preserved in the nanoparticles. Figure 3.24 shows a cryogenic field emission scanning electron microscopy (cryo-FESEM—a technique that complements cryo-TEM and small-angle x-ray scattering SAXS) image of a discrete particle with the same underlying tortuous structure entirely consistent with the bulk cubic phase, from which the particles are prepared, and closely resembling the mathematical description of nodal surface representation (Rizwan et al. 2007). Such dispersed particles are termed *cubosome*, *hexasomes,* after the name of the corresponding liquid-crystalline bulk phase, or *isasomes* as a collective name, which comes from internally self-assembled particles or *somes*, which has been coined by Yaghmur and coworkers (Yaghmur et al. 2006b).

Aqueous dispersions of reverse bicontinuous cubic phase, Q_{II} (Gustafsson et al. 1997, Larsson 2000, Spicer et al. 2001, de Campo et al. 2004, Barauskas et al. 2005); reverse hexagonal phase, H_{II} (Yaghmur et al. 2005, Dong et al. 2006); reverse discontinuous cubic phase, I_{II} (Nakano et al. 2002, Yaghmur et al. 2006b); and sponge phases, L_3 (Barauskas et al. 2005, 2006) are good examples for dispersions of

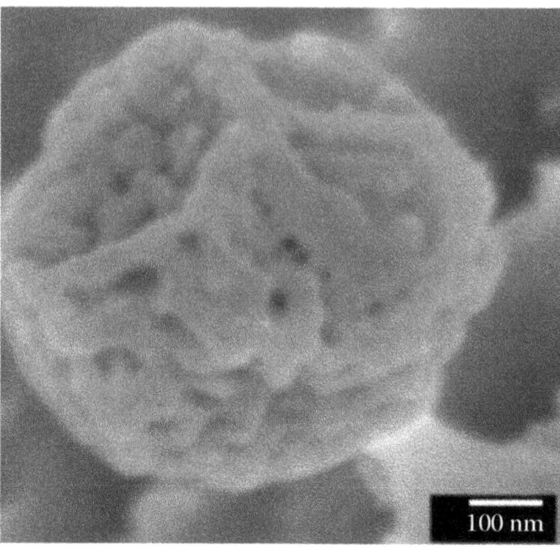

FIGURE 3.24 Cryo-FESEM image of a particle with an internal structure consistent with the reverse cubic bicontinuous phase. (Reprinted from *Micron*, 38, Rizwan, S.B., Dong, Y.D., Boyd, B.J., Rades, T., and Hook, S., 478–485, Copyright 2007, with permission from Elsevier.)

particles with a size of few nanometers and distinctive internal well-ordered periodic structure. Neither of these liquid-crystalline phases, however, is able to form stable dispersions in water because large portions of the hydrophobic sections are exposed to water. Therefore, aqueous dispersions are typically produced by high-energy mechanical fragmentation or ultrasonic irradiation of a bulk phase in the presence of dispersing agents—typically amphiphilic polymers—to efficiently cover the outer surface of the dispersed particles.

Typical lipids forming reverse phases are unsaturated long-chain monoglycerides such as glyceryl monooleate (GMO). Since the pioneering studies of Larsson and coworkers on the formation and characterization of GMO-based dispersions (Larsson 1989, Gustafsson et al. 1996, 1997), various nanostructured particles based on other lipids have been described (Abraham et al. 2005, Johnsson et al. 2005, Boyd et al. 2006, Dong et al. 2006, Yaghmur and Glatter 2008). The most suitable techniques for characterization of the morphology and internal structure of the dispersed particles are SAXS and cryo-TEM. Figures 3.25 and 3.26 show collections of cryo-TEM images taken from dispersions of particles with internal morphologies.

The stabilization of the dispersions in their large majority is performed by one of the copolymers of the PEO-PPO-PEO family, namely, $(EO)_{99}$-$(PO)_{67}$-$(EO)_{99}$. Presumably, the copolymer has no significant impact on the confined nanostructure as it is accommodated at outermost particle surface. According to some authors, however, due to its penetration into the oil-free cubic phase, the copolymer can change the symmetry thus influencing the internal structure of the dispersions (Yaghmur et al. 2006b). The internal structure can be modulated by varying the lipid composition and temperature conditions. The structural transitions induced are reversible and identical to those observed in the corresponding nondispersed phases coexisting with an excess of water (Yaghmur et al. 2005, 2006a).

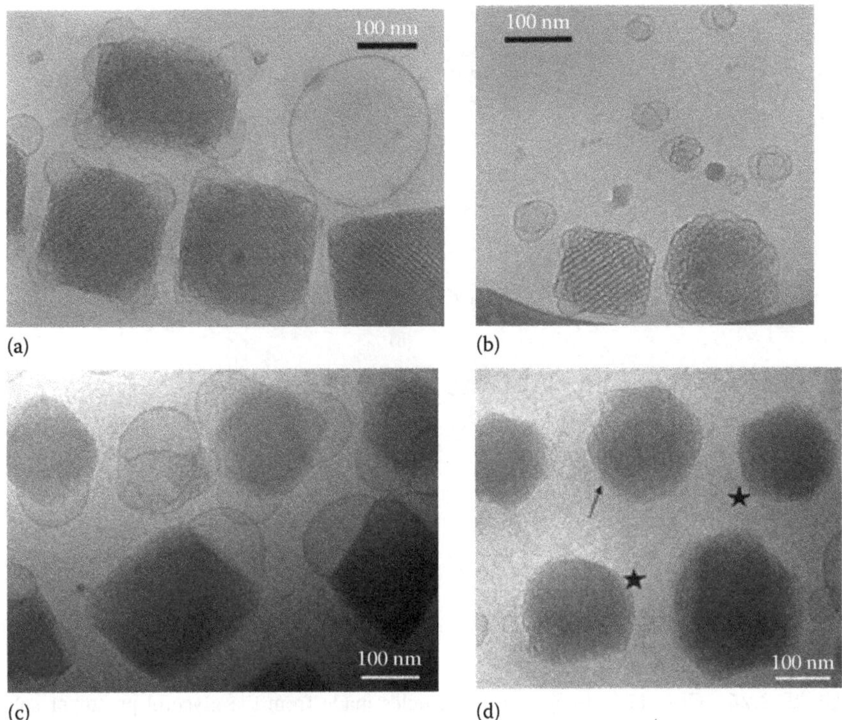

FIGURE 3.25 Cryo-TEM micrographs of particles with internal cubic and hexagonal morphologies. The sample in (a) was prepared by microfluidization at 80°C of a mixture of GMO (92 parts) and EO_{99}-PO_{67}-EO_{99} (8 parts) in excess water and quenching at 25°C. (b) The sample of the same composition in (a) prepared by sonication and the micrograph was taken 5 weeks after sonication. Bar = 100 nm. (Reprinted with permission from Gustafsson, J., Ljusberg-Wahren, H., Almgren, M., and Larsson, K., *Langmuir*, 12, 4611–4613. Copyright 1996, American Chemical Society.) Cryo-TEM micrographs of cubosomes at 25°C (c) and hexasomes at 55°C (d). The dispersions consisted of 4.625 wt.% monolinolein, 0.375 wt.% EO_{99}-PO_{67}-EO_{99}, and 95 wt.% water were prepared by ultrasonication. The arrow and stars denote an internal hexagonal symmetry and curved striations, respectively. (Reprinted with permission from de Campo, L., Yaghmur, A., Sagalowicz, L., Leser, M.E., Watzke, H., and Glatter, O., *Langmuir*, 20, 5254–5261. Copyright 2004, American Chemical Society.)

Dispersions by means of microfluidization are typically made by forming a coarse dispersion by mixing a homogeneous sample of melted GMO and stabilizing copolymer with an excess water (93–99 w/w%) followed by microfluidization at 80°C and slow cooling at room temperature. In this way, the cubosomes crystallized slowly inside the original emulsion droplets. The content of the stabilizing PEO-PPO-PEO copolymer is typically below 10 w/w%. When properly prepared, the cubosomes formed in this system are particles of roughly cubic shape with about 200 nm side length that stay stable for months (Gustafsson et al. 1996). Typical aggregates found after microfluidization are shown in Figure 3.25a. Besides a minor population of vesicles, the dominating aggregate morphology is faceted particles.

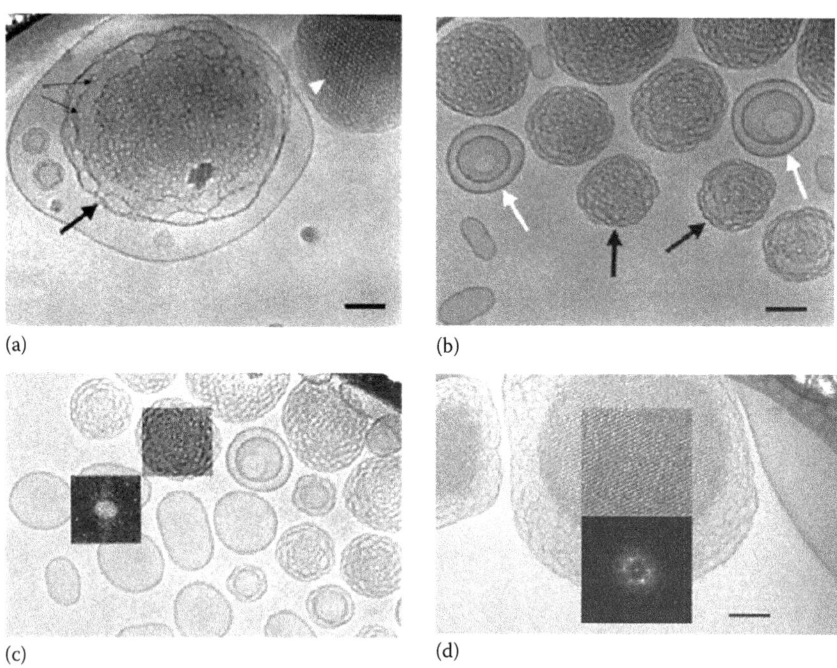

FIGURE 3.26 Cryo-TEM images from particles made from the glycerol precursor solutions. DDP$(EO)_{92}$/GMO at 0.01 (a), 0.048 (b), and 0.05 (c) molar ratios. The arrowhead in (a) shows particles with a periodic internal structure, compatible with a bicontinuous cubic phase. White arrows in (b) indicate particles with deep invaginations. The black arrows point out interlamellar attachments, in side on view (thick arrows) and face on view (thin arrows). The FFT insert in (c) reveals a disordered, but bicontinuous, L_3, interior. (d) Particles found in the excess water of a bulk gel of GMO containing 7 wt.% of $(DDGG)_4(EO)_{114}$. The bar represents 100 nm. (a and b: Reprinted with permission from Rangelov, S. and Almgren, M., *J. Phys. Chem. B*, 109, 3921–3929. Copyright 2005, American Chemical Society; c and d: Almgren, M. and Rangelov, S., *J. Disp. Sci. Technol.*, 27, 599–609, 2006. Reprinted with permission from Taylor & Francis Group.)

Vesicular structures were often found attached to them, usually at the corners. An increase in the proportions of copolymer from 8 to 11 w/w% leads to an increased amount of vesicles in the dispersion but did not significantly change the structure or internal morphology of the particles. If instead the same mixture is dispersed by sonication at room temperature, which involves the highest input of energy, much smaller "folded vesicles" are formed initially, and only slowly, some of these particles evolve into well-shaped cubosomes (Figure 3.25b). De Campo et al. have demonstrated for the first time that the internal structure of the dispersed particles can be tuned by temperature in a reversible way (de Campo et al. 2004). Upon increasing temperature, the internal structure underwent a transition from cubic via hexagonal to fluid isotropic L_{II} phase as documented by SAXS. The samples were prepared by means of ultrasonication from a binary monolinolein/water system stabilized by EO_{99}-PO_{67}-EO_{99}. Figure 3.25c and b show micrographs of particles with an internal bicontinuous cubic structure, cubosomes, at 25°C (Figure 3.25c) that transformed into particles exhibiting an internal hexagonal symmetry, hexasomes, at 55°C (Figure 3.25d). The surface vesicles attached on the cubosomes is a notable feature.

Traditionally, stabilization of *isasomes* is achieved by $(EO)_{99}$-$(PO)_{67}$-$(EO)_{99}$ and very few other stabilizers have been reported. It has recently been discovered that another copolymer of the same family—$(EO)_{132}$-$(PO)_{50}$-$(EO)_{132}$, which is characterized by a slightly shorter PPO block and higher (80%) PEO content—is superior to $(EO)_{99}$-$(PO)_{67}$-$(EO)_{99}$ (Chong et al. 2011). An interesting approach is that of Driever and coworkers who have shown that cubosomes stabilized by $(EO)_{99}$-$(PO)_{67}$-$(EO)_{99}$ can be incorporated into a polymeric layer-by-layer system while maintaining their internal structure (Driever et al. 2011). In the following, we give some examples of alternative polymeric stabilizers and techniques to produce discrete internally self-organized nanoparticles.

Rangelov and Almgren have applied a liquid precursor method to obtain particles with an internal structure (Rangelov and Almgren 2005, Almgren and Rangelov 2006). The liquid precursor consisted of solutions of GMO and different PEG-based copolymers bearing blocks of lipid-mimetic anchors (see Figure 3.14 for the chemical structure of the lipid-mimetic anchors) in copolymer to GMO molar ratios <0.1 in glycerol. The glycerol solutions were heated to 60°C and added dropwise to 10-fold excess of water under vigorous stirring thus avoiding the input of high energy via microfluidization or ultrasonication. Upon hydration of the liquid precursors, nanosized (from ca. 80 to 130 nm in radii, according to DLS) particles were produced. Cryo-TEM (some micrographs are shown in Figure 3.26) evidenced formation of polymorph dispersed particles. A fraction of them exhibited highly ordered interior, compatible with a bicontinuous cubic phase, and less-ordered outer regions (Figure 3.26a and b). Such particles coexisted with other particles seemingly formed from another bicontinuous isotropic phase, L_3 (Figure 3.26c). The latter, together with arrays of interlamellar attachments that were found to consist of the interior of a third group of particles with an internal structure, were considered ancestors of the cubic structures. A particulate phase was also found in the excess water of bulk phases. Particles with a disordered outer region that surrounded an inner highly ordered region of the type shown in Figure 3.26d were abundant (Almgren and Rangelov 2006). The fast Fourier transform (FFT) insert shows an evident hexagonal motif.

Other techniques to obtain cubic particles, involving preparation of liquid or dry precursors, have been described elsewhere (Spicer et al. 2001, 2002). These techniques also avoid the high-energy dispersion of cubic gels. The "hydrotrope" process allowed the creation of concentrated liquid precursor, in particular, solution of the lipid in ethanol, that spontaneously produced cubosomes upon hydration with water containing a stabilizing polymer (Spicer et al. 2001). The dry-powder precursors were produced by spray drying of GMO encapsulated in starch or dextran (Spicer et al. 2002). Cryo-TEM images in Figure 3.27 reveal nanostructured particles with large unfolded bilayers on their surfaces. One limitation of this technique is the large proportion of polymer required for encapsulation (75% and 60% w/w for starch and dextran, respectively) that will limit the amount of active material that can be incorporated for subsequent delivery.

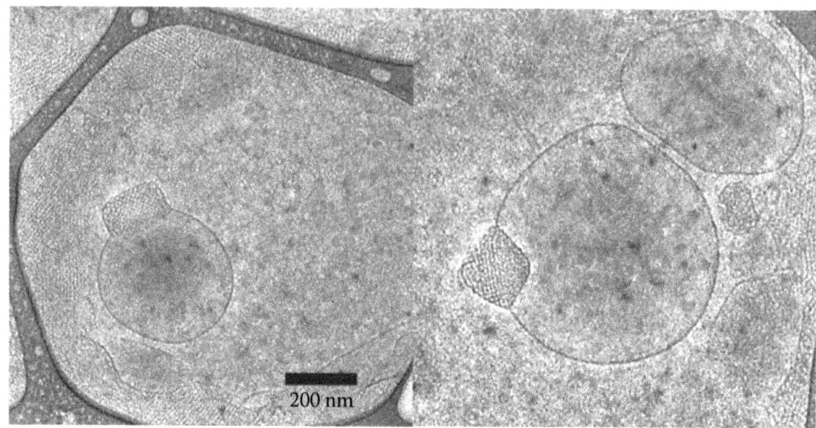

FIGURE 3.27 Cryo-TEM photographs of cubosome particles (with large surface vesicles) that form upon hydration of the starch–monoolein powders. Also visible on the edges of the grid surrounding the particles are regions of cubic liquid-crystalline gel that are not dispersed into discrete cubosomes. (With kind permission from Springer Science+Business Media: *J. Nanoparticle Res.*, 4, 2002, 297, Spicer, P.T., Small, W.B., Lynch, M.L., and Burns, J.L.)

Sterically stabilized cubosomic nanoparticles have been prepared by fragmentation of PEGylated cubic lipid phase (Angelov et al. 2012). The latter was formed from GMO and its PEGylated derivative (GMO-PEG2000) in molar ratios favoring the formation of reverse liquid-crystalline structures. The amphiphilic polymer considerably affected the interfacial curvature of the cubic membrane and, under agitation, contributed to the fragmentation of the cubic lattice into nanoparticles. In fact, the purpose of the work was to investigate the entrapment of protein molecules in the cubosomic nanoparticles. The authors studied α-chymotrypsin A and suggested that the protein molecules were entrapped in the interior of the PEGylated cubosomes via a "nanopocket" mechanism. The protein was able to essentially stabilize the cubosomic particles, which display well-defined inner organization of nanochannels in their freeze-fracture planes.

A specially synthesized charged block copolymer (see Figure 3.28 for its structural formula and molecular characteristics (Stubenrauch et al. 2007)) has been used in quantities up to 4 wt.% to stabilize reverse lyotropic phases based on phytantriol (Dulle and Glatter 2012). By using charged stabilizers, the authors aimed at creating a highly negative zeta-potential of the particles, which could be exploited as a means to control their adsorption onto charged surfaces. Compared to other stabilizers such as the traditionally used $(EO)_{99}$-$(PO)_{67}$-$(EO)_{99}$ as well as low molecular weight surfactant, SDS, and silica particles, the newly synthesized copolymers gave the highest increase in surface charge of the particles without changing their internal structure.

3.4 POLYMER–POLYMER HYBRID STRUCTURES

The size, morphology, and properties of copolymer assemblies are typically controlled through changes in molecular parameters, copolymer chemistry, and functionality via synthetic manipulations. The high kinetic stability of these assemblies,

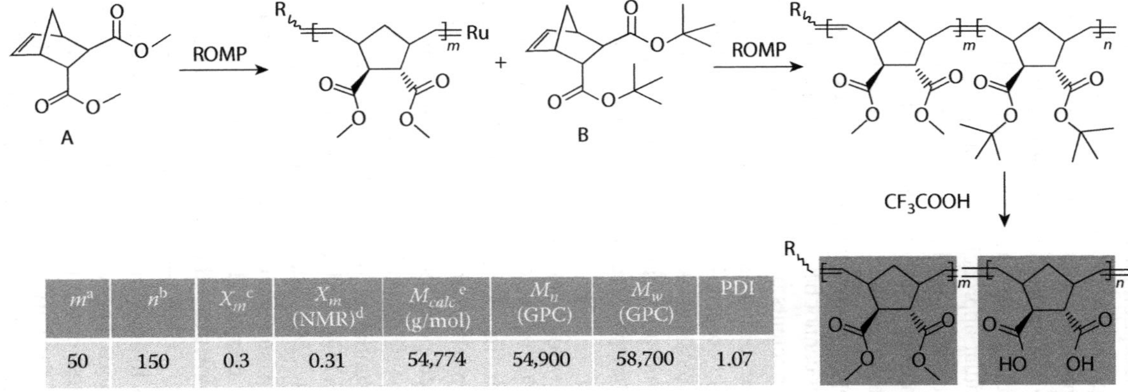

m^a	n^b	$X_m{}^c$	X_m (NMR)[d]	$M_{calc}{}^e$ (g/mol)	M_n (GPC)	M_w (GPC)	PDI
50	150	0.3	0.31	54,774	54,900	58,700	1.07

FIGURE 3.28 Reaction pathway for the synthesis of a block copolymer of endo,exo[2.2.1]bicyclo-2-ene-5,6- dicarboxylic acid dimethyl ester (monomer A) and endo,exo[2.2.1]bicyclo-2-ene-5,6-dicarboxylic acid di-tert-butyl ester (monomer B) and molecular characteristics of the copolymer. a and b—theoretical numbers of lyophobic and lyophilic units, respectively. c—theoretical ratio of monomer A to monomer B. d—A/B ratio, determined from the ^{1}H NMR spectrum. e—calculated molecular weight. (Reprinted with permission from Dulle, M. and Glatter, O., *Langmuir*, 28, 1136–1141. Copyright 2012, American Chemical Society.)

compared to those formed by surfactants and lipids, is a frequently cited reason for their appeal in various applications, including delivery (Allen et al. 1999, Kataoka et al. 2001, Torchilin 2001). The nanostructures they form are typically kinetically trapped, metastable, and frequently sensitive to the way of preparation, which expands the type of self-assembled structures that can be achieved and, in fact, provides an additional level of morphological control. However, control not only over morphology but also over properties of the self-assemblies can be achieved by using mixtures of different amphiphiles thus creating hybrid structures. Polymer–surfactant and polymer/lipid hybrid structures are considered in the previous two sections of this chapter. Due to the greater structural, functional, and chemical variability of the polymers, the modular approach of mixing different (amphiphilic co) polymers offers additional opportunities and broadens the possibilities to control the morphology, properties, and behavior of polymer/polymer hybrid structures. In this section, we make no attempt to thoroughly cover the literature on this topic. Instead, we highlight modulation of properties, formation of new structures, or tuning and remodeling the existing ones that cannot be achieved when the copolymers are used alone or in combination with surfactants or lipids. Hybrid polymer–polymer structures may result from various types of interactions and forces (see Figure 3.1). Some of the latter, however, are typically observed in nonaqueous solutions, and in spite of their ability to facilitate formation of self-assembled structures and excitement from phenomenological, theoretical, and practical points of view, these interactions and forces as well as the structures they produce will be left out of consideration or just briefly mentioned in this chapter.

The group of relevant polymers to start may certainly be polymers that are commercially available such as PEO-PPO-PEO copolymers. The latter have been on the market for more than 50 years in wide ranges of molecular weights and ratios between hydrophilic and hydrophobic components. These copolymers are probably the most extensively studied polymeric nonionic surfactants that find wide application in fields as diverse as medicine and petroleum industry. Interestingly, the first anticancer micellar formulation that reached clinical trials is based on mixed micelles of $(EO)_2(PO)_{31}(EO)_2$ and $(EO)_{100}(PO)_{65}(EO)_{100}$ in which doxorubicin is loaded (Danson et al. 2004, Batrakova and Kabanov 2008). Another system that exhibits potential in anticancer drug delivery is reported elsewhere (Lee et al. 2011a). The authors combined two PEO-PPO-PEO copolymers with the same length of the PPO moieties—the lamellar-forming $(EO)_5(PO)_{68}(EO)_5$ and micellar-forming $(EO)_{20}(PO)_{68}(EO)_{20}$—to produce self-assembled structures with spherical morphology and hydrodynamic diameters less than 200 nm. The systems, in different weight ratios between the constituting copolymers, were characterized with CMCs located in the middle of the CMCs of the individual copolymers, high solubilization capacity due to $(EO)_5(PO)_{68}(EO)_5$, and high stability due to $(EO)_{20}(PO)_{68}(EO)_{20}$. Folic-acid-functionalized $(EO)_{20}(PO)_{68}(EO)_{20}$/$(EO)_{100}(PO)_{65}(EO)_{100}$ mixed micelles encapsulating paclitaxel have been developed and tested in vitro and in vivo (Zhang et al. 2011). $(EO)_{100}(PO)_{65}(EO)_{100}$ is considerably more hydrophilic than $(EO)_5(PO)_{68}(EO)_5$. Both copolymers individually form small spherical micelles, and accordingly, the size of the mixed micelles was also low—about 20 nm in diameter. In another paper (Kulthe et al. 2011),

$(EO)_{20}(PO)_{68}(EO)_{20}$ has been used with another hydrophobic copolymer of this family—$(EO)_3(PO)_{43}(EO)_{43}$—which has a shorter PPO block. In contrast to the systems with $(EO)_5(PO)_{68}(EO)_5$ (see preceding text), DLS revealed formation of very small particles—about 20 nm. Their stability was somewhat dependent on the composition: stable dispersions were obtained in the interval of $(EO)_3(PO)_{43}(EO)_{43}$/$(EO)_{20}(PO)_{68}(EO)_{20}$ weight ratios from 0.1/1.0 to 0.5/3.0. The mixed micelles showed high entrapment efficiency, loading capacity, and sustained release profile for aceclofenac.

As seen, the copolymer pairs considered so far are of the same, ABA, chain architecture. The copolymers differ in the length of the individual blocks and some of them in their propensity to form particular self-assembled structures. A systematic study on variations of size, size distribution, and particle morphology upon mixing a polymersome-forming copolymer, $(DDGG)_8(EO)_{45}$, with related diblock copolymers of different phase propensity is presented elsewhere (Rangelov 2006). In the interval in which the size of the particles did not change, the vesicular structure was preserved. In that interval, the structure of the protective layer and the grafting density of the PEG chains can be varied by varying the content and the type of the guest copolymer. A cartoon of a protective layer of high grafting density consisting of different in length PEG chains is presented in Figure 3.29. Such a layer is expected to be more effective in providing steric stabilization and, in particular, resistance against opsonization and phagocytosis.

Honda et al. (2010) have shown that the co-assembly of linear and cyclic copolymer allows for tuning the cloud points (CPs) to any temperature between those of the homoassemblies. They used a triblock copolymer of reverse, that is, BAB, chain architecture, poly(butyl acrylate)-PEO-poly(butyl acrylate) and its cyclized

FIGURE 3.29 Schematic representation of a monolayer of vesicles composed of $DDP(EO)_{92}$ and $(DDGG)_8(EO)_{45}$. The circles with the two narrow curved lines represent the lipid-mimetic anchors. The wider curved lines represent the PEG chains. (Reprinted with permission from Rangelov, S., *J. Phys. Chem. B*, 110(9), 4256–4262. Copyright 2006, American Chemical Society.)

analogue. Both copolymers formed flowerlike spherical micelles with hydrodynamic diameters ~20 nm and exhibited CPs that differed by more than 40°C. Information about the size of the mixed micelles was not provided; however, it was shown that the CPs were influenced and can be controlled by controlling the mixing ratios between the two copolymers.

To obtain hybrid copolymer–copolymer structures, however, it is not necessary that the copolymers are of the same chain architecture or all constituent blocks are chemically identical. The pairs may contain only one type of chemically identical blocks or may be composed of moieties of entirely different nature. A variety of combinations such as AB-BC, ABA-BC, AB-CD, ABA-BCB, and so forth may be easily found in the literature. Rangelov et al. have described composite particles of a block copolymer of PO and ethoxyethyl glycidyl ether, which has been found to possess lower critical solution temperature (LCST) properties at temperatures below 20°C (Dimitrov et al. 2004), with $(EO)_{13}(PO)_{30}(EO)_{13}$ or $(EO)_{26}(PO)_{40}(EO)_{26}$ at weight ratios in the interval from 1:0.1 to 1:10 (Rangelov et al. 2005). The hybrid aggregates, composed of copolymers of ABA and BCB chain architectures, exhibited tunable temperature behavior in that they rearranged at a certain temperature that was composition dependent. The rearrangement was related with dissociation of the large (50–60 nm) composite particles initially formed and may have potential applications in areas such as delivery and release of active substances. In an early paper, Yang et al. (1996) investigated the secondary association of micelles. By this time, the approach, in the sense that it involved mixed micelles formed in aqueous solution from a diblock copolymer $(BO)_8(EO)_{41}$ and a triblock copolymer $(BO)_{12}(EO)_{76}(BO)_{12}$, is considered as new. A 50:50 wt.% mixture of the two copolymers in dilute aqueous solution formed well-separated mixed micelles in which the triblock copolymer macromolecules were in looped conformation, that is, with both poly(butylene oxide) (PBO) blocks in the same micellar core. At higher concentrations, micellar clusters were formed. They were bridged by the triblock copolymer macromolecules that adopted unlooped, extended conformation with PBO blocks anchored in different micellar cores.

Biodegradable nanoparticles for optical imaging, again with participation of a PEO-PPO-PEO copolymer, have been developed by blending $(EO)_{78}(PO)_{30}(EO)_{78}$ with chitosan-modified poly(L-lactide-*co*-ε-caprolactone) (Ranjan et al. 2011). Note that the two copolymers do not possess chemically identical moieties. The particles loaded with indocyanine green were spherical in shape with dimensions 146–260 nm and increasing ζ-potential from −41.6 to +25.3 mV depending on chitosan content. Stable aqueous dispersions have been obtained by adsorption of copolymers of EO and PO on agglomerates of micelles formed by an entirely hydrophobic triblock copolymer polystyrene-poly(hydrogenated butadiene)-polystyrene (Munk et al. 1998). The dispersions were obtained by injection of a micellar solution of polystyrene-poly(hydrogenated butadiene)-polystyrene in dioxane containing appropriate quantities of an EO-PO block copolymer into excess water. When the micellar solution was injected into water, the dioxane diffused from the micellar shells; the shells collapsed but for some period of time remained swollen enough to allow interpenetration with shells of the neighboring micelles. The resulting particles were 55–75 nm in radii depending on the type and quantity of the stabilizing copolymer and molar masses of hundreds of millions. They have a compartmentalized structure consisting of a hydrophobic interior of polystyrene (PS) in which domains of poly(hydrogenated butadiene) were distributed. These structures were stabilized by a thin layer of macromolecules of the EO-PO block copolymer anchored on the surface via its PPO block. The aqueous complexation of an amphiphilic block copolymer AB and a double hydrophilic block copolymer CD is a straightforward approach to compartmentalize micelles as shown by Lutz and Laschewsky (2005). The AB copolymer contained a hydrophobic block of poly(*n*-butyl acrylate) and a polyanionic segment of poly(sodium 2-acrylamido-2-methylpropanesulfonate), whereas the CD copolymer was composed of cationic and nonionic blocks of poly(*N*,*N*,*N*-trimethylaminoethyl acrylate chloride) and poly[oligo(ethylene glycol) acrylate], respectively. Their complexation led to a core–shell–corona morphology, where the intermediate shell region was formed from the polyionic complex between the cationic and anionic blocks.

Micelle evolution due to a collision/fusion/fission process has been observed in a binary mixture of spherical micelles, formed from a PEE-*b*-PEO diblock copolymer, and segmented wormlike micelles, formed from a μ-PEE-PEO-poly(perfluoropropylene oxide) miktoarm star copolymer (Li et al. 2006). The long segmented wormlike micelles initially fused with the spherical micelles, followed by fission to progressively shorter micelles, which finally evolved into unique multicompartment micelles referred to as "hamburger" micelles (Figure 3.30). Note that here the binary mixture is composed by *preformed* micelles; a binary blend of the miktoarm star terpolymer with the diblock copolymer in the bulk *prior* to micellization did not generate new intermediate micelle structures, presumably because of phase separation between the two copolymers during the film-casting process.

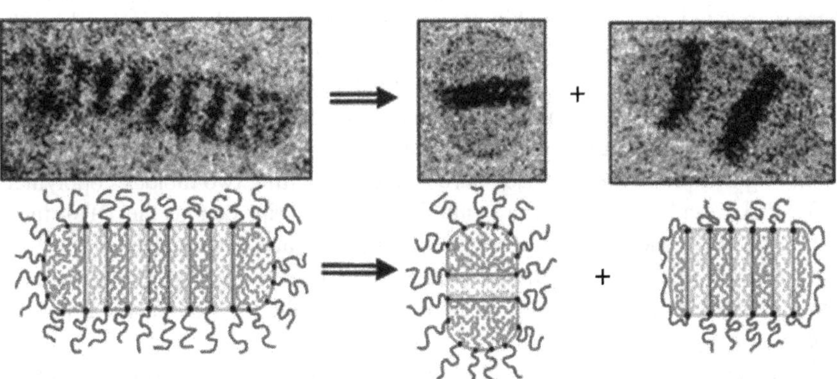

FIGURE 3.30 Evolution in multicompartment micelles ("hamburger" micelles) from a binary mixture of micelles of different morphology. Segmented wormlike micelles were formed from a μ-PEE-PEO-poly(perfluoropropylene oxide) miktoarm star copolymer. Spherical micelles were formed from a PEE-*b*-PEO diblock copolymer. Dark gray, light gray, and black correspond to poly(perfluoropropylene oxide), PEE, and PEO, respectively. (Reprinted with permission from Li, Zh., Hillmyer, M.A., and Lodge, T.P., *Macromolecules*, 39(2), 765–771. Copyright 2006, American Chemical Society.)

The polyester-based polymersomes, which have thus far been made from PEO-PCL or PEO-PLA, show promise as hydrolysis-triggered controlled-release vesicles. By blending with inert PEO-PBD in different mole-percent ratios, the encapsulant release can be controlled on time scales from hours to weeks (Ahmed and Discher 2004). Poration in these polymersomes was shown to occur as the hydrophobic polyester block is hydrolytically scissioned, progressively generating an increasing number of pore-preferring copolymers in the membrane. The rate of release rose linearly with the molar ratio of degradable copolymer (PEO-polyester) to nondegradable (PEO-PBD) copolymer. Porous polymersomes intended for magnetic resonance imaging (see preceding text) have been prepared through co-assembly of PEO-PBD and poly(ethylene oxide)-poly(ε-caprolactone) (PEO-PCL) (Cheng et al. 2009). Subsequent hydrolysis of the PCL block resulted in vesicles of highly porous membrane which exhibited improved permeability to water flux and a large capacity to store chelated Gd within the aqueous lumen.

Another approach to generate membrane permeability involves incorporation of a stimuli-responsive block copolymer in the membrane of polymersomes consisting of a conventional amphiphilic block copolymer, PEO-*b*-PS (Kim et al. 2009a). The stimuli-responsive block copolymers, PEO-*b*-poly(styrene boronic acid), contained boronic acid residues which imparted both pH and sugar sensitivity. Polymersomes composed entirely of the latter copolymers disassembled at higher pH and/or presence of sugar, whereas formation of holes was induced in the membranes of the polymersomes composed of the conventional and stimuli-responsive copolymers. The same approach can be used to modify the polymersome's surface as recently shown by Massignani et al. (2009) for a pair of polymersome-forming copolymers, poly(2-methacryloxyethyl phosphorylcholine)-*b*-poly (2-(diisopropylamino) ethyl methacrylate) (PMPC-*b*-PDPA) and PEG-PDPA. Nanometer-sized polymersomes were formed upon mixing the two copolymers at different ratios (Figure 3.31). As seen, the two copolymers were phase segregated, which resulted in patchy polymersomes with surface topology that can be controlled by the ratio between the copolymers.

Zhang et al. (2010a) report on vesicles with an asymmetric membrane, composed of two diblock copolymers with a common hydrophobic block. The preparation procedure and structure of the polymersomes are presented in Figure 3.32. The vesicles containing dextran in their interior were formed by adding two diblock copolymers, PEO_{45}-PCL_{30} and PCL_{66}-$dextran_{22}$ (numbers in subscript denote the DPs), into a dextran-in-PEO aqueous two-phase system. The inner and outer leaflets of the membrane are composed of dextran-PCL and PEO-PCL, respectively. The asymmetric bilayer created a different chemical environment of the interior to which proteins were effectively encapsulated.

Self-assembled structures can be prepared by blending of copolymers bearing oppositely charged blocks (moieties). A distinctive feature here is that one or both copolymers may be water soluble, that is, taken separately, they do not self-associate. By a proper selection of the constituent block, both the morphology and the properties of the self-assembled structures can be controlled. De Santis and coworkers describe for the first time in a recent paper that PIC micelles formed by co-assembly of oppositely and permanently charged poly(sodium 2-acrylamido-2-methylpropanesulfonate)-block-poly(*N*-isopropylacrylamide)

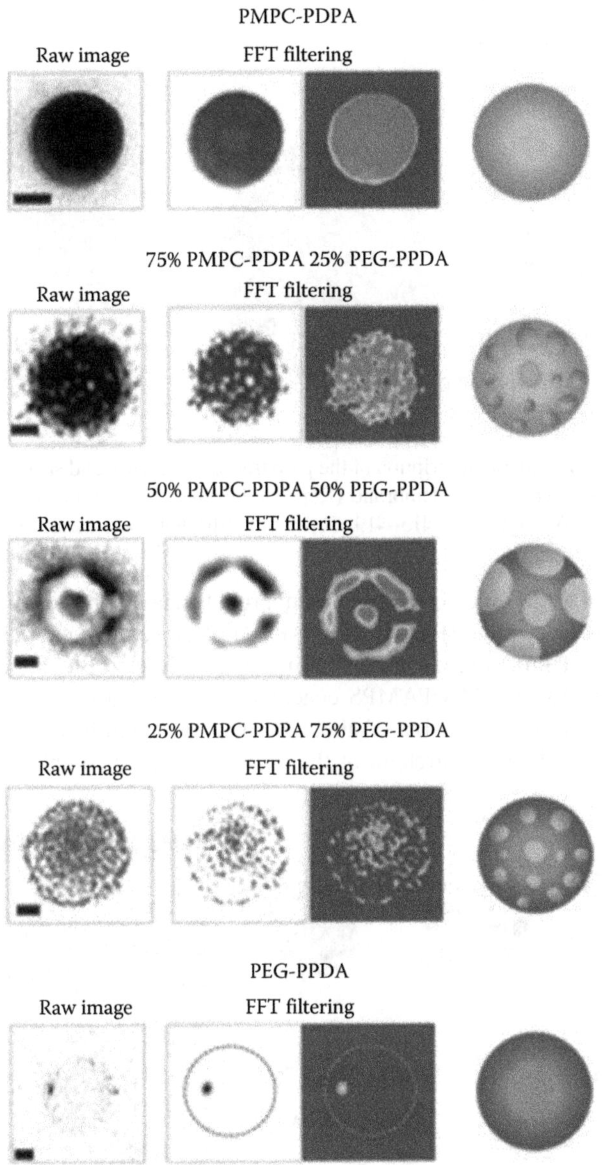

FIGURE 3.31 Nanometer-sized polymersomes imaged by TEM and analyzed using FFT filtering. Phase segregation generated by binary different ratio mixtures of poly(2-methacryloxyethyl phosphorylcholine)-*b*-poly(2-(diisopropylamino) ethyl methacrylate) (PMPC-*b*-PDPA) and PEG-*b*-PDPA copolymers forming patchy polymersomes that displays surface domains highlighted by selective staining of the PMPC chain by phosphotungstic acid (selective for ester groups). *Scale bar*=50 nm. (Massignani, M., LoPresti, C., Blanazs, A., Madsen, J., Armes, S.P., Lewis, A.L., and Battaglia, G.: *Small*. 2009. 5(21). 2424–2432. Copyright Wiley-VCH Verlag GmbH & Co. KGaA. Reproduced with permission.)

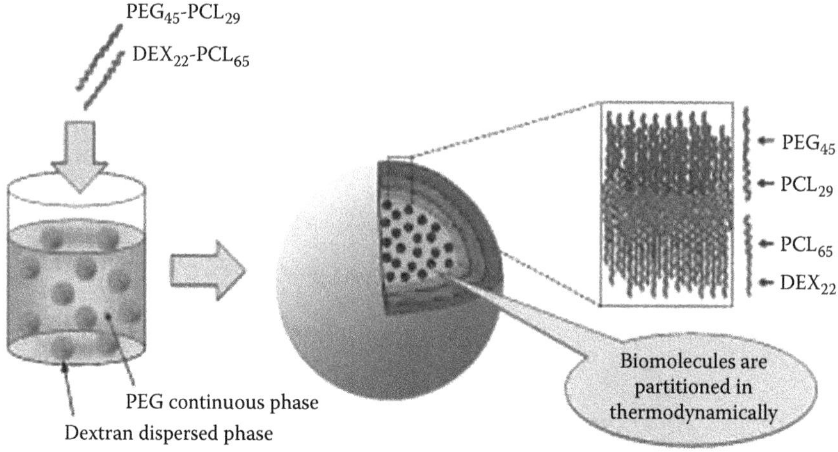

FIGURE 3.32 Schematic description of the preparation procedure and structure of polymersomes of asymmetrical bilayer membrane. (Reprinted from *J. Control. Release*, 147(3). Zhang, Y., Wu, F., Yuan, W., and Jin, T. 413–419, Copyright 2010, with permission from Elsevier.)

(PAMPS-*b*-PNIPAM) and poly[(3-acrylmidopropyl)-trimethylammonium chloride]-block-poly(ethylene oxide) (PAMPTMA-*b*-PEO) block copolymers (De Santis et al. 2010). At room temperature, spherical almost monodisperse PIC micelles, consisting of a mixed PAMPTMA/PAMPS coacervate core and a mixed PEO/PNIPAM corona, were formed, with size of about 80–110 nm (Figure 3.33). The micelles were sensitive to the ionic strength of the solutions as they completely dissociated

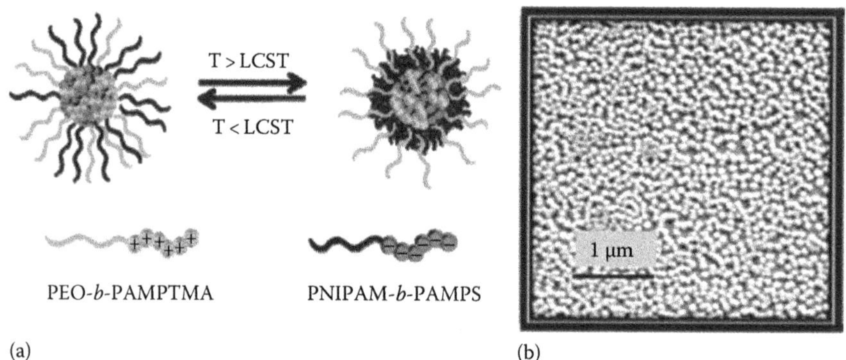

FIGURE 3.33 Formation of PIC micelles by co-assembly of oppositely and permanently charged poly(sodium 2-acrylamido-2-methylpropanesulfonate)-block-poly (*N*-isopropylacrylamide), PAMPS-*b*-PNIPAM, and poly[(3-acrylmidopropyl)-trimethylammonium chloride]-block-poly(ethylene oxide) (PAMPTMA-*b*-PEO) block copolymers and their temperature-induced transition from core–corona to core–shell–corona structure (a). SEM image of the particles (b). (Reprinted with permission from De Santis, S., Ladogana, R.D., Diociaiuti, M., and Masci, G., *Macromolecules*, 43(4), 992–2001. Copyright 2010, American Chemical Society.)

in the presence of NaCl at concentrations above 0.4–0.6 M. PNIPAM segments at 34°C–37°C gave rise to a temperature-induced transformation of the familiar core–corona structure into a compartmentalized core–shell–corona one: fully interconnected and continuous collapsed PNIPAM shell was formed, with PEO chains forming channels across the PNIPAM membrane (see Figure 3.33). It was demonstrated that the properties of the PIC micelles could be controlled by the DPs of the individual blocks. For example, the longer polyelectrolyte chains gave rise to bigger micelles, which were more stable with respect to the ionic strength. The transition temperature was modulated by the PNIPAM chain length, whereas PEO chains, as long as 114 repeating units, were necessary to effectively stabilize PIC micelles both below and above LCST of PNIPAM.

In contrast, the morphology of the C3Ms described elsewhere (Voets et al. 2006) was disklike. The particles were formed spontaneously in water from two water-soluble diblock copolymers—poly(acrylic acid)-*block*-poly(acryl amide) (PAA_{42}-*b*-$PAAm_{417}$) and poly(2-methylvinylpyridinium iodide)-*block*-poly(ethylene oxide) ($P2MVP_{42}$-*b*-PEO_{446}). The combination of associative and segregative phase separation resulted in formation of disk-shaped *Janus* particles with a complex coacervate core (C3) and an asymmetric water-swollen corona which was microphase separated into two distinct faces.

An interesting case of hybrid structures that combine a diblock copolymer, poly(*N*-methyl-2-vinyl pyridinium iodide)-*b*-PEO (Figure 3.34a), and a supramolecular coordination polymer to eventually produce C3M has been described elsewhere (Yan et al. 2008). The supramolecular coordination polymer was formed spontaneously from small bisligand molecules (Figure 3.34b) and zinc ions at a 1:1 M ratio (Vermonden et al. 2003). As every coordination center carries two elementary negative charges ($4COO^- + Zn^{2+}$), the coordination complex can be viewed as a negatively charged polyelectrolyte existing in the form of either small rings or long chains depending on concentration (Figure 3.34c). The systems were studied by DLS, cryo-TEM, and SAXS (Yan et al. 2007a,b, 2008). The size and morphology of the particles varied with the charge ratio and concentration. While they were spherical in the case of charge neutrality (Figure 3.34d and e), many wormlike and spherocylindrical objects were observed in coexistence with the spherical ones when the positive charges were in excess (Figure 3.34f and g).

Coordination bonds are involved in the formation of supramolecular graft copolymers, the side chains of which are composed of PEG linked to a poly(methyl methacrylate) (PMMA) backbone via ruthenium (II)–terpyridine complexes (Gohy et al. 2003). The synthetic pathway for preparation of different copolymers with varying numbers of PEG chains and length of the backbone is presented in Figure 3.35. Polydisperse spherical micelles were detected in aqueous solution by DLS, TEM, and atomic force microscopy (AFM).

The incorporation of various multi[(porphinato)zinc(II)]-based supermolecular fluorophores (see Figure 3.36 for the chemical structures of some of them) in the membranes of polymersomes formed from a PEO_{30}-PBD_{46} diblock copolymer can be mentioned as a representative example of the hybrid emissive vesicles (Ghoroghchian et al. 2005). Depending on their chemical structure, the fluorophores can be located in the membrane–aqueous interface (compound A, Figure 3.36) or in the hydrophobic

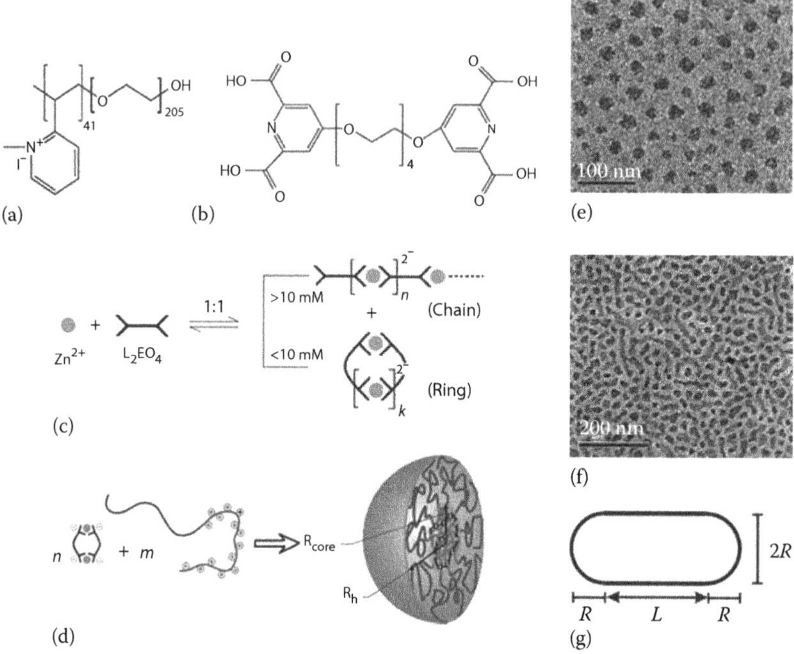

FIGURE 3.34 Structures of the diblock copolymer, poly(*N*-methyl-2-vinyl pyridinium iodide)-*b*-PEO (a), bisligand (b), coordination complex of a 1:1 Zn–bisligand molar ratio (c), and formation of spherical PIC micelles (d). Cryo-TEM images at charge neutrality (e) and in excess of positive charges (f). An illustration of a spherocylinder of radius R and length L + 2R (g). (a–d: Yan, Y., Harnau, L., Besseling, N.A.M., Keizer, A.D., Ballauff, M., Rosenfeldt, S., and Cohen Stuart, M.A., *Soft Matter*, 4, 2207–2212, 2008. Reproduced by permission of The Royal Society of Chemistry; e and f: Reprinted with permission from Yan, Y., Besseling, N.A.M., de Keizer, A., Fokkink, R., Drechsler, M., and Cohen Stuart, M.A., *J. Phys. Chem. B*, 111, 11662. Copyright 2007, American Chemical Society; g: Reprinted with permission from Yan, Y., Besseling, N.A.M., de Keizer, A., Marcelis, A.T.M., Drechsler, M., and Cohen Stuart, M.A., *Angew. Chem., Int. Ed.*, 46, 1807. Copyright 2007, American Chemical Society.)

core of the bilayer membrane (compound B, Figure 3.36). The polymersomes can be prepared in desired dimensions—for example, giant, micrometer-sized vesicles or small (<300 nm) unilamellar polymersomes—by applying the common preparation methods. Fluorescence energy modulation over a broad spectrum domain of the visible and near-infrared (600–900 nm) was demonstrated. The polymersomal structures highlighted that the nature of intramembranous polymer-to-fluorophore contacts depended on the position and identity of the porphyrins' ring substituents.

One of the co-assembling copolymers of the PIC micelles of Yuan and coworkers is a graft copolymer—polyethylenimine-graft-poly(*N*-vinylpyrrolidone) (Yuan et al. 2011). Upon interactions in aqueous solution with an oppositely charged block copolymer, poly(*N*-vinylpyrrolidone)-block-poly(2-acrylamido-2-methyl-1-propanesulphonic acid), it formed stable PIC micelles that exhibited spherical morphology and dimension ~140 nm. A pH-dependent release profile of a model

PEG-tpyRu(III)Cl_3

EtOH
N-ethylmorpholine
reflux 4 h
NH_4PF_6

1a–c

2a–c

a: $m = 2; n = 34$
b: $m = 4; n = 35$
c: $m = 5; n = 51$

FIGURE 3.35 Synthetic pathway for preparation of supramolecular graft copolymers. (Gohy, J.-F., Hofmeier, H., Alexeev, A., and Schubert, U.S.: *Macromol. Chem. Phys.* 2003. 204. 1524–1530. Copyright Wiley-VCH Verlag GmbH & Co. KGaA. Reproduced with permission.)

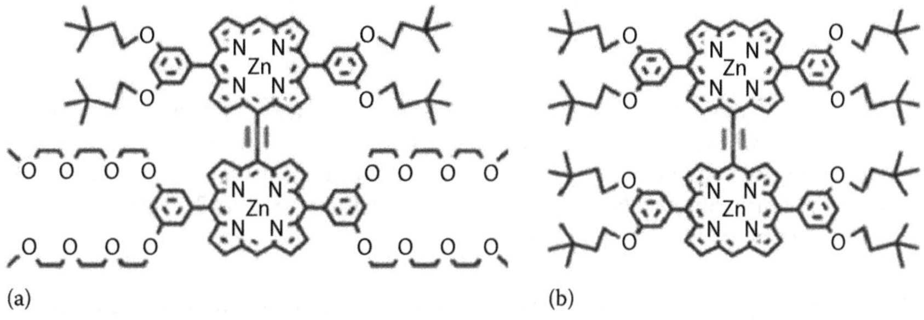

FIGURE 3.36 Chemical structures of meso-to-meso ethynyl-bridged bis[(porphyrinato) Zn(II)] supermolecular fluorophores. The fluorophores can be located at the interface (a) or in the core (b) of the bilayer membrane. (Reprinted with permission from supporting information of Ghoroghchian, P.P., Frail, P.R., Susumu, K., Park, T.-H., Wu, S.P., Uyseda, H.T., Hammer, D.A., and Therien, M.J., *J. Am. Chem. Soc.*, 127, 15388–15390. Copyright 2005, American Chemical Society.)

drug, incorporated into the cores of the micelles via electrostatic interactions, was observed.

Binding a homopolymer to a specific block or moiety of a given copolymer can also induce micelle formation, stabilize the micelles, induce morphological transitions, and modulate the properties of the self-assembled structures. As shown

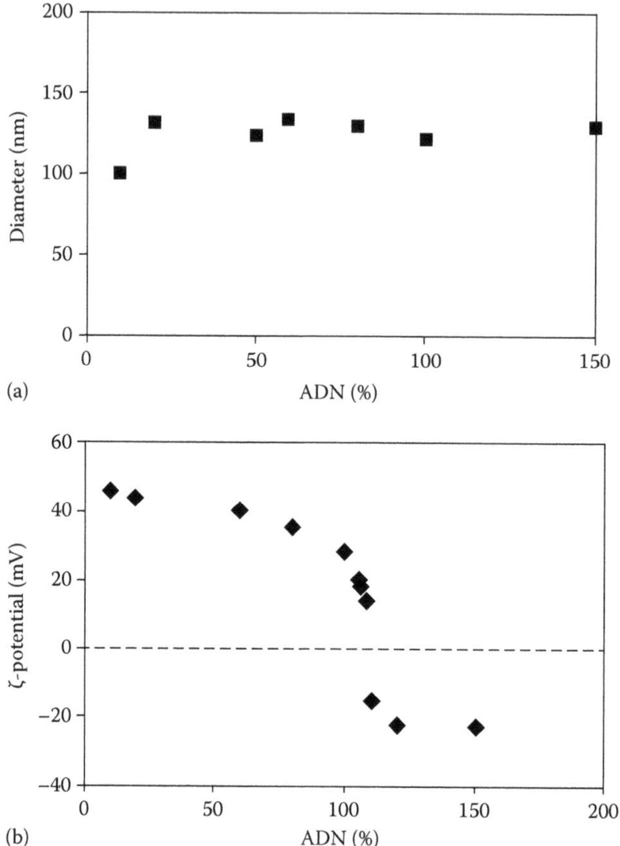

FIGURE 3.37 Variations of hydrodynamic diameters (a) and ζ-potential (b) of mixed micelles of PMAA/PS-*b*-PMAPTAC-*b*-PEO at different ADNs. The concentration of PS-*b*-PMAPTAC-*b*-PEO is fixed at 0.1 g/L for both experiments. (Liu, J., Yoneda, A., Liu, D., Yokoyama, Y., Yusa, Sh., and Nakashima, K., *Can. J. Chem.*, 88, 208–216, 2010. Reprinted with permission from NRC Research Press.)

earlier for the polystyrene-*b*-poly(3-(methacryloylamino)propyltrimethylammonium chloride)-*b*-poly(ethylene oxide) (PS-*b*-PMAPTAC-*b*-PEO) triblock copolymers (see Figure 3.9 for the chemical formula), the addition of a negatively charged substance, SDS, substantially decreased the CMC, indicating that the micelle formation is induced at lower concentrations and caused shrinkage of the segments in the shell region resulting in overall reduction of particle dimensions and macrophase separation at ca. 60% of apparent ADN (Liu et al. 2010). Figure 3.37 shows the variations of the hydrodynamic diameters and ζ-potential of the particles with ADN, which is proportional to the content of poly(methacrylic acid) (PMAA) added. Nearly uniform micelles with dimensions of about 120 ± 5 nm were obtained in a wide interval of PMAA contents, and in contrast to the case of SDS (see preceding text), neither morphological changes from an extended to shrunken conformation nor macrophase separation were observed even at ADN as high as 150%. The differences could be attributed to the polymer nature of the PMAA and to the hydrophilicity of the residual carboxyl groups present in the methacrylic acid after neutralization. The ζ-potential of the particles continuously decreased to 0 mV at about 100% ADN (Figure 3.37b) due to gradual neutralization of the positive charges during the binding of the oppositely charged PMAA. Noteworthy, the addition of more PMAA caused a further decrease in the ζ-potential indicating excess negative charges in the mixed micelles, which implied for another type of interactions between the triblock copolymer and PMAA. According to the authors, this could be attributed to hydrophobic interactions between PMAA and the PS-*b*-PMAPTAC-*b*-PEO or hydrogen bonding between PMAA and the PEO block (Liu et al. 2010).

For different types of interactions in binary mixtures of a nonionic–cationic diblock copolymer, poly(ethylene oxide-*block*-2-(diethylamino)ethyl methacrylate) (PEO-*b*-PDEA) and an anionic polyelectrolyte, PMAA, have been reported a bit earlier (Weaver et al. 2003). Depending on the solution pH, the different interactions accordingly led to the formation of different types of self-assembled structures. At pH>8.5, PMAA did not participate in the micelle formation and only the familiar core–corona micelles with hydrophobic PDEA cores and with the PEO chains located in the corona were observed. PIC micelles with charge-compensated PDEA/PMAA cores and PEO coronas are obtained at pH 6–8.5, whereas colloidal structures due to hydrogen bonding between the PEO and PMAA chains were formed below pH 3.

In a number of studies, Rotello and coworkers applied the concept of molecular recognition for creation of supramolecular assemblies. Unprecedented formation of vesicles by mixing random or block copolymers lacking defined head groups but containing complementary diaminopyridine and thymine moieties has been reported (Ilhan et al. 2000, Drechsler et al. 2002, Thibault et al. 2006, Uzun et al. 2010). The vesicles were termed recognition-induced polymersomes (RIPs) (Ilhan et al. 2000). For their formation, specific three-point hydrogen bonding interactions are required (Figure 3.38). The RIPs are typically formed in organic solvents (chloroform) and occasionally in water, when one of the blocks is PEG (Uzun et al. 2010), but are invariably micrometer in size.

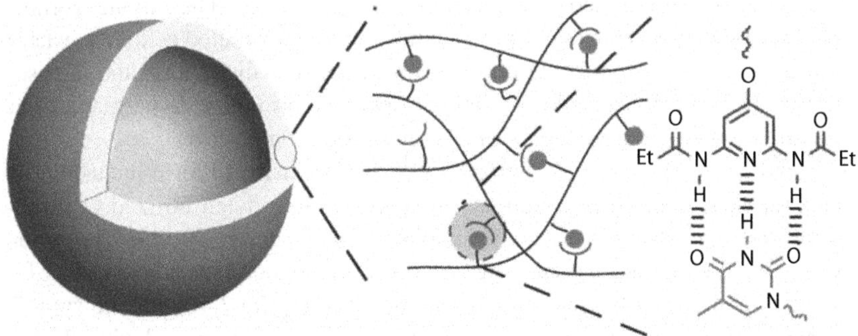

FIGURE 3.38 Formation of RIPs by specific hydrogen bonding between polymers containing diaminopyridine and thymine moieties. (Thibault, R.J., Uzun, O., Hong, R., and Rotello, V.M.: *Adv. Mater.* 2006. 18. 2179–2183. Copyright Wiley-VCH Verlag GmbH & Co. KGaA. Reproduced with permission.)

A modular system with which mixing and matching hydrophobic and hydrophilic blocks is possible with resultant control over the size, morphology, and surface functionality has been created by utilizing noncovalent coiled-coil interactions between complementary peptides. The latter, denoted E and K, are composed of Ac-G(E I A A L E K)$_3$-NH$_2$ with $M_n = 2380$ g/mol and Ac-(K I A A L K E)$_3$G-NH$_2$ and $M_n = 2378$ g/mol, respectively. Peptide E is sequence specific; it folds into a stable coiled-coil dimer with peptide K due to their amino acid sequence designs, and the E/K pair has been shown to retain its coiled-coil binding upon conjugation of macromolecules such as short PS, PEO, or poly(γ-benzyl L-glutamate) (PBLG) to E and/or K (Robson Marsden et al. 2008, 2009, 2010, Yano et al. 2008). PS$_9$-E forms spherical micelles in aqueous solution, and via coiled-coil folding with peptide K or K-PEO$_{77}$, the corona can be altered, resulting in larger spherical micelles or rodlike micelles (Robson Marsden et al. 2009). Polymersomes and disklike micelles with a range of sizes, membrane thickness, and surface functionality were formed upon noncovalent complexation of PBLG-E with K or PEO-K (Figure 3.39). Rather uniform membrane thicknesses that slightly increased with increasing size of the hydrophilic moieties were observed for the systems with the shortest PBLG block (Figure 3.39a through c). They were in remarkably close accordance with the calculated bilayer thicknesses, implying that the rigid hydrophobic PBLG rods are able to assemble into very well-defined bilayers through coupling to the water-soluble peptide rods. Rather peculiar spherical objects that withstand the drying process were observed for the noncovalent complexes with longer PBLG blocks (Figure 3.39d). As seen, the spherical particles contain aqueous interiors and have very bulky membranes of variable thickness (~68 ± 22 nm). The range of membrane thicknesses likely arises from the polydispersity of molecular lengths in this sample.

Noncovalent interactions between a homopolymer and either preformed aggregates or a copolymer followed by co-assembly also present useful attributes for production of new properties of the resulting self-assembled hybrid structures. An ionic complexation between fibrillar and spherical colloidal objects has been presented elsewhere (Wang et al. 2011). The resulting nanocomposites were prepared via adsorption of cationic micelles based on PBD-*b*-quaternized poly(dimethylaminoethyl methacrylate) on carboxymethylated nanofibrillated cellulose. The cationic corona of the block copolymer micelles served as a binder to nanofibrillated cellulose, whereas the hydrophobic rubbery micellar cores facilitated energy dissipation and nanoscale lubrication between the cellulose domains under deformation. Synergistic effects were observed with promoted work of fracture in one composition.

The well-known host–guest interactions of cyclodextrins to form inclusion complexes with various (hydrophobic macro) molecules have been utilized to produce temperature-responsive nanosized assemblies (Zhang et al. 2010b). The authors designed and synthesized a double hydrophilic block copolymer composed of PEG and a block containing β-cyclodextrin moieties (see Figure 3.40a for the synthesis route), which formed spontaneously spherical assemblies when mixed with PNIPAM in water (Figure 3.40b). The driving force for the formation of these assemblies was found to be the inclusion complexation interaction between the hydrophobic cavity of β-cyclodextrin and the isopropyl group of PNIPAM. These assemblies exhibited loose cores since only a portion of the isopropyl groups participated in the host–guest

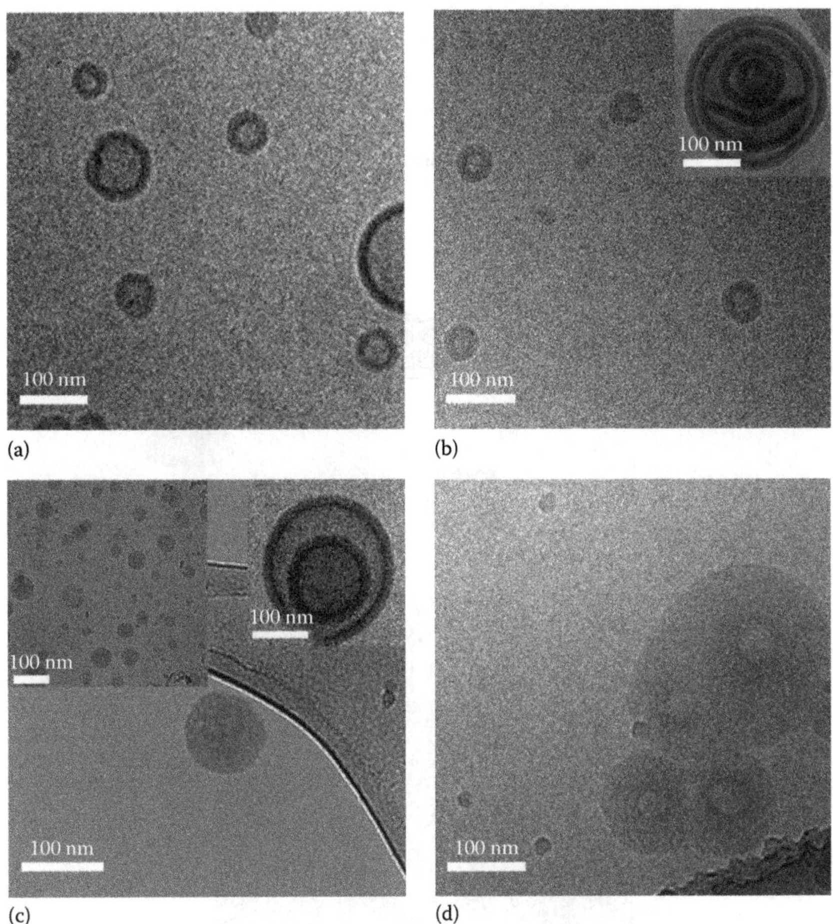

FIGURE 3.39 Cryo-TEM images of (a) $PBLG_{36}$-E, (b) $PBLG_{36}$-E/K (with MLV inset), (c) $PBLG_{36}$-E/K-PEG (with bilamellar vesicle and bicelles inset), and (d) $PBLG_{100}$-E/K-PEG. Scale bars = 100 nm ([total peptide] = 1500 μM, PBS). (Reprinted with permission from Robson Marsden, H., Handgraaf, J.-W., Nudelman, F., Sommerdijk, N.A.J.M., and Kros, A., *J. Am. Chem. Soc.*, 132, 2370–2377. Copyright 2010, American Chemical Society.)

interactions, while the rest of them were free to respond to temperature variations. The β-cyclodextrin functioned as multiple cross-linking points to hold the PNIPAM chains together. This particular structure made the size of the self-assemblies dependent on the weight ratio PNIPAM to PEG-*b*-poly(β-cyclodextrin) and sensitive to external temperature variations.

The special stereochemistry of polylactides has been used to provide access to noncovalent stabilization via stereocomplexation of two identical diblock copolymers containing enantiomerically pure polylactide chains—poly(L-lactide) and poly(D-lactide) (Wolf et al. 2010). The hydrophilic block of these copolymers is composed of poly(1,3-dihydroxypropyl methacrylate). Taken separately the two copolymers formed small spherical micelles with an approximate diameter of 20–30 nm (Figure 3.41a and b).

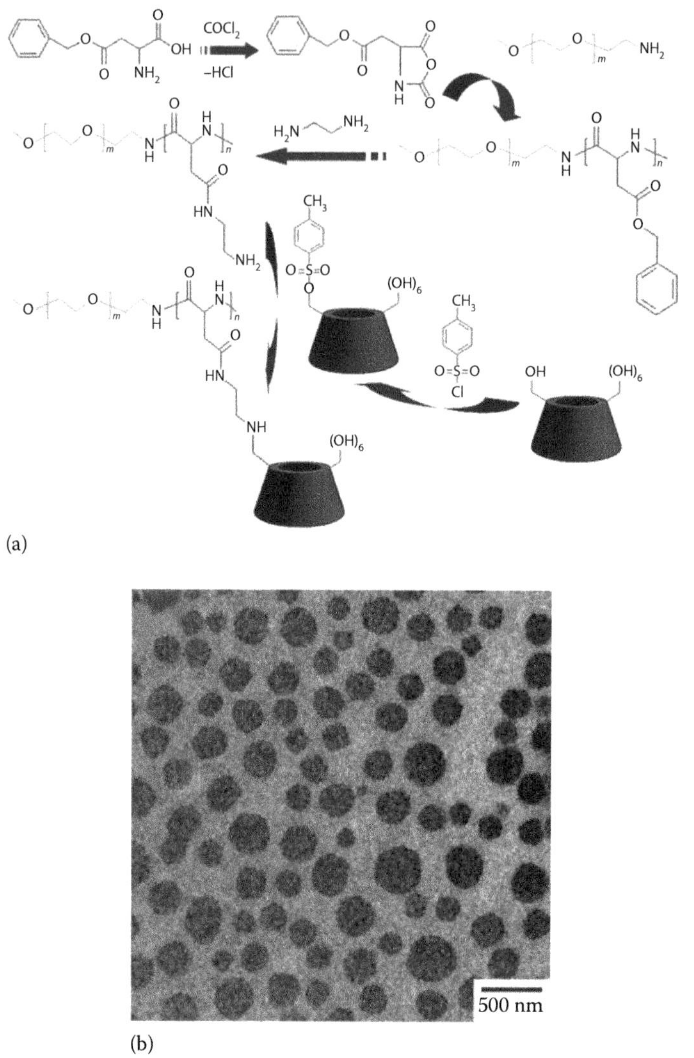

(a)

(b)

FIGURE 3.40 (a) Synthesis of PEG-*b*-poly(β-cyclodextrin) with $m=114$ and $n=13$ according to the ^{1}H NMR analysis. (b) TEM image of assemblies based on PNIPAM (number average molecular weight of 1000) and PEG-*b*-poly(β-cyclodextrin) at 1:10 weight ratio. (Zhang, J., Feng, K., Cuddihy, M., Kotov, N.A., and Ma, P.X., *Soft Matter*, 6(3), 610–617, 2010. Reproduced by permission of The Royal Society of Chemistry.)

While the influence of the stereocomplexation (mixture of the two copolymers at 1:1 with regard to enantiomeric polylactide blocks) on the CAC was very small, the aggregate morphology was strongly affected as large vesicles (polymersomes) with diameters in the 600–1400 nm range were formed (Figure 3.41c1 and c2). The authors argued that the stereocomplex-induced crystallization promoted formation of planar assemblies and hence formation of polymersome-like structures.

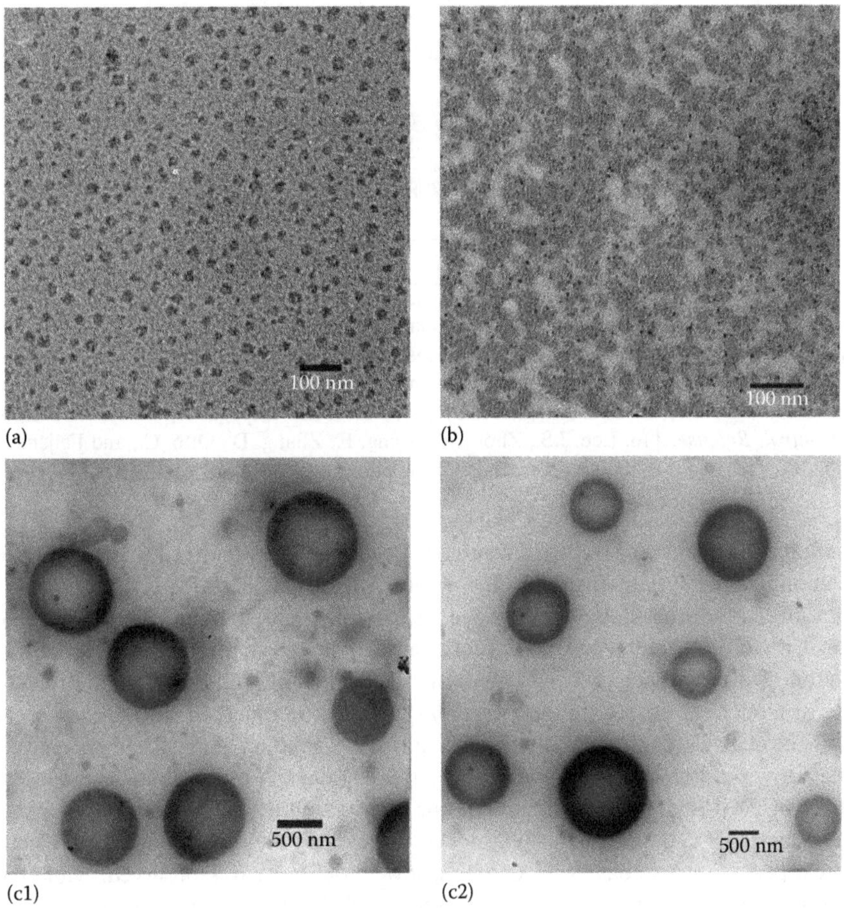

FIGURE 3.41 TEM images of micellar aggregates with an average diameter of 20–30 nm formed from poly(1,3-dihydroxypropyl methacrylate)$_{47}$-*b*-poly(L-lactide)$_{26}$ (a), poly(1,3-dihydroxypropyl methacrylate)$_{47}$-*b*-poly(D-lactide)$_{28}$ (b), and large polymersomes with a diameter between 600 and 1400 formed from a 1:1 mixture of the two copolymers under identical conditions (c1 and c2). (Reprinted with permission from Wolf, F.K., Hofmann, A.M., and Frey, H., *Macromolecules*, 43, 3314–3324. Copyright 2010, American Chemical Society.)

HomoPEO has been shown to affect the colloidal stability of polymersome dispersions prepared from PEO$_{16}$-*b*-PBD$_{22}$ (Smart et al. 2008) by triggering either slow or rapid aggregation depending on the PEO molecular weight. A critical molecular weight of 4000 (four times larger than the molar mass of the PEO block of the polymersome-forming block copolymer) was found, above which the polymersomes underwent rapid aggregation. Interestingly, the depletion forces were ruled out as a possible reason for aggregation. Instead, the presence of PEO seemed to drive the formation of assemblies that are normally stable at higher concentrations most probably as a result from adsorption on the polymersome surface and linking the polymersomes in a way similar to its interactions with SDS micelles. Thus, the added PEO

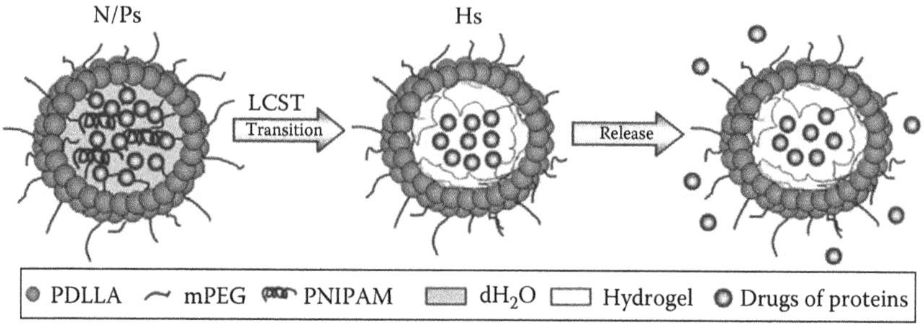

FIGURE 3.42 Schematic 2D cross-sectional illustration of formation of membrane-associated hydrogel layer influencing the release rate of incorporated agents. N/Ps and Hs refer to PNIPAM-containing polymersomes and hydrogel-containing (i.e., at temperatures above the LCST of PNIPAM, e.g., 37°C) polymersomes, respectively. (Reprinted from *J. Control. Release*, 146, Lee, J.S., Zhou, W., Meng, F., Zhang, D., Otto, C., and Feijen, J., 400–408, Copyright 2010, with permission from Elsevier.)

disrupted the membrane in such a way as to create a need for the vesicles to aggregate to minimize the system energy, as the authors suggested. The effects of homopolymers that are encapsulated in the aqueous interior of the polymersomes could be rather versatile. Lee and coworkers, for example, have recently reported on the formation of a membrane-associated hydrogel layer that strongly reduces the release rate of model substances, in this particular case, dextran marked with a fluorescent dye (Lee et al. 2010). The polymersomes, in which a solution of PNIPAM was encapsulated, were prepared from PEO-*b*-poly(D,L-lactide). At a temperature above the LCST of PNIPAM, phase separation in the interior of the polymersomes took place and a hydrogel and an aqueous phase were formed (Figure 3.42). Sustained release over a period of 4 weeks at 37°C with a low initial burst was observed, while the release from the polymersomes not containing PNIPAM was completed in 6 days.

3.5 HYBRID STRUCTURES FORMED UPON INTERACTION OF POLYMERS WITH PROTEINS AND PEPTIDES

The examples addressed to encapsulation of proteins and enzymes in polymeric self-assembled structures and polymersomes, in particular, are numerous. In the following, a few examples are given to demonstrate the versatility of self-assembled structures prepared from both commercially available and specially synthesized amphiphilic copolymers in providing protection and shielding of proteins and enzymes while maintaining their functions. The encapsulation can additionally benefit from the extended circulation kinetics and controlled-release properties of the self-assembled structures.

Napoli and coworkers have presented oxidation-responsive vesicles, composed of a symmetric triblock copolymer, PEO-*b*-poly(propylene sulfide)-*b*-PEO, with DPs of the constituent blocks of 16, 50, and 16, respectively, in which oxidant-generating enzyme, glucose oxidase, was incorporated (Napoli et al. 2004). Following an extravesicular addition of a substrate, glucose, to which the membrane is permeable, an oxidant, H_2O_2, was generated intravesicularly, which converted the hydrophobic poly(propylene sulfide) block into more hydrophilic sulfoxides and sulfones, thus inducing polymersome destabilization. The enzyme-catalyzed production of H_2O_2 and the possible changes associated with the oxidation of the poly(propylene sulfide) moieties as well as cryo-TEM micrograph of the polymersomes are presented in Figure 3.43. The materials and the approach may have utility in drug delivery and detection and sensing of biological analytes.

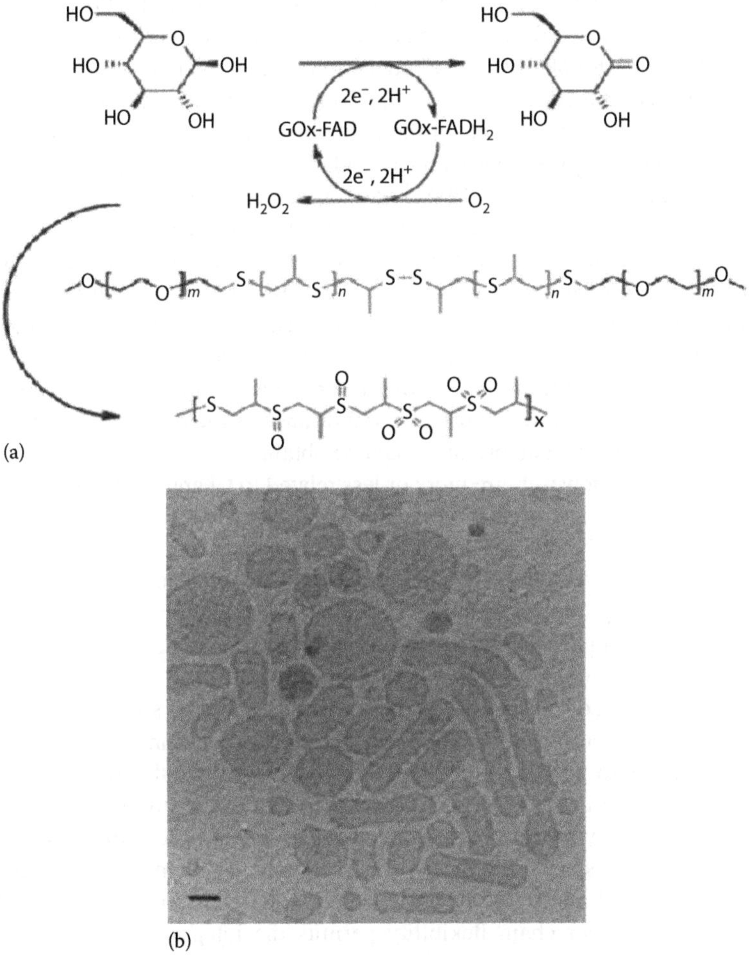

FIGURE 3.43 (a) Glucose-oxidase-catalyzed conversion of β-D-glucose into gluconolactone with production of H_2O_2 in the presence of oxygen. Below the symmetric PEG-*b*-poly(propylene sulfide)-*b*-PEG block copolymer under study and a possible repeating unit of poly(propylene sulfide) after oxidation by H_2O_2, $m = 16$, $n = 25$. (b) A cryo-TEM micrograph of the polymersomes formed in buffer after extrusion (the bar represents 100 nm). (Reprinted with permission from Napoli, A., Boerakker, M.J., Tirelli, N., Nolte, R.J.M., Sommerdijk, N.A.J.M., and Hubbell, J.A., *Langmuir*, 20, 3487–3491. Copyright 2004, American Chemical Society.)

Arifin and Palmer have demonstrated the ability of PEO-PBD polymersomes to encapsulate bovine hemoglobin without perturbing its affinity for oxygen mimicking the red blood cells' capability, thus generating alternative in vivo oxygen therapeutics (Arifin and Palmer 2005). Encapsulation of insulin in neutral and biologically stable polymersomes, based on the same copolymer, has been shown to provide a promising method to increase the therapeutic efficiency by maintaining the protein structure (Christian et al. 2008). Vesicles based on neutral diblock (PEO-PBD, PEO-PCL) as well as charged triblock (PEO-PCL-PAA) copolymers have shown abilities to encapsulate large globular proteins (Kishimura et al. 2007, Wittemann et al. 2007, Lee et al. 2001). Polymersomes formed from poly(2-methacryloxyethyl phosphorylcholine)-*b*-poly (2-(diisopropylamino) ethyl methacrylate), mentioned earlier, have been studied for delivery of primary human antibodies in order to both achieve immunolabeling and exploit the antibodies' therapy potential (Massignani et al. 2010). Superoxide dismutase, an antioxidant enzyme, has been encapsulated in the vesicular cavity of PMOXA-PDMS-PMOXA polymersomes, the membrane of which is oxygen permeable (Axthelm et al. 2008); the enzyme remained functional in neutralizing superoxide radicals in situ and exhibited prolonged lifetime. A cascade reaction has been sequentially catalyzed by three enzymes—glucose oxidase, encapsulated in the vesicular cavity of polymersomes based on a PS-polyisocyanopeptide diblock copolymer; horseradish peroxidase inserted in the membrane; and *Candida antarctica lipase B* in the outside medium (Vriezema et al. 2007). The positional assembly of the enzymes was of primary importance as the removal of any enzyme disrupted the cascade and no product could be obtained.

These examples, however, are more or less related to Chapter 5 and together with many others are considered in more details there. What we seek in this chapter is to demonstrate how the hybrid systems produce new properties and functions, how they are designed to match application demands, how one can obtain control over the nanostructures produced as a result of co-assembly, and how novel membrane composites with unique selectivity and permeability are created. Perhaps the most important breakthrough in creation of polymer–protein hybrid structures is the successful incorporation of functional proteins in polymer bilayers since this opened the doors for development of nanoreactors. In general, incorporation of synthetic or biological channels into polymer membranes has been considered extremely difficult, because the bilayer thickness is much larger than that of the liposomes. Membrane proteins are naturally optimized for biological membranes and therefore are too small to completely span the bilayer of thicker polymer membranes. Simulations using coarse-grained molecular dynamics recently performed (Srinivas et al. 2005) suggested that polymer chain flexibility permits the integration of proteins with small membrane-spanning domains, but the flexibility of the hydrophilic chains can partially block the functional pore of the membrane proteins, resulting in decreased functionality compared to when the proteins are incorporated in more natural lipid membranes. However, the pioneering work of Meier et al. (2000), who integrated membrane proteins into "black" block copolymer membranes, proved that it is possible to incorporate proteins into hyperthick membranes without disturbing their functionality. Diverse channel proteins, such as FhuA, OmpF, Tsx, and enzymes, have been inserted in polymer membranes and shown to remain functional in such environment (Meier et al. 2000, Nardin et al. 2001, Ye and van den Berg 2004, Ranquin et al. 2005, Onaca et al. 2006). The main advantage is that it is possible via protein engineering to design functionalized protein channels that function as an on/off switch. Thus, the membrane permeability is controlled by bringing substrates to the inside and transporting products to the outside medium. Polymersome nanoreactors based on PMOXA-PDMS-PMOXA triblock copolymers are presented in Figure 3.44. In a series of papers, Meier and coworkers demonstrated successful incorporation of the channel protein OmpF and encapsulation of enzymes and proteins (Nardin et al. 2000, 2001, Ranquin et al. 2005, Mecke et al. 2006, Nallani et al. 2006, Ben-Haim et al. 2008). Figure 3.44d shows a more sophisticated nanoreactor which was realized by incorporating channel protein LamB that served as a receptor for the λ-phage virus. The λ-phage virus recognizes its receptor, binds to the polymersomes, and injects its DNA through the channel to the inside of the vesicle (Graff et al. 2002). PMOXA-PDMS-PMOXA polymersomes loaded with phosphate anions have been used as scaffolds for biomineralization when alamethicin, an ion

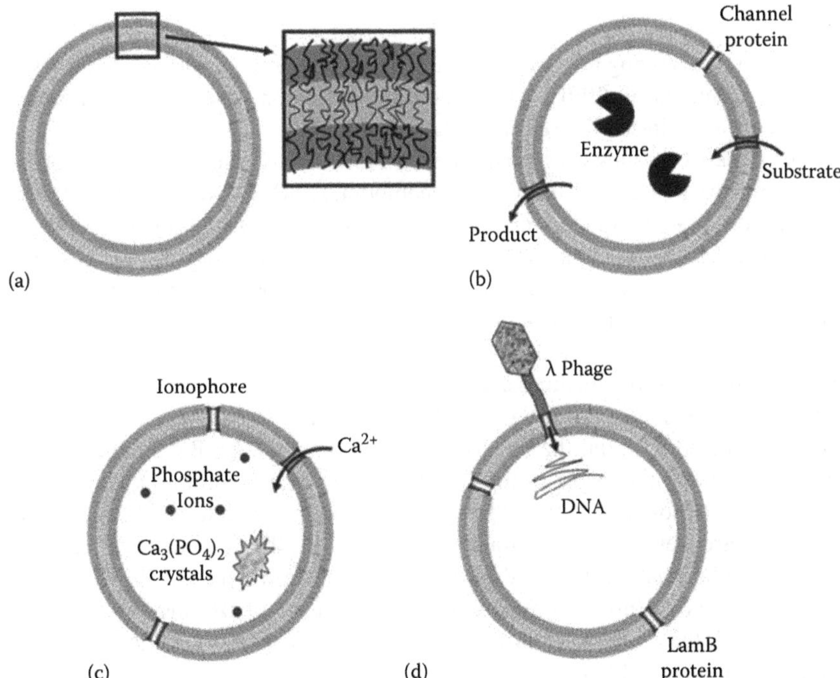

FIGURE 3.44 Schematic representation of polymer nanoreactors. (a) Cross section of triblock copolymer vesicle. (b) Polymersome with encapsulated enzyme and membrane-embedded channel protein. In the case described in the text, the substrate entering the vesicle is ampicillin, and the product of the hydrolysis is ampicillinoic acid. (c) Polymersome with embedded ionophores allows Ca^{2+} ions to enter the vesicle where they react with phosphate ions to form calcium phosphate crystals. (d) The LamB protein serves as a receptor to the λ-phage virus that can inject its DNA through the channel into the polymersome. (Mecke, A., Dittrich, C., and Meier, W., *Soft Matter*, 2, 751, 2006. Reproduced by permission of the Royal Society of Chemistry.)

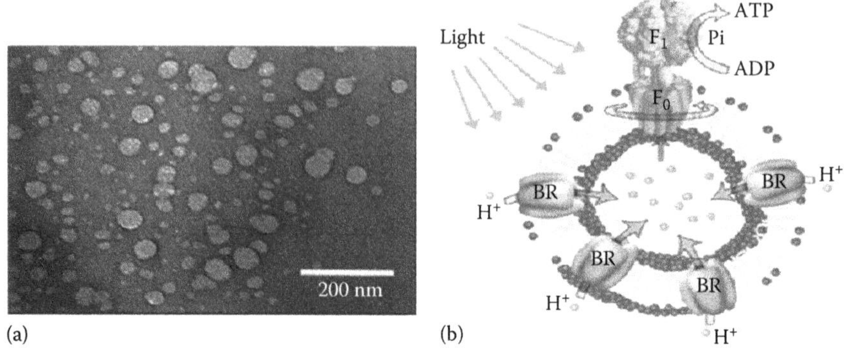

FIGURE 3.45 TEM micrograph (a) and schematic representation (b) of polymersomes based on a poly(2-ethyl-2-oxazoline)-*b*-poly(dimethylsiloxane)-*b*-poly(2-ethyl-2-oxazoline) triblock copolymer reconstituted with both BR and F_0F_1-ATP synthase. (Reprinted with permission from Choi, H.-J. and Montemagno, C.D., *Nano Lett.*, 5, 2538–2542. Copyright 2005a, American Chemical Society.)

channel protein allowing for cation (calcium) transport, is reconstituted in the vesicle membrane (Sauer et al. 2001). After a certain incubation time, calcium phosphate crystals were seen inside the vesicles (Figure 3.44c).

Most of the works, however, focus on measurements of a single-channel protein's functionality after its reconstitution in the membrane. Choi and Montemagno reported on incorporation of two membrane proteins for adenosine triphosphate (ATP) production (Choi and Montemagno 2005). They used a poly(2-ethyl-2-oxazoline)-*b*-poly(dimethylsiloxane)-*b*-poly(2-ethyl-2-oxazoline) triblock copolymer that yielded polymersomes with a narrow size distribution, a mean diameter of 37 ± 12.2 nm from TEM (Figure 3.45a), and a wall thickness of about 4 nm—that is, rather similar to the typical lipid bilayer thickness. The authors achieved light-driven ATP biosynthesis by simultaneous reconstitution of bacteriorhodopsin (BR) and ATP synthase. When illuminated by light, the reconstituted in the membrane BR pumped protons from the outside to the inside of the vesicles thus creating a proton gradient across the membrane (Choi et al. 2005). ATP synthase, also incorporated in the membrane, uses the electrochemical proton gradient generated by BR to synthesize ATP from adenosine diphosphate (ADP) and inorganic phosphate (Figure 3.45b). Essentially the same authors developed a method for reconstitution of BR and F_0F_1-ATP synthase by which the direction of proton pumping and the ATP production profile can be altered (Choi et al. 2006).

Onaca and coworkers have introduced a reduction-triggered release switch based on an engineered FhuA channel variant (Onaca et al. 2008). The polymersomes were formed from PMOXA-PDMS-PMOXA with molecular weight of 20,000. They were spherical objects, characterized by a hydrodynamic radius from DLS of 105 nm and aggregation number from SLS of about 5400 macromolecules per vesicle. This reduction-triggered and sterically controlled-release system for transmembrane control of the chemical fluxes has been developed by chemical modification of the lysine residues in the channel FhuA Δ1–160 with either a pyridyl or a biotinyl label.

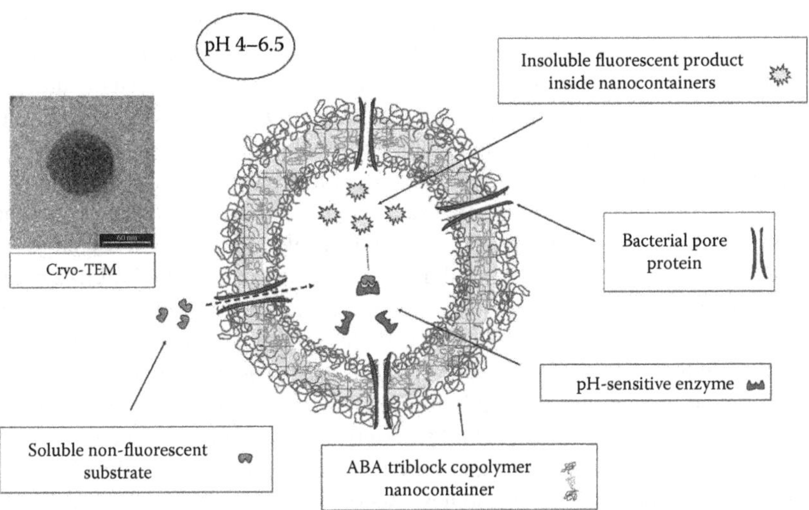

FIGURE 3.46 2D outline and visualization of the nanoreactor system: The nanoreactors are based on a synthetic triblock copolymer membrane functionalized with bacterial ompF pore proteins that make intact, size-selective channels for passive diffusion across the membrane. Encapsulated acid phosphatase enzyme processes a nonfluorescent substrate into an insoluble, fluorescent reaction product at pH 4–6.5. Visualization of the final nanoreactor was done by cryo-TEM, where a distinct core is surrounded by a fine halo, corresponding to the polymer membrane. (Reprinted with permission from Broz, P., Driamov, S., Ziegler, J., Ben-Haim, N., Marsch, S., Meier, W., and Hunziker, P., *Nano Lett.*, 6, 2349. Copyright 2006, American Chemical Society.)

Activation of an encapsulated phosphatase in polymersomes after change in pH has been reported elsewhere (Broz et al. 2006). To achieve this, a pH switchable nanoreactor based on polymersomes formed from the same PMOXA-PDMS-PMOXA triblock copolymer was built. The polymersomes were equipped with bacterial transmembrane OmpF proteins, reconstituted in the membrane, and the pH-sensitive enzyme, acid phosphatase, encapsulated in the vesicle lumen. Schematic representation of the nanoreactor together with cryo-TEM and confocal microscopy images is shown in Figure 3.46. The substrate processing was switchable at pH 4–6.5. By variation of the pH of the surrounding solution, the nanoreactor was able to change its state of activity as demonstrated by producing a water-insoluble fluorescent dye inside the polymeric vesicle. The bacterial pores that were integrated into the polymer membrane remained functional (even though the polymer membrane is three to four times thicker than a biological, lipid-based membrane) and allowed the passive diffusion of both protons for activity control of the encapsulated acid phosphatase, as well as the diffusion of the nonfluorescent substrate into the nanoreactor (Broz et al. 2006).

Other vesicle-forming copolymers have also been used to construct nanoreactors and nanocapsules. Furthermore, not only were channel proteins reconstituted to produce pores or to induce selective transport through the membrane but also other proteins and peptides. Some of them may as well provide target specificity, stimuli responsiveness, and triggering properties upon activation of the systems. Stable protein polymersomes have been generated by a *triggered templated assembly* method (Li et al. 2010). In the first step of the method, vesicles, which form the template material, were formed from the commercially available $(EO)_5(PO)_{68}(EO)_5$ triblock copolymer in the presence of fully water-soluble protein copolymers denoted CS^XS^XC. The latter were prepared in large quantities by a biosynthetic approach retaining the amino acid sequences and length. Two such protein copolymers (twin compounds) were prepared, in which the C blocks are collagen-like; they carry few charges and are water soluble at all pH. The central motive S^XS^X has a number of silklike repeats separated by a chargeable amino acid X, which differ for the twin compounds (histidine, CS^HS^HC; glutamic acid, CS^ES^EC). If the groups are charged (neutral pH), the S block has no secondary structure and is water soluble (Figure 3.47). In the absence of the charge, however, folding takes place and β-rolls or in some cases β-sheets form through intramolecular hydrogen bonding. These secondary structure elements have hydrophobic faces and they insert into the template, thus leading to the formation of protein polymersomes. According to the authors, the method is versatile in the sense that various biologically active materials can be incorporated into the protein polymersome by selection of the triggering method. For instance,

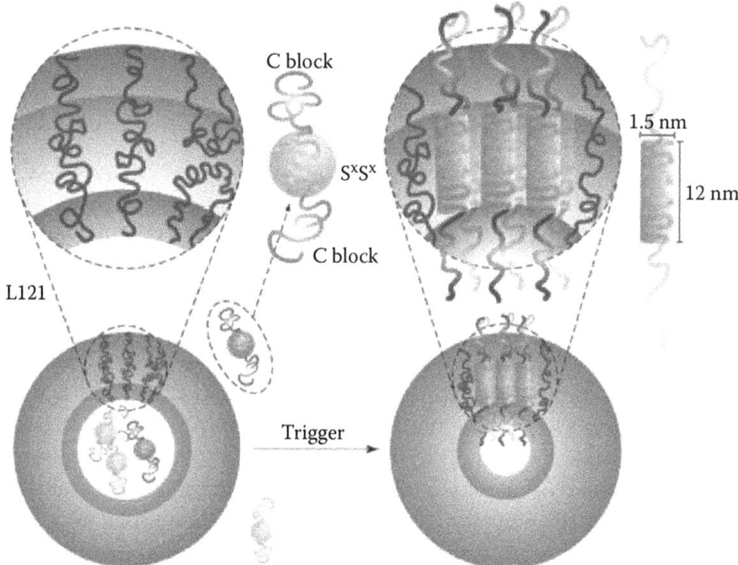

FIGURE 3.47 Triggered templated assembly of protein polymersomes. $(EO)_5(PO)_{68}(EO)_5$ (denoted L121) vesicles (light gray core with thin dark gray corona) surrounded by triblock peptide copolymer CS^XS^XC (S and C blocks indicated). After a trigger, the X groups (spherical dots) become uncharged. The S blocks adopt a β-sheet secondary structure, the hydrophobicity of which drives the insertion of the protein polymers into the capsule. (Li, F., de Wolf, F.A., Marcelis, A.T.M., Sudholter, E.J.R., Cohen Stuart, M.A., and Leermakers, F.A.M.: *Angew. Chem. Int. Ed.* 2010. 49. 9947–9950. Copyright Wiley-VCH Verlag GmbH & Co. KGaA. Reproduced with permission.)

the positively charged protein copolymer CS^HS^HC can be triggered by addition of negatively charged polyelectrolytes, such as siRNA, which then makes a stable biocompatible gene container. In this case, the biologically active species is also responsible for the stability of the capsules and gives the capsules a cooperative release mechanism (Li et al. 2010).

Stable incorporation and active proton transport for porous structures assembled from either dendritic dipeptide, (6Nf-3,4-3,5)12G2-CH_2-Boc-L-Tyr-L-Ala-Ome, or dendritic diester, (R)-4Bp-3,4-dm8G1-COOMe, incorporated into polymersomes have been reported elsewhere (Kim et al. 2009b). The self-assembly of the helical porous structures from the dipeptide and diester is schematically shown in Figure 3.48, left panel. Giant (5–20 μm) polymersomes were spontaneously formed upon hydration of a thin copolymer film, whereas small (~200 nm) unilamellar polymersomes were prepared by sonication of aqueous dispersions and extrusion through 200 nm pore-size filters. A PEO-PBD diblock copolymer (molar masses of 1400 and 2400 Da, respectively) and a triblock PMOXA-PDMS-PMOXA copolymer (molar masses of 2200, 5000, and 2200, respectively) were used for preparation of polymersomes. The dendritic dipeptide and diester assembled in the PEO-PBD polymersomes but not in those prepared from PMOXA-PDMS-PMOXA, most probably due to their higher solubility in PBD than in PDMS. In addition, the triblock copolymer may adopt both membrane-spanning and membrane-looping conformations, which might interfere with orderly channel assembly. The pores were of diameters ~14 to 15 Å and provided an extremely hydrophobic interior, which, in turn, facilitated the transport of water and ions with high rate and selectivity. Relatively large membrane loading levels (up to 5 and 11 mol% for the dipeptide and diester, respectively) were achieved, while the vesicular structure was retained (see Figure 3.48, right panel).

3.6 HYBRID STRUCTURES FORMED UPON INTERACTIONS OF POLYMERS WITH OLIGO- AND POLYNUCLEOTIDES

Hybrid structures between polymers and oligo- and polynucleotides have been mainly created and investigated for the development of polymer-based gene delivery systems. Similarly to the hybrid structures considered in the previous sections, upon interactions of mostly synthetic (co)polymers with DNA, RNA, and antisense oligonucleotides, new structures are formed and new properties are produced. By rational selection of existing natural and synthetic (co)polymers and proper design of novel polymers, specific properties of the hybrid structures could be rendered. The fundamental design criteria for any synthetic gene delivery system includes the ability to (1) neutralize the negatively charged phosphate backbone of DNA to prevent charge repulsion against the anionic cell surface, (2) condense the bulky structure of DNA to appropriate length scales for cellular internalization, and (3) protect the DNA from both extracellular and intracellular nuclease degradation (Schaffer and Lauffenburger 1998, Lechardeur et al. 1999, Abdelhady et al. 2003). At first glance, these criteria look easily achievable especially in the light of significant progress in polymer science, in particular the synthetic versatility afforded by the polymer chemistry enabling the creation of a diverse array of polymers ranging in size, architecture, and chemical composition. However, the effective gene delivery faces

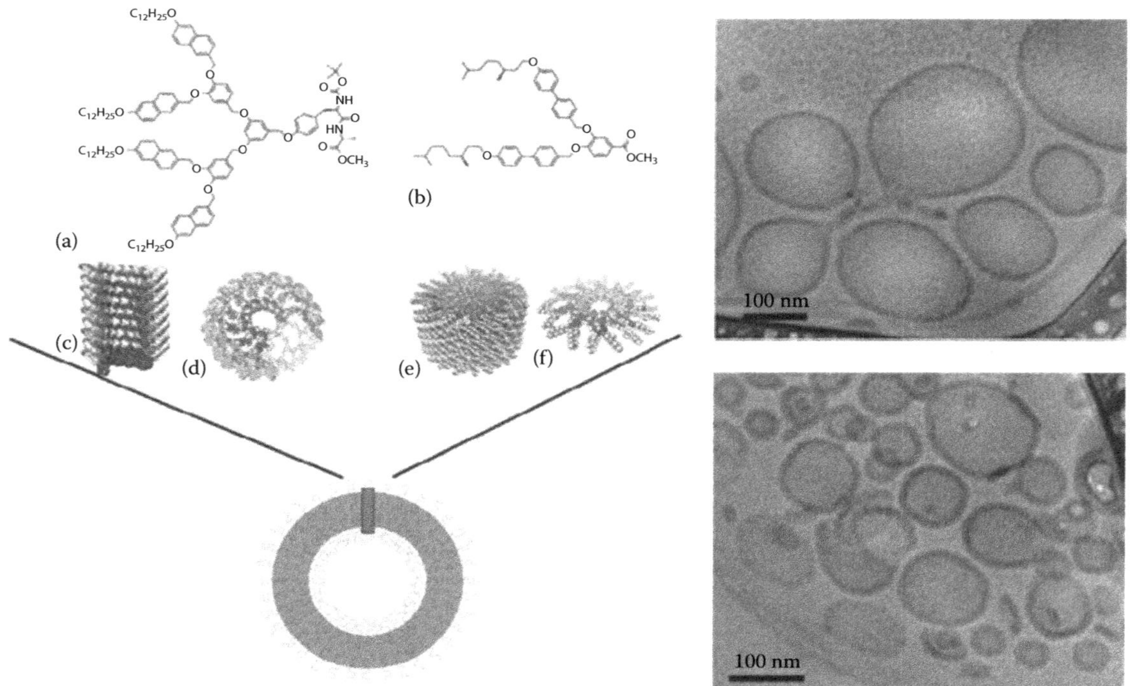

FIGURE 3.48 Left panel: schematics of chemical structures and self-assembly of dendritic dipeptide, (6Nf-3,4-3,5)12G2-CH_2-Boc-L-Tyr-L-Ala-Ome (a,c,d), and dendritic ester, (R)-4Bp-3,4-dm8G1-COOMe (b,e,f) in polymersome membranes. (c,e) Self-assembled pore structures with side view and (d,f) top view for dendritic dipeptide and dendritic ester, respectively. Right panel: cryo-TEM images for dendritic dipeptide (top)- and dendritic diester (bottom)-incorporated polymersomes prepared from PEO-PBD diblock copolymer. (Kim, A.J., Kaucher, M.S., Davis, K.P., Peterca, M., Imam, M.R., Christian, N.A., Levine, D.H., Bates, F.S., Percec, V., and Hammer, D.A.: *Adv. Funct. Mater.* 2009b. 19. 2930–2936. Copyright Wiley-VCH Verlag GmbH & Co. KGaA. Reproduced with permission.)

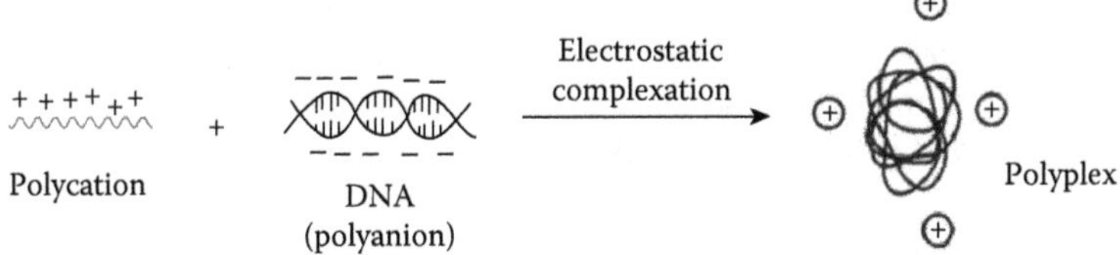

FIGURE 3.49 Schematics of polyplex formation via electrostatic interactions between DNA and a polycation.

various biological barriers that should be overcome, issues such as biocompatibility, systematic stability in the bloodstream, minimal induction of toxic effects, as well as structural and processing considerations. Very frequently, properties designed to overcome one barrier or to fulfill a certain demand can directly hinder the ability of the system to overcome other barriers or fulfill other demands. Obviously, creation of polymer-based gene delivery systems hinges upon a critical balance between various structural, physical, and chemical properties.

In nature, the negatively charged DNA macromolecule forms complexes with positively charged proteins resulting in formation of nanosized aggregates (Burlingame et al. 1985, Luger et al. 1997). Similarly, DNA can form complexes with other positively charged species. Many of the polymeric vectors developed so far have exploited the anionic nature of DNA to drive the complexation via electrostatic interactions (Figure 3.49). The complexes thus formed are termed *polyplexes*, in analogy with lipoplexes—complexes formed between DNA and cationic lipids. Polycations commonly used in the capacity to form complexes with DNA are PEI, polylysine, polyamidoamine, as well as natural polymers such as chitosan, gelatin, dextran, and pullulan. All of these synthetic and natural occurring polymers possess amino groups. At least a portion of them are protonated at neutral pH to enable electrostatically driven self-assembly with DNA or other oligo- or polynucleotides. At a sufficient nitrogen to phosphate (N/P) charge ratio, the polymer can condense DNA to sizes compatible with cellular uptake while providing steric protection from nuclease degradation (Bielinska et al. 1997).

The main synthetic polymer that is used extensively to prepare polyplexes is PEI (Godbey et al. 1999a,b, Wightman et al. 2001, Neu et al. 2005, Pack et al. 2005, Chumakova et al. 2008, Hosseinkhani et al. 2008). PEI is commercially available in both linear and branched chain topology and different molecular weights. It is often considered to be the gold standard for the polymer-based vectors for DNA delivery. The branched PEI has proven to be more efficient as a gene delivery agent and is now most commonly used although its relatively high cytotoxicity is an issue. Increased molecular weight corresponds to higher transfection but also increased toxicity issues; therefore, a balance must be found.

Polypropyleneimine dendrimers are also commercially available. Zinselmeyer and coworkers have recently evaluated polypropyleneimine dendrimers of different generations (from 1 to 5) as gene delivery systems (Zinselmeyer et al. 2002). The authors concluded that the second generation combined a sufficient level of DNA binding with a low level of cell cytotoxicity to give it optimum in vitro gene transfer activity. Gebhart and Kabanov evaluated a panel of polycations including branched and linear PEIs, poly[*N*-ethyl-4-vinyl pyridinium bromide], polyamido-amine dendrimer, polypropyleneimine dendrimer, and a conjugate of a PEO-PPO-PEO copolymer and PEI having a graft–block copolymer architecture (Gebhart and Kabanov 2001).

Poly(L-lysine) and poly(D-lysine) are other synthetic polymers employed in polyplex formation (Capan et al. 1999a,b, Berry et al. 2001, Meilander et al. 2003, Trentin et al. 2005) and are also frequently used as a control in studies to demonstrate the advantages and disadvantages of other polyplexes. They are less toxic than PEI but still their toxicity is an issue. In general, the reduction of the positive charges has been found to increase the transfection efficiency of the polyplexes most probably due to weakening of the polymer/DNA interactions resulting in easier dissociation of DNA from its vector (Erbacher et al. 1997, Kimura et al. 2002, Gabrielson and Pack 2006). Polylysine segment is a constituent moiety of a number of block and graft copolymers. Eldred and coworkers have studied the effects of side chain configuration and backbone spacing in the gene delivery of lysine-derived cationic polymers (Eldred et al. 2005), whereas Ahn and coworkers reported on the synthesis of biodegradable multiblock copolymers of polylysine and PEG (Ahn et al. 2004). Also biodegradable are the PEG-co-poly(L-lysine)-graft-histidine multiblock copolymers prepared from the same group (Bikram et al. 2004). They were of the type $(AB)_n$ and consisted of short PEG and poly(L-lysine) blocks. The former imparted steric stabilization and also introduced biodegradable ester bond linkages. *N,N*-Dimethylhistidine in varying molar ratios was coupled to the ε-amines of poly(L-lysine) thus producing PEG-co-poly(L-lysine)-graft-histidine multiblock copolymers. Its role was to improve the endosome-disrupting capabilities of the copolymers. The copolymers were able to condense plasmid DNA into nanostructures with dimensions between 150 and 200 nm and surface potential of ~4–45 mV and protect the latter from endonuclease digestion for about 2 h.

A diblock copolymer comprising of thermally sensitive block of PNIPAM ($M_n = 3500$) and polycationic block of poly(L-lysine) (DP = 46) has been shown to form polyplexes in cationic to phosphate group (N/P) molar ratios from 0.6:1 to 5:1 (Dimitrov et al. 2011). Stable polyplexes with maximum DNA condensation and particle diameter around 60 nm were obtained at N/P = 5:1 (Figure 3.50).

Conjugated and chemically modified polylysines have been reported as well. These are glycosylated and lactosylated polylysines for gene transfer and nuclear translocation in cystic fibrosis airway epithelial cells (Hashida et al. 1998, Klink et al. 2001, 2003, Grosse et al. 2002) and polylysine conjugated to antibody for targeted gene delivery to leukemia T cells (Suh et al. 2001). Folate receptor-mediated and integrin-mediated gene delivery has been achieved using folate-PEG-poly(L-lysine) and RGD-oligolysine peptide conjugates (Mislick et al. 1995, Harbottle et al. 1998, Cho et al. 2005).

Linear polyamidoamines are known to exhibit good DNA-binding characteristics and are considered as effective gene transfer agents. Linear polyamidoamine has been used in conjunction with stabilizing trehalose to complex DNA and evaluate its usefulness as a gene carrier when released from a scaffold (Heyde et al. 2007). Dendritic polyamidoamines have also been applied to polyplex formation

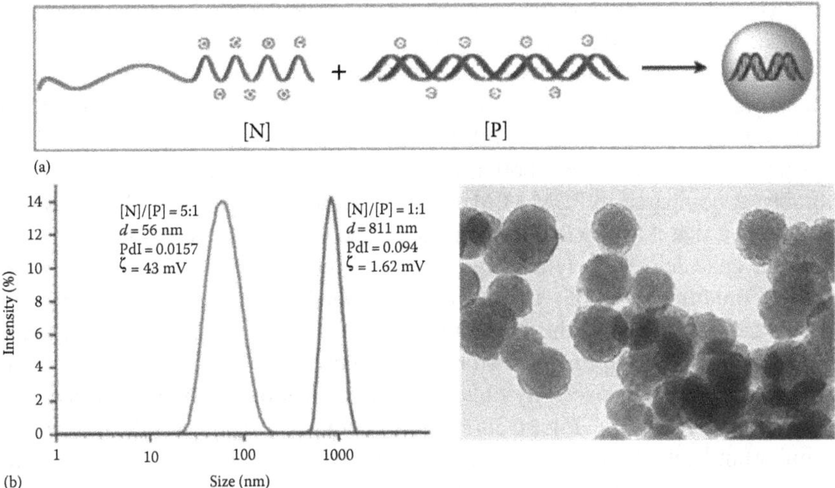

FIGURE 3.50 (a) Schematics of polyplex formation between DNA (salmon sperm, 2000 bp) and PNIPAM-poly(L-lysine) diblock copolymer. (b) Size distribution, diameter, and ζ-potential values of polyplexes formed at different N/P molar ratios. TEM image of the polyplex at N/P = 5:1. (Dimitrov, I.V., Petrova, E.B., Kozarova, R.G., Apostolova, M.D., and Tsvetanov, Ch.B., *Soft Matter*, 7(18), 8002, 2011, Supporting Information. Reproduced by permission of The Royal Society of Chemistry.)

(Bielinska et al. 2000, Fu et al. 2007). The size and charge of dendrimers are typically dependent on the generation. Particularly for the dendritic polyamidoamines, the sixth generation has been reported as the most efficient at transfection (Haensler et al. 1993), though other generations have been used as well. Compared to other agents (polylysine, PEI), polyamidoamines exhibit lower cytotoxicity; even, no cytotoxic effects have been observed for polyamidoamine polyplexes used to transfect cells on a film (Bielinska et al. 2000). Partially degradable polyamidoamine dendrimers are commercially available and have been reported to be 50-fold (!) more efficient at gene transfer than the nondegradable analogues (Tang et al. 1996).

Another representative of the rich collection of polycations that is known for nearly two decades is poly(2-(dimethylamino)ethyl methacrylate) (PDMAEMA) (Cherng et al. 1996, van de Wetering et al. 1997, 1998, Funhoff et al. 2005, Dubruel et al. 2006). PDMAEMA is water soluble and able to bind DNA in a wide interval of ratios. At a polymer/DNA ratio above 2, positively charged polyplexes are formed with a size around 200 nm (van de Wetering et al. 1998). Their transfection efficiency was severalfold higher than those of dextran- and poly(L-lysine)-based polyplexes, although all polyplexes had about the same physicochemical characteristics indicating that the differences in the transfection efficiency can be ascribed to differences in the structural properties of the transfectant (van de Wetering et al. 1997). Transfection efficiency was also studied as a function of PDMAEMA molecular weight. Polymers with molecular weight >300 kDa were better transfection agents in both COS-7 and OVCAR-3 cells than low molecular weight polymers. DLS measurements showed that high molecular weight polymers were able to condense DNA effectively resulting in particles of 170–210 nm. In contrast, when plasmid DNA was incubated with low molecular weight PDMAEMA, large (~1.0 μm) complexes were formed (van de Wetering et al. 1997). Copolymers of DMAEMA with methyl methacrylate, ethoxytriethylene glycol methacrylate, or *N*-vinylpyrrolidone also acted as transfection agents (van de Wetering et al. 1998). However, only for the copolymer containing the latter monomer (54 mol%), both transfection efficiency and cytotoxicity changed favorably as compared to PDMAEMA homopolymer. In a recent study, it has been demonstrated that it is possible to covalently attach an endosome-disruptive peptide to cationic gene delivery polymers with preservation of its membrane destabilization activity (Funhoff et al. 2005). Efforts have been made to develop materials and systems with ability to escape the endosomal–lysosomal pathway (see more in Chapter 5). In this particular chapter, the membrane-disrupting protein was derived from the influenza virus. It was covalently linked to different cationic polymethacrylates and polymethacrylamides (see Figure 3.51 for the chemical structures of the monomers), and importantly, it retained its membrane-disruptive properties. DLS and ζ-potential measurements revealed formation of relatively small (100–250 nm) and positively charged (15–25 mV) polyplexes upon the interaction with DNA (Funhoff et al. 2005).

Researchers have also explored the possibilities of using natural polymers for preparation of nonviral vectors. Chitosan, dextran, gelatin, and pullulan have emerged as promising candidates because of their nontoxicity, biocompatibility, biodegradability, favorable physicochemical properties, and ease of chemical modification. In a recent review, Prabaharan and Mano have examined the advances in the application of chitosan and its derivatives to nonviral gene delivery and have given an overview of transfection studies that use chitosan as a transfection agent (Prabaharan and Mano 2004). When using high molecular weight (100–400 kDa) chitosans, problems related with aggregation, low solubility under physiological conditions, and high viscosity at concentrations used in vivo, slow dissociation and degradation have been observed. On the contrary, chitosans of molecular weights of less than 10 kDa form weak complexes, resulting in physically unstable polyplexes. Those of moderate molecular weights, for example, 50 kDa, seem to be promising materials. Mao et al. reported on preparation of chitosan–DNA nanoparticles by a complex coacervation method (Mao et al. 2001). At N/P ratios between 3 and 8, particles of narrow

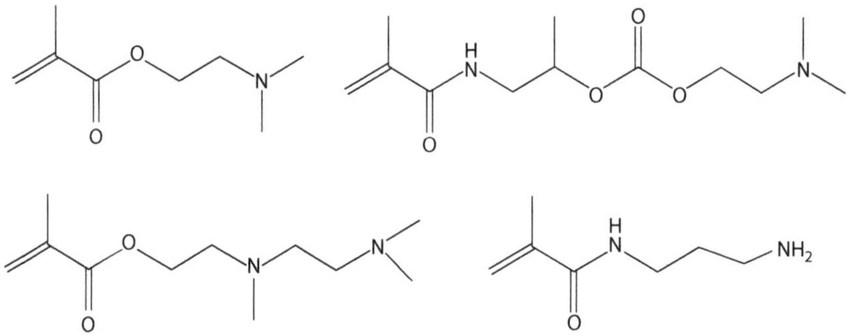

FIGURE 3.51 Chemical structures of various monomers used to prepare cationic polymethacrylates and polymethacrylamides with covalently linked membrane-disrupting peptide.

size distribution and dimensions of about 100–250 nm were obtained (Figure 3.52). They were composed of 35.6 and 64.4 wt.% of DNA and chitosan, respectively, and were slightly positive (ζ potential from +12 to + 18 mV) at pH 6 and nearly neutral at pH 7.2 (Figure 3.52).

By incorporation of poly(propyl acrylic acid)—a polymer specifically designed to disrupt lipid bilayer membrane within a sharply defined pH range—in a chitosan gene carrier, the release of plasmid DNA from the endosomal compartment was enhanced (Kiang et al. 2004). The optimal content of poly(propyl acrylic acid) in the ternary complex was 10 μg, corresponding to a 1:1 weight ratio with DNA. At this dose, the polyplex was characterized with an average diameter of approximately 400 nm and exhibited net negative surface charge of −17 mV. The chemical modification of chitosans and grafting with synthetic polymers offer additional possibilities and advantages. Lee and coworkers have described the synthesis of thiolated chitosan which is known to possess enhanced mucoadhesiveness and cell penetration properties (Lee et al. 2007). Relatively small (from 75 to 120 nm in diameter) particles of thiolated chitosan/DNA were formed. Depending on the weight ratios of chitosan to DNA, the ζ-potential ranged from +2.2 to +19.7 mV. At a weight ratio of 2.5:1, the polyplexes exhibited physical stability and protection against DNase I digestion. Oligo-chitosan (M_n = 3400) grafted with short PEI chains is reported elsewhere (Wong et al. 2006). The results indicated that all amines of chitosan were grafted with short (M_n = 206) PEI chains. The complete retardation of DNA migration at an N/P ratio of 2.5:1 evidenced the good DNA condensation capability of PEI-g-chitosan. Good DNA-binding ability, reduced cytotoxicity, and high transfection efficiency have been reported for chitosan grafted with macrocyclic polyamines on the C-2 and the C-6 position (Li et al. 2011). The grafting positions were important as the C-2-grafted chitosan exhibited higher transfection efficiency but was more cytotoxic than the C-6-grafted one. To improve the transfection efficiency of chitosan, Deng and coworkers have prepared a dendronized chitosan derivative by a copper-catalyzed azide alkyne cyclization reaction of propargyl focal point polyamidoamine dendron with 6-azido-6-deoxy-chitosan (Deng et al. 2011). Figure 3.53 depicts schematics of the synthesis. Better solubility and buffering capacity compared to unmodified chitosan as well as enhanced transfection efficiency in some cases and lower cytotoxicity compared to PEI were reported. Though the chitosan (modified or not) particles loaded with DNA frequently exhibit high transfection efficiency, the latter is accompanied by strong immunological reaction of cells. In order to reduce the latter, the DNA/chitosan nanoparticles have been encapsulated in scaffolds based on PLGA (Nie and Wang 2007, Nie et al. 2009).

Other natural polymers, gelatin and dextran, have been cationized by introducing spermine to carboxyl groups and by means of reductive amination of oxidized dextran, respectively (Hosseinkhani et al. 2006a,b). Nonviral vectors for gene delivery from 3D scaffolds were prepared and both systems successfully transfected cells in vitro. Nagane et al. (2009) by using *N,N′*-carbonyldiimidazole activation method (Hermanson 1996) have cationized pullulan of M_w = 47,300 by introducing 14.5 mol% of spermine. The cationized pullulan readily formed PICs with plasmid DNA. The apparent size and ζ-potential of the discrete polyplex particles were 236 ± 51 nm and 14.0 ± 1.0 mV, respectively. By applying reverse transfection technique, consisting of

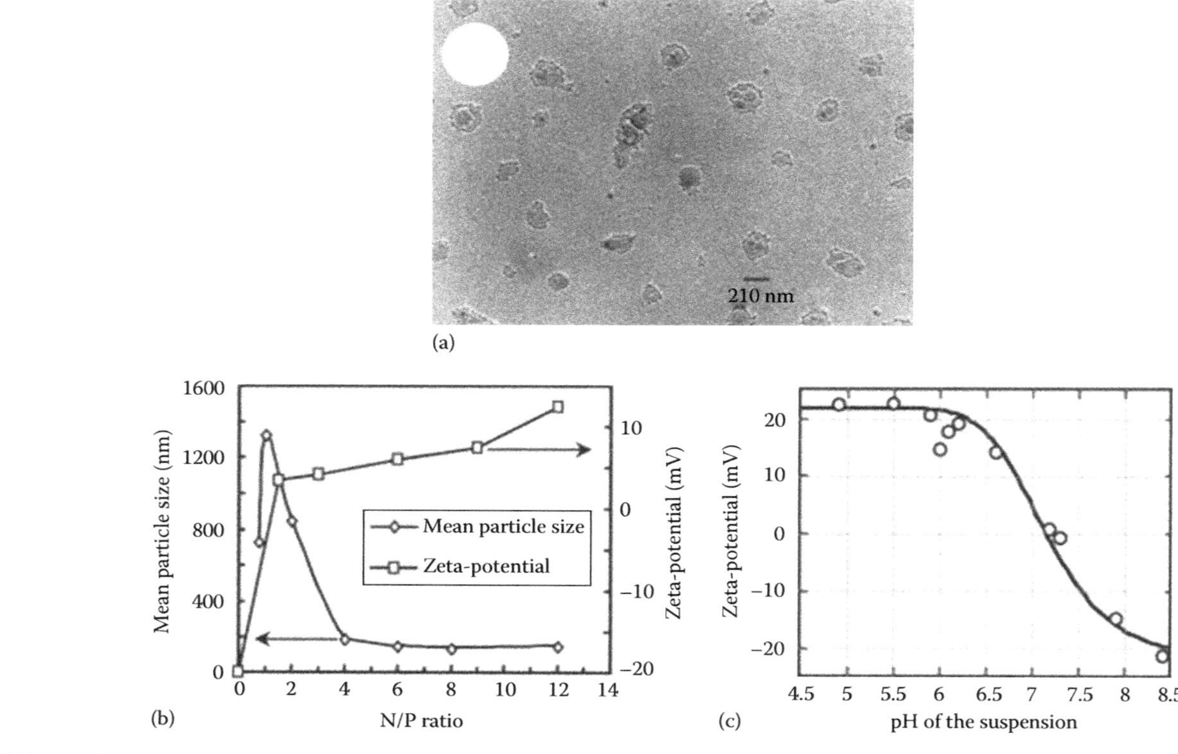

FIGURE 3.52 Physical characterization of chitosan–DNA nanoparticles. (a) SEM image of nanoparticles. (b) Effect of N/P ratio on the mean size (◇) and zeta-potential (◻) of the nanoparticles. (c) Effect of the pH of the nanoparticle suspension on the zeta-potential of the nanoparticles. (Reprinted from *J. Control. Release*, 70(3), Mao, H.Q., Roy, K., Truong-Le, V.L., Janes, K.A., Lin, K.Y., Wang, Y., August, J.T., and Leong, K.W., 399–421, Copyright 2001, with permission from Elsevier.)

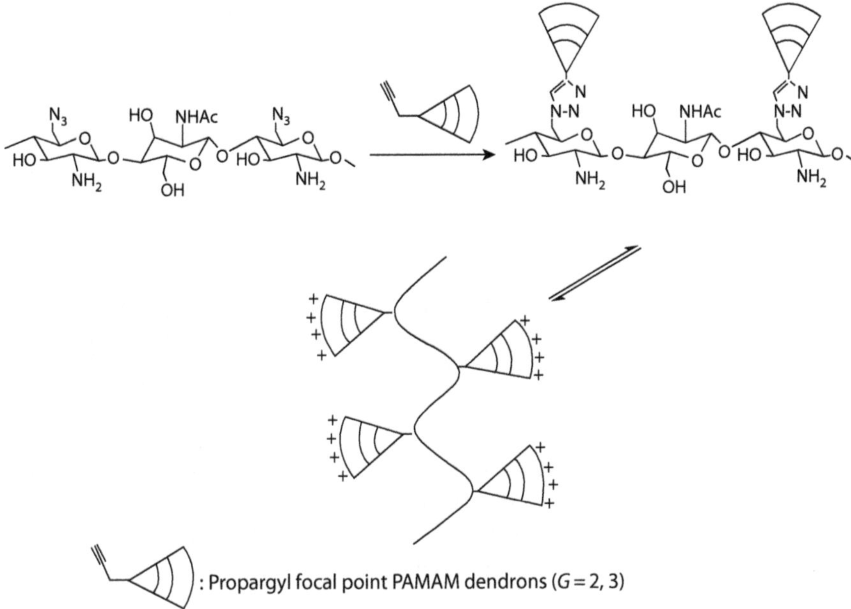

FIGURE 3.53 Schematics of the synthesis of a dendronized chitosan derivative via a copper-catalyzed azide alkyne cyclization reaction of propargyl focal point polyamido-amine (PAMAM) dendron with 6-azido-6-deoxy-chitosan. (Reprinted with permission from Deng, J., Zhou, Y., Xu, B., Mai, K., Deng, Y., and Zhang, L.-M., *Biomacromolecules*, 12(3), 642–649. Copyright 2011, American Chemical Society.)

attachment of the nonviral vector to the base of tissue culture plastic by an artificial cell adhesion protein before the addition of the cells, the transfection was significantly enhanced (Okazaki et al. 2007, Nagane et al. 2009).

Besides the application of polymers that are readily available (both synthetic commercial products and natural polymers), other directions in which research is progressing to meet the demands of nonviral gene delivery are the design of new polymers and copolymers as well as self-assembled systems able to carry genes. A series of poly(glycidyl methacrylate)s, both linear and starlike, with molecular weights typically below 30 kD have been modified with amines such as methylethylamine, 2-amino-1-butanol, and 4-amino-1-butanol (Gao et al. 2011). The amino poly(glycidyl methacrylate)s were able to bind antisense oligonucleotide (a single strand of DNA) in an interval of N/P ratios from 0.5 to 3, which resulted in formation of spherical nanosized complexes with dimensions of 100–300 nm. The authors showed that the increased charge ratio and a synergistic effect of hydrogen bonding in the systems contributed to the increased stability of the complexes, which prevented the incorporated antisense oligonucleotide from nuclease degradation.

An interesting approach toward gene silencing (switching off of a gene by the machinery of the cells) has been suggested elsewhere (Lee et al. 2011b). These authors reported on the synthesis of hybrid conjugates composed of biodegradable PLGA that was conjugated to the 3′ end of small interfering RNA (siRNA) via a disulfide

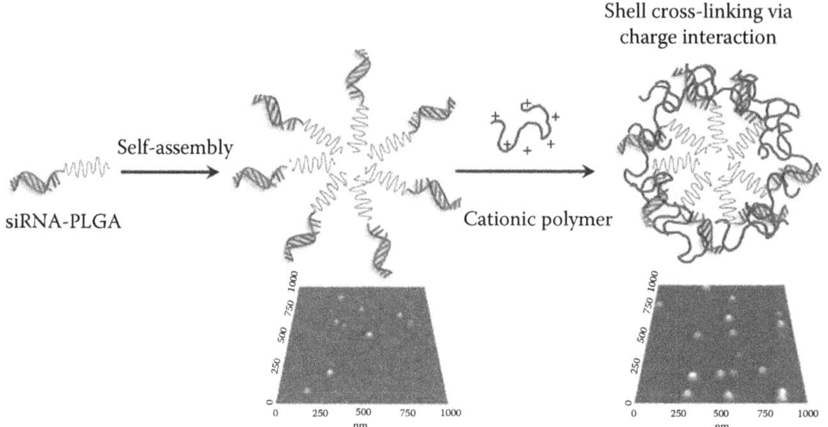

FIGURE 3.54 Schematic representation of a self-assembled structure formed from siRNA-PLGA conjugates and its polyplex with linear PEI. Lower panel: AFM images of the structures. (Reprinted from *J. Control. Release*, 152(1), Lee, S.H., Mok, H., Lee, Y., and Park, T.G., 152–158, Copyright 2011b, with permission from Elsevier.)

bond. By hydrophobic interactions of PLGA blocks, the siRNA-PLGA conjugates spontaneously self-assembled to form small (~20 nm) core–corona-type micelles (Figure 3.54). Hybrid micellar structures were formed via ionic complexation upon the addition of linear PEI. These cationic siRNA-PLGA/PEI micelles (Figure 3.54) were slightly larger (~30 nm) than the initial ones and exhibited superior intracellular uptake and enhanced gene silencing compared to the simple siRNA-PEI polyplexes.

Addition of polynucleotides to preformed micelles has also been described and it presents an efficient method to prepare stable polyplexes. Convertine and coworkers have shown that the addition of siRNA to the micellar solutions of their "diblock" copolymers poly(DMAEMA)-*b*-poly[(butyl methacrylate)-*co*-(DMAEMA)-*co*-(propylacrylic acid)] (see Figure 2.30) did not significantly change the particle dimensions (Convertine et al. 2010). The anionic siRNA was completely bound to the cationic corona of the micelles at N/P ratios of 4:1 and higher. Bridging between the positively charged micelles was not observed. The binding did not neither perturb the stability of the particles nor compromise the pH-responsive destabilizing properties of the copolymer. Moreover, the micellar structure appeared to be important for RNA binding; as noted by the authors, at the CMC and below, the copolymer chains began to form less-ordered aggregates and dissociated, which was expected to reduce cationic charge density and therefore RNA binding. In an earlier paper, complexes between DNA and micelles of cationic diblock copolymers, polystyrene-*b*-poly(quaternized 2-vinylpyridine), have been investigated (Talelli and Pispas 2008). Polyplexes in a wide range of P/N ratios were formed upon the addition of DNA to micellar solutions of these copolymers. Precipitation was observed at P/N ratios close to unity; however, beyond the precipitation region, that is, at P/N ratios >2.5 polyplexes with narrow size distribution, constant dimensions of about 120 nm in diameter, and negative ζ-potential of −15 mV were reproducibly formed. The picture of the polyplexes deduced from the data was of discrete spherical objects consisting

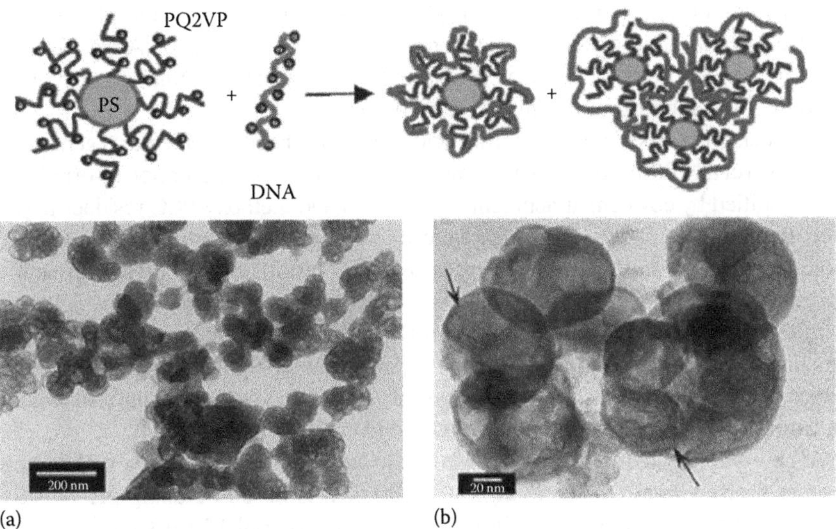

FIGURE 3.55 Schematic representation (top) and TEM micrographs (bottom) of multimicellar polyplex formed from polystyrene-*b*-poly(quaternized 2-vinylpyridine) micelles and DNA at a P/N ratio of 3. (a) and (b) represent TEM images at different magnification. The arrows in (b) denote individual block copolymer micelles inside the polyplex particles. (Top: Talelli, M. and Pispas, S.: *Macromol. Biosci.* 2008. 8. 960. Copyright Wiley-VCH Verlag GmbH & Co. KGaA. Reproduced with permission; bottom: Haladjova, E., Rangelov, S., Tsvetanov, Ch., and Pispas, S.. *Soft Matter*, 8(10), 2884–289, 2012. Reproduced by permission of the Royal Society of Chemistry.)

of several individual copolymer micelles wrapped with DNA chains as shown in Figure 3.55. Note that in contrast to the previous cases, here the negative charges (i.e., DNA) are in excess. The overall negative ζ-potential is beneficial as far as the ability of the polyplex to avoid the interactions with the plasma proteins, ultimately leading to sequestration from the blood stream, is concerned. Preparation of polymeric nanocapsules with DNA entrapped in their interior also benefitted from the DNA excess in the polyplex particles. In a subsequent paper (Haladjova et al. 2012), the latter were coated by a cross-linked polymer shell, and conditions were found to simultaneously destroy the polyplex, disintegrate the micelles, and remove the block copolymer macromolecules thus leaving DNA entrapped in polymeric nanocapsules.

Encapsulation of both DNA fragments and plasmid DNA in cationic diblock copolymer vesicles has been reported by Korobko and coworkers (Korobko et al. 2005, 2006). The encapsulation was achieved by applying a single emulsion technique using polybutadiene-*b*-poly(*N*-methyl-4-vinyl pyridinium) diblock copolymer. For this purpose, an aqueous DNA solution was emulsified in an organic solvent (toluene) and stabilized by the amphiphilic diblock copolymer. The PBD block formed an interfacial brush, whereas the cationic poly(*N*-methyl-4-vinyl pyridinium) block formed complexes with DNA. A subsequent change of the quality of the organic solvent resulted in the collapse of the brush and the formation of a capsule. The capsules were subsequently dispersed in aqueous medium to form vesicles and stabilized with an osmotic agent in the external phase. Inside the vesicles, the plasmid was compacted in a liquid-crystalline fashion. It was shown that the DNA was released from the vesicles once the osmotic pressure dropped below 10^5 N/m^2 or if the ionic strength of the supporting medium exceeded 0.1 M.

Polymersomes based on poly(amino acids) have also been considered as gene carriers (Brown et al. 2000). Amino acid homopolymers, poly(L-lysine) or poly(L-ornithine), were modified by covalent attachment of palmitoyl and methoxy PEG residues to produce amphiphilic copolymers which were found to self-associate in the presence of cholesterol into vesicles. By virtue of methoxy PEG residues, they bore a near neutral ζ-potential. The polymeric vesicles condensed DNA at a polymer to DNA weight ratio of 5:1 or greater, and the polymeric vesicle–DNA complexes improved gene transfer to human tumor cell lines in comparison to the parent homopolymers despite the absence of receptor-specific ligands and lysosomotropic agents. In vivo studies have shown that gene transfer can be achieved in the lungs and liver (Brown et al. 2003).

A new paradigm for DNA encapsulation and intracellular delivery has been described elsewhere (Lomas et al. 2007, 2008). It is based on using pH-sensitive poly(2-(methacryloyloxy)ethyl-phosphorylcholine)-*co*-poly(2-(diisopropylamino) ethyl methacrylate) (PMPC-PDPA) diblock copolymers. They were reported (Du et al. 2005) to form stable vesicles (polymersomes) at physiological pH. Due to protonation of the tertiary amine groups on the PDPA chains, which switched from hydrophobic at physiological pH to hydrophilic (a weak cationic polyelectrolyte) at mildly acidic conditions, dissociation of the polymersomes at around pH 5–6 to form molecularly dissolved copolymer chains was observed. The pH sensitivity of a diblock copolymer of composition $PMPC_{25}$-$PDPA_{70}$ was exploited for the formation of the vesicles and simultaneous encapsulation of DNA. As shown in Figure 3.56, firstly, the copolymer was dissolved in mildly acidic aqueous solution and then DNA plasmid was added at pH 6.0. At this pH, the PMPC-PDPA chains were molecularly dissolved and protonated. When protonated, the cationic PDPA chains interacted strongly with the anionic phosphate groups on the DNA. As the pH was increased further to 7.5, self-assembly into polymersomes occurred and some of the DNA plasmid was encapsulated. An entrapment efficiency of around 20% was determined after purification of the DNA-loaded polymersomes by preparative GPC. The observation of ζ-potential values of zero when the solution pH is greater than the p*K*a of the PDPA block suggested that there was no DNA bound to the *outside* of the polymersomes. At neutral pH, DNA was relatively weakly bound, which did not affect the vesicle formation: DLS studies indicate that empty PMPC-PDPA vesicles exhibited a similar particle size distribution to that of the vesicles prepared in the presence of DNA, and the TEM analysis clearly showed that vesicles are the predominant colloidal aggregate morphology. In contrast, the strong DNA–copolymer electrostatic interactions that occurred in mildly acidic solution led to the formation of DNA–copolymer aggregates. TEM images (not shown) confirmed the formation of well-defined quasi-tubular aggregates with lengths ranging from 100 to 200 nm and tube diameters of 15 to 20 nm. Schematics of the two morphologies are presented in Figure 3.56. Both of them ensured protection of DNA regardless of pH of the environment. The sharp pH-driven transition occurring within just one pH unit makes these polymersomes

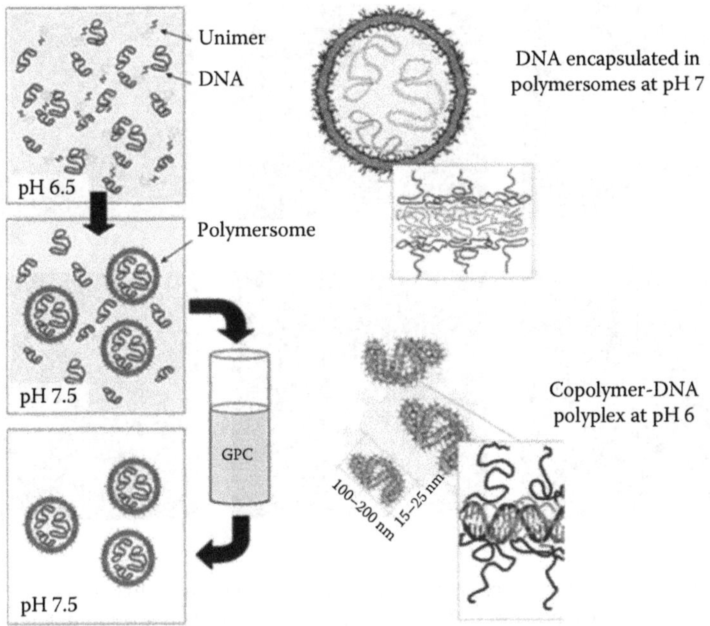

FIGURE 3.56 (Left) Schematic presentation of DNA encapsulation using poly(2-(methacryloyloxy)ethyl-phosphorylcholine)-*co*-poly(2-(diisopropylamino)ethyl methacrylate) (PMPC–PDPA) diblock copolymer polymersomes. (Right) Schematics of morphologies observed at different pH. (Lomas, H., Canton, I., MacNeil, S., Du, J., Armes, S.P., Ryan, A.J., Lewis, A.L., and Battaglia, G.: *Adv. Mater.* 2007. 19. 4238–4243. Copyright Wiley-VCH Verlag GmbH & Co. Reproduced with permission.)

very attractive for intracellular delivery applications, which are considered in more details in Chapter 5.

Similar is the approach of Kim and coworkers to load siRNA and antisense oligonucleotide in polymersomes prepared from various biodegradable, nonionic copolymers—PEG-*b*-PCL, PEG-*b*-PLA, and their blends with the inert PEG-*b*-PBD (Kim et al. 2009c). Copolymer solutions in DMSO were added slowly to oligonucleotide solutions in deionized water and vortexed briefly. The DMSO was extracted via extensive dialysis of the mixed solution in cold PBS during which the vesicles were formed. The free oligonucleotides were removed via a second dialysis step, and finally, 100 nm vesicles were generated via extrusion through nanoporous filters. Before size downing, the polymersomes were sufficiently large to visualize the encapsulation of fluorescent antisense oligonucleotide (Figure 3.57). The inset in this figure shows edge-bright intensity profile for polymersome membrane and the filled lumen profile for antisense oligonucleotide, indicating a lack of membrane interaction. The absence of interactions was to be expected considering the nonionic nature of the constituent blocks of the copolymers. In spite of this, loading efficiencies of up to 30% for siRNA and at least 20% for antisense nucleotide were quantified that appeared similar to those of oligonucleotides in nanocomplexes with the commercially available cationic lipid delivery vehicle Lipofectamine 2000 and

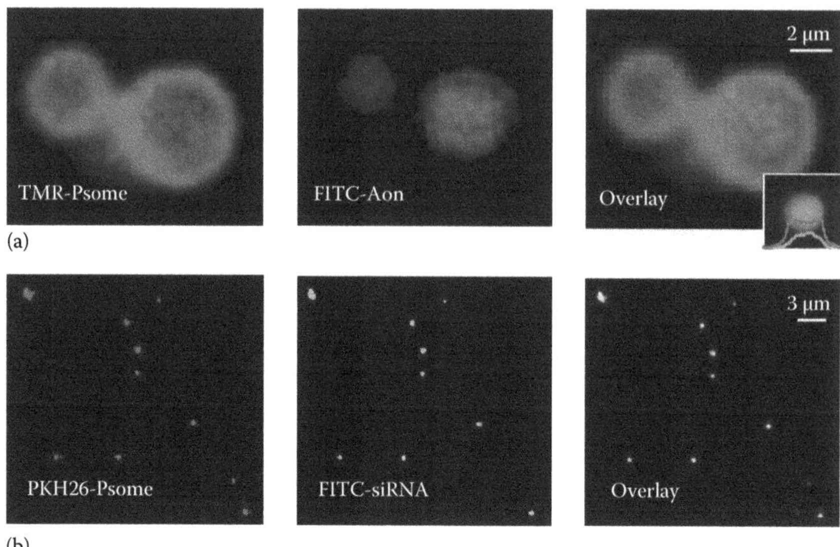

FIGURE 3.57 (a) Fluorescence microscope images of antisense oligonucleotide (middle) loaded in biodegradable polymersomes based on PEG-*b*-PCL (left) after formation of LUVs and the overlay image (right). The lack of membrane interaction was evidenced by the edge-bright intensity profile for polymersome membrane and the filled lumen profile for the antisense oligonucleotide shown in the inset of the overlay image. (b) Fluorescence images of siRNA (middle) loaded into fluorescently labeled nanosized polymersomes (left) based on PEG-*b*-PLA and the overlay image (right). (Reprinted from *J. Control. Release*, 134, Kim, Y., Tewari, M., Pajerowski, J.D., Cai, Sh., Sen, Sh., Williams, J., Sirsi, Sh., Lutz, G., and Discher, D.E., 132–140, Copyright 2009c, with permission from Elsevier.)

DNA in the previously mentioned PMPC-PDPA polymersomes. Because of the lack of interactions between the copolymer and oligonucleotides, these structures are not further considered in the present section. However, more details of their encapsulation capacity, controlled release, cellular uptake, and efficient in vivo delivery are presented in Chapter 5.

Research has been carried out to solve the common dilemma for all gene carrier systems, namely, simultaneous fulfillment of (at least) two opposite requirements: (1) tight complex formation, favorable for cellular uptake and evasion of DNA degradation, and (2) complex dissociation or loose complex formation, favorable for transcription (Figure 3.58). The concept is based on introduction of stimuli-sensitive moieties in the copolymer to control the formation of the polyplex and its dissociation at specific site and time thus optimizing the complex status at each intracellular process. The first demonstration of the control of transfection efficiency by light using a nonviral gene vector has been done by Nagasaki and coworkers who synthesized a cationic L-lysine-modified polyazobenzene dendrimer (Nagasaki et al. 2000). The azo moiety of the dendrimer showed trans-to-cis isomerization upon irradiation by UV light. UV irradiation of the plasmid DNA complex into the cytoplasm caused a 50% increase in the transfection efficiency,

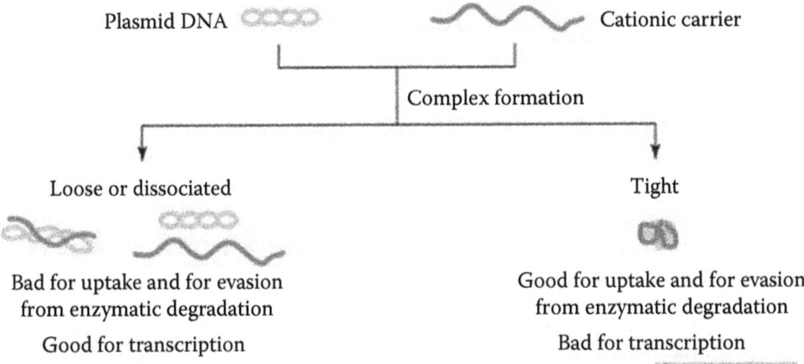

FIGURE 3.58 The dilemma of the DNA–polymer complex: just how tightly should DNA be complexed? In a cationic DNA–polymer complex, the common dilemma is that these systems must fulfill two opposite requirements simultaneously: (1) tight complex formation, favorable for cell uptake and evasion of DNA degradation, and (2) complex dissociation or loose complex formation, favorable for transcription by RNA polymerase. (Reprinted from *Drug Disc. Today*, 7(7), Yokoyama, M., 426. Copyright 2002, with permission from Elsevier.)

compared with the negative control. A decrease of the cationic charge density on the dendrimer surface was considered to contribute to the transfection increase by promoting DNA release from the complex.

The first photochemical-internalization-mediated gene delivery in vivo has been reported by Nishiyama et al. (2005). This has been achieved by using a ternary complex composed of a core containing DNA packaged with cationic peptides and enveloped in the anionic dendrimer phthalocyanine, which provides the photosensitizing action. The ternary complex showed more than 100-fold photochemical enhancement of transgene expression in vitro with reduced photocytotoxicity.

Besides light, the temperature is also applicable as a stimulus. Light has the advantage over temperature with respect to the site precision, whereas temperature is the stimulus of choice in terms of available depth from the surface. In principle, thermoresponsive polymers with LCSTs below body temperature can be used to deliver tightly condensed DNA complexes to cells. Once inside the cells, an externally applied temperature reduction to below the LCST induces the relaxed, extended-chain conformation to result in DNA release (Figure 3.59). Kurisawa and coworkers have designed a temperature-responsive polymeric gene carrier system composed of three components (Kurisawa et al. 2000): a temperature-responsive block, a cationic block, and a hydrophobic block. The cationic block was based on PDMAEMA. It interacted with DNA to ensure high DNA-binding efficiency. The phase transition temperature was adjusted by the incorporation of the hydrophobic block; it was based on poly(*n*-butyl methacrylate). Satisfactory control over the formation and dissociation of the polyplex in response to temperature variations was achieved for a copolymer with DPs of 81, 8, and 11 for PNIPAM, PDMAEMA, and poly(*n*-butyl methacrylate), respectively. The phase transition of the copolymer was 21°C and did not change in the presence of DNA. Its transfection efficiency was enhanced on lowering the temperature.

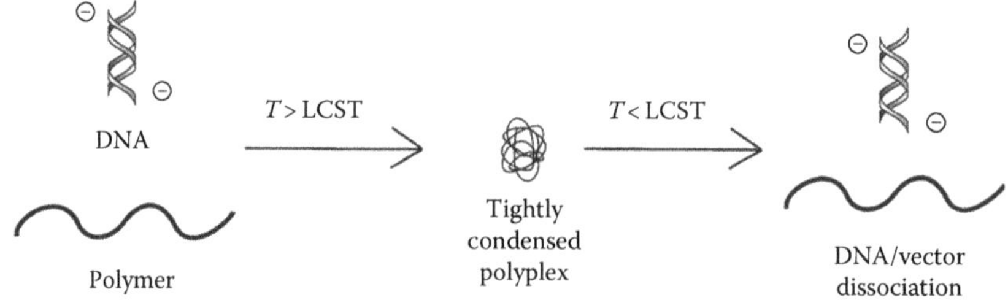

FIGURE 3.59 Schematic illustration of thermoresponsive polymers as gene vectors. Polymers with an LCST below body temperature assume a collapsed conformation at *T*>LCST to form tightly condensed polyplexes, which promotes cellular uptake. Upon reducing the temperature to *T*<LCST (e.g., from an externally applied source), the polymer assumes a relaxed, linear conformation to enable DNA/vector dissociation. (Reprinted from *Prog. Polym. Sci.*, 32, Wong, Sh.Y., Pelet, J.M., and Putnam, D., 799–837. Copyright 2007, with permission from Elsevier.)

Sun et al. (2005) have reported on controllable gene expression of chitosan-based vectors. They synthesized a carboxyl-terminated random copolymer of NIPAm and vinyl laurate which was coupled to chitosan. The copolymer exhibited LCST of ca. 26°C, but normally it self-associated to form micelle-like structures with an average diameter of 118 nm. Globular complexes with DNA in a wide N/P range were formed. Their size was dependent on the interplay of hydrophobic and electrostatic interactions between DNA and copolymer and reached a maximum of 148 nm at 1:1. At lower charge ratio, the self-aggregation from hydrophobic alkyl chains of vinyl laurate caused a slight increase in the size of complex. Whereas at higher charge ratio, more compact complexes reaching a value of 67 nm at N/P of 15:1 were formed as a result from the strong electrostatic attraction. Further, the authors showed that the affinity between DNA and copolymer could be tuned by varying surrounding temperature. At temperatures above LCST, tight and stable complexes were formed due to the collapse of PNIPAM chains. Reducing temperature below LCST facilitated the dissociation of the polyplex, resulting in the exposure of more DNA molecules. The level of gene expression could be thermally controlled and increased by lowering incubation temperature to temperatures below the LCST, for example, 20°C. Note that similar to the PNIPAM-*b*-PDMAEMA-*b*-poly(*n*-butyl methacrylate) triblock copolymer (Kurisawa et al. 2000), this copolymer also contained three types of moieties though not organized in a block architecture—thermosensitive, cationic, and hydrophobic—responsible for three different functions of the vector. A rather similar effect, however, has been achieved with a series of thermosensitive copolymers—trimethyl chitosan-*g*-PNIPAM (Mao et al. 2007). An LCST of 32°C was measured regardless of the grafting ratios. Upon mixing with DNA, polyplexes with spherical morphology (Figure 3.60) were formed with sizes ranging from 200 to 900 nm depending on the N/P ratio. It was shown that the copolymers exhibited stronger ability to bind DNA at 40°C (above LCST) when PNIPAM was collapsed. Again, improvement of gene transfection efficiency was reported following incubation of the cultured cells at 25°C (below LCST) for a while.

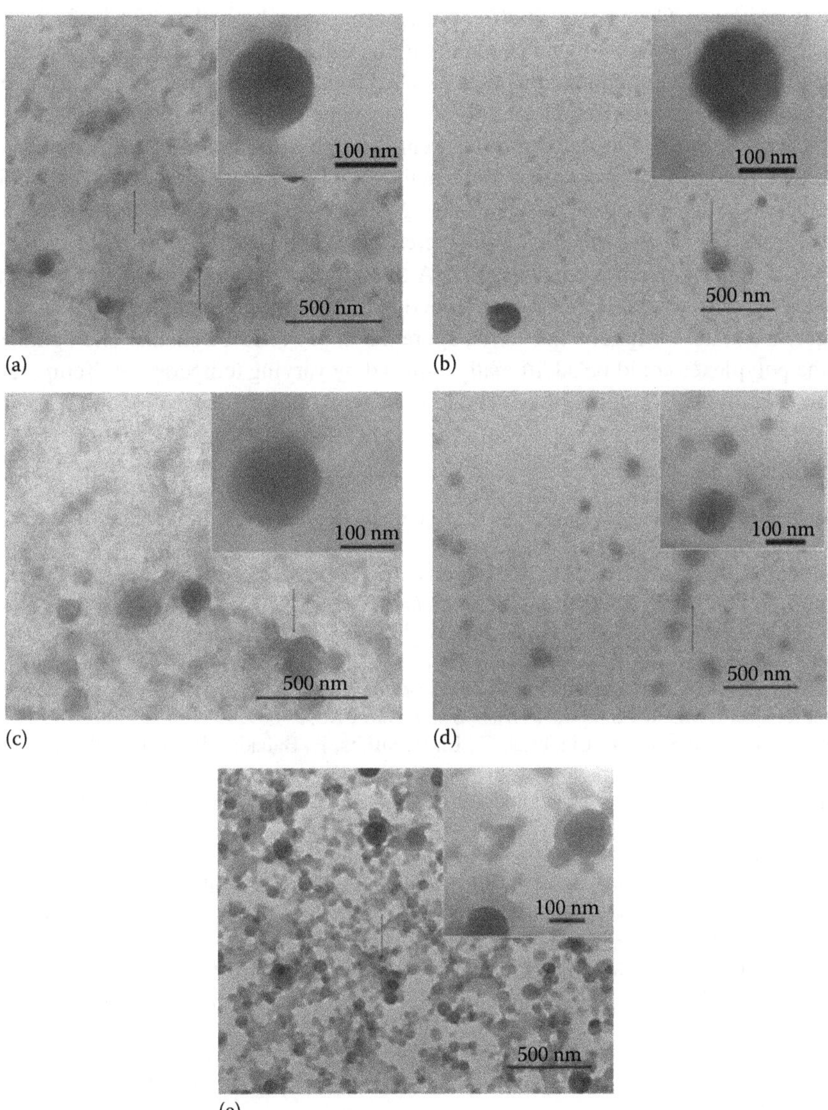

FIGURE 3.60 TEM images of (a) trimethyl chitosan/DNA particles dried at 37°C, (b) trimethyl chitosan-*g*-PNIPAM26/DNA particles dried at 20°C, (c) trimethyl chitosan-*g*-PNIPAM26/DNA dried at 37°C, (d) trimethyl chitosan-*g*-PNIPAM60/DNA dried at 20°C, and (e) trimethyl chitosan-*g*-PNIPAM60/DNA dried at 37°C. The insets are higher magnification of the corresponding images. All N/P ratios are 10. (Reprinted from *Biomaterials*, 28(30), Mao, Zh., Ma, L., Yan, J., Yan, M., Gao, Ch., and Shen, J., 4488–4500. Copyright 2007, with permission from Elsevier.)

In another paper, the synthesis of a thermosensitive bioconjugate prepared by coupling hydroxyl-terminated PNIPAM with the cell-penetrating peptide, poly(L-arginine) (Cheng et al. 2006). The bioconjugate exhibited an LCST of 35.2°C, which was slightly higher than that of pure PNIPAM. No evidence for self-association was provided. The complexation with DNA did not influence the LCST; however, globular structures with dimensions 50–120 nm, depending on the bioconjugate/DNA ratio, were formed. The size variation went through a minimum at a bioconjugate/DNA wt. ratio of 10/1. According to the authors, at lower mass ratio, the electrostatic interaction between the bioconjugate and DNA was weaker, which resulted in formation of relatively loose structures. With increasing vector content, more of the cationic poly(L-arginine) moieties condensed DNA to a more compact entity; that is why the particle size showed a decreasing trend and reached a minimum at a 10/1 ratio. The larger size at 20/1 might be due to the aggregation of PNIPAM chains. The tightness of the polyplexes could be additionally adjusted by varying temperatures. Temporary cooling below LCST of the bioconjugate was favorable for gene expression.

REFERENCES

Abdelhady, H. G., Allen, S., Davies, M. C., Roberts, C. J., Tendler, S. J., Williams, P. M. 2003. *Nucleic Acids Res.* 31(14):4001.

Abraham, T., Hato, M., Hirai, M. 2005. *Biotechnol. Progr.* 21:255–262.

Ahmed, F., Discher, D. E. 2004. *J. Control. Release* 96(1):37–53.

Ahn, C. H., Chae, S. Y., Bae, Y. H., Kim, S. W. 2004. *J. Control. Release* 97(3):567–574.

Alexander, S. 1977. *J. Phys. (Paris)* 38:983.

Allen, C., Maysinger, D., Eisenberg, A. 1999. *Colloids Surf. B* 16(1–4):3–27.

Almgren, M., Rangelov, S. 2006. *J. Disp. Sci. Techn.* 27:599–609.

Almgren, M., van Stam, J., Lindblad, C., Li, P., Stillbs, P., Bahadur, P. 1991. *J. Phys. Chem.* 995:5677.

Angelov, B., Angelova, A., Papahadjopoulos-Sternberg, B., Hoffmann, S., Nicolas, V., Lesieu, S. 2012. *J. Phys. Chem. B* 116(26):7676–7686.

Arifin, D. R., Palmer, A. F. 2005. *Biomacromolecules* 6:2172–2181.

Axthelm, F., Casse, O., Koppenol, W. H., Nauser, T., Meier, W., Palivan, C. 2008. *J. Phys. Chem. B* 112:8211–8217.

Bangham, A. D., Standish, M. M., Watkins, J. C. 1965. *J. Mol. Biol.* 13:238–252.

Barauskas, J., Johnsson, M., Tiberg, F. 2005. *Nano Lett.* 5:1615–1619.

Barauskas, J., Misiunas, A., Gunnarsson, T., Tiberg, F., Johnsson, M. 2006. *Langmuir* 22:6328–6334.

Batrakova, E. V., Kabanov, A. V. 2008. *J. Control. Release* 130:98–106.

Belsito, S., Bartucci, R., Montesano, G., Marsh, D., Sportelli, L. 2000. *Biophys. J.* 78:1420–1430.

Ben-Haim, N., Broz, P., Marsch, S., Meier, W., Hunziker, P. 2008. *Nano Lett.* 8:1368.

Bergstrand, N., Edwards, K. 2004. *J. Colloid Interface Sci.* 276(2):400–407.

Berkovich, A. K., Lukashev, E. P., Melik-Nubarov, N. S. 2012. *Biochim. Biophys. Acta Biomembr.* 1818:375–83.

Berry, M., Gonzalez, A. M., Clarke, W., Greenlees, L., Barrett, L., Tsang, W., Seymour, L., Bonadio, J., Logan, A., Baird, A. 2001. *Mol. Cell Neurosci.* 17:706–716.

Bielinska, A. U., Kukowska-Latallo, J. F., Baker, R. 1997. *Biochim. Biophys. Acta* 1353:180–190.

Bielinska, A. U., Yen, A., Wu, H. L., Zahos, K. M., Sun, R., Weiner, N. D., Baker Jr., J. R., Roessler, B. J. 2000. *Biomaterials* 21:877–887.

Bikram, M., Ahn, C., Chae, S., Lee, M., Yockman, J., Kim, S. 2004. *Macromolecules* 37(5):1903–1916.

Binder, W. H. 2008. *Angew. Chem. Int. Ed.* 47:3092–3095.

Boyd, B. J., Whittaker, D. V., Khoo, S. M., Davey, G. 2006. *Int. J. Pharm.* 318:154–162.

Bronich, T. K., Ouyang, M., Kabanov, V. A. Eisenberg, A., Szoka, Jr., F. C., Kabanov. A. V. 2002. *J. Am. Chem. Soc.* 124(40):11872–11873.

Bronich, T. K., Popov, A. M., Eisenberg, A., Kabanov, V. A., Kabanov, A. V. 2000. *Langmuir* 16(2):481–489.

Bronstein, L. M., Chernyshov, D. M., Timofeeva, G. I., Dubrovina, L. V., Valetsky, P. M., Khokhlov, A. R. 2000a. *J. Colloid Interface Sci.* 230(1):140–149.

Bronstein, L. M., Chernyshov, D. M., Timofeeva, G. I., Dubrovina, L. V., Valetsky, P. M., Obolonkova, E. S., Khokhlov, A. R. 2000b. *Langmuir* 16(8):3626–3632.

Bronstein, L. M., Chernyshov, D. M., Vorontsov, E., Timofeeva, G. I., Dubrovina, L. V., Valetsky, P. M., Kazakov, S., Khokhlov, A. R. 2001. *J. Phys. Chem. B* 105(38):9077–9082.

Brown, M. D., Gray, A. I., Tetley, L., Santovena, A., Rene, J. Schaetzlein, A. G., Uchegbu, I. F. 2003. *J. Control. Release* 93:193.

Brown, M. D., Schaetzlein, A., Brownlie, A., Jack, V., Wang, W., Tetley, L., Gray, A. I., Uchegbu, I. F. 2000. *Bioconj. Chem.* 11(6):880–891.

Broz, P., Driamov, S., Ziegler, J., Ben-Haim, N., Marsch, S., Meier, W., Hunziker, P. 2006. *Nano Lett.* 6:2349.

van der Burgh, S., de Keizer, A., Cohen Stuart, M. A. 2004. *Langmuir* 20(4):1073–1084.

Burke, S. E., Eisenberg, A. 2001. *Langmuir* 17(26):8341–8347.

Burlingame, R. W., Love, W. E., Wang, B. C., Hamlin, R., Ngyyen, H. X., Moudrianakis, E. N. 1985. *Science* 228:546.

de Campo, L., Yaghmur, A., Sagalowicz, L., Leser, M. E., Watzke, H., Glatter, O. 2004. *Langmuir* 20:5254–5261.

Capan, Y., Woo, B. H., Gebrekidan, S., Ahmed, S., DeLuca, P. P. 1999a. *J. Control. Release* 60:279–286.

Capan, Y., Woo, B. H., Gebrekidan, S., Ahmed, S., DeLuca, P. P. 1999b. *Pharm. Res.* 16:509–513.

Chen, D., Li, Zh., Yu, Ch., Shi, Y., Zhang, Zh., Tu, B., Zhao, D. 2005. *Chem. Mater.* 17(12):3228–3234.

Cheng, N., Liu, W., Cao, Zh., Ji, W., Liang, D., Guo, G., Zhang, J. 2006. *Biomaterials* 27:4984–4992.

Cheng, Zh., Thorek, D. L. J., Tsourkas, A. 2009. *Adv. Func. Mater.* 19(23):3753–3759.

Cheng, Zh., Tsourkas, A. 2008. *Langmuir* 24:8169–8173.

Cherng, J. Y., van de Wetering, P., Talsma, H., Crommelin, D. J., Hennink, W. E. 1996. *Pharm. Res.* 13(7):1038–1042.

Cheyne, R. B., Moffitt, M. G. 2005. *Langmuir* 21:5453–5460.

Chieng, Y. Y., Chen, S. B. 2009. *J. Phys. Chem. B* 113:14934–14942.

Cho, K. C., Kim, S. H., Jeong, J. H., Park, T. G. 2005. *Macromol. Biosci.* 5(6):512–519.

Choi, H.-J., Germain, J., Montemagno, C. D. 2006. *Nanotechnology* 17:1825–1830.

Choi, H.-J., Lee, H., Montemagno, C. D. 2005. *Nanotechnology* 16:1589.

Choi, H.-J., Montemagno, C. D. 2005. *Nano Lett.* 5:2538–2542.

Chong, J. Y. T., Mulet, X., Waddington, L. J., Boyd, B. J., Drummond, C. J. 2011. *Soft Matter* 7:4768–4777.

Christian, D. A., Cai, S., Bowen, D. M., Kim, Y. H., Pajerowski, J. D., Discher, D. E. 2008. *Polym. Preprints* 49:1075.

Chumakova, O. V., Liopo, A. V., Andreev, V. G., Cicenaite, I., Evers, B. M., Chakrabarty, S., Pappas, T. C., Esenaliev, R. O. 2008. *Cancer Lett.* 261:215–225.

Convertine, A. J., Diab, C., Prieve, M., Paschal, A., Hoffman, A. S., Johnson, P. H., Stayton, P. S. 2010. *Biomacromolecules* 11:2904–2911.

Cox, J. K., Yu, K., Eisenberg, A., Lennox, R. B. 1999. *Phys. Chem. Chem. Phys.* 1:4417–4421.
Danson, S., Ferry, D., Alakhov, V., Margison, J., Kerr, D., Jowle, D., Brampton, M., Halbert, G., Ranson, M. 2004. *Br. J. Cancer* 90:2085–2091.
De Santis, S., Ladogana, R. D., Diociaiuti, M., Masci, G. 2010. *Macromolecules* 43(4): 1992–2001.
Demina, T., Grozdova, I., Krylova, O., Zhirnov, A., Istratov, V., Frey, H., Kautz, H., Melik-Nubarov, N. 2005. *Biochemistry* 44:4042–4054.
Deng, J., Zhou, Y., Xu, B., Mai, K., Deng, Y., Zhang, L.-M. 2011. *Biomacromolecules* 12(3):642–649.
Dimitrov, I. V., Petrova, E. B., Kozarova, R. G., Apostolova, M. D., Tsvetanov, Ch. B. 2011. *Soft Matter* 7(18):8002.
Dimitrov, Ph., Rangelov, S., Dworak, A., Tsvetanov, Ch. 2004. *Macromolecules* 37: 1000–1008.
Drechsler, U., Thibault, R. J., Rotello, V. M. 2002. *Macromolecules* 35:9621–9623.
Driever, C. D., Mulet, X., Johnston, A. P. R., Waddington, L. J., Thissen, H., Caruso, F., Drummond, C. J. 2011. *Soft Matter* 7:4257.
Dong, Y. D., Larson, I., Hanley, T., Boyd, B. J. 2006. *Langmuir* 22:9512–9518.
Du, J., Tang, Y., Lewis, A. L., Armes, S. P. 2005. *J. Am. Chem. Soc.* 127:17982.
Dubruel, P., Schacht, E. 2006. *Macromol. Biosci.* 6(10):789–810.
Dulle, M., Glatter, O. 2012. *Langmuir* 28:1136–1141.
Edwards, K., Johnsson, M., Karlsson, G., Silvander, M. 1997. *Biophys. J.* 73:258–266.
Eldred, S. E., Pancost, M. R., Otte, K. M., Rozema, D., Stahl, S. S., Gellman, S. H. 2005. *Bioconjug. Chem.* 16(3):694–699.
Erbacher, P., Roche, A. C., Monsigny, M., Midoux, P. 1997. *Biochim. Biophys. Acta* 1324(1):27–36.
Feitosa, E., Winnik, F. M. 2010. *Langmuir* 26:17852–17857.
Firestone, M. A., Wolf, A. C., Seifert, S. 2003. *Biomacromolecules* 4:1539–1549.
Francis, R., Skolnik, A. M., Carino, S. R., Logan, J. L., Underhill, R. S., Angot, S., Taton, D., Gnanou, Y., Duran, R. S. 2002. *Macromolecules* 35:6483–6485.
Frey, S. L., Zhang, D., Carignano, M. A., Szleifera, I., Lee, K. Y. C. 2007. *J. Chem. Phys.* 127:114904–114912.
Fu, H.-L., Cheng, S.-X., Zhang, X.-Z., Zhuo, R.-X. 2007. *J. Control. Release* 124:181–188.
Funhoff, A. M., van Nostrum, C. F., Lok, M. C., Kruijtzer, J. A., Crommelin, D. J., Hennink, W. E. 2005. *J. Control. Release* 101(1–3):233–246.
Gabriel, G. J., Pool, J. G., Som, A., Dabkowski, J. M., Coughlin, E. B., Muthukumar, M., Tew, G. N. 2008. *Langmuir* 24:12489–12495.
Gabrielson, N. P., Pack, D. W. 2006. *Biomacromolecules* 7(8):2427–2435.
Ganguly, R., Aswal, V. K., Hassan, P. A., Gopalakrishnan, I. K., Kulshreshtha, S. K.2006. *J. Phys. Chem. B* 110:9843.
Gao, H., Lu, X., Ma, Y., Yang, Y., Li, J., Wu, G., Wang, Y., Fan, Y., Ma, J. 2011. *Soft Matter* 7(19):9239–9247.
Gau-Racine, J., Lal, J., Zeghal, M., Auvray, L. 2007. *J. Phys. Chem. B* 111:9900–9907.
Gebhart, C. L., Kabanov, A. V. 2001. *J. Control. Release* 73(2–3):401–416.
de Gennes, P. G. 1980. *Macromolecules* 13:1069.
de Gennes, P. G. 1987. *Adv. Colloid Interface Sci.* 27:189.
Ghoreishi, S. M., Fox, G. A., Bloor, D. M., Holzwarth, J. F., Wyn-Jones, E. 1999. *Langmuir* 15:5474.
Ghoroghchian, P. P., Frail, P. R., Susumu, K., Park, T.-H., Wu, S. P., Uyseda, H. T., Hammer, D. A., Therien, M. J. 2005. *J. Am. Chem. Soc.* 127:15388–15390.
Godbey, W. T., Wu, K. K., Mikos, A. G. 1999a. *J. Control. Release* 60:149–160.
Godbey, W. T., Wu, K. K., Mikos, A. G. 1999b. *J. Biomed. Mater. Res.* 45A:268–275.

Goddard, E. D., Ananthapadmanabhan, P. K. (Eds.). 1993. *Interactions of Surfactants with Polymers and Proteins*, Boca Raton, FL, CRC Press.
Gohy, J.-F., Hofmeier, H., Alexeev, A., Schubert, U. S. 2003. *Macromol. Chem. Phys.* 204:1524–1530.
Gohy, J. F, Varshney, S. K, Antoun, S., Jerome, R. 2000. *Macromolecules* 33(25):9298–9305.
Gouveia, L. M., Grassl, B., Muller, A. J. 2009. *J. Colloid Interface Sci.* 333:152–163.
Graff, A., Sauer, M., Van Gelder, P., Meier, W. 2002. *Proc. Natl. Acad. Sci. USA* 99:5064.
Grosse, S., Tremeau-Bravard, A., Aron, Y., Briand, P., Fajac, I. 2002. *Gene Ther.* 9(15): 1000–1007.
Gustafsson, J., Ljusberg-Wahren, H., Almgren, M., Larsson, K. 1996. *Langmuir* 12:4611–4613.
Gustafsson, J., Ljusberg-Wahren, H., Almgren, M., Larsson, K. 1997. *Langmuir* 13:6964–6971.
Haensler, J., Szoka, F. C. 1993. *Bioconjug. Chem.* 4:372–379.
Haladjova, E., Rangelov, S., Tsvetanov, Ch., Pispas, S. 2012. *Soft Matter* 8(10):2884–2889.
Harada, A., Kataoka, K. 1995. *Macromolecules* 28(15):5294–5299.
Harbottle, R. P., Cooper, R. G., Hart, S. L., Ladhoff, A., McKay, T., Knight, A. M., Wagner, E., Miller, A. D., Coutelle, Ch. 1998. *Hum. Gene Ther.* 9(7):1037–1047.
Hashida, M., Takemura, S., Nishikawa, M., Takakura, Y. 1998. *J. Control. Release* 53(1–3):301–310.
Hecht, E., Hoffmann, H. 1994. *Langmuir* 10:86.
Hecht, E., Hoffmann, H. 1995. *Colloids Surf.* 96:181.
Hecht, E., Mortensen, K., Gradzielski, M., Hoffmann, H. 1995. *J. Phys. Chem.* 99:4866.
Hermanson, G. T. 1996. *Bioconjugate Techniques*. San Diego, CA: Academic Press.
Heyde, M., Partridge, K. A., Howdle, S. M., Oreffo, R. O. C., Garnett, M. C., Shakesheff, K. M. 2007. *Biotechnol. Bioeng.* 98:679–693.
Hofs, B., Voets, I. K, de Keizer, A., Cohen Stuart, M. A. 2006. *Phys. Chem. Chem. Phys.* 8(36):4242–4251.
Honda, S., Yamamoto, T., Tezuka, Y. 2010. *J. Am. Chem. Soc.* 132:10251–10253.
Hong, S., Leroueil, P. R., Janus, E. K., Peters, J. L., Kober, M.-M., Islam, M. T., Orr, B. G., Baker, J. R., Banaszak Holl, M. M. 2006. *Bioconjugate Chem.* 17:728–734.
Hosseinkhani, H., Azzam, T., Kobayashi, H., Hiraoka, Y., Shimokawa, H., Domb, A. J., Tabata, Y. 2006a. *Biomaterials* 27:4269–4278.
Hosseinkhani, H., Hosseinkhani, M., Gabrielson, N. P., Pack, D. W., Khademhosseini, A., Kobayashi, H. 2008. *J. Biomed. Mater. Res.* 85A:47–60.
Hosseinkhani, H., Yamamoto, M., Inatsugu, Y., Hiraoka, Y., Inoue, S., Shimokawa, H., Tabata, Y. 2006b. *Biomaterials* 27:1387–1398.
Hristova, K., Kenworthy, A., McIntosh, T. J. 1995. *Macromolecules* 28:7693–7699.
Hristova, K., Needham, D. 1995. *Macromolecules* 28:991–1002.
Ilhan, F., Galow, T. H., Gray, M., Clavier, G., Rotello, V. M. 2000. *J. Am. Chem. Soc.* 122:5895–5896.
Israelachvili, J. 1992. *Intermolecular and Surface Forces*, 2nd edn. San Diego, CA: Academic Press.
Jansson, J., Schillen, K., Nilsson, M., Soderman, O., Fritz, G., Bergmann, A., Glatter, O. 2005. *J. Phys. Chem. B* 109:7073.
Jansson, J., Schillen, K., Olofsson, G., Silva, R. C. D., Loh, W. 2004. *J. Phys. Chem. B* 108:82.
Johnsson, M., Barauskas, J., Tiberg, F. 2005. *J. Am. Chem. Soc.* 127:1076–1077.
Johnsson, M., Hansson, P., Edwards, K. 2001. *J. Phys. Chem. B* 105:8420–8430.
Johnsson, M., Silvander, M., Karlsson, G., Edwards, K. 1999. *Langmuir* 15:6314–6325.
Kabanov, A. V., Bronich, T. K., Kabanov, V. A., Yu K., Eisenberg, A. 1996. *Macromolecules* 29(21):6797–6802.
Kabanov, A. V., Bronich, T. K., Kabanov, V. A., Yu, K., Eisenberg, A. 1998. *J. Am. Chem. Soc.* 120:9941–9942.

Kaschiev, D., Exerowa, D. 1983. *Biochim. Biophys. Acta* 732:133–145.

Kataoka, K., Harada, A., Nagasaki, Y. 2001. *Adv. Drug Delivery Rev.* 47(1):113–131.

Kelarakis, A., Castelletto, V., Krysmann, M. J., Havredaki, V., Viras, K., Hamley, I. W. 2008. *Langmuir* 24:3767–3772.

Kenworthy, A. K., Simon, S. A., McIntosh, T. J. 1995. *Biophys. J.* 68:1903–1920.

Kepczynski, M., Jamroz, D., Wytrwal, M., Bednar, J., Rzad, E., Nowakowska, M. 2012. *Langmuir* 28:676–688.

Kiang, T., Bright, C., Cheung, C. Y., Stayton, P. S., Hoffman, A. S., Leong, K. W. 2004. *J. Biomater. Sci. Polym. Ed.* 15(11):1405–1421.

Kim, A. J., Kaucher, M. S., Davis, K. P., Peterca, M., Imam, M. R., Christian, N. A., Levine, D. H., Bates, F. S., Percec, V., Hammer, D. A. 2009b. *Adv. Funct. Mater.*19:2930–2936.

Kim, K. T., Cornelissen, J. J. L. M., Nolte, R. J. M., Van Hest, J. C. M. 2009a. *Adv. Materials* 21:2787–2791.

Kim, Y., Tewari, M., Pajerowski, J. D., Cai, Sh., Sen, Sh., Williams, J., Sirsi, Sh., Lutz, G., Discher, D. E. 2009c. *J. Control. Release* 134:132–140.

Kimura, T., Yamaoka, T., Iwase, R., Murakami, A. 2002. *Macromol. Biosci.* 2(9):437–446.

Kishimura, A., Koide, A., Osada, K., Yamasaki, Y., Kataoka, K. 2007. *Angew. Chem. Int. Ed.* 46:6085.

Klink, D., Chao, S., Glick, M. C., Scanlin, T. F. 2001. *Mol. Ther.* 3(6):831–841.

Klink, D., Yu, Q. C., Glick, M. C., Scanlin, T. 2003. *Mol. Ther.* 7(1):73–80.

Kono, K., Igawa, T., Takagishi, T. 1997. *Biochim. Biophys. Acta* 1325:143–154.

Korobko, A. V., Backendorf, C., van der Maarel, J. R. C. 2006. *J. Phys. Chem. B* 110:14550.

Korobko, A. V., Jesse, W., van der Maarel, J. R. C. 2005. *Langmuir* 21:34–42.

Kostarelos, K., Kipps, M., Tadros, Th. F., Luckham, P. F. 1998a. *Colloids Surf. A* 136:1.

Kostarelos, K., Luckham, P. F., Tadros, Th. F. 1995. *J. Liposome Res.* 5:117.

Kostarelos, K., Luckham, P. F., Tadros, Th. F. 1998b. *J. Chem. Soc. Faraday Trans.* 94:2159.

Kostarelos, K., Tadros, Th. F., Luckham, P. F. 1999. *Langmuir* 15:369.

Kriz, J., Dybal, J., Dautzenberg, H. 2001. *J. Phys. Chem. A* 105(31):7486–7493.

Kulthe, S. S., Inamdar, N. N., Choudhari, Y. M., Shirolikar, S. M., Borde, L. C., Mourya, V. K. 2011. *Colloids Surf. B: Biointerfaces* 88(2):691–696.

Kumbhakar, M. 2007. *J. Phys. Chem. B* 111:14250–14255.

Kurisawa, M., Yokoyama, M., Okano, T. 2000. *J. Control. Release* 69:127–137.

Langevin, D. 2009. *Adv. Colloid Interface Sci.* 147–148:170–177.

Larsson, K. 1989. *J. Phys. Chem.* 93:7304.

Larsson, K., 2000. *Curr. Opin. Colloid Interface Sci.* 5:64–69.

Lawrence, J. M. 1994. *Chem. Soc. Rev.* 23:417.

Lechardeur, D., Sohn, K. J., Haardt, M., Joshi, P. B., Monck, M., Graham, R. W., Beatty, B., Squire, J., O'Brodovich, H., Lukacs, G. L. 1999. *Gene Ther.* 6(4):482–497.

Lee, D., Zhang. W., Shirley, S., Kong, X., Hellermann, G., Lockey, R., Mohapatra, S. S. 2007. *Pharm. Res.* 24(1):157–167.

Lee, E. S., Oh, Y. T., Youn, Y. S., Nam, M., Park, B., Yun, J., Kim, J. S., Song, H.-T., Oh, K. T. 2011a. *Colloids Surf. B: Biointerfaces* 82(1):190–195.

Lee, J. C., Bermudez, H., Discher, B. M., Sheehan, M. A., Won, Y. Y., Bates, F. S., Discher, D. E. 2001. *Biotechnol. Bioeng.* 73:135.

Lee, J. S., Zhou, W., Meng, F., Zhang, D., Otto, C., Feijen, J. 2010. *J. Control. Release* 146:400–408.

Lee, S. H., Mok, H., Lee, Y., Park, T. G. 2011b. *J. Control. Release* 152(1):152–158.

Li, B., Xu, H., Li, Z., Yao, M. F., Xie, M., Shen, H., Shen, S., Wang, X. S., Jin, Y. 2012. *Int. J. Nanomed.* 7:187–197.

Li, Ch., Tian, H., Rong, N., Liu, K., Liu, F., Zhu, Y., Qiao, R., Jiang, Y. 2011. *Biomacromolecules* 12(2):298–305.

Li, F., de Wolf, F. A., Marcelis, A. T. M., Sudhölter, E. J. R., Cohen Stuart, M. A., Leermakers, F. A. M. 2010. *Angew. Chem. Int. Ed.* 49:9947–9950.

Li, Y., Kwak, J. C. T. 2004. *Langmuir* 20:4859–4866.

Li, Y., Xu, R., Couderc, S., Bloor, D. M., Holzwarth, J. F., Wyn-Jones, E. 2000. *Langmuir* 16:10515.

Li, Y., Xu, R., Couderc, S., Bloor, D. M., Wyn-Jones, E., Holzwarth, J. F. 2001. *Langmuir* 17:183.

Li, Zh., Hillmyer, M. A., Lodge, T. P. 2006. *Macromolecules* 39(2):765–771.

Lindblom, G., Rilfors, L. 1989. *Biochim. Biophys. Acta* 988:221.

Liu, J., Yoneda, A., Liu, D., Yokoyama, Y., Yusa, Sh., Nakashima, K. 2010. *Can. J. Chem.* 88:208–216.

Logan, J. L., Masse, P., Gnanou, Y., Taton, D., Duran, R. S. 2005. *Langmuir* 21:7380–739.

Lomas, H., Canton, I., MacNeil, S., Du, J., Armes, S. P., Ryan, A. J., Lewis, A. L., Battaglia, G. 2007. *Adv. Mater.* 19:4238–4243.

Lomas, H., Massignani, M., Abdullah, K. A., Canton, I., Lo Presti, C., MacNeil, S., Du, J., Blanazs, A., Madsen, J., Armes, S. P., Lewis, A. L., Battaglia, G. 2008. *Faraday Discuss.* 139:143–159.

Lu, Q., Bazuin, C. G. 2005. *Nano Lett.* 5(7):1309–1314.

Luger, K., Mader, A. W., Richmond, R. K., Sargent D. F., Richmond, T. J. 1997. *Nature* 389:251.

Lutz, J. F., Laschewsky, A. 2005. *Macromol. Chem. Phys.* 206:813–817.

Mao, H. Q., Roy, K., Truong-Le, V. L., Janes, K. A., Lin, K. Y., Wang, Y., August, J. T., Leong, K. W. 2001. *J. Control. Release* 70(3):399–421.

Mao, Zh., Ma, L., Yan, J., Yan, M., Gao, Ch., Shen, J. 2007. *Biomaterials* 28(30):4488–4500.

Marsh, D., Bartucci, R., Sportelli, L. 2003. *Biochim. Biophys. Acta* 1615:35–59.

Massignani, M., Lomas, H., Battaglia, G. 2010. *Adv. Polym. Sci.* 229:115–154.

Massignani, M., LoPresti, C., Blanazs, A., Madsen, J., Armes, S. P., Lewis, A. L., Battaglia, G. 2009. *Small* 5(21):2424–2432.

Mecke, A., Dittrich, C., Meier, W. 2006. *Soft Matter* 2:751.

Meier, W., Nardin, C., Winterhalter, M. 2000. *Angew. Chem. Int. Ed.* 39:4599.

Meilander, N. J., Pasumarthy, M. K., Kowalczyk, T. H., Cooper, M. J., Bellamkonda, R. V. 2003. *J. Control. Release* 88:321–331.

Menger, F. M., Davydov, D. A., Yaroslavova, E. G., Rakhnyanskaya, A. A., Efimova, A. A., Ermakov, Y. A., Yaroslavov, A. A.2009. *Langmuir* 25:13528–13533.

Menger, F. M., Yaroslavov, A. A., Melik-Nubarov, N. S. 2006. *Acc. Chem. Res.* 10:702–710.

Menger, F. M., Yaroslavov, A. A., Sitnikova, T. A., Rakhnyanskaya, A. A., Yaroslavova, E. G., Davydov, D. A., Burova, T. V., Grinberg, V. Y., Shi, L. 2009. *J. Am. Chem. Soc.* 131: 1666–1667.

Menger, F. M., Yaroslavov, A. A., Sybachin, A. V., Kesselman, E., Schmidt, J., Talmon, Y. Rizvi, S. A. A. 2011. *J. Am. Chem. Soc.* 133:2881–283.

Mislick, K. A., Baldeschwieler, J. D., Kayyem, J. F., Meade, T. J. 1995. *Bioconjug. Chem.* 6(5):512–515.

Momekova, D., Momekov, G., Rangelov, S., Storm, G., Lambov, N. 2010. *Soft Matter* 6:591–601.

Momekova, D., Rangelov, S., Lambov, N., Karlsson, G., Almgren M. 2008. *J. Disper. Sci. Technol.* 29(8):1106–1113.

Momekova, D., Rangelov, S., Yanev, S., Nikolova, E., Konstantinov, S., Romberg, B., Storm, G., Lambov, N. 2007. *Eur. J. Pharm. Sci.* 32:308–317.

Munk, P., Rangelov, S., Tuzar, Z. 1998. *Int. J. Polym. Analysis. Charact.* 4:435–446.

Nagane, K., Kitada, M., Wakao, S., Dezawa, M., Tabata, Y. 2009. *Tissue Eng. Part A* 15:1655–1665.

Nagasaki, T. Atarashi, K., Makino, K., Noguchi, A., Tamagaki, S. 2000. *Mol. Cryst. Liq. Cryst.* 345:227–232.
Nakamura, K., Shikata, T. 2003. *Macromolecules* 36:9698–9700.
Nakamura, K., Shikata, T. 2004. *Macromolecules* 37:8381–8388.
Nakamura, K., Shikata, T. 2007. *J. Phys. Chem. B* 111:12411–12417.
Nakamura, K., Shikata, T., Takahashi, N., Kanaya, T. 2005. *J. Am. Chem. Soc.* 127: 4570–4571.
Nakamura, K., Yamanaka, K., Shikata, T. 2003. *Langmuir* 19:8654–8660.
Nakano, M., Teshigawara, T., Sugita, A., Leesajakul, W., Taniguchi, A., Kamo, T., Matsuoka, H., Handa, T. 2002. *Langmuir* 18:9283–9288.
Nallani, M., Benito, S., Onaca, O., Graff, A., Lindemann, M., Winterhalter, M., Meier, W., Schwaneberg, U. 2006. *J. Biotechnol.* 123:50.
Nam, J., Beales, P. A., Vanderlick, T. K. 2011. *Langmuir* 27:1–6.
Napoli, A., Boerakker, M. J., Tirelli, N., Nolte, R. J. M., Sommerdijk, N. A. J. M., Hubbell, J. A. 2004. *Langmuir* 20:3487–3491.
Nardin, C., Thoeni, S., Widmer. J., Winterhalter. M., Meier, W. 2000. *Chem. Commun.* 15:1433–1435.
Nardin, C., Widmer, J., Winterhalter, M., Meier, W. 2001. *Eur. Phys. J. E* 4:403.
Neu, M., Fischer, D., Kissel, T. 2005. *J. Gene Med.* 7(8):992–1009.
Ngo, A. T., Cosa, G. 2010. *Langmuir* 26:6746–6754.
Nie, H., Ho, M.-L., Wang, C.-K., Wang. C.-H., Fu, Y.-C. 2009. *Biomaterials* 30:892–901.
Nie, H., Wang, C.-H. 2007. *J. Control. Release* 120:111–121.
Nikolova, A. N., Jones, M. N. 1996. *Biochim. Biophys. Acta* 1304:120–128.
Nishiyama, N., Iriyama, A., Jang, W.-D., Miyata, K., Itaka, K., Inoue, Y., Takahashi, H. et al. 2005. *Nat. Mater.* 4:934–941.
Oikonomou, E., Bokias, G., Kallitsis, J. K., Iliopoulos, I. 2011. *Langmuir* 27:5054–5061.
Okazaki, A., Jo, J., Tabata, Y. 2007. *Tissue Eng.* 13:245–251.
Onaca, O., Nallani, M., Ihle, S., Schenk, A., Schwaneberg, U. 2006. *Biotechnol. J.* 1:795.
Onaca, O., Sarkar, P., Roccatano, D., Friedrich, T., Hauer, B., Grzelakowski, M., Guven, A., Fioroni, M., Schwaneberg, U. 2008. *Angew. Chem. Int. Ed.* 47:7029–7031.
Orwick, M. C., Judge, P. J., Procek, J., Lindholm, L., Graziadei, A., Engel, A., Grobner, G., Watts, A. 2012. *Angew. Chem. Int. Ed.* 51:4653–4657.
Pack, D. W., Hoffman, A. S., Pun, S., Stayton, P. S. 2005. *Nat. Rev. Drug Discov.* 4:581–593.
Panmai, S., Prud'homme, R. K., Peiffer, D. G., Jockusch, S., Turro, N. J. 2002. *Langmuir* 18:3860–3864.
Papahadjopoulos, D., Allen, T. M., Gabizon, A., Mayhew, E., Matthay, K., Huang, S. K., Lee, K. D., Woodle, M. C., Lasic, D. D., Redemann, C. 1991. *Proc. Natl. Acad. Sci. USA* 88:11460–11464.
Patton, J. S., Carey, M. C. 1979. *Science* 204:145.
Patton, J. S., Vetter, R. D., Hamosh, B., Borgstrom, B., Lindstrom, M., Carey, M. C. 1985. *Food Microstruct.* 4:29.
Peleshanko, S., Jeong, J., Gunawidjaja, R., Tsukruk, V. V. 2004. *Macromolecules* 37:6511–6522.
Pergushov, D. V., Remizova, E. V, Feldthusen, J., Zezin, A. B, Müller, A. H. E, Kabanov, V. A. 2003. *J. Phys. Chem. B* 107(32):8093–8096.
Piculell, L., Egermayer, M., Sjöström, J. 2003. *Langmuir* 19:3643–3649.
Pispas, S. 2011a. *Soft Matter* 7:474–482.
Pispas, S. 2011b. *Soft Matter* 7:8697–8701.
Pispas, S., Hadjichristidis, N. 2003. *Langmuir* 19:48–54.
Pispas, S., Sarantopoulou, E. 2007. *Langmuir* 23:7484–7490.
Prabaharan, M., Mano, J. F. 2004. *Drug Deliv.* 12(1):41–57.
Rangelov, S. 2006. *J. Phys. Chem. B* 110(9):4256–4262.
Rangelov, S., Almgren, M. 2005. *J. Phys. Chem. B.* 109:3921–3929.
Rangelov, S., Dimitrov, Ph., Tsvetanov, Ch. 2005. *J. Phys. Chem. B* 109:1162–1167.

Rangelov, S., Edwards, K., Almgren, M., Karlsson, G. 2003. *Langmuir* 19:172–181.
Rangelov, S., Petrova, E., Berlinova, I., Tsvetanov, Ch. 2001. *Polymer* 42:4483–4491.
Ranjan, A., Zeglam, K., Mukerjee, A., Thamake, S., Vishwanatha, J. K. 2011. *Nanotechnology* 22(29), Art. No 295104.
Ranquin, A., Versées, W., Meier, W., Steyaert, J., Van Gelder, P. 2005. *Nano Lett.* 5:2220.
Rinaudo, M., Quemeneur, F., Pépin-Donat, B. 2012. *Int. J. Polym. Analysis Charact.* 17:1–10.
Rizwan, S. B., Dong, Y. D., Boyd, B. J., Rades, T., Hook, S. 2007. *Micron* 38:478–485.
Robson Marsden, H., Elbers, N. A., Bomans, P. H. H., Sommerdijk, N. A. J. M., Kros, A. 2009. *Angew. Chem. Int. Ed.* 48:2330–2333.
Robson Marsden, H., Handgraaf, J.-W., Nudelman, F., Sommerdijk, N. A. J. M., Kros, A. 2010. *J. Am. Chem. Soc.*132:2370–2377.
Robson Marsden, H., Korobko, A. V., van Leeuwen, E. N. M., Pouget, E. M., Veen, S. J., Sommerdijk, N. A. J. M., Kros, A. 2008. *J. Am. Chem. Soc.*130:9386–9393.
Rodriguez, C., Sanchez-Dominguez, M., Sarac, B., Rogac, M., Shrestha, R. G., Shrestha, L. K., Varade, D., Ghosh, G., Aswal, V. K. 2010. *Colloid Polym. Sci.* 288:739–751.
Rodriguez, C., Shrestha, L. K., Lopez-Quintela, M. A. 2007. *J. Colloid Interface Sci.* 312:108–113.
Rodriguez, M., Xue, J., Gouveia, L. M., Muller, A. J., Eduardo Saez, A., Rigolini, J., Grassl, B. 2011. *Colloids Surf. A: Physicochem. Eng. Aspects* 373:66–73.
Rojas, M., Muller, A. J., Saez, A. E. 2010. *J. Colloid Interface Sci.* 342:103–109.
Ruysschaert, T., Sonnen, A. F. P., Haefele, T., Meier, W., Winterhalter, M., Fournier, D. 2005. *J. Am. Chem. Soc.* 127:6242–6247.
Sakaguchi, N., Kojima, C., Harada, A., Kono, K. 2008. *Bioconjug. Chem.* 19:1040–1048.
Sanchez-Moreno, P., Ortega-Vinuesa, J. L., Martin-Rodriguez, A., Boulaiz, H., Marchal-Corrales, J. A., Peula-Garcia, J. M. 2012. *Int. J. Molecular Sci.* 13:2405–2424.
Santore, M. M., Discher, D. E., Won, Y. Y., Bates, F. S., Hammer, D. A. 2002. *Langmuir* 18:7299–7308.
Sauer, M., Haefele, T., Graff, A., Nardin, C., Meier, W. 2001. *Chem. Commun.* 23:2452–2454.
Schaffer, D. V., Lauffenburger, D. A. 1998. *J. Biol. Chem.* 273(43):28004–28009.
Schulz, M., Olubummo, A., Binder, M., W. H. 2012. *Soft Matter* 8:4849–4864.
Scriven, L. E. 1976. *Nature* 263:123.
Shin, J., Shum, P., Thompson, D. H. 2003. *J. Control. Release* 91:187–200.
Sikor, M., Sabin, J., Keyvanloo, A., Schneider, M. F., Thewalt, J. L., Bailey, A. E., Frisken, B. J. 2010. *Langmuir* 26:4095–4102.
Smart, T. P., Fernyhough, C., Ryan, A. J., Battaglia, G. 2008. *Macromol. Rapid Commun.* 29:1855–1860.
Spicer, P. T., Hayden, K. L., Lynch, M. L., Ofori-Boateng, A., Burns, J. L. 2001. *Langmuir* 17:5748.
Spicer, P. T., Small, W. B., Lynch, M. L., Burns, J. L. 2002. *J. Nanoparticle Res.* 4:297.
Srinivas, G., Discher, D. E., Klein, M. L. 2005. *Nano Lett.* 5:2343.
Stubenrauch, K., Fritz-Popovski, G., Ingolic, E., Grogger, W., Glatter, O., Stelzer, F., Trimmel, G. 2007. *Macromolecules* 40:4592.
Sun, Sh., Liu, W., Cheng, N., Zhang, B., Cao, Zh., Yao, K., Liang, D., Zuo, A., Guo, G., Zhang, J. 2005. *Bioconj. Chem.* 16:972–980.
Suh, W., Chung, J.-K., Park, S.-H., Kim, S. W. 2001. *J. Control. Release.* 72:171–178.
Taboada, P., Castro, E., Mosquera, V. 2005. *J. Phys. Chem. B* 109:23760.
Talelli, M., Pispas, S. 2008. *Macromol. Biosci.* 8:960.
Tan, Y., Srinivasan, Sh., Choi., K.-Sh. 2005. *J. Am. Chem. Soc.*127(10):3596–3604.
Tang, M. X., Redemann, C. T., Szoka, F. C. 1996. *Bioconjug. Chem.* 7:703–714.
Thibault, R. J., Uzun, O., Hong, R., Rotello, V. M. 2006. *Adv. Mater.*18:2179–2183.
Torchilin, V. 2001. *J. Control. Release* 73(2–3):137–172.
Trentin, D., Hubbell, J., Hall, H. 2005. *J. Control. Release* 102:263–275.

Tribet, C., Sebai, S. C., Cribier, S., Karimi, A., Massotte, D. 2010. *Langmuir* 26:14135–14141.

Tribet, C., Vial, F. 2008. *Soft Matter* 4:68–81.

Tribet, C., Vial, F., Oukhaled, A. G., Auvray, L. 2007. *Soft Matter* 3:75–78.

Uzun, O., Sanyal, A., Jeong, Y., Rotello, V. M. 2010. *Macromol. Biosci.*10:481–487.

Vermonden, T., van der Gucht, J., de Waard, P., Marcelis, A. T. M., Besseling, N. A. M., Sudholter, E. J. R., Fleer, G. J., Cohen Stuart, M. A. 2003. *Macromolecules* 36:7035.

Voets, I. K., de Keizer, A., Cohen Stuart, M. A. 2009. *Adv. Colloid Interface Sci.* 147–148: 300–318.

Voets, I. K., de Keizer, A., Cohen Stuart, M. A., Justynska, J., Schlaad, H. 2007. *Macromolecules* 40(6):2158–2164.

Voets, I. K., de Keizer, A., De Waard, P., Frederik, P. M., Bomans, P. H. H., Schmalz, H., Walther, A., King, S. M., Leermakers, F. A. M., Cohen Stuart, M. A. 2006. *Angew. Chem. Int. Ed.* 45(40):6673–6676.

de Vries, R., Cohen Stuart, M. A. 2006. *Curr. Opin. Colloid Interface Sci.* 11(5):295–301.

Vriezema, D. M., Garcia, P. M. L., Oltra, N. S., Natzakis, N. S., Kuiper, S. M., Nolte, R. J. M., Rowan, A. E., van Hest, J. C. M. 2007. *Angew. Chem. Int. Ed.* 46:7378

Wang, M., Olszewska, A., Walther, A., Malho, J.-M., Schacher, F. H., Ruokolainen, J., Ankerfors, M. et al. 2011. *Biomacromolecules* 12(6):2074–2081.

Wang, Y., Han, P., Xu, H., Wang, Z., Zhang, X., Kabanov, A. V. 2010. *Langmuir* 26:709–715.

Wang, Y. J., Kho, K., Cheow, W. S., Hadinoto, K. 2012. *Int. J. Pharm.* 424:98–106.

Waschinski, C. J., Barnert, S., Theobald, A., Schubert, R., Kleinschmidt, F., Hoffmann, A., Saalwaechter, K., Tiller, J. C. 2008. *Biomacromolecules* 9:1764–1771.

Weaver, J. V. M., Armes, S. P., Liu, S. 2003. *Macromolecules* 36:9994.

van de Wetering, P., Cherng, J. Y., Talsma, H., Crommelin, D. J., Hennink, W. E. 1998. *J. Control. Release* 53(1–3):145–53.

van de Wetering, P., Cherng, J., Talsma, H., Hennink, W. 1997. *J. Control. Release* 49(1):59–69.

Whitten, D. G., Ding, L., Chi, E. Y., Chemburu, S., Ji, E., Lopez, G. P., Schanze, K. S. 2009. *Langmuir* 25:13742–13751.

Wightman, L., Kircheis, R., Rossler, V., Carotta, S., Ruzicka, R., Kursa, M., Wagner, E. 2001. *J. Gene Med.* 3(4):362–372.

Wittemann, A., Azzam, T., Eisenberg, A. 2007. *Langmuir* 23:2224.

Wolf, F. K., Hofmann, A. M., Frey, H. 2010. *Macromolecules* 43:3314–3324.

Wong, K., Sun, G., Zhang, X., Dai, H., Liu, Y., He, C., Leong, K. W. 2006. *Bioconjug. Chem.* 7(1):152–158.

Wong, Sh. Y., Pelet, J. M., Putnam, D. 2007. *Prog. Polym. Sci.* 32:799–837.

Woodle, C. 1995. *Adv. Drug Delivery Rev.*16:249.

Woodle, M. C., Lasic, D. D. 1992. *Biochim. Biophys. Acta* 1113:171.

Woodle, M. C., Newman, M. S., Martin, F. J. 1992. *Int. J. Pharm.* 88:327.

Wu, G., Khant, H. A., Chiu, W., Lee, K. Y. C. 2009. *Soft Matter* 5:1496–1503.

Wu, G., Lee, K. Y. C. 2009. *J. Phys. Chem. B* 113:15522–15531.

Yaghmur, A., de Campo, L., Sagalowicz, L., Leser, M. E., Glatter, O. 2005a. *Langmuir* 21:569–577.

Yaghmur, A., de Campo, L., Sagalowicz, L., Leser, M. E., Glatter, O. 2006b. *Langmuir* 22:9919–9927.

Yaghmur, A., de Campo, L., Salentinig, S., Sagalowicz, L., Leser, M. E., Glatter. O. 2006. *Langmuir* 22:517–521.

Yaghmur, A., Glatter, O. 2008. *Adv. Colloid Interface Sci.* 147–148:333–342.

Yan, Y., Besseling, N. A. M., de Keizer, A., Fokkink, R., Drechsler, M., Cohen Stuart, M. A. 2007a. *J. Phys. Chem. B* 111:11662.

Yan, Y., Besseling, N. A. M., de Keizer, A., Marcelis, A. T. M., Drechsler, M., Cohen Stuart, M. A. 2007b. *Angew. Chem. Int. Ed.* 46:1807.

Yan, Y., Harnau, L., Besseling, N. A. M., Keizer, A. D., Ballauff, M., Rosenfeldt, S., Cohen Stuart, M. A. 2008. *Soft Matter* 4:2207–2212.
Yang, Xh., Yang, Y.-W., Zhou, Z.-K., Attwood, D., Booth, C. 1996. *J. Chem. Soc. Faraday Trans.* 92(2):257–265.
Yano, Y., Yano, A., Oishi, S., Sugimoto, Y., Tsujimoto, G., Fujii, N., Matsuzaki, K. 2008. *ACS Chem. Biol.* 3:341–345.
Yaroslavov, A. A., Efimova, A. A., Kostenko, S. N. 2012. *Polym. Sci. Series A* 54(4):264–269.
Yasuda, S., Townsend, D., Michele, D. E., Favre, E. G., Day, S. M., Metzger, J. M. 2005. *Nature* 436:1025–1029.
Ye, J., van den Berg, B. 2004. *EMBO J.* 23:3187.
Yeh, Y.-B., Chen, B.-Ch., Lin, H.-P., Tang. Ch.-Y. 2006. *Langmuir* 22(1):6–9.
Yokoyama, M. 2002. *Drug Disc. Today* 7(7):426.
Yuan, J., Luo, Y., Gao, Q. 2011. *J. Microencapsul.* 28(2):93–98.
Yuba, E., Kojima, C., Harada, A., Tana, Watarai, Sh., Kono, K. 2010. *Biomaterials* 31:943–951.
Zhang, J., Feng, K., Cuddihy, M., Kotov, N. A., Ma, P. X. 2010b. *Soft Matter* 6(3):610–617.
Zhang, W., Shi, Y., Chen, Y., Ye, J., Sha, X., Fang, X. 2011. *Biomaterials* 32(11):2894–2906.
Zhang, Y., Wu, F., Yuan, W., Jin, T. 2010a. *J. Control. Release* 147(3):413–419.
Zhao, J., Zhang, G., Pispas, S. 2009. *J. Phys. Chem. B* 113:10600–10606.
Zheng, Y., Davis, H. T. 2000. *Langmuir* 16:6453–6459.
Zhu, J., Yu, H., Jiang, W. 2005. *Macromolecules* 38:7492–7501.
Zinselmeyer, B. H., Mackay, S. P., Schatzlein, A. G., Uchegbu, I. F. 2002. *Pharm. Res.* 19(7):960–967.

4 Hybrid Polymeric Nanoparticles Containing Inorganic Nanostructures

4.1 INTRODUCTION

The synthesis and property evaluation of inorganic nanoparticles (NPs) is a mature and ongoing research direction both in the material science and nanoscience fields. Combining properties of NPs with those of polymers, aiming at the production of hybrid multifunctional NPs and nanostructures, presents several advantages and opportunities. Hybrid NPs and nanostructures dispersed in liquids will be the focus of this chapter. The field is enormous and is developing rapidly. It is difficult to really mention every effort. Some general relevant reviews on different aspects of this fascinating research direction can be found in the reference section. In an effort to discuss basic concepts and recent developments in the field, we present and discuss some selected examples.

4.2 SOME GENERAL FEATURES OF INORGANIC NANOPARTICLES

Metal-containing inorganic NPs continue to attract significant scientific attention due to their unique properties that differ from those of the bulk material (Zhu et al. 2004, Merican et al. 2007). These specific properties are due to their nanoscale dimensions that give rise to quantum size effects. Noble metal NPs, like those of Au, Ag, and Pt, have been extensively studied due to their potential utilization in many applications such as catalysis (Manea et al. 2004), sensors (Shipway et al. 2000), electronics (Thomas and Kamat 2003), photonics (Tan et al. 2006), and medicine/bioimaging (Huang et al. 2007, Neoh 2011). Their photophysical and catalytic properties depend and are strongly related to NP size, shape, internal morphology, and interparticle distances and interactions. Noble metal NPs have brilliant colors due to their surface plasmon resonance (SPR). Gold NPs (AuNPs) usually have a red color, while silver NPs (AgNPs) are yellow. Their color depends on the size and shape of the metal NPs and the dielectric constant of the medium surrounding them. Changes in the size or the shape of metal NPs lead in color changes as well, due to the oscillation of the electrons in the conductive band (Eustis and El-Sayed 2006). This oscillation is usually in the visible region for AgNPs and AuNPs and is observed as the SPR absorption, while the SPR for Pt NPs is not in the visible region. The shape of the SPR band is strongly related to the size and shape of the metal NPs. Small spherical AuNPs and AgNPs display an SPR at around ~520 and ~430 nm, respectively

(Douglas et al. 2008). Anisotropic metallic NPs have two or more SPR bands, the so-called the transverse and the longitudinal band (Malikova et al. 2002).

Usually the synthesis of noble metal NPs can be performed by simple reduction of metal precursors in the presence or absence of additives for further stabilization of the produces metal NPs. The main goal is to achieve control over the size, shape, and size distribution of the resulting NPs via controlling the parameters that influence the nucleation and growth during NP formation. The use of additives as ligands and stabilizing agents (they can be surfactants or polymers) contributes to the control of the growth process and simultaneously to the colloidal stability (inhibition of agglomeration) of the obtained NPs.

Semiconductor NPs are also interesting for their photophysical/optical properties. The adsorption and photoluminescence (PL) spectra of these nanoclusters highly depend on the nature of the inorganic cluster, as well as on the size (Wang and Herron 1991, El-Sayed 2001). A number of colloidal chemistry routes have been proposed for the synthesis of well-defined semiconductor NPs. They depend on the nature of semiconductor to be synthesized (e.g., metal sulfides vs. metal oxides). In this case, passivation of the cluster surface may be needed in order to alleviate defects on the nanocrystal surface and to enhance the photophysical properties of the particle. Additionally, capping of the semiconductor NPs with appropriate ligands increases their chemical and colloidal stability and also imparts stealth properties to the NPs, allowing their utilization in medicinal and bioimaging applications.

Inorganic NPs possessing magnetic properties have also attracted considerable interest in the last 10 years due to a number of possible applications. Especially, iron oxide NPs (IONPs) are investigated intensively for applications in the field of nanomedicine due to their unique physical properties (De et al. 2008, Jun et al. 2008). Their high surface area to volume ratios and superparamagnetism are useful characteristics for utilization in magnetic resonance imaging (MRI), drug and gene delivery, tissue engineering, and bioseparation (Laurent et al. 2008). IONP can be classified into two separate categories according to their size: (1) superparamagnetic iron oxide (SPIO) NPs with a mean particle diameter of 50–100 nm and (2) ultrasmall superparamagnetic iron oxide (USPIO) NPs with a size sometimes well below 50 nm. Both are composed of ferrite nanocrystallites of magnetite (Fe_3O_4) or maghemite (γ-Fe_2O_3) (Laurent et al. 2008, Gao et al. 2009); they are biocompatible and can be easily synthesized. The biodistribution of these IONPs can be altered by the application of an external magnetic field, achieving a short of active, nonchemical targeting, thus avoiding direct contact of any medical device with the inner parts of the body. IONPs may also have potential applications in hyperthermia therapy since they can heat up under the influence of a localized high-frequency magnetic field, achieving in this way destruction of surrounding infected tissue.

Several chemical methods have been proposed for the synthesis of IONPs. The coprecipitation process is the simplest and most widely employed route. In this approach, IONPs are prepared by aging a stoichiometric mixture of ferrous and ferric salts in aqueous media, in the absence of oxygen, under basic conditions yielding magnetite NPs. Iron in Fe_3O_4 is an unstable oxidation state; therefore, the material can quickly transform to maghemite in air or under acidic conditions in the absence of oxygen. The main advantage of this method is that it can produce relatively large amounts of material, with good control over particle size (in the range 2–20 nm) and shape. Control can be achieved by adjusting the pH, ionic strength, and concentration of the reaction solution. Another advantage is the possibility for in situ NP surface functionalization if the coprecipitation process is performed in the presence of additives, such as low molecular weight organic compounds as ligands or polymers (Boyer et al. 2010b). The magnetic properties of the produced IONPs may also be tailored via the coprecipitation conditions (e.g., the saturation magnetization varies with the concentration of NaCl used in the synthesis in the range 63–71 emu/g for magnetite).

A second method is based on the high-temperature decomposition of organometallic precursors (e.g., $Fe(CO)_5$) in organic solvents. This method also offers improved control over the size and shape of IONPs (Peng et al. 2006). The IONP size can be tuned in the range 3–19 nm by altering the precursor utilized and decomposition temperature. The yield of spherical NPs is also enhanced using the decomposition process, providing some control over the shape of IONPs. The IONPs produced by the decomposition route are coated with hydrophobic compounds to facilitate stabilization in organic solvents; however, this reduces their solubility in aqueous biological fluids. The magnetic properties of IONPs can be tuned by the incorporation of other metals, such as cobalt, nickel, and manganese, into the starting material (Sun et al. 2004).

Concerning the magnetic properties of the resulting IONPs, this can also be influenced by reaction temperature. At low temperature, IONPs self-organize in solution and all magnetic spins align, resulting in ferromagnetism. When the temperature is sufficiently high, above a certain value, thermal energy can disrupt spin alignment. Above this temperature (the blocking temperature), the NPs lose their magnetic properties. The application of an external magnetic field is then required, which induces the spontaneous reorganization of magnetic spin directions and re-magnetizes the IONPs. If the magnetic field is removed, the particles lose their spin alignment, liberating heat. Therefore, ferromagnetism is essential for the synthesis of stable colloidal IONP suspensions, as the nonalignment of spin limits the interparticle attraction that would otherwise occur. Very small IOPNs exhibit superparamagnetism. The saturation magnetization, M_s, defined by the alignment of all magnetic spins in a sample decreases with IONP size (Jun et al. 2005). The magnetic spins of atoms close to the surface are less well organized than for atoms in the bulk of the particle near the core. This phenomenon is referred to as "spin canting." In addition, high crystallinity results in higher saturation magnetization. To improve M_s, the addition of dopants, such as manganese, has been reported (Sun et al. 2004). Magnetite (M_s=92 emu/g) has a higher M_s value than maghemite (78 emu/g) for a similar particle size (Gupta and Gupta 2005).

Silica microparticle and NPs are widely used in a number of applications, for example, as fillers in plastics and paints, as stationary phases for chromatographic separations, as catalyst supports, and in electronics and sensors. This is because silica NPs are easy to prepare, even at large scale, with precise control over the size, size distribution, surface functionalities, and internal morphology. Additionally, silica is generally characterized as a safe material for biomedical applications (Burns et al. 2006, Radhakrishman et al. 2006, Piao et al. 2008). The synthesis of spherical monodispersed silica NPs via the hydrolysis and condensation of silicon alkoxides in mixtures of alcohol and water with ammonia acting as the catalyst is known for many years. Simple variation in the composition of the mixture can result in NPs having sizes in the range of 20 nm–2 μm. Smaller, nevertheless uniform, silica NPs can be produced by the water-in-oil reverse microemulsion method. Versatile chemical functionalization methodologies have been developed and allow for alteration of the surface functionalities on silica NPs. The production of mesoporous silica NPs, with the ability to encapsulate various compounds and nanomaterials, has opened new directions for the utilization of silica NPs in medicinal and bioimaging applications.

By closing this section, it should be noted that besides spherical inorganic NPs composed of only one type of material, several unique NP overall morphologies and internal architectures have been reported so far. These include, for example, gold nanorods and nanocubes; core/shell structures, where the core and the shell are composed of different inorganic material; alloy NPs, produced by mixing different metal precursors and incorporating different elements within the same NP; and dumbbell-shaped mixed NPs with chemically segregated structures.

4.3 POLYMERS AS LIGANDS FOR NANOPARTICLE FORMATION AND STABILIZATION

Amine-terminated polymers have been widely used for the preparation of metal NPs via reduction of metal precursors, as well as for the stabilization of the obtained metal NPs. Amines are used as reducing and stabilizing agents in AuNPs formation, and this also applies to polymers carrying amine groups. However, it is not clear how the chemical or structural properties of these species are responsible for their reducing character (Newman and Blanchard 2006). It is known that the electrons transferred from the amine groups to the metal ion provoke its reduction to zero valence state and therefore induce the growth of metal NPs. The formation of an Au–amine complex for the protection of Au NPs was reported by Aslam et al. (2004).

Water-soluble, amine-terminated poly(ethylene oxide) (PEO) (PEO-NH_2) was mixed with $HAuCl_4$ and heated at 100°C, in order to produce AuNPs of good colloidal stability with an average diameter of ca. 16 nm (Iwamoto et al. 2003). A complex between the $AuCl_4^-$ and the $-NH_2$ group of the PEO-NH_2 was formed, resulting to the reduction of the gold ions and NP formation. It is worth mentioning that the utilization of a hydroxyl-terminated PEO (PEO-OH) did not result in the formation of AuNPs.

Branched poly(ethylenimine) (PEI) was utilized as both the reducing and the stabilizing agent for the formation of Au NPs at room temperature (Chen et al. 2007a, Kim et al. 2008). Varying the amount of PEI initially used, both the size and the optical properties of the PEI-capped AuNPs could be controlled, leading to the formation of AuNPs with excellent colloidal stability and amine surface groups (Figure 4.1).

Dendrimers, such as those based on poly(propyleneimine) (PPI), have also been used for the formation of AuNPs. NP synthesis has been facilitated by heating the solution containing the third-generation PPI dendrimer, without the additional step of introducing any extra reducing agent (Sun et al. 2003).

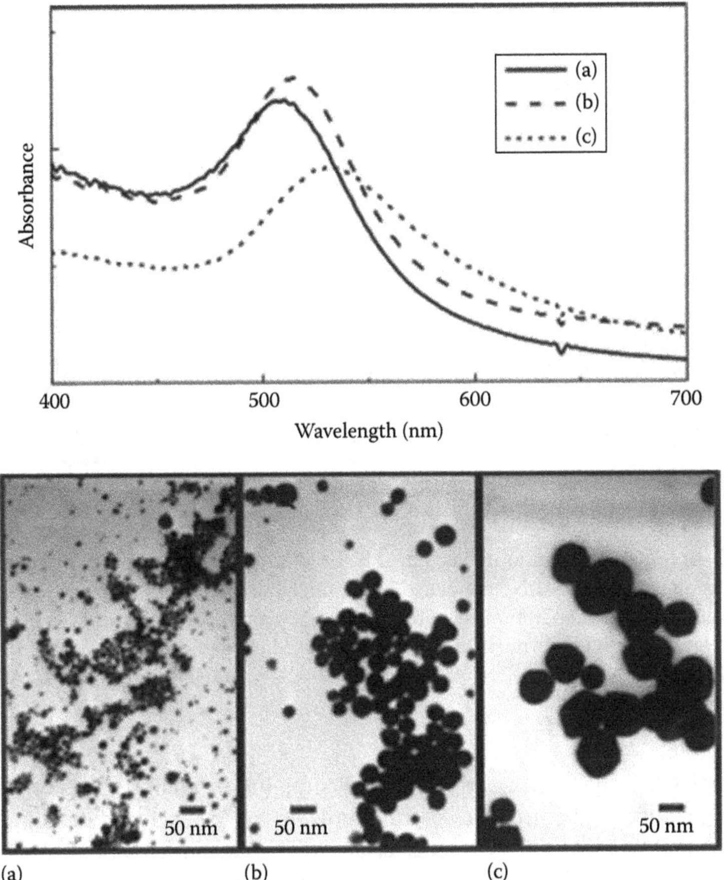

FIGURE 4.1 Synthesis of PEI-AuNPs: UV-vis spectra and corresponding TEM images taken from mixtures containing 25 mL of 1.4 mM aqueous $HAuCl_4$ solution, and (a) 0.9, (b) 0.7, and (c) 0.5 mL of a 1%(w/w) aqueous PEI solution were stirred for 16 h at room temperature. (Reprinted with permission from Kim, K., Bong Lee, H., Won Lee, J., Kun Park, H., and Soo Shin, K., *Langmuir*, 24, 7178–7183, 2008. Copyright 2008, American Chemical Society.)

Preformed AuNPs were covalently conjugated onto the periphery of thiol terminally functionalized thermosensitive poly(*N*-isopropylacrylamide) (PNIPAM) unimolecular micelles in order to fabricate satellite-like nanostructures (Xu et al. 2007). These hybrid nanostructures displayed high stability due to the covalent linkage between the AuNPs and the thiol end groups. The satellite-like nanostructures were thermoresponsive in the sense that they are able to shrink and swell reversibly in response to changes in solution temperature. Consequently, the spatial distances between AuNPs attached at the unimolecular micelles surface could be tuned, and the latter effected changes in the optical properties (adsorption spectrum) of the hybrid NPs (Figure 4.2).

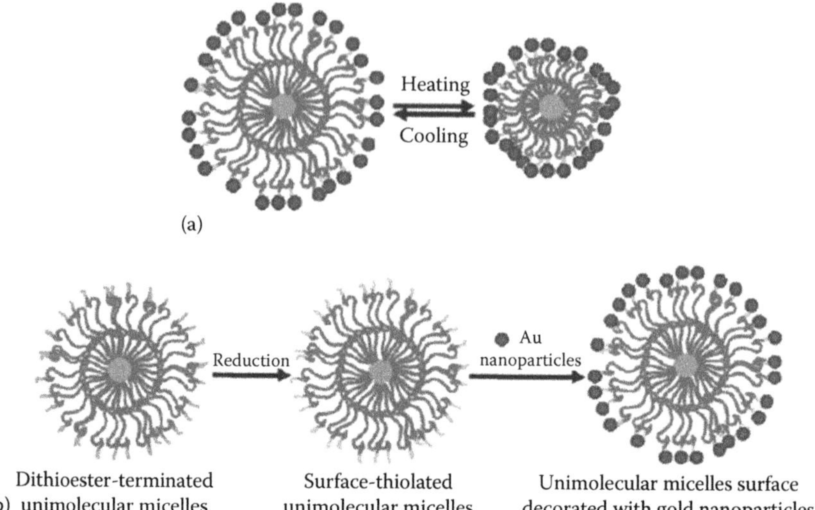

FIGURE 4.2 (a) Hybrid unimolecular PNIPAM micelles exhibiting thermo-tunable spatial distance between AuNPs attached at the micelle surface and (b) schematic illustration of the two-step preparation of hybrid unimolecular micelles surface decorated with AuNPs. (Reprinted with permission from Xu, H., Xu, J., Jiang, X., Zhu, Z., Rao, J., Yin, J., Wu, T., Liu, H., and Liu, S., *Chem. Mater.*, 19, 2489–2495, 2007. Copyright 2007, American Chemical Society.)

Bio-polyelectrolytes, like carboxylated dextran (of different degrees of chemical functionalization), have been successfully utilized for the surface functionalization of iron oxide magnetic NPs (Wagner et al. 2004, Bakandritsos et al. 2008, 2011, Wotschadlo et al. 2009). Hybrid NPs are usually larger than the original iron oxide particles (in the 100 nm range) since self-organization in this case results in structures where several IONPs are incorporated within a supramolecular polymeric nano-object, for reasons related to the stability of the self-assembled structures. Stable colloidal dispersions of the biocompatible hybrid NPs are obtained, using mixing protocols, which in term show promise as biocompatible MRI contrast agents due to the magnetic properties of the inorganic component. Low molecular weight polyethylamine, hydrophobically modified with alkyl groups, was used as the phase transfer agent for iron oxide nanoclusters (Liu et al. 2011a). The cationic groups of PEI bind siRNA, leading to the formation of hybrid magnetic NPs that can be used in gene delivery applications.

Polymeric components can be used as ligands for semiconductor NPs (quantum dots [QDs]). These include charged homopolymers, thiolated polyethylene glycol (PEG) chains, PAMAM dendrimers, engineered peptides for targeting, as well as thiolated proteins (Mattoussi et al. 2000, Akerman et al. 2002, Potapova et al. 2003, Ding et al. 2004, Wall and Himmel 2004, Huang and Tomalia 2005, Pan et al. 2006, Wisher et al. 2006, Hezinger et al. 2008). Mixed ligands on the same QD, composed of PEG and designed peptides, introduce targeting/recognition protein-like moieties and stealth function at the same time. PEI is able for multidentate binding to semiconductor NPs through weak coordination of the amine groups (Nann 2005). PEI enhances the water solubility of the NPs and mediates the complexation of the hybrid NPs with nucleic acids. However, the use of PEI was found to also enhance the photooxidation of the semiconductor NPs. Poly(2-dimethylaminoethyl methacrylate) can be also utilized (Wang et al. 2006) with similar results regarding solubility and colloidal stabilization in biological fluids. The next generation of polymeric multidentate ligands for semiconductor NPs is presented in the development of phosphine-containing polymers (Kim and Bawendi 2003, Kim et al. 2005a). This type of polymers presents chemical and binding similarities with several low molecular weight ligands used in the preparation of QDs. Additionally, they do not compromise the PL properties of the NPs, like amine-containing polymer do. In a similar fashion, polyether-based dendrons and dendrimers with aryl phosphine groups at the focal point have been used as ligands for the stabilization of semiconductor NPs (Huang and Tomalia 2006).

4.4 SYNTHESIS OF INORGANIC NANOPARTICLES IN BLOCK COPOLYMER MICELLES AND OTHER POLYMERIC NANOSTRUCTURES IN SOLUTIONS

It has been emphasized that NP characteristics are influenced by the formation route leading to NP creation, as well as by the nature of the matrix or the ligands surrounding the NP (in the most general sense, the environment around the NP). Therefore, it is certain that in order to exploit the NPs' properties, for example, in device fabrication, their formation should be morphologically controlled (Shenhar et al. 2005). As far as the surface chemistry and colloidal stability of metal NPs are concerned, one should bear in mind that usually inorganic nanoparticles are stabilized by an organic medium/ligand in order to prevent their agglomeration (Filali et al. 2005) (Figure 4.3). At the same time, this "organic layer" provides additional properties to the system, such as chemical functionality (Ofir et al. 2008), that influences to a large extent the targeting ability and affects the biodistribution and pharmacokinetics of the hybrid NPs in bioapplications, the affinity toward other NPs or surfaces and the responsive properties of the nanostructure toward the environment, and consequently the self-assembly properties or deposition directionality of the NPs.

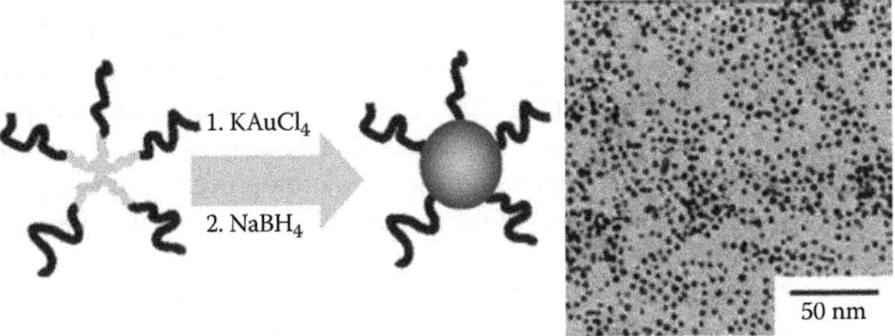

FIGURE 4.3 (PEO-*b*-PCL)$_5$ star-block copolymers as templates for the synthesis of AuNPs. (Reprinted with permission from Filali, M., Meier, M.A., Schuber, U.S., and Gohy, J.F., *Langmuir*, 21, 7995–8000, 2005. Copyright 2005, American Chemical Society.)

Thus, the organic ligand, which in the cases presented in this chapter is a polymeric substance (amphiphilic copolymers or polyelectrolytes in most cases), dictates in several ways the properties of the inorganic NP, in the neat state or in solution/dispersed state in organic or aqueous solvents. These ways depend on the macromolecular architecture, the chemical functionality, and the chemical composition of the polymeric ligand. In many instances, reported polymers induce the assembly of the metal precursors into well-defined metal NPs, in terms of size and morphology. The in situ formed NPs are therefore incorporated into composite/hybrid materials with the macromolecular ligands significantly influencing and directing NP ordering within the polymeric matrix.

Since the utilized polymer facilitates the nucleation and growth of metal NPs, it directly influences the size, the size distribution, the shape, and the morphology of NPs. This influence is caused by the interactions between certain polymer connected chemical moieties and the inorganic species, frequently leading to the formation of small particle sizes and narrow size distributions of the resulting metal NPs. Amphiphilic diblock copolymers tend to self-assemble in the presence of selective solvents in order to form micelles, consisting of a dense hydrophobic core and a soluble hydrophilic corona as has been discussed previously. Metal NPs can be in situ synthesized by being either embedded inside the dense core or coordinated at the corona block, leading to the formation of well-defined NPs of the desired size and shape. More elaborate localizations can be achieved when a triblock terpolymer or other more architecturally complex macromolecules are utilized, that is, at the inner shell of a core/shell/corona micelle formed by a triblock terpolymer.

The size and the shape of the metal NPs are also influenced by the reduction method introduced to the colloidal material. Small metal NPs are formed with rapid reducing agents, since this process favors the simultaneous nucleation of the metal NPs, leading to a "raspberrylike" architecture of the hybrid micelles. On the other hand, larger metal NPs are observed when employing a less effective agent, which favors slow nucleation leading to hybrid micelles with a "cherrylike" morphology (Antonietti et al. 1995, Mayer 2001, Forster 2003).

A general strategy for the formation of metal NPs inside block copolymer micelles involves the interaction of the metal precursor salt with one of the blocks of the copolymer. This largely pinpoints the spatial localization of primary nuclei within the micelles that eventually will develop to metal NPs, as a result of the following reduction step that will transform precursor species into metal NPs. The reduction leads to nucleation and growth of the metal NPs, resulting in inorganic aggregates (Forster 2003, Fustin et al. 2006). The role of precursor block copolymer chemical interactions, as well as conformational characteristics and rearrangements of the copolymer blocks, is influential to the formation process. Some representative examples of such approaches are presented in the following in order to better illustrate the general protocol.

Möller et al. (Mössmer et al. 2000) have synthesized AuNPs inside the core of poly(styrene-*b*-2-vinylpyridine) (PS-*b*-P2VP) micelles. The PS-*b*-P2VP block copolymer forms reverse micelles, when dissolved in toluene (a selective solvent for P2VP), consisting of a P2VP core and a poly(styrene) (PS) corona block. The $HAuCl_4$ was used as the metal precursor, and when added, it was preferentially dissolved into the core of the micelles, due to protonation of the 2VP groups. Reduction was achieved by hydrazine. N_2H_2 was also taken up into the core of the micelles due to its polar character, leading to the formation of well-defined, spherical AuNPs with an average size of ca. 9 nm. The self-assembly behavior of the hybrid PS-*b*-P2VP/Au colloids during the preparation procedure was followed by means of static and dynamic light scattering (DLS), giving valuable insights into the formation process and the structural rearrangements of the nanosystem.

Reverse micelles of poly(styrene-*b*-4-vinylpyridine) ((PS-*b*-P4VP)$_n$) star-block copolymers with several block copolymer arms were used as nanoreactors in order to prepare spherical AuNPs (Li et al. 2006). (PS-*b*-P4VP)$_n$ star-block copolymers were synthesized with different lengths of the P4VP blocks, the block that interacts with the NP precursor, and were dissolved in toluene. Micelles consisting of a P4VP core and a petallike shell of PS were formed as evidenced by light scattering and microscopic techniques. Addition of $HAuCl_4$ led to protonation of 4VP segments and to the formation of a polyionic block complexed with gold atoms. As a result of the protonation/complexation, an increase in the size of the micellar cores was observed, mainly due to the electrostatic repulsion among the polyionic blocks. However, small AuNPs of ca. 2 nm having spherical morphology were formed regardless the length of the 4VP block, indicating that the nucleation and growth process were independent of the molecular weight and the conformation of the P4VP block.

Other metal NPs, such as the ones composed of Pt and Pd, were prepared within PS-*b*-PEO and P2VP-*b*-PEO micelles, relying on a similar coordination/reduction mechanism (Bronstein et al. 1999, 2000a). In the case of PS-*b*-PEO copolymers, which do not carry any charge, the cationic surfactant cetylpyridinium bromide (CPC) was incorporated leading to the formation of mixed block copolymer/surfactant aggregates. Ion exchange of the surfactant counterions by $PtCl_6^{2-}$ and $PdCl_4^{2-}$ ions resulted in the incorporation of metal ion precursor within the micellar structures. Subsequent reduction of the metal-containing PS-*b*-PEO/CPC/MXn hybrid assemblies using $NaBH_4$ and molecular hydrogen resulted in the formation of metal NPs mainly located within the block copolymer/surfactant mixed assemblies. In the case of P2VP-*b*-PEO copolymers in aqueous media, the incorporation of metal salts induced micellization even at low pH values where P2VP blocks, which are the blocks that can coordinate to the metal ions, are protonated and no micelles are formed in the absence of the metal precursors. The structural characteristics of block copolymer/salt assemblies were significantly influenced by the type of salt used. This in turn was influential to the size and the morphology of the produced PtNPs and palladium NPs (PdNPs).

Pd metal NPs were also synthesized in block copolymers using hydrophobic metal precursors (Wang et al. 2008a). A toluene solution of $Pd(PPh_3)_4$ was mixed in a micellar solution of poly(styrene-*b*-*N,N*-dimethylacrylamide) (PS-*b*-PDMA) copolymer in isopropanol. $Pd(PPh_3)_4$ was taken by the PS cores due to its incompatibility with isopropanol. Aging of the solutions for a period of 3 months resulted in the formation of Pd metal NPs through dissociation of the ligands. Transmission electron microscopy (TEM) images showed the presence of 1–3 nm–sized metal NPs with well-developed crystalline lattices. The rate of metal NP formation could be tuned through temperature and the concentration of the metal precursors.

The pH- and temperature-sensitive poly(methacrylic acid-*b*-*N*-isopropylacrylamide) (PMAA-*b*-PNIPAM) copolymer was utilized for the one-pot synthesis of AuNPs with the core of the formed micelles (Nuopponen and Tenhu 2007). It was shown that the structural properties of the polymeric aggregates can be adjusted by variation of both the pH and the temperature, resulting in significant changes of their size, depending on the physicochemical parameters of the system. Shifts in the maximum of the SPR band of AuNPs were observed by changing the pH and the temperature (Figure 4.4), due to stretching/collapse of the PMAA blocks, which results in an increase/decrease of the distances between neighboring AuNPs (Li et al. 2008). Therefore, the photophysical properties of the system can be regarded as being responsive to physicochemical changes of the surrounding medium.

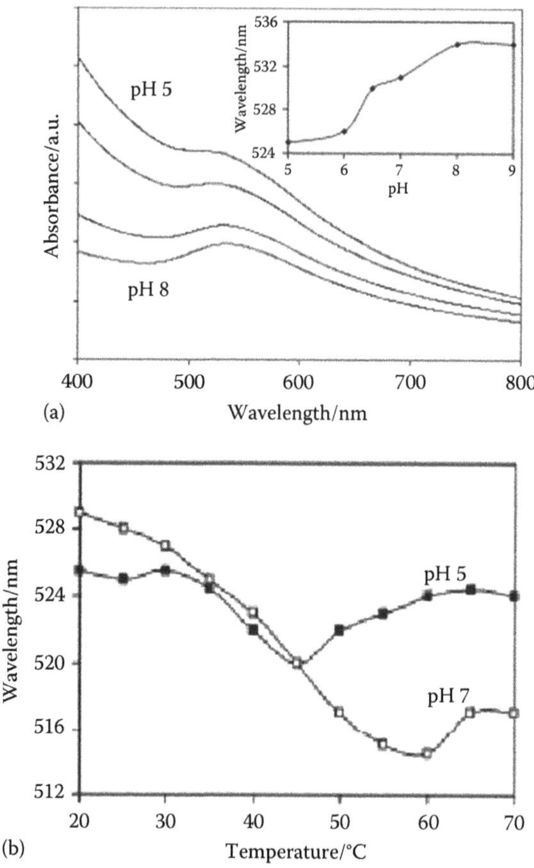

FIGURE 4.4 Au/PMAA-*b*-PNIPAM hybrid NPs. (a) UV-vis spectra as a function of pH. Turbidity increases as pH is decreased; pH decreases from bottom to top (pH 8, 7, 6, and 5). The SPR band undergoes a blue shift (534–525 nm) when pH is decreased. (b) The λ_{max} of the SPR as a function of temperature at two pHs (7.0 and 5.0). (Reprinted with permission from Nuopponen, M. and Tenhu, H., *Langmuir*, 23, 5352–5355, 2007. Copyright 2007, American Chemical Society.)

AgNPs have been in situ synthesized inside the core of poly(isoprene-*b*-acrylic acid) (PI-*b*-PAA) block copolymer micelles (Meristoudi et al. 2008). PI-*b*-PAA copolymer was dissolved in tetrahydrofuran (THF), a solvent selective for polyisoprene (PI) blocks, leading to the formation of micellar aggregates with an average diameter of ca. 68 nm, with a PI corona and a poly(acrylic acid) (PAA) core. Addition of $AgNO_3$ into the copolymer solution and subsequent reduction of Ag^+ cations to Ag^o led to the formation of large hybrid aggregates of 194 nm diameter. It was postulated that Ag^+ cations interact strongly with the carboxylic acid groups of the acrylic acid units, creating physical cross-links between block copolymer chains, which in turn promote the formation of micellar-like aggregates. However, the AgNPs prepared had good colloidal stability, implying that the copolymer acted as a good colloidal stabilizer. Nevertheless, smaller micellar aggregates (~66 nm) were observed after addition of $HAuCl_4$, instead of $AgNO_3$, to the PI-*b*-PAA/THF solution, presumably due to the weak coordination interactions among the acrylic acid units and the $AuCl_4^-$ ions. The study gives some guidelines regarding the nature of interactions and the effects of block copolymer chemical structure and of the structural characteristics of copolymer nanoassemblies that can be active and may influence structure formation and colloidal stability in hybrid metal NPs/block copolymer systems.

Alexandridis et al. have reported the synthesis of AuNPs in the presence of block copolymers and in the absence of an externally added reducing agent (Alexandridis and Sakai 2004a,b, 2005a,b, 2006a). In this particular case, the block copolymer that hosts the metal precursors in aqueous solution can simultaneously induce the reduction toward NPs. In their earlier study, a poly(ethylene oxide)-*b*-poly(propylene oxide) (PEO-*b*-PPO) diblock copolymer has been utilized as the reducing, stabilizing, and morphogenic agent for the one-pot synthesis of AgNPs and AuNPs (Alexandridis and Sakai 2006a) (Figure 4.5). The same group (Alexandridis and Sakai 2005a) later reported that the size and shape of AuNPs formed in aqueous solutions, at ambient temperature, can be efficiently controlled by varying the concentration metal ions, as well as the concentration and the type of the block copolymer

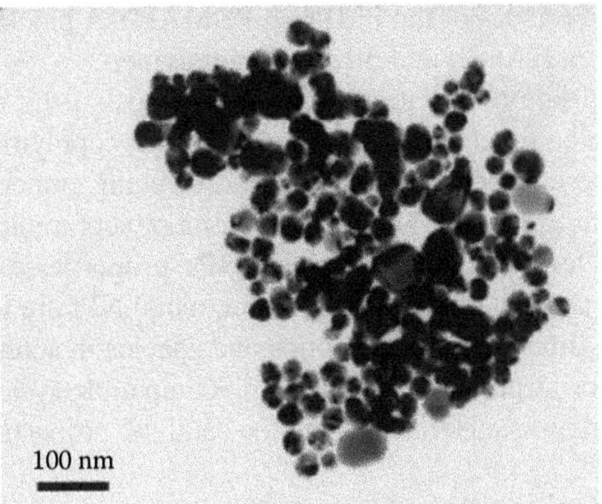

FIGURE 4.5 TEM images of spherical AgNPs synthesized in 10wt.% $PO_{19}EO_{33}PO_{19}$ formamide solutions. (Reprinted with permission from Alexandridis, P. and Sakai, T., *Chem. Mater.*, 18, 2577–2583, 2006a. Copyright 2006, American Chemical Society.)

(in this case, Pluronic PEO-PPO-PEO triblock copolymers of different compositions and molecular weights have been used alternatively to diblock copolymers). It was proposed that main-chain ether polymers, with hydroxyl terminal functionalities, can form pseudo-crown ether cavities in the presence of ions (Mill and Longenbeger 1995). Metal ions become bound in these cavities and are further reduced to NPs. By increasing the molecular weight and the concentration of the polyether, an increase in the number of existing cavities is taking place, resulting in an enhancement of the reduction of the precursor metal ions to metal NPs.

AgNPs were formed in situ in aqueous solutions of the double-hydrophilic block copolymer poly(ethylene oxide-*b*-methacrylic acid) (PEO-*b*-PMAA) with no additional reducing agent, ultraviolet irradiation, or other electrochemical method (Zhang et al. 2001). The PMAA blocks were coordinated with the Ag^+ ions, due to the presence of carboxylic acid groups, while the PEO block acted as the stabilizing agent of the formed hybrid assemblies. At the same time, PEO blocks promoted the gradual reduction of the metal ions to AgNPs, in a manner analogous to that reported for Pluronic copolymers. The AgNPs were gradually transformed to silver nanowires during aging of the solution. Adsorption spectra and TEM observations as a function of time indicated that nanowire formation was accomplished through an increase of AgNPs, accompanied by adhesion of the initial hybrid assemblies. This resulted in directed aggregation growth, forming initially silver nanowires with coarse surfaces, and finally (after more than 50 h), Ag nanowires with smooth surfaces were observed.

The class of poly[(2-methacryloyloxy)ethyl phosphorylcholine-*b*-(2-dimethylamino)ethyl methacrylate] (PMPC-*b*-PDMAEMA) block copolymers was successfully utilized for the in situ synthesis of high-quality AuNPs, without the need of any external reducing agent (Yuan et al. 2006). The poly(2-(methacryloyloxy)ethyl phosphorylcholine) (PMPC) block acts as the water-soluble, ionic stabilizing agent, while the tertiary amine groups of poly(*N,N*-dimethylaminoethyl methacrylate (PDMAEMA) induced the reduction of $AuCl_4^-$ precursors to AuNPs. It was demonstrated that the size and the shape of the AuNPs could be controlled by changing the molecular characteristics of the PMPC-*b*-PDMAEMA block copolymer and the relative ratios between $HAuCl_4$ and the block copolymer.

Following appropriate preparation protocols and using the right block copolymers, metal NPs can be fixed at the corona of the block copolymer micelles or similar core/shell polymeric superstructures, instead of being embedded into the dense micellar cores. Fixing the metal NPs at the corona of supramolecular nano-objects provides advantages for further exploitation of NPs' properties, such as the catalytic, optical, and biological ones. In this arrangement, metal NPs are more accessible and free to interact with other functional groups and species in solution (i.e., in ligand-exchange reactions or targeting protocols in biocompatible systems [Sperling et al. 2008]) and with particular substrates, when the catalytic properties of the metal NPs are of interest.

The formation of an AuNPs within a core/shell/corona mixed micelle composite structure in organic solvents was first proposed by Shi (Chen et al. 2007b). In particular, mixed micelles consisting of a PEO corona, a AuNPs/poly(4-vinylpyridine) (Au/P4VP) inner shell, and a PS core were prepared. First, micelles consisting of a PS core and a P4VP corona were prepared, and $HAuCl_4$ was preferentially coordinated at the 4VP outer block. Addition of poly(ethylene oxide-*b*-4-vinylpyridine) copolymers led to the formation of a mixed core/shell/corona micelle due to the coordination and mixing of the P4VP blocks of the P4VP-*b*-PEO copolymer with the outer Au/P4VP corona of the existing PS-*b*-P4VP micelles. Finally, after NP formation, a multifunctional hybrid nanostructure composed of a PS core, a solvent swollen Au/P4VP shell, which is very sensitive to the solution pH, and a biocompatible, nonionic PEO water-soluble corona was created.

In a similar approach, core/shell/corona Au/micelle hybrid nanostructures with a responsive smart hybrid shell were prepared (Chen et al. 2008). Micelles composed of a polystyrene core and a poly(ethylene glycol) and poly(4-vinylpyridine) mixed shell were formed in acidic water through self-assembly of a PEG-*b*-PS-*b*-P4VP triblock terpolymer. Addition of $HAuCl_4$ and reduction to zerovalent Au led to the formation of a multicompartment Au/micelle hybrid nanostructure composed of a PS core, a hybrid shell of P4VP/Au/PEG, and a PEG corona. It was postulated that in aqueous solutions, the inner shell is swollen at different degrees, depending on the solution pH, due to the differences in P4VP protonation, Au/P4VP coordination, and hydrophilicity of the compartment. Under basic conditions, when P4VP blocks are water-insoluble, channels through the shell are produced, due to the presence of hydrophilic PEO chains, connecting the core with the outer environment. It is expected that such a topological arrangement and responsiveness of the nanostructure will modulate and fine-tune the catalytic properties of AuNPs (or the catalytic properties of other metal NPs that can be produced by this process).

In another case, AuNPs were formed in the corona of poly[(*p*-*tert*-butylstyrene)-*b*-sodium(sulfamate/carboxylate)isoprene] (PtBS-*b*-SCPI) block copolymers micelles (Meristoudi and Pispas 2009). Formation of AuNPs was achieved in the absence of any reducing agent, since it was observed that the amine groups of the corona SCPI chains were able to promote reduction of the metal precursor (Figure 4.6), as it was demonstrated by TEM observations. The kinetics of NP formation were followed by UV-vis spectroscopy measurements. Heating of the solutions resulted in speeding up the reduction process. The hybrid PtBS-*b*-SCPI/AuNPs micelles were able to complex with lysozyme forming multicomponent/multifunctional hybrid inorganic–(bio) organic NPs.

The formation of AuNPs within core/shell/corona polymeric NPs was mediated by the triblock copolymer poly[(ethylene glycol)-*b*-(4-vinylpyridine)-*b*-(*N*-isopropylacrylamide)] (PEG-*b*-P4VP-*b*-PNIPAM) (Zheng et al. 2006) in another case. This triblock terpolymer is thermosensitive and pH responsive, due to the presence of PNIPAM and P4VP blocks, respectively. Gold ions are complexed with the P4VP middle block that forms the shell of the polymeric micelles, resulting in the preparation of size and morphologically controlled AuNPs, by varying the pH or the temperature of the aqueous solution. The resultant AuNPs/block copolymer hybrid nanostructures are thermoresponsive as well and retain their colloidal stability over a long period of time (Figure 4.7).

Metal NPs attached to the shell of block copolymer micelles have been also prepared by micellar structural inversion, due to a change in the solvent, from micelles containing already formed NPs in their cores (Mayer and Mark 1996, Zhao and

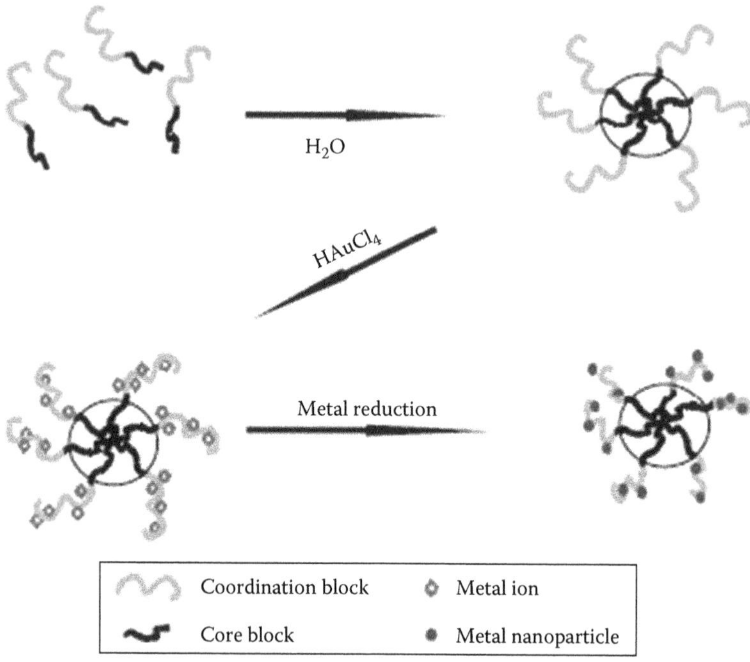

FIGURE 4.6 AuNP formation in the corona of PtBS-*b*-SCPI amphiphilic block copolymer micelles in the absence of an additional reducing agent. (Reprinted from *Polymer*, 50, Meristoudi, A. and Pispas, S., 2743–2751, Copyright 2009, with permission from Elsevier.)

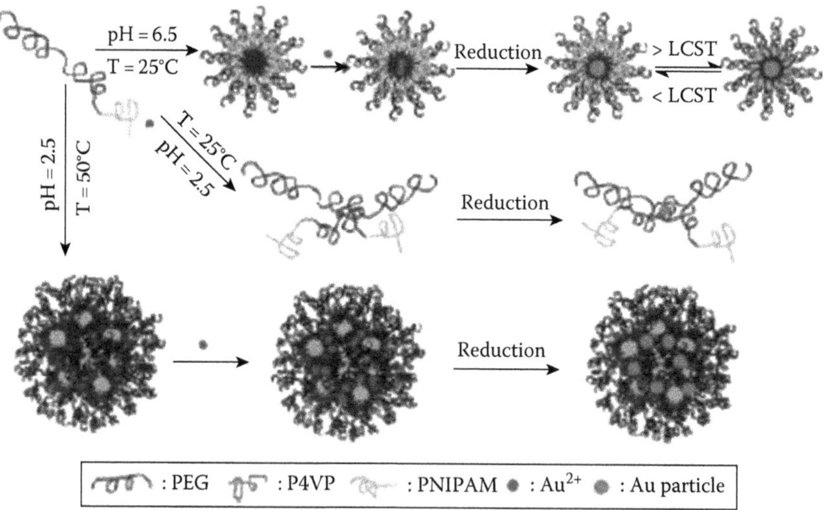

FIGURE 4.7 Thermoresponsive and pH-responsive micellization of PEG_{110}-*b*-$P4VP_{35}$-*b*-$PNIPAM_{22}$ and synthesis of the discrete AuNPs, gold–polymer core/shell NPs, and AuNP clusters. (Reprinted with permission from Zheng, P., Jiang, X., Zhang, X., Zhang, W., and Shi, L., *Langmuir*, 22, 9393–9396, 2006. Copyright 2006, American Chemical Society.)

Douglas 2002). AuNPs were prepared in situ in the core of PS-*b*-P4VP copolymer in chloroform (Hour et al. 2007) by the precursor complexation/reduction method. The authors have elegantly demonstrated that the gradual switching of the solvent composition, by addition of methanol, to a mixture containing primarily alcohol (methanol/chloroform 9:1) resulted in a core/shell reversal, thus creating spherical hybrid block copolymer micelles with a PS core and a P4VP/AuNPs corona.

Another strategy for fixing metal NPs on the corona of block copolymer micelles is the ligand-exchange reaction, as proposed by Eisenberg et al. (Azzam et al. 2008). In this particular case, preformed AuNPs and PdNPs were anchored on the corona of poly[(ethylene oxide)-*b*-styrene-*b*-4-vinylpyridine] (PEO-PS-P4VP) triblock terpolymer micelles. AuNPs and PdNPs were initially stabilized by tetraoctylammonium bromide (TAOB) as the ligand and were mixed with the triblock copolymer solution. A ligand-exchange reaction occurred, replacing the TAOB molecules from the metal surface, due to strong coordination forces between the 4-vinylpyridine block and the metal species. Spontaneous self-assembly of the nanostructure resulted in micelle formation having the NPs located within the corona of the micelles, that is, in the P4VP domain.

Giant hybrid compound block copolymer micelles encapsulating AuNPs were prepared by the use of poly(styrene-*b*-2-vinyl pyridine) and poly(isoprene-*b*-styrene) block copolymers mixtures (Mantzaridis and Pispas 2008). The AuNPs were in situ synthesized inside the cores of the PS-*b*-P2VP block copolymer micelles in toluene. The resulting hybrid polymeric/metal NPs micelles were mixed with a toluene solution of a PI-*b*-PS diblock copolymer having a high molecular weight, and the solvent was allowed to evaporate. Redissolution of the solid mixture in heptane, a solvent selective for PI blocks, allowed the encapsulation of PS-*b*-P2VP/AuNPs hybrid micelles inside the core of PI-*b*-PS giant micelles, due to the compatibility of the PS core of the PI-*b*-PS micelles with the PS corona of the PS-*b*-P2VP/AuNPs micelles.

In several cases, bimetallic colloids have been formed within block copolymer aggregates, either by simultaneous (Bronstein 2000a) or successive (Alexandridis and Sakai 2006b) reduction of both metal precursor compounds. Usually, core/shell bimetallic NPs are produced by successive reduction, while simultaneous reduction yields alloy NPs. Bimetallic colloids display enhanced properties compared to monometallic NPs, due to the combination of the properties of both metals and those of the polymer, as well as to their unusual morphology and variable composition (Chatterjee and Jewrajka 2007). PdNPs and bimetallic PdAu, PdPt, and PdZn NPs were synthesized in PS-*b*-P4VP block copolymer micelles in order to exploit their catalytic properties (Bronstein et al. 2000b, Sulman et al. 2009). It was found that the catalytic capability of the bimetallic NPs formed in this way was enhanced compared to that of Pd alone. This was attributed to an increase in the number of active centers on the surface of the bimetallic particles, due to changes of the electronic structure and the surface geometry of Pd, derived from the incorporation of the second metal (i.e., Au, Pt, Zn) as observed by Fourier transform infrared (FTIR) spectroscopy and x-ray photoelectron spectroscopy (XPS). Besides acting as a nanoreactor, block copolymers provide solubility and colloidal stability to the hybrid NPs.

Silver–gold alloy NPs have been synthesized by the use of PDMAEMA-*b*-PMMA-*b*-PDMAEMA triblock copolymer nanoreactors (Chatterjee and Jewrajka 2007). Simultaneous reduction of $AgNO_3$ and $HAuCl_4$ resulted in the formation of homogeneous alloy NPs, stabilized by the copolymer chains, and not to a mixture of Ag and Au NPs, as it was evidenced by UV-vis spectroscopy. However, Alexandridis and Sakai (2006b) utilized the successive reduction of $AgNO_3$ and $AuCl_4$, which led to the synthesis of Ag-Au core/shell NPs of different morphologies.

Similar preparation protocols can also be applied for the synthesis of semiconductor or magnetic NPs within amphiphilic block copolymers self-assembled nanostructures. Semiconductor NPs display size-dependent properties, which facilitate their potential applications, especially in nonlinear optical materials (Sun and Riggs 1999). In this respect, Eisenberg et al. were among the first to study systematically the effects of block copolymer characteristics on the size of CdS NPs formed in the cores of poly(styrene-*b*-acrylic acid) (PS-*b*-PAA) diblock copolymers micelles in organic solvents (Moffitt and Eisenberg 1995a, Moffitt et al. 1995b). By UV-vis and TEM, observations revealed that the size of spherical CdS NPs could be varied along with the size of the PAA ionic cores of the micelles, allowing for tuning the diameter of the NPs in the range 3–5 nm. Neutralization of remaining PAA core segments increased the stability of the hybrid colloids and allowed several precipitation and redispersion cycles in organic solvents, without loss of the photophysical properties of the semiconductor NPs. It was also possible to increase the size of preformed CdSNPs by adding new Cd ions and treatment with H_2S in a second step.

Utilizing another preparation protocol, CdS NPs were prepared in the core of PS-*b*-P2VP block copolymer micelles (Zhao et al. 2001). Cadmium acetate was used as the metal precursor, and it was added into the block copolymer solution in THF, a good solvent for both blocks. The complexation of Cd^{2+} ions with the P2VP blocks induced micelle formation, with a 2VP core loaded with Cd^{2+}, where the ion/2VP segment coordination clusters are acting as nonpermanent cross-links, and a PS corona. Addition of H_2S gas resulted in CdS NP formation. The produced NPs displayed good colloidal stability, due to the interaction between the 2VP units and the surface of the CdS NPs and the stabilizing effect of the PS corona chains. It was shown that the size of the CdS NPs could be modified by changing the 2VP:Cd^{2+} molar ratio.

CdS NPs embedded within the corona of block copolymer micelles could be formed by changing the preparation protocol mentioned earlier (Zhao and Douglas 2002). The compound micelles consisting of a PS corona and a 2VP core loaded with CdS NPs can undergo solubilization in water at low pH, resulting in a collapse of the PS chains and protonation of the 2VP units. Thus, a new hybrid micelle is created with a PS core and a P2VP corona decorated with randomly distributed CdS NPs, which are still interacting with the ionic segments of the corona. Alternatively, CdS NPs can be in situ prepared at the corona of the PS-*b*-P2VP micelles as well. In this method, after the induced micellization of PS-*b*-P2VP in THF, due to the presence of Cd^{2+} ions, acidic water has been introduced leading to the formation of micelles composed of a collapsed PS core and a hybrid P2VP/Cd^{2+} corona. Semiconductor CdS NPs coordinated at the corona of the PS-*b*-P2VP inverted micelles are prepared by addition of either a Na_2S solution or a H_2S gas.

In several attempts, CdS NPs were successfully synthesized at the corona of various amphiphilic block copolymers, consisting of the same hydrophilic block and different hydrophobic blocks (Mandal and Chatterjee 2007). Characterization techniques have shown that the size, the shape, and the size distribution of the CdS NPs were similar, regardless the nature of the block copolymer used. The reason for this observation lies on the fact that NP formation occurs at the hydrophilic corona of the micelles, and thus, their properties are independent of the nature of the hydrophobic core that does not participate in any way in the formation of the semiconductor NPs. However, in each case investigated, 5% (v/v) of a nonaqueous solvent was added during CdS NPs formation, which is believed to have influenced the nucleation and growth of the inorganic NPs.

CdS and CdSe NPs were also prepared in amphiphilic block copolymer matrixes in organic solvents (Gatsouli et al. 2007). Poly(sulfonated styrene-*b*-*tert*-butylstyrene) (SPS-*b*-PtBS) and poly(styrene-*b*-sulfonated isoprene) (PS-*b*-SPI), of different molecular weights and compositions, were the two block copolymer families utilized for the formation and stabilization of the CdS and CdSe NPs. Cd^{2+} ions addition in block copolymer solutions in good solvents for the two blocks led to the induction of micellar-like aggregates. The hybrid block copolymer–based colloids displayed good stability, due to the strong binding between the Cd^{2+} ions and the sulfonate groups of the copolymers, which directed the NP formation in the core of the micellar aggregates, after reaction with S^{2-} and Se^{2-} anions. PS- and PI-soluble chains provided stabilization of the assembly. The NPs' size was strongly dependent on the Cd^{2+}/SO_3H ratio, leading to the formation of CdS NPs with larger size than those of CdSe for the same Cd^{2+}/SO_3H ratio.

Triblock copolymers have been also employed for the synthesis of semiconductor NPs. Eisenberg et al. utilized a specially synthesized poly(ethylene oxide-*b*-styrene-*b*-acrylic acid) (PEO-*b*-PS-*b*-PAA) triblock copolymer for the in situ formation of CdS QDs (Duxin et al. 2005) and the development of hybrid copolymer/NPs assemblies. Different morphologies for the CdS-containing micellar aggregates were obtained depending on the preparation conditions. In all cases, inverse internal aggregate structures were obtained, consisting of a PAA core neutralized with Cd^{2+}, a PS inner shell, and a PEO corona that were obtained in THF. Four different overall aggregate morphologies were observed, (1) a worm-shaped nanoassembly, where the CdS QDs were surrounded by the PAA block (Figure 4.8); (2) a multicore micelle with spherical PS nanodomains surrounded by the PEO chains, where the CdS QDs are partially enveloped by the PAA chains; (3) a nanostructure where the CdS NPs were located on the surface of the PS cores; and (4) a fourth nanoaggregate consisting of spherical single-core micellar aggregates.

Other metals, metal oxide, and mixed metal NPs embedded in the polymeric matrix of block copolymer nanoassemblies display unusual optical, electronic, and magnetic properties, and some examples of their preparation within self-organized block copolymer colloids should be presented in this chapter.

Magnetic Co NPs having different sizes and shapes were synthesized in poly(stryrene-*b*-4-vinylpyridine) block copolymer micelles (Platonova et al. 1997). The properties of the Co particles could be adjusted from nonmagnetic to superparamagnetic and ferromagnetic by changing the reaction conditions. Similar methods

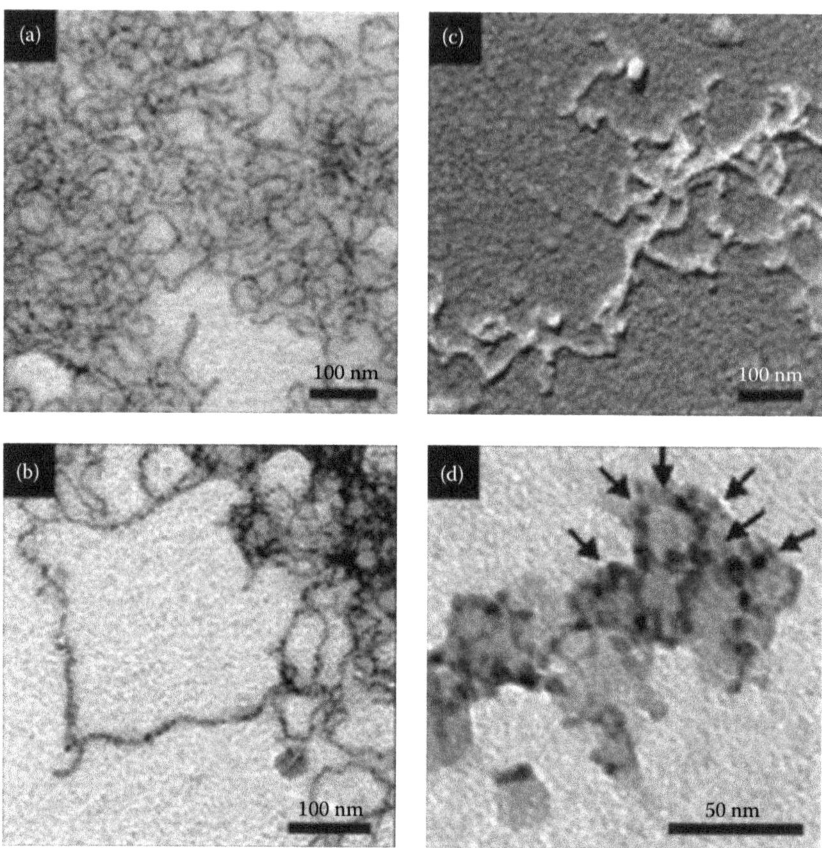

FIGURE 4.8 (a–c) Wormlike assemblies of PEO(45)-*b*-PS(150)-*b*-PAA(108) triblock copolymer, containing cadmium sulfide QDs in their core. The neutralization ratio $nCd^{2+}/2nAA^{-}$ is 2. (a, b) As prepared sample. (b) A close-up of (a), displaying the spatial distribution of QDs within the polymer aggregates. (c) Platinum–carbon replica at an angle of 45°. Negative picture (the shadow is dark). The average rod width is 7 nm. (d) PS core micelles in water, surrounded by CdS particles. The arrows are indicating the presence of some of the QDs. (Reprinted with permission from Duxin, N., Liu, F., Vali, H., and Eisenberg, A., *J. Am. Chem. Soc.*, 127, 10063–10069, 2005. Copyright 2005, American Chemical Society.)

were used for the synthesis of Co and Co–Co oxide NPs in PS-*b*-P2VP block copolymer micelles (Boyen et al. 2003, Diana et al. 2003). Recently, iron oxide and Co-doped IONPs were synthesized in PEO-*b*-PMAA aggregates (Bronstein et al. 2008). The block copolymer–stabilized magnetic NPs formed display remarkable solubility in water and aqueous buffers and high stability over periods of years.

Magnetite NPs were prepared in situ in the presence of the double-hydrophilic block copolymer poly[sodium (sulfamate/carboxylate)isoprene-*b*-ethylene oxide], (PSCI-*b*-PEO), following an one-step precipitation method based on the slow oxidation of $Fe(OH)_2$ by atmospheric oxygen in alkaline environment. Initially, hydrous $FeCl_2$ was added to an aqueous solution of the copolymer under vigorous stirring and a N_2 blanket, and the temperature was raised to 80°C. The $Fe(OH)_2$ gel was formed by addition of KOH, and the mixture was allowed to be oxidized by atmospheric oxygen. The resulting IONPs had a diameter ca. 15 nm. The hybrid nanoassemblies carried a negative surface charge and a biocompatible PEO shell, while they showed high saturation magnetization up to 67.7 emu/g (Basina et al. 2009).

Liu et al. have utilized cross-linked cylindrical polymer assemblies resulting from a linear poly(styrene-*b*-2-(cinnamoyl methacrylate)-*b*-*tert*-(butyl acrylate) (PS-*b*-PCMA-*b*-PtBA) triblock terpolymer for magnetic NP production. The polymeric assemblies were obtained via photo-cross-linking of the terpolymer in the solid state and subsequent redispersion in THF. The polymeric nanofibers were converted to nanotubes having interior carboxylic acid groups, as a result of the hydrolysis of the PtBA blocks into PAA blocks. Coordination of iron ions to the PAA segments followed by alkali addition produced γ-Fe_2O_3 NPs inside the terpolymer nanotubes (Yan et al. 2001, 2004, Liu et al. 2003). The same group utilized the nanotubes formed by another water-soluble triblock quaterpolymer, namely, poly[glyceryl methacrylate-*b*-[2-cinnamoyloxyethcinnamoyloxyethyl methacrylate-*co*-(2-hydroxyethyl methacrylate)]-*b*-(*tert*-butyl acrylate)] (PGMA-*b*-P(CEMA-*co*-HEMA)-*b*-PtBA), in order to prepare Pd/Ni alloy NPs (Yan et al. 2005). First, the nanotubes of the triblock were self-assembled in water, and the PtBA blocks were hydrolyzed to PAA. Palladium loading of the nanotubes was performed via addition of $PdCl_2$ and complexation of carboxylate groups with Pd^{2+} ions. Subsequent reduction with $NaBH_4$ produced PdNPs within the polymeric nanotubes. Addition of nickel ions led to the palladium-catalyzed electroless deposition of Ni inside the nanotubes (Figure 4.9).

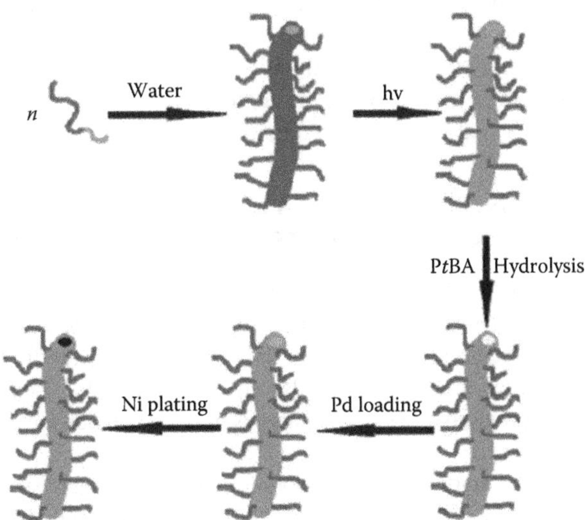

FIGURE 4.9 Illustration of the formation and chemical processing of a PGMA-*b*-P(CEMA-HEMA)-*b*-PtBA cylindrical aggregate for alloy inorganic NP formation through consecutive metal loading. (Reprinted with permission from Yan, X.H., Liu, G.J., Haussler, M., and Tang, B.Z., *Chem. Mater.*, 17, 6053–6059, 2005. Copyright 2005, American Chemical Society.)

In this way, triblock copolymer/Pd/Ni nanofibers were obtained, exhibiting saturation magnetization of c.a. 12.5 emu/g of Ni.

Manners et al. demonstrated that magnetic nanowires could be prepared from cylindrical nanoassemblies of poly(ferrocenyldimethylsilane-*b*-2-vinyl pyridine) (PFS-*b*-P2VP) diblock copolymers (Wang et al. 2009a). Due to the semicrystalline nature of the poly(ferrocenyldimethylsilane) (PFS) blocks, stable assemblies were produced in 2-propanol, with a PFS inner core and a P2VP corona. These could be easily transferred into water via an intermediate quaternization step on the P2VP blocks. The outer corona interacted electrostatically with preformed dextran-functionalized magnetite NPs creating hybrid magnetic nanoassemblies.

Cylindrical polymeric brushes, that is, macromolecules having densely grafted side polymer chains on a linear main chain, have been used as templates for the preparation of magnetic IONPs and hybrid nanoassemblies. Brushes with poly(*n*-butyl acrylate-*b*-sodium acrylate) diblock side chains, where the poly(sodium acrylate) blocks are directly connected to the main chain, were prepared by Muller et al., by polymerization of macromonomers, hydrolysis/neutralization of protected carboxylate groups in a second step (Zhang et al. 2004). Coordination of iron ions with the carboxylate groups, in organic solvents, resulted in hybrid brushes. The iron ions were converted to γ-Fe_2O_3 NPs under alkaline conditions, which were further aligned and fused into 1D objects by the action of the polymeric template.

Poly(ethylene oxide-*b*-ethylenimine) (PEO-*b*-PEI) and poly(ethylenimine-*b*-ethylene oxide-*b*-ethylenimine) (PEI-*b*-PEO-*b*-PEI) double-hydrophilic diblock and triclock linear copolymers were used as the catalysts and the templates for the production of low polydispersity silica NPs (Kind et al. 2010). Tetraethylorthosilicate (TEOS) was used as the SiO_2 precursor. Its addition to slightly acidic aqueous solutions of the copolymers resulted in the formation of near monodisperse silica NPs with diameter ca. 30 nm, after rather long reaction time (~1 week). This was possible due to the catalytic action of PEI amine functionalities on TEOS hydrolysis. Aggregation of the NPs was observed. The degree of aggregations as well as the size of NPs assemblies formed was influenced by the solution pH.

4.5 HYBRID NANOSTRUCTURES FROM PREFORMED INORGANIC NANOPARTICLES AND BLOCK COPOLYMERS

An alternative route for the production of hybrid inorganic–polymeric NPs is to combine preformed inorganic NPs and polymer chains. In this respect, block copolymers present the greater potential in the formation and stabilization of hybrid NPs, although homopolymers have been also utilized as ligands for inorganic NPs as has been discussed previously. Several examples of this approach have been reported in the literature.

In order to prepare magnetic hybrid inorganic–polymeric nanostructures, which could be more useful in biomedical applications, the group of Muller (Xu et al. 2010a) have used a polymeric brush with poly(methacrylic acid-*b*-oligoethylene glycol methacrylate) (PMAA-*b*-POEGMA) double-hydrophilic block copolymers as the side chains. In this case, PMAA chains were directly attached to the backbone chain, forming the core of the molecular brush. Preformed magnetite NPs (ca. 10 nm in diameter) were added to the aqueous solution of the brush and were coordinated to the PMAA core of the brushes. This process resulted in a self-assembled hybrid nanostructure with linearly aligned IONPs within its core.

A linear amphiphilic random terpolymer containing hydrophilic free acrylic acid segments and hydrophobically modified acrylic acid segments (reacted with octylamine and isopropylamine) (Luccardini et al. 2006) has been employed for the water solubilization and stabilization of CdSe/ZnS QDs, originally stabilized with trioctylphosphine oxide (TOPO) ligands. CdSe/ZnS NPs were dispersed in a chloroform solution of the terpolymer. After evaporation of the organic solvent, the solid mixture was dispersed in water. The hybrid NPs thus formed displayed a narrow size distribution and negative surface charge and retained the photophysical properties of the inorganic component.

Lecommandoux and coworkers have reported the preparation of hybrid micelles from amphiphilic poly(butadiene-*b*-glutamic acid) (PB-*b*-PGA) diblock copolymers in aqueous media, incorporating hydrophobically surface-modified magnetic γ-Fe_2O_3 NPs (Lecommandoux et al. 2005, 2006a,b). Hybrid micelles were produced by mixing an aqueous solution of the PB-*b*-PGA copolymer and a dichloromethane solution of the IONPs (Lecommandoux et al. 2005, 2006a) (Figure 4.10). The obtained dispersions were stable with time. Characterization of the resulting nanostructures by DLS and small angle neutron scattering (SANS) showed that the hybrid nanosystems were larger than the original block copolymer micelles (in the 330–430 nm range, compared to 60–70 nm for the original block copolymer micelles, incorporating ca. 10^4 NPs per aggregate). This also indicated that the incorporation of IONPs modifies the organization of block copolymer chains in solution. Nevertheless, the hybrid nanostructures had a spherical shape, and they incorporated several IONPs (of ca. 8 nm diameter each) in their volume. The higher volume fraction of ferrofluid achieved inside the micellar self-assembled structures reached 45%.

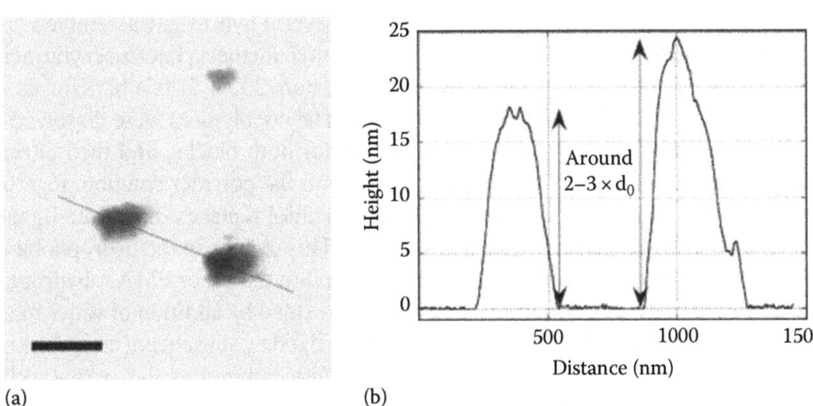

FIGURE 4.10 (a) AFM picture of PB48-*b*-PGA114 copolymer micelles loaded with NPs; scale bar, 400 nm. (b) Height profile along the line shown on the left image. (Reprinted with permission from *J. Magn. Magn. Mater.*, 300, Lecommandoux, S., Sandre, C.O.F., Rodriguez-Hernandez, J., and Perzynski, R., 71–74, Copyright 2006, with permission from Elsevier.)

In a relevant study reported by the same group (Lecommandoux et al. 2006b), different ferrofluids were utilized in conjunction with PB-*b*-PGA block copolymer micelles. The structural properties of the hybrid entities obtained in each case, as well as the dispersibility and the stability of the dispersions, were found to depend on the composition of the copolymer, the nature of the ferrofluid, and the pH of the medium. The dispersions were found to be responsive to external magnetic fields, as well as to changes in the solution pH and ionic strength due to the presence of the polypeptide block.

Nasongkla and coworkers have used maleimide end-functionalized poly(ethylene glycol-*b*-D,L-lactide) (PEG-*b*-PLA) amphiphilic block copolymers in order to encapsulate hydrophobically surface-modified IONPs, together with the anticancer drug doxorubicin (DOX) (Nasongkla et al. 2006). The hybrid magnetic NPs acted as nanocarriers of DOX, having also targeting functionalities for cancer treatment, due to the covalent attachment of cRGD factor on the hybrid micelle surface through the maleimide functionalities of the PEG corona.

Poly(ethylene oxide-*b*-dimethylaminoethyl methacrylate) (PEO-*b*-PDMAEMA) block copolymers, synthesized by a combination of anionic ring-opening polymerization (ROP) and atom transfer radical polymerization (ATRP), were electrostatically grafted on AuNPs (Miyamoto et al. 2008). Copolymers with a different length of PDMAEMA block have been utilized for the stabilization of the AuNPs. The hybrid NPs prepared under acidic conditions showed aggregation after several hours of preparation. On the contrary, the Au NPs/PEO-*b*-PDMAEMA prepared under strong alkaline conditions (pH > 10) was stable. This behavior was attributed to the protonation characteristics of the PDMAEMA blocks. It was proposed that at pH > 10, the PDMAEMA blocks interact with the NP surface via coordination of their tertiary amine groups (and not electrostatically) since at this pH, they are completely deprotonated. It is worth noting that the dispersions containing NPs stabilized with copolymers having short PDMAEMA blocks were more stable, due mainly to the steric effects of the solvated PEO coronas.

Taton and coworkers reported the synthesis of several hybrid self-assembled NPs composed of AuNPs and PS-*b*-PAA or poly(methyl methacrylate-*b*-acrylic acid) (PMAA-*b*-PAA) block copolymers (Kang and Taton 2003, 2005a,b, Kim et al. 2007, Laicera et al. 2007, Shibasaki et al. 2009). The copolymers were dissolved in *N,N*-dimethylformamide (DMF), a good solvent for both blocks, and then citrate-functionalized AuNPs in DMF were introduced into the polymer solution, together with a small amount of 1-dodecanethiol. Dodecanethiol replaces the citrate ligands and renders the AuNP surface more hydrophobic. This change in the hydrophobicity of the AuNP surface allows for their incorporation into the PS or PMAA hydrophobic cores of the block copolymer micelles that are formed by addition of water in the DMF solution. The assembled hybrid micelles were fixed by subsequent cross-linking of the PAA coronas using 2,2′-(ethylenedioxy)bis(ethylamine) as the cross-linking agent and 1-(3-dimethylaminopropyl)-3-ethyl-carbodiimide methiodide as the coupling reagent. An aqueous solution of the hybrid micelles was obtained by dialysis that removed DMF and excess reagents. TEM observations revealed that hybrid micelles containing multiple NPs were obtained when the smaller 4 nm AuNPs were utilized whereas it was possible to obtain block copolymer micelles containing only one-metal NP when AuNPs with diameters larger than 10 nm were used. In the former case, the size destruction of the hybrid colloids was broad. Interestingly the thickness of the polymer shell surrounding a single large AuNP could be fine-tuned by changing the size of the metal NP and the ratio of metal NP to copolymer and by altering the length ratio of the two blocks. It was found that (1) the thickness of the shell increased by increasing the size of the AuNPs (at constant polymer to Au particle surface ratio); (2) shell thickness increased by decreasing the weight ratio of AuNPs, at constant polymer concentration, due to the decrease in the available surface area for block copolymer adsorption; and (3) shell thickness increased when the length of the PS block was increased relative to the PAA block.

Silica NPs were coated with poly(ethylenimine-*g*-ethylene glycol) (PEI-*g*-PEO) graft copolymer through electrostatic interactions (Thierry et al. 2008). The negative-charged surface of silica particles allows adsorption of the positively charged PEI backbone on their surface, while the PEG side chains confer steric stabilization, stealth properties, and increased solubility in water. It is characteristic to note that the hybrid NP coating with proteins, as well as NP aggregation, was minimal when the silica/polymer NPs were incubated in buffer solutions containing human serum albumin and lysozyme.

More complex hybrid nanoassemblies were reported by Winnik et al. (Wang et al. 2007, Zhang et al. 2010a). They utilized PS-*b*-P4VP block copolymer micelles in order to encapsulate poly(3-hexylthiophene) (P3HT) chains in their cores taking advantage of the insolubility of PS and P3HT in the surrounding medium. Different QDs were then successfully connected to the corona of the micelles due to favorable interactions with the P4VP corona blocks. This sequential assembly provided a spatial organization between the "active" components (i.e., light adsorbing and fluorescent QDs and P3HT conductive chains), which however were still able to interact with each other, as steady-state fluorescence spectroscopy experiments indicated.

It is interesting to mention the preparation of hybrid colloidal particles based on the concurrent self-assembly of gold nanowires (Au NWs) and amphiphilic block copolymers in water (Xu et al. 2010). Initially the PS-*b*-PAA block copolymer was dissolved in a DMF/THF/H_2O ternary solvent mixture, and a solution of oleylamine-stabilized Au NWs was added. THF was used as the common organic solvent for the solubilization of Au NWs in the mixed solution and together with DMF promoted the swelling of the PS block in order to accommodate the hydrophobically functionalized Au NWs. Dilution of the aged mixed solution with water and subsequent centrifugation steps resulted in an aqueous dispersion of the hybrid colloidal nanostructures. TEM images indicated the presence of hybrid colloids containing a single Au NW in a springlike coiled state with 5–10 loops. This was a result of the constrains imposed on the nanowires by the micellization process of the block copolymer chains, which formed the shells of the metal wires in the unfriendly aqueous environment.

4.6 HYBRID POLYMERSOMES

The aforementioned examples clearly demonstrate the great variety and different approaches toward the formation of hybrid assemblies based on presynthesized inorganic NPs and block copolymers. Another methodology that is also of significant interest regards the incorporation of inorganic NPs within preorganized block copolymer nanostructures. Encapsulation of inorganic NPs into relatively simple block copolymer assemblies with a more or less compact structure, like spherical block copolymer micelles, is less challenging, due to the relatively easy ways of obtaining such structures and the relative structural stability of the polymeric micelles. In the examples discussed in the previous section, it was shown how such structures can be generally obtained and characterized in terms of structure and properties. The incorporation of inorganic NPs into block copolymer vesicles (also known as polymersomes) is starting to attract special attention, due to the unique structural and functional properties of polymersomes. Although polymersomes are generally considered more stable than the analogous vesicular structures formed by low molecular weight surfactants and lipids, the compartmentalization of such structures, as a result of the existence of a delicate polymeric membrane, may lead to a number of phenomena after interaction with inorganic NPs, including membrane collapse and global morphological transformations of the nanostructure. On the other hand, the nanocompartmentalized structure of polymersomes opens opportunities for different placement of the inorganic NPs. In principle, NPs can be placed inside the polymeric membrane or can bound to the outer or inner surface of the polymersome bilayer or can be simply encapsulated within the central solvent reservoir of the polymersome. All these possibilities can lead to the formation of several different multifunctional and responsive hybrid nanoassemblies. Such nanostructures could be utilized in several application fields related to medicine and nanotechnology, like catalysis in constrained geometries, controlled/targeted/triggered drug and gene delivery, and selective/responsive nanostructures for (bio)sensor applications. The application can be dictated by or selected according to the nature of the incorporated NPs and their synergistic effects with the polymeric part of the nanostructure, in addition to the properties of the polymer component. For a successful construction of such elegant hybrid nanostructures, the size, shape, and surface characteristics of the inorganic NPs, mode of mixing, and NP formation protocol, as well as the possible interactions being active with the block copolymer chains constituting the polymersome membrane, are important factors that have to be taken into account and will determine the formation and the final structural and functional properties of the resulting hybrid nanostructure. Some examples that are discussed in the following illustrate the methodologies and nanomorphologies that can be achieved in such mixed systems.

Armes and coworkers have constructed polymersomes through the self-assembly of a pH-responsive, hydrolytically self-cross-linkable copolymer, poly[ethylene oxide-*b*-(2-(diethylamino)ethyl methacrylate-stat-3-(trimethoxysilyl) propyl methacrylate], [PEO-*b*-P(DEA-stat-TMSPMA)], in THF/water mixtures (Du and Armes 2005a). DEA segments are pH responsive due to the presence of tertiary amine groups, which can also coordinate to inorganic ions. The TMSPMA segments offer the possibility for stabilizing the nanostructures formed in solutions via cross-linking, while the PEO chains ensure solubility of the nanostructures. The characteristics of the polymersomes could be affected and tuned by a number of parameters including initial terpolymer concentration, water content, and terpolymer composition. After polymersomes were formed, with the P(DEA-stat-TMSPMA) blocks forming the

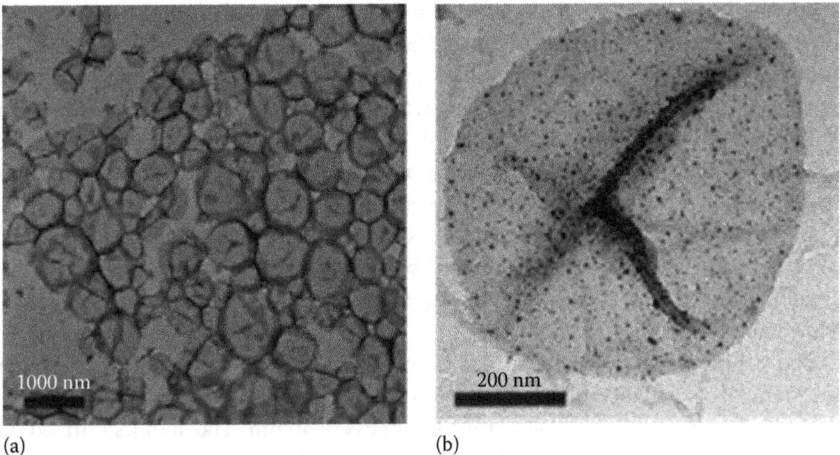

FIGURE 4.11 TEM images of (a) vesicles prepared using the PEO_{43}-*b*-P(DEA_{40}-stat-$TMSPMA_{40}$) copolymer in 1:2 v/v THF/water at an initial copolymer concentration of 40.0 g/L and (b) the same vesicles decorated with AuNPs located solely within the vesicle walls. (Reprinted with permission from Du, J. and Armes, S.P., *J. Am. Chem. Soc.*, 127, 12800–12801, 2005a. Copyright 2005, American Chemical Society.)

inner part of the membrane bilayer, addition of $HAuCl_4$ resulted in protonation of the DEA segments. Reduction of the Au(III) ions with $NaBH_4$ produced AuNPs with size in the 4 nm range. These NPs were found, by TEM observations, to be incorporated within the membrane of the initial polymersomes (Figure 4.11). Although this NP synthesis was utilized for staining the polymersomes for subsequent imaging of their structure, it can also lead to the production of novel gold colloidal catalysts supported on block copolymer vesicles. The pH-sensitive microenvironment and the nanoconfinement of the NPs may also result in fine-tuning of their catalytic properties.

The same group has also reported the synthesis of AuNPs within the walls of polymersomes made of the biocompatible zwitterionic block copolymer poly[2-(methacryloyloxy)ethyl phosphorylcholine-*b*-2-(diisopropylamino)ethyl methacrylate] (PMPC-*b*-PDPA) (Du et al. 2005b). The PMPC block is biocompatible and water soluble in the whole pH range, due to its zwitterionic character, while the poly(2-(diisopropylamino)ethyl methacrylate (PDPA) block confers pH sensitivity to the membrane walls. The tertiary amine groups of PDPA were also used as coordinating sites for Au(III) precursors, which were subsequently reduced to Au NPs. These two examples demonstrate that it is possible to form inorganic NPs in situ within preformed polymersome membranes, under appropriate conditions, and by the right choice of polymeric components.

Binder and coworkers have reported a very systematic study on the incorporation of preformed Au and CdSe NPs into poly(butadiene-*b*-ethylene oxide) (PB-*b*-PEO) polymersomes (Binder et al. 2007). They used NPs with hydrophilic and hydrophobic surfaces. Liposomes were also investigated, but we focus on the results related to polymersomes. It was found that hydrophobically modified inorganic NPs could be incorporated within the polymersome membranes, presumably through hydrophobic interactions with the hydrophobic inner PB blocks of the bilayers. However, stable hybrid polymersomes could not be obtained above a critical amount of NPs. In the case of hydrophilic inorganic NPs, these were found to be evenly distributed within the inner reservoir and the walls of the polymersomes. The authors observed several more specific phenomena due to the incorporation of hydrophobic NPs in the system, like reduction in the polymersome's size and size distribution, a large number of empty polymersomes, and budding effects. It should be mentioned that stabilization of the nanostructures was performed by producing an outer shell of SiO_2 via a sol–gel coating process, thus in reality creating complex hybrid organic–inorganic hollow nanostructures.

Highly fluorescent CdS/ZnS-based, core–multi-shell QDs with a diameter ca. 6 nm were also successfully incorporated within the central region of PB-PEO polymersomes, having a polymer membrane thickness of 16 nm. The authors utilized the film rehydration method supported by sonication and heating at temperatures higher than ambient. They also incorporated dyes within the bilayer using the cosolvent method (Mueller et al. 2008).

Morphological transitions were observed when it was attempted to embed hydrophobically surface-modified magnetic IONPs into the bilayers of PS-*b*-PAA polymersomes (Hickey et al. 2010). TEM images showed the formation of IONP-loaded micelles with increasing amounts of the NPs. The authors proposed a budding mechanism for the transition triggered by the aggregation of IONPs. However, incorporation of magnetic NPs into block copolymer vesicles may present several specific advantages, especially for nanomedicinal applications, and therefore, it is currently pursued by a number of research groups.

In terms of intrinsic physicochemical properties and added or synergistic characteristics, magnetic NPs should be viewed as interesting components of hybrid polymersomes. It is worth emphasizing three of them. (1) Magnetic NPs are generally used as contrast agents in diagnostics involving MRI protocols. (2) Drug and gene nanocarriers containing magnetic NPs can be moved to the infected/targeted organs by application of external magnetic field gradients. This macroscopic property allows for considerable active targeting efficiency, in the microscale and nanoscale, without the need for implementing specific, tedious, and costly chemical functionalization schemes on the selected nanocarrier. (3) By using ultrasmall superparamagnetic IONPs as the inorganic functional component of multifunctional nanocarriers, tumors can be thermo-ablated by hyperthermia, making use of the heating up effect of the NPs under the influence of a localized high-frequency magnetic field. The latter mechanism can also be exploited as a triggering stimulus for releasing an encapsulated drug. In this case, the nanocarrier should be designed in such a way that it can undergo a structural transition because of changes in the local temperature around it. Such temperatures can be the main-chain transition, T_m, from gel to a fluid state for a lipid membrane in the case of liposomes or a LCST transition for a thermosensitive polymer chain (e.g., PNIPAM) that can be part of the polymeric component. This specific approach is named magneto-chemotherapy, due to the synergistic effects of the different components of the nanocarrier, and is starting to gain attention for overriding several drug delivery drawbacks in the nanomedicine field.

The principles and assets described so far concerning the incorporation of magnetic NPs into vesicles, in relation to drug delivery applications, have been already demonstrated with liposomes (Chen et al. 2010, Pradhan et al. 2010), and surely can be extended successfully to polymersomes with added advantages.

Nanosystems based on hybrid polymersomes with embedded magnetic NPs (presynthesized independently, in the absence of the block copolymer component) have been developed and characterized in terms of structure and function by the Lecommandoux group (Lecommandoux et al. 2005, 2006a,b). IONPs having a hydrophobic coating and with a diameter close to 8 nm were loaded into the hydrophobic part of the vesicles membrane, formed by poly(butadiene-*b*-L-glutamic acid) (PB-*b*-PGA) block copolymers. The thickness of the hydrophobic membrane was estimated to 14 nm via SANS measurements, which correlates well with the diameter of the NPs incorporated inside the membrane. Geometrical factors may be important in such cases and should be taken into consideration for the construction of the hybrid nanostructures. Anisotropy 2D SANS measurements in the presence of a magnetic field demonstrated that it was possible to modify the shape of the vesicles membrane even by the use of rather low external magnetic fields (Lecommandoux et al. 2005). This is an additional factor that can influence the structure and properties of the nanostructures, and in this way, their functionality for biomedical drug delivery oriented applications. The deformation of vesicles under a magnetic field turns out to be an interesting way to trigger the release of entrapped molecules. This is due to the possible formation of pores within the membrane, as a result of the tension created during the deformation. However, the authors did not present any experimental data that could prove the occurrence of such a process.

Along the same lines, Forster and coworkers successfully incorporated magnetic NPs into polymersomes made from the self-assembly of poly(isoprene-*b*-ethylene oxide) block copolymers (Krack et al. 2008). They found that the IONPs were located at the PI/PEO interface. This could be due to the larger size of the inorganic NPs utilized (ca. 14 nm diameter) compared to the membrane thickness (ca. 12 nm), indicating once again the crucial role of size matching between inorganic NPs diameter and membrane thickness. As a result of the interfacial placement of IONPs, extensive bridging of adjacent bilayers was observed, via cryo-TEM imaging, leading to the formation of multi-lamellar vesicles (onion like). The loading capability of the polymeric nanostructures in respect to the incorporated IONPs was high enough in order to induce a large magnetophoretic mobility of the hybrid vesicles as it was evidenced by video microscopy, and this characteristic is significant for biorelated applications of these nanostructures.

Recently, Lecommandoux et al. reported on the preparation of hybrid polymersomes with the incorporation of hydrophobically modified maghemite (γ-Fe_2O_3) NPs in the membrane of poly(trimethylene carbonate-*b*-L-glutamic acid) (PTMC-*b*-PGA) block copolymer vesicles (Sanson et al. 2011). They have utilized a nanoprecipitation process, which involved the addition of aqueous buffer in the block copolymer or block copolymer/IONPs solutions in DMSO or THF, under stirring. Besides choice of organic solvent, duration of buffer injection, as well as order of addition, was important in controlling the size and polydispersity of polymersomes. The resulting hybrid vesicles had a size in the 100–400 nm range and a considerably narrow size distribution and were highly magnetic (it was possible to incorporate up to 70%w/w of magnetic NPs inside the polymeric structures). More importantly, their shape could be deformed under a relatively low-intensity magnetic field (ca. 0.1 T), and their magnetophoretic mobility was high enough to allow their use for magnetic field directed location in vivo. These magnetic polymersomes exhibited a long-term colloidal stability and showed an important contrast enhancement in MRI experiments with a particularly low (subnanomolar) detection limit. DOX was also incorporated in the hybrid polymersomes via the nanoprecipitation protocol. The release rate of encapsulated DOX could be enhanced twice via the application of an external RF oscillating magnetic field at 500 kHz. This was made possible, presumably, by the melting of the semicrystalline PTMC hydrophobic blocks, which in turn was a result of the local hyperthermia effect taking place only at the scale of the vesicles membranes, due to the nanoscale placement of the IONPs.

Another type of hybrid polymersomes was reported by the group of Eisenberg. Metal NPs stabilized by the block copolymer coronas were incorporated into block copolymer vesicles via a mixing protocol that resulted in the co-assembly of the components (Mai and Eisenberg 2010). In this particular work, PEO-*b*-PS-*b*-PAA hybrid block copolymer micelles having Pb ions in their PAA coronas (which act as cross-links between the PAA chains making the corona more rigid) and AuNPs with end-grafted PS-*b*-PAA chains (via the PS end on the AuNP surface) were incorporated within the polymersome membranes of PS-*b*-PEO diblock copolymers (Figure 4.12). This was possible due to the compatibility of the PS blocks present on the AuNPs or in the cores of the hybrid PEO-*b*-PS-*b*-PAA(Pb) micelles and the PS blocks present in the walls of the PS-*b*-PEO vesicles. These expectations were also confirmed by TEM observations that showed the spatial placement of the inorganic components within the inner domains of the vesicles' membrane (NPs were actually located in the central 10%–20% region of the polymersome walls), where polymeric hydrophilic components were seen in the outer rims of the walls

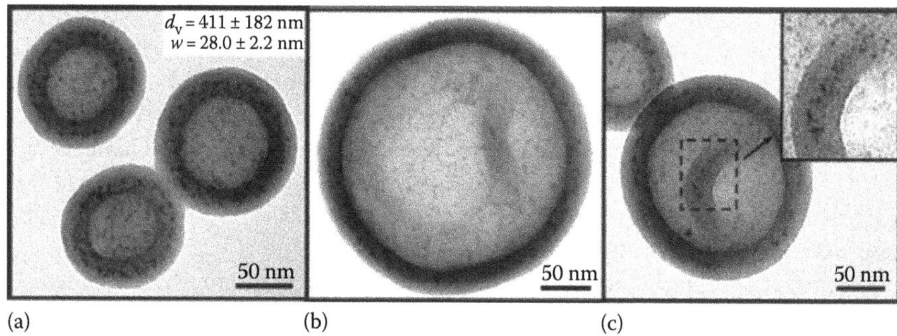

FIGURE 4.12 TEM micrographs of the vesicles with incorporated Pb-based inorganic NPs prepared from the combined solution of the $PS2_{35}$-*b*-PEO_{45} copolymer (0.5 wt.%) and the PEO_{45}-*b*-PS_{155}-*b*-$P(APb)_{25}$ micelles (0.5 wt.%). (a) Small vesicles, (b) a large vesicle, and (c) an indented vesicle; the magnified indentation of the vesicle is shown in the inset. (Reprinted with permission from Mai, Y. and Eisenberg, A., *J. Am. Chem. Soc.*, 132, 10078–10084, 2010. Copyright 2010, American Chemical Society.)

on both sides of the membrane, that is, no particles were found on the inner or the outer interface of the membrane.

Closing this section, it is important to refer to another type of hybrid polymersomes: those formed by the self-organization of amphiphilic gold nanocrystals (spherical NPs and nanorods) surface functionalized by two chemically different end-grafted polymers of hydrophilic and hydrophobic nature (Song et al. 2011). These amphiphilic polymer-functionalized gold nanocrystals are analogous to block copolymers. The vesicular structures are formed due to the amphiphilicity and the geometrical packing requirements imposed by the "molecular" structure of the nanocrystals on the self-assembled structures. Plasmonic properties are thus imprinted in the hybrid vesicles through the structure of the building blocks. The nanostructures are responsive in the sense that the disruption of the plasmonic vesicles can be triggered by stimulus mechanisms inherent to either the grafted polymers or the gold nanocrystal. The synthesis of these grafted NPs in the next section since it involves the use of surface-initiated polymerization (SIP) methodologies.

4.7 SURFACE-MODIFIED INORGANIC NANOPARTICLES BY GRAFTED POLYMER CHAINS

Modification of surfaces with polymer chains has received great attention in an effort to tune the wettability, biocompatibility, and friction of the surfaces, as well as to improve their resistance to corrosion. Surface modification by grafted polymer chains offers possibilities for creating responsive surfaces (also termed "smart surfaces") since the properties and the response of the attached polymer layer may be triggered by external stimuli, like temperature, pH, ionic strength, and UV light. Block copolymers and polyelectrolytes can be successfully utilized for this purpose due to their stimuli-responsive properties stemming from their chemical structure and design. Inorganic NP surfaces can be functionalized/modified by tethering onto or from polymer chains (Advincula 2003, Neoh and Kang 2011). As has been noted in several cases metal, semiconductor and magnetic NPs are interesting due to their electronic, optical, magnetic, and catalytic properties, and their combination with polymeric components enhances the properties and their application potential. Surface functionalization of inorganic NPs facilitates their dispersion in organic and aqueous media and tailors the properties of their surfaces in terms of interactions with their environment. Interactions are particularly important in the case of NPs in biological fluids, like the bloodstream, when they are used as drug delivery nanovehicles or as contrast agents for imaging or as hyperthermia agents. In these cases, the polymeric layer should also provide stealth properties to the NPs in order to avoid their "detection" as foreign bodies by the macrophages and other cells of the reticuloendothelial immune defense system of the body. Escape from detection increases the circulation time of the NP in the bloodstream and prolongs the beneficial action of the NPs. Therefore, by following surface modification methodologies, hybrid inorganic–polymeric NPs are created, which present combined and in most cases designed properties, suitable for advanced applications.

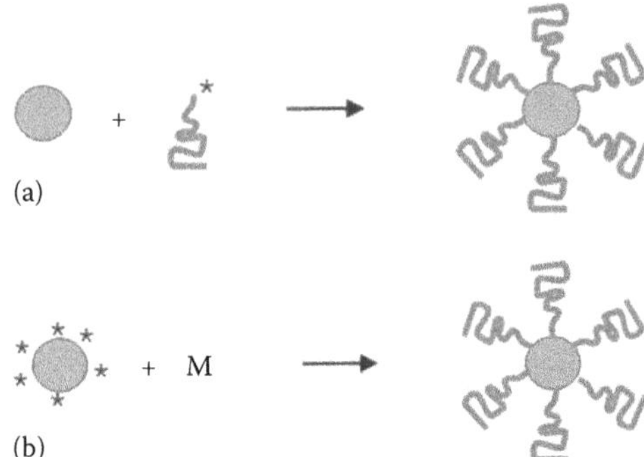

FIGURE 4.13 "Grafting to" (a) and "grafting from" (b) approaches for the surface functionalization of NPs with polymer chains.

There are two general approaches for the chemical functionalization of NPs surfaces with polymer chains. In the "grafting to" approach, end-reactive polymers can react with functional groups on the surface, immobilizing the polymer chain on the surface by one of its ends, thus producing a polymer brush around the inorganic NP (Figure 4.13). The end-reactive polymer chain can be a "living" polymer that is created in the course of a living/controlled polymerization scheme, and the reactive end is the one used for the polymerization of the monomer(s) and at the same time ensures effective reaction with the NP's surface. Alternatively, a polymer with an end-functional group can be created via a polymerization and a post-polymerization reaction sequence, the end-functional group being chosen such as to be able to react with the surface groups of the NP.

The "grafting to" approach is a rather straightforward and simple technique, in the sense that it requires the formation of an end-reactive polymer chain in solution, which can be isolated and characterized independently, and in a second step, its reaction with the surface takes place. The polymer chain in principle can have one of all different architectures: linear, branched, graft, block structure, etc. However, there are some disadvantages in this methodology: (a) High grafting densities cannot be achieved, mainly due to the steric hindrance exercised by the initially grafted chains toward the new incoming ones. (b) Brush thickness is directly related and limited by the molecular weight of the polymer utilized in the reaction solution. "Grafting from" methodologies involve the creation of active sites on the surface, from which monomer present in the reaction solution in contact with the surface can be polymerized, forming the desired polymer layer/brush. In this case, diffusion of a low molecular weight monomer toward the active chain ends is required in order to extend the tethered chains, which is much easier than polymer chain diffusion. This approach has also been termed surface initiated polymerization (SIP) (Advincula 2003, Edmondson et al. 2004) and has received much attention in recent years, due to its great versatility in producing a variety of polymer brushes and surfaces modified with a variety of polymeric overlayers.

It has been demonstrated that controlled/living polymerizations can be initiated by initiator molecules bound on curved colloidal NP and planar solid surfaces. By choosing the appropriate functionalization reactions, essentially all surfaces can be functionalized by initiators, and by judicious choice of conditions of the polymerization, a wide range of monomer functionalities can be incorporated in a polymeric brush/layer in a controllable manner. All known solution living/controlled polymerization mechanisms have been employed for the preparation of tethered homopolymer and block copolymer chains (Advincula 2006, Brittain et al. 2006, Buchmeiser 2006, Tsujii et al. 2006). The living/controlled characteristics of such polymerization systems allow for complete utilization of essentially all the advantages of a traditional solution living/controlled polymerization reaction scheme, including molecular weight control, low polydispersity, synthesis of compositionally uniform block copolymers, polyelectrolyte functionalities, and polymer brushes with end functionalities, together with the ability to control brush density. However, it seems that controlled radical polymerization schemes are more successful in the preparation of thick and dense polymer brushes on surfaces. This may be a consequence of the kinetics of radical polymerization processes, although sometimes sacrificial amounts of initiator, present in solution state, are necessary to achieve controlled polymerizations and high molecular weights. This may require additional purification and manipulation steps after brush synthesis due to the possible formation of homopolymers in solution as well as to incomplete consumption of monomer(s).

The literature is rich in examples where hybrid inorganic–polymeric NPs are produced via surface modification of preformed inorganic NPs with grafted polymer chains. In the following, we present some relevant works starting with the ones based on the "grafting to" methodology and continuing with works based on the "grafting from" technique.

Thiol groups have been widely used for the attachment of small molecules and polymers on AuNPs. PEG homopolymers with -SH end groups have been grafted on AuNPs and nanorods for increasing their biocompatibility and circulation time in body fluids (Niidome et al. 2006, 2009, Kim and Taton 2007, Akiyama et al. 2009). The length of PEG-grafted chains was found to play a crucial role in the circulation time and biodistribution of gold nanorods (Akiyama et al. 2009). Changes in the molecular architecture of the grafted chains were also found to significantly influence the circulation time of gold nanorods. Gold nanorods grafted with branched PEGs had a larger circulation time compared to nanorods grafted with linear chains (Tong et al. 2009). The effects may be related to the steric shielding of the different macromolecular components, since a poor shielding effect allows protein binding (i.e., enhanced opsonization) and increased phagocytosis by RES. Surface grafting of PEG chains using thiol chemistry has been also reported for gold nanoshells (Kah et al. 2009).

Surface grafting of preformed AuNPs with thermosensitive random copolymers has been also demonstrated (Boyer et al. 2009a). Copolymers based on di(ethylene glycol) ethyl ether acrylate] (DEGA), a monomer that forms a thermoresponsive homopolymer, and oligoethylene glycol ethyl ether acrylate (OEGA) were prepared by reversible addition–fragmentation chain transfer polymerization (RAFT) copolymerization. P(DEGA-*co*-OEGA) copolymers with different compositions were obtained. The trithiocarbonate end group of these copolymer shows high affinity for gold surfaces, and therefore, it was used for the surface grafting of the copolymers on the AuNPs. The copolymers conferred both thermosensitivity and antifouling properties to the hybrid NPs as it was deduced from the decreased adsorption of bovine serum albumin onto the particles. In an extension of this approach, two types of copolymers thermosensitive and charged ones were co-grafted on the AuNPs (Boyer et al. 2010a).

Block copolymer chains have been also grafted on AuNPs. Eisenberg et al. reported the synthesis of AuNPs carrying PS-*b*-PAA block copolymer chains through a thioacetate ester end group placed at the PS terminus of the copolymer chain (Mai and Eisenberg 2010). Therefore, these hybrid NPs can be viewed as star-block copolymers having a AuNP core, a PS inner shell, and a PAA corona, which renders them water soluble.

The hydroxyl groups present on the surface of iron oxide magnetic NPs can be utilized for the covalent fixation of a number of polymer chains having suitable end groups that can react with hydroxyl functionalities (Boyer et al. 2010b). In such cases, PEG is the polymer of choice mainly due to the stealth properties and the water solubility that imparts to the NPs. End functionalities may include trimethoxysilane groups, carboxyl groups, dopamine, primary amine groups, cysteine, and phosphonic acid groups (Zhang et al. 2002, Xu et al. 2004, Zhang and Zhang 2005, Sun et al. 2006, Boyer et al. 2009b, 2010b). Reaction conditions can be well controlled in most cases, and the grafting density of polymer chains on the NP surface can be varied through changes in the molar ratios of the surface/polymer end groups.

Other polymers can be also utilized. α-Phosphonic acid-ω-dithiopyridine-functionalized poly(oligoethylene glycol acrylate) (POEGA) was synthesized via RAFT by the use of a suitable trithiocarbonate chain transfer agent bearing a dimethyl phosphonate group. The dimethyl phosphonate group was transformed to phosphonic acid group via reaction with bromosilane, while the trithiocarbonate group was transformed to pyridyl disulfide group via aminolysis, after the polymerization step. The phosphonic acid group provided strong anchoring of the POEGA chains on IONPs surface, while the pyridyl disulfide group was used for the conjugation of biomolecules (Boyer et al. 2009b). Following a similar reaction scheme, phosphonic acid–terminated poly(dimethylamino ethyl acrylate) (PDMAEA) chains were synthesized and used in conjunction with POEGA chains of the same end functionality for the preparation of IONPs bearing both types of polymeric chains grafted on their surface. The POEGA chains were used for screening the positive charges of the PDMAEA chains in order to increase biocompatibility of the hybrid NPs. The positively charged PDMAEA chains can be utilized for the complexation of DNA, RNA, or oligonucleotides in an effort to create hybrid multifunctional NPs for gene delivery applications. Silica NPs can be easily functionalized with PEG chains following similar approaches as the ones discussed for IONPs (Feng et al. 2009, He et al. 2010).

Grafting from methodologies has also been utilized for end-grafting polymer chains on inorganic NPs. Silica NPs are the most frequently used NPs in order to demonstrate the feasibility of controlled polymerization surface-initiated methodologies for the polymer functionalization of NPs.

ATRP initiators, including 2-(4-chloromethylphenyl) ethyl) dimethylethoxysilane (CPTS), (3-(2-bromoisobutyryl) propyl) dimethylethoxysilane (BPDS), and (3-(2-bromopropionyl) propyl) dimethylethoxysilane (BIDS), were immobilized onto silica NPs via reaction of the alkoxysilane moieties of the initiators molecules with the surface hydroxyl groups of the particles (Von Werne and Patten 1999). Styrene and methyl methacrylate (MMA) were then polymerized giving silica NPs grafted with PS and poly(methacrylic acid) (PMMA) chains. A less than 100% initiating efficiency was observed. Polymerizations were controlled at least for the smaller 75 nm particles. For larger particles (ca. 300 nm), the addition of free initiator or deactivator was necessary in order to achieve a controlled polymerization. Grafting density of initiator molecules was found to increase by increasing the size of the silica NPs.

Armes and coworkers utilized surface-initiated ATRP for attaching hydrophilic polymer chains on the surface of silica NPs (Wang et al. 1999, Wang and Armes 2000). Polymerizations took place in aqueous media and poly(oligo(ethylene glycol) methacrylate) (POEGMA) and poly(2-(*N*-morpholino)ethyl methacrylate) (PMEMA) thermosensitive homopolymer chains could be fixed on the particle surface. The hybrid NPs showed temperature-dependent aggregation/precipitation depending on the nature of the surface grafted polymer. POEGMA-grafted silica particles were water soluble for temperatures in the 25°C–60°C range, while PMEMA-grafted ones showed aggregation at temperatures higher than 34°C (the temperature corresponding to the LCST of the PMEMA chains).

The SIP of sodium 4-styrenesulfonate, sodium 4-vinyl benzoate, 2-(dimethylamino)-ethyl methacrylate (DMAEMA), and 2-(diethylamino)ethyl methacrylate (DEAEMA) in protic solvents, from suitably functionalized silica NPs, allowed the attachment of polyelectrolyte chains (Chen et al. 2003). The grafted polyelectrolytes provided electrosteric stabilization to the NPs at the pH range where polymer chains are negatively or positively charged, depending on the chemical nature of the chain. Thus, hybrid NPs with pH-sensitive colloidal stability were formed.

Grafting poly(*n*-butyl acrylate) (PnBA) chains on silica NPs provided NPs dispersible in organic solvents, while the low T_g of the PnBA shell led to the formation of hybrid films with desirable elastomeric properties (Carrot et al. 2001). Organic solvent dispersible silica NPs were prepared by grafting from long PMMA chains through ATRP methodologies (Boettcher et al. 2000).

Matyjaszewski et al. have reported the synthesis of hybrid NPs by the use of surface-initiated ATRP of styrene and several (meth)acrylates from colloidal silica particles functionalized with 2-bromoisobutyrate groups (Pyun et al. 2003). They presented a detailed study on the kinetics of polymerization under such conditions. Styrene, *n*-butyl acrylate, and MMA monomers were employed in order to obtain hybrid NPs having surface-attached PS-*b*-PnBA, PMMA-*b*-PnBA, and PnBA-*b*-PMMA block copolymer chains with controlled molecular weights and compositions and low molecular weight distributions. Characterization of the grafted chains was possible after hydrolysis of the silica cores by hydrofluoric acid. TEM observations revealed the core/shell structure of the resulting NPs.

Silica particles functionalized with thermosensitive and photoresponsive block copolymer chains have been reported by Liu et al. (Wu et al. 2009). Densely grafted thermoresponsive PNIPAM brushes with inner and outer layers selectively labeled with fluorescence resonance energy transfer (FRET) donors, 4-(2-acryloyloxyethylamino)-7-nitro-2,1,3-benzoxadiazole (NBDAE) dye molecules, and photoswitchable acceptors, 10-(2-methacryloxyethyl)-30,30-dimethyl-6-nitro-spiro(2*H*-1-benzo-pyran-2,20-indoline) (SPMA) molecules, respectively, were prepared by surface-initiated sequential ATRP (Figure 4.14). Due to the majority of *N*-isopropylacrylamide segments, the P(NIPAM-*co*-NBDAE)-*b*-P(NIPAM-*co*-SPMA) brushes were found to collapse in the temperature range of 20°C–37°C. The FRET process between NBDAE and SPMA groups was observed after UV irradiation of the aqueous dispersion of hybrid NPs, which induces the transformation of SPMA moieties in the outer layer of the polymer brushes from nonfluorescent spiropyran (SP) form to the fluorescent merocyanine (MC) form. The FRET efficiency was found to be coupled to the thermally induced conformational changes of PNIPAM-labeled chains. In particular, it could be effectively tuned by the thermoinduced collapse/swelling of P(NIPAM-*co*-NBDAE)-*b*-P(NIPAM-*co*-SPMA) brushes through changes in the relative distances between donor and acceptor moieties fixed in the respective layers of the polymer brushes. In a next step, when the hybrid NP dispersion was irradiated with visible light again after UV irradiation, the MC form of SPMA moieties reverts back to the nonfluorescent SP form, leading to a seizure of the FRET process. It was postulated that such hybrid silica/polymer NPs can serve as sensitive ratiometric nanoscale fluorescent thermometers with a turn-on/turn-off ability, based on the right conditions of light irradiation and temperature of the system.

Muller and coworkers used surface-initiated self-condensing ATRP for the fixation of a hyperbranched polymer on silica NPs (Mori et al. 2002). An α-bromoester-type initiator carrying a chlorosilane end group was covalently attached on the NPs. An acrylic-type AB* inimer, 2-(2-bromopropionyloxy)ethyl acrylate (BPEA), containing a polymerizable double bond and an initiating 2-bromopropionyl group were then polymerized under self-condensing ATRP conditions. Well-defined surface attached were obtained. The use of *tert*-butyl acrylate as a comonomer in the polymerization step resulted, after hydrolysis, in the formation of hybrid silica/polymer NPs carrying hyperbranched water-soluble PAA chains.

Zhao and coworkers succeeded in the preparation of silica NPs with two chemically different surface-attached chains, namely, PS and PAA (Li et al. 2005a). A dual functional initiator able for promoting ATRP and nitroxide-mediated polymerization was attached on the NPs through a chlorosilane functionality. First, poly(*t*-butyl acrylate) chains were grown by ATRP, followed by the growth of PS chains by NMP in a controlled manner. PtBA chains were hydrolyzed to PAA. Thus, hydrophilic and hydrophobic chains were simultaneously attached on the same inorganic particle, as proven by NMR characterization. The obtained hybrid particles are amphiphilic, and the mixed brushes can reorganize in different solvent environments.

RAFT polymerization techniques were also used for the grafting of polymer chains onto silica NPs but admittedly to a lesser extend compared to ATRP techniques. Benicewicz et al. utilized a RAFT agent carrying a silane group that allowed its covalent fixation on the surface of silica NPs (Li and Benicewiez 2005b). The surface-functionalized NPs were used for the polymerization of styrene and *n*-butyl acrylate. In this way, silica NPs having grafted polystyrene, PnBA homopolymers, and PS-*b*-PnBA diblock copolymer chains were prepared, under controlled polymerization conditions.

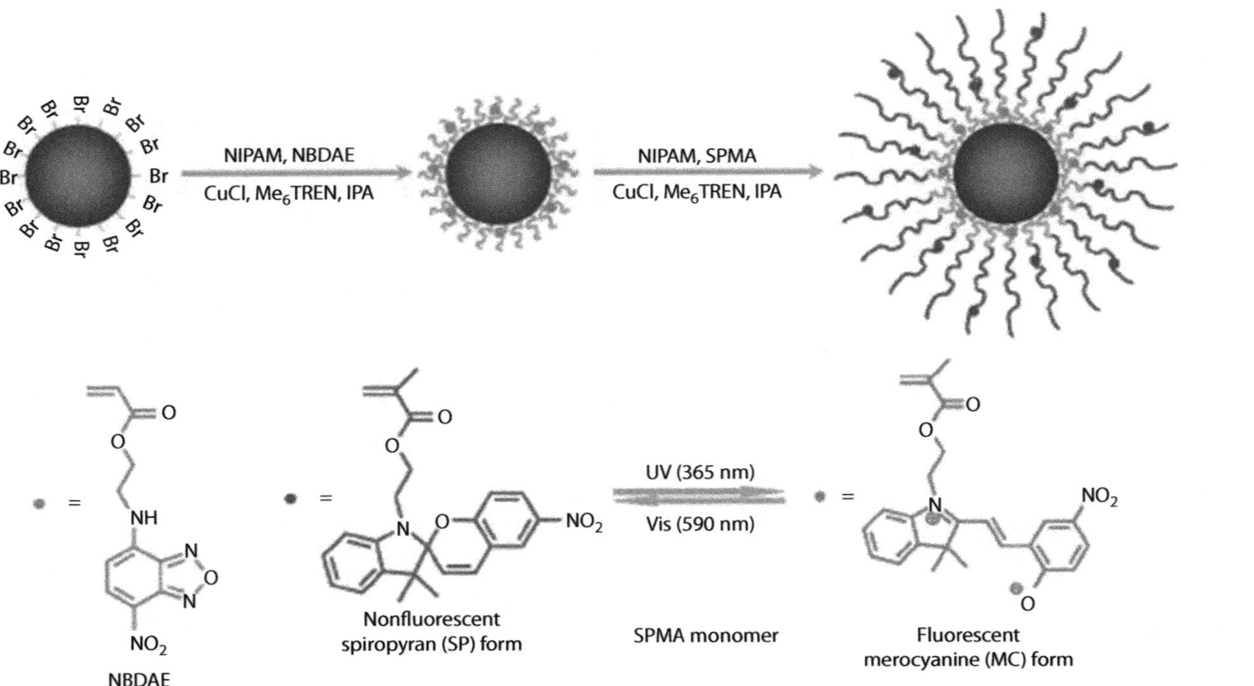

FIGURE 4.14 ATRP synthesis of hybrid silica NPs coated with thermoresponsive P(NIPAM-*co*-NBDAE)-*b*-P(NIPAM-*co*-SPMA) brushes with the inner and outer layers selectively labeled with FRET donor (NBDAE) and photoswitchable acceptor (SPMA) residues, respectively. (Reprinted with permission from Wu, T., Zou, G., Hu, J., and Liu, S., *Chem. Mater.*, 21, 3788–3798, 2009. Copyright 2009, American Chemical Society.)

Surface-initiated strategies for grafting polymers on inorganic NPs based on nitroxide-mediated polymerization have also been reported. The alkoxyamine initiator, *N*-*tert*-butyl-*N*-(1-diethylphosphono-2,2-dimethylpropyl)nitroxide (DEPN), was functionalized by a silane group in order to be covalently immobilized on the surface of silica NPs, up to a relatively high grafting density (Bartholome et al. 2003). Then the NMP of styrene was performed, in the presence of free alkoxyamine initiator in order to achieve a better controlled polymerization process. PS chains with predicted molecular weights and relatively narrow size distributions were grafted on the silica NPs.

Polymer surface-functionalized silica NPs were prepared by surface-initiated anionic polymerization techniques (Advincula et al. 2002, Zhou et al. 2002). In this case, diphenylethylene molecules were grafted on the silica NPs, and then they were activated by BuLi in order to produce surface anionic initiating moieties that could be used for the anionic polymerization of styrene, isoprene, and other monomers. Production of surface-attached block copolymers was also possible.

ROP initiated from the surface of silica NPs was utilized for grafting biocompatible and biodegradable poly(ε-caprolactone) (PCL) chains on them (Carrot et al. 2002). The surface of silica NPs was functionalized with a trimethoxysilane agent carrying an amine functionality, which then was activated by an aluminum alkoxide in order to promote polymerization. Well-defined hybrid NPs were obtained that could be dispersed in organic solvents for the PCL shell.

A "grafting through" approach was employed by Shipp and coworkers in order to produce hybrid silica/polymer core/shell NPs (Chinthamanipeta et al. 2008). The surface of the silica core was functionalized with 3-(methacryloxypropyl) dimethyl chlorosilane, introducing vinyl double bonds on the surface. RAFT polymerization of MMA produced PMMA-grafted chains on the silica NPs. In these hybrid NPs, most probably each PMMA chain has several anchors on the surface.

In another approach, silica NPs surface functionalized by ATRP initiators were produced by the co-condensation reaction of [3-(2-bromoisobutyryl) propyl] triethoxysilane and [3-(2-bromoisobutyryl)-propyl] ethoxydimethylsilane, tetraethoxysilane. The reaction yields silica NPs having diameters in the range 10–300 nm and 2-bromoisobutyryl surface functionalities that were utilized as ATRP initiating sites. Styrene, *tert*-butyl acrylate, and methyl acrylate were successfully polymerized to produce hybrid NPs with grafted homopolymer chains (Radhakrishnan et al. 2008).

Brittain and Ranjan employed a specially designed RAFT chain transfer agent, which allowed the utilization of both RAFT polymerization and click chemistry, in the surface functionalization of silica NPs (Ranjan and Brittian 2007). They were able to modify the surface with polystyrene and polyacrylamide chains. In the first step, azide-functionalized NPs were prepared, and then the PS and PAcA homopolymer chains carrying alkyne functionalities, due to the use of the special RAFT agent in the polymerization step, were attached to the surface via click reaction. Using a similar RAFT–click chemistry combination protocol, the same authors were able to graft poly(styrene-*b*-methyl methacrylate) diblock copolymer chains onto silica NPs (Ranjan and Brittain 2008). The initially silica grafted PS-based macro-CTA agent was used in the polymerization of MMA, giving silica NPs with an inner shell of PS blocks and an outer shell of PMMA blocks (Figure 4.15).

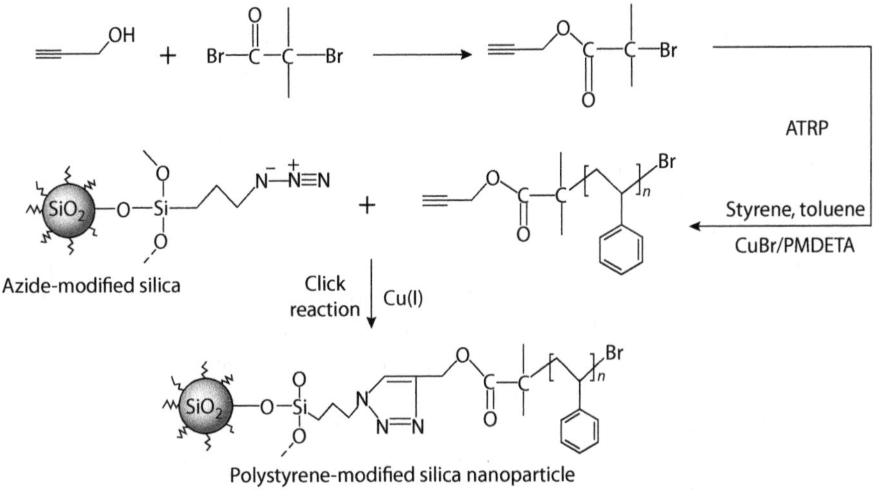

FIGURE 4.15 Surface modification of silica particles using a combination of ATRP and click chemistry. (Reprinted with permission from Ranjan, R. and Brittain, W.J., *Macromolecules*, 40, 6217–6223, 2007. Copyright 2007, American Chemical Society.)

PS chains were grafted from TiO_2 NP surfaces using the ATRP technique (Wang et al. 2010). Initially TiO_2 NPs were coated with a trimethoxysilane containing a 2-bromo-2-methylpropyl group. Then polymerization of styrene took place in anisole in the presence of CuBr and PMDETA. The molecular weight of the grafted chains was well controlled, and the molecular weight distribution was narrow. TEM images revealed the core/shell structure of the hybrid NPs.

Thermoresponsive hybrid gold/PNIPAM NPs were prepared by surface-initiated ATRP of NIPAM (Kim et al. 2005b). The AuNPs surfaces were functionalized by the disulfide initiator $[BrC(CH_3)_2COO(CH)_{11}S]_2$ and were utilized as multifunctional initiating centers for the polymerization of the monomer. Well-separated hybrid NPs with a AuNP core carrying PNIPAM brushes were observed by TEM. SIP of NIPAM was also carried out in the presence of ethylene diacrylate as a cross-linker, at variable ratios, leading to the synthesis of AuNPs with cross-linked PNIPAM shells. The authors observed that the cross-linking affected the shrinkage of the corona as temperature was increased through the LCST transition of the PNIPAM chains, compared to the case of non-cross-linked brushes. Brushes with cross-links were less contracted at higher temperatures.

AuNPs stabilized with a thiol ligand bearing an amine end group were used for the ROP of β-benzyl-L-aspartate *N*-carboxy anhydride (BLA-NCA) (Prabaharan et al. 2009), giving AuNPs functionalized with polyaminoacid chains. The amine end groups of PBLA were used for the covalent attachment of PEG chains. In this way, AuNPs carrying PBLA-*b*-PEG chains were produced. The hydroxyl end groups of the PEG blocks were transformed to folate acid groups for active targeting, while DOX was covalently attached to the side groups of PBLA, after an intermediate functionalization step with hydrazine. The multistep synthetic route resulted in multifunctional hybrid NPs with biofunctional targeting moieties and anticancer drug as a cargo.

Poly(oligoethylene glycol methacrylate) (POEGMA) chains were fixed on IONPs via an SIP methodology (Hu et al. 2006). Initially the 4-(chloromethyl)phenyl trichlorosilane ATRP initiator was immobilized on the surface of the NPs via reaction of the trichlorosilane moiety with the surface hydroxyl groups. Then OEGMA monomer was polymerized by the surface-bound initiator moieties through copper-catalyzed ATRP. The hybrid NPs thus obtained were well dispersible in aqueous media, and the presence of PEGMA thick layer reduced the interaction with macrophages.

Hyperbranched polyglycerol (HPG) was also grafted on the surface of magnetic IONPs by SIP of glycidol. HPG is biocompatible and protein resistant like linear PEG (Siegers et al. 2004, Wang et al. 2008b). Two different polymerization approaches were followed (Khan and Huck 2003, Wang et al. 2008, 2009b). In the first approach, 3-mercaptopropyltrimethoxy silane was fixed on the NP surface, followed by deprotonation of the mercaptan group with sodium methoxide. Then ROP of glycidol took place in toluene and elevated temperature. Due to the structure and reactivity of the monomer and the active alkoxide groups, a hyperbranched polymer (HPG) surface layer is formed, which comprises ca. 13%wt of the final material. The hybrid NPs were well dispersed in water and resistant to protein adsorption (like fibrinogen, lysozyme, and γ-globin) at physiological pH. In the second approach, the hydrophobic ligands of IONPs (oleic acid (OA) and oleylamine) were substituted with 6-hydroxy caproic acid. After treatment with aluminum isopropoxide, the surface hydroxyl groups initiated the polymerization of glycidol giving HPG-coated magnetic NPs with excellent solubility and colloidal stability in water as well as with low cytotoxicity.

The preparation of biofunctional hybrid NPs with fluorescent/magnetic properties and polymer stabilizing grafted chains was reported by Zhu and coworkers (Liu et al. 2011b). Iron(III)-mediated ATRP with activators generated by electron transfer (AGET ATRP) was used for the "grafting from" polymerization of the fluorescent monomer 9-(4-vinylbenzyl)-9*H*-carbazole (VBK) from magnetic ferroferric oxide NPs. $FeCl_3.6H_2O$ was the catalyst, *tris*(3,6-dioxaheptyl)amine (TDA-1) was used as the ligand, and ascorbic acid (AsAc) was used as the reducing agent. The initiator for ATRP was covalently attached on the magnetic NPs with ligand exchange with 3-aminopropyltriethoxysilane (APTES) and then esterification with 2-bromoisobutyryl bromide. After polymerization, well-defined core/shell NPs (Fe_3O_4@PVBK) were obtained with a magnetic core and a fluorescent shell (PVBK). In the next step of the synthetic process, well-dispersed bifunctional NPs stabilized with block copolymer chains (Fe_3O_4@PVBK-*b*-P(PEGMA)) were obtained in water, via consecutive AGET ATRP of the hydrophilic monomer poly(ethylene glycol) methyl ether methacrylate (PEGMA) (Figure 4.16). The chemical composition of the hybrid NPs' surface, at different surface modification stages, was monitored with FTIR spectroscopy. The magnetic and fluorescent properties of the NPs were evaluated and found to be useful for biomedical applications. In particular, the Fe_3O_4@PVBK-*b*-P(PEGMA) NPs showed effective imaging ability in enhancing the negative contrast in MRI experiments.

In another work, the same group presented the synthesis of hybrid water-soluble multifunctional NPs with thermoresponsive, magnetic, and fluorescent properties (Li et al. 2011). The hybrid Fe_3O_4@SiO_2-PNIPAM NPs were prepared via surface-initiated RAFT polymerization, using fluorescent RAFT agent–functionalized silica-coated

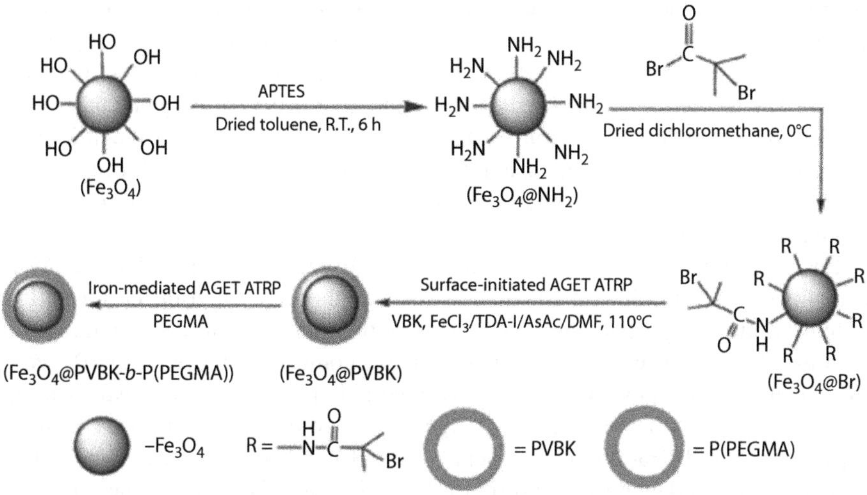

FIGURE 4.16 Synthesis route for the polymer-grafted magnetic NPs (Fe_3O_4@PVBK and Fe_3O_4@PVBK-*b*-P(PEGMA)) with core/shell structure. (Reprinted with permission from Liu, G., Xie, J., Zhang, F., Wang, Z.Y., Luo, K., Zhu, L., Quan, Q.M., Niu, G., Lee, S., Ai, H., and Chen, X.Y., *Small*, 7, 2742–2749, 2011. Copyright 2011, American Chemical Society.)

magnetic NPs as the chain transfer agent in the SIP of *N*-isopropylacrylamide. The systematic characterization of the particles obtained by a number of techniques revealed their structure and demonstrated their combined physicochemical properties, as well as their high potential for utilization in fluorescence and MRI biomedical applications.

CdTe QDs possessing surface functional groups, as a result of their preparation scheme, were used for the surface-initiated anionic polymerization of glycidol (Zhou et al. 2009). An HPG layer was formed on the surface of CdTe NP also in this case, improving their water solubility, biocompatibility, and survivability in biological media, without compromising the fluorescence properties of the semiconductor NPs. The less well-controlled surface-initiated ROP of ethylene oxide (EO), in the presence of aluminum isopropoxide, from surface hydroxyl groups of silica NPs resulted also in biocompatible silica/PEG hybrid NPs containing up to 40 wt.% grafted PEO chains (Joubert et al. 2005).

4.8 HYBRID POLYMERIC NANOPARTICLES VIA (MINI)EMULSION POLYMERIZATION

Emulsion and miniemulsion polymerization–based methodologies have been frequently utilized in order to prepare hybrid polymeric NPs. These composite latex particles, with sizes in the submicrometer scale, are composed of organic and inorganic domains organized into well-defined core/shell, multinuclear, raspberrylike, or armored morphologies. Particular emphasis is placed on the synthetic strategies for fabrication of these colloidal materials. The synthetic methodologies described in the literature regarding the synthesis of such hybrid colloids can be categorized into two main synthetic approaches. The first involves the polymerization of organic monomers in the presence of preformed inorganic particles. In the second approach, inorganic nanomaterials are synthesized in the presence of preformed polymer latexes. The techniques allow for the incorporation of different inorganic nanoscale species, for example, silica, iron oxide, pigments, clays, QDs, and metallic NPs into polymeric particles. Some examples will be provided in the following focusing on giving an overview of the synthetic methods as well as the new trends and approaches in the field.

Inverse miniemulsion polymerization of 2-hydroxyethyl methacrylate (HEMA) with silver tetrafluoroborate ($AgBF_4$) as the lipophobe resulted in the formation of hybrid latex polymeric particles incorporating the silver salt. Salt loadings as high as 13%, in respect to the disperse phase, could be achieved. AgNPs were subsequently formed by a gas phase in situ reduction of $AgBF_4$ using hydrazine as the reducing agent. The formation of AgNPs was confirmed by UV-vis spectroscopy and x-ray diffraction experiments. Hybrid nanoparticles sizes were determined by DLS and electron microscopy. Overall hybrid NP sizes in the range ca. 250 nm were obtained. In particular, TEM and atomic force microscopy (AFM) imaging revealed the morphology of the hybrid NPs, showing that AgNPs were distributed on the surface of PHEMA NPs (Cao et al. 2011) (Figure 4.17). The formation of raspberrylike hybrid particles was attributed to the slow diffusion of hydrazine to the dispersed phase and the fast reduction of silver particles by hydrazine. The effects related to the type and amount of cosolvent, salt content, and type of surfactant used for NP stabilization, on the hybrid NP characteristics and colloidal stability during the reduction process, were investigated in detail. Relatively better colloidal stability could be obtained in the dispersions with lower molecular weight P(E/O)-PEO copolymers as stabilizers, when using a relatively greater amount of $AgBF_4$ salt in the initial steps of particle synthesis and a suitable amount of cosolvent. The size of the AgNPs was relatively larger when ethylene glycol was utilized as the cosolvent compared to the dispersions with water. The size of the AgNPs was found to increase slightly with increasing the amount of cosolvent. The authors proposed that such hybrid particles could be used for antibacterial coatings in medical applications. Utilizing similar synthetic approaches of the same group succeeded in encapsulating different hydrophilic metal salts such as tetrafluoroborates of iron(II), cobalt(II), nickel(II), copper(II), and zinc(II) and nitrates of cobalt(II), nickel(II), copper(II), zinc(II), and iron(III) and cobalt(II) chloride into PHEMA-based polymeric NPs. In this case, some differences in the distribution of the inorganic component within the hybrid nanoparticles were observed for iron salts, as indicated by TEM imaging (Cao et al. 2010).

Hawker and van Berkel reported the incorporation of gold, AgNPs, and $MnFe_2O_4$ NPs (of ca. 10 nm in diameter) and gold nanorods within larger spherical NPs of poly(divinylbenzene) (PDVB NPs of ca. 100 nm in diameter) via miniemulsion polymerization (van Berkel and Hawker 2010). It was found that grafting of the AuNPs with low molecular weight polystyrene thiols enhanced the encapsulation of NPs within the PDVB NPs. TEM images showed that metal NPs were distributed within the PDVB particles.

Composite NPs of PMMA, PS, poly(lauryl methacrylate) (PLMA), and (PBA-*co*-PMMA) incorporating different amounts and types of hydrophobic rare earth clusters (Ln = Y, Pr, Eu, Sm, Nd) were prepared by miniemulsion polymerization (Hauser et al. 2011)]. DLS and TEM observations revealed that the resulting cluster-polymer hybrid NPs were spherical in shape with a narrow size distribution. The successful encapsulation reached more than 1000 mg/L of the hydrophobic Ln clusters (as total cluster concentration dispersed in water). The study of the photophysical properties of the dispersions showed that water was successfully restrained from the vicinity of the lanthanide clusters. Furthermore, a very efficient energy transfer from the ligand and polymeric unit to Eu^{3+}ions could be observed in the

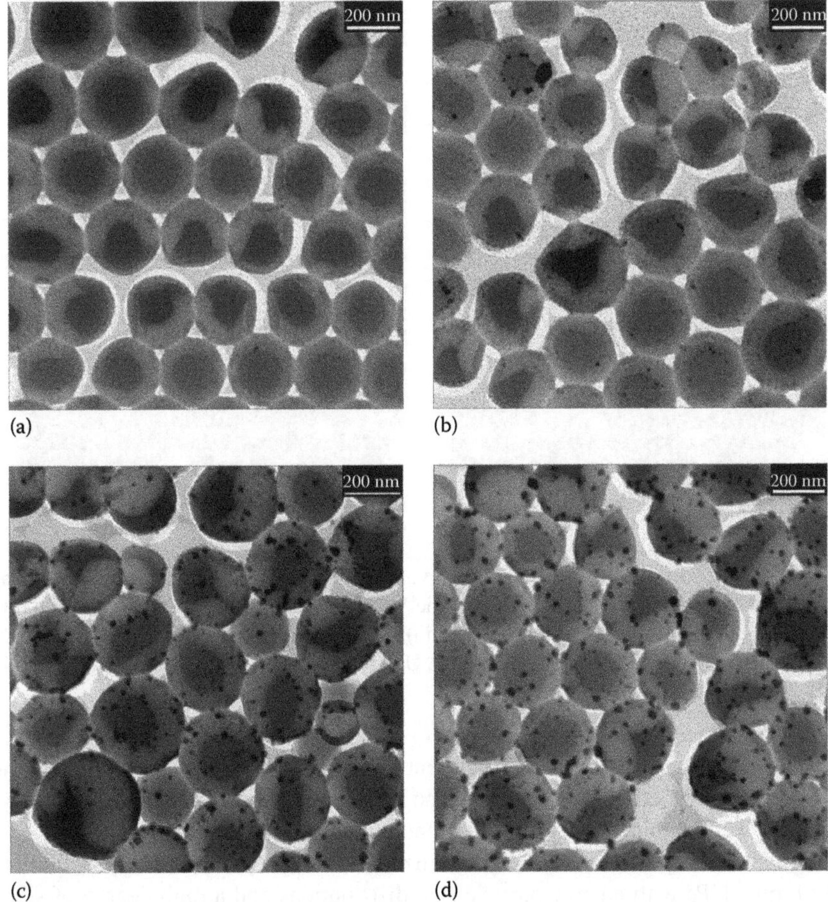

FIGURE 4.17 Morphological evolution of Ag/PHEMA hybrid particles during the reduction process of AgBF4 by hydrazine via gas-phase diffusion (a, 15; b, 30; c, 60; d, 120; e, 180; f, 240; g, 300; and h, 360 min). (Reprinted with permission from Cao, Z., Walter, C., Landfester, K., Wu, Z., and Ziener, U., *Langmuir*, 27, 9849–9859, 2011. Copyright 2011, American Chemical Society.)

(continued)

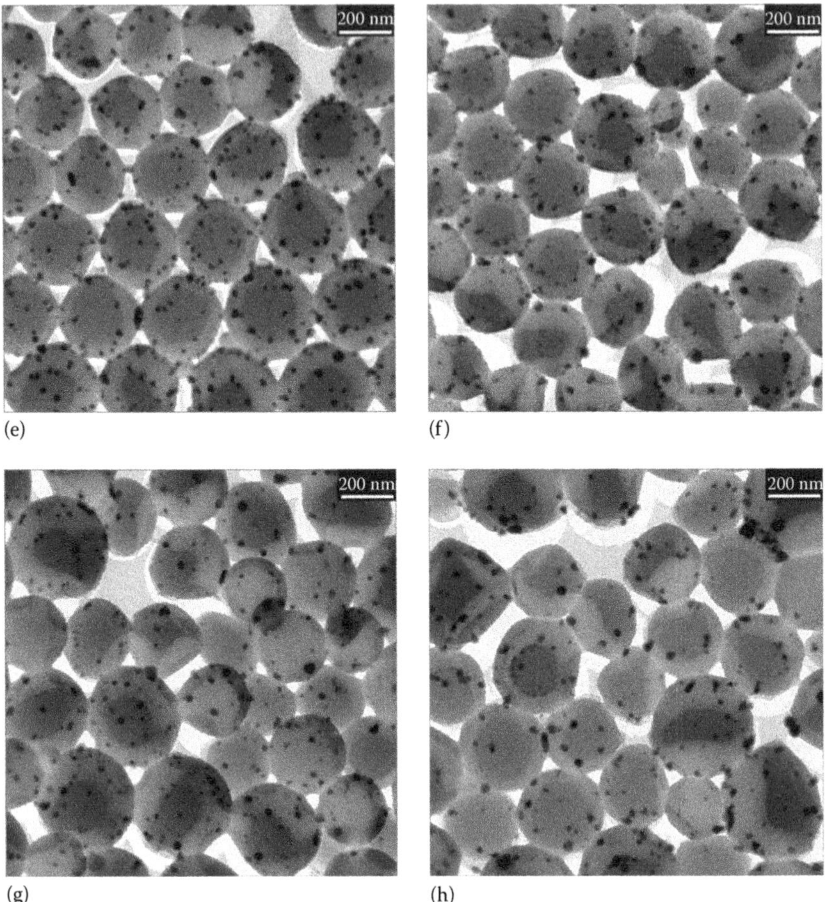

FIGURE 4.17 (continued) Morphological evolution of Ag/PHEMA hybrid particles during the reduction process of AgBF4 by hydrazine via gas-phase diffusion (a, 15; b, 30; c, 60; d, 120; e, 180; f, 240; g, 300; and h, 360 min). (Reprinted with permission from Cao, Z., Walter, C., Landfester, K., Wu, Z., and Ziener, U., *Langmuir*, 27, 9849–9859, 2011. Copyright 2011, American Chemical Society.)

dispersions. By making use of the different glass transition temperatures (T_g) of the utilized polymer matrix, monolayers of the NPs and efficient luminescent thin polymer films were obtained by spin coating techniques.

Miniemulsion polymerization was utilized for the preparation of hybrid silica/polystyrene NPs with narrow particle size distributions and a high degree of silica encapsulation (Costoyas et al. 2009). Preformed silica NPs were utilized. The effects of adding surface modifiers, the size of silica NPs, the ratio styrene monomer/silica, the surfactant concentration, and the presence of ethanol in the reaction mixture, on the properties of the hybrid NPs, were investigated. A synergistic effect was observed using OA together with 3-(trimethoxysilyl)propyl methacrylate (TPM) in the compatibilization step between the organic phase (monomer) and inorganic

NPs (silica). Mononuclear and multinuclear eccentric core/shell hybrid NPs were obtained depending on the experimental conditions utilized.

A one-step method was reported by Wu et al. in order to synthesize PS/TiO_2 nanocomposite particles via miniemulsion polymerization methodologies (Wu et al. 2010). Styrene monomer and acetylacetone (Acac)-chelated tetra-*n*-butyl titanate (TBT) were confined in the miniemulsion droplets with the aid of the cationic surfactant cetyltrimethylammonium bromide (CTAB) and hexadecane (HD) as the co-stabilizer. It was postulated that during the polymerization of styrene, TBT diffused toward the surface of the miniemulsion droplets owing to its hydrophilicity and yielded TiO_2 particles via an in situ sol–gel process. Owing to the electrostatic interactions between the positively charged CTAB and negatively charged Ti-OH groups, TiO_2 particles were coated onto the PS cores to form PS/TiO_2 nanocomposite particles having the inorganic component on their surface. The morphology of the hybrid NPs was confirmed by TEM.

Topfer and Schmidt-Naake (2007) prepared two types of inorganic NPs, consisting of silica and titania, via basic Stoeber synthesis, sized ca. 80 and 115 nm, respectively. These inorganic particles were utilized as core material into a miniemulsion copolymerization system after having been surface functionalized with trimethoxysilyl propyl methacrylate (MPTMS). The miniemulsion monomer reaction system consisted of styrene and 2-hydroxyethylmethacrylate and styrene sulfonic acid (SSA) or aminoethyl methacrylate hydrochloride (AEMA) as comonomers in varying compositions. The use of polar comonomer was found to enhance inorganic–polymeric component compatibility and inorganic NP encapsulation. The size, size distribution, and morphology of the resulting hybrid NPs were investigated by scanning electron microscopy (SEM) and DLS measurements. The composition and thermal properties of materials were studied by photoacoustic FTIR (PA-FTIR) spectroscopy, elemental analysis, thermogravimetric analysis (TGA), and DSC. Hybrid nanoparticles in the size range 100–500 nm were obtained.

The synthesis of hybrid silica/polymer hollow microspheres via the double in situ miniemulsion polymerization was reported by Zhang et al. (2010b). In order to explain the formation of hollow structures, the authors proposed the following mechanism: the organic monomers MMA and γ-(trimethoxysilyl)propyl methacrylate (MPS), as well as the SiO_2 precursor tetraethoxysilane (TEOS), were restricted in the miniemulsion monomer droplets. As the polymerization of organic monomers proceeded, the low miscibility between the polymeric phase and TEOS resulted in microphase separation. The compressed TEOS was pushed out from the initially formed PMMA particles as nodules at high MPS content, or it could be accumulated at one internal side of PMMA particles at low MPS content. By subsequent hydrolysis/condensation reactions under basic conditions, TEOS nodules turned into SiO_2 particles, located on the shells of the growing polymer particles, that is, the TEOS phase transformed into silica. The MPS-modified SiO_2 particles were thus chemically bonded onto the surfaces of polymer particles, owing to the copolymerization of MPS and MMA. TEM images of the nanoparticles were in general agreement with the proposed NP formation process and expected NP morphology (Figure 4.18).

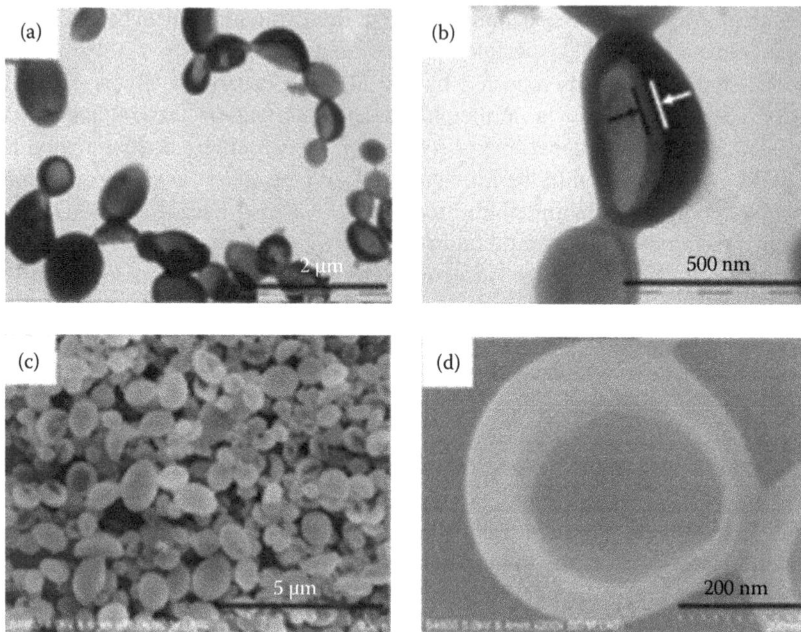

FIGURE 4.18 (a–d) Morphologies of hybrid microspheres with mass ratio of MMA/TEOS/MPS/SDS = 14/6/0/0.15: TEM of (a) low-magnification and (b) magnified images and SEM of (c) low-magnification and (d) magnified images. (Reprinted with permission from Zhang, J., Yang, J., Wu, Q., Wu, M., Liu, N., Jin, Z., and Wang, Y., *Macromolecules*, 43, 1188–1190, 2010b. Copyright 2010, American Chemical Society.)

More complex hybrid NPs were prepared via a Pickering miniemulsion polymerization process. Specifically, positively charged hybrid NPs, having a raspberrylike morphology, consisting of a polystyrene core and an alumina-coated silica shell, were formed in a surfactant-free system, via the radical miniemulsion copolymerization of styrene and different comonomers (i.e., acrylic acid, methacrylic acid, and acrylamide) by using a cationic silica sol as the sole emulsifier (Schrade et al. 2011). The influence of different parameters including pH of the dispersion, comonomer content, and the amount and size of silica NPs on the colloidal stability of the systems was investigated in detail. It was found that stable dispersions could be obtained by using 1–4 wt.% of the acidic comonomers, at low pH. By increasing the content of the silica particles (of ca. 22 nm size), the hybrid particle size in the dispersion could be decreased in a size range from 750 to 300 depending on the acidic comonomer utilized. TEM and SEM images revealed the presence of alumina-coated silica particles on the surface of the polymeric matrix.

An intriguing morphology of hybrid NPs was recently achieved via emulsion polymerization methods. Tri-layer poly(methacrylic acid-*co*-ethylene glycol dimethacrylate)/silica/poly(ethylene glycol dimethacrylate) (P(MAA-*co*-EGDMA)/SiO_2/PEGDMA) and P(MAA-*co*-EGDMA)/SiO_2/polydivinylbenzene hybrid microspheres were prepared by distillation–precipitation polymerization of ethylene glycol dimethacrylate (EGDMA) and divinylbenzene (DVB) in the presence of 3-(methacryloxy)propyl trimethoxysilane (MPS)-modified P(MAA-*co*-EGDMA)/SiO_2 microspheres as the seeds (Ji et al. 2009). The polymerization of EGDMA and DVB was performed in neat acetonitrile with AIBN as initiator, in order to coat the MPS-modified P(MAA-*co*-EGDMA)/SiO_2 seeds through the capture of EGDMA and DVB oligomer radicals with the aid of vinyl groups on the surface of modified seeds. No stabilizer or surfactant was utilized. Monodisperse P(MAA-*co*-EGDMA)/SiO_2 core/shell microspheres were synthesized by forming a layer of silica onto P(MAA-*co*-EGDMA) microspheres, via a sol–gel process. These nanospheres were further grafted by MPS, incorporating the reactive vinyl groups onto the surface, to be used as the seeds for the construction of hybrid microspheres with the tri-layer structure. Hollow poly(ethylene glycol dimethacrylate) (PEGDMA) and PDVB microspheres with free movable P(MAA-*co*-EGDMA) core were subsequently obtained after the selective etching of the silica mid-layer from the tri-layer hybrid microspheres in hydrofluoric acid (Figure 4.19). The morphology and structure of the tri-layer polymer hybrids and the corresponding hollow polymer microspheres with movable P(MAA-*co*-EGDMA) core were verified by TEM, FTIR, and XPS (Figure 4.20).

Fluorescent polystyrene particles have been synthesized via miniemulsion polymerization. The synthetic scheme involved the incorporation of preformed CdSe/ZnS core/shell QDs (QDs) into the PS particles in the course of styrene polymerization (Joumaa et al. 2006). QDs were coated with organic ligands in order to improve compatibility with the polymeric matrix. The influence of QD concentration, QD coating (either TOPO-coated or vinyl functionalized), and surfactant concentration on the polymerization kinetics and the PL properties of the resulting hybrid NPs has been studied. Experimental results indicate that polymerization kinetics were not altered by the presence of QDs, irrespective of their surface coating. Hybrid NPs with sizes ranging from 100 to 350 nm, depending on surfactant concentration, and having narrow size distributions were obtained. The fluorescence intensity of the hybrid NPs increased with the number of incorporated TOPO-coated QDs. A slight red shift of the emission maximum was observed and was correlated with phase separation between PS matrix material and QDs. It was noted that such a phase separation should have occurred during the polymerization and results in localization of the QDs in the vicinity of the particle/water interface. Hybrid NPs with TOPO-coated QDs displayed higher fluorescence intensity compared to those functionalized with the vinyl moiety. The obtained fluorescent NPs should find a variety of applications in biotechnology.

Improved miniemulsion polymerization methodologies were utilized for the formation of PS/PMMA core/shell spherical NPs carrying highly fluorescent CdS/ZnS-coated CdSe QDs in their core (Fleischhaker and Zentel 2005). For the synthesis of the PS core, a modified miniemulsion polymerization has been used. In this step, the inorganic QDs were incorporated into the core particles. The presynthesized CdSe QDs embedded in the core were coated with the higher band gap semiconductor materials CdS and ZnS in a successive ion layer adsorption reaction (SILAR) keeping up the light-emitting properties of the QDs during the integration process. The PMMA shell was prepared by a newly developed core/shell polymerization

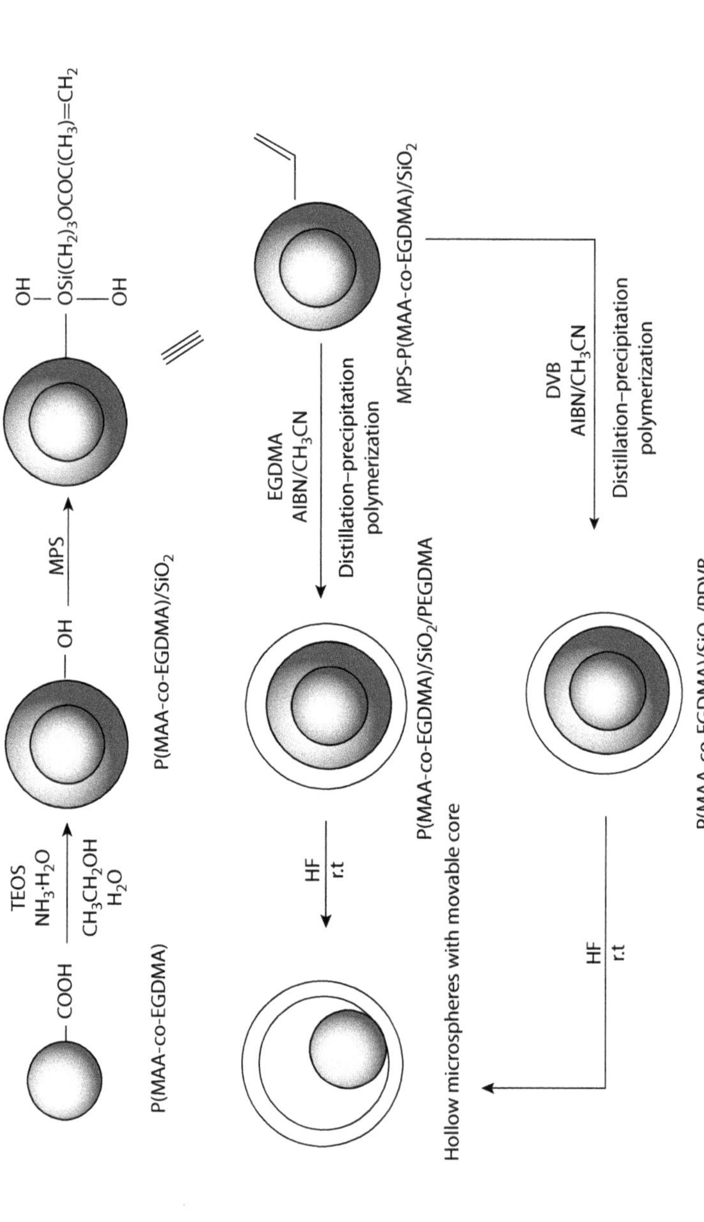

FIGURE 4.19 Preparation of P(MAA-*co*-EGDMA)/silica/polymer tri-layer hybrid particles and the corresponding hollow polymer microspheres with movable P(MAA-*co*-EGDMA)cores. (Reprinted with permission from *Polymer*, 50, Ji, H., Wang, S., and Yang, X., 133–140, Copyright 2009, with permission from Elsevier.)

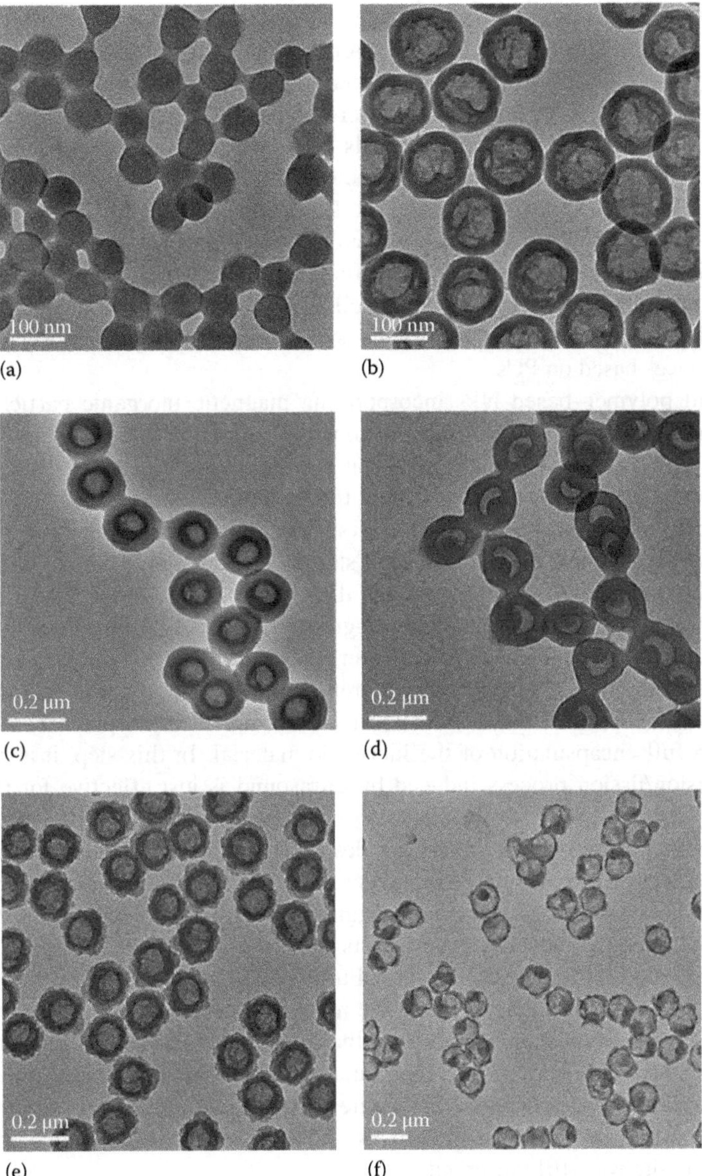

FIGURE 4.20 TEM micrographs of P(MAA-*co*-EGDMA)/silica/polymer tri-layer hybrid particles and the corresponding hollow polymer microspheres with movable P(MAA-*co*-EGDMA) cores: (a) P(MAA-*co*-EGDMA) NPs, (b) MPS-modified P(MAA-*co*-EGDMA)/silica core/shell microspheres, (c) P(MAA-*co*-EGDMA)/silica/PEGDMA tri-layer hybrid microspheres (EGDMA 0.5 mL), (d) hollow PEGDMA microspheres with movable P(MAA-*co*-EGDMA) cores, (e) P(MAA-*co*-EGDMA)/silica/PDVB tri-layer hybrid microspheres (DVB 0.5 mL), and (f) hollow PDVB microspheres with movable P(MAA-*co*-EGDMA) cores. (Reprinted with permission from *Polymer*, 50, Ji, H., Wang, S., and Yang, X., 133–140, Copyright 2009, with permission from Elsevier.)

using the PS cores as seed particles. The size of the core/shell colloids could be controlled by the injected volume of MMA monomer comprising the shell, as well as by the amount of core particles used in reaction leading to the formation of the shell. Hybrid spheres of different diameters were prepared and subsequently were self-assembled into colloidal photonic crystals (PCs) with photonic stop bands located in the visible range of the electromagnetic spectrum. A modifying influence of the photonic band structure of the PC on the PL of the embedded QDs was observed by angular-dependent fluorescence measurements. The controlled combination of electronic confinement, originating from the QDs, and photon confinement, due to the periodic dielectric structure of the colloidal crystal, as it has been realized in this work presents a huge platform for the design and construction of novel optoelectronic devices based on PCs.

Hybrid polymer–based NPs incorporating magnetic inorganic particles were also synthesized by the aid of miniemulsion polymerization as reported by Ramirez and Landfester (2003). The encapsulation of high amounts of magnetite NPs into polystyrene particles was achieved by a three-step preparation scheme involving two miniemulsion processes. In the first step, OA-coated magnetite particles were dispersed in octane. In the second step, magnetite aggregates in water were obtained in a miniemulsion process, via the use of sodium dodecyl sulfate (SDS) as surfactant. In this way, magnetite aggregates covered by an OA/SDS bilayer were obtained. In the third step, the dispersion containing the bilayer-coated magnetite aggregates was mixed with a styrene monomer miniemulsion, and a second miniemulsion process, a so-called ad-miniemulsification process, was used to obtain full encapsulation of the inorganic material. In this step, it is believed that a fusion/fission process induced by ultrasound is just effective for the pure monomer droplets, whereas the monomer-coated magnetite aggregates stay intact. In this way, all monomer droplets are destroyed and nucleated onto the magnetite aggregates to form a surface monomer film. After polymerization, polystyrene nanoparticles encapsulating magnetite aggregates were obtained. TGA, preparative ultracentrifugation, and TEM experiments showed that up to 40% magnetite could be encapsulated in the polymer matrix. Furthermore, as indicated by TEM imaging, NPs had a high homogeneity of the magnetite content. This is in agreement with the proposed mechanism active in the third preparation step. Magnetometry measurements revealed that the encapsulated iron oxide aggregates still consist of separated superparamagnetic magnetite particles, due to the OA original layer. Therefore, after the encapsulation process, 60% of the magnetization compared to bulk magnetite was still preserved.

Magnetic/polymer hybrid NPs were prepared by Ramos and Forcada in a single-step miniemulsion process, developed to achieve better control of the morphology of the magnetic nanocomposite particles (Ramos and Forcada 2011). They followed a surfactant-free miniemulsion polymerization scheme using styrene as the monomer, HD as the hydrophobe, and potassium persulfate as an initiator in the presence of OA-modified magnetite NPs. The effect of the type of cross-linker used [divinylbenzene and bis-[2-(methacryloyloxy)ethyl] phosphate (BMEP)] together with the effect of the amount of an aid stabilizer (dextran) on size, particle size distribution, and morphology of the hybrid NPs synthesized was studied. The mixture of different surface modifiers produces hybrid nanocolloids with various morphologies: from a typical core/shell structure composed of an agglomerated magnetite core surrounded by a polymer shell (when the hydrophobic DVB was used as the cross-linker) to a homogeneous morphology where the magnetite particles are uniformly distributed throughout the entire nanocomposite (when BMEP was used for cross-linking, enhancing the polymer/magnetite compatibility). Hybrid NPs had low size distributions in all cases. Dextran was found to decrease the size of the hybrid NPs. A synergistic effect of dextran and cross-linker was observed. Core/shell particles with non-agglomerated magnetite particles in the core were formed in the simultaneous presence of dextran and DVB.

Delgado and coworkers reported an interesting method to prepare composite colloidal NPs, consisting of a magnetic core (magnetite) and a biodegradable polymeric shell (poly(ethyl-2-cyanoacrylate) or PE-2-CA) (Arias et al. 2001). The method is based on an anionic polymerization procedure initiated/promoted by bases present in the polymerization medium under emulsion conditions. Presynthesized magnetite NPs were added to the miniemulsion containing the PE-2CA monomer. During the polymerization and as the polymeric phase increases, magnetite NPs were encapsulated into the polymer matrix giving hybrid magnetic NPs. NPs with a diameter ca. 145 nm were obtained. The authors stated that the heterogeneous structure of the particles can confer both magnetic field responsiveness and potential applicability as a biodegradable drug carrier. Their studies showed that the hybrid NPs display an intermediate behavior between that of pure magnetite particles and PE-2-CA spheres, respectively.

Multifunctional hybrid NPs incorporating magnetic NPs and fluorescent dyes were also reported (Holzapfel et al. 2006). Polystyrene particles encapsulating magnetite NPs (ca. 10–12 nm) in a hydrophobic poly(styrene-*co*-acrylic acid) shell were synthesized by a three-step miniemulsion process. A high amount of magnetic iron oxide was incorporated by this process (typically 30%–40% w/w). As a second functional component, a fluorescent dye was also integrated in order to obtain multifunctional NPs. Polymerization of styrene in the presence of the functional components resulted in hybrid NPs with sizes in the range 45–70 nm. It was possible to surface functionalize NPs and to vary the amount of carboxyl groups on the surface by copolymerization of styrene with the hydrophilic acrylic acid monomer. Magnetic measurements evaluated the inherent magnetization properties of the nanoparticles. For biomedical evaluation, the NPs were incubated with different cell types. The introduction of carboxyl groups on the particle surfaces enabled the uptake of NPs as demonstrated by the detection of the fluorescent signal by fluorescent-activated cell sorter (FACS) and laser scanning microscopy. However, the quantity of iron within the cells was lower than that which is required for most biomedical applications (e.g., in MRI). A further increase of NP uptake could be accomplished by transfection agents like poly(L-lysine) or other positively charged polymers. Lysine has also been covalently bound to particle surface by coupling to the carboxyl groups. In the latter case, the amount of internalized iron was even higher than that with the NPs carrying a transfection agent physically adsorbed on their surface. These NPs were found to be clustered in endosomal compartments.

4.9 HYBRID NANOPARTICLES BY LBL APPROACHES AND HYBRID POLYMERIC (POLYELECTROLYTE) NANOCAPSULES

Polymeric multilayers can be prepared onto inorganic NP surfaces using the layer-by-layer (LbL) self-assembly technique, as it is done for planar surfaces (Decher 1997, Srivastava and Kotov 2008). The LbL technique is mostly applied with polyelectrolytes where alternating layers of oppositely charged polyelectrolytes are deposited on the surface due to electrostatic attraction. After each layer formation, overcharging of the surface occurs allowing for the deposition of the next layer. The technique is versatile and provides great flexibility for the formation of polymer multilayers with variable architecture, thickness, morphology, and surface characteristics, as well as for the incorporation of non-polymeric components (Srivastava and Kotov 2008). Obviously, this can be another strategy for the preparation of hybrid inorganic–polymeric NPs utilizing preformed inorganic NPs as the cores and suitably chosen polymers as the outer layers.

Based on the aforementioned context, AuNPs, having negative surface charges from their synthesis, were used as substrates for the deposition of a cationic polyethylenimine (PEI) hydrophilic layer, due to electrostatic interactions. The second layer was composed of negatively charged poly(acrylic acid)-*g*-poly(oligoethylene glycol methylester acrylate) (PAA-*g*-POEGMEA). The hybrid NPs were stable in phosphate buffer for 14 h, due to the steric stabilization of neutral hydrophilic and biocompatible POEGMEA chains (Bousquet et al. 2010).

In another work, five alternating layers of cationic poly(allylamine hydrochloride) (PAH) and anionic poly(styrene sulfonate) (PSS) were deposited on AuNPs, with PAH being the last outer layer (Schneider et al. 2009). These NPs were stable in water but presented aggregation when the ionic strength of the aqueous medium was increased to physiological conditions. The authors deposited an outer functional layer of a terpolymer composed of three monomers: (1) *N*-(2-hydroxypropyl) methacrylamide, for increased water solubility and resistance to opsonization by serum proteins; (2) *N*-methacryloyl-glycyl-glycyl thiazolidine-2-thione, for covalent attachment of the terpolymer layer to the PAH outer layer via aminolysis; and (3) *N*-methacryloyl-glycyl-glycyl-D,L-phenylalanyl-leucyl-glycyl DOX, a conjugate of DOX that can release the anticancer drug via enzymatic degradation of the oligopeptide spacer. The deposition of the terpolymer layer increased the solubility of the hybrid NPs under physiological conditions and increased the stealth properties of the nanostructures.

Alternatively, the core material can be a polymeric latex particle, and the surface multilayer can be formed by the deposition of charged polymer components and inorganic NPs. This leads to the formation of a hybrid coating on the latex particle. In these preparation protocols, excess components of LbL assembly are usually removed by intermediate and repeated centrifugation steps, due to the large difference in mass and size of the polymeric/inorganic NP components and the coated core particle. This step is analogous to intermediate washing steps performed in the case of LbL polyelectrolyte assembly on solid planar surfaces.

Following this construction principles, Caruso and coworkers reported the coating of PS submicrometer size latex particles with a hybrid coating of a polyelectrolyte multilayer and an outer layer of CdTe and CdTe(S) semiconductor NPs (Rogach et al. 2000). First, a PAH layer was formed on the negatively charged surface of the PS latex particles by electrostatically driven adsorption of the cationic polyelectrolyte. Two consecutive alternative layers of PSS and PAH were then deposited on the first layer, forming an LbL coating of five layers with the outermost layer being PAH. Then CdTe and CdTe(S) semiconductor NPs, with sizes in the range 2.5–5 nm of narrow size distribution, coated with 1-mercapto-2,3-propanediol and 1,2-dimercapto-3-propanol were deposited on the PAH layer. It was postulated that the –OH and –SH groups of the QDs ligands enhanced the interactions of the inorganic nanocrystal with the $-NH_3^+$ groups of the PAH layer. The QDs had size-dependent electronic transitions and strong tunable PL in the range 530–680 nm. TEM observations demonstrated the successful synthesis of the hybrid NPs. The hybrid spherical colloids were subsequently used for the construction of 3D colloidal crystals.

In another study (Wang et al. 2002), the same groups reported the preparation of PS latex particles coated with polyelectrolyte/semiconductor NPs multilayers of more complex architecture. LbL assembly was utilized in order to form these complex multilayers composed of several groups of alternating PAH/PSS layers, separated by layers of CdTe semiconductor NPs (~4 nm in diameter). The CdTe nanocrystals were coated with thioglycolic acid ligands form their synthesis. The carboxylic acid groups of the ligands could interact with the positively charged amine groups of PAH. Each CdTe layer was separated with a PAH/PSS/PAH polyelectrolyte triple layer in order to regenerate a positively charged surface for the deposition of the next CdTe layer. It was found that the deposition of these triple polyelectrolyte layers between the inorganic nanocrystal layers could result in more uniform NP coatings compared with a single polyelectrolyte interlayer. The final outermost layer deposited on the modified latex surface was a layer of anti-immunoglobulin g (anti-IgG), which imparts specific biofunctionality to the surface of the hybrid particle. The protein layer was separated from the outermost CdTe layer by a PAH/PSS/PAH/PSS tetra layer, since it was known from previous studies that anti-IgG retains its biological activity when it is deposited onto such layers (Figure 4.21).

Hybrid PS latex particles coated with alternating multilayers of polyelectrolytes and magnetic magnetite NPs were also prepared by utilizing the LbL technique (Caruso et al. 2001). Negatively surface charged PS latexes of 270 nm diameter were coated with cationic poly(diallyldimethylammonium chloride) (PDADMAC) chains, anionic PSS chains, and Fe_3O_4 NPs, 8–12 nm in diameter, stabilized with tetrabutylammonium hydroxide. The thickness and the architecture of the LbL assembled multilayer could be well controlled by the number the adsorption cycles and the deposition sequence. TEM observations indicated that more uniform layer could be obtained when three alternating PDADMAC/PSS layers were deposited between the magnetite layers. The applications of these hybrid NPs in drug delivery and diagnostics were discussed.

The LbL technique can be also utilized for the preparation of nano- and microcapsules comprising nonsupported (hollow) polyelectrolyte multilayers (Decher 1997, Sukhorukov et al. 1998, De Koker et al. 2007, Skirtach et al. 2007).

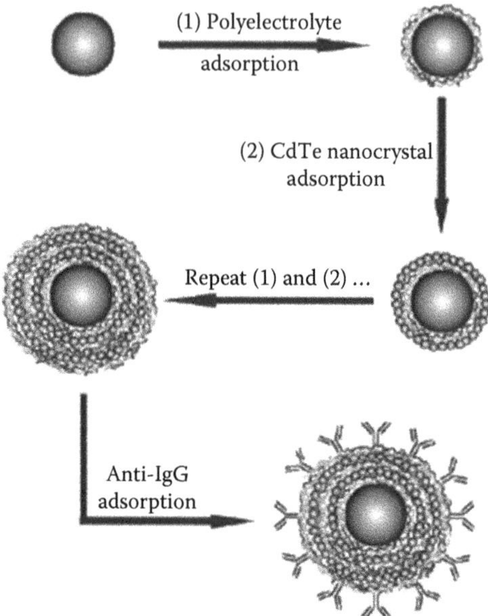

FIGURE 4.21 Procedure used to prepare CdTe QD–microsphere bioconjugates via LbL assembly. (Reprinted with permission from Wang, D., Rogach, A.L., and Caruso, F., *Nano Lett.*, 2, 857–861, 2002. Copyright 2002, American Chemical Society.)

The methodology is based on the consecutive deposition of oppositely charged polyelectrolytes on a complementary charged spherical NP that acts as a core structure, in an analogous way to the examples discussed previously. The new feature is that at the end of the deposition process the core is removed leaving unsupported stable polyelectrolyte multilayers in solution, which are now actually hollow nanocapsules. The thickness and internal structure of the nanocapsule walls can again be controlled via alterations in the number of polyelectrolyte adsorption steps, and the adsorption sequence followed. Thicknesses in the few tens of nanometers to several micrometers range can be achieved, with the overall size of the capsules being varied in the range of tens of nanometers to several micrometers (De Koker et al. 2007, Skirtach et al. 2007). The latter parameter is largely defined by the size of the initial core particle. Deposition of other building blocks, for example, inorganic NPs, can be also achieved in the intermediate steps resulting in the preparation of hybrid polyelectrolyte capsules.

Skirtach et al. constructed PAH/PSS capsules by polyelectrolyte LbL deposition on positively charged melamine formaldehyde core particles (Skirtach et al. 2004). AgNPs were synthesized within the polyelectrolyte multilayers from silver nitrate precursor, coordinated on the multilayer, by the reduction method. Dissolution of the melamine formaldehyde core in 0.1 M HCl resulted in hollow PAH/PSS capsules encapsulating AgNPs. Irradiation of the hybrid capsules with a continuous wave laser at 830 nm resulted in modification of the capsule wall integrity. Cutting through, deformation and ablative destruction of the capsules were observed depending on the laser power employed each time. The same group reported the preparation of hybrid capsules with walls composed of PAH/PSS multilayers and a gold sulfide core/gold shell NPs layer, encapsulating PSS/rhodamine-labeled PSS in their interior (Skirtach et al. 2005). Laser irradiation promoted the release of the encapsulated polymer chains. These experiments indicate the great potential of hybrid polymeric capsules for controlled disruption by external laser radiation and tunable liberation of encapsulated materials.

Caruso et al. also constructed hybrid capsules based on PAH and PSS polyelectrolytes and AuNPs by using LbL techniques (Angelatos et al. 2005). In this case, nine polyelectrolyte layers were created with the outer layer being PSS, giving negative charges to the resulting capsules after dissolution of the melamine formaldehyde core. Then fluorescein isothiocyanate–labeled dextran chains were loaded into the capsules by lowering the pH at 5. This decrease in pH increased the permeability of the nanocapsules walls allowing encapsulation of dextran. By increasing the pH at 9, labeled dextran chains were entrapped into the capsules due to lower permeability of the walls at this pH value. Excess non-entrapped dextran was removed by centrifugation. Next, AuNPs, with an average diameter of 6 nm and a 4-(dimethylamino) pyridine stabilizing shell, were incorporated into the nanocapsules walls through electrostatic interactions (Figure 4.22). These complex hybrid capsules encapsulating dextran were found to be responsive to laser radiation due to the presence of AuNPs in their walls. Absorption of light by AuNPs increased the local temperature of the walls and their permeability, resulting in the release of dextran macromolecules. In the same study, it was shown that it was possible to functionalize the outer surface of melamine formaldehyde core PAH/PSS shell NPs with lipid bilayers. On these

FIGURE 4.22 (a) SEM image of magnetic hollow spheres prepared by exposing PE3-coated PS particles to five adsorption cycles of Fe_3O_4 NPs and PDADMAC, followed by calcinations at 500°C. (b) TEM image of the same sample showing the hollow nature of the particles. The inset shows the regularity of the wall structure. The scale bar in the inset represents 100 nm. (Reprinted with permission from Caruso, F., Spasova, M., Susha, A., Giersig, M., and Caruso, R.A., *Chem. Mater.*, 13, 109–116, 2001. Copyright 2001, American Chemical Society.)

new lipid-based surface, antibodies were noncovalently attached (via electrostatic interactions), which in turn could bound fluorescently labeled secondary antibodies. Although the second LbL procedure has not been extended to hybrid nanocapsules, there is no obvious reason that forbids the biofunctionalization of hybrid nanocapsules or other hybrid core/shell NPs by following similar LbL protocols.

Rogach and coworkers utilized melamine formaldehyde spherical core particles for the construction of hybrid polymeric capsules by LbL assembly of PAH and PSS polyelectrolytes incorporating two different types of inorganic NPs (Gaponik et al. 2003). In the polyelectrolyte walls, thioglycolic acid negatively charged CdTe semiconductor NPs and citric acid–stabilized Fe_3O_4 magnetic NPs, also carrying a negative surface charge, were incorporated due to electrostatic interactions with the positive charges of the PAH chains. In this way, hybrid polymeric capsules with PL and magnetic properties were prepared, allowing their use as luminescence markers and their external manipulation with a magnetic field. The author demonstrated, by in vitro experiments, the increased uptake of the capsules by cells under the application of a localized and directing magnetic field, due to the convenient monitoring of the process under a fluorescence microscope (Zebli et al. 2005). They concluded that such hybrid capsules would be useful for targeted delivery of therapeutic agents encapsulated in them.

Polymeric hybrid capsules were prepared by the use of solely biocompatible polymers that were prepared in water by colloidal LbL technique (Gaponik et al. 2004). The biocompatible materials from natural sources included the polyelectrolytes alginic acid sodium salt, protamine sulfate, dextran sulfate, and chitosan. The core material used for the construction of the capsules was $MnCO_3$ near monodisperse particles that could be easily dissolved after the deposition process by decreasing the pH of the aqueous dispersion, by simple addition of HCl. The architecture of the capsules' layers, including all four biopolyelectrolytes, ensured stability of the capsules. These microcapsules were labeled with water-soluble thiol-stabilized CdTe NPs emitting in the visible range and with $Cd_xHg_{1-x}Te$ or HgTe NPs that emit in the near-IR electromagnetic spectrum. Simultaneous incorporation of magnetic Fe_3O_4 NPs was also achieved allowing external manipulation of the capsules by a magnetic field. The inorganic components carried a negative surface charge as a result of the chemical structure of the ligands used in their preparation. It is interesting to note that incorporation of the NPs was done after preparation of the capsules employing a simple mixing procedure of the capsules and NPs solutions, followed by centrifugation for removal of free NPs. The photophysical properties of the inorganic NPs were not altered due to their incorporation into the polyelectrolyte membranes, indicating low levels of NP aggregation. The luminescence efficiency of the inorganic NPs encapsulated in the polyelectrolyte capsules remained stable, under physiological conditions (for weeks in the case of CdTe-labeled capsules and at least for a month for CdHgTe and dropped by 80% for HgTe, because of the shift of the luminescence band outside the water transmission window). Biocompatible microcapsules labeled with $Cd_xHg_{1-x}Te$ NCs emitting at 750–1200 nm range might be of special interest for monitoring the drug delivery processes and biomedical applications related to hyperthermia due to the magnetic properties of the inorganic components (Figure 4.23).

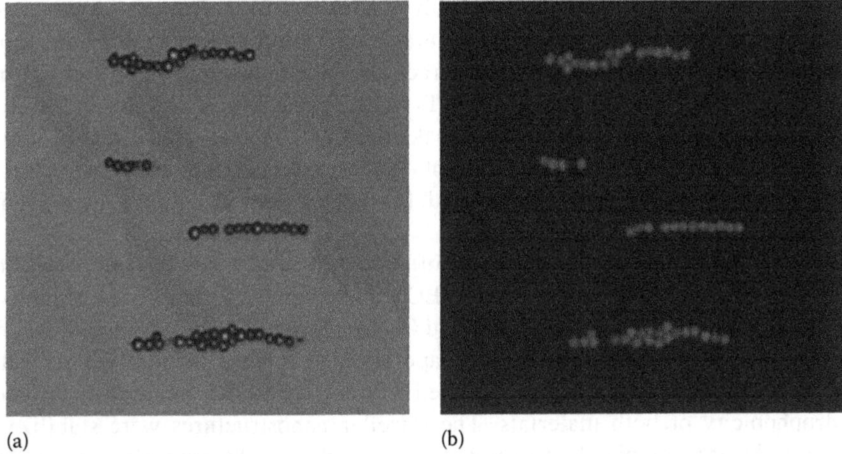

FIGURE 4.23 Microcapsules simultaneously loaded with luminescent semiconductor and magnetic oxide NPs are aligned in magnetic field. The images were obtained with a confocal laser scanning microscope TCS Leica operating in transmission (a) and in a luminescence excitation wavelength 476 nm) mode (b), respectively. Capsule diameter is 5.6 μm in all cases. (Reprinted with permission from Gaponik, N., Radtchenko, I.L., Sukhorukov, G.B., and Rogach, A.L., *Langmuir*, 20, 1449–1452, 2004. Copyright 2004, American Chemical Society.)

4.10 HYBRID NANOPARTICLES INCORPORATING BLOCK COPOLYMERS AND CARBON NANOMATERIALS

Besides inorganic NPs like metallic, semiconductor, or metal oxide NPs, carbon nanostructures (including fullerenes, carbon nanotubes (CNTs), carbon nanohorns (CHNs)) have been utilized as components of hybrid NPs and nanostructures in conjunction with polymers and especially with block copolymers. This is a relatively new area of research triggered by the discovery of several novel carbon nanomaterials that has revolutionized the field of material science in the last 10–15 years. Interest in these kinds of hybrid colloids is stemming from the need to solubilize the carbon nanostructures, which are scarcely soluble in organic solvents and most importantly in aqueous media. By achieving this, it then becomes possible to completely utilize the optical, conductive, and mechanical properties of the carbon-based nanomaterials for biomedical applications, as well as for the preparation of advanced, complex, and functional NPs and nanostructured materials for, for example, (bio)sensor and light harvesting–related applications or novel drug delivery nanosystems (Tasis et al. 2006, Kim et al. 2012).

Some examples of hybrid nanostructures combining carbon nanomaterials and block copolymers are discussed in order to illustrate the possibilities in this field of research. We focus our attention to hybrid colloids based on block copolymers since the flexibility in construction functional colloids and the number of structural variations are by far more extended. Several additional examples can be found in the literature employing homopolymers. In the case of CNTs, specialty homopolymers include poly(vinyl pyrrolidone), PSS, poly(metaphenylene vinylene),

PEO, polystyrene, polyisoprene (PI), polybutadiene (PB), poly(methyl methacrylate), and poly(dimethylsiloxane) (Baskaran et al. 2005 and references therein). Weak noncovalent and nonspecific interactions (like π–π interactions and CH–π interactions) between polymers and CNTs result in polymer adsorption and wrapping around the CNTs. This in turn breaks the CNT bundles and results in solubilization of the CNTs in organic solvents good for the polymer chains. Polymer/CNT nanocomposites can be produced from these hybrid colloids by solvent evaporation (Baskaran et al. 2005).

Hybrid micellar NPs composed of C_{60} fullerene and the amphiphilic poly(styrene-*b*-ethylene oxide) (PS-*b*-PEO) block copolymer have been obtained by initially dissolving the copolymer and C_{60} in toluene, then by addition of water in the solid mixed film formed by evaporation of toluene (Mountrichas et al. 2007a). In this way, C_{60} was encapsulated within the PS core domains due to the hydrophobicity of both materials. The micellar nanostructures were stabilized in water by the soluble PEO chains as indicated by light scattering and SEM measurements. The obtained stable aqueous dispersions showed optical limiting and nonlinear optical properties under visible nanosecond laser excitation due to the presence of fullerene particles. In another study (Wang et al. 2008a), C_{60} molecules have been incorporated at the core/corona interface of poly(styrene-*b*-dimethylaminoethyl methacrylate) (PS-*b*-PDMAEMA) or PS-*b*-PDMA block copolymer micelles. This spatial placement of C_{60} allows for a better exposure to water and acts favorably for photosensitization of the nanostructure. In fact, when the hybrid colloids were irradiated with red light, large amounts of active oxygen species were produced, indicating that these colloids can be utilized as photosensitizers in photodynamic therapy of cancer. The photoactivities of the hybrid self-assembled NPs could be varied through changes in the amount of C_{60} incorporated in the micelles.

The block copolymer poly(styrene-*b*-(sulfamate/carboxylate)isoprene) (PS-*b*-SCPI) was utilized for the solubilization of CNTs in water and the subsequent formation of hybrid colloidal NPs, comprising the amphiphilic block copolymer and the elongated carbon nanostructures. The particular copolymer can be physisorbed on the surface of the nanotubes, through hydrophobic interactions between the phenyl rings of the PS block and the nonpolar CNT outer walls, without disturbing the nanotube carbon atoms network. On the other hand, the SCPI block keeps the system in solution, due to electrosteric repulsion forces (Mountrichas et al. 2007b,c). Interestingly, the same block copolymer has been also used for the solubilization of CNH, that is, spherical aggregates of carbon nanostructure with a hornlike morphology, leading to stable water-soluble nanohybrids through an identical solubilization scheme (Mountrichas et al. 2009a). It is worth emphasizing that the optical properties as well as the mechanical and conductive properties of the carbon nanomaterials are retained due to preservation of the carbon network, since no chemical reaction with the block copolymer is involved in either case.

Following the successful solubilization of the carbon-based nanomaterials, the functional groups of the polyelectrolyte block SCPI of the physisorbed block copolymer, decorating the carbon nanostructures, were utilized for the template synthesis of either semiconductor or metallic NPs. The synthesis (Mountrichas et al. 2007b)

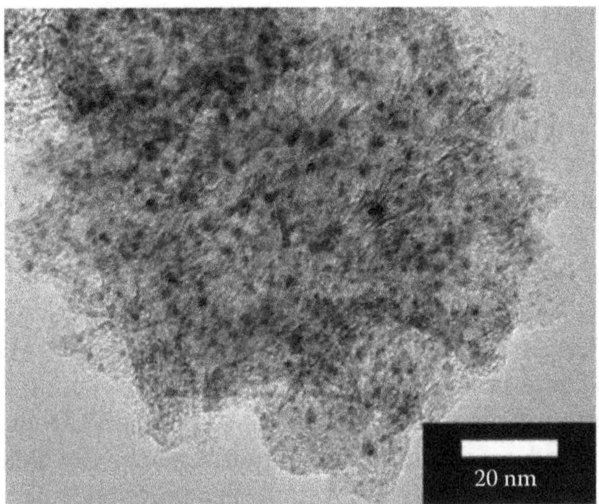

FIGURE 4.24 High-resolution TEM image of hybrid NPs composed of CNHs/PS-*b*-PSCI copolymer/AuNPs. AuNPs loaded on the CNHs/copolymer ensemble are identified as dark spots. (Reprinted with permission from Mountrichas, G., Ichihashi, T., Pispas, S., Yudasaka, M., Iijima, S., and Tagmatarchis, N., *J. Phys. Chem. C*, 113, 5444–5449, 2009a. Copyright 2009, American Chemical Society.)

and the photophysical properties (Mountrichas et al. 2009b) of CdS NPs on CNT/PS-*b*-SCPI hybrids have been studied, showing that the semiconductor NPs are actually placed in the polyelectrolyte block domain, because of electrostatic interaction between the cadmium salt precursor and the chemically modified SCPI block. These multifunctional hybrid NPs were utilized as the active material in the construction of a photoelectrochemical cell (Mountrichas et al. 2009b). By utilizing a similar preparation protocol, the synthesis of gold metallic NPs has been demonstrated at the periphery of CNH decorated with the PS-*b*-PSCI copolymer (Mountrichas et al. 2009a) (Figure 4.24). Again the SCPI blocks worked as the functional domains for the growth of Au NPs creating complex NPs composed of carbon nanomaterial, block copolymer chains, and metallic NPs in water. It was demonstrated that the hybrid nanostructures possess the combined properties of each component.

In a similar fashion, CNTs were noncovalently modified with PI-*b*-PAA amphiphilic block copolymer chains, through physisorption, which resulted in the formation of water dispersible CNT/PI-*b*-PAA hybrid nano-objects with a PAA outer shell (Tasis et al. 2007). The PAA corona chains were utilized as the molecular templates for the biomimetic crystallization of calcium carbonate, due to the presence of the PAA carboxylate groups, which play the role of anchoring groups that interact with the growing crystals. This study further demonstrates the possibilities for potential utilization of such functional hybrid nanostructures in material synthesis and a variety of applications.

Hybrid NPs composed of carbon nanostructures and homopolymers or block copolymers have been also prepared by "grafting from" and "grafting to" polymerization methodologies, similar to the strategies discussed in the section on hybrid NPs via SIP. A variety of polymers have been combined with a large number of carbon nanomaterials. The disadvantage of these approaches is that the carbon network is disturbed by the covalent fixation of polymeric chains. Although solubilization of carbon nanostructures is achieved, mechanical, optical, and conductive properties of the carbon nanomaterials may be, however, compromised.

Baskaran et al. reported the covalent functionalization of multiwalled CNTs (MWNTs) using surface-initiated ATRP. The carboxylic groups bound on the surface of MWNTs were used for the attachment of the ATRP initiator molecules. Polymerization of MMA and styrene resulted in MWNTs grafted with PMMA and PS chains. The success of the functionalization methodology was verified by FTIR, Raman, TGA, and TEM measurements on the hybrids. The hybrid nanostructures were soluble in a variety of solvents (Baskaran et al. 2004). The same authors also reported the covalent attachment of PS and PEO chains on MWNTs via a surface-initiated anionic polymerization scheme. Covalent attachment of precursor anions, namely, 4-hydroxyethyl benzocyclobutene (BCB-EO) and 1-benzocyclobutene-1′-phenylethylene (BCB-PE), was possible through Diels–Alder cycloaddition at 235°C (Sakellariou et al. 2008). BCB-EO groups were used for the polymerization of EO after reaction with potassium triphenylmethane to produce reactive alkoxides. BCB-PE groups were used for the polymerization of styrene after reaction with *sec*-butyllithium. The living PS chains were also utilized for the polymerization of a second monomer (isoprene) leading to the formation of PS-*b*-PI block copolymer chains attached on the walls of MWNTs. Initially insoluble MWNTs were gradually dissolved in the reaction medium as polymerization proceeded, and after termination, polymer-grafted MWNTs dispersible in organic solvents were obtained. The hybrid nanostructures were characterized in detail. More recently, the same group showed that single-walled nanotubes (SWNTs) and MWCNTs could be functionalized with a titanium alkoxide catalyst through a Diels–Alder cycloaddition reaction. The catalyst-functionalized CNTs were used for the surface-initiated titanium-mediated coordination polymerizations of L-lactide (L-LA), ε-caprolactone (ε-CL), and *n*-hexyl isocyanate (HIC) employing the “grafting from” technique. In this way, CNTs carrying biocompatible and/or biodegradable polymer chains were prepared (Priftis et al. 2009). The grafting was supported by NMR, FTIR, Raman, TEM, and TGA characterization of the hybrid nanostructures, which were found to be soluble in organic solvents, due to the presence of the polymer chains. Suppression of PLA and PCL crystallization was observed in the solid state.

CNTs carrying two chemically different polymer chains were prepared by the grafting from technique (Priftis et al. 2009). A Diels–Alder cycloaddition reaction of substituted benzocyclobutanes was employed in order to functionalize MWNTs with two different initiators, which could be used for ROP of L-LA and ε-CL, and for the ATRP polymerization of MMA and PS, respectively. Following CNT functionalization with the precursor initiator moieties, the simultaneous SIPs of different monomers followed, under appropriate conditions, leading to the formation of Janus-type polymer/MWNT nanostructures. Nanostructures of the types MWNT-*g*-(PCL/PS), MWNT-*g*-(PCL/PMMA), MWNT-*g*-(PLA/PS), and MWNT-*g*-(PLA/PMMA) were obtained as revealed by NMR, FTIR, Raman,

TEM, and TGA characterization. These Janus hybrid nanostructures were found to enrich the interface between two immiscible organic solvents for the two different grafted chains.

Mountrichas et al. have reported the preparation of polymer-grafted nanohorns by an anionic polymerization "grafting to" scheme (Mountrichas et al. 2007d). Living PI and poly(styrene-*b*-isoprene) chains, produced by *s*-BuLi initiation, were grafted onto purified CNHs in benzene solutions under vacuum conditions, in order to ensure stability of the active anionic chain ends. TEM observations together with FITR, Raman, and TGA characterization proved the successful synthesis of the hybrid core/shell-like nanostructures. The hybrids were soluble in nonpolar organic solvents for the polymeric chains. The same group reported the attachment of PEO chains on CNHs utilizing a "grafting to" scheme (Mountrichas et al. 2009c). Monohydroxyl end-functionalized PEO was coupled to the carboxylic acid groups present on the surface of CNHs, after their transformation to alkyl chlorides, creating a covalent link between the PEO chains and the carbon nanostructures. The hybrid colloids thus obtained were soluble in water. The particular NPs could be incorporated in a poly(*p*-hydroxyl styrene) (PHOS) polymer matrix due to hydrogen bonding interactions between PEO and PHOS segments.

In a more straightforward approach, Mountrichas et al. synthesized CNHs with grafted PMAA chains by conventional bulk radical polymerization. In situ formation of radicals in the CNH/AIBN/MAA system allowed for creation of radicals onto the carbon network. These sites initiated the polymerization of MAA forming PMAA-decorated CNHs. The attachment of PMAA chains provided water solubility to the hybrid nanostructure. The polyelectrolyte chains were further utilized as templates for the formation of AuNPs from metal precursors through reduction. TEM images showed the existence of rather uniform AuNPs on the surface of CNHs, with the PMAA corona acting as the glue between the inorganic components of the hybrid nanoensemble and rendering the nanostructures with water solubility (Mountrichas et al. 2010).

REFERENCES

Advincula, R. C. 2003. *J. Disper. Sci. Technol.* 24:343–361.

Advincula, R. 2006. *Adv. Polym. Sci.* 197:107–136.

Advincula, R., Zhou, Q. G., Park, M., Wang, S. G., Mays, J. W., Sakellariou, G., Pispas, S., Hadjichristidis, N. 2002. *Langmuir* 18:8672–8684.

Akerman, M.E., Chan, W. C., Laakkonen, W. P., Bhatia, S. N., Ruoslahti, E. 2002. *PNAS* 99:12617–12621.

Akiyama, Y., Mori, T., Katayama, Y., Niidome, T. 2009. *J. Control. Release* 139:81–84.

Alexandridis, P., Sakai, T. 2004a. *Langmuir* 20:8426–8432.

Alexandridis, P., Sakai, T. 2004b. *Polym. Mater. Sci. Eng.* 91:939–945.

Alexandridis, P., Sakai, T. 2005a. *Nanotechnology* 16:S344–S350.

Alexandridis, P., Sakai, T. 2005b. *J. Phys. Chem. B* 109:7766–7772.

Alexandridis, P., Sakai, T. 2006a. *Chem. Mater.* 18:2577–2583.

Alexandridis, P., Sakai, T. 2006b. *Mater. Lett.* 60:1983–1987.

Angelatos, A. S., Radt, B., Caruso, F. 2005. *J. Phys. Chem. B* 109:3071–3076.

Antonietti, M., Wenz, E., Bronstein, L., Seregina, M. 1995. *Adv. Mater.* 7:1000–1005.

Arias, J. L., Gallardo, V., Gomez-Lopera, S. A., Plaza, R. C., Delgado, A. V. 2001. *J. Control. Release* 77:309–321.

Aslam, M., Fu, L., Vijayamohanan, K., Dravid, V. P. 2004. *J. Mater. Chem.* 14:1795–1799.

Azzam, T., Bronstein, L., Eisenberg, A. 2008. *Langmuir* 24:6521–6527.

Bakandritsos, A., Mattheolabakis, G., Chatzikyriakos, G., Szabo, T., Tzitzios, V., Kouzoudis, D., Couris, S., Avgoustakis, K. 2011. *Adv. Funct. Mater.* 21:1465–1475.

Bakandritsos, A., Psarras, G. C., Boukos, N. 2008. *Langmuir* 24:11489–11496.

Bartholome, C., Beyou, E., Bourgeat-Lami, E., Chaumont, P., Zydowicz, N. 2003. *Macromolecules* 36:7946–7952.

Basina, G., Mountrichas, G., Devlin, E., Boukos, N., Niarchos, D., Petridis, D. Pispas, S., Tzitzios, V. 2009. *J. Nanosci. Nanotechnol.* 9:4753–4759.

Baskaran, D., Mays, J. W., Bratcher, M. S. 2004. *Angew. Chem. Int. Ed.* 43:2138–2142.

Baskaran, D., Mays, J. W., S. Bratcher, M. S. 2005. *Chem. Mater.* 17:3389–3397.

Binder, W. H., Sachsenhofer, R., Farnik, D., Blaas, D. 2007. *Phys. Chem. Chem. Phys.* 9:6435–6441.

Boettcher, H., Hallensleben, M. L., Nub, S., Wurm, H. 2000. *Polym. Bull.* 44:223–228.

Bousquet, A., Boyer, C., Davis, T. P., Stenzel, M. H. 2010. *Polym. Chem.* 1:1186–1195.

Boyen, H. G., Kastle, G., Zurn, K., Herzog, T., Weigl, F., Ziemann, P., Mayer, O., Jerome, C., Moeller, M., Spatz, J. P., Garnier, M. G., Oelhagen, P. 2003. *Adv. Funct. Mater.* 13:359–363.

Boyer, C., Bulmus, V., Priyanto, P., Teoh, W. Y., Amal, R., Davis, T. P. 2009b. *J. Mater. Chem.* 19:111–123.

Boyer, C., Whittaker, M. R., Bulmus, V., Liu, J., Davis, T. P. 2010b. *NPG Asia Mater.* 2:23–30.

Boyer, C., Whittaker, M. R., Chuah, K., Liu, J., Davis, T. P. 2010a. *Langmuir* 26:2721–2730.

Boyer, C., Whittaker, M. R., Luzon, M., Davis, T. P. 2009a. *Macromolecules* 42:6917–6926.

Brittain, W. J., Boyes, S. G., Granville, A. M., Baum, M., Mirous, B. K., Akgun, B., Zhao, B., Blickle, C., Foster, M. D. 2006. *Adv. Polym. Sci.* 198:125–147.

Bronstein, L. M., Chernyshov, D. M., Timofeeva, G. I., Valetsky, L. V., Obolonkova, E. S., Khokhlov, A. R. 2000b. *Langmuir* 16:3626–3632.

Bronstein, L. M., Chernyshov, D. M., Volkov, I. O., Ezernitskaya, M. G., Valetsky, P. M., Matveeva, V. G., Sulman, E. M. 2000a. *J. Catal.* 196:302–308.

Bronstein, L. M., Kostylev, M., Shtykova, E., Vlahu, T., Huang, X., Stein, B. D., Bykov, A., Remmes, N. B., Baxter, D. V., Svergun, D. I. 2008. *Langmuir* 24:12618–12623.

Bronstein, L. M, Sidorov, S. N., Valetsky, P. M., Hartmann, J., Cölfen, H., Antonietti, M. 1999. *Langmuir* 15:6256–6260.

Buchmeiser, M. R. 2006. *Adv. Polym. Sci.* 197:137–171.

Burns, A., Ow, H., Weisner, U. 2006. *Chem. Soc. Rev.* 35:1028–1042.

Cao, Z., Walter, C., Landfester, K., Wu, Z., Ziener, U. 2011. *Langmuir* 27:9849–9859.

Cao, Z., Wang, Z., Herrmann, C., Landfester, K., Ziener, U. 2010. *Langmuir* 26:18008–18015.

Carrot, G., Diamanti, S., Manuszak, M., Charleux, B., Vairon, J. P. 2001. *J. Polym. Sci., Part A: Polym. Chem.* 39:4294–4299.

Carrot, G., Rutot-Houze, D., Pottier, A., Degee, P., Hilborn, J., Dubois, P. 2002. *Macromolecules* 35:8400–8406.

Caruso, F., Spasova, M., Susha, A., Giersig, M, Caruso, R. A. 2001. *Chem. Mat.* 13:109–116.

Chatterjee, U., Jewrajka, S. K. 2007. *J. Colloid Interface Sci.* 313:717–723.

Chen, C. C., Hsu, C. H., Kuo, P. L. 2007a. *Langmuir* 23:6801–6806.

Chen, X., An, Y., Zhao, D., He, Z., Cheng, J., Shi, L. 2008. *Langmuir* 15:8198–8204.

Chen, X., Liu, Y., An, Y., Lu, J., Li, J., Xiong, D., Shi, L. 2007b. *Macromol. Rapid Commun.* 28:1350–1355.

Chen, X., Randall, D. P., Perruchot, C., Watts, J. F., Patten, T. E., von Werne, T., Armes, S. P. 2003. *J. Colloid Interface Sci.* 257:56–62.

Chen, Y., Bose, A., Bothun, G. D. 2010. *ACS Nano* 4:3215–3218.
Chinthamanipeta, P. S., Kobukata, S., Nakata, H., Shipp, D. A. 2008. *Polymer* 49:5636–5642.
Costoyas, A., Ramos, J., Forcada, J. 2009. *J. Polym. Sci. Part A: Polym. Chem.* 47:935–948.
De, M., Ghosh, P. S., Rotello, V. M. 2008. *Adv. Mater.* 20:4225–4234.
De Koker, S., De Geest, B. G., Cuvelier, C., Ferdinande, L., Deckers, W., Hennink, W. E., De Smedt, S., Mertens, N. 2007. *Adv. Funct. Mat.* 17:3754–3763.
Decher, G. 1997. *Science* 277:1232–1237.
Diana, F. S., Lee, S. H., Petroff, P.M., Kramer, E. J. 2003. *Nano Lett.* 3:891–895.
Ding, S.-Y., Rumbles, G., Lones, M., Tucker, M. P., Nedeljkovic, J., Simon, M. N., Wall, J. S., Himmel, M. E. 2004. *Macromol. Mater. Eng.* 289:622–628.
Douglas, F., Yañez, R., Ros, J., Marin, S., De la Escosura-Muñiz, A., Alegret, S., Merkoci, A. 2008. *J. Nanopart. Res.* 10:97–104.
Du, J., Armes, S.P. 2005a. *J. Am. Chem. Soc.* 127:12800–12801.
Du, J., Tang, Y., Lewis, A. L., Armes, S. P. 2005b. *J. Am. Chem. Soc.* 127:17982–17983.
Duxin, N., Liu, F., Vali, H., Eisenberg, A. 2005. *J. Am. Chem. Soc.* 127:10063–10069.
Edmondson, S., Osborne, V. L., Huck, W. T. S. 2004. *Chem. Soc. Rev.* 33:14–22.
El-Sayed, M. A. 2001. *Acc. Chem. Res.* 34:257–264.
Eustis, S., El-Sayed, M. A. 2006. *Chem. Soc. Rev.* 35:209–217.
Feng, L., Wang, Y., Wang, N., Ma, Y. 2009. *Polym. Bull.* 63:313–327.
Filali, M., Meier, M. A., Schuber, U. S., Gohy, J. F. 2005. *Langmuir* 21:7995–8000.
Fleischhaker, F., Zentel, R. 2005. *Chem. Mater.* 17:1346–1351.
Forster, S. 2003. *J. Mater. Chem.* 13:2671–2688.
Fustin, C. A., Colard, C., Filali, M., Guillet, P., Duwez, A. S., Meier, M. A. R., Schubert, U. S., Gohy, J. F. 2006. *Langmuir* 22:6690–6695.
Gao, J., Gu, H., Xu, B. 2009. *Acc. Chem. Res.* 42:1097–1104.
Gaponik, N., Radtchenko, I. L., Gerstenberger, M. R., Fedutik, Y. A., Sukhorukov, G. B., Rogach, A. L. 2003. *Nano Lett.* 3:369–372.
Gaponik, N., Radtchenko, I. L., Sukhorukov, G. B. Rogach, A. L. 2004. *Langmuir* 20:1449–1452.
Gatsouli, K. D., Pispas, S., Kamitsos, E. I. 2007. *J. Phys. Chem. C* 111:15201–15209.
Gupta, A. K., Gupta, M. 2005. *Biomaterials* 26:3995–3102.
Hauser, C. P., Thielemann, D. T., Adlung, M., Wickleder, C., Roesky, P. W., Weiss, C. K., Landfester, K. 2011. *Macromol. Chem. Phys.* 212:286–296.
He, Q., Zhang, J., Shi, J., Zhu, Z., Zhang, L., Bu, W., Guo, L., Chen, Y. 2010. *Biomaterials* 31:1085–1092.
Hezinger, A. F. E., Tebmar, J., Gopferich, A. 2008. *Eur. J. Pharm. Biopharm.* 68:138–152.
Hickey, R. J., Sanchez-Gaytan, B. L., Cui, W., Composto, R. J., Fryd, M., Wayland, B. B., Park, S. J. 2010. *Small* 6:48–51.
Holzapfel, V., Lorenz, M., Weiss, C. K., Schrezenmeier, H., Landfester, K., Mailander, V. 2006. *J. Phys.: Condens. Matter* 18:S2581–S2594.
Hou, G., Zhu, L., Chen, D., Jiang, M. 2007. *Macromolecules* 40:2134–2140.
Hu, F. X., Neoh, K. G., Cen, L., Kang, E. T. 2006. *Biomacromolecules* 7:809–816.
Huang, B., Tomalia, D. A. 2005. *J. Luminesc.* 111:215–223.
Huang, B., Tomalia, D. A. 2006. *Inorg. Chim. Acta* 359:1951–1966.
Huang, X., Jain, P. K., El-Sayed, I. H., El-Sayed, M. A. 2007. *Nanomedicine* 2:681–688.
Iwamoto, M., Kuroda, K., Zaporojtchenko, V., Hayashi, S., Faupel, F. 2003. *Eur. Phys. J. D* 24:365–371.
Ji, H., Wang, S., Yang, X. 2009. *Polymer* 50:133–140.
Joubert, M., Delaite, C., Bourgeat-Lami, E., Dumas, P. 2005. *Macromol. Rapid Commun.* 26:602–607.
Joumaa, N., Lansalot, M., Theretz, A., Elaissari, A., Sukhanova, A., Artemyev, M., Nabiev, I., Cohen, J. H. M. 2006. *Langmuir* 22:1810–1816.

Jun, J.-W., Huh, Y.-M., Choi, J., Lee, J.-H., Song, H.-T., Kim, S., Yoon, S., Kim, K.-S., Shin, J.-S., Suh, J.-S., Cheon, J. 2005. *J. Am. Chem. Soc.* 127:5732–5733.
Jun, Y.-W., Lee, J.-H., Cheon, J. 2008. *Angew. Chem. Int. Ed.* 47:5122–5127.
Kah, J. C. Y., Wong, K. Y., Neoh, K. G., Song, J. H., Fu, J. W. P., Mhaisalkar, S., Olivo, M., Sheppard, C. J. R. 2009. *J. Drug Target.* 17:181–193.
Kang, Y., Taton, T. A. 2003. *J. Am. Chem. Soc.* 125:5650–5651.
Kang, Y., Taton, T. A. 2005a. *Angew. Chem. Int. Ed.* 44:409–412.
Kang, Y., Taton, T. A. 2005b. *Macromolecules* 38:6115–6121.
Khan, M., Huck, W. T. S. 2003. *Macromolecules* 36:5088–5093.
Kim, B.-S., Taton, T. A. 2007. *Langmuir* 23:2198–2202.
Kim, D., Park, S., Lee, J. H., Jeong, Y. Y, Jon, S. 2007. *J. Am. Chem. Soc.* 129:7661–7665.
Kim, D. J., Kang, S. M., Kong, B., Kim, W.-J., Paik, H., Choi, H., Choi, I. S. 2005a. *Macromol. Chem. Phys.* 206:1941–1946.
Kim, K., Bong Lee, H., Won Lee, J., Kun Park, H., Soo Shin, K. 2008. *Langmuir* 24:7178–7183.
Kim, S., Bawendi, M. G. 2003. *J. Am. Chem. Soc.* 125:14652–14653.
Kim, S. W., Kim, T., Kim, Y. S. 2012. *Carbon* 50:3–33.
Kim, S.-W., Kim, S., Tracy, J. B., Jasanoff, A., Bawendi, M. G. 2005b. *J. Am. Chem. Soc.* 127:4556–4557.
Kind, L., Shkilnyy, A., Schlaad, H., Meier, W., Taubert, A. 2010. *Colloid Polym. Sci.* 288:1645–1650.
Krack, M., Hohenberg, H., Kornowski, A., Lindner, P., Weller, H., Forster, S. 2008. *J. Am. Chem. Soc.* 130:7315–7320.
Laicera, C. S. T., Mrozeka, R. A., Taton, T. A. 2007. *Polymer* 48:1316–1328.
Laurent, S., Forge, D., Port, M., Roch, A., Robic, C., Elst, L. V., Muller, R. N. 2008. *Chem. Rev.* 108:2064–2110.
Lecommandoux, S., Sandre, C. O. F., Rodriguez-Hernandez, J., Perzynski, R. 2005. *Adv. Mater.* 17:712–718.
Lecommandoux, S., Sandre, C. O. F., Rodriguez-Hernandez, J., Perzynski, R. 2006a. *J. Magn. Magn. Mater.* 300:71–74.
Lecommandoux, S., Sandre, C. O. F., Rodriguez-Hernandez, J., Perzynski, R. 2006b. *Prog. Solid State Chem.* 34:171–179.
Li, C., Benicewicz, B. C. 2005b. *Macromolecules* 38:5929–5935.
Li, D., He, Q., Yang, Y., Möhwald, H., Li, J. 2008. *Macromolecules* 41:7254–7259.
Li, D., Sheng, X., Zhao, B. 2005a. *J. Am. Chem. Soc.* 127:6248–6252.
Liu, G., Xie, J., Zhang, F., Wang, Z. Y., Luo, K., Zhu, L., Quan, Q. M., Niu, G., Lee, S., Ai, H., Chen, X. Y. 2011b. *Small* 7:2742–2749.
Liu, G. J., Yan, X. H., Li, Z., Zhou, J. Y., Duncan, S. 2003. *J. Am. Chem. Soc.* 125:14039–14045.
Liu, J., He, W., Zhang, L., Zhang, Z., Zhu, J., Yuan, L., Chen, H., Cheng, Z., and Zhu, X. 2011a. *Langmuir* 27:12684–12692.
Li, J., Shi, L., An, Y., Li, Y., Chen, X., Dong, H. 2006. *Polymer* 47:8480–8487.
Li, Q., Zhang, L., Bai, L., Zhang, Z., Zhu, J., Zhou, N., Cheng, Z., Zhu, X. 2011. *Soft Matter* 7:6958–6966.
Luccardini, C., Tribet, C., Vial, F., Marchi-Artzner, V., Dahan, M. 2006. *Langmuir* 22:2304–2310.
Mai, Y., Eisenberg, A. 2010. *J. Am. Chem. Soc.* 132:10078–10084.
Malikova, N., Pastiza-Santos, I., Schierhorn, M., Kotov, N. A., Liz-Marzan, L. M. 2002. *Langmuir* 18:3694–3699.
Mandal, D., Chatterjee, U. 2007. *J. Chem. Phys.* 126:134507–134513.
Manea, F., Houillon, F. B., Pasquato, L., Scrimin, P. 2004. *Angew. Chem. Int. Ed.* 43:6165–6169.
Mantzaridis, C., Pispas, S. 2008. *Macromol. Rapid Commun.* 29:1793–1796.
Mattoussi, H., Mauro, J. M., Goldman, E. R., Anderson, G. P., Sundar, V. C., Mikulec, F. V., Bawendi, M. G. 2000. *J. Am. Chem. Soc.* 122:12142–12150.

Mayer, A. B. R. 2001. *Polym. Adv. Technol.* 12:96–102.
Mayer, A. B. R., Mark, J. E. 1996. *Polym. Prepr. (Am. Chem. Soc., Polym. Chem. Div.)* 74:459–460.
Merican, Z., Schiller, T. L., Hawker, C. J., Fredericks, P. M., Blakey, I. 2007. *Langmuir* 23:10539–10545.
Meristoudi, A., Pispas, S. 2009. *Polymer* 50:2743–2751.
Meristoudi, A., Pispas, S., Vainos, N. 2008. *J. Polym. Sci. Part B: Polym. Phys.* 46:1515–1524.
Mill, G., Longenbeger, L. 1995. *J. Phys. Chem.* 99:475–481.
Miyamoto, D., Oishi, M., Keitaro, K. K., Nagasaki, Y. 2008. *Langmuir* 24:5010–5017.
Moffitt, M., Eisenberg, A. 1995a. *Chem. Mater.* 7, 1178–1184.
Moffitt, M., McMahon, L., Pessel, V., Eisenberg, A. 1995b. *Chem. Mater.* 7:1185–1192.
Mori, H., Seng, D. C., Zhang, M., Muller, A. H. E. 2002. *Langmuir* 18:3682–3687.
Mössmer, S., Spatz, J. P., Möller, M., Aberle, T., Scmidt, J., Burchard, W. 2000. *Macromolecules* 33:4791–4797.
Mountrichas, G., Ichihashi, T., Pispas, S., Yudasaka, M., Iijima, S., Tagmatarchis, N. 2009a. *J. Phys. Chem. C* 113:5444–5449.
Mountrichas, G., Pispas, S., Ichihasi, T., Yudasaka, M., Iijima, S., Tagmatarchis, N. 2010. *Chem. Eur. J.* 16:5927–5933.
Mountrichas, G., Pispas, S., Tagmatarchis, N. 2007b. *Small* 3:404–407.
Mountrichas, G., Pispas, S., Tagmatarchis, T. 2007d. *Chem. Eur. J.* 13:7595–7599.
Mountrichas, G., Pispas, S., Xenogiannopoulou, E., Aloukos, P., Couris, S. 2007a. *J. Phys. Chem. B* 111:4315–4319.
Mountrichas, G., Sandanayaka, A. S. D., Economopoulos, S. P., Pispas, S., Ito, O., Hasobe, T., Tagmatarchis, N. 2009b. *J. Mater. Chem.* 19:8990–8998.
Mountrichas, G., Tagmatrchis, N., Pispas, S. 2007c. *J. Phys. Chem. B* 111:8369–8372.
Mountrichas, G., Tagmatarchis, N., Pispas, S. 2009c. *J. Nanosci. Nanotechnol.* 9:3775–9.
Mueller, W., Koynov, K., Fischer, K., Hartmann, S., Pierrat, S., Baschee, T., Maskos, M. 2008. *Macromolecules* 42:357–361.
Nann, T. 2005. *Chem. Commun.* 13:1735–1736.
Nasongkla, N., Bey, E., Ren, J., Ai, H., Khemtong, C., Guthi, J. S., Chin, S. F., Sherry, A. D., Boothman, D. A., Gao, J. 2006. *Nano Lett.* 6:1427–1430.
Neoh, K. G., Kang, E. T. 2011. *Polym. Chem.* 2:747–759.
Newman, J. D. S., Blanchard, G. J. 2006. *Langmuir* 22:5882–5888.
Niidome, T., Akiyama, Y., Yamagata, M., Kawano, T., Mori, T., Niidome, Y., Katayama, Y. 2009. *J. Biomater. Sci. Polym. Ed.* 20:1203–1215.
Niidome, T., Yamagata, M., Okamoto, Y., Akiyama, Y., Takahashi, H., Kawano, T., Katayama, Y., Niidome, Y. 2006. *J. Control. Release* 114:343–347.
Nuopponen, M., Tenhu, H. 2007. *Langmuir* 23:5352–5357.
Ofir, Y., Samanta, B., Rotello, V. M. 2008. *Chem. Soc. Rev.* 37:1814–1820.
Pan, B., Gao, F., He, R., Cui, D., Zhang, Y. 2006. *J. Colloid Interface Sci.* 297:151–156.
Peng, S., Wang, C., Xie, J., Sun, S. 2006. *J. Am. Chem. Soc.* 128:10676.
Piao, Y., Burns, A., Kim, J., Wiesner, U., Hyeon, T. 2008. *Adv. Funct. Mater.* 18:3745–3758.
Platonova, O. A., Bronstein, L. M., Solodovnikov, S. P., Yanovskaya, I. M., Obolonkova, E. S., Valetsky, P. M., Wenz, E., Antonietti, M. 1997. *Colloid Polym. Sci.* 275:426–431.
Potapova, I., Mruk, R., Prehl, S., Zentel, R., Basche, T., Mews, A. 2003. *J. Am. Chem. Soc.* 125(2003):320–321.
Prabaharan, M., Grailer, J. J., Pilla, S., Steeber, D. A., Gong, S. 2009. *Biomaterials* 30:6065–6075.
Pradhan, P., Giri, J., Rieken, F., Koch, C., Mykhaylyk, O., Doblinger, M., Banerjee, R., Bahadur, D., Plank, C. 2010. *J. Control. Release* 142:108–121.
Priftis, D., Petzetakis, N., Sakellariou, G., Pitsikalis, M., Baskaran, D., Mays, J. W., Hadjichristidis, N. 2009. *Macromolecules.* 42:3340–3346.

Priftis, D., Sakellariou, G., Baskaran, D., Mays, J. W., Hadjichristidis, N. 2009. *Soft Matter* 5:4272–4278.

Pyun, J., Jia, S., Kowalewski, T., Patterson, G. D., Matyjaszewski, K. 2003. *Macromolecules* 36:5094–5104.

Radhakrishnan, B., Constable, A. N., Brittain, W. J. 2008. *Macromol. Rapid Commun.* 29:1828–1833.

Radhakrishnan, B., Ranjan, R., Brittain, W. J. 2006. *Soft Matter* 2:386–396.

Ramirez, L. P., Landfester, K. 2003. *Macromol. Chem. Phys.* 204:22–31.

Ramos, J., Forcada, J. 2011. *Langmuir* 27:7222–7230.

Ranjan, R., Brittain, W. J. 2007. *Macromolecules* 40:6217–6223.

Ranjan, R., Brittain, W. J. 2008. *Macromol. Rapid Commun.* 29:1104–1110.

Rogach, A. L., Susha, A., Caruso, F., Sukhorukov, G., Kornowski, A., Kershaw, S., Mohwald, H., Eychmüller, A., Weller, H. 2000. *Adv. Mater.* 12:333–337.

Sakellariou, G., Ji, H., Mays, J. W., Baskaran, D. 2008. *Chem. Mater.* 20:6217–6230.

Sanson, C., Diou, O., Ibarboure, E., Soum, A., Brulet, A., Miraux, S., Sandre, O., Lecommandoux, S. 2011. *ACS Nano* 5:1122–1140.

Schneider, G. F., Subr, V., Ulbrich, K., Decher, G. 2009. *Nano Lett.* 9:636–642.

Schrade, A., Mikhalevich, V., Landfester, K., Ziener, U. 2011. *J. Polym. Sci. Part A: Polym. Chem.* 49:4735–4746.

Shenhar, R., Norsten, T. B., Rotello, V. M. 2005. *Adv. Mater.* 17:657–663.

Shibasaki, Y., Kim, B.-S., Young, A. J., McLoon, A. L., Ekker, S. C., Taton, T. A. 2009. *J. Mater. Chem.* 19:6324–6327.

Shipway, A. N., Katz, E., Willner, I. 2000. *Phys. Chem.* 1:18–25.

Siegers, C., Biesalski, M., Haag, R. 2004. *Chem.-Eur. J.* 10:2831–2838.

Skirtach, A. G., Antipov, A. A., Shchukin, D. G., Sukhorukov, G. B. 2004. *Langmuir* 20:6988–6992.

Skirtach, A. G., De Geest, B. G., Mamedov, A., Antipov, A. A., Kotov, N. A., Sukhorukov, G. B. 2007. *J. Mat. Chem.* 17:1050–1054.

Skirtach, A. G., Dejugnat, C., Braun, D., Susha, A. S., Rogach, A. L., Parak, W. J., Mohwald, H., Sukhorukov, G. B. 2005. *Nano Lett.* 5:1371–1377.

Song, J. B., Cheng, L., Liu, A. P., Yin, J., Kuang, M., Duan, H. W. 2011. *J. Am. Chem. Soc.* 133:10760–10763.

Sperling, R. A., Rivera-Gill, P., Zhang, F., Zanella, M., Parak, W. J. 2008. *Chem. Soc. Rev.* 37:1896–1902.

Srivastava, S., Kotov, N. A. 2008. *Acc. Chem. Res.* 41:1831–1841.

Sukhorukov, G. B., Donath, E., Lichtenfeld, H., Knippel, E., Knippel, M., Budde, A., Mohwald, H. 1998. *Coll. Surf. A* 137:253–266.

Sulman, E. M., Matveeva, V. G., Sulman, M. G., Demidenko, G. N., Valetsky, P. M., Stein, B., Mastes, T., Bronstein, L. M. 2009. *J. Catal.* 262:150–156.

Sun, C., Sze, R., Zhang, M. 2006. *J. Biomed. Mater. Res., Part A* 78A:550–557.

Sun, S., Zeng, H., Robinson, D. B., Raoux, S., Rice, P. M., Wang, S. X., Li, G. 2004. *J. Am. Chem. Soc.* 126:273–279.

Sun, X., Jiang, X., Dong, S., Wang, E. 2003. *Macromol. Rapid Commun.* 24:1024–1028.

Sun, Y., Riggs, J. E. 1999. *Int. Rev. Phys. Chem.* 18:43–49.

Tan, Y., Qian, W., Ding, S., Wang, Y. 2006. *Chem. Mater.* 18:3385–3391.

Tasis, D., Pispas, S., Galiotis, C., Bouropoulos, N. 2007. *Mater. Lett.* 61:5044–5046.

Tasis, D., Tagmatarchis, N., Bianco, A., Prato, M. 2006. *Chem. Rev.* 106:1105–1136.

Thierry, B., Zimmer, L., McNiven, S., Finnie, K., Barbe, C., Griesser, H. J. 2008. *Langmuir* 24:8143–8150.

Thomas, K. G., Kamat, P. V. 2003. *Acc. Chem. Res.* 36:888–893.

Tong, L., He, W., Zhang, Y., Zheng, W., Cheng, J. X. 2009. *Langmuir* 25:12454–12459.

Topfer, O., Schmidt-Naake, G. 2007. *Macromol. Symp.* 248:239–248.

Tsujii, Y., Ohno, K., Yamamoto, S., Goto, A., Fukuda, T. 2006. *Adv. Polym. Sci.* 197:1–45.
Van Berkel, K. Y., Hawker, C. J. 2010. *J. Polym. Sci. Part A: Polym. Chem.* 48:1594–1606.
Von Werne, T., Patten, T. E. 1999. *J. Am. Chem. Soc.* 121:7409–7410.
Wagner, K., Kautz, A., Röder, M., Schwalbe, M., Pachmann, K., Clement, J. H., Schnabelrauch. M. 2004. *Appl. Organomet. Chem.* 18:514–519.
Wall, J. S., Himmel, M. E. 2004. *Macromol. Mater. Eng.* 289:622–628.
Wang, D., Rogach, A. L., Caruso, F. 2002. *Nano Lett.* 2:857–861.
Wang, H., Patil, A. J., Liu, K., Petrov, S., Mann, S., Winnik, M. A., Manners, I. 2009b. *Adv. Mater.* 21:1805–1809.
Wang, L., Neoh, K. G., Kang, E. T., Shuter, B., Wang, S. C. 2009a. *Adv. Funct. Mater.* 19:2615–2622.
Wang, M., Felorzabihi, N., Guerin, G., Haley, J. C., Scholes, G. D., Winnik, M. A. 2007. *Macromolecules* 40:6377–6384.
Wang, M., Oh, J. K., Dykstra, T. E., Lou, X., Scholes, G. D., Winnik, M. A. 2006. *Macromolecules* 39:3664–3672.
Wang, S., Zhou, Y., Yang, S., Ding, B. 2008b. *Colloids Surf. B* 67:122–126.
Wang, W., Cao, H., Zhu, G., Wang, P. 2010. *J. Polym. Sci. Part A: Polym. Chem.* 48:1782–1790.
Wang, X. S., Armes, S. P. 2000. *Macromolecules* 33:6640–6645.
Wang, X. S., Lascelles, S. F., Jackson, R. A., Armes, S. P. 1999. *Chem. Commun.* 8:1817–1820.
Wang, X.-S., Metanawin, T., Zheng, X.-Y., Wang, P. Y., Ali, M., Vernon, D. 2008a. *Langmuir* 24:9230–9232.
Wang, Y., Herron, N. 1991. *J. Phys. Chem.* 95:525–532.
Wisher, A. C., Bronstein, I., Chechik, V. 2006. *Chem. Commun.* 15:1637–1639.
Wotschadlo, J., Liebert, T., Heinze, T., Wagner, K., Schnabelrauch, M., Dutz, S., Müller, R. et al. 2009. *J. Magn. Magn. Mater.* 321:1469–1474.
Wu, T., Zou, G., Hu, J., Liu, S. 2009. *Chem. Mater.* 21:3788–3798.
Wu, Y., Zhang, Y., Xu, J., Chen M., Wu, L. 2010. *J. Colloid Interface Sci.* 343:18–24.
Xu, C. J., Xu, K. M., Gu, H. W., Zheng, R. K., Liu, H., Zhang, X. X., Guo, Z. H., Xu, B. 2004. *J. Am. Chem. Soc.* 126:9938–9939.
Xu, H., Xu, J., Jiang, X., Zhu, Z., Rao, J., Yin, J., Wu, T., Liu, H., Liu, S. 2007. *Chem. Mater.* 19:2489–2495.
Xu, J., Wang, H., Liu, C., Yang, Y., Chen, T., Wang, Y., Wang, F., Liu, X., Xing, B., Chen, H. 2010. *J. Am. Chem. Soc.* 132:11920–11922.
Xu, Y., Yuan, J., Fang, B., Drechsler, M., Mullner, M., Bolisetty, S., Ballauff. M., Muller, A. H. E. 2010a. *Adv. Funct. Mater.* 20:4182–4186.
Yan, X. H., Liu, G. J., Haussler, M., Tang, B. Z. 2005. *Chem. Mater.* 17:6053–6059.
Yan, X. H., Liu, G. J., Li, Z. 2004. *J. Am. Chem. Soc.* 126:10059–10063.
Yan, X. H., Liu, G. J., Liu, F. T., Tang, B. Z., Peng, H., Pakhomov, A. B., Wong, C. Y. 2001. *Angew. Chem. Int. Ed.* 40:3593–3597.
Yuan, J. J., Schmid, A., Armes, S. P. 2006. *Langmuir* 22:11022–11028.
Zebli, B., Susha, A. S., Sukhorukov, G. B., Rogach, A. L., Parak, W. J. 2005. *Langmuir* 21:4262–4265.
Zhang, D., Qi, L., Ma, J., Cheng, H. 2001. *Chem. Mater.* 13:2753–2759.
Zhang, J., Yang, J., Wu, Q., Wu, M., Liu, N., Jin, Z., Wang, Y. 2010b. *Macromolecules* 43:1188–1190.
Zhang, M., Estournes, C., Bietsch, W., Muller, A. H. E. 2004. *Adv. Funct. Mater.* 14:871–878.
Zhang, M., Wang, M., He, S., Qian, J., Saffari, A., Lee, A., Kumar, S., Hassan, Y., Guenther, A., Scholes, G., Winnik, M. A. 2010a. *Macromolecules* 43:5066–5074.
Zhang, Y., Kohler, N., Zhang, M. 2002. *Biomaterials* 23:1553–1561.
Zhang, Y., Zhang, J. 2005. *J. Colloid Interface Sci.* 283:352–357.
Zhao, H., Douglas, E. P. 2002. *Chem. Mater.* 14:1418–1423.

Zhao, H., Douglas, E. P., Harrison, B. S., Schanze, K. S. 2001. *Langmuir* 17:8428–8434.
Zheng, P., Jiang, X., Zhang, X., Zhang, W., Shi, L. 2006. *Langmuir* 22:9393–9396.
Zhou, L., Gao, C., Xu, W., Wang, X., Xu, Y. 2009. *Biomacromolecules* 10:1865–1874.
Zhou, Q., Wang, S., Fan, X., Advincula, R. 2002. *Langmuir* 18:3324–3329.
Zhu, L., Zheng, X., Liu, X. 2004. *J. Colloid Interface Sci.* 273:155–161.

5 Biological Applications of Polymeric Nanoparticles

The preparation and physicochemical properties of several distinct classes of polymeric nanoparticles have been presented in Chapters 2 through 4. This chapter focuses on present and potential biological applications of polymer-containing nanoparticles and nanoassemblies with emphasis given to drug and gene delivery nanocarriers, imaging and diagnostics, and others. The general ideas and principles that allow the use of different polymeric nanoparticles according to the specific application are also briefly discussed. The focus is placed on polymeric nanostructures and nanocolloids that their potential applications have been demonstrated already to a significant degree. There are a number of excellent reviews discussing the applications of polymeric nanoparticles, especially of block copolymer micelles and other self-assembled structures, on drug delivery (Kwon and Forrest 2006, Wiradharma et al. 2009, Du et al. 2010, Horgan et al. 2010, Kedar et al. 2010, Mishra et al. 2010, Ponta and Bae 2010, Tyrrell et al. 2010, Miyata et al. 2011, Xiong et al. 2011, Wagner 2012).

5.1 DELIVERY OF LOW-MOLECULAR-WEIGHT DRUGS

Several pharmaceutical compounds, which show therapeutic activity toward a number of diseases and especially toward cancer tumors, are hydrophobic, having very low solubility in aqueous media. Some of them also show side effects, like nonspecific cytotoxicity, so their administration should be kept within certain concentration windows (Langer 1998). The use of specifically designed polymeric nanocarriers for the solubilization, transfer, and delivery/release of hydrophobic drugs is a rapidly developing field in nanomedicine. Among such nanodelivery systems, polymeric micelles formed by amphiphilic block copolymers stand as a highly interesting and promising class of self-assembled nanocarriers (Kwon and Forrest 2006, Miyata et al. 2011, Xiong et al. 2011). The interest is primarily due to the great versatility of block copolymer–based micelles through the flexible manipulation of their chemical structure and properties via synthetic chemistry. In such nanocarrier systems, the hydrophobic core plays the role of the drug-containing compartment, and the micellar shell acts as the interface toward the surrounding biological media providing stabilization, stealth, and targeting properties. In principle, other parts of the micellar carrier, like the micellar corona and the core/corona interface, can be utilized for drug entrapment inside block copolymer micelles (Figure 5.1).

There are several aspects of the structure of amphiphilic block copolymer (AmBC) micelles that should be considered and optimized in order to obtain an advanced nanocarrier system (Xiong et al. 2011). The designing principles should

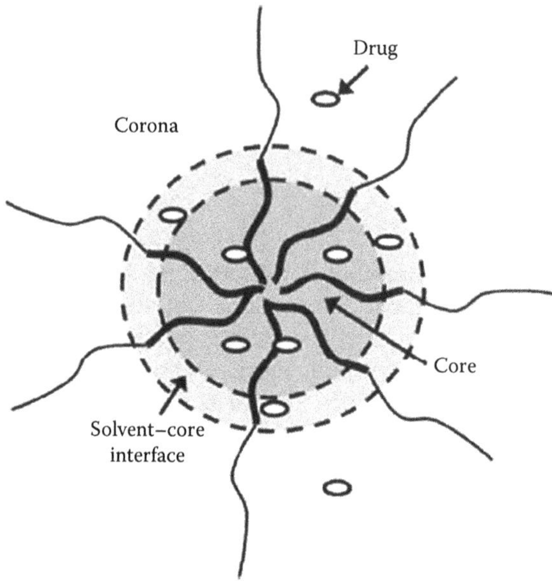

FIGURE 5.1 Illustration of the three regions for drug solubilization in a block copolymer micelle. (Reprinted from *Prog. Polym. Sci.*, 35, Tyrrell, Z.L., Shen, Y., and Radosza, M., 1128–1143, Copyright 2010, with permission from Elsevier.)

also take into account the chemical nature of the drug to be carried. Incorporation of a drug into the nanocarrier can be achieved by physical entrapment of the pharmaceutical, for example, into the hydrophobic core of the micelle, or via chemical conjugation of the drug onto one of the blocks of the copolymer. In the latter case, stability of the polymer–drug bond should be such that it allows release of the drug under certain conditions, usually met at the desired point of delivery (Figure 5.2). This approach is strongly connected to the requirement that a specific drug nanocarrier should be prepared for a specific drug, something that cannot always be achieved, since it requires specific chemistry schemes and it is usually laborious, time consuming, and cost inefficient. However, several examples of drug conjugation on the block copolymer chains can be found in the literature.

One of the early studies on micelle-forming block copolymer–drug conjugates for tumor therapy has been reported by Kataoka and coworkers (Bae et al. 2003). They investigated the physicochemical properties and biodistribution of poly(ethylene oxide-*b*-aspartate) block copolymer–adriamycin conjugates (PEO-*b*-PAsp [ADR]) in murine colon adenocarcinoma 26 (C-26) tumor-bearing mice after intravenous injection. Long circulation times in blood and a parallel reduced uptake in the major organs of the reticuloendothelial system (i.e., liver and spleen) were observed. This was consistent with the *in vivo* persistence of a micellar core/shell structure in which the block copolymer–drug conjugate has been shown to adopt in aqueous media with the PAsp(ADR) blocks forming the core and the PEO chains the micellar corona. Enhanced accumulation of the micellar PEO-*b*-PAsp (ADR) conjugate in tumors evident after 24 h (ca. 10% dose per g tumor), relative to free ADR (ca. 0.90% dose

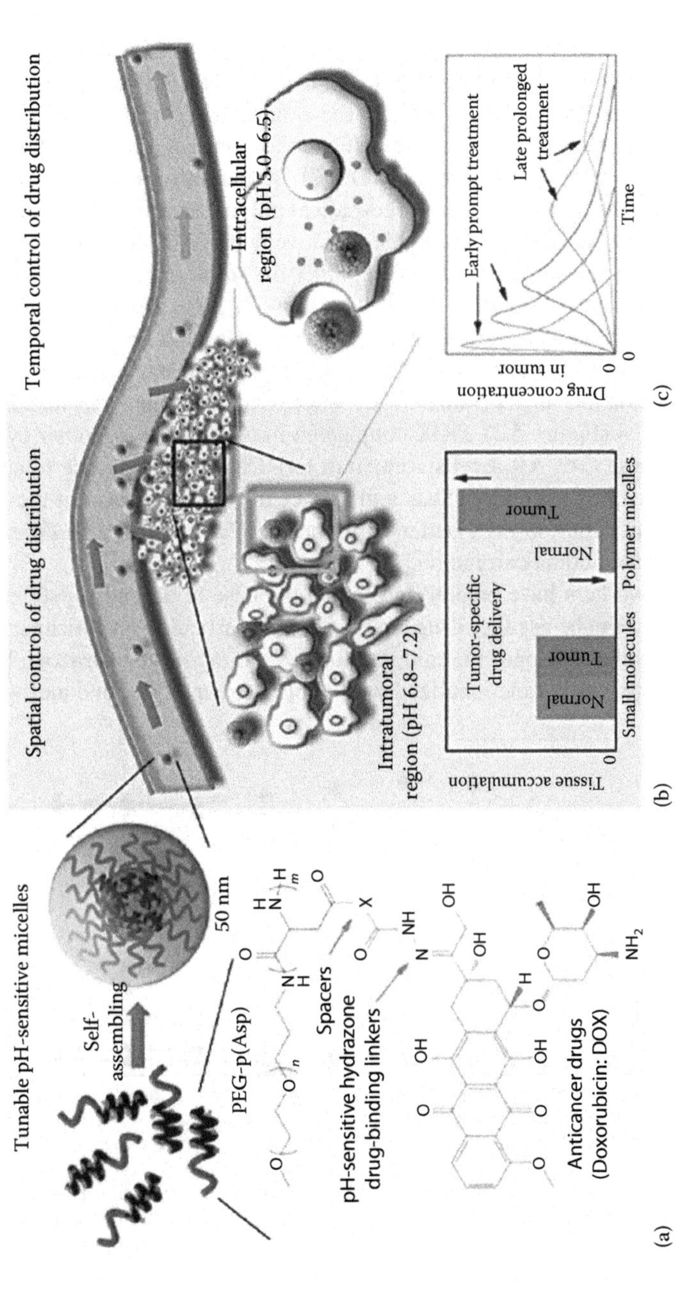

FIGURE 5.2 Desired spatial (a,b) and temporal (c) control of drug delivery to tumor microenvironment using tunable block polymer micelles with conjugated drug. (With kind permission from Springer Science + Business Media: *Pharm. Res.*, 27, 2010, 2330–2342, Ponta, A. and Bae, Y.)

per g tumor), was demonstrated. Further, peak levels of PEO-PAsp(ADR) conjugate in the heart were lower than for free ADR.

It has been reported by Hennink and coworkers that core cross-linked biodegradable polymeric micelles composed of poly(ethylene glycol)-*b*-poly[*N*-(2-hydroxypropyl)methacrylamide-lactate] (PEG-*b*-P(HPMAm-Lac_n)) diblock copolymers show prolonged circulation in the bloodstream upon intravenous administration and enhanced tumor accumulation through the enhanced permeation and retention (EPR) effect. In order to fully exploit the EPR effect for drug targeting, the group developed a doxorubicin methacrylamide (DOX-MA) derivative that was covalently incorporated into the micellar core by free-radical polymerization. The structure of the DOX derivative was susceptible to pH-sensitive hydrolysis, enabling controlled release of the drug in acidic conditions (in either the intratumoral environment or the endosomal vesicles) (Talelli et al. 2010). Between 30% and 40% w/w of the added drug was covalently entrapped, and the micelles with covalently attached DOX had an average diameter of 80 nm. The entire drug payload was released within 24 h incubation at pH 5 and 37°C, whereas only around 5% release was observed at pH 7.4 (Figure 5.3). DOX-conjugated micelles showed higher cytotoxicity in B16F10 and OVCAR-3 cells compared to DOX-MA, likely due to cellular uptake of the micelles via endocytosis and intracellular drug release in the acidic organelles. The micelles showed better antitumor activity than free DOX in mice bearing B16F10 melanoma carcinoma.

Gong and coworkers have recently presented the case of an end-functionalized ABA triblock that can be used for drug conjugation. In particular, an anticancer drug was covalently connected onto the middle block by pH-sensitive hydrazone bonds. Moreover, the polymer is functionalized with a folate group at one end and with an

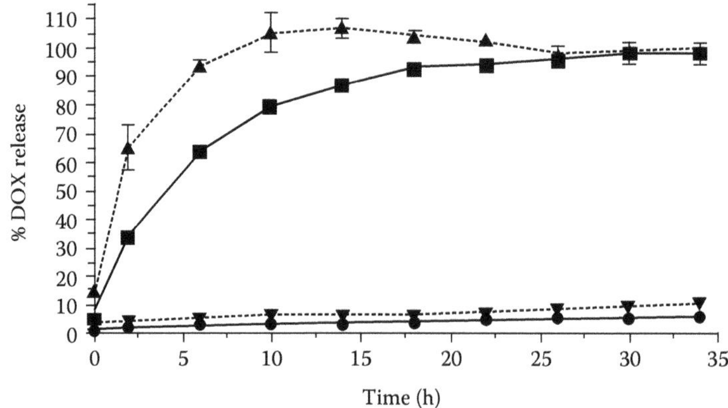

FIGURE 5.3 Release of core-covalently bound DOX from methacrylated mPEG-*b*-p((50% HPMAm-Lac1)-*co*-(50% HPMAm-Lac2)) cross-linked micelles at pH 5 (■) and pH 7.4 (●) and 37°C and of DOX from the DOX-MA monomer containing the acid-sensitive hydrazone linker at pH 5 (▲) and pH 7.4 (▼) at 37°C. (Reprinted from *J. Control. Release*, 151, Talelli, M., Rijcken, C.J.F., Oliveira, S., van der Meel, R., van Bergen en Henegouwen, P.M.P., Lammers, T., van Nostrum, C.F., Storm, G., and Hennink, W.E., 183–192, Copyright 2011, with permission from Elsevier.)

acrylate group at the other. The drug-conjugated triblock could self-assemble into stable vesicles in aqueous solution. The folate groups have been found to be located at the outer vesicle layer, providing active tumor targeting, while acrylate groups were located at the inner layer, able to provide stability through cross-linking. Moreover, magnetic nanoparticles could be encapsulated into the aqueous core of the stable vesicles, allowing ultrasensitive magnetic resonance imaging (MRI) detection. The described system can offer desirable drug release, stability, and excellent tumor-targeting activity in line with MRI ability, suitable for diagnosis purposes (Yang et al. 2010), as will be discussed in the following sections.

Well-defined hyperbranched double-hydrophilic block copolymer of poly(ethylene oxide)-hyperbranched-polyglycerol (PEO-hb-PG) was developed as an efficient drug delivery nanocarrier for the hydrophobic anticancer drug DOX (Lee et al. 2012). The authors demonstrated that the hydrophilic hyperbranched PEO-hb-PG copolymer formed micellar structures after conjugation with DOX. The drug was chemically linked to the copolymer by pH-sensitive hydrazone bonds, resulting in pH-responsive controlled release of DOX. Structural characterization of the macromolecular drug conjugate with dynamic light scattering (DLS), atomic force microscopy (AFM), and transmission electron microscopy (TEM) showed that the material assembled in spherical core–shell-type micelles with an average diameter of ca. 200 nm. Studies on the pH-responsive release of DOX and *in vitro* cytotoxicity experiments revealed the stimuli-responsive and controlled drug delivery character of the nano-system. Taking into account the several beneficial features of hyperbranched double-hydrophilic block copolymers, such as enhanced biocompatibility, increased water solubility, and drug-loading efficiency, as well as the improved clearance of the copolymer after drug release, the authors proposed that double-hydrophilic block copolymers are able to provide versatile platforms for the development of efficient drug delivery systems for effective treatment of cancer.

The approach of physical entrapment allows for greater flexibility in nanodrug formulation, since the same micellar nanocarrier can, in principle, be used for the encapsulation of a number of different hydrophobic drugs. However, also in this case, certain design aspects should be considered and completed successfully.

The first aspect concerns drug loading into the micellar nanocarrier. Drug loading should be as a high as possible, since in this way the amount of drug to be delivered can be maximized and therapeutic drug levels can be sustained for a longer time upon systemic administration. The physicochemical affinity of the pharmaceutical compound toward the hydrophobic core of the micelles plays a crucial role. The higher is the affinity, the higher should be the loading ability of the micellar nanocarrier. A measure of the aforementioned affinity is the Flory–Huggins interaction parameter (χ_{sp}); the lower the value of the parameter, the higher the affinity between the micellar core and the drug. The interaction parameter is dramatically affected by the characteristics of the drug, such as hydrophobicity, polarity, or degree of ionization in respect to the same parameters of the micellar core (Tan et al. 2010). Therefore, a systematic chemical design for preparing the desired core blocks should be followed.

Lavasanifar and coworkers have demonstrated both theoretically and experimentally that the use of block copolymer micelles with different hydrophobic blocks as the cores leads to changes of the micellar drug capacity. The Flory–Huggins interaction parameter was calculated by the group contribution method between a drug, cucurbitacin, and three different block copolymers based on PCL-*b*-PEO copolymers, namely, regular PCL-*b*-PEO copolymer and those with poly(ε-caprolactone) (PCL) blocks functionalized with benzyl and cholesteryl pendant groups. It was found that the pair with the lower χ_{sp}, that is, the cucurbitacin/cholesteryl–PCL-*b*-PEO pair, was also the system showing the highest drug loading. It is noteworthy that the release profile of the three systems was found to be influenced not only by the interaction parameter but also by the core viscosity (Molavi et al. 2008, Mahmud et al. 2009). Chemically modified PCL-*b*-PEO copolymers were also investigated for the encapsulation of amphotericin B (AmB) in block copolymer micelles. Pendant benzyl, carboxyl, stearyl, palmitoyl, and cholesteryl groups were attached on the PCL hydrophobic block. The drug-loading capacity was found to increase in the order cholesteryl < stearyl < palmitoyl < carboxyl. In the carboxyl group case, hydrogen bond formation may be also active and leads to the greater increase in drug loading compared to the precursor block copolymer micelles (Falamarzian et al. 2010a,b). In another work, micelles based on PEO-*b*-PLAsp block copolymers were utilized as drug encapsulating agents. The poly(L-aspartic acid) blocks were chemically modified with fatty acid groups increasing the hydrophobicity of this block in respect to the benzyl modified precursor block copolymer. AmB was again the drug that was encapsulated in all block copolymers investigated (Lavasanifar et al. 2002a). Results showed that the incorporation of fatty acid groups in the core-forming block increased substantially the encapsulation efficiency of the micellar nanocarriers.

In one of the early and successful studies on encapsulation of hydrophobic anticancer drugs into block copolymer micelles, the concept of physicochemical affinity between the micellar core and the drug was utilized (Kataoka et al. 2000). DOX was loaded into micelles formed by poly(ethylene glycol)-*b*-poly(*b*-benzyl-L-aspartate) (PEG-*b*-PBLA) block copolymer following an oil/water (O/W) emulsion methodology. Rather high drug-loading levels were achieved (ca. 15–20 w/w%). This was correlated to the low water solubility of DOX and the possible interactions with the benzyl side groups of PBLA segments through π–π stacking. Drug-loaded micelles had a narrow size distribution and diameters in the range 50–70 nm. It was observed that the release process for DOX from the micelles was comprised of two steps. First, a rapid release of the drug took place followed by a slow and long-lasting release step. Lowering of the media pH from 7.4 to 5.0 resulted in an acceleration of DOX release. The observed pH-sensitive release of DOX from the block copolymer micelles was connected to the protonation of the amine group of the drug, which is expected to result in an increase of its water solubility. Blood circulation of DOX was substantially improved due to its encapsulation into the micellar nanocarrier. DOX loaded in the PEG-*b*-PBLA micelles showed a considerably higher antitumor activity compared to free DOX against mouse C26 tumor by intravenous injection, indicating the potential of PEG-*b*-PBLA/DOX nanoensemble to act as a long-circulating carrier system for modulated drug delivery. Related studies have been reported by the Kataoka group even earlier (Kwon et al. 1994, 1995).

Along the same lines, camptothecin and paclitaxel (PTX) were efficiently solubilized into poly(isoprene-*b*-ethylene oxide) micelles. This was the first time that this particular type of block copolymer micelles was proposed as nanocarriers for a hydrophobic drug. Polyisoprene block is highly hydrophobic and the presence of C=C bonds allows for π–π interactions with several drugs carrying unsaturated groups. It has a low T_g and also shows chemical structure similarities with natural terpenoids. Solubilization of the drugs resulted in an increase in the size of the micelles, while no change was observed in the surface potential of the micelles. Micellar suspensions of these drugs containing up to 30 μg of the drug per mL of final solution could be prepared. Micelles loaded with drugs were stable for more than 2 weeks (Levchenco et al. 2007).

The affinity between drug and polymeric carrier can be increased by introducing specific interactions between the micellar core and the drug. Specific interactions can be enhanced by, for example, hydrogen bonding. Tan et al. have described the synthesis of a block copolymer with urea pendant groups. The existence of urea groups not only changes the physical properties of the polymer but also allows the increased loading of an anticancer drug, due to hydrogen bonding, without considerable changes on the micellar size. Furthermore, experimental results have indicated that, even though the copolymer is nontoxic, the drug-loaded micelles were able to destroy cancer cells (Tan et al. 2010).

Poly(β-lactam-isoprene-*b*-ethylene oxide) copolymers have been recently utilized for the solubilization of the novel hydrophobic anticancer drug curcumin (Gardikis et al. 2010). It was expected that the hydrogen-bonding interaction between lactam groups of the copolymer and curcumin would enhance curcumin encapsulation. The size of the drug-loaded copolymer micelles was found to be smaller than the size of the empty micelles. Drug encapsulation efficiency and loading capacity increased by an increase in the poly(β-lactam-isoprene) block in the copolymer. Encapsulation efficiency reached 78% and maximum loading capacity was ca. 2.5 mg of drug/mL of solution. Empty and loaded micelles were found substantially stable for a period of nearly 2 months. *In vitro* release of the drug was slower for the micelles possessing the highest poly(β-lactam-isoprene) content, suggesting the determining role of block copolymer characteristics on the release profile of the drug. The cytotoxic activity of the drug was also improved.

The interaction between block copolymers and pharmaceuticals can be also enhanced through electrostatic interactions, that is, when a charged block copolymer and a complementary charged drug molecule are paired. Typically, the addition of an oppositely charged drug molecule leads to the formation of a water-insoluble polyion complex between the drug and the charged block, while the second block (usually a neutral PEG/PEO chain) ensures the solubility of the complex. Following this concept, Li et al. reported the complexation of the cationic drugs dibucaine, tetracaine, and procaine with anionic poly(methacrylic acid-b-ethylene oxide) (PMAA-*b*-PEO) copolymers to give complex micelles incorporating the ionic drugs in their cores (Li et al. 2003). The block copolymer–drug pairs self-assembled to form micelle-like nanoaggregates comprised of a core of neutralized polyions (PMAA/drug complex) surrounded by the PEO corona. It was found that the properties of the aggregates strongly depended on the hydrophobicity of the anesthetics and that besides electrostatic interactions, hydrophobic interactions as well play an important role. The aggregates induced by dibucaine were found to have a highly viscous interior, whereas those induced by tetracaine had a fluid interior, due to the lesser hydrophobic effect of the latter drug. It was observed that procaine did not induce aggregate formation due to its low hydrophobicity. All these characteristics are expected to influence significantly drug encapsulation and release in such systems that show good potential as nanocarriers for ionic drugs.

More recently, dibucaine has been successfully incorporated into the inner part of the corona of amphiphilic poly(ethylene oxide-*b*-sodium 2-(acrylamido)-2-methyl-1-propanesulfonate-*b*-styrene) (PEO-*b*-PAMPS-*b*-PS) triblock copolymer micelles through electrostatic interactions by the same group (Bastakoti et al. 2010). Spherical micelles with zero surface charge and a stealth PEO corona were formed for up to 100% incorporation of the drug (in respect to the anionic sites of the corona chains), indicating that the amount of incorporated drug can be modulated by the length of the charged block.

The preparation of micelles with a complex core containing a drug molecule was achieved by complex formation between the model cationic drug, diminazene diaceturate (DIM), and a series of novel diblock copolymers, that is, carboxymethyldextran-poly(ethylene glycols) (CMD-PEGs) (Soliman and Winnik 2008). Micellar properties were found to be dependent on the ionic charge density (or degree of substitution, DS) of the CMD block and the ratio, [+]/[−], of positive charges of the drug to the negative charges of the copolymers. Micelles were formed at [+]/[−] = 2 and incorporated up to 64% by weight DIM. Their sizes were in the nanometer range (R_h values from 36 to 50 nm were observed) depending on the molecular weight and DS of CMD-PEG. The critical aggregation concentration was also found to depend on DS. The micelles were stable within the $4 < \text{pH} < 11$ range. A more detailed investigation at pH 5.3 proved that the micelles were not destabilized for an ionic strength increase up to almost 0.4 M NaCl in the case of CMD-PEG copolymers of high DS. Micelles of CMD-PEG of low DS (ca. 30%) disintegrated in solutions containing more than 0.1 M NaCl. Sustained *in vitro* DIM release was observed for micelles of CMD-PEG of high DS ([+]/[−] = 2) (Figure 5.4).

Besides physicochemical affinity, other parameters can simultaneously influence the loading of a drug into a polymeric micellar nanocarrier. In particular, the solubility of a drug molecule into the hydrophobic compartment of a block copolymer aggregate is also highly influenced by the chosen incorporation method. A systematic encapsulation study of two drugs into aggregates of a series of block copolymers with variable hydrophobicity has been implemented following two distinct encapsulation procedures, namely, dialysis and the O/W emulsion method. The results indicate the great influence of both the interaction parameter between different drugs and copolymer chains on drug loading, as well as that of the followed encapsulation protocol (Sant et al. 2004).

Another study on a similar context was reported recently by Harada and coworkers (Harada et al. 2011). They incorporated the anticancer agent, camptothecin, into poly(ethylene glycol)-*b*-poly(aspartic acid-*co*-benzyl aspartate) (PEG-P(Asp(Bzl))) polymeric micelle carriers by using two different solvents (tetrafluoroethylene [TFE] and chloroform) using a solvent-evaporation drug-incorporation process. They have observed significant differences in the drug-incorporation behavior, in the morphologies of the incorporated drug and the loaded block copolymer

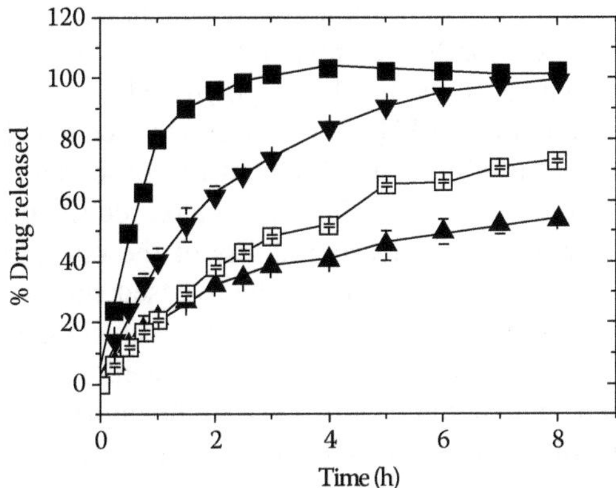

FIGURE 5.4 Release of DIM evaluated by the dialysis bag method from (■) DIM alone in tris–HCl, 25 mM [NaCl] = 150 mM and pH 7.4; (▼) DIM/85-CMD40-PEG140 micelles, [+]/[−] = 2, in 25 mM tris–HCl, [NaCl] = 150 mM and pH 7.4; (▲) DIM/85-CMD40-PEG140 micelles, [+]/[−] = 2, in 25 mM tris–HCl [NaCl] = 0 mM and pH 5.3; and (⊞) DIM/30-CMD68-PEG64 at [+]/[−] = 2, in tris–HCl, 25 mM [NaCl] = 0 mM and pH 5.3. (Reprinted from *Int. J. Pharmacol.*, 356, Soliman, G.M. and Winnik, F.M., 248–258, Copyright 2008, with permission from Elsevier.)

micelles, and in the pharmacokinetic behavior of the nanocarriers by using the two different solvents. The block copolymer micelles, which were prepared with TFE as the incorporation solvent, exhibited more stable circulation in the bloodstream than those prepared with chloroform. This contrast indicates a novel technological perspective regarding the drug incorporation into polymeric micelle carriers. Morphological analysis of the inner core, using AFM in conjunction with fluorescence anisotropy measurements, revealed the presence of a directed alignment of the drug molecules and camptothecin crystals in the micelle inner core. These observations were correlated with the observed pharmacokinetic behavior. This was also the first report on the morphologies of the incorporated drug into the cores of block copolymer micelles.

Architecture of the block copolymer chain may also influence drug loading. Zhang et al. have reported on the comparison of drug-loading capacities between hyperbranched copolymers and linear analogues. The experimental data reveal that the hyperbranched architecture is superior in terms of both drug-loading and entrapment efficiency. Moreover, the hyperbranched copolymer had a more sustained drug-release behavior (Zhang et al. 2010).

Controlled release of the encapsulated drugs is also of great importance in the design of a nanocarrier system. In many cases, drug-loading and drug-release processes are closely interrelated and mutually influenced procedures, in the sense that the way the drug was incorporated into a micellar nanocarrier would influence its release out of the polymeric nanostructure. For example, the use of a common solvent for drug encapsulation may influence the viscosity of the micellar core where the drug is incorporated, and this may also influence drug diffusion out of the core. More importantly, triggered release of encapsulated drugs becomes more important especially in cases that the external stimulus is related to the specific characteristics of the targeted organ/tissue for drug administration within the body.

In a rather informative study involving poly(ethylene oxide)-*b*-poly(*N*-hexyl stearate-L-aspartamide) (PEO-*b*-PHSA) block copolymer micelles incorporating AmB in their cores (Lavasanifar et al. 2002b), it was observed that the increase in the hexyl group substitution of the PHSA block increased the incorporation of the specific drug but decreased the releasing ability of the micellar nanocarriers. In an analogous case study, PTX was found to be released slower from PEO-*b*-PCL copolymers containing benzyl groups on the PCL block compared to the original nonfunctionalized PEO-*b*-PCL block copolymer micelles. The authors claimed that the observed behavior should be attributed to the higher rigidity of the benzyl containing blocks resulting in a greater viscosity of the micellar cores (Shahin and Lavasanifar 2010).

Triggered release as a response to pH changes is highly useful for the release of anticancer drugs in the acidic microenvironment of tumor cells. Acidic microenvironment of tumor cells results from the increased aerobic and anaerobic glucose metabolism taking place inside such cells. Additionally, cancer tumors are known to have an increased temperature, in comparison to healthy tissues, due also to their increased metabolic rate. In any case, drug levels should be kept within certain limits in order to achieve the desired therapeutic levels. It has been shown that cancer cells are more sensitive to exposure to low drug levels for longer time periods than the case of exposure to high drug levels for shorter periods. Amphiphilic block copolymers of PEG as the first block and a second block composed of hydrophobic methacrylate monomers and methacrylic acid (MAA) were utilized for the solubilization of the poorly water-soluble model drugs, indomethacin, fenofibrate, and progesterone. Drug loadings of <6% and 6%–14% w/w were achieved by the dialysis and emulsion methods, respectively. Evaluation of progesterone release *in vitro* has demonstrated that the drug release from the block copolymer micelles increased when the pH of the release medium was raised from 1.2 to 7.2 (Sant et al. 2004).

pH-responsive drug conjugate micelles have been also proposed as a tool for achieving triggered site-specific release of the carried drug. Usually the drug is chemically bonded to the core-forming block via an acid-labile bond. This bond should be stable at physiological pH and should be easily cleaved at the acidic pH of a tumor's extracellular space or in its cell endosomes, allowing for the release of the therapeutic agent. Poly(ethylene oxide)-*b*-poly(L-lactic acid) copolymers having DOX linked on the PLLA terminus via hydrazone bond were investigated and showed effective drug release at acidic pH. DOX has been also linked to poly(aspartic acid) blocks via the same hydrazone linker, and pH-sensitive drug release was also observed in this case (Bae et al. 2003). In an analogous study, PTX was attached to the poly(aspartic acid) block of a PEO-*b*-PAsp block copolymer formerly functionalized with levulinic acid moieties. The hydrazone-type linker was cleavable at low pH and the drug could be released in a controlled way. However, the preparation of mixed micelles containing levulinic acid and 4-acetyl benzoic acid moieties with attached DOX showed a clearly moderated release profile for the drug compared to the original pure micelles. So, blending of block copolymer–drug conjugates can be another way of fine tuning the release of a drug in such cases (Alani et al. 2010).

Frechet and coworkers have utilized an acid-labile cyclic benzylidene acetal linker for the chemical conjugation of drugs on the poly(aspartic acid) block of PEO-*b*-poly(aspartic acid) block copolymers (Gillies and Frechet 2005). The authors found that this type of linkers was rapidly hydrolyzed at pH 5, showing a half-life of ca. 60 min at normal body temperature (37°C). Hydrolysis of the acetal linkers resulted in cleavage of the hydrophobic groups of the copolymer chains that in turn led to the disruption of the copolymer micelles and the subsequent release of the carried hydrophobic drugs. In a similar fashion, but taking advantage of the protonation effect, encapsulated hydrophobic drugs can be released from the nanocarrier by the dissociation of the micelles due to an increase in copolymer solubility. Poly(ethylene glycol)-*b*-poly(histidine) (PEG-*b*-PHis) copolymers form stable, narrow-sized micelles at physiological pH because of the hydrophobicity of the PHis block. Therefore, DOX could be encapsulated into the PHis core of the micelles at loading levels close to 20% by weight. When the drug-loaded micelles were present in the lower extracellular pH of the tumor or into the endosomal compartments of tumor cells, they were disintegrated, due to protonation of histidine groups, leading to the release of DOX (Lee et al. 2003, 2005).

In a more recent study, a series of synthetic block copolymers consisting of poly(2-hydroxyethyl methacrylate) (PHEMA) and PHis, showing biocompatibility and membranolytic activity, were investigated as pH-sensitive drug carrier for tumor targeting (Johnson et al. 2012). DOX was successfully encapsulated in the nanosized micelles and delivered under different pH conditions. DOX release was investigated *in vitro* according to the pH of the surrounding medium. DOX was released from the micelles in a controlled and sustained manner and the release rate of DOX was accelerated at low pH 5.5 compared to pH 7.4. The empty micelles could be effectively internalized by human embryonic kidney 293T cells and HCT116 cells. The viability of both cell types was higher than 80% at a micelle concentration range 1–50 μg/mL. The drug-loaded PHEMA-*b*-PHis micelles could also be effectively internalized by HCT116 human colon carcinoma cells. It was found that they slowly released the encapsulated DOX molecules, showing effective cell proliferation inhibition compared to free DOX, specifically at an acidic environment.

Nanosized drug carriers with stealth properties and long circulation times are preferentially accumulated to tumors through the EPR effect, due to the differences in the microstructure/physiology of the tumor tissues compared to normal ones. In order to effectively drive the nanocarriers toward the specific sites for release of their cargo, targeting capabilities should be introduced onto the nanocarriers. This is more frequently achieved by the introduction of targeting chemical groups/moieties at the free end of the corona-forming block of the diblock copolymer forming the micellar nanocarrier. Targeting ligands are helpful in attaching the drug nanovehicles to the target cells or tissues and for enhancing crossing of the cell membrane. Various ligands can be utilized depending on the target and may include small organic molecules as chain-end connected chemical groups, carbohydrates, peptides, and antibodies.

In the category of small organic molecules targeting moieties, folic acid (FA) groups are very commonly used because its receptor is overexpressed in many human cancer cells. Folate-PEO-*b*-PCL micelles (the folate group being attached to the free end of the PEG block) loaded with PTX exhibited distinctly higher cytotoxicity for two types of human adenocarcinoma cells compared to normal human fibroblast cells (Park et al. 2005). A mixed block copolymer micellar system containing folate-PEO-*b*-PLGA and PEO-*b*-PLGA-DOX conjugate chains, having one DOX molecule attached to the poly(D,L-lactic-*co*-glycolic acid) block, was more effective than DOX alone in terms of cytotoxicity toward tumor cells *in vitro* (Yoo and Park 2004).

Complex amino acid–based core cross-linked star-block copolymers, that is, (poly(L-lysine)$_{arm}$poly(L-cystine)$_{core}$) copolymers, bearing peripheral allyl functionalities, which were further functionalized with a poly(ethylene glycol)-folic acid (PEG-FA) conjugate at the periphery, forming FA-PEG$_{arm}$-(poly(L-lysine)$_{arm}$poly(L-cystine)$_{core}$) stars, were evaluated as drug carriers for targeted drug delivery *in vitro*. Confocal microscopy and flow cytometry revealed that the functionalized stars could be internalized into MDA-MB-231 cells, while cytotoxicity studies indicate that the stars are nontoxic to cells at concentrations of up to 50 μg/mL (Figure 5.5). These results show that these amino acid–based star polymers can be utilized in targeted drug delivery applications including chemotherapy (Sulistio et al. 2011).

Aiming at developing block copolymer micellar nanocarriers with targeting and pH-responsive drug release, Armes and coworkers have presented a series of FA corona–functionalized block copolymer micelles with pH-responsive blocks. They found that two different drugs can be loaded in the respective micelles forming well-defined nanocarriers at physiological pH. However, due to dissociation (or not) of the polymeric micelles, the release profile of the drug was found to be strongly depended on the environmental pH value (Licciardi et al. 2006).

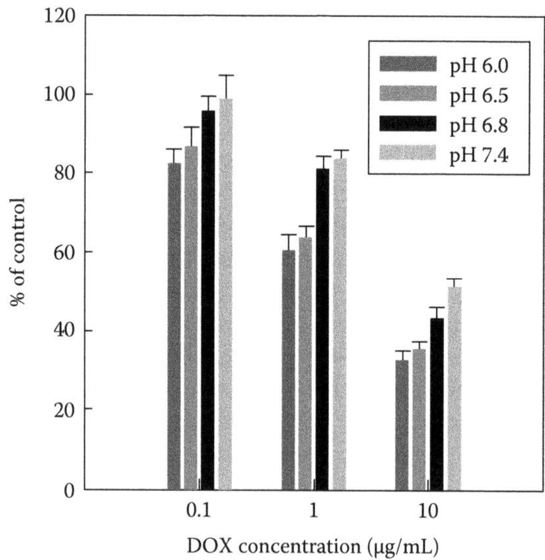

FIGURE 5.5 Dose-dependent antitumor activity of DOX-loaded p(HEMA)$_{25}$-*b*-p(His)$_{15}$ micelles to HCT-116 cells after 48 h incubation. (Johnson, R.P., Jeong, Y.-I., Choi, E., Chung, C.-W., Kang, D.H., Oh, S.-O., Suh, H., and Kim, I.: *Adv. Funct. Mater.* 2012. 22. 1058–1068. Copyright Wiley-VCH Verlag GmbH & Co. KGaA. Reproduced with permission.)

Folate-functionalized polymeric micelles with pH-triggered drug-release properties have been also presented by Yung and coworkers. In particular, the use of mixed micelles composed of two different copolymers has been proposed. One of the copolymers is suitable for the drug loading, while the other copolymer offers pH sensitivity and therefore determines the desired drug-release profile. Notably, the experimental results indicate that by tuning the ratio between the block copolymers utilized for micelle formation, the desirable physicochemical characteristics can be achieved for the nanocarrier system (Zhao et al. 2010).

The use of carbohydrates as active targeting ligands on the surface of block copolymer micelles has been also reported (Nagasaki et al. 2001, Jule et al. 2003). Glucose, galactose, mannose, and lactose can be utilized for this purpose since the targeting groups can bind to glycol receptors of cells. It has been demonstrated that lectins, ConA- and RCA-1-type receptors, can recognize mannose and β–D-glucose residues.

Polycaprolactone-g-dextran (Gal-PCL-g-Dex-FITC) polymers carrying galactose groups for active targeting and fluorescein isothiocyanate (FITC) groups for fluorescence labeling were self-assembled into stable, spherical micelles in aqueous medium and in serum. The anti-inflammatory drug prednisone acetate was loaded in the polymeric micelles through hydrophobic interactions. The *in vitro* drug-release investigations showed that the galactosylated micelles could be selectively recognized by and subsequently accumulate in HepG2 cells. It was demonstrated that the relative uptake of the micelles by liver was much higher than the other tissues, indicating that the galactosylated micelles have great potential as a liver-targeting drug carrier, as a result of their design and structure (Wu et al. 2009) (Figure 5.6).

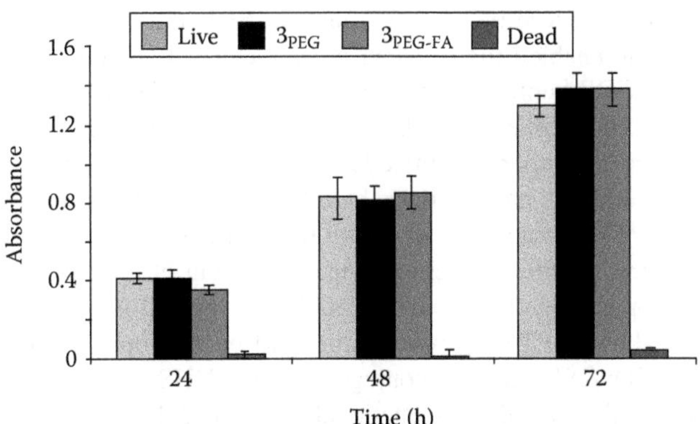

FIGURE 5.6 Viability of breast cancer cells (MDA-MB-231) after incubation with FA-PEG$_{arm}$-(poly(L-lysine)$_{arm}$poly(L-cystine)$_{core}$) stars, 3PEG-FA and 3PEG stars at 5.0 μg/mL for 24, 48, and 72 h, as assessed by MTS assays. Absorbance is proportional to MTS reduction by metabolically active cells. (Reprinted with permission from Sulistio, A., Lowenthal, J., Blencowe, A., Bongiovanni, M.N., Ong, L., Gras, S.L., Zhang, X., and Qiao, G.G., *Biomacromolecules*, 12, 3469–3477, 2011. Copyright 2011, American Chemical Society.)

Small peptide sequences and proteins have been also utilized for introducing targeting properties to polymeric nanoparticles and nanoassemblies. PEO-*b*-PCL block copolymer micelles having a cyclic pentapeptide (cyclic-Arg-Gly-Asp-d-Phe-Lys, cRGDfK) attached to the free ends of the PEO blocks, and carrying the hydrophobic drug DOX in their PCL core, were prepared following a multistep synthetic protocol (Nasongkla et al. 2004). Such a peptide, which contains the RGD sequence, can recognize $\alpha_v\beta3$-integrins that are overexpressed on the surface of tumor cells or the angiogenic endothelial cells of the tumor vasculature. First, maleimide-terminated PEO-*b*-PCL copolymers were synthesized by a well-established end functionalization reaction on the –OH terminus of PEO blocks. After preparation of the micelles, the cRGDfK peptide was attached on the micelle surface via reaction of its thiol group with the maleimide groups of the copolymer chains. A ca. 76% yield of the coupling reaction was achieved. Nevertheless, confocal laser scanning microscopy (CLSM) observations indicated that the drug-loaded micelles with the targeting peptide could be accumulated into human Kaposi's sarcoma tumor endothelial SLK cells in a 30 times greater extend compared to non-peptide-modified micelles.

In another work, acetal-terminated-PEO-*b*-PCL block copolymers were used for the formation of micelles in aqueous media. Then the acetal group on the free PEO terminus was transformed to an aldehyde group by hydrolysis under acidic conditions. The aldehyde group on the micellar surface was utilized for the attachment of the GRGDS targeting peptide by a Schiff-type reaction. Fluorescence spectroscopy/microscopy experiments showed that the uptake of GRGDS carrying micelles by mouse melanoma B16-F10 cells was almost five times greater than that of the PEO-PCL micelles that did not carry the targeting peptide moiety (Xiong et al. 2007). In an analogous study, an acetal-PEO-*b*-poly(α-carboxyl-ε-caprolactone) (ac-PEO-*b*-PCCL) block copolymer was used for the chemical conjugation of DOX through the α-carboxyl groups of the PCCL block forming an amide group between the polymer chain and the drug. The terminal acetal group was again transformed to an aldehyde, and this end group enabled the covalent attachment of RGD-containing peptides, including the GRGDS peptide, in order to functionalize the micellar surface. The RGD-functionalized micelles with the conjugated DOX presented higher toxicity toward B16-F10 cells compared to micelles without the peptide (Xiong et al. 2008).

Polymeric micelle-like nanoparticles based on PEGylated poly(trimethylene carbonate) (PEG-*b*-PTMC) and decorated with the cyclic RGD peptide were prepared for active targeting to integrin-rich cancer cells (Jiang et al. 2011a). The amphiphilic diblock copolymer, α-carboxyl poly(ethylene glycol)-poly(trimethylene carbonate) (HOOC-PEG-*b*-PTMC), was synthesized by ring-opening polymerization (ROP). The c(RGDyK) ligand was conjugated to the NHS-activated PEG terminus of the copolymer. The c(RGDyK)-functionalized PEG-*b*-PTMC micellar nanoparticles were loaded with PTX using the emulsion/solvent-evaporation technique, and they were found to have nanometer-scale sizes. Cellular uptake of c(RGDyK)-PEG-*b*-PTMC/PTX nanoassemblies was found to be higher than that of the non-functionalized micelles containing PTX, due to the integrin-mediated endocytosis effect, taking place in the former case as a result of the presence of the targeting functionalities. *In vitro* cytotoxicity, cell apoptosis, and cell cycle arrest studies were also employed in order to assess the efficiency of the nanodelivery systems and

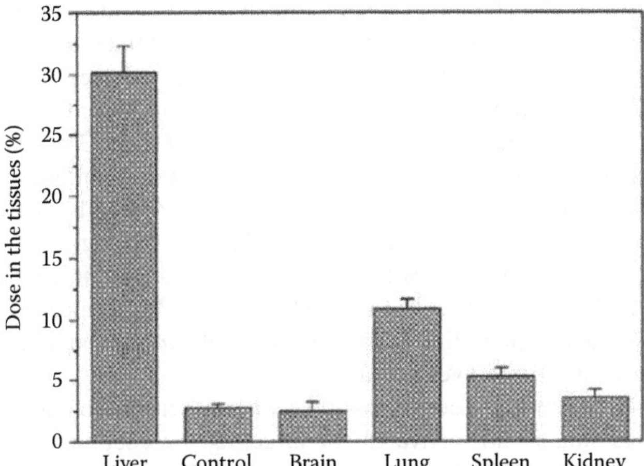

FIGURE 5.7 Overall uptake of Gal-PCL-g-Dex-FITC micelles into different tissues after 2 h injection. The uptake of PCL-g-Dex-FITC micelles in liver was used as a control. (Reprinted from *Biomaterials*, 30, Wu, D.Q., Lu, B., Chang, C., Chen, C.S., Wang, T., and Zhang, Y.Y., 1363–1369, Copyright 2009, with permission from Elsevier.)

revealed that the c(RGDyK)-PEG-*b*-PTMC/PTX exhibited significantly stronger *in vitro* anti-angiogenic activity than PEG-*b*-PTMC/PTX and also Taxol. A pharmacokinetic study in rats demonstrated that the micellar nanoparticles significantly enhanced the bioavailability of PTX than Taxol, by increasing the blood circulation time of the nanocarrier. Enhanced accumulation of c(RGDyK)-PEG-*b*-PTMC/PTX in tumor tissue *in vivo* xenograft tumor-bearing model was also observed, by real-time fluorescence imaging, highlighting the high specificity and efficiency of the functionalized micellar nanoparticles for tumor active targeting (Figure 5.7).

Since it has been realized that transferring receptors are considerably elevated on various types of cancer cells, transferrin itself was proposed as a potential targeting protein to be conjugated to the surface of block copolymer micelles carrying hydrophobic anticancer drugs. Following this concept, transferrin, a serum glycoprotein of 80 kDa molecular weight, was attached to the corona of PEO-*b*-PEI polyion complex micelles, with a PEO corona and a core composed of the electrostatic complex of polyethylenimine (PEI) and phosphorothioate antisense oligonucleotides (ODNs), via an avidin/biotin structure. Fluorescent-labeled transferrin-conjugated micelles were found to accumulate with a high efficiency into resistant human oral epidermoid carcinoma cells (Vinogradov et al. 1999). Some additional examples pertaining to the delivery of relatively higher molecular weight nucleotides using block copolymer micelles and other polymeric nanoparticles containing targeting moieties will be presented in the following sections.

Monoclonal antibodies or their Fab fragments have also shown promise as targeting ligands on the surface of block copolymer micelles. Kabanov et al. have developed immunomicelles based on Pluronic P-85 micelles, to which a murine polyclonal antibody against α_2-glycoprotein was chemically attached on the micellar corona, in an effort to deliver the neuroleptic agent haloperidol to the brain

(Kabanov et al. 2005). The monoclonal antibody C225 against epidermal growth factor receptors (EGFRs) was attached to the PEG terminus of a poly(ethylene glycol)-*b*-poly(L-glutamic acid) (PEG-*b*-PLG) block copolymer. The attachment was accomplished by reaction of the sulfhydryl group introduced to C225 with the vinylsulfone group on the PEG end of the copolymer. The PLG block was chemically conjugated to DOX. The micelles carrying the C225 antibody were selectively bound to human vulvar squamous carcinoma A431 cells that overexpress EGFR, in contrast to non-modified micelles. Receptor-mediated micelle uptake was found to occur within 5 min where the non-modified DOX micellar conjugates required ca. 24 h to be internalized into cells. The modified micelles were found to be more effective in inhibiting growth of A431 cells in comparison to free DOX after a 6 h exposure period (Vega et al. 2003).

PEG-*b*-P(HPMAm-Lac$_n$) core cross-linked thermosensitive biodegradable polymeric micelles suitable for active tumor targeting, by coupling the anti-EGFR (EGFR) EGa1 nanobody to their surface as targeting ligand, were developed by Hennink and coworkers (Talelli et al. 2011). In the first synthetic step, PEG was functionalized with *N*-succinimidyl 3-(2-pyridyldithio)-propionate (SPDP) to yield a PDP–PEG-*b*-P(HPMAm-Lac$_n$) block copolymer. Micelles composed of 80% PEG-*b*-P(HPMAm-Lac$_n$) and 20% PDP–PEG-*b*-P(HPMAm-Lac$_n$) were prepared from mixtures of the two copolymers, and then lysozyme (utilized as a model protein) was modified with *N*-succinimidyl-S-acetylthioacetate, deprotected with hydroxylamine hydrochloride and subsequently coupled to the micellar surface. The micellar conjugates were characterized using SDS-PAGE and gel permeation chromatography showing the success of the synthetic protocol. Following an analogous conjugation step, the EGa1 nanobody was coupled to PEG/PDP–PEG micelles. Conjugation was successful as demonstrated by Western blot and dot blot analysis. Rhodamine-labeled EGa1-micelles showed substantially higher binding, as well as uptake by EGFR-overexpressing cancer cells (A431 and UM-SCC-14C) than rhodamine-labeled micelles without the targeting ligand. No binding of the nanobody-functionalized micelles was observed to EGFR-negative cells (3T3) as well as to 14C cells in the presence of an excess of free nanobody. The latter result demonstrates that the binding of the nanobody micelles was indeed through interaction with the EGF receptor. Although these micelles did not contain any active pharmaceutical compound, the presented studies illustrate the ability of the particular ligand to act as a valuable and specific moiety for tumor targeting of block copolymer micellar nanocarriers.

Nucleic acid ligands (aptamers) are another class of ligands potentially well suited for the selective targeting of drug-encapsulated controlled-release polymer particles in a cell- or tissue-specific manner. Bioconjugates composed of controlled-release polymer nanoparticles and aptamers were synthesized from a poly(lactic acid)-*b*-poly(ethylene glycol) (PLA-*b*-PEG) copolymer with a terminal carboxylic acid functional group at the PEG free end (PLA-*b*-PEG-COOH) and were examined in terms of their efficiency for targeted delivery to prostate cancer cells. Rhodamine-labeled dextran was encapsulated within the micellar nanoparticles as a model drug. The particular nanoparticles showed several desirable characteristics including (1) substantial negative surface charge (corresponding to a surface zeta-potential of ca. −50 mV), which may minimize nonspecific interaction with the negatively charged nucleic acid aptamers; (2) carboxylic acid groups on the particle surface for modification via covalent conjugation to amine-modified aptamers; and (3) presence of PEG chains on the particle surface, which is expected to increase circulating half-life, while contributing to decreased uptake in nontargeted cells. Nanoparticle–aptamer bioconjugates contained RNA aptamers that bind to the prostate-specific membrane antigen, a well-known prostate cancer tumor marker that is overexpressed on prostate acinar epithelial cells (Farokhzad et al. 2004). The authors demonstrated that these bioconjugates could efficiently target and be incorporated into the prostate LNCaP epithelial cells, which express the prostate-specific membrane antigen protein. A 77-fold increase in binding was observed for the aptamer-functionalized micellar nanoparticles compared to reference nanoparticles with no specific binding ligands. In contrast to LNCaP cells, the uptake of the aptamer-functionalized block copolymer nanoparticles was not enhanced in cells that do not express the prostate-specific membrane antigen protein.

It is evident from the former discussion that for the construction of a polymeric nanocarrier with a success potential for drug delivery applications, a number of basic functionalities and responsive properties should be taken into consideration and be physically accommodated into the nanostructure. This holds true in particular for those based on block copolymers for the delivery of small drug molecules. However, some design and construction principles are universal and should be considered also in the case of delivery of large molecules although some fine tuning should always take place. In the following, we continue with the presentation of several elucidating examples of such designed nanocarriers for small drugs that give a better picture of the breadth of the field and some of the advances accomplished so far.

A rather complex block copolymer, poly(*N*-isopropyl acrylamide-*co*-*N*,*N*-dimethylacrylamide-*co*-2-aminoethyl methacrylate)-*b*-poly(10-undecenoic acid) (P(NIPAM-*co*-DMAAm-*co*-AEMA)-*b*-PUA), containing four chemically different segments, each one carrying a specific function, was synthesized and studied in the context of an advanced drug nanocarrier (Liu et al. 2007). FA was conjugated to the hydrophilic random copolymer corona-forming block, through the amine group in AEMA units. The copolymer self-assembled into micelles in aqueous media, which exhibited pH-induced responsiveness and temperature sensitivity. The micelles had a rather small size and a well-defined PUA core, while the folate-targeting moieties were concentrated in the corona. The anticancer drug, DOX, was encapsulated into the micelles by a membrane dialysis method. DOX release was pH dependent, being faster at low pH (as is the case of endosomes and lysosomes). Therefore, it was concluded that DOX could be readily released from the micelles into the cell nucleus after being incorporated into cells. The IC50 value of DOX-loaded micelles with folate-targeting groups against folate receptor–expressing 4T1 and KB cells was much lower than that of the DOX-loaded micelles without folate (3.8 vs. 7.6 mg/L for 4T1 cells and 1.2 vs. 3.0 mg/L for KB cells). *In vivo* experiments conducted in a 4T1 mouse breast cancer model demonstrated that DOX-loaded micelles had a longer blood circulation time than free DOX ($t_{1/2}$ was 30 and 140 min, respectively). In addition, the micelles delivered an increased amount of DOX to the tumor when compared to free DOX.

Micelles formed by specially designed amphiphilic 6-arm star-block copolymers with PCL inner blocks and zwitterionic poly(2-methacryloyloxyethyl phosphorylcholine) outer blocks (6sPCL-*b*-PMPC) have been utilized for the encapsulation and delivery of PTX (Tu et al. 2011). The star copolymers synthesized by a combination of ROP and atom transfer radical polymerization (ATRP) formed spherical micelles with PCL hydrophobic cores. Micelle size increased by 30%–80% after incorporation of PTX. Star-block micelles were labeled with FITC in order to follow their internalization profile into cancer cells. Fluorescence microscopy observations confirmed that the loaded micelles have been efficiently internalized by tumor cells. It was possible to directly visualize the micelles within tumor cells using TEM, and it was concluded that the 6sPCL-*b*-PMPC drug-loaded micelles were more efficiently uptaken by tumor cells compared to diblock PCL-*b*-PEG micelles. The 6sPCL-*b*-PMPC micelles carrying PTX showed much higher cytotoxicity against HeLa cells than PCL-*b*-PEG micelles loaded with the same drug, a result that was attributed to the higher efficiency of cellular uptake of the particular micelles.

Dendron-like poly(ε-benzyloxycarbonyl-L-lysine)/linear PEO block copolymer micelles encapsulating DOX were tested for their ability to carry and release the drug under different pH environmental conditions (Xu and Dong 2012). Drug loading in the range 5%–10%wt were achieved depending on the composition of the copolymers. The *in vitro* experiments indicated that the drug-loaded micelles showed a triphasic drug-release profile at aqueous pH 7.4 or 5.5 and at 37°C. The authors argued that the initial fast drug-release behavior was probably due to the large surface area of nanoparticles, and the subsequent slower drug release was mainly controlled by the drug diffusion out of the nanoparticles. The nanoparticles sustained a longer drug-release period for about 2 months. The drug-release profile at aqueous pH 5.5 similar to endosome and/or lysosome compartments inside cells was slightly faster, which was due to the relatively increased aqueous solubility of DOX at mildly acidic pH.

Multicomponent micellar nanocarriers based on ABC miktoarm block copolymers (where A = PEG, B = PCL, and C = a short triphenylphosphonium bromide residue) were used for the delivery of coenzyme Q10 (CoQ10) in mitochondria (Sharma et al. 2012). The micellar nanoparticles displayed sizes of 25–60 nm and a drug-loading capacity reaching 60 wt%. The nanoassemblies were stable in solution for more than 3 months. The authors attributed the extraordinarily high CoQ10 loading capacity of the miktoarm micelles to the good chemical compatibility between CoQ10 and the PCL arm of the copolymer. Confocal microscopy observations of the fluorescently labeled block copolymer analogue, together with the mitochondria-specific vital dye label, showed that the nanocarriers could actually reach the mitochondria within the cells. The high CoQ10 loading efficiency allowed testing of loaded micelles within a broad concentration range and provided evidence for CoQ10 effectiveness in two different experimental paradigms: oxidative stress and inflammation. The experimental results obtained suggested that the miktoarm-based carrier could effectively deliver CoQ10 to mitochondria without loss of drug effectiveness. This particular work shows that controlled and targeted delivery of low molecular weight drugs to intracellular organelles is possible through the use of specially designed block copolymer micellar nanocarriers.

Amphiphilic, shell cross-linked, knedel-like (SCK) polymer nanoparticles were studied as potential nanocarriers of DOX (Lin et al. 2011a). The main goals were to investigate the effects of the core and shell dimensions on the loading and release of DOX as a function of the solution pH. SCKs were constructed from poly(acrylic acid)-*b*-polystyrene (PAA-*b*-PS) amphiphilic diblock copolymers having different relative block lengths and overall degrees of polymerization. A series of different SCK nanoparticle samples with hydrodynamic diameters from 14 to 30 nm were prepared and studied. Block copolymer compositions employed gave SCKs with ratios of shell to core volumes ranging from 0.44 to 2.1. The SCKs were capable to sustain drug loadings in the range 1500–9700 DOX molecules per particle, with larger numbers of DOX molecules encapsulated within the larger core SCKs. The volume occupied by the PAA shell relative to the volume occupied by the polystyrene core was found to correlate inversely with the diffusion-based release of DOX. SCKs having smaller cores and higher acrylic acid corona volume to styrene core volume ratios showed lower final extents of release. Higher final extents of release and faster rates of release were observed for all DOX-loaded SCK particles at pH 5.0 versus pH 7.4, respectively, ca. 60% versus 40% at 60 h, a promising characteristic for enhanced DOX delivery within tumors and cancer cells (Figure 5.8). Quantitative determination of the kinetics of release was made by fitting the release profile data to the Higuchi model. The release rate constants determined ranged from 0.0431 to 0.0540 $h^{-1/2}$ at pH 7.4 and 0.106 to 0.136 $h^{-1/2}$ at pH 5.0 for SCKs, compared to the non-cross-linked block copolymer micelle analogues that exhibited rate constants for release of DOX of 0.245 and 0.278 $h^{-1/2}$ at pH 7.4 and 5.0, respectively.

Nanoparticles or less well-defined micellar aggregates formed by random copolymers have been investigated in parallel for use as delivery nanovehicles for small

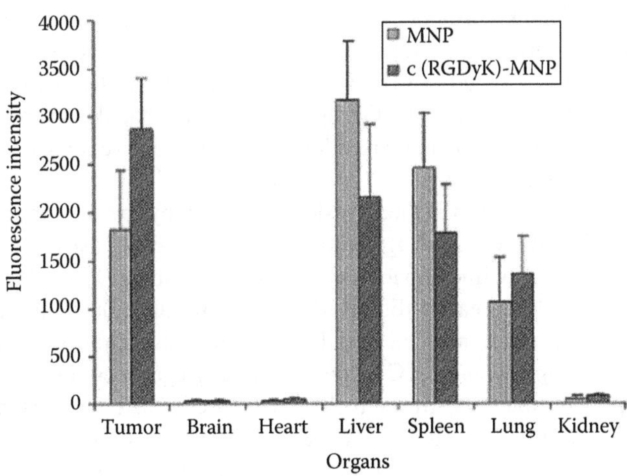

FIGURE 5.8 Fluorescence intensity of Dir-loaded MNP and c(RGDyK)-MNP in various organs. (Reprinted from *Biomaterials*, 32, Jiang, X., Sha, X., Xin, H., Chen, L., Gao, X., Wang, X., Law, K., Gu, J., Chen, Y., Jiang, Y., Ren, X., Ren, Q., and Fang, X., 9457–9469, Copyright 2011a, with permission from Elsevier.)

drug molecules. Ulbrich and coworkers utilized *N*-(2-hydroxypropyl)methacrylamide (HPMA) random copolymers with covalently bonded cholesteryl groups and their conjugates with DOX (Chytil et al. 2012). DOX and cholesterol derivatives were bound by pH-sensitive hydrazone bonds to the main polymer chain. The presence of the hydrophobic cholesteryl groups results in the formation of aggregates in aqueous media, and the covalently bonded drug resides in the hydrophilic corona of the assemblies. DOX could be released from the aggregates at endosomal pH (pH=5) due to hydrolysis of the covalent bonds. The rate of hydrolysis was strongly dependent on the microenvironment around the polymer–drug bond. Although DOX could be released at first, it was observed that the high molecular weight supramolecular polymer carrier was disintegrated very slowly, forming relatively short polymer fragments that are small enough to be eliminated from the body by glomerular filtration.

A random copolymer composed of thermosensitive *N*-isopropyl acrylamide, hydrophilic *N,N*-dimethylacrylamide, and pH-responsive undecenoic acid segments, P(NIPAM-*co*-DMAAm-*co*-UA), has been assembled in micelles in aqueous media, at pH 7.4 and at 37°C, and DOX has been encapsulated into the copolymer nanoassemblies (Soppimath et al. 2005, 2007). NIPAM and DMAAM units were forming the corona of the micelles. Protonation of the UA units at lower pH (e.g., 5 or 6.6) made them more hydrophobic and resulted in a destabilization of the micelles at 37°C and release of DOX. Release of the anticancer drug was found to be slow at pH 7.4, but it was increased at lower pH values due to the deformation of the core/shell structure. Attachment of cholesterol groups on the UA units made the copolymer more hydrophobic, and PTX could be now easily encapsulated into the micelles. PTX release was found to depend on environmental pH and temperature as in the case of DOX-loaded P(NIPAM-*co*-DMAAm-*co*-UA) micelles without cholesterol groups. Comparison with thermoresponsive drug-loaded block copolymer micelles showed that pH sensitivity played a major role in the extent of internalization of the micellar nanocarriers. FA groups were also attached for targeting on the particular copolymer. Active targeting increased considerably DOX uptake from 4T1 mouse breast cancer cells compared to drug-loaded copolymer micelles without targeting functionalities. The IC50 value of DOX against KB cells that also overexpress folate receptors was also found to be increased for the case of folate-conjugated copolymer micelles encapsulating the drug.

Besides self-assembled micelles, vesicles formed by block copolymers have been successfully utilized as drug nanocarriers in several cases. Xu et al. have presented the case of a linear triblock copolymer, namely, PEO-PAA-PNIPAM. This copolymer could be easily dissolved molecularly at room temperature, but it was self-assembled into nanosized vesicles upon increasing the solution temperature (temperature higher than 32°C). The formed aggregates could be stabilized by cross-linking at the interface through the PAA block. Next, the encapsulation of biologically interesting molecules took place at relatively high loading efficiency. It was demonstrated that the formed carrier was stable against a number of environmental changes, like temperature, salinity, and dilution. However, dissociation was observed under reductive conditions, leading to the conclusion that such systems can be particularly useful for intracellular delivery of biopharmaceuticals (Xu et al. 2009). Another example of complex nanocarrier has been presented by

Wang and coworkers. A triblock copolymer with a cross-linkable middle block was used for the release of an entrapped anticancer drug into the cell cytoplasm. The triblock copolymer forms onion-type micelles upon dissolution in aqueous media, while cross-linking can be realized at the shell, through disulfide bonds. The formed aggregate was stable during circulation and it was found to be effective in terms of drug delivery, under the reductive intracellular environment, where cross-linking is damaged (Wang et al. 2010).

Hollow core spherical micelles formed by stereocomplexation of AB_2 miktoarm copolymers of the types poly(ethylene glycol)-[poly(L-lactide)]$_2$ (PEG-(PLLA)$_2$) and poly(ethylene glycol)-[poly(D-lactide)]$_2$ (PEG-(PDLA)$_2$) and nanofibers formed by stereocomplexation between poly(ethylene glycol)-*b*-poly(D-lactide) (PEG-*b*-PDLA) and poly(ethylene glycol)-*b*-poly(L-lactide) (PEG-*b*-PLLA) were used for the encapsulation of PTX (Tan et al. 2009). Drug loading of 12 and 40%wt were achieved, respectively, showing the effect of nanostructure morphology on the loading capacity of the nanostructure in respect to hydrophobic drugs. Release of PTX was continuous in both cases without an initial rapid increase of the released drug.

5.2 DELIVERY OF MACROMOLECULAR DRUGS

5.2.1 Gene Delivery

Gene therapy refers to treatment of diseases by modifying gene expression within specific cells. It has the potential to shift the way numerous human diseases such as sickle cell anemia, HIV, Parkinson's disease, Huntington's disease, and Alzheimer's disease to which a genetic identity has been given are managed and treated. The concept behind is simple—exogenous delivery of therapeutic genes to the body to treat genetic-based diseases. Engineered viruses were the first gene delivery agents. Some of the earliest clinical trials produced promising results. However, although viral therapies have proven to be efficient, they appeared quite risky to the patient (Jooss et al. 1998, Marshall 1999, Boyce 2001, Check 2002) as severe side effects were made strikingly evident. The documented dangers of using the efficient recombinant viruses as carriers have motivated the exploration for safer, less pathogenic, and immunogenic gene delivery alternatives including lipid-based vectors, chemically modified viruses, inorganic materials, and polymer-based gene delivery systems. As stated elsewhere (Wong et al. 2007), in addition to the potential safety benefits, such nonviral systems offer greater structural and chemical versatility for manipulating physicochemical properties, vector stability upon storage and reconstitution, and a larger gene capacity compared to their viral counterparts. In the following, we focus on the polymer-based systems that have been used as nonviral carriers for complexing and intracellular delivery of genetic material (DNA, RNA, other nucleotides).

5.2.1.1 Biological Barriers to Gene Delivery

The current understanding of the various biological barriers that face efficient gene delivery has been discussed in a recent review (Wong et al. 2007). Figure 5.9 schematically represents the whole process of gene delivery—from DNA packaging to gene expression—as well as all barriers and obstacles that should be

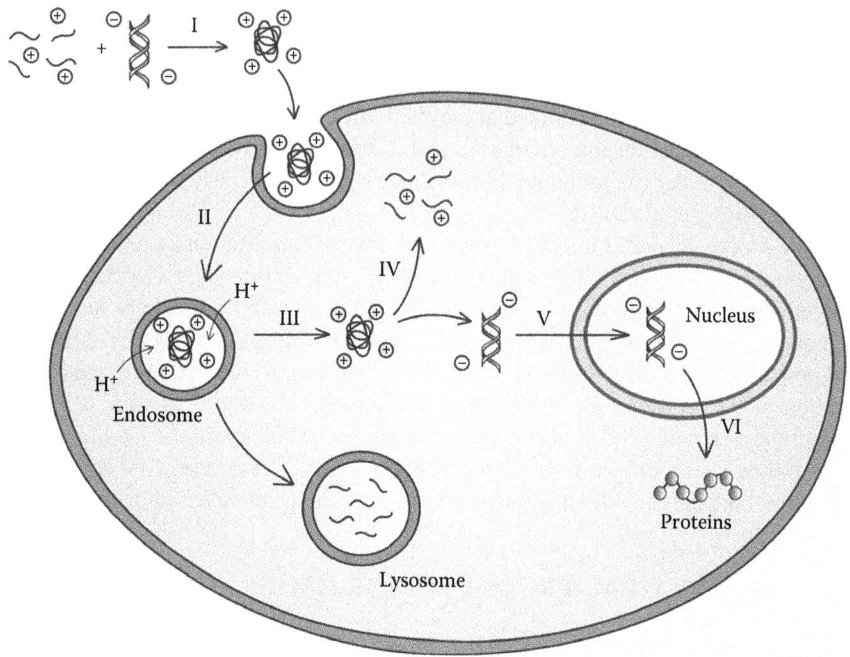

FIGURE 5.9 Barriers to gene delivery. (Reprinted from *Prog. Polym. Sci.*, 32, Wong, S.Y., and Pelet, J.M., Putnam, D., 799–837, Copyright 2007, with permission from Elsevier.)

overcome. The fundamental criteria for any synthetic gene delivery system require the ability to (I) package therapeutic genes, (II) gain entry into cells, (III) escape the endolysosomal pathway, (IV) effect DNA/vector release, (V) traffic through the cytoplasm and into the nucleus, (VI) enable gene expression, and, last but not least, remain biocompatible (Figure 5.9).

The packaging of DNA is the first stage of the gene delivery process. It is typically achieved by three main strategies—electrostatic interactions, encapsulation, and adsorption—aiming at (1) neutralizing the negatively charged phosphate backbone of DNA to prevent charge repulsion against the anionic cell surface, (2) condensing the bulky structure of DNA to appropriate length scales for cellular internalization, and (3) protecting the DNA from both extracellular and intracellular nuclease degradation (Schaffer and Lauffenburger 1998, Lechardeur et al. 1999, Abdelhady et al. 2003). The polymeric vectors developed to date have exploited the anionic nature of DNA and other nucleotides to drive complexation via electrostatic interactions. As a result, a polyplex is formed, in which the DNA is condensed but remains active and capable of transfecting cells. The size of polyplexes is typically in the range 30–500 nm. Their characteristics can be controlled through the proper selection of the (co)polymer and through the N/P ratio. A variety of polyplexes of DNA, RNA, and other ODNs with natural and synthetic (co)polymers is described in Section 3.6. It has been documented that positively charged particles are taken up by cells faster than negatively or neutrally charged ones. In addition, the excess positive charge results in a final positive zeta-potential for the polyplex and the plasmid is protected by the polymer. That is why a surplus of positive charge or polymer added to the polyplex is beneficial as far as gene delivery is concerned.

Another possible way to protect genes from enzymatic degradation is encapsulation within nano- or microspherical structures. The latter are typically prepared from biodegradable polymers such as polyphosphoesters, polyphosphazenes, poly(β-amino ester)s, poly(lactic acid), poly(lactide-*co*-glycolide), and polyorthoesters, which, owing to the presence of hydrolytically unstable linkages, can degrade to shorter oligomeric and monomeric components. The degradation kinetics as well as the size of the DNA-encapsulated complex can be controlled by varying polymer properties and method of particle formation (Santos et al. 1999, Panyam and Labhasetwar 2003, Hedley 2005, Jang and Shea 2006). Drawbacks related to exposure to high shear stress, organic solvent and extreme temperatures, low encapsulation efficiency and DNA bioavailability, and potential DNA degradation have been frequently reported (Ando et al. 1999, Walter et al. 1999, Wang et al. 1999, Fu et al. 2000). A "hybrid" approach is the chemical conjugation of cationic surfactants or polymers to the surface of biodegradable particles. DNA can be electrostatically bound to the latter, thus enhancing its availability for immediate release (Kasturi et al. 2005, Munier et al. 2005, O'Hagan et al. 2006).

Cellular entry is the first obstacle encountered by the polyplex to be overcome. The strategies developed to surmount this barrier have mimicked the endocytosis—the process by which cells internalize nutrients, bacteria, etc., from the extracellular space. The endocytic uptake is known to occur by at least five pathways—phagocytosis, macropinocytosis, clathrin-mediated endocytosis, caveolae-mediated endocytosis, and clathrin- and caveolae-independent endocytosis (Conner and Schmid 2003). All of them share the common uptake mode of enclosing the internalized polyplexes within transport vesicles derived from the plasma membrane. Differences arise, however, in the subsequent intracellular processing and routing of internalized polyplexes, which consequently affect the extent to which the polyplexes can deliver their genetic cargo efficiently (Wong et al. 2007). The cellular entry can ideally be achieved by targeted uptake via receptor-mediated endocytosis and size exclusion–mediated phagocytosis. The former consists of attaching of targeting ligands to the vectors, which induce endocytosis upon binding to their cognate surface receptors. Table 5.1 lists some examples of the most commonly used ligands that have provided selective targeting to various cell types. Preferential internalization of DNA-encapsulated microspheres (1–10 μm size range) by phagocytic cells is the core of the size exclusion–mediated phagocytosis. Such a strategy has been adopted for DNA vaccine applications where antigen-presenting cells of the immune system are the target cells (Wang et al. 2004, O'Hagan et al. 2006).

The nonspecific uptake, that is, in the absence of targeting strategies, is another possible way to gain entry in the cells. It involves ionic interactions with membrane-bound proteoglycans, lipophilic interactions with phospholipid membrane, and cell-penetrating peptide (CPP)-mediated uptake. Proteoglycans are composed of a membrane-associated core protein from which a chain of sulfated or carboxylated glycosaminoglycans (GAGs) extend into the extracellular space (Hardingham and Fosang 1992). These highly anionic GAG units determine much of the interactions

TABLE 5.1
Endogenous Ligands, Carbohydrates, Antibodies, and CPP Employed to Promote Cellular Uptake

Endogenous Ligands	Carbohydrates	Antibodies	CPP
Transferrin	Galactose	Anti-cd3	HIV Tat
TGF-α	Mannose	Anti-EGF	Penetratin
Folate	Lactose	Anti-HER2	Transportan
FGF		Anti-pIgR	Polyarginine
EGF		Anti-JL-1	
VEGF			
Asialoorosomucoid			
RGD peptide			

between the cell surface and the extracellular macromolecules and are responsible for the overall negative charge of the plasma membrane (Yanagishita and Hascall 1992); they are believed to play a central role in the endocytic uptake of many non-targeted, positively charged gene delivery systems (Mislick and Baldeschwieler 1996, Kichler et al. 2006). It is important to consider, however, that differences in proteoglycan distribution between cell types can lead to varying degrees of internalization from one cell type to another and, in addition, the nonspecificity of this mode of interaction can lead to indiscriminate uptake and undesirable gene expression by unanticipated cell populations (Wong et al. 2007).

Polyplexes have been able to induce cellular entry through interactions between lipophilic residues bound to the vector and the phospholipid layer of the cell membrane (Takigawa and Tirrell 1985). In particular, increased transfection efficiency of branched PEI modified with long lipophilic chains has been reported elsewhere (Thomas and Klibanov 2002). The results of these authors indicate that even the position of the substituents (primary or tertiary amines of PEI) can have effects on the extent of this interaction and thus transfection efficiencies. CPPs are shown in Table 5.1. Typically they are positively charged, amphiphilic in nature, and composed of 5–40 amino acids. They are able to simultaneously bind DNA and serve as cell-penetrating agents. The mechanism of their action is still unclear. The largely accepted hypotheses include (1) the formation of peptide-lined pores within the membrane, (2) direct penetration through the membrane and into the cytoplasm, (3) transient uptake into a membrane-bound micellar structure that inverts to release the CPP and its genetic cargo inside the cytosol, and (4) the induction of endocytosis (El-Andaloussi et al. 2005).

After internalization, the polyplex is believed to be contained within an endocytic vesicle. It is then either (1) recycled back to the cell surface; (2) sorted to acidic, degradative vesicles (e.g., lysosomes); or (3) delivered to an intracellular organelle (e.g., Golgi apparatus, endoplasmic reticulum) (Khalil et al. 2006). The intracellular itinerary that the endocytic vesicles follow depends on the uptake pathway (i.e., phagocytosis, macropinocytosis, clathrin-mediated endocytosis, caveolae-mediated endocytosis, and clathrin- and caveolae-independent endocytosis), which in turn is dependent on a number of factors including cell line, polyplex type, and the conditions under which the polyplex is formulated (Rejman et al. 2005, von Gersdorff et al. 2006). The clathrin-mediated endocytosis is currently the most characterized pathway. Within this pathway, the polyplex is sequestered in endosomal vesicles and shuttled through the endolysosomal pathway. The *escape from endolysosomal pathway* is of paramount importance to avoid enzymatic degradation within lysosomal compartment; otherwise, the gene delivery fails at this stage. Strategies to escape lysosomal degradation have exploited the influx of protons by incorporating pH-sensitive moieties into vector designs. These are protonatable amines such as linear and branched PEI; PDMAEMA; polyamidoamine; endosomal viral or synthetic release proteins such as melittin, influenza virus hemagglutinin, diphtheria toxin, GALA, KALA, and JTS-1; and alkylated carboxylic acids such as poly(propylacrylic acid), poly(ethylacrylic acid), and poly(ethylacrylate-*co*-acrylic acid). The ultimate effect of these vector components is the disruption of the membrane of the endolysosomal vesicles leading to subsequent release of the vesicles' content in the cytosol.

Before the double-layer membrane surrounding the nucleus is reached, the polyplexes (or naked DNA) that have escaped from the endolysosomal pathway must be trafficked through the cytosol. The cytosolic environment is both physically and metabolically hostile to the polyplexes. The mesh-like structure of the cytoskeleton can severely impede the diffusion of both naked DNA greater than 250 bp with an extended linear length of approximately 85 nm (Lukacs et al. 2000, Dauty and Verkman 2005) and polyplexes that are typically on the order of 150–200 nm in size. In addition, nucleolytic enzymes ready to degrade unprotected nucleic acids are interspersed among the microtubules, intermediate filaments, and microfilaments that build the network-like structure of the cytoskeleton. Strategies have been elaborated to shorten the residence time within the cytosol and promote transport toward and into nucleus. They are typically based on nuclear import machinery naturally utilized by cells to transport proteins into nucleus. Cytosolic proteins that are destined for the nucleus contain a distinct amino acid sequence known as nuclear localization signal (NLS), which is recognized by import proteins (importins). The latter direct their subsequent transport and shuttle them through the nuclear pore complexes that perforate the nuclear membrane (Pante and Kann 2002). The NLSs are short, cationic peptide segments and can be directly used to interact and condense DNA (Kichler et al. 2000). They can be also attached to a polymer vector that is subsequently complexed with gene material (Moffatt et al. 2006, Talsma et al. 2006). NLSs used to promote cytosolic transport and nuclear import are, for example, human T-cell leukemia virus (HTLV) type 1, M9, HIV type-1 viral protein R (Vpr), and SV40 large T antigen. The limitations of this approach are related to the size and type of DNA used (i.e., linear, plasmid), the method of NLS incorporation (i.e., covalent conjugation to DNA, electrostatic complexation with DNA, or conjugation to a polymer vector), the type of NLS peptide employed, the number of NLSs incorporated, and the type of polymer vector used (e.g., liposomes, PEI) (Bremner et al. 2004, Bergen and Pun 2005). Various natural or recombinant proteins, for example, protamine, histone, and high-mobility group proteins, contain NLS sequences and a net positive charge. They fall into two broad categories—histones and nonhistone proteins. The former are involved with DNA binding and condensation as well as regulating transcription and cell cycle progression (Haberland and Bottger 2005, Kaouass et al. 2006), whereas the latter function similarly and support various DNA-related activities including transcription, replication, and recombination (Kaouass et al. 2006). Carbohydrates such as lactose, mannose, and *N*-acetylglucosamine can also mediate cytosolic transport and nuclear import: similar to the NLS-based transport, the carbohydrate-mediated transport is an energy- and signal-dependent process that imports the cargo into the nucleus through nuclear pores (Wong et al. 2007). In spite of some conflicting reports (Klink et al. 2001, Monsigny et al. 2004, Grosse et al. 2006), the relative ease of synthesis, inherent biocompatibility, and reduced immunogenic concerns associated with carbohydrate-based moieties have contributed to its advantageous utility in gene vector design. Polyplexes may also gain nuclear entry by taking advantages of nuclear membrane breakage that occurs during cell mitosis (Brunner et al. 2000, Pack et al. 2005, Wong et al. 2007). This approach, however, is somewhat limited since many applications of nonviral gene delivery target nondividing or slow-dividing cells (Parker et al. 2007).

The final step of the journey of the exogenous genes is to *express their encoded therapeutic proteins*. Genome insertion has been attempted by incorporating DNA transposons within the delivery system. Transposons are naturally occurring DNA sequences capable of enzymatically excising themselves out of the one chromosomal locus and reinserting itself into another locus (Yant et al. 2000). An important limitation of transposon-mediated genome integration is the random insertion into the host genome leading to inadvertent gene disruption and unwanted side effects. Therefore, systems providing a certain degree of site specificity have been explored. These are integrase enzymes derived from bacteriophages and hybrid systems comprising a transposable element and a DNA sequence recognition element. Transgene insertion occurs through their ability to recognize and dock the transgenic plasmid to a known genome locus (Wong et al. 2007). Furthermore, both physical and chemical stimuli have been explored to provide an additional level of control over transcription.

5.2.1.2 Polymer-Based Nonviral Gene Carriers

In Section 3.6, we refer to a number of hybrid co-assembled nanoparticles, the constituents of which are a synthetic or natural (co)polymer and oligo- or polynucleotides. They are discussed from the point of view of structure, morphology, and dimensions and, mainly, from the point of view of new properties produced as a result of the interactions between the two (or more in some cases) species. In the following, these and similar structures are discussed from the perspective of gene therapy, though gene delivery via polyplexes has some issues to overcome such as low transfection, nonspecific short-term delivery, rapid clearance, and toxicity concerns and has to achieve a significant presence in a clinical capacity based on the small percentage of nonviral vectors currently used in clinical trials (Wong et al. 2007, O'Rorke et al. 2010). The examples given as follows as well as those in Section 3.6 demonstrate how research is progressing to meet the demands of nonviral polymer-based gene delivery. When possible, information about the functional effects of a particular structural feature and mechanisms that govern structure–function relationships and, in turn, gene delivery is provided.

Nonviral transfection systems based on polymer vectors have been periodically reviewed by various authors (Gebhart and Kabanov 2001, Wong et al. 2007, Mintzer and Simanek 2009, O'Rorke et al. 2010, Tamboli et al. 2011, Kesharwani et al. 2012, Mastrobattista and Hennink 2012). Typically, polyplexes between polycations and DNA (RNA or other ODNs) have been evaluated with respect to their effectiveness, toxicity, and cell type dependence. Not only have commercially available polycations such as linear and branched PEIs, polyamidoamine dendrimers, and linear and dendritic polypropyleneimines been examined but also other polymer vectors. These are polymethacrylates, carbohydrate-based polymers (chitosan, dextran, cyclodextrins, polyglycoamidoamine), biodegradable polymers (poly(4-hydroxy-1-proline ester), poly[α-(4-aminobutyl)-1-glycolic acid], poly(amino ester), phosphorus-containing polymers), and polypeptides (Tat-based peptides, antennapedia homeodomain peptide, MPG peptide, transportan peptide). These systems exhibit activity higher or comparable to "standard" lipid-based transfection reagents. Other studies demonstrate good potential of structurally diverse polymers and polyplex systems as transfection reagents with low cytotoxicity.

5.2.1.2.1 Linear (Co)Polymers

Zheng and coworkers have prepared hydrophobically modified 1.8 kDa PEI conjugates with lipoic acid (6,8-dithiooctanoic acid) using carbodiimide chemistry (Zheng et al. 2011). The DNA-binding ability was impaired by lipoylation and much smaller (84–183 nm) particles were formed. The MTT assays demonstrated that all PEI–lipoic acid conjugates and polyplexes were essentially nontoxic to HeLa and 293T cells up to a tested concentration of 50 μg/mL and an N/P ratio of 80/1, respectively. The *in vitro* gene transfection studies in HeLa and 293T cells showed that lipoylation of 1.8 kDa PEI markedly boosted its transfection activity.

A library of 39 strictly linear PEG-PEI diblock copolymers has been synthesized (Bauhuber et al. 2012). The copolymers were composed of PEG moieties with fixed molar mass of 2, 5, or 10 kDa, whereas the PEI molar mass ranged from 1.5 to 10.8 kDa. They were expected to build small and stable polyplexes, as the two blocks were clearly segregated, which implied that the PEG might not sterically counteract the interaction between the nucleic acid and PEI. The results indicated that the PEG domain had a greater influence on the physicochemical properties of the polyplexes than PEI. A PEG content higher than 50% led to small (<150 nm), nearly neutral polyplexes with favorable stability. The transfection efficacy of these polyplexes was significantly reduced compared to the PEI homopolymer but was restored by the application of the corresponding degradable copolymer, which involved a redox triggerable PEG domain.

Quaternized poly[3,5-bis(dimethylaminomethylene)-p-hydroxyl styrene] homopolymer (QNPHOS), having two permanently charged cationic sites per monomer unit as well as its block copolymer with PEO (QNPHOS-*b*-PEO), were studied as small interfering RNA (siRNA) and plasmid DNA carriers (Varkouhi et al. 2012). Comparisons with the standard transfectant PDMAEMA, in terms of nucleic acid–binding strength, gene silencing, and transfection activities of the complexes, were made. It was shown that siRNA complexes based on QNPHOS and QNPHOS-*b*-PEO dissociate in the presence of a fourfold higher heparin concentration than necessary to destabilize PDMAEMA-based complexes. Under the same conditions, complexes of DNA and QNPHOS or QNPHOS-*b*-PEO did not show any dissociation, in contrast to PDMAEMA polyplexes. The DNA polyplexes based on QNPHOS or QNPHOS-*b*-PEO did not show transfection activity, which was mainly ascribed to their high physicochemical stability. On the other hand, siRNA complexes based on QNPHOS and QNPHOS-*b*-PEO were found to show a low cytotoxicity and an improved siRNA delivery and high gene-silencing activity, even higher than those based on PDMAEMA. The authors concluded that the observed behavior must be due to the excellent binding characteristics of QNPHOS and QNPHOS-*b*-PEO toward siRNA, which in turn can be correlated with the presence of two permanently charged cationic groups per monomer unit in the QNPHOS chains. Based on the results of this study, it can be concluded that formation of strong siRNA nanosized complexes with polymers containing double charges per monomer is advantageous.

In a recent study, oligomers and polymers of *N*-ethyl pyrrolidine methacrylamide as well as its copolymers with DMAAm have been synthesized (Velasco et al. 2011). Cell viability and proliferation after contact with polymer and polyplexes were studied using 3T3 fibroblasts, and the systems showed an excellent biocompatibility at 2 and 4 days. Transfection studies were performed with plasmid Gaussia luciferase (GLuc) kit and were found that the highest transfection efficiency in serum-free was obtained with oligomers from the P/N ratio of 1/6 to 1/10.

Polyphosphonium polymers can be considered as an efficient and nontoxic alternative to polyammonium carriers. A study allowing for direct comparison of DNA binding and gene transfection of phosphonium-containing macromolecules and their respective ammonium analogues has been performed (Hemp et al. 2012a). Conventional free-radical polymerization of quaternized 4-vinylbenzyl chloride monomers afforded phosphonium- and ammonium-containing homopolymers for gene transfection experiments of HeLa cells. DNA gel shift assays and luciferase expression assays revealed that the phosphonium-containing polymers bound DNA at lower charge ratios and displayed improved luciferase expression relative to the ammonium analogues. The triethyl-based vectors for both cations failed to transfect HeLa cells, whereas tributyl-based vectors successfully transfected HeLa cells similar to SuperFect demonstrating the influence of the alkyl substituent lengths on the efficacy of the gene delivery vehicle. Cellular uptake of Cy5-labeled DNA highlighted successful cellular uptake of triethyl-based polyplexes, showing that intracellular mechanisms presumably prevented luciferase expression. Endocytic inhibition studies demonstrated the caveolae-mediated pathway as the preferred cellular uptake mechanism for the delivery vehicles examined. The authors noted that changing the polymeric cation from ammonium to phosphonium enables an unexplored array of synthetic vectors for enhanced DNA binding and transfection that may transform the field of nonviral gene delivery (Hemp et al. 2012a).

In another paper, a series of polymers based on poly(ethoxytriethylene glycol acrylate) bearing ammonium or phosphonium quaternary groups have been prepared (Ornelas-Megiatto et al. 2012). The triethylphosphonium polymer showed transfection efficiency up to 65% with 100% cell viability, whereas the best result obtained for the ammonium analogue reached only 25% transfection with 85% cell viability. The nature of the alkyl substituents on the phosphonium cations (*tert*-butyl,

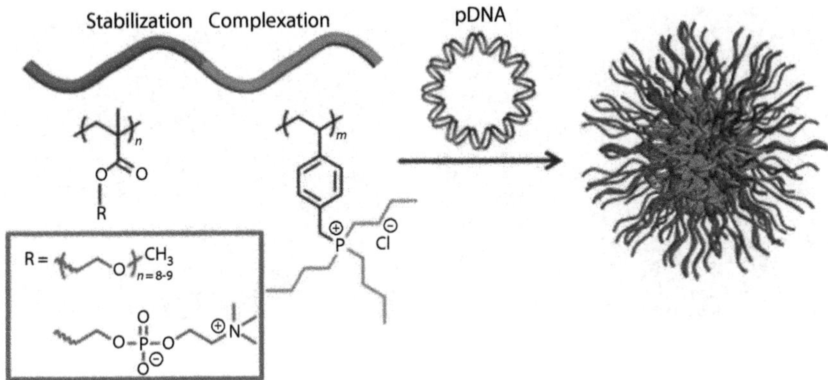

FIGURE 5.10 Phosphonium-containing diblock copolymers for enhanced colloidal stability and efficient nucleic acid delivery. The stabilizing blocks were of molecular weight M_n=25,000 g/mol. By subsequent chain extension, phosphonium-containing blocks with DPs of 25, 50, or 75 were synthesized. (Reprinted with permission from Hemp, S.T., Smith, A.E., Bryson, J.M., Allen, M.H., and Long, T.E., *Biomacromolecules*, 13(8), 2439–2445, 2012. Copyright 2012, American Chemical Society.)

hydroxypropyl, phenyl were studied) was shown to have an important influence on the transfection efficiency and toxicity of the polyplexes. The results presented by the authors showed that the use of positively charged phosphonium groups was a worthy choice to achieve a good balance between toxicity and transfection efficiency in gene delivery systems (Ornelas-Megiatto et al. 2012).

Phosphonium-based diblock copolymers for nonviral gene delivery have been synthesized by reversible addition fragmentation transfer (RAFT) (Hemp et al. 2012b). They were composed of a stabilizing block of either oligo(ethylene glycol$_9$) methyl ether methacrylate or 2-(methacryloxy)ethyl phosphorylcholine and a phosphonium-containing cationic block of 4-vinylbenzyltributylphosphonium chloride, which induced electrostatic complexation with DNA (Figure 5.10). All block copolymers exhibited low cytotoxicity (>80% cell viability) and generated stable polyplexes with hydrodynamic radii between 100 and 200 nm. The cellular uptake by COS-7 and HeLa cells and, consequently, the transfection in these cell lines were reduced, whereas serum transfection in HepaRG cells, which are a predictive cell line for *in vivo* transfection studies, showed successful transfection using all diblock copolymers with luciferase expression.

A micellar delivery system based on "diblock" copolymers of poly(DMAEMA)-*b*-poly[(butyl methacrylate)-*co*-(DMAEMA)-*co*-(propylacrylic acid)] (see Figure 2.30) was described in the previous chapters. As noted, the copolymer was composed of a polycationic block of PDMAEMA, which mediated siRNA binding and a second block of dimethylaminoethyl methacrylate (DMAEMA) and propylacrylic acid in roughly equimolar ratios with butyl methacrylate (BMA), which was pH sensitive and endosome releasing (Convertine et al. 2010). Delivery systems were formed upon binding siRNA to the corona of the preformed micelles at theoretical +/– charge ratios of 4/1 and greater. Their ability to deliver siRNA through

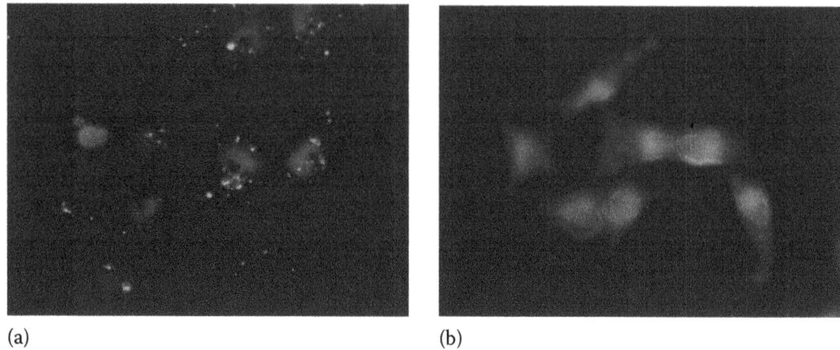

FIGURE 5.11 Polymer-enhanced intracellular delivery of FAM-labeled siRNA. Representative images illustrating (a) punctate staining (dots of strong contrast) in the samples treated with lipofectamine/siRNA complexes alone and (b) dispersed fluorescence within the cytosol following delivery of poly(DMAEMA)-*b*-poly[(butyl methacrylate)-*co*-(DMAEMA)-*co*-(propylacrylic acid)]/siRNA complexes. Samples were treated for 15 min with 25 nM FAM-siRNA and prepared for microscopic examination following DAPI nuclear staining (areas of low contrast). (Reprinted with permission from Convertine, A.J., Diab, C., Prieve, M., Paschal, A., Hoffman, A.S., Jonhson, P.H., and Stayton, P.S., *Biomacromolecules*, 11, 2904–2911, 2010. Copyright 2010, American Chemical Society.)

the endosomal pathway into the cytoplasm was investigated by conducting mRNA knockdown experiments against GAPDH. Transfection experiments in HeLa cells were conducted at siRNA concentrations between 12.5 and 100 nM under high serum conditions. Flow cytometry analysis of cell uptake properties indicated enhancement of the knockdown potential of the micellar system: an uptake in 90% of cells and a threefold increase in siRNA per cell compared to a standard lipid transfection agent. The fluorescence micrographs (Figure 5.11) showed that labeled siRNA is dispersed throughout the cellular cytoplasm in contrast to the commercial reagent that was primarily punctate.

Also mentioned in Chapter 3 are the amino poly(glycidyl methacrylate)s. These were shown to form stable complexes with antisense ODNs that prevented the latter from nuclease degradation. In addition, the polyplexes of linear and starlike poly(glycidyl methacrylate)s of M_n in the range 15–20 kDa modified with methylethylamine as well as linear poly(glycidyl methacrylate) modified with 4-amino-1-butanol exhibited higher transfection efficiency *in vitro* compared to the “golden” standard of PEI 25 k (Gao et al. 2011a).

The potential of a series of triblock copolymers for messenger RNA (mRNA)-based strategies has recently been demonstrated (Cheng et al. 2012). The materials were composed of a cationic DMAEMA segment to mediate mRNA condensation, a hydrophilic poly(ethylene glycol) methyl ether methacrylate (PEGMA) segment to enhance stability and biocompatibility, and a pH-responsive endosomolytic copolymer of diethylaminoethyl methacrylate (DEAEMA) and BMA designed to facilitate cytosolic entry. They were able to condense mRNA into 86–216 nm particles, and the polyplexes formed from polymers with the PEGMA segment in the center of the polymer chain displayed the greatest stability to heparin displacement; they were associated with the highest transfection efficiencies in two immune cell lines, RAW 264.7 macrophages (77%) and DC2.4 dendritic cells (50%). Transfected DC2.4 cells were shown to be capable of subsequently activating antigen-specific T cells.

The efficacy of using a polysorbitol-based osmotically active transporter (PSOAT) system (see Figure 5.12a for the chemical formula of the copolymer) for siRNA delivery and its specific mechanism for cellular uptake to accelerate targeted gene silencing has been reported (Islam et al. 2012). The authors found that PSOAT functioned via a caveolae-mediated uptake mechanism due to its hyperosmotic activity. An example of silencing effect of PSOAT/siLuc and PSOAT/siGFP in A549 cells is demonstrated in Figure 5.12a, right panel: luciferase- or green fluorescent protein (GFP)-expressing A549 cells at 80% confluence were transfected with polymer/siRNA polyplexes. Graphical presentation of the caveolae-mediated endocytosis is shown in Figure 5.12b. Detailed description of the different stages and events of the caveolae-mediated uptake mechanism is given by the authors (Islam et al. 2012): (1–2) PSOAT/siRNA complex recognizes and binds to the sorbitol-transporting channel (STC) on the extracellular membrane and creates a hyperosmotic environment. (3–5) The osmotic pressure–sensitive PSOAT/siRNA complex induces caveolin (Cav)-1 expression and selectively stimulates caveolae-mediated endocytosis. This event subsequently generates COX-2 expression under osmotic stress. At this stage, Cav-1 expression augments and a mature caveolae structure is formed where COX-2 might work on disrupting the joint surface of particle deposition and accelerate caveolae-mediated endocytosis. (6) Selective caveolae endocytosis allows the PSOAT/siRNA complex containing caveolae endosome (caveosome) to avoid lysosomal fusion. (7–8) The endosome containing PSOAT/siRNA complex swells and eventually bursts due to the proton sponge effect of LPEI, which allows the complex to escape into the cytosol. (9) Due to the nature of degradable linkages, PSOAT degradation occurs and siRNAs are released. (10–12) The released siRNA recognizes and breaks down the target mRNA at the posttranscriptional level through an RNAi mechanism.

The transfection and local tissue distribution of plasmid DNA using different cationic polymers have been analyzed after intradermal injection in mice (Palumbo et al. 2012). Cationic polymers with distinct chemical structures: branched PEI, linear poly(2-aminoethyl methacrylate) (PAEM), and diblock copolymer PEG-*b*-PAEM were selected to study the structure–function relationship of the polymer carriers in the context of *in vivo* administration (Figure 5.13). The authors found that naked DNA dispersed quickly in hours within the skin and showed limited colocalization with APCs. On the other hand, polyplexes formed depots at the injection site and persisted for days to engage and transfect skin cells. PEGylated polyplexes, in particular, possessed superior stability against aggregation than non-PEGylated polyplexes, and they disseminated well in the skin that promoted interaction with both the antigen-presenting cells and dermal fibroblasts. These findings provide *in vivo* evidence to support the use of PEGylated polymer carriers for DNA vaccine delivery and suggest possible approaches to further improve polymer design.

In a preceding paper (Ji et al. 2011), a comprehensive study has been conducted to evaluate the colloidal properties of PAEM/plasmid DNA polyplexes, the uptake and subcellular trafficking of polyplexes in antigen-presenting dendritic cells,

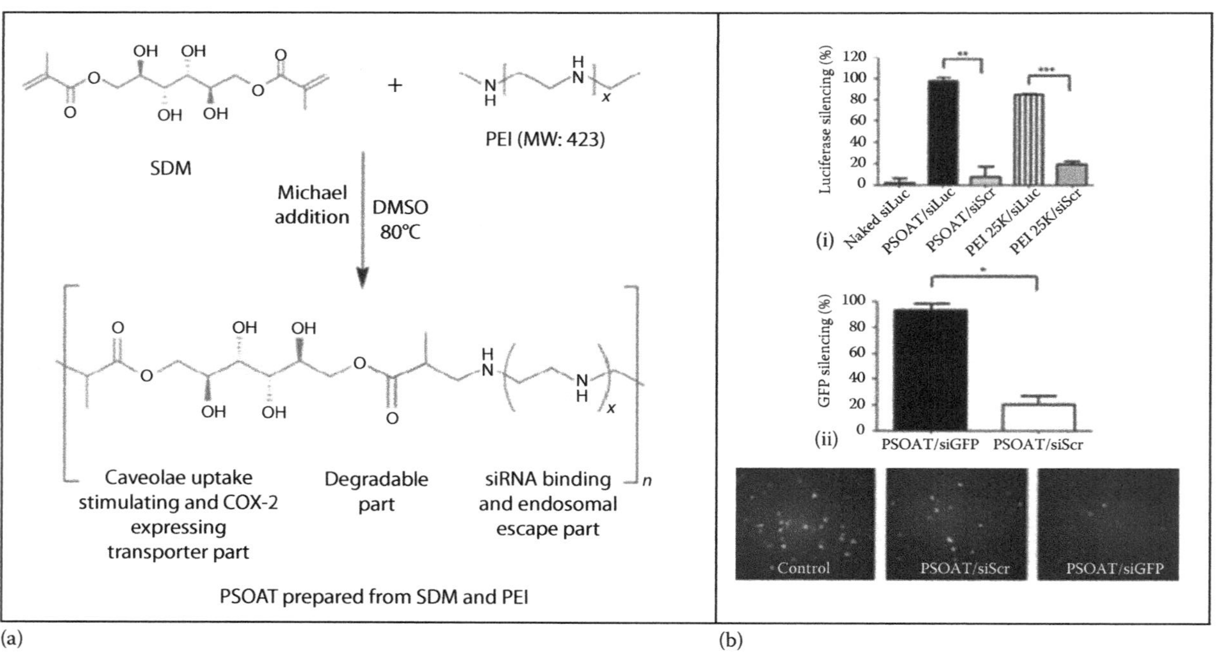

FIGURE 5.12 (a) Left panel: proposed reaction scheme for the synthesis of PSOAT from sorbitol dimethacrylate (SDM) and PEI. PSOAT was prepared through a Michael addition reaction in DMSO at 80°C. (b) Right panel: silencing efficacy of PSOAT/siLuc and PSOAT/siGFP in A549 cells. (i) Luciferase silencing by PSOAT/siLuc and PEI 25K/siLuc complexes compared to the respective scrambled siRNA (siScr) group. (ii) GFP silencing efficacy of PSOAT/siGFP complexes compared to the respective siScr group at a siRNA concentration of 100 pmol/L ($n=3$, error bar represents SD) ($^{*}P<0.05$; $^{**}P<0.01$; $^{***}P<0.001$, one-way ANOVA) and fluorescent images of control, PSOAT/siScr- and PSOAT/siGFP-treated A549 cells (magnification: 10×). (Reprinted from *Biomaterials*, 33(34), Islam, M.A., Shin, J.-Y., Firdous, J., Park, T.-E., Choi, Y.-J., Cho, M.-H., Yun, C.-H., and Cho, C.-S., 8868–8880, Copyright 2012, with permission from Elsevier.)

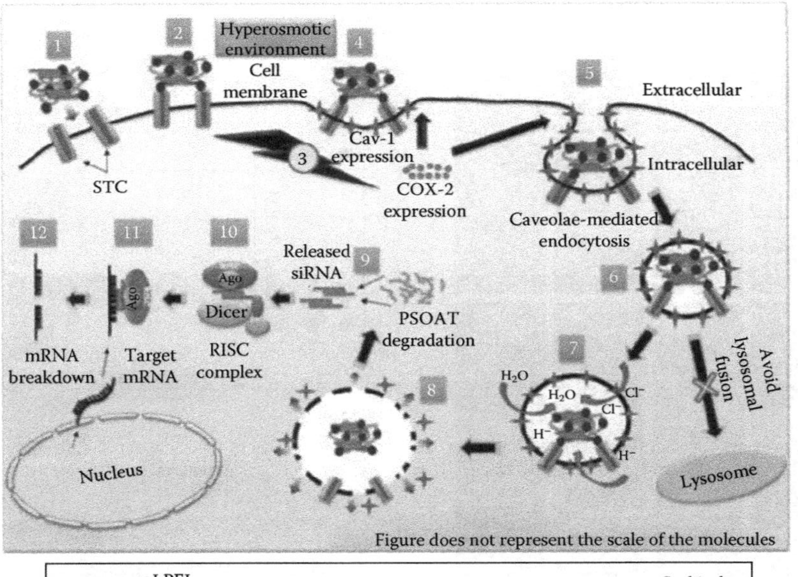

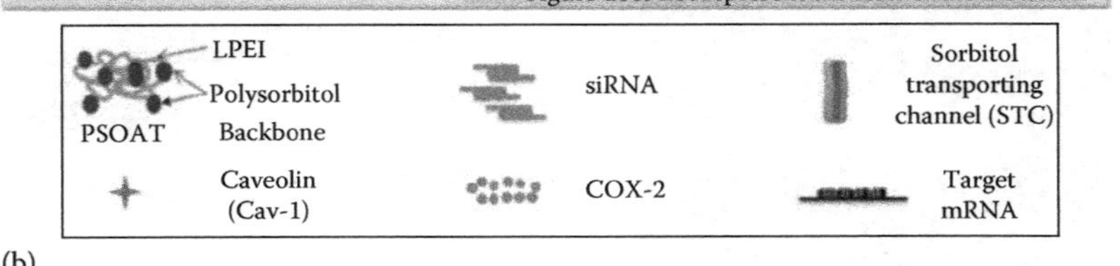

(b)

FIGURE 5.12 (continued) (b) Schematic representation on the function of PSOAT. Graphical presentation shows selective caveolae-mediated endocytosis of the PSOAT/siRNA complex following hyperosmotic pressure that induces caveolin (Cav-1) and cyclooxygenase (COX-2) expression, which accelerates the efficacy of the transporter for RNAi silencing. (Reprinted from *Biomaterials*, 33(34), Islam, M.A., Shin, J.-Y., Firdous, J., Park, T.-E., Choi, Y.-J., Cho, M.-H., Yun, C.-H., and Cho, C.-S., 8868–8880, Copyright 2012, with permission from Elsevier.)

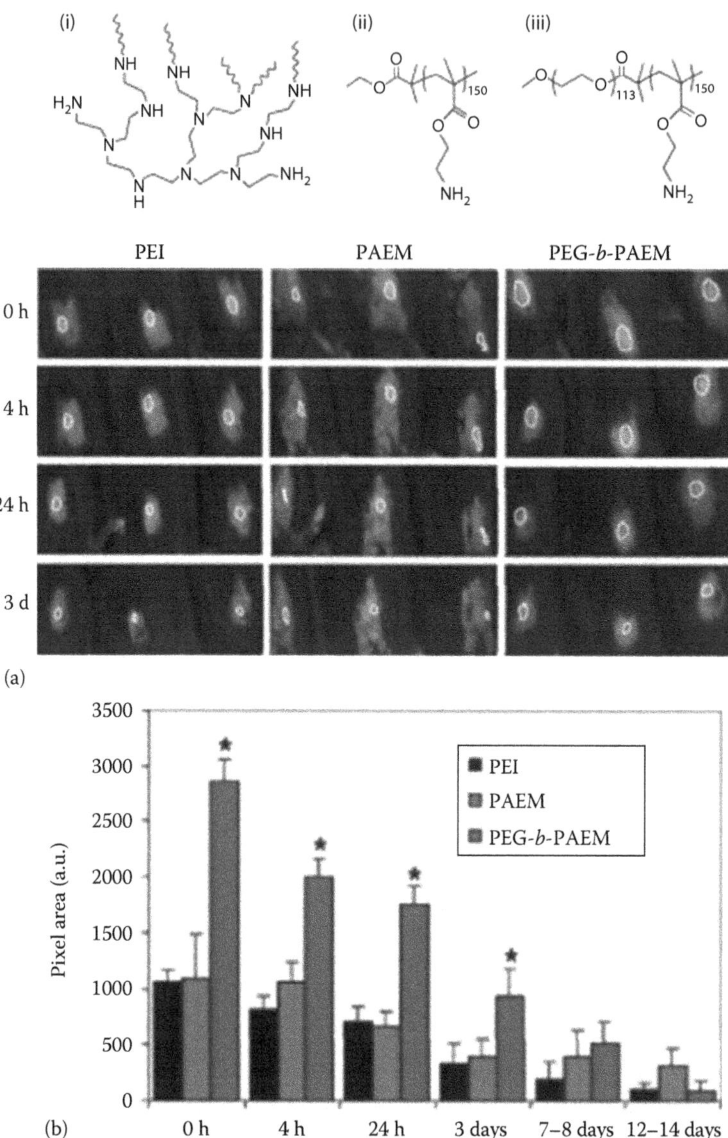

FIGURE 5.13 (a) Chemical structures of (i) branched PEI, (ii) linear PAEM, and (iii) diblock copolymer PEG-*b*-PAEM. (b) Tissue distribution of polyplexes in live animals after intradermal injection. Three mice were each injected with polyplexes containing 10 μg of Cy3-labeled DNA in the right hind quadriceps region and were imaged together. Plasmid distribution in three injected mice was shown at indicated time points as marked by a white outline (A), and the area of signal was quantified (B). *PEGylated polyplexes showed statistically larger area of spreading than PEI- and PAEM-based polyplexes (t-test, $p < 0.001$). (Reprinted from *J. Control. Release*, 159(2), Palumbo, R.N., Zhong, X., Panus, D., Han, W., Ji, W., and Wang, C., 232–239, Copyright 2012, with permission from Elsevier.)

and the biological performance of PAEM as a potential DNA vaccine carrier. PAEM of different chain length (45, 75, and 150 repeating units) showed varying strength in condensing plasmid DNA into narrowly dispersed nanoparticles with very low cytotoxicity. Longer polymer chains resulted in higher levels of overall cellular uptake and nuclear uptake of plasmid DNA, but shorter polymer chains favored intracellular and intranuclear release of free plasmid from the polyplexes.

Packaging of plasmid DNA within both rod- and sphere-shaped polyplex micelles has been investigated, with a focus on DNA rigidity and folding (Osada et al. 2012). Selective preparation of rod or spherical structures was accomplished by modulating the polylysine segment length of the PEG–polylysine block copolymers. The correlation of these packaging structures to gene expression efficiency was determined both *in vitro* and *in vivo*, with improved gene expression resulting from folded plasmid DNA demonstrated in cultured cells as well as in skeletal muscle following intravenous injection. The authors concluded that the enhanced gene expression demonstrated by folded plasmid DNA packaging may be attributed to its nuclease tolerability and transcription-active nature. Results obtained for regularly folded plasmid DNA polyplex micelles were in stark contrast with collapsed plasmid DNA polyplex micelles, probably due to limited transcription efficiency, clearly showing that controlled packaging of plasmid DNA is crucial for achieving effective gene transfer (Osada et al. 2012).

5.2.1.2.2 Nonlinear (Co)Polymers

Besides molecular weight and composition, the chain architecture of gene delivery vectors plays a critical role in DNA complexation and, hence, gene expression. Generally, the nonlinear (graft, branched, starlike) polymer exhibits superior gene delivery abilities compared to their linear analogues.

Reversibly hydrophobilized 10 kDa PEI based on rapidly acid-degradable acetal-containing hydrophobe has been designed for nontoxic and highly efficient nonviral gene transfer (Liu et al. 2011a). Water-soluble PEI derivatives with average 5, 9, and 14 units of pH-sensitive 2,4,6-trimethoxybenzylidene-tris(hydroxymethyl)ethane (TMB-THME) hydrophobe per molecule, denoted as PEI-g-$(TMB\text{-}THME)_n$, were readily obtained by treating 10 kDa PEI with varying amounts of TMB-THME-nitrophenyl chloroformate. Polyplexes of smaller size (100–170 nm) and higher surface charges (+25 to +43 mV) were obtained. *In vitro* transfection experiments performed at N/P ratios of 10/1 and 20/1 in HeLa, 293T, HepG2, and KB cells using plasmid pGL3 expressing luciferase as the reporter gene showed that reversibly hydrophobilized PEIs had superior transfection activity to 25 kDa PEI control. For example, polyplexes of PEI-g-$(TMB\text{-}THME)_{14}$ showed about 235-fold and 175-fold higher transfection efficiency as compared to 10 kDa PEI in HeLa cells in serum-free and 10% serum media, respectively, which were approximately 7-fold and 16-fold higher than 25 kDa PEI formulation at its optimal N/P ratio under otherwise the same conditions. Hydrophobic modification of 10 kDa PEI enhances its DNA condensation ability and cellular interactions, while reversal of hydrophobic modification in endosomes facilitates intracellular release of DNA (Figure 5.14).

Star copolymers have been developed as siRNA carriers by Matyjaszewski and coworkers (Cho et al. 2011). The authors utilized ATRP methodologies and an

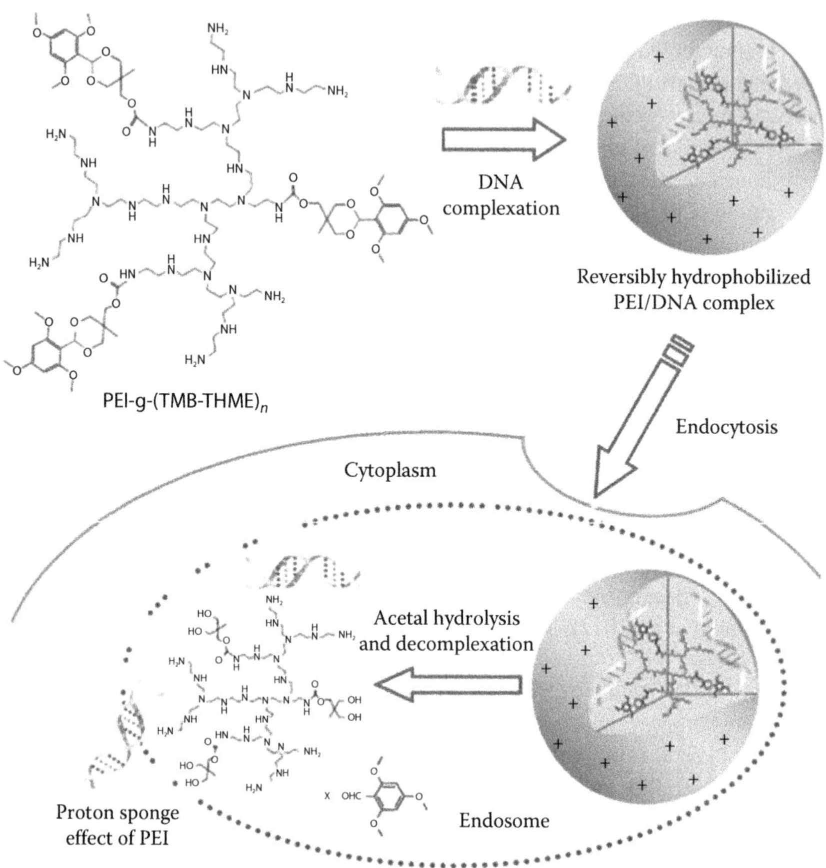

FIGURE 5.14 Illustration on reversibly hydrophobilized 10 kDa PEI for efficient intracellular delivery and release of DNA. TMB-THME denotes 2,4,6-trimethoxybenzylidene-tris(hydroxymethyl)ethane. (Reprinted from *Biomaterials*, 32(34), Liu, Z., Zheng, M., Meng, F., and Zhong, Z., 9109–9119, Copyright 2011a, with permission from Elsevier.)

"arm-first" approach for the synthesis of the copolymers having PEG arms and a degradable cationic core of PEGMA macromonomer, 2-(dimethylamino)ethyl methacrylate (DMAEMA), and a disulfide dimethacrylate (acting as the cross-linker). The star polymers had a diameter of ca. 15 nm in aqueous solutions and were found to be degraded under redox conditions by glutathione treatment into individual polymeric chains, due to cleavage of the disulfide cross-linker. Culture tests with mouse calvarial preosteoblast-like cells, embryonic day 1, subclone 4 (MC3T3-E1.4) indicated that the star copolymers were biocompatible, with more than 80% cell viability after 48 h of incubation even at high concentration of the copolymer (800 μg/mL). Surface charge of the nanoassemblies formed by electrostatic interaction with siRNA at varying N/P ratios confirmed the formation of star copolymer/siRNA complexes. Confocal microscopy and flow cytometry measurements showed the cellular uptake of the nanocomplexes in MC3T3-E1.4 cells after 24 h of incubation.

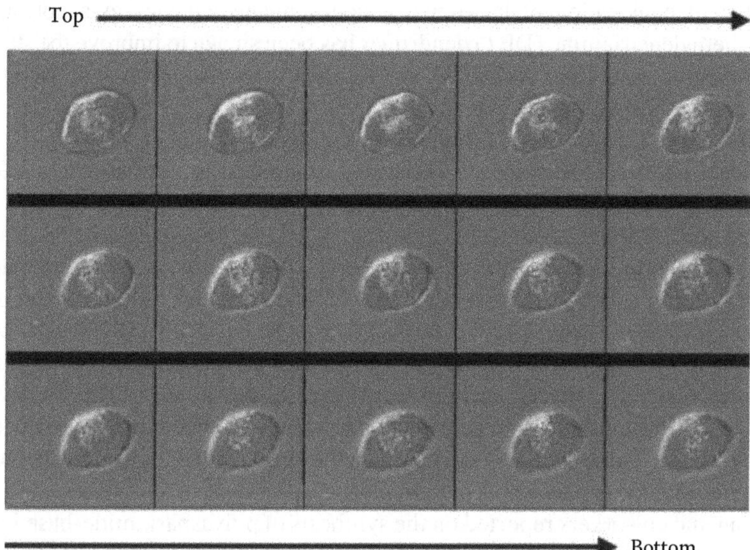

FIGURE 5.15 Optical section z series of A2780 human ovarian cancer cells incubated with siRNA complex with polyamidoamine-PEG-poly(L-lysine) copolymer. (Reprinted with permission from Patil, M.L., Zhang, M., and Minko, T., *ACS Nano*, 22, 1877–1887, 2011. Copyright 2011, American Chemical Society.)

A copolymer composed of dendritic polyamidoamine, PEG, and poly(L-lysine) has been designed for siRNA delivery (Patil et al. 2011). Each moiety played a specific role: the polyamidoamine dendrimer worked as a proton sponge and participated in the endosomal escape and cytoplasmic delivery of siRNA; PEG rendered nuclease stability, whereas poly(L-lysine) provided primary amines to form polyplexes with siRNA and also acted as penetration enhancer. The copolymer provided excellent cellular uptake by A2780 human ovarian cancer cells and ensured homogeneous and uniform distribution of the polyplex in different cellular layers from the top of the cell to the bottom as evidenced by Figure 5.15.

Poly(cyclooctene-graft-oligolysine)s, having a comblike architecture, have been prepared by ring-opening metathesis polymerization of the corresponding oligopeptide-substituted cyclic olefin monomers (Parelkar et al. 2011). The reconfiguration of polylysine into short oligolysine grafts, strung from a hydrophobic polymer backbone, gave transfection reagents greatly superior to polylysine, despite having the identical cationic functional groups. Altering the oligolysine graft length modulated DNA–polymer interactions and transfection efficiency, while incorporating the PKKKRKV heptapeptide (the Simian virus SV40 large T-antigen nuclear localization sequence) pendent groups onto the polymer backbone led to even greater transfection efficiency over the oligolysine-grafted structures. The relative strength of the polymer/DNA complex was key to the transfection performance, as judged by serum stability and PicoGreen analysis. Moreover, the polyplexes exhibited low cytotoxicity, contributing to the therapeutic promise of these reagents.

Chemical conjugation of hydrophobic deoxycholate moieties to a polyamido-amine-diethylenetriamine (DET) dendrimer has been shown to improve the stability of the resulting polyplexes with DNA against ionic strength (Jeong et al. 2011). The transfection efficiency of the latter is higher than that of the polyplex formed from the unmodified dendrimer, but its cytotoxicity remains the same.

Cationic polymeric amphiphiles composed of a poly(L-lactide) middle block and two flanking dendritic poly(L-lysine) moieties of the second generation have been prepared and characterized with respect to CAC, self-aggregation and plasmid DNA-binding affinities, and the effects of poly(L-lactide) block length on particle size, morphology, and ζ-potential (Zhu et al. 2011). Toxicities of these amphiphiles and their polyplexes were assayed by MTT with HeLa, SMMC-7721 and COS-7 cells, and COS-7 cell luciferase and eGFP gene transfection efficacies with these amphiphiles as the delivery carriers were investigated as well. The copolymers exhibited low toxicities. The polyplex nanoparticles were shown to form via two different aggregation mechanisms depending on the length of the middle poly(L-lactide) block (Figure 5.16).

Zhang and coworkers reported on the synthesis of polyaspartamide-based disulfide containing brushed PEI derivatives, $P(Asp\text{-}Az)_X$-SS-PEI (Zhang et al. 2012a).

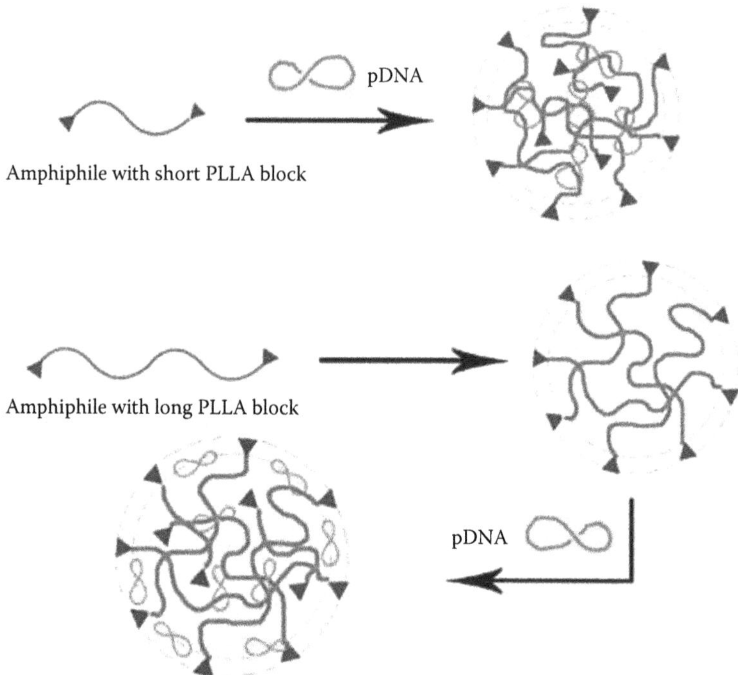

FIGURE 5.16 Schematic presentation of polyplex structures depending on the length of the middle poly(L-lactide) block of dendritic poly(L-lysine)-linear poly(L-lactide)-dendritic poly(L-lysine) symmetric polymeric amphiphiles. (Reprinted with permission from Zhu, C., Zheng, M., Meng, F., Mickler, F. M., Ruthardt, N., Zhu, X., and Zhong, Z., *Biomacromolecules*, 13(3), 769–778, 2012. Copyright 2012, American Chemical Society.)

They were reduction sensitive and able to condense DNA into small positive nanoparticles. *In vitro* experiments revealed that the reducible P(Asp-Az) $_X$-SS-PEI not only had much lower cytotoxicity but also posed high transfection activity (both in the presence and absence of serum) as compared to the control nondegradable 25 kDa PEI.

The influence of graft densities of PCL-*b*-PEG chains on physicochemical properties, DNA complexation, and transfection efficiency of amphiphilic copolymers prepared by grafting PCL-*b*-PEG chains onto hyperbranched PEI has been studied (Zheng et al. 2012a). It was found that the transfection efficiencies of these copolymers increased at first toward an optimal graft density ($n=3$) and then decreased. The buffer-capacity test showed almost exactly the same tendency as transfection efficiency. Cytotoxicity depended on the collective effect of PEG molecular weight and graft density of PCL-*b*-PEG chains. The cytotoxicity, zeta-potential, affinity with DNA, stability of the polyplexes, and critical micellization concentration (CMC) values were reduced strongly and regularly with increasing graft density. Increasing the excess of polymer over DNA was shown to result in a decrease of the observed particle size to 100–200 nm. In an earlier paper, essentially the same authors have created a library of PEI-graft-(PCL-block-PEG) copolymers to establish structure–function relationships for siRNA delivery (Liu et al. 2011b). It was found that longer PEG chains, longer PCL segments, and higher graft density beneficially affected the stability and formation of polyplexes and reduced the zeta-potential of siRNA polyplexes. Significant siRNA-mediated knockdown was observed for hyperbranched PEI25k-(PCL900-mPEG2k)$_1$ at N/P 20 and 30, implying that the PCL hydrophobic segment played a very important role in siRNA transfection.

Zhong and coworkers have grafted PEO with 1.8 kDa branched PEI and prepared copolymers with varying compositions (Zhong et al. 2012). The copolymers were able to effectively condense DNA into small (80–245 nm) particles with moderate positive (+7.2 to +24.1 mV) surface charges. Reduced cytotoxicity (MTT assay in 293T cells, cell viability >80%) with increasing PEO molecular weight and decreasing PEI graft densities was observed. PEO(13k)-g-10PEI (the numbers correspond to molecular weight and grafting densities, respectively) polyplexes formed at an N/P ratio of 20/1, which were essentially nontoxic (100% cell viability), displayed over three- and fourfold higher transfection efficiencies in 293T cells than 25 kDa PEI standard under serum-free and 10% serum conditions, respectively. CLSM studies using Cy5-labeled DNA confirmed that these PEO-g-PEI copolymers could efficiently deliver DNA into the perinuclei region as well as into nuclei of 293T cells at an N/P ratio of 20/1 following 4 h transfection under 10% serum conditions. Enhanced gene delivery efficacy and decreased cytotoxicity of polyplexes of branched 25 kDa PEI grafted with PEO-PPO-PEO copolymers with varying molecular weights and PEO contents have been reported (Liang et al. 2011). The (PEO-PPO-PEO)-g-PEI copolymers showed lower cytotoxicity in three different cell lines (HeLa, MCF-7, and HepG2) than PEI 25 kDa. pGL3-lus was used as a reporter gene, and the transfection efficiency was *in vitro* measured in HeLa cells.

The role of boronic acid moieties in polyamidoamines has been investigated in the application of these polymers as gene delivery vectors (Piest and Engbersen 2011). The polymers contained 30% of phenylboronic acid side chains and exhibited improved polyplex formation abilities with plasmid DNA since smaller and more polydisperse polyplexes were formed as compared to their non-boronated counterparts. The transfection efficiency was approximately similar to that of commercial PEI; however, the polyplexes showed increased cytotoxicity most probably caused by increased membrane-disruptive interactions.

Lin and Engbersen have recently demonstrated that a one-pot Michael addition reaction using an aminoterminated PEG as a comonomer in an equimolar mixture of bisacrylate (*N,N′*-cystaminebisacrylamide) and primary amine monomers (4-amino-1-butanol) is a versatile approach to obtain PEGylated polyamidoamines with disulfide linkages (Lin and Engbersen 2011b). The copolymers condensed DNA into nanoscaled PEGylated polyplexes (<250 nm) with near neutral (2–5 mV) or slightly positive (9–13 mV) surface charge that remained stable in 150 mM buffer solution over 24 h. The PEGylated polyplexes showed very low cytotoxicity in MCF-7 and NIH 3T3 cells and induced appreciable transfection efficiencies in the presence of 10% serum, although those were lower than those of the corresponding unPEGylated complexes. The notably lower transfection was likely caused by limited endosomal escape. In another paper, the cellular dynamics of the polyplexes was studied *in vitro* on retinal pigment epithelium cells (Vercauteren et al. 2011). It was shown that these net cationic polyplexes required a charge-mediated attachment to the sulfate groups of cell-surface heparan sulfate proteoglycans in order to be efficiently internalized. In addition, the involvement of defined endocytic pathways in the internalization of the polyplexes in ARPE-19 cells by using a combination of endocytic inhibitors, RNAi depletion of endocytic proteins, and live-cell fluorescence colocalization microscopy, was assessed. The authors found that the polyplexes entered RPE cells via both flotillin-dependent endocytosis and PAK1-dependent phagocytosis-like mechanism. The capacity of polyplexes to transfect cells was, however, primarily dependent on a flotillin-1-dependent endocytosis pathway.

5.2.1.2.3 Stimuli-Sensitive Systems, Targeting

Clinical therapies have long used control of some environmental parameters at the therapeutic target area. Thermoresponsive nonviral gene carriers based on PNIPAM have been shown to enhance their transfection efficiency upon applying hypothermia, since DNA is assumed to be released as PNIPAM undergoes a globule to coil conformational transition on cooling below its lower critical solution temperature (LCST) (Kurisawa et al. 2000, Sun et al. 2005, Lavigne et al. 2007, Turk et al. 2007). However, contradictory observations, that is, enhanced transfection obtained by raising the temperature from below to above the LCST, have been reported (Bisht et al. 2006, Zintchenko et al. 2006). The enhanced transfection above the LCST has been attributed to increased cation numbers in the endosomal compartment resulting from polyplex aggregation (Zintchenko et al. 2006), whereas the former authors attributed this observation to greater cationic presentation on the surface of polyplexes. Recombinant, protein-based polymer analogues of PNIPAM are the elastin-like polypeptides (ELPs). They are a class of genetically engineered, thermoresponsive polymers that consist of VPGXG pentapeptide monomers, where X is a guest residue that can be any amino acid except proline (Meyer and Chilkoti 1999). The ELPs are water soluble, biocompatible, and thermally responsive with an inverse temperature phase transition

(Urry 1997). Preparation and evaluation of a hybrid recombinant material that possess a thermoresponsive ELP segment and a DET-modified poly-L-aspartic acid segment have been described elsewhere (Chen et al. 2012a). The reaction scheme is presented in Figure 5.17. The polyplexes formed by the copolymer and pGL4 plasmid were of dimensions between 90 and 100 nm in diameter and neutral even at N/P>2. The latter indicated that the ELP block was able to effectively shield the positive charges. The polyplexes also showed appreciable transfection efficiency with low cytotoxicity. The polyplexes practically retained the thermal phase transition behavior conferred by the copolymer; however, they exhibited a two-step transition process: an irreversible primary aggregation below the copolymer's transition temperature, followed by a reversible secondary aggregation at temperatures about the transition temperature. The aggregation is critical for thermal targeting, as particles aggregate at the disease site in response to external heating, which facilitates the particles deposition. Both the copolymer and the polyplexes aggregated at elevated temperatures (~52°C). Although this temperature is not clinically relevant, it may be modulated by changing the ELP block composition or molecular weight. Therefore, the authors concluded the thermal responsiveness of this copolymer can be used for thermal targeting of therapeutic pDNA to hyperthermic disease sites.

Thermosensitive properties and sharp phase transition behavior have been observed for two types of copolymers—random and block–random copolymers of poly(NIPAM-*co*-hydroxyethyl methacrylate-*co*-DMAEMA) (Shen et al. 2012). Cell viability assays indicated low cytotoxicity of block–random copolymers and almost no cytotoxicity of the random copolymers that was ascribed to the presence of low positive charges along copolymer backbones. The random copolymers were used as DNA delivery vectors to deliver GFP gene into HEK293T cells. The transfection efficiency was evaluated by GFP protein expression, which was affected by the charge content and transfection operation temperature. Lowering the temperature to 25°C during the transfection course improves transfection efficiency, which results from the polyplexes dissociation at the temperature below the LCST (Shen et al. 2012).

Poly[oligo(ethylene glycol) methacrylate] is another thermosensitive polymer exhibiting LCST properties. Zhang and coworkers have exploited its properties to prepare a family of thermosensitive cationic copolymers that contain branched PEI 25 K as the cationic segment and poly[methoxy(diethylene glycol) methacrylate-*co*-oligo(ethylene glycol) methacrylate] as the thermosensitive block (Zhang et al. 2011). All of these copolymers were able to condense DNA and formed polyplexes with diameters of 150–300 nm and zeta-potentials of 7–32 mV at N/P ratios between 12 and 36. The key factor for shielding the positive surface charge of the polyplexes and protecting them against protein adsorption was the length of thermosensitive block. Lower cytotoxicity and the best transfection performance were achieved with copolymers with, respectively, a higher content and longest chains of the thermosensitive moieties.

Ternary polyplexes comprising plasmid DNA, PEG-based block cationomer PEG-*b*-{*N*-[*N*-(2-aminoethyl)-2-amnoethyl]aspartamide} (PEG-*b*-P[Asp(DET)], and poly(ethylene oxide)-*b*-poly(propylene oxide)-*b*-poly(ethylene oxide)-*b*-P[Asp(DET)] (P(EPE)-*b*-P[Asp(DET)]) have been prepared by Lai and coworkers

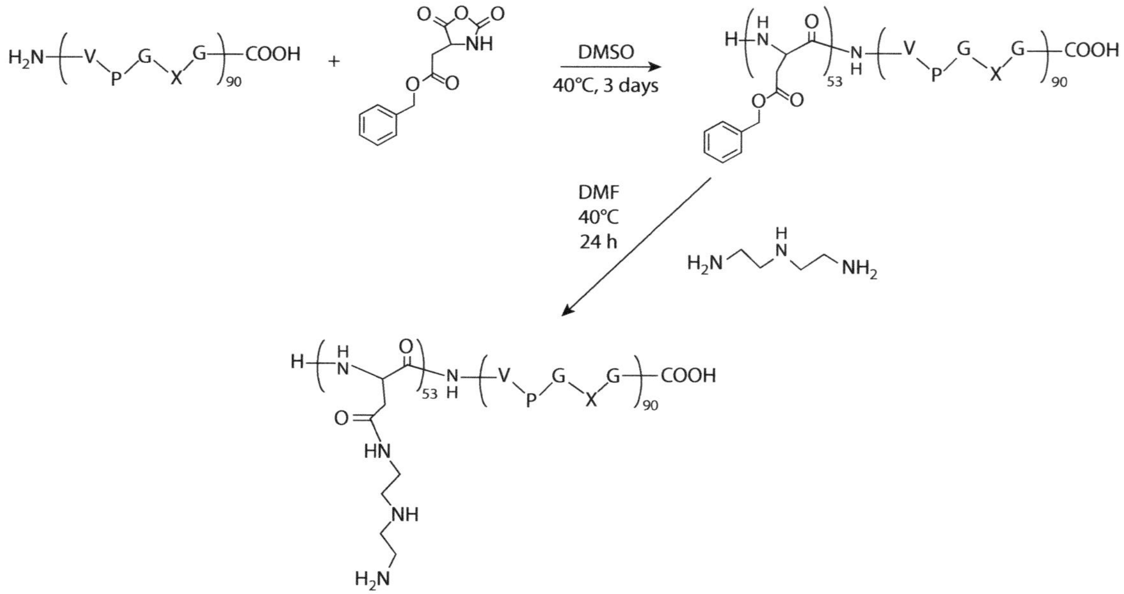

FIGURE 5.17 Reaction scheme for the synthesis of a diblock copolymer possessing a thermoresponsive ELP block and a DET-modified poly-L-aspartic acid block. (Reprinted from *Int. J. Pharm.*, 427(1), Chen, T.-H., Bae, Y., Furgeson, D.Y., and Kwon, G.S., 105–112, Copyright 2012a, with permission from Elsevier.)

aiming at maintaining adequate transfection efficiency and solving stability issues (Lai et al. 2012). The ternary complexes exhibited improved stability against salt-induced aggregation, although the gene delivery ability dropped with increasing amount of PEG-*b*-P[Asp(DET)]. Reducible ternary complexes, prepared by substituting P(EPE)-*b*-P[Asp(DET)] with a corresponding copolymer possessing a redox potential–sensitive disulfide linkages, P(EPE)-SS-P[Asp(DET)], achieved higher transfection compared to the nonreducible polyplexes while exhibiting comparable stability.

Sensitive to light are polyplexes constructed via host–guest interactions between β-cyclodextrin and azobenzene (Li et al. 2012a). The dePEGylation of the complex upon light irradiation facilitated the release of DNA and its entry into the nucleus (Figure 5.18), which resulted in efficient transfection.

Magnetic gene vectors for targeting gene delivery have been described elsewhere (Zheng et al. 2012b). They were prepared by surface modification of Fe_3O_4

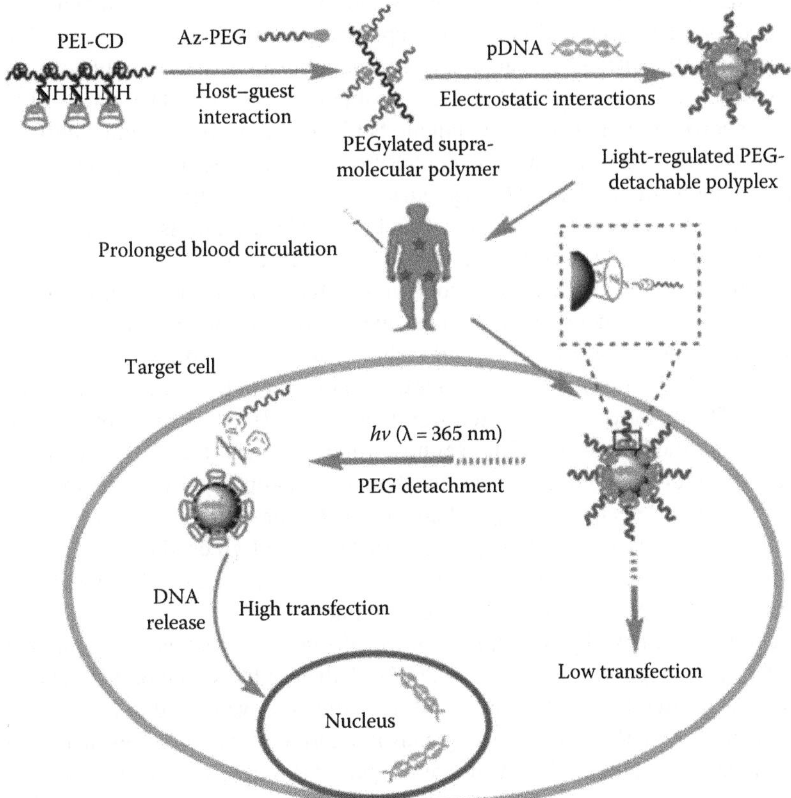

FIGURE 5.18 Light-regulated host–guest interaction as a new strategy for intracellular PEG-detachable polyplexes to facilitate nuclear entry. PEI, CD, Az, and PEG correspond to PEI, β-cyclodextrin, azobenzene, and PEG, respectively. (Li, W., Wang, Y., Chen, L., Huang, Z., Hu, Q., Ji, J., *Chem. Commun.*, 48(81), 10126–10128, 2012a. Reproduced by permission of The Royal Society of Chemistry.)

nanoparticles with carboxymethyl dextran and PEI and used to deliver GFP gene into BHK21 cells. The transfection efficiency and gene expression efficiency of those transfected with a magnet were much higher than that of standard transfection. Sun and coworkers have prepared PEI-decorated magnetic nanoparticles and determined their potency for efficiently complexing and delivering DNA *in vitro* with the help of a magnetic field (Sun et al. 2012). PEI was associated with PAA-bound superparamagnetic iron oxide (PAAIO) through electrostatic interactions (PEI-PAAIO). PEI-PAAIO formed stable polyplexes with pDNA in the presence and absence of 10% fetal bovine serum (FBS) and could be used for *magnetofection*. The effect of a static magnetic field on the cytotoxicity, cellular uptake, and transfection efficiency of PEI-PAAIO/pDNA was evaluated with and without 10% FBS. Magnetofection efficacy in HEK 293T cells and U87 cells containing 10% FBS was significantly improved in the presence of an external magnetic field.

Other magnetic targeting gene delivery systems have been described by Han and coworkers (Han et al. 2011). The systems prepared by modification of magnetosomes with polyamidoamine dendrimers and Tat peptides were designed to cross the blood–brain barrier and deliver therapeutic genes to brain cancerous tissues. Transfection efficiencies of Tat–magnetosome–polyamidoamine polyplexes with pGL-3 were studied using U251 human glioma cells *in vitro*. The results showed that the incorporation of external magnetic field and Tat peptides could significantly improve transfection efficiency of delivery system. Furthermore, biodistribution *in vivo* demonstrated that Tat–magnetosome–polyamidoamine could efficiently transport across the blood–brain barrier and assemble at brain tissue of rat detected by single-photon emission computed tomography (SPECT).

Branched PEI–hyaluronic acid copolymers have been successfully prepared using carbodiimide chemistry and subsequently functionalized with mannose to improve the transfection efficiency, cell viability, and cell specificity in macrophages (Mahor et al. 2012). Formation of polyplexes at polymer to DNA weight ratio ≥2 was observed. The nanohybrids exhibited significantly ($P<0.05$) lower cytotoxicity than that of unmodified branched PEI. Mannose functionalization of these nanohybrids showed specificity for both murine and human macrophage-like cell lines RAW 264.7 and human acute monocytic leukemia cell line (THP1), respectively, with a significant level ($P<0.05$) of expression of GLuc and green fluorescent reporter plasmids. Internalization studies indicate that a mannose-mediated endocytic pathway is responsible for this higher transfection rate.

The use of hydroxyethyl starch for controlled shielding/deshielding of polyplexes has recently been reported (Noga et al. 2012). Hydroxyethyl starch of different molar masses was grafted to PEI, and the resulting conjugates were used to generate polyplexes with the luciferase-expressing plasmid DNA pCMVluc. Deshielding was tested *in vitro* by ζ-potential measurements and erythrocyte aggregation assay upon addition of α-amylase to the hydroxyethyl starch–decorated particles. The addition of α-amylase led to gradual increase in the zeta-potential of the nanoparticles over 0.5–1 h and to a higher aggregation tendency for erythrocytes due to the degradation of the hydroxyethyl starch—coat and exposure of the polyplexes' positive charge. *In vitro* transfection experiments were conducted in two cell lines ± α-amylase in the culture medium. The amylase-treated hydroxyethyl starch–decorated complexes showed up to 2 orders of magnitude higher transfection levels compared to the untreated hydroxyethyl starch–shielded particles, while α-amylase had no effect on the transfection of PEG-coated or uncoated polyplexes.

Reduction-sensitive reversibly shielded DNA polyplexes based on PDMAEMA-SS-PEG-SS-PDMAEMA triblock copolymers have been prepared and shown highly promising for nonviral gene transfection (Zhu et al. 2012). They were prepared by RAFT polymerization with controlled compositions of 6.6-6-6.6 and 13-6-13 kDa. The resulting polyplexes (diameters < 120 nm, 0 to +6 mV) showed excellent colloidal stability against 150 mM NaCl. In the presence of 10 mM of dithiothreitol, however, they were rapidly deshielded, unpacked, and DNA was released. Notably, *in vitro* transfection studies showed that reversibly shielded polyplexes afforded up to 28 times higher transfection efficacy as compared to stably shielded control under otherwise the same conditions. Confocal laser scanning microscope studies revealed that reversibly shielded polyplexes efficiently delivered and released pDNA into the perinuclei region as well as nuclei of COS-7 cells (Zhu et al. 2012).

Folate-conjugated tercopolymers have been shown to deliver gene materials into target tumor cell via folate receptor–mediated endocytosis (Liu et al. 2012a). The copolymers were based on PEI-g-PCL-*b*-PEG conjugated with folate. They were able to condense DNA completely at N/P ratio >2 and polyplexes of N/P ratio 10 with sizes of about 120 nm, and positive zeta-potentials were selected for biological evaluations due to their stability. An enhancement of both cellular uptake of PEI-g-PCL-*b*-PEG-Fol/pDNA polyplexes and their transfection efficiency was observed in folate receptor–overexpressing cells in comparison to unmodified PEI-g-PCL-*b*-PEG.

An attractive target for tumor-targeting therapies is the EGFR. A gene delivery system based on PEGylated linear PEI has been developed (Schafer et al. 2011). Peptide sequences directly derived from the human EGF molecule enhanced transfection efficiency with concomitant EGFR activation. Only the EGFR-binding peptide GE11, which has been identified by phage display technique, showed specific enhancement of transfection on EGFR-overexpressing tumor cells including glioblastoma and hepatoma but without EGFR activation. In a clinically relevant orthotopic prostate cancer model, intratumorally injected GE11 polyplexes were superior in inducing transgene expression when compared with untargeted polyplexes. Induction of tumor-selective iodide uptake and therapeutic efficacy of ^{131}I in a hepatocellular carcinoma (HCC) xenograft mouse model, using the same polyplexes, has been reported in another paper (Klutz et al. 2011).

Other authors have isolated and covalently linked to the distal end of PEG (3500)-PEI(25 kDa) an anti-DF3/Mucin1 (MUC1) nanobody with high specificity for the MUC1 antigen, which is an aberrantly glycosylated glycoprotein overexpressed in tumors of epithelial origin (Sabeqzadeh et al. 2011). The resulting macromolecular conjugate successfully condensed plasmids coding a transcriptionally targeted truncated-Bid (tBid) killer gene under the control of the cancer-specific MUC1 promoter. The engineered polyplexes exhibited favorable physicochemical characteristics for transfection and dramatically elevated the level of Bid/tBid expression in both MUC1 overexpressing caspase-3-deficient (MCF7 cells) and caspase-3-positive (T47D and SKBR3) tumor cell lines and, concomitantly, induced considerable cell death.

A newly developed delivery vector has been designed to impart bioreducibility for greater intracellular pDNA release, higher serum stability, and efficient complexing ability by incorporating disulfide linkage, PEG, and low molecular weight PEI, respectively (Son et al. 2011). RVG peptide as a targeting ligand for neuronal cells was used to deliver genes to mouse brain overcoming the blood–brain barrier. The physiochemical properties of the polyplexes formed with plasmid DNA, their cytotoxicity, and the *in vitro* transfection efficiency on Neuro2a cell were studied prior to the successful *in vivo* study. *In vivo* fluorescence assay substantiated the permeation of the plasmid DNA–loaded polymeric vector through the blood–brain barrier.

Nam and coworkers have described the development of a cardiomyocyte-targeting nonviral gene carrier and anti-apoptotic candidate siRNAs to inhibit cardiomyocyte apoptosis (Nam et al. 2011). The newly synthesized bioreducible polymers were based on cystamine bisacrylamide-diaminohexane and incorporated a cardiomyocyte-targeting peptide (PCM) and the CPP HIV Tat 49–57 (Figure 5.19, left panel A). Gel retardation results demonstrated that peptide-modified polymers could effectively condense anionic genes at a weight ratio over 5 (Figure 5.19, left panel B). As shown in Figure 5.19 (right panel), the cells treated with polyplexes of polymer modified by both Tat and PCM peptides, whose plasmid DNA was labeled with YOYO-1 iodide, showed a more intense green signal in the cytoplasm with an additional 2 h of treatment. These results demonstrate that conjugation with the Tat peptide increases transfection efficiency, while addition of the PCM peptide results in a further increase in transfection efficiency in H9C2 cardiomyocytes. In addition, *in vitro* cytotoxicity by MTT assay demonstrated good cell viability after transfection using the PCM-CD-Tat polyplexes, which verified that PCM conjugation facilitated cardiomyocyte targeting, while Tat peptide conjugation facilitated gene delivery into the cells without significant cytotoxicity.

5.2.1.2.4 Sugars, Amino Acids, Glycopolymers, Peptides, Hybrids, Polysaccharides

Wu and coworkers have focused on the chain flexibility of novel arginine-based polycations with incorporated ethylene glycol units (Figure 5.20) in the gene delivery efficiency and the relationships between polyplex size, ζ-potential, and transfection efficiency (Wu et al. 2012). The polymers were synthesized by the solution polycondensation reaction of the *p*-toluenesulfonic acid salt of L-arginine diester from oligoethylene glycol and di-*p*-nitrophenyl esters of dicarboxylic acids. The transfection results obtained from luciferase and GFP assays from a wide range of cell lines, primary cells, and stem cells showed that the introduction of flexible segments was beneficial for the improvement of the transfection efficiency, which was comparable or better than that of commercial transfection reagents, but at a much lower cytotoxicity.

In a recent work, Aldawsari and coworkers have demonstrated that the conjugation of amino acids such as arginine, lysine, and leucine to the third generation of polypropyleneimine dendrimers led to an enhanced antiproliferative activity of the polyplexes *in vitro* (Aldawsari et al. 2011). *In vivo*, the intravenous administration of amino acid–bearing polypropyleneimine dendrimer polyplexes resulted in a

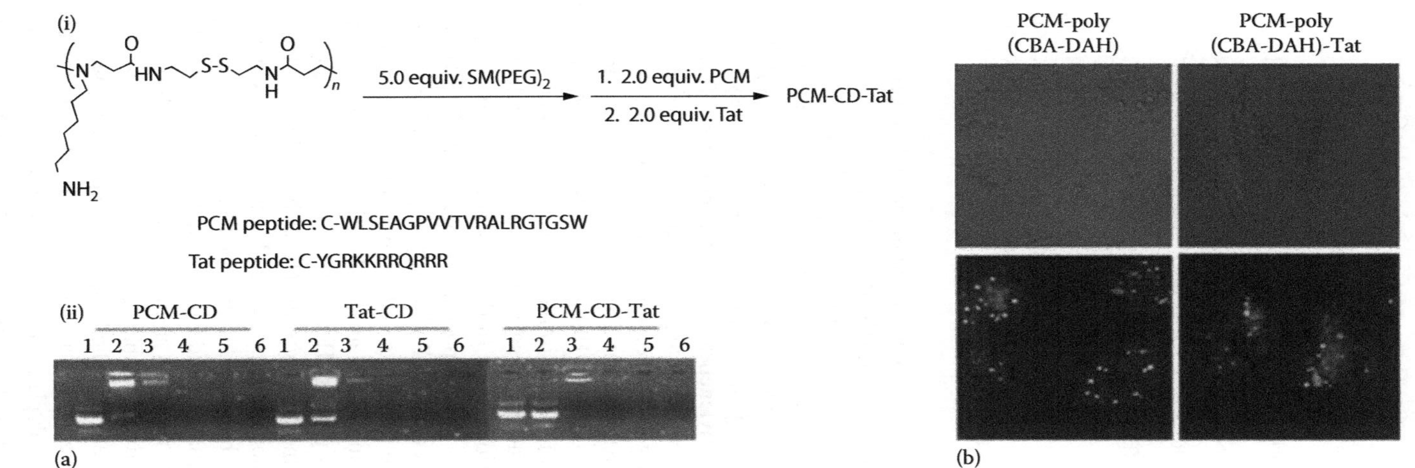

FIGURE 5.19 (a): (i) Schematic diagram of a cystamine bisacrylamide-diaminohexane polymer incorporating a PCM and the CPP HIV Tat 49–57. (ii) Agarose gel electrophoresis of polymers complexed with plasmid DNA at the various weight ratios of polymer/plasmid DNA=0, 1, 5, 10, 20, and 40 (lanes 1, 2, 3, 4, 5, and 6, respectively). (b) Intracellular trafficking of YOYO-1 iodide-labeled plasmid DNA polyplex in H9C2 cells incubated with PCM-conjugated or PCM and Tat-conjugated polymer using confocal microscopy after a 2 h treatment. (Reprinted from *Biomaterials*, 32(22), Nam, H.Y., Kim, J., Kim, S., Yockman, J. W., Kim, S.W., and Bull, D.A., 5213–5222, Copyright 2011, with permission from Elsevier.)

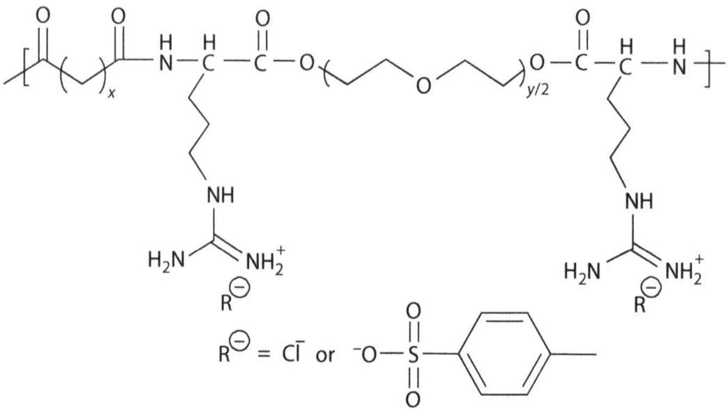

FIGURE 5.20 Chemical formula of L-arginine-based polycations. (Wu, J., Yamanouchi, D., Liu, B., and Chu, C.-C., *J. Mater. Chem.*, 22(36), 18983–18991, 2012. Reproduced by permission of The Royal Society of Chemistry.)

significantly improved tumor gene expression, with the highest gene expression level observed after treatment with lysine-modified polyplex.

Anderson and coworkers have recently demonstrated that very subtle structural changes in polymer chemistry, such as end groups, can yield significant differences in the biological delivery efficiency and transgene expression of polymers used for plasmid DNA delivery (Anderson et al. 2012). These authors synthesized six analogues of a trehalose-pentaethylenehexamine glycopolymer (Tr4) that contained (1A) adamantane, (1B) carboxy, (1C) alkynyl-oligoethyleneamine, (1D) azido trehalose, (1E) octyl, or (1F) oligoethyleneamine end groups and evaluated the effects of polymer end group chemistry on the ability of these systems to bind, compact, and deliver plasmid DNA to cultured HeLa cells. All polymers studied were able to bind and compact plasmid DNA at similarly low N/P ratios and form polyplexes. The effects of the different end group structures were most evident in the polyplex internalization and transfection assays in the presence of serum as determined by flow cytometry and luciferase gene expression, respectively. The Tr4 polymers end capped with carboxyl groups (1B) (N/P=7), octyne (1E) (N/P=7), and oligoethyleneamine (1F) (N/P=7) were taken into cells as polyplex and exhibited the highest levels of fluorescence, resulting from labeled plasmid. Similarly, the polymers end functionalized with carboxyl groups (1E at N/P=7), octyl groups (1E at N/P=15), and in particular oligoethyleneamine groups (1F at N/P=15) yielded dramatically higher reporter gene expression in the presence of serum (Anderson et al. 2012).

RAFT polymerization has been used to prepare a series of diblock glycopolymers composed of 2-deoxy-2-methacrylamido glucopyranose and the primary amine–containing *N*-(2-aminoethyl) methacrylamide (Smith et al. 2011). The polyplexes with plasmid DNA were found to be stable against aggregation in the presence of salt and serum over the 4 h time period studied. Delivery experiments were performed *in vitro* to examine the cellular uptake, transfection efficiency, and cytotoxicity of the glycopolymer/plasmid DNA polyplexes in cultured HeLa cells: the diblock copolymer with the shortest *N*-(2-aminoethyl) methacrylamide block was found to be the most effective. The gene knockdown efficacy (delivery of siRNA to U-87 glioblastoma cells) of the diblock copolymer with the longest *N*-(2-aminoethyl) methacrylamide block was found to be similar to commercial transfecting agents.

Reilly and coworkers have found that histone H3 peptides (peptides containing transcriptionally activating modifications) conjugated to PEI can effectively bind and protect plasmid DNA (Reilly et al. 2012). The H3/PEI hybrid polyplexes did not compromise the cell viability. They were found to transfect a substantially larger number of CHO-K1 cells *in vitro* compared to both polyplexes that were formed with only the H3 peptides and those that were formed with only PEI at the same total charge ratio. Transfections with the endolysosomal inhibitors chloroquine and bafilomycin A1 indicated that the H3/PEI hybrid polyplexes exhibited slower uptake and a reduced dependence on endocytic pathways that trafficked to the lysosome, indicating a potentially enhanced reliance on caveolar uptake for efficient gene transfer (Reilly et al. 2012).

A panel of HPMA-oligolysine copolymers with varying peptide length and polymer molecular weight has been built using RAFT polymerization (Johnson et al. 2011). The panel was screened for optimal DNA binding, colloidal stability in salt, high transfection efficiency, and low cytotoxicity. Increasing polyplex stability in phosphate buffered saline (PBS) correlated with increasing polymer molecular weight and decreasing peptide length. Copolymers containing chains of 5 and 10 (K_5 and K_{10}) oligocations transfected cultured cells with significantly higher efficiencies than copolymers of K_{15}.

By exploiting the self-assembly properties and net positive charge of ELP, hollow spheres of sizes of 100, 300, 500, and 1000 nm have been fabricated (Dash et al. 2011). The microbial transglutaminase cross-linking provided robustness and stability to the hollow spheres while maintaining surface functional groups for further modifications. The resulting hollow spheres showed a higher loading efficiency of plasmid DNA by using polyplex (~70 μg plasmid DNA/mg of hollow sphere) than that of self-assembled ELP particles and demonstrated controlled release triggered by protease and elastase. Moreover, polyplex-loaded hollow spheres showed better cell viability than polyplex alone and yielded higher luciferase expression by providing protection against endosomal degradation.

Hyperbranched cationic glycopolymers have been prepared by RAFT using 2-aminoethyl methacrylamide (AEMA) and 3-glucoamidopropyl methacrylamide (GAPMA) or 2-lactobionamidoethyl methacrylamide (LAEMA) as shown in Figure 5.21 (Ahmed and Narain 2012). The hyperbranched polymers were synthesized with varying molecular weights and compositions, and their gene expressions were evaluated in two different cell lines. The copolymers formed complexes with β-galactosidase plasmid at varying molar ratios. Regardless of the molecular weight, the polyplexes produced in water were about 70–200 nm in diameter (Figure 5.21). The high polydispersities of ζ-potential also observable in Figure 5.21 may be associated with polydisperse nature of the hyperbranched copolymers (M_w/M_n in the range 1.70–11.2!). All copolymers of high molecular weight (50–60 kDa) showed very low gene expression, regardless of composition and polymer/plasmid ratio used. The copolymers of molecular weights 5–30 kDa exhibited high gene expression (Figure 5.21), along with high cell viability. Furthermore, the cellular uptake and gene expression were studied in two different cell lines in the presence

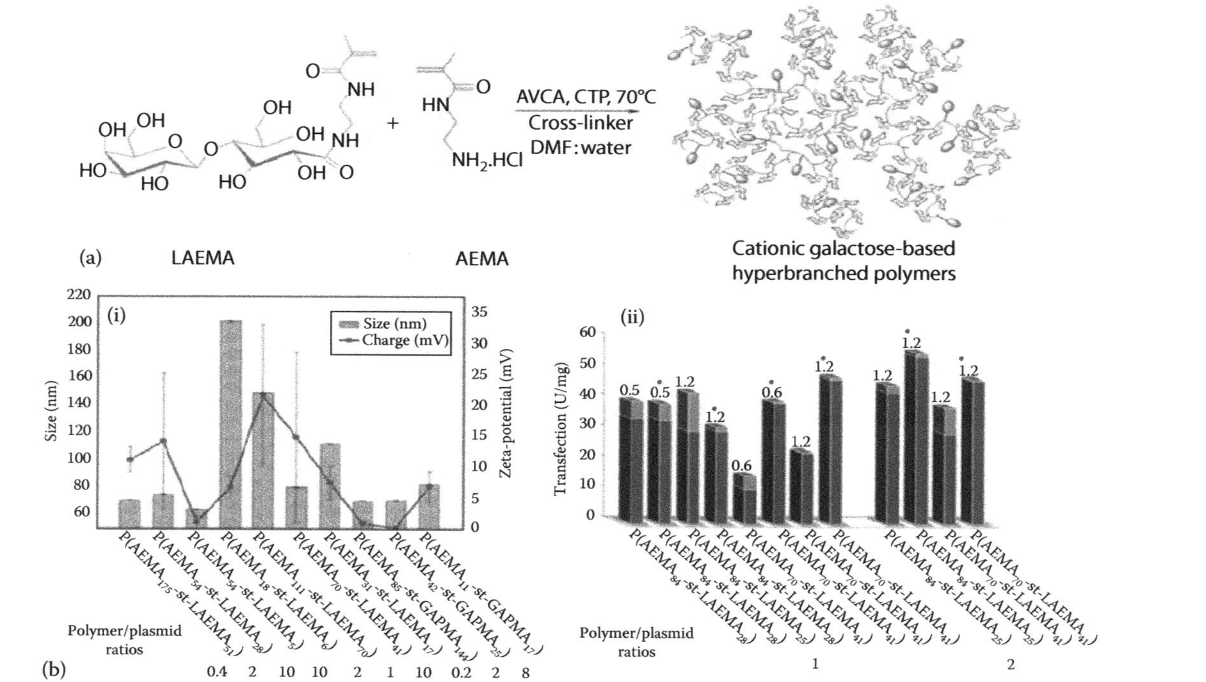

FIGURE 5.21 (a) Synthesis of hyperbranched cationic glycopolymers by RAFT. LAEMA, AEMA, AVCA, and CTP correspond to LAEMA, AEMA, 4,4′-azobis(4-cyanovaleric acid), and 4-cyanopentanoic acid dithiobenzoate, respectively. (b): (i) DLS and zeta-potential data for the hyperbranched glycopolymer-DNA polyplexes. All samples are prepared in deionized water at varying polymer/DNA molar ratios. (ii) Gene expression of galactose-based polymers of varying molecular weights, in the presence (*) and absence of serum using Hep G2 cells. Gene expression is evaluated using β-galactosidase assay at DNA doses 0.6 and 1.2 μg and varying polymer/plasmid molar ratios as shown. (Reprinted from *Biomaterials*, 33(15), Ahmed, M. and Narain, R., 3990–4001, Copyright 2012, with permission from Elsevier.)

of lectins. It was found that polyplexes–lectin conjugates showed enhanced cellular uptake *in vitro*; however, their gene expression was cell line- and lectin-type dependent (Ahmed and Narain 2012).

The same authors have synthesized a library of cationic glycopolymers of predetermined molar masses (3–30 kDa) and narrow polydispersities using RAFT polymerization (Ahmed and Narain 2011). The polymers contained pendant sugar moieties and differed from each other in their architecture (block vs. random), molecular weights, and monomer ratios (carbohydrate to cationic segment). It was shown that the aforementioned parameters can largely affect the toxicity, DNA condensation ability, and gene delivery efficacy of these polymers. For instance, random copolymers of high degree of polymerization were found to be the ideal vector for gene delivery purposes. They showed lower toxicity and higher gene expression in the presence and absence of serum, as compared to the corresponding diblock copolymers. The effect of serum proteins on random copolymer–based and diblock copolymer–based polyplexes and hence gene delivery efficacy was studied as well: the diblock copolymer–based polyplexes showed lower interactions with serum proteins, lower cellular uptake, and lower gene expression in both Hep G2 and HeLa cells in comparison to the polyplexes based on the random copolymers.

Cyclodextrin-modified PEIs with different cyclodextrin grafting levels have been synthesized (Li et al. 2011). The polycation containing an average of 15 cyclodextrin moieties per PEI chain could protect DNA completely above N/P ratio of 2. The particle sizes of these polyplexes were about 120 nm. Excellent stability in physiological conditions, probably due to the hydration shell of cyclodextrin, was observed at N/P ratios of 8 and 10. Uptake inhibition experiments indicated that PEI/DNA polyplexes were internalized by HEK293T cells by both clathrin-mediated endocytosis and caveolae-mediated endocytosis. The route of caveolae-mediated endocytosis was significantly promoted after cyclodextrin modification; hence, the cell uptake and transfection efficiency of polyplexes were significantly improved for HEK293T cells. However, the uptake and transfection efficiency in HepG2 cells were similar for the PEI/DNA polyplexes and the polyplexes of cyclodextrin-modified PEI, probably due to the lack of endogenous caveolins. A β-cyclodextrin-PEI-based polymer has been modified with the Tat peptide for plasmid DNA delivery to placenta mesenchymal stem cells (Lai et al. 2011). Nanoparticles (~150 nm) were formed with DNA at the optimal N/P ratio of 20/1. The conjugation of the Tat peptide onto PEI β-cyclodextrin was demonstrated to improve the transfection efficiency in cells after 48 and 96 h of post-transfection incubation. The viability of the cells was shown to be over 80% after 5 h of treatment and 24 h of posttreatment incubation.

Water-soluble chitosan-graft-(polyethylenimine-β-cyclodextrin) cationic copolymers have been synthesized via reductive amination between oxidized chitosan and low molecular weight PEI-modified β-cyclodextrin (Ping et al. 2011). The synthetic procedure is presented in Figure 5.22a. The polycations exhibited good ability to condense both plasmid DNA and siRNA into compact and spherical nanoparticles. Their gene transfection activity showed improved performance in comparison with native chitosan in HEK293, L929, and COS-7 cell lines. The pendant β-cyclodextrin moieties of the copolymers allowed the supramolecular PEGylation through self-assembly with adamantyl-modified PEG (Figure 5.22b), which significantly improved

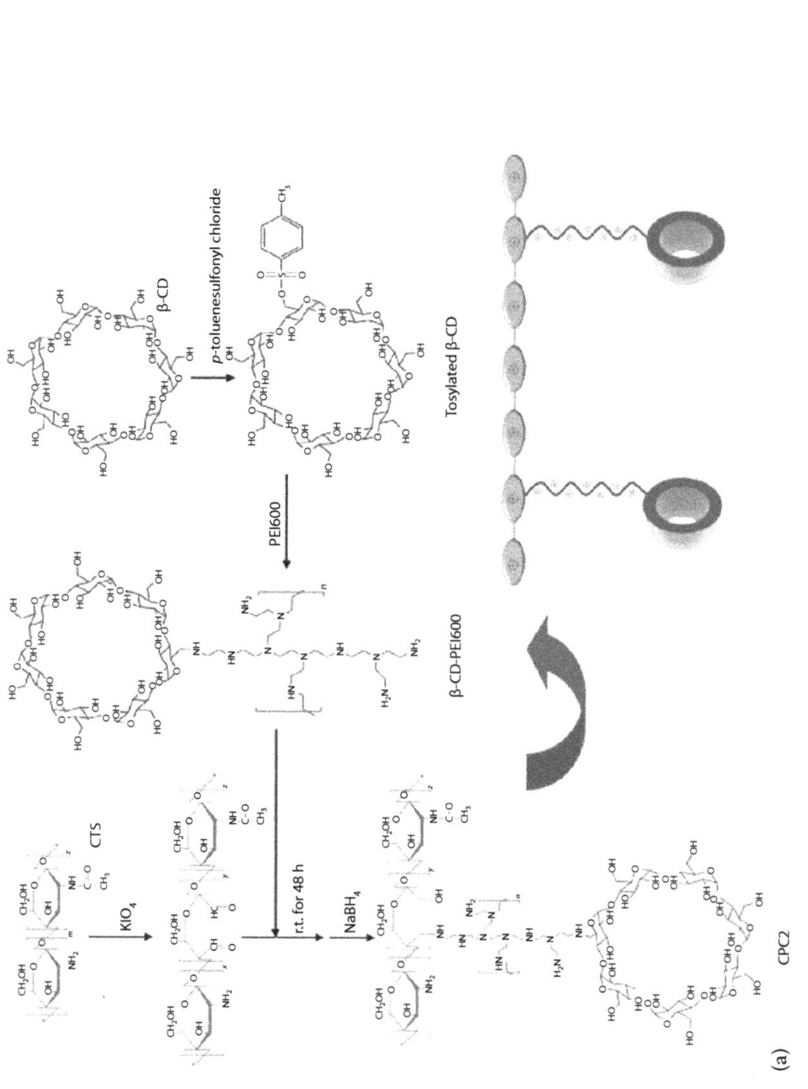

FIGURE 5.22 (a) Synthetic procedure for the preparation of chitosan-graft-(polyethylenimine-β-cyclodextrin) cationic copolymers. (Reprinted from *Biomaterials*, 32(32), Ping, Y., Liu, C., Zhang, Z., Liu, K. L., Chen, J., and Li, J., 8328–8341. Copyright 2011, with permission from Elsevier.)

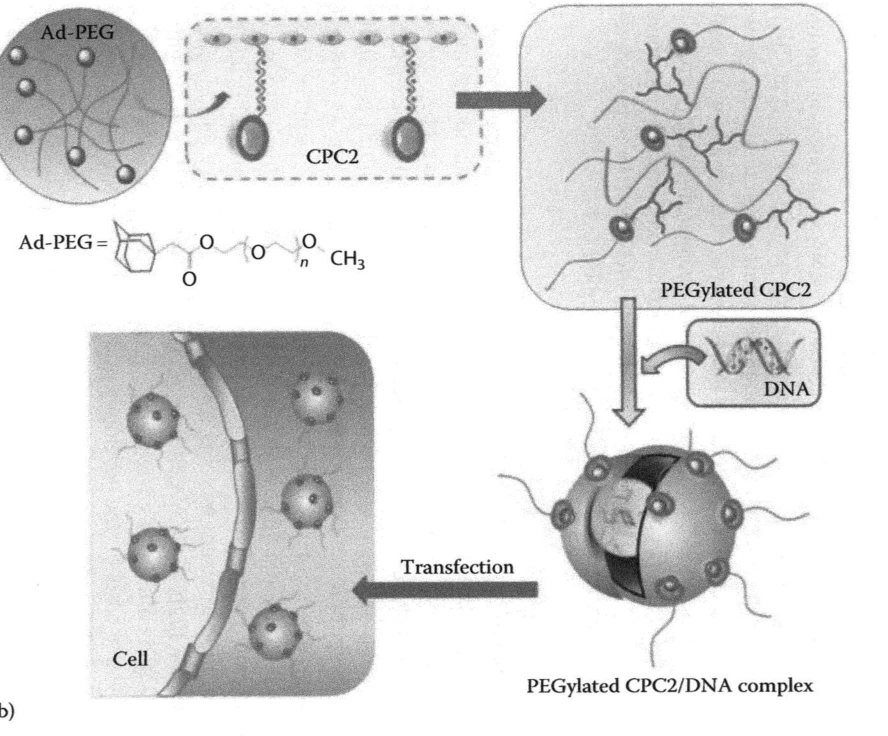

FIGURE 5.22 (continued) (b) Illustration of PEGylation through supramolecular self-assembly between pendant β-cyclodextrin moieties and adamantyl-modified PEG and the formation of supramolecular PEGylated copolymer/DNA complex. CTS, chitosan; β-CD, β-cyclodextrin; Ad-PEG5k, adamantyl-modified PEG; CPC, chitosan-graft-(polyethylenimine-β-cyclodextrin) copolymer. (Reprinted from *Biomaterials*, 32(32), Ping, Y., Liu, C., Zhang, Z., Liu, K. L., Chen, J., and Li, J., 8328–8341. Copyright 2011, with permission from Elsevier.)

their stability under physiological conditions. The supramolecular PEGylated polyplexes showed decreased transfection efficiency in all tested cell lines; however, their silencing efficiency in HEK293 and L929 cells was higher (up to 84%) compared to that of commercial agents.

(Coixan polysaccharide)-graft-PEI-folate has been prepared as an effective vector for *in vitro* and *in vivo* tumor-targeted gene delivery (Jiang et al. 2011b). The cytotoxicity of the copolymers was significantly lower than that of PEI 25 kDa and close to that of PEI 1200. The *in vitro* transfection, tested in both FR-positive cells (C6 and HeLa cells) and FR-negative cells (A549 cells), showed a high targeting specificity and good gene transfection efficiency in FR-positive cells.

Arima and coworkers have prepared complexes of siRNA with polyamidoamine dendrimer of the third generation conjugated with α-cyclodextrin (Arima et al. 2011). siRNA appeared to be well protected in the complexes against degradation by serum. The complexes showed negligible cytotoxicity and hemolytic activity and were found to deliver fluorescent-labeled siRNA to cytoplasm, not nucleus, after transfection in NIH3T3-luc cells. The authors concluded that the dendritic polyamidoamine conjugated with α-cyclodextrin could be potentially used as a siRNA carrier to provide the RNAi effect on endogenous gene expression with negligible cytotoxicity.

5.2.1.2.5 Ternary Complexes

Loosening of the polyplex, weakening of the adsorption of serum proteins, and improving of cellular uptake have been considered as important factors leading to high transfection efficiency of DNA ternary complexes. They have been specifically investigated for such complexes composed of DNA, 25 kDa PEI, and different types of biocompatible polyanions—polysaccharides and polypeptides containing carboxyl groups or sulfonic acid groups such as heparin sodium salt, alginic acid sodium salt, poly(aspartic acid), and PLG (Wang et al. 2012a). The results obtained indicated that the low pK_a and flexible structure of the polyanions tended to loosen the compact DNA polyplexes. The ternary polyplexes exhibited lower binding affinities and less adsorption to serum proteins compared with the binary ones. In addition, they maintained high levels of cellular uptake and intracellular accumulation in serum-containing medium that correlated with their high transfection efficiency. These results provide a basis for the development of polyanion/DNA/polycation ternary polyplexes gene delivery.

Complexes designed to increase the stability of mRNA, to improve transfection efficiency, and to reduce the cytotoxicity have been prepared (Debus et al. 2011). These are composite polyplexes with mRNA consisting of PEI and PEI-PEG copolymers. Stable complexes were formed even at low N/P ratios. Most of them showed small particle dimensions (<200 nm) and positive zeta-potentials of +20–+30 mV. Polyplexes with mRNA Luc and blends of low molecular weight PEI (5 kDa) and PEI (25 kDa)-PEG (20 kDa) block copolymer showed protein expression as high as polyplexes with PEI (25 kDa). Moreover, luciferase expression was significantly higher than that obtained with one of the components alone.

Integration of homo-catiomer, poly{*N*′-[*N*-(2-aminoethyl)-2-aminoethyl]aspartamide}, PAsp(DET), into PEGylated polyplex micelles prepared from the corresponding block copolymers with PEG, (PEG)-*b*-PAsp(DET), and DNA has been attempted aiming at enhancing the cell transfection efficiency (Chen et al. 2012b). *In vivo* anti-angiogenic tumor suppression evaluations validated the feasibility of the approach. The integration promoted gene transfection to the affected cells via systemic administration. The loaded anti-angiogenic gene remarkably expressed in the tumor site, thereby imparting significant inhibitory effect on the growth of vascular endothelial cells, ultimately leading to potent tumor growth suppression. In a subsequent paper, it has been shown that the formulation of both homo PAsp(DET) and copolymer (PEG)-*b*-PAsp(DET) in the polyplex can help to improve gene therapy for the respiratory system because both effective PEG shielding of polyplexes and functioning of PAsp(DET) polycations to enhance endosomal escape were achieved (Uchida et al. 2012).

Copolymers of HPMA and methacrylamido-functionalized oligo-L-lysine peptide monomers with either a nonreducible 6-aminohexanoic acid linker or a reducible 3-[(2-aminoethyl)dithiol]propionic acid linker have been prepared and shown able to form polyplexes with DNA (Shi et al. 2012). The copolymer containing the reducible linkers was less efficient at transfection than the nonreducible polymer and was prone to flocculation in saline and serum-containing conditions but was also not cytotoxic at charge ratios tested. Optimal transfection efficiency and toxicity were attained with mixed 1:1 formulation of copolymers.

Ternary complexes formed by mixing (1) chitosan with lipoplex of *N*-[1-(2,3-dioleyloxy)propyl]-*N*,*N*,*N*-trimethylammonium chloride (DOTAP) and plasmid DNA and (2) DOTAP with chitosan/plasmid DNA polyplex have been prepared (Wang et al. 2012b). The DOTAP/chitosan/plasmid DNA complexes were in compacted spheroids and irregular lump of larger aggregates in structure, while the short rodlike and toroid-like and donut shapes were found in chitosan/DOTAP/pDNA complexes by AFM. The transfection efficiency of the *lipopolyplexes* showed higher GFP gene expression than lipoplex and polyplex controls in Hep-2 and HeLa cells and luciferase gene expression two- to threefold than lipoplex control and 70- to 120-fold than polyplex control in Hep-2 cells. The intracellular trafficking was examined by CLSM. Rapid plasmid DNA delivery to the nucleus enhanced by chitosan was achieved after 4 h transfection.

A facile approach to constructing ternary gene carriers by adding hyaluronic acid to preformed PEI/DNA polyplexes has been demonstrated (Wang et al. 2011). Spherical particles with diameter about 250 nm and shifting of ζ-potential from positive to negative were observed upon the formation of hyaluronic acid–shielded polyplexes. Although the electrostatic complexation was loosened, DNA disassembly was not observed. The stability of PEI/DNA/hyaluronic acid ternary polyplexes in physiological condition was improved, and the cytotoxicity was reduced. Comparing with PEI/DNA polyplexes, the uptake and transfection efficiency of hyaluronic acid–shielded polyplexes was lower for HEK293T cells probably due to the reduced adsorptive endocytosis, whereas it was higher for HepG2 cells due to HA receptor–mediated endocytosis.

By adding mannosylated and histidylated liposomes to mRNA-PEGylated histidylated polylysine polyplex, ternary lipopolyplexes have been formed (Perche et al. 2011a). The lipopolyplexes enhanced the transfection of dendritic cells *in vivo* and the anti-B16F10 melanoma vaccination in mice. In a subsequent paper, it has been shown by confocal microscopy study that with DC2.4 cells expressing Rab5-EGFP or Rab7-EGFP, DNA uptake occurred through clathrin-mediated endocytosis. The transfection of DC2.4 cells with mannosylated and histidylated lipopolyplexes containing DNA encoding luciferase gene gave luciferase activity two to three times higher than with non-mannosylated ones. In contrast to the latter, it was inhibited by 90% in the presence of mannose (Perche et al. 2011b).

Amine-terminated generation 5 polyamidoamine dendrimers have been utilized as templates to synthesize gold nanoparticles (Figure 5.23). The dendrimer-entrapped gold nanoparticles were prepared in different Au atom/dendrimer molar ratios (25:1–100:1) and were used as nonviral gene delivery vectors (Shan et al. 2012). Gel retardation assay, AFM imaging, and DLS experiments demonstrated that dendrimer-entrapped gold nanoparticles were able to effectively compact pDNA to form polyplexes with a smaller size when compared to the initial polyamidoamine dendrimers without gold nanoparticles entrapped. The results clearly showed that dendrimer-entrapped gold nanoparticles with an appropriate composition (Au atom/dendrimer molar ratio = 25:1) enabled enhanced gene delivery, with a gene transfection efficiency more than 100 times higher than that of the dendrimers without nanoparticles entrapped (Figure 5.23). The authors believed that the entrapment of gold nanoparticles within dendrimer templates helped preserve 3D spherical shape of dendrimers, enabling high compaction of DNA to form smaller particles and consequently resulting in enhanced gene delivery.

A two-step process for preparation of ternary *magnetoplexes* has recently been described (Liu et al. 2011c). The sixth generation of polyamidoamine dendrimer modified with superparamagnetic nanoparticles was introduced to PEI/DNA polyplexes. The ternary *magnetoplexes* exhibited enhanced transfection efficiency in COS-7, 293T, and HeLa cells when a magnetic field was applied. Importantly, time-resolved and dose-resolved transfection indicated that high-level transgene expression was achievable with a relatively short incubation time and low DNA dose

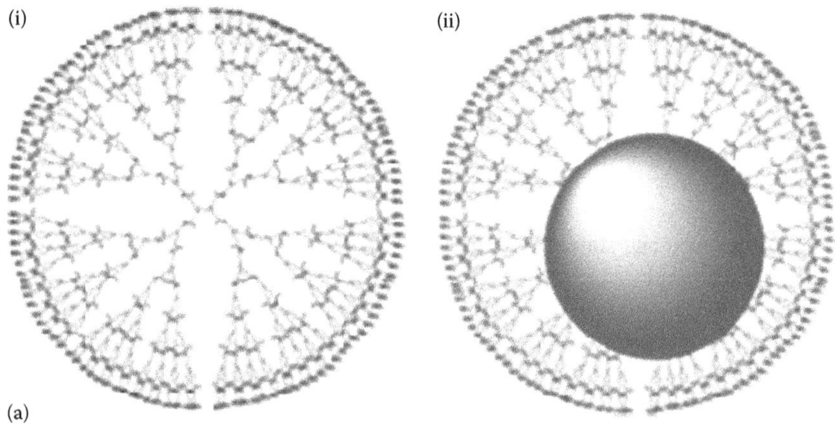

FIGURE 5.23 (a) Schematic illustration of the structures of (i) aminoterminated polyamidoamine dendrimers of the fifth generation and (ii) dendrimer-entrapped gold nanoparticles.

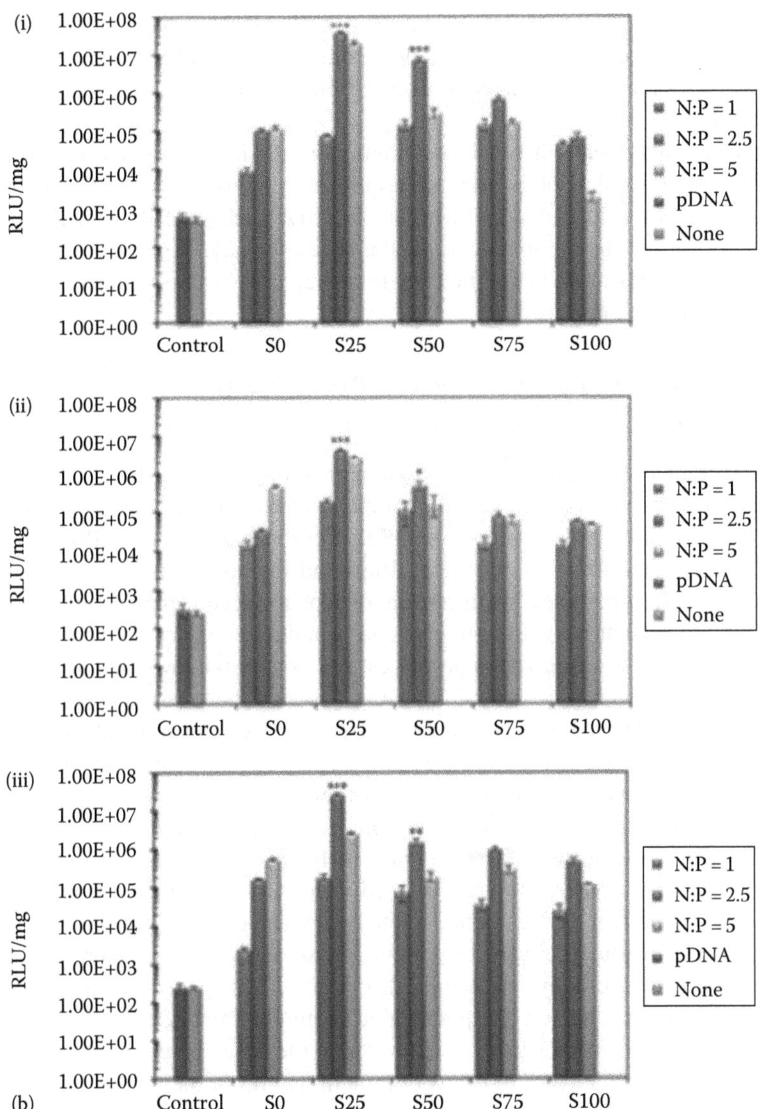

FIGURE 5.23 (continued) (b) Luciferase gene transfection efficiency of dendrimer-entrapped gold nanoparticles/DNA polyplexes determined in (i) HeLa, (ii) COS-7, and (iii) 293T cells at the N/P ratios of 1:1, 2.5:1, and 5:1, respectively. Transfection was performed at a dose of 1 μg/well of DNA (mean ± SD, $n = 3$). Cells without treatment (None) and cells treated with vector-free pDNA (pDNA) were used as controls. Statistical differences between Au DENPs (S25, S50, S75, and S100, respectively) versus G5.NH_2 dendrimers (S0) at an N/P ratio of 2.5:1 were compared and indicated with (*) for $p < 0.05$, (**) for $p < 0.01$, and (***) for $p < 0.001$, respectively. (Reprinted from *Biomaterials*, 33(10), Shan, Y., Luo, T., Peng, C., Sheng, R., Cao, A., Cao, X., Shen, M., Guo, R., Tomas, H., and Shi, X., 3025–3035, Copyright 2012, with permission from Elsevier.)

when *magnetofection* was employed. Further evidence from Prussian blue staining, quantification of cellular iron concentration, and cellular uptake of Cy-3-labeled DNA demonstrated that the magnetic field could quickly gather the magnetoplexes to the surface of target cells and consequently enhance the uptake of magnetoplexes by the cells.

Somewhat untypical but worth mentioning are the hybrid polyplexes composed of DNA and polyhedral oligomeric silsesquioxanes (siloxane materials that have a cage-like structure) loaded with hydrophobic drugs such as PTX, which is encapsulated within the hydrophobic core of the silsesquioxane (Loh et al. 2011). They exhibited superior transfection efficiency in human breast cancer cells than the non-drug-loaded polyplexes.

5.2.2 Delivery and Encapsulation of Proteins and Enzymes: Enzymatic Nanoreactors

Protein therapy is a promising strategy to fight protein-deficiency diseases by using *in vitro* produced proteins to intracellularly replace or complement the faulty ones (De Duve and Wattiaux 1966, Barton et al. 1991, Rohrbach and Clarke 2007, Christian et al. 2009). Administration of protein drugs and therapeutic peptides poses some additional issues as compared with conventional pharmaceutical compounds because of the high molecular weight of proteins and peptides and their relatively short half-life in plasma due to opsonization processes. It is now well known that protein delivery can be significantly improved by using designed nanocarriers with targeting abilities. Polymer-based nanoparticles of different structure can be utilized for this purpose since, because of their subcellular size, they can cross the fenestration of the vascular epithelium and penetrate tissues. Furthermore, nanocarriers can be confined at the location of choice by conjugation to molecules that strongly bind to the target cells. Enzymatic conversions can take place in the interior of such nanocarriers, and their membranes or outer shells can be used to confine and tune reaction pathways. Despite the significant progress made in the design and engineering of nanoparticles tailored to the targeted delivery of proteins, these nanocarriers seldom succeed in delivering proteins directly inside the cell cytosol. These nanosystems also have a significant drawback, namely, the potential to be highly immunogenic (Solaro 2008). Therefore, extensive efforts have been devoted to the development of polymer-based nanocarriers for the delivery of drugs of protein nature. In the following, we give a brief overview of polymer-based nanocarriers of proteins and polypeptides that have been constructed from synthetic and/or natural building segments using the self-assembly and co-assembly approaches. Liposomes, capsules solely based on proteins (viral capsides), microspheres formed by precipitation polymerization or polymerization-induced phase separation, layer-by-layer capsules, as well as unimolecular dendrimer-based containers are not covered or are just occasionally mentioned.

Polymeric protein delivery systems generally rely on the formation of micrometer- or nanometer-size polymer particles containing protein cargo. In these delivery systems, the delivery particles are synthesized by assembling polymer and protein cargo through noncovalent interactions such as electrostatic interactions, hydrogen bonding, and hydrophobic interactions. Proteins can be adsorbed onto polymer particles or dendrimers. Alternatively, when polymers with heterogeneous molecular structure are used, they can co-assemble with protein molecules through appropriate moieties via electrostatic or hydrophobic interactions or other noncovalent interactions (e.g., hydrogen bonding). Other sections of the polymer may serve to provide a stabilizing effect of the complex in solution. Particularly common, although limited to anionic proteins, is the formation of protein–polymer complexes from block copolymers containing a cationic block and a PEG block.

Proteins can be readily complexed to the hydrophobic moieties of amphiphilic copolymer through hydrophobic interactions. For example, cholesterol-bearing pullulan or its derivative was assembled with proteins, resulting in the formation of cationic nanogels with diameters less than 50 nm (Ayame et al. 2008). In this case, hydrophobic domains formed by the cholesterol moieties on the polymer chains served as complexing sites for the proteins. Enhanced cellular uptake, as well as retained intracellular enzymatic activity, was reported. Effective endosomal release of the proteins was also observed 18 h after transduction.

Interactions different from electrostatic forces or hydrophilic/hydrophobic interactions can also be utilized to form polymer–protein complexes. These are affinity binding, which has the advantage of minimizing protein denaturation and well-controlled complex structure. However, the approach is limited to proteins that are able to fuse with their binding ligands (Du et al. 2012). A good example of this approach are the PEI–glutathione conjugates that bind with glutathione S-transferase-fused proteins and induce cellular uptake in mammalian cells (Murata et al. 2008).

As other delivery systems, the protein delivery by polymers also faces a dilemma: On one side, the weak noncovalent interactions make the complexes vulnerable to dissociation, while strong interactions may alter the protein structure and even cause denaturation, on the other side. The encapsulation of proteins or enzymes in the interior or corona of polymeric micelles or in the lumen of polymersomes represents a possible approach to find the right balance between the magnitude of the interactions and the stability of the loaded carriers.

Lee and coworkers have demonstrated the use of polyion complexes for antibody delivery into cytoplasm. The antibody alone cannot penetrate into the cytoplasm. However, the complexation with a carefully designed block copolymer, consisting of a water-soluble PEO block and a polyelectrolyte block, can facilitate the endosomal delivery. In the described system, the formed polyion complex was stable at pH 7.4, but it could be dissociated at pH 5.5. This behavior led to the targeted release of the antibody into the cytoplasm, where the pH is lower (Lee et al. 2010).

Bulmus and coworkers have reported on novel pH-responsive polymeric nanocarriers for the enhanced cytoplasmic delivery of enzyme susceptible drugs, such as proteins and peptides as well as antisense ODNs (Bulmus et al. 2003). A new functionalized monomer, namely, pyridyl disulfide acrylate, was specifically synthesized and incorporated into an amphiphilic copolymer containing also MAA and butyl acrylate segments with the aid of free-radical polymerization techniques. This resulted in a glutathione- and pH-sensitive, membrane-disruptive terpolymer with functional groups, which allowed thiol-containing molecules to be readily conjugated. In this way, the polymeric nanocarriers posed two main functions: (a) pH-dependent, endosomal membrane disruption and escape into the cytoplasm, followed by (b) reaction of disulfide-conjugated drug with glutathione, a normal constituent of the cytoplasm of cells, causing release of the drug from the nanocarrier. Chemical conjugation and/or ionic complexation with oligopeptides (or antisense ODNs) was performed in order to load the corresponding macromolecular drugs on the nanocarrier. Hemolytic activity at low pH remained high even after the conjugation/complexation with oligopeptides. The terpolymer itself showed no toxicity even at high concentrations, as determined with tests involving mouse 3T3 fibroblasts and human THP-1 macrophage-like cells. Uptake of the radiolabeled polymer and enhanced cytoplasmic delivery of FITC-ODN was also studied in THP-1 macrophage-like cells.

Poly(ethylene glycol) dimethacrylate (PEGDMA) and MAA-based nanoparticles and microparticles have been prepared by emulsion polymerization. Particles of different size were obtained by using different concentration of sodium lauryl sulfate as the emulsifying agent. The particles were further evaluated as carrier systems for the oral delivery of insulin (Tomar et al. 2011). Insulin loading efficiency of the particles was found to be directly proportional to the particle size and inversely proportional to the acid content of the particles. *In vitro* insulin release studies from insulin-loaded particles were performed by simulating the gastrointestinal tract conditions using HPLC. At pH 2.5, the release of insulin from polymeric particles was observed in the range of 5%–8%, while a significantly higher release (20%–35%) was observed at pH 7.4 during the first 15 min of *in vitro* release. Larger size particles (of sizes ca. 8.3 μm) showed the highest efficiency to reduce the blood glucose level in diabetic rabbits.

Two types of blend formulations, poly(lactic-*co*-glycolic acid) (PLGA)/linear PEO-PPO-PEO and PLGA/x-shaped PEO-PPO, have been studied as model nanocarriers for insulin in potential oral administration (Santander-Ortega et al. 2009). The results showed that the interactions with the digestive enzymes were considerably reduced in the blend formulations. The net charge of the encapsulated protein showed a clear effect in the final size of the nanoparticles, while the encapsulation efficiency was controlled by the polyoxyethylene derivative. The authors concluded that the carriers formed with encapsulated insulin in PLGA/linear PEO-PPO-PEO particles are capable, at least *in vitro*, to overcome the gastrointestinal barrier.

He and coworkers have recently focused on the size-dependent oral absorption mechanism of polymeric nanoparticles loaded with protein drugs (He et al. 2012). The authors prepared rhodamine B–labeled carboxylated chitosan–grafted nanoparticles with particle sizes of 300, 600, and 1000 nm and similar ζ-potentials (~−35 mV), which were loaded with FITC-labeled bovine serum albumin. The smallest particles demonstrated elevated intestinal absorption, as mechanistically evidenced by higher mucoadhesion in rat ileum, release amount of the payload into the mucus layer, Caco-2 cell internalization, transport across Caco-2 cell monolayers and rat ileum, and systemic biodistribution after oral gavage. Peyer's patches could play a role in the mucoadhesion of nanoparticles, resulting in their close association with the intestinal absorption of nanoparticles.

For insulin-loaded nanoparticles prepared using trimethyl chitosan chloride modified with a targeting peptide (CSKSSDYQC) has been reported elsewhere (Jin et al. 2012). The authors demonstrated the effects of goblet cell–targeting nanoparticles on the oral absorption of insulin *in vitro*, *ex vivo*, and *in vivo*. The article is mainly focused on the targeting, permeation, uptake, and internalization of the particles and showed that the orally administrated peptide-modified nanoparticles produced a better hypoglycemic effect with a 1.5-fold higher relative bioavailability compared with unmodified ones.

Chitosan–carrageenan nanoparticles have shown promising properties as carriers of therapeutic macromolecules (Grenha et al. 2010). By using these natural marine-derived polymers, nanoparticles in the 350–650 nm size range and positive zeta-potentials of 50–60 mV were obtained in a hydrophilic environment, under very mild conditions, avoiding the use of organic solvents or other aggressive technologies for their preparation. They exhibited a noncytotoxic behavior in biological *in vitro* tests performed using L929 fibroblasts, which is critical regarding the biocompatibility of those carriers. Using ovalbumin as model protein, nanoparticles evidenced loading capacity varying from 4% to 17% and demonstrated excellent capacity to provide a controlled release for up to 3 weeks.

Core/corona nanoparticles based on various monomers, prepared by emulsion polymerization methods, have been utilized by Akagi and coworkers as therapeutic peptide carriers (Akagi et al. 2007). Salmon calcitonin (sCT) was used as the model peptide drug. After the oral administration of mixtures of sCT and nanoparticles to rats, it was found that the blood ionized calcium concentration decreased indicating that the use of nanoparticles as sCT carriers enhanced peptide adsorption through the gastrointestinal tract. sCT absorption was larger for nanoparticles with PNIPAM chains on their surfaces (Sakuma et al. 1997). Nanoparticles with PMAA or poly(vinyl amine) chains on the surface showed a smaller effect. A further enhancement of nanoparticle peptide delivery efficiency was observed for nanoparticles with polystyrene cores and coronas of mixed PNIPAM and poly(vinyl amine) chains (Sakuma et al. 2002).

As shown in the previous chapters, the polymersomes have an ideal architecture that can be used as a scaffold for precise positioning of biomacromolecules. By encapsulation of proteins, incorporation of membrane proteins in the polymeric bilayer, and surface conjugation with, for example, cellular signaling molecules, the polymersomes can be applied for various purposes. The first encapsulation experiments have been performed with proteins such as myoglobin, hemoglobin, and albumin via the addition of the solid block copolymer polybutadiene (PBD)-PEG to an aqueous solution of the desired solute, after which the mixture was incubated for a day (Lee et al. 2001). The efficiency of this method varied considerably with the type of the solute, ranging from about 5% for albumin to more than 50% for myoglobin.

Pang and coworkers have created biodegradable polymersomes for brain delivery of peptides (Pang et al. 2008). They used the thiols present in the monoclonal antibody OX26 to decorate PEG-PCL polymersomes with maleimide functions on their surface. The antibody was shown able to initiate endogenous receptor–mediated transcytosis of the polymersomes across the blood–brain barrier. The polymersomes should degrade with time to release their contents. Biodegradable and biocompatible polymersomes as oxygen carriers have been developed as well (Arifin and Palmer 2005, Rameez et al. 2008). They were based on PEG-PCL and PEG-PLA in which bovine or human hemoglobin was encapsulated. By varying the hemoglobin concentration or the block copolymer concentration, the aggregate size distribution and encapsulation efficiency were tuned. Due to the higher membrane thickness, the oxygen affinity of the polymersome-encapsulated hemoglobin was lower than that of red blood cells but still consistent with what is required for efficient oxygen delivery in the systematic circulation.

Albumin and tetanus toxoid have been successfully encapsulated by solvent evaporation in PLA nanoparticles of sizes about 100–120 nm (Soppimath et al. 2001). In order to retain the secondary and tertiary structures of the protein drugs, a slow and intermittent sonication at low temperatures was applied. Bovine serum albumin and immuno-γ-globulin (IgG) encapsulated in nanoparticles prepared from PEO-PLGA using solvent diffusion method exhibited encapsulation efficiency of 58.9% and slow *in vitro* release rate. However, the problem of protein hydration, pH reduction derived from polymer degradation, and presence of the hydrophobic interfaces in PLGA-based delivery devices led to protein inactivation or irreversible aggregation inside PLGA or PLA nanoparticles (Santander-Ortega et al. 2010). These problems have been prevented by incorporation of stabilizers such as PEO and its derivatives. It is noteworthy that the type of the PEO derivative and pH of internal aqueous phase were the most important factors influencing BSA protein encapsulation and release kinetics (Yadav et al. 2011). Degradation and release characteristics of polyester particles can be improved by the incorporation of polyoxyethylene derivatives with different hydrophilic–lipophilic balance (Santander-Ortega et al. 2010). The extended half-life of bovine serum albumin encapsulated in PEG-PLGA nanoparticles was found to alter the protein biodistribution in rats compared to that of the same protein loaded in PLGA nanoparticles (Li et al. 2001). Similar results have been reported for tetanus toxoid encapsulated in PLA and PLA-PEG nanoparticles of comparable dimensions (137–156 nm) but differing in their hydrophobicity (Simone et al. 2007). In addition, the PLA-PEG nanoparticles led to greater penetration of tetanus toxoid into the blood circulation and in lymph nodes than PLA-encapsulated tetanus toxoid (Vila et al. 2002). The results suggested that PLA nanoparticles suffered an immediate aggregation upon incubation with lysozyme, whereas the PEG-coated nanoparticles remained totally stable (Simone et al. 2009). The antibody levels elicited following in administration of PEG-coated nanoparticles were significantly higher than those corresponding to PLA nanoparticles (Vila et al. 2004).

Porous Tat-functionalized polymersomes prepared by co-assembly of polystyrene$_{40}$-block-poly[l-isocyanoalanine(2-thiophen-3-ylethyl)amide]$_{50}$ and polystyrene-block-poly(ethylene glycol)-oxanorbornadiene (Figure 5.24) have been described elsewhere (van Dongen et al. 2010). The polymersomes had an average diameter of about 114 nm with no obvious size variation between polymersomes with different protein types and contents. Efficient cellular uptake of polymersomes loaded with GFP by a variety of cell lines was observed. After their uptake into cells, a considerable fraction of the GFP signal colocalized with the acidic vesicles; however, almost equally large population of GFP-containing punctuate structures retained a neutral pH. Furthermore, the same polymersomes loaded with horseradish peroxidase functioned as nanoreactors inside cells. They maintained their activity to a much higher degree than what was reported for free horseradish peroxidase trafficked to lysosomes (van Dongen et al. 2010).

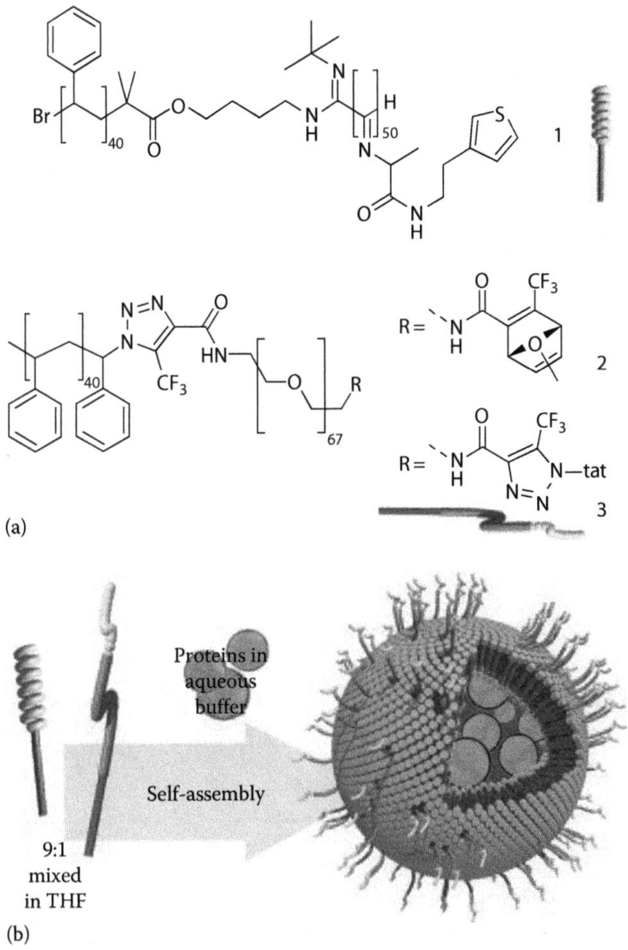

FIGURE 5.24 (a) Structures and representations of polystyrene$_{40}$-block-poly[l-isocyanoalanine(2-thiophen-3-ylethyl)amide]$_{50}$ (1) and polystyrene-block-poly(ethylene glycol)-oxanorbornadiene (2). (b) Formation of Tat-functionalized polymersomes loaded with proteins by co-assembly of 1 and 3 in 9:1 ratio. (van Dongen, S.F.M., Verdurmen, W.P.R., Peters, R.J.R.W., Nolte, R.J.M., Brock, R., and van Hest, J.C.M.: *Angew. Chem. Int. Ed.* 2010. 49. 7213–7216. Copyright Wiley-VCH Verlag GmbH & Co. KGaA. Reproduced with permission.)

The potential of polymeric vesicles as a light-triggered delivery system has been studied (Cabane et al. 2011). The polymersomes were prepared by self-assembly of a photocleavable amphiphilic block copolymer composed of PAA and poly(methyl caprolactone) as hydrophilic and hydrophobic blocks, respectively, linked by O-nitrobenzyl photocleavable segment (Figure 5.25). The vesicles disintegrated upon UV irradiation and rearranged into small micelle-like structures, thus releasing their payload. Graphical summary of the mechanism of polymersome degradation is presented in Figure 5.25. Enhanced GFP was encapsulated in the polymersomes with encapsulation efficiency of 22% and releases upon reorganization of the polymer chains.

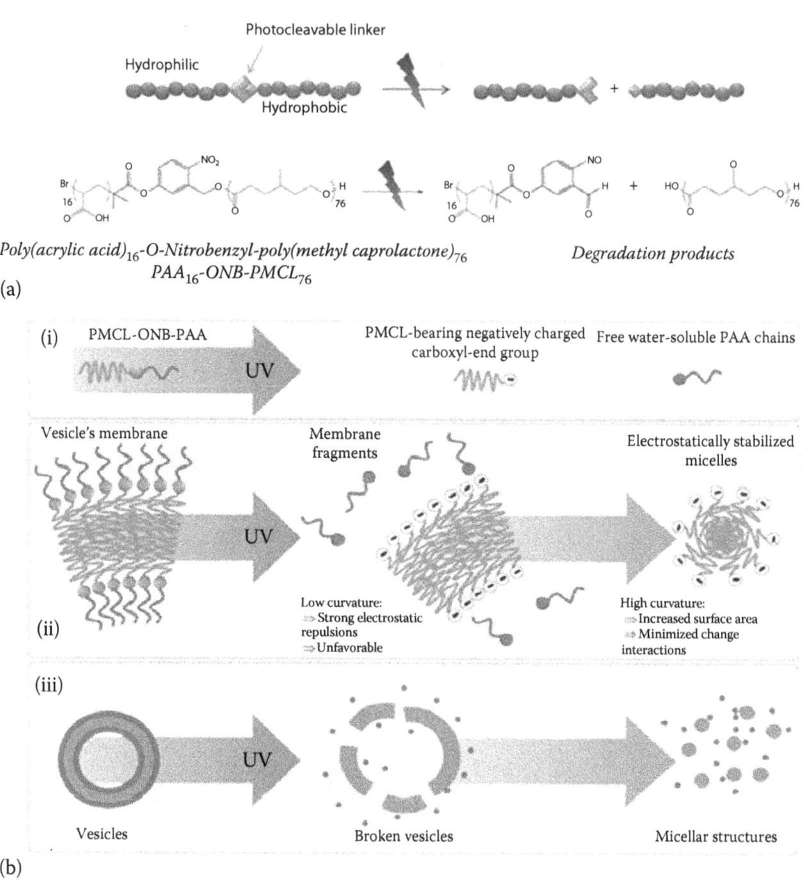

FIGURE 5.25 (a) Schematic view of the amphiphilic photocleavable block copolymer, chemical structure of the poly(methyl caprolactone)-O-nitrobenzyl-poly(acrylic acid) diblock copolymer, and its degradation products upon UV irradiation. (b) Graphical summary depicting the proposed mechanism of polymersome degradation. (i) Molecular level (the diblock copolymer is rapidly cleaved upon UV irradiation), (ii) supramolecular level (as a result of chain scission, packing of the PMCL chains forming the membrane is progressively destabilized and evolves into a more favorable arrangement), and (iii) aggregate morphology (from vesicles to broken vesicles to stabilized micellar structures). (Cabane, E., Malinova, V., Menon, S., Palivan, C.G., and Meier, W., *Soft Matter*, 7(19), 9167–9176, 2011. Reproduced by permission of The Royal Society of Chemistry.)

PIC micelles based on PEG-poly(aspartic acid) copolymers with biological polyions or enzymes such as lysozyme and trypsin have been formed (Harada and Kataoka 1998, Yuan et al. 2005). By derivatization of the aspartate moiety with citraconic anhydride using ethylene diamine as a linker, a copolymer that was stable at neutral and basic pH but degraded at acidic pH was formed (Lee et al. 2007). The derivatized copolymer formed PIC micelles that could contain lysozyme in their cores and degraded rapidly when brought into an acidic environment (Figure 5.26).

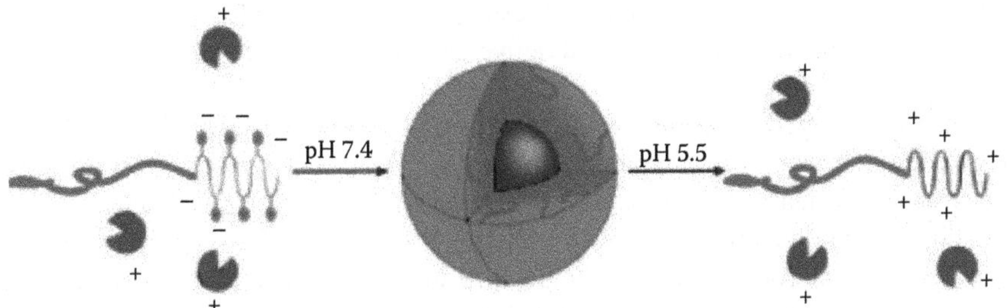

FIGURE 5.26 Schematic representation of the micellar assembly of the PEG-poly[(*N'*-citraconyl-2-aminoethyl)aspartamide] diblock copolymer with lysozyme and its degradation with time at acidic pH. (Reprinted with permission from Lee, Y., Bae, Y., Hiki, S., Ishii, T., and Kataoka, K., *J. Am. Chem. Soc.*, 129, 5362, 2007. Copyright 2007, American Chemical Society.)

Intercellular adhesion molecule-1 (ICAM-1)-targeted nanocarriers have been used to deliver *acid α-glucosidase* into cells to address the specific enzyme deficiency in Pompe's disease (Hsu et al. 2012). ICAM-1 is a protein involved in inflammation and overexpressed on most cells under pathological conditions. The deficiency of acid α-glucosidase causes Pompe's disease, which leads to excess glycogen storage throughout the body, mainly in the liver and striated muscle. To improve delivery of acid α-glucosidase, the latter was coupled to polymer nanocarriers (average dimensions ~180 nm), which were coated with an antibody specific to ICAM-1. Fluorescence microscopy showed specific targeting to cells, with efficient internalization and lysosomal transport, enhancing glycogen degradation over nontargeted acid α-glucosidase. Radioisotope tracing in mice demonstrated enhanced acid α-glucosidase accumulation in all organs, including Pompe targets. In an earlier paper, markedly enhanced delivery in brain, kidney, heart, liver, lung, and spleen of mice was reported (Hsu et al. 2011); TEM showed attachment and internalization into vascular endothelium. Fluorescence microscopy proved targeting, endocytosis, and lysosomal transport of the loaded particles in macro- and microvascular endothelial cells and a marked enhancement of globotriaosylceramide degradation. The lysosomal accumulation of the latter in multiple tissues, due to deficiency of α-galactosidase, causes Fabry disease. Therefore, this ICAM-1-targeting strategy may help improve the efficacy of therapeutic enzymes for Fabry disease and Pompe's disease.

Encapsulation of antioxidant enzymes in PEG-coated liposomes has been shown to increase the antioxidant enzyme bioavailability and enhance protective effects in animal models. Micelles based on PEO-PPO-PEO copolymers showed even more potent protective effect (Hood et al. 2011). Such nanocarriers protected encapsulated antioxidant enzymes from proteolysis and improved delivery to the target cells, such as the endothelium lining the vascular lumen. A recent work of Simone and coworkers has been focused on protective encapsulation of the potent antioxidant enzyme, catalase, by filamentous polymer nanocarriers (Simone et al. 2009). The authors maintained the same molecular weight ratio of PEG to PLA in a series of PEG-*b*-PLA diblock copolymers and varied the total copolymer molecular weight from about 10,000 to 100,000 g/mol. All diblock copolymers formed filamentous particles upon processing, which encapsulated active enzyme. The latter proved resistant to protease degradation.

A nanogel formed by self-assembly of a conjugate of heparin and PEO-PPO-PEO block copolymer has been prepared (Choi et al. 2011). Heparin possesses advantageous structural features such as negatively charged structure and presence of specific binding domains for proteins. The nanogel solution of the conjugate exhibited high stability, hydrodynamic size below 100 nm, and ability to encapsulate both small molecules and proteins. Circular dichroism and gel electrophoresis demonstrated that the stability of monoclonal antibodies (3D8 scFv) encapsulated into the nanogel had been maintained.

Advanced protein delivery systems for treatment of lysosomal storage diseases have recently been described (Giannotti et al. 2011). Polyelectrolyte complexes between trimethyl chitosan and the lysosomal enzyme α-GAL via self-assembly and ionotropic gelation were prepared. The particles were characterized with an average particle size below 200 nm, polydispersity index <0.2, and a protein loading efficiency of about 65%. The polyelectrolyte nanoparticles were stable and active under physiological conditions and able to release the enzyme at acidic pH, as demonstrated by *in situ* AFM (Figure 5.27). They were further functionalized with Atto 647N and their cellular uptake and fate were tracked using high-resolution fluorescence microscopy. In contrast to their precursor, the polyelectrolyte complexes were efficiently internalized by human endothelial cells and mostly accumulated in lysosomal compartments.

Bovine serum albumin has been encapsulated in levan-based nanocarriers (Sezer et al. 2011). Microbial levans are biopolymers produced from sucrose-based substrates

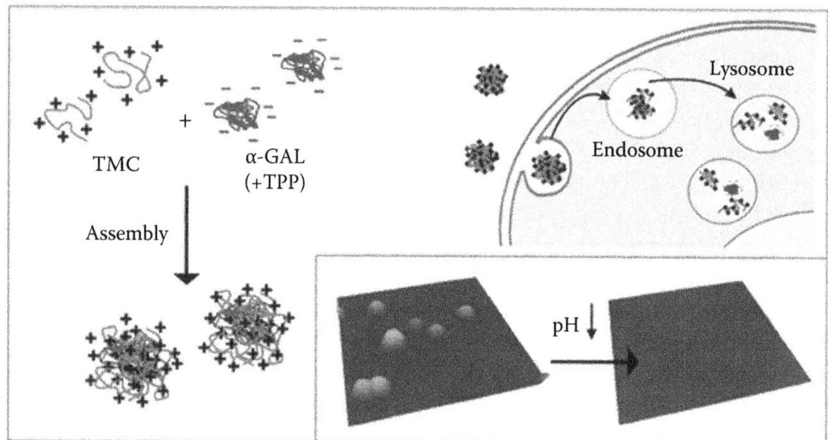

FIGURE 5.27 Schematic representation of the formation of polyelectrolyte complexes between trimethyl chitosan and α-GAL, their cellular uptake and accumulation in lysosomal compartments, and the release of the enzyme at acidic pH. (Reprinted with permission from Giannotti, M.I., Esteban, O., Oliva, M., Garcia-Parajo, M.F., and Sanz, F., *Biomacromolecules*, 12(7), 2524–2533, 2011. Copyright 2011, American Chemical Society.)

by a variety of microorganisms. For this study, levan produced by Halomonas sp. was used. Encapsulation efficiency 49.3%–71.3%, particle dimensions in the 200–537 range, and slightly positive ζ-potentials were reported.

5.3 APPLICATIONS IN IMAGING AND DIAGNOSTICS

Bioimaging techniques have been widely applied in modern biomedical field in order to make disease diagnosis more accurate and rapid (Li et al. 2010, Koo et al. 2011, Parveen et al. 2012). Due to their significant potential for disease analysis and diagnosis at the molecular and cellular level, imaging technologies, including techniques like optical imaging, MRI, and ultrasonic imaging, combined with bioimaging and molecular cell biology, have been developed to meet the demands of modern biomedical research. Through these noninvasive methods, the organs' and cell functions *in vivo* can be inspected, and the molecular interactions can be monitored on line. Because the signal between the normal and pathological organs is too low to be detected, fine medical images cannot be readily obtained usually without the use of functional and assisting chemicals and molecular nanodevices designated as the "imaging agent" or "imaging probe." Such agents have been developed for enhancing the image contrast and spatial resolution, which in turn can significantly improve the accuracy of diagnosis. Research on suitable imaging agents therefore became a significant research area in life sciences, medical sciences, and material sciences. With the development of nanotechnology, novel nanoimaging agents have been developed. Nanoparticles, including polymeric, inorganic, and hybrid nanoparticles with sizes lower than 100 nm, have unique properties that can be fine tuned and utilized for bioimaging applications. Generally some basic properties/functionalities should be incorporated into the nanoparticle in order to function properly as a contrast agent: the nanoparticle should possess (1) an inorganic or organic core/active material that can interact with incoming radiation for imaging enhancement, (2) a hydrophilic shell for improving the stability of the nanoconstruct within the probed medium/site, and (3) functional outer ligands that can offer targeting properties to the nanoparticulate imaging agent. The nature of each component may vary according to the technique implemented for imaging as well as the application environment.

Broadly speaking, the field of biomedical imaging can be divided into several categories based upon the electromagnetic spectrum utilized in each case, that is, magnetic resonance, optical/near infrared (NIR), and ionizing radiation (x-rays and γ-rays) (Pansare et al. 2012). Imaging based on ionizing radiation generally refers to the detection of high-frequency emissions from radioactive elements such as the gamma-ray emitters ^{111}In or ^{99}mTc or the passage of x-rays through the body. The main technologies involved in this category are positron emission tomography (PET), SPECT, and x-ray computed tomography (CT). MRI tends to operate on the other end of the spectrum in the MHz frequency range, relying upon contrast agents such as gadolinium or superparamagnetic iron oxide nanoparticles (SPIONs) to modify the relaxivity of water molecules for providing soft tissue contrast. Long-wavelength and NIR imaging has been developed due to the desire to perform whole animal/body and deep tissue imaging. NIR imaging applications are also growing rapidly due to the availability of targeted biological agents for diagnosis and basic medical research that can be imaged *in vivo*. The wavelength range of 650–1450 nm falls in the region of the spectrum with the lowest absorption in tissue and therefore enables the deepest tissue penetration. The rapid growing interest in multimodal imaging, which involves simultaneous delivery of actives, targeting, and imaging, requires nanoparticles or supramolecular assemblies. Well-defined and smart-designed nanoparticles for diagnostics also have advantages in increasing circulation time and increased imaging brightness relative to single-molecule imaging agents. This has led to rapid advances in nanocarriers for long-wavelength, NIR imaging. Some examples pertaining to polymer-based nanoparticles for bioimaging will be discussed in the following.

Ghoroghchian and coworkers utilized PBD-*b*-PEO block copolymer vesicles in water in order to encapsulate multi[(porphinato)zinc(II)]-based supermolecular fluorophores. This led to the formation of emissive polymersomes with a rich photophysical diversity enabling emission energy modulation over a broad spectral domain (Ghoroghchian et al. 2005). They have shown that by controlling polymer-to-fluorophore noncovalent interactions, the bulk photophysical properties of these soft, supramolecular, optical materials could be finely tuned allowing utilization of the nanoassemblies in NIR bioimaging protocols.

Prud'homme and coworkers reported the preparation of hybrid nanoparticles for NIR imaging composed of rare earth ion–doped phosphors nanocrystals ($NaYF_4$:Yb^{3+},Er^{3+}) surface modified by PAA homopolymers and poly(ethylene glycol)-*b*-poly(caprolactone) (PEG-*b*-PCL), poly(ethylene glycol)-*b*-poly(lactic-*co*-glycolic acid) (PEG-*b*-PLGA), and poly((ethylene glycol)-*b*-lactic acid) (PEG-*b*-PLA) amphiphilic block copolymers (Budijono et al. 2010). Initially hexagonal phase nanophosphors were prepared using one-step cothermolysis utilizing oleic acid (OA) and trioctylphosphine (TOP) ligands. Then direct ligand exchange with PAA and using an amphiphilic copolymer encapsulation via flash nanoprecipitation took place (Akbulut et al. 2009). Both surface modification routes produced colloidally stable dispersions in water and in buffers and serum media (Figure 5.28). The PEG block confers biocompatibility to the nanostructures and ensures long circulation times in the bloodstream. These polymer-modified upconverting nanophosphors provide promising new nanomaterials also for photodynamic therapy (PDT) applications.

An MRI contrast agent based on polymeric nanoparticles was developed by conjugation of gadolinium (Gd) chelate groups onto the biocompatible poly(L-lactide-*b*-ethylene glycol) (PLA-*b*-PEG) block copolymers (Chen et al. 2011). The MRI contrast agent was targeted to liver. PLA-*b*-PEG copolymers were conjugated with diethylenetriaminepentaacetic acid (DTPA). The PLA-*b*-PEG-DTPA nanoparticles were prepared by the solvent diffusion method, and then Gd was loaded onto the nanoparticles due to complexation with the DTPA moieties on the surface of the PLA-*b*-PEG-DTPA nanoparticles. The mean size of the nanoparticles was ca. 269 nm. The relaxivity of the Gd-labeled polymeric nanoparticles was determined against the conventional contrast agent Magnevist. Compared to Magnevist, the Gd-labeled PLA-PEG nanoparticles showed significant enhancement on imaging signal intensity. The T_1 and T_2 relaxivities per Gd atom of the Gd-labeled nanoparticles were found to be 18.865 and 24.863 mM^{-1} s^{-1} at 3 T, respectively. In addition, the signal intensity *in vivo* was stronger comparing with the Gd-DTPA contrast agent

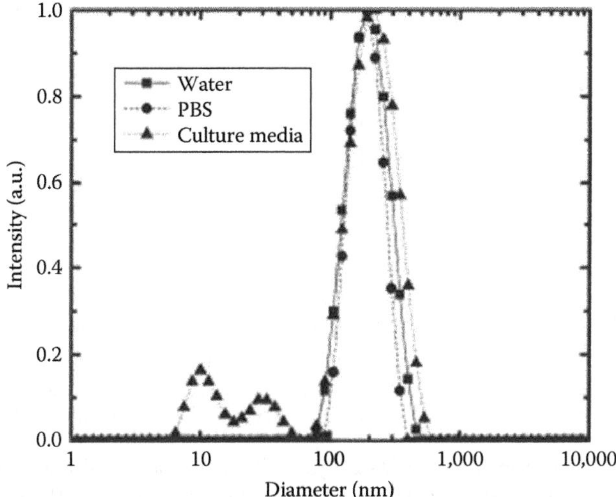

FIGURE 5.28 Size distributions of PEG-*b*-PLA-protected rare earth ion–doped phosphors nanocrystals in water (■), PBS (●), and culture media (▲). The peaks at 10 and 30 nm are proteins in the FBS culture media. (Reprinted with permission from Budijono, S.J., Shan, J., Yao, N., Miura, Y., Hoye, T., Austin, R.H., Ju, Y., and Prud'homme, R. K., *Chem. Mater.*, 22, 311–318, 2010. Copyright 2010, American Chemical Society.)

alone, and the T_1 weight time was lasting for 4.5 h. The distribution of nanoparticles *in vivo* was evaluated in rats, and it was found to be also enhanced compared to Magnevist. The liver-targeting efficiency of the Gd-labeled PLA-*b*-PEG-based nanoparticles in rats was 14.6 comparing with Magnevist injection. Therefore, the Gd-labeled nanoparticles showed the potential as targeting molecular MRI contrast agent for further clinical utilization.

A polymeric micellar system was utilized for creating a novel MRI contrast agent by Shiraishi et al. (2009). The block copolymer PEG-*b*-poly(L-lysine) was used as the micelle-forming copolymer, as well as the matrix for the incorporation of MRI-active gadolinium ions through polymer-linked chelating moieties, namely, DOTA. The DOTA moieties were chemically linked to the primary amine groups of the lysine segments in the copolymer. The functionalized PEG-*b*-poly(L-lysine-DOTA) block copolymer could form micelles in aqueous media in the absence of Gd ions. The polymeric micelle structure was maintained even after partial gadolinium chelation (ca. 40%) to the existing DOTA moieties. The prepared polymeric micelle MRI contrast agent was injected into a mouse tail vein at a dose of 0.05 mmol Gd/kg and was found to exhibit stable blood circulation. A considerable amount (6.1 ± 0.3% of ID/g of the polymeric micelle MRI contrast agent) was found to accumulate at solid tumors 24 h after intravenous injection by means of the EPR effect alone (Figure 5.29). An MRI analysis revealed that the signal intensity of the tumor was enhanced by two-fold via the use of the new hybrid polymeric nanoparticle contrast agent.

A similar approach based on mixed micellar polymeric systems was also reported for the preparation of nanoparticle MRI contrast agents (Nakamura et al. 2006).

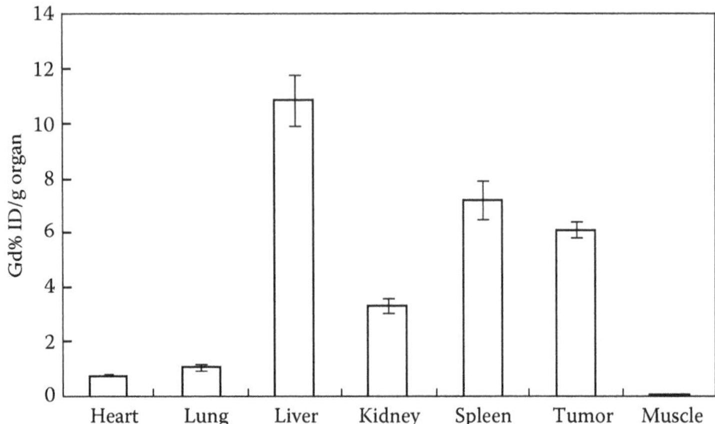

FIGURE 5.29 Biodistribution of PEG-P(Lys-DOTA-Gd) micelle 24 h after injection at a dose of 0.05 mmol Gd/kg. (Reprinted from *J. Control. Release*, 136, Shiraishi, K., Kawano, K., Minowa, T., Maitani, Y., and Yokoyama, M., 14–20, Copyright 2009, with permission from Elsevier.)

Polymeric micelles that were formed from cationic polymers (polyallylamine or protamine) and anionic block copolymers (poly(ethylene glycol)-*b*-poly(aspartic acid) derivative) could bind Gd ions, creating hybrid mixed polymeric micelles. These micelles were found to provide high contrasts in MRI experiments by shortening the T_1 longitudinal relaxation time of protons of water. The PEG-*b*-PAsp block copolymer with bound Gd ions showed high relaxivity (T_1-shortening ability) values (from 10 to 11 mol^{-1} s^{-1}), while the mixed polymeric micelles incorporating Gd ions exhibited low relaxivity values (from 2.1 to 3.6 mol^{-1} s^{-1}). The authors concluded that these findings illustrate the feasibility of the novel MRI micellar contrast agent that selectively provides high contrasts at solid tumor sites owing to a dissociation of the micelle structures, while selective delivery to the tumor sites is achieved in the polymeric micelle form.

Iron oxide (IO) magnetic nanoparticles embedded into polymeric nanoparticles and micelles have been also tested as MRI contrast agents. Ujiie et al. reported the preparation of IO nanoparticles in the presence of poly(ethylene glycol)-*b*-poly(4-vinylbenzylphosphonate) (PEG-*b*-PVBP) block copolymer (Ujiie et al. 2011). The magnetic nanoparticles were obtained through alkali coprecipitation of iron salts in the presence of the copolymers. In this way, the surface of the particles was coated with a relatively immobilized PEG layer with a high density of PEG chains (referred as PEG-protected iron oxide nanoparticles, PEG-PIONs). The PEG chains were bound to the IO surface via multipoint anchoring of the phosphonate groups present in the PVBP block. The surface density of PEG chains in the PEG-PIONs was varied through variation of the [PEG-*b*-PVBP]/[iron salts] feed–weight ratio in the coprecipitation reaction. PEG-PIONs prepared at an optimal feed–weight ratio in this study showed a hydrodynamic diameter of ca. 50 nm. The PEG-PIONs could be dispersed in PBS that contains 10% serum without any change in their hydrodynamic diameters over a period of 1 week, indicating that PEG-PIONs possess

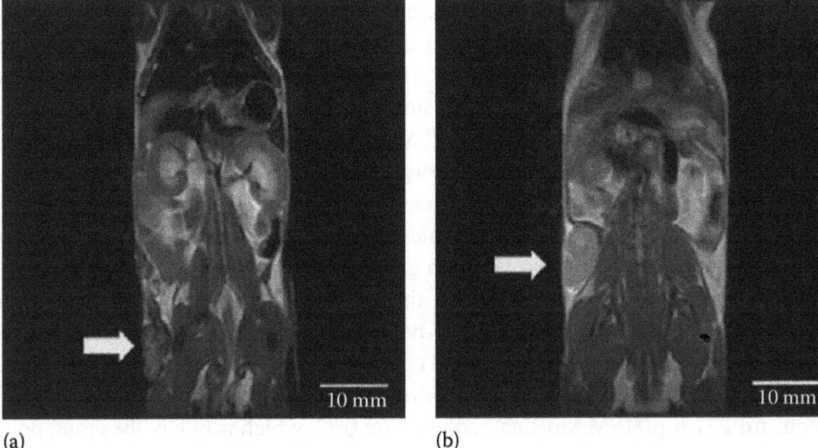

FIGURE 5.30 T_2-weighted images of C-26 tumor-bearing mice at 24 h postinjection of 84 mg Fe kg^{-1} of (a) PEG-PION4 and (b) control. The arrow denotes tumor site. (Reprinted from *Colloids Surf. B*, 88, Ujiie, K., Kanayama, N., Asai, K., Kishimoto, M., Ohara, Y., Akashi, Y., Yamada, K., Hashimoto, S., Oda, T., Ohkohchi, N., Yanagihara, H., Kita, E., Yamaguchi, M., Fujii, H., and Nagasaki, Y., 771–778, Copyright 2011, with permission from Elsevier.)

high dispersion stability under *in vivo* physiological conditions as well as excellent anti-biofouling properties. PEG-PIONs showed a long blood circulation time and significant tumor accumulation (more than 15% ID/g of tumor) without the aid of any surface targeting ligand in mouse tumor models (Figure 5.30). It was observed that the majority of the PEG-PIONs accumulated in the tumor in 96 h after administration, whereas those in normal tissues were smoothly eliminated in the 96 h period. This observation illustrates the enhancement of tumor selectivity in the localization of PEG-PIONs due to the physicochemical characteristics of the nanoparticles. The performance characteristics of such hybrid nanoparticles should promote their use as MRI contrast agents.

On a similar context, a series of pH-responsive polymeric micelles based on a pH-responsive poly(β-amino ester)/(amido amine) block and PEG were developed to act as carriers for the delivery/accumulation of magnetic IO (Fe_3O_4) nanoparticles that respond rapidly to an acidic tumor environment for effective MRI of tumors (Gao et al. 2011b). At physiological conditions, the Fe_3O_4 nanoparticles could be well encapsulated into the poly(β-amino ester)/(amido amine) core of the polymeric micelles, due to hydrophobic interactions, shielded by the PEG coronal shell. In an acidic tumor environment, the pH-responsive block, carrying ionizable tert-amino groups on its backbone, can become protonated and therefore soluble. In this case, the hydrophobic Fe_3O_4 nanoparticles are released, thus enhancing their accumulation into targeting tumors. The Fe_3O_4-loaded polymeric micelles were tested in a disease rat model of cerebral ischemia that produces acidic tissue, due to its pathological condition, in respect to the ability of this MRI probe as a pH-triggered agent, using a 3.0 T MRI scanner. It was observed that the IO particles were gradually accumulated in the brain ischemic area, indicating that the pH-triggered MRI probe may be effective for targeting the acidic environment and for diagnostic imaging of pathological tissues. The same group utilized a similar system where the red fluorescent dye sulforhodamine 101 (SR101) was conjugated to the poly(β-amino ester)/(amido amine) block allowing for simultaneous MRI and optical imaging of pathological tissues (Gao et al. 2010). CLSM observations demonstrated the cellular uptake of SR101-labeled, Fe_3O_4-loaded polymeric micelles by breast cancer cells.

Hydrophobic OA-coated SPIONs (diameter 5–10 nm) were encapsulated into biodegradable thermosensitive polymeric micelles resulting in a system fulfilling the requirements for systemic administration and bioimaging (Talelli et al. 2009). The micelles were composed of amphiphilic, thermosensitive, and biodegradable block copolymers of poly(ethylene glycol)-*b*-poly[*N*-(2-hydroxypropyl)methacrylamide dilactate] (mPEG-*b*-p(HPMAm-Lac_2)). The encapsulation was performed by addition of one volume of SPIONs in tetrahydrofuran (THF) to nine volumes of a cold aqueous mPEG-*b*-p(HPMAm-Lac_2) solution (at 0°C, which is below the cloud point of the polymer), followed by rapid heating of the resulting mixture to 50°C, in order to induce micelle formation. Hybrid nanoparticles with ca. 200 nm diameter and rather low polydispersity were obtained. TEM analysis demonstrated that clusters of SPIONs were present in the core of the block copolymer micelles. A maximum loading of 40% was reported, while MRI scanning of the samples demonstrated that the SPION-loaded micelles had high r_2 and r_2^* relaxivities. The r_2^* values were determined to be at least twofold higher than the r_2 values, confirming the clustering of the SPIONs in the micellar core. The particles showed excellent stability under physiological conditions for 1 week, even in the presence of FBS. The authors proposed that due to the stability of the hybrid micelles together with their ease of preparation and their nanoscale size, these systems are highly suitable for image-guided drug delivery.

Polymeric nanoparticles carrying fluorine atoms have been also explored in MRI applications since they have large structural design potential when compared to traditional systems such as emulsions and solutions of smaller molecules. There is generally a growing interest in the development of different fluorinated compounds capable of being tracked *in vivo* using ^{19}F-MRI techniques. By the use of commercially available ^{19}F-MRI coils, ^{19}F imaging is easily achievable in the clinical setting. When the ^{19}F image is superimposed on the familiar ^1H density image, the location of a fluorinated compound or a fluorinated particle can be determined in a noninvasive manner. ^{19}F is an attractive tracking nucleus for MRI due to its high sensitivity and the absence of a confounding ^{19}F background signal within the body. The requirements for a successful ^{19}F-MRI tracking agent include that it should have high fluorine content; the fluorine nuclei should have appropriate NMR properties, for example, sufficient mobility; and preferably there should be a single peak in the ^{19}F-NMR spectrum. Such compounds have the potential to provide high image intensity without the need for selective excitation sequences.

Core/shell micelles from well-defined diblock copolymers, with acrylic acid hydrophilic blocks and hydrophobic blocks composed of partially fluorinated acrylate and methacrylate monomers synthesized using ATRP, were evaluated as potential ^{19}F-MRI agents (Peng et al. 2009a). The diblock copolymers spontaneous self-assembled into stable micelles, in aqueous solutions or in mixed organic/aqueous solvents, with diameters from approximately 20 to 45 nm. These micelles had a fluorine-rich core that provides a strong signal for MRI. The observed MRI image intensities were related to the NMR longitudinal and transverse relaxation times and were found to depend on polymer structure and method of micelle formation. Two distinct T_2 relaxation times were observed and measured. By comparison of expected MRI image intensities with those observed experimentally, it was concluded that methacrylate-based polymers show systematically lower signal intensity than acrylate polymers. This was related to the presence of a population of nuclear spins having very short T_2 relaxation times that generally cannot be detected under high-resolution NMR and MRI conditions.

A number of hyperbranched fluoropolymers were synthesized and their micelles in water were studied as potential ^{19}F-MRI agents (Du et al. 2008a). A hyperbranched starlike core was first prepared via atom transfer radical self-condensing vinyl (co) polymerization (ATR-SCVCP) of 4-chloromethyl styrene (CMS), lauryl acrylate (LA), and 1,1,1-tris(4′-(2″-bromoisobutyryloxy)phenyl)ethane (TBBPE). The polymerization gave a small core that served as a macroinitiator. Trifluoroethyl methacrylate (TFEMA) and *tert*-butyl acrylate (tBA) in different ratios were then "grafted" from the core to give hyperbranched starlike polymers with M_n of about 120 kDa and polydispersity values of about 1.6–1.8, having statistical copolymer chains as arms. After acidic hydrolysis of the *tert*-butyl ester groups, amphiphilic, hyperbranched starlike polymers with M_n of about 100 kDa were obtained. These branched copolymers were used for micelle formation in aqueous solutions, giving micelles with TEM-measured diameters ranging from 3 to 8 nm and DLS-measured hydrodynamic diameters from 20 to 30 nm. These micelles gave a narrow, single resonance as determined by ^{19}F-NMR spectroscopy, with a half width of approximately 130 Hz (Figure 5.31). The T_1 and T_2 relaxation times of the micellar systems were about 500 and 50 ms, respectively, and were not significantly affected by the composition and the overall size of the micelles. ^{19}F-MRI phantom images of these fluorinated

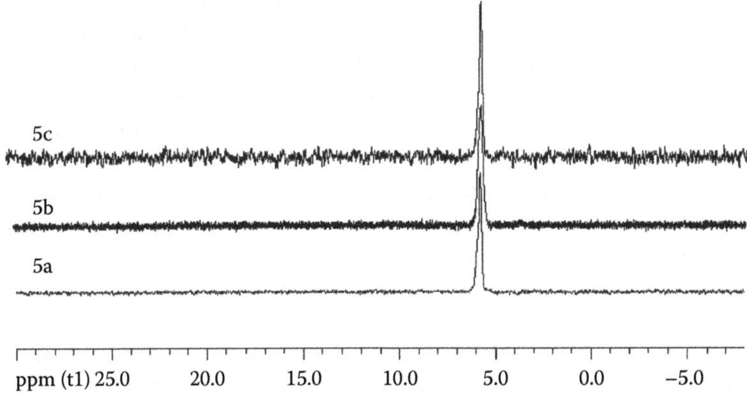

FIGURE 5.31 19F-NMR spectra of hyperbranched fluoropolymer micelles. (Reprinted with permission from Du, W., Nystrom, A.M., Zhang, L., Powell, K.T., Li, Y., Cheng, C., Wickline, S.A., and Wooley, K.L., *Biomacromolecules*, 9, 2826–2833, 2008a. Copyright 2008, American Chemical Society.)

micelles were also acquired. These images demonstrated that the particular fluorinated polymeric micelles may be useful as novel ^{19}F-MRI agents for utilization in a variety of biomedical applications.

Fluorinated nanoparticles, constructed by amphiphilic block copolymers based on precursor copolymers of styrene (PS) and 2,3,4,5,6-pentafluorostyrene (PPFS), as well as from block copolymers with tBA segments, that is, (PtBA)-b-PS-*co*-PPFS, were evaluated as potential MRI contrast agents (Nystrom et al. 2009). Reversible addition–fragmentation chain transfer polymerization was successfully employed in order to synthesize the aforementioned copolymers with control over molecular weight and composition and with narrow polydispersity. It was found that the copolymerization of styrene and PFS allowed for the preparation of gradient copolymers with opposite levels of monomer consumption, depending on the feed ratio. Conversion to amphiphilic block copolymers, namely, PAA-*b*-(PS-*co*-PPFS), was achieved by removing the protecting groups via hydrolysis, followed by chemical linking of with monomethoxy PEG chains for the enhancement of hydrophilicity and water solubility. Subsequently solution-state assembly and intramicellar cross-linking afforded SCK block copolymer nanoparticles with fluorinated segments. These final fluorinated nanoparticles (having hydrodynamic diameters of ca. 20 nm) were studied as potential MRI contrast agents based on the existing ^{19}F nuclei. It was found that packing of the hydrophobic fluorinated polymers into the core domain restricted the mobility of the chains and prohibited ^{19}F-NMR spectroscopy when the particles were dispersed in water without an organic cosolvent. The authors succeeded in encapsulating perfluoro-15-crown-5-ether (PFCE) into the polymeric micelles with good uptake efficiency, in order to increase fluorine content and mobility of core chains. Nevertheless, they observed that it was necessary to swell the core with a good solvent (dimethyl sulfoxide [DMSO]) to increase the mobility sufficiently enough to observe the ^{19}F-NMR signal of the PFCE blocks. These studies elucidate some aspects on the chemical design and properties of fluorinated polymeric nanoparticles with the potential to be used as ^{19}F-MRI contrast agents.

In another approach, a series of partly fluorinated polyelectrolytes were evaluated for their applicability as corona-forming components in ^{19}F-MRI-active nanoparticles in aqueous solutions (Nurmi et al. 2010). The polyelectrolytes were statistical and block copolymers of TFEMA and DMAEMA. The statistical copolymers were directly dissolved in water, whereas the block copolymers assembled into micellar nanoparticles with poly-TFEMA cores and poly(TFEMA-*co*-DMAEMA) coronas in aqueous media. The ^{19}F-spin–lattice (T_1) and spin–spin (T_2) relaxation times and ^{19}F image intensities of solutions of the polymers were measured and related to polymer chemical structure, as well as the chain conformation of the coronas of the micellar nanoparticles in aqueous solutions. The ^{19}F-NMR T_2 relaxation times were found to be highly indicative of the ^{19}F imaging performance of the dispersed nanomaterials. It was concluded that maintaining sufficient mobility of the ^{19}F nuclei was important for obtaining images of high intensity. ^{19}F mobility could be increased by preventing their aggregation in water as it was done in the case of statistical copolymers or by incorporating fluorine-containing segments within the corona of the micellar nanoparticles, in the case of block copolymers, where electrostatic repulsion between monomer units was present.

5.4 POLYMERIC NANOPARTICLES FOR PHOTODYNAMIC THERAPY

PDT is a treatment that utilizes the combined action of photosensitizer molecules and a specific light source for various types of cancer (Dolmans et al. 2003). In comparison to chemotherapy, PDT can be regarded as a noninvasive therapy and in principle induces less harmful effects to the healthy tissues. Photosensitizer molecules utilized in PDT act as strong absorbing agents, in a wide spectrum of wavelengths, and convert oxygen to highly reactive oxygen species (ROS) that in turn induce damage to tumor cells. It has been shown that photodynamic therapies destroy tumors through three main mechanisms: (1) have a direct effect on cancer cells leading to cell death such as apoptosis and necrosis (Pogue et al. 2001), (2) produced ROS damage the vascular and subsequent deprivation of oxygen and nutrients to the tumor (Krammer 2001), and (3) have several effects on the immune system causing an immunosuppressive response to the tumor (van Duijnhoven et al. 2003). In order to increase therapeutic efficiency of the photosensitizer in cancer therapy, a suitable carrier is needed for enhancing the usually poor solubility of such agents and to improve targeting. A variety of nanocarriers have been proposed and developed for PDT, including polymer-photosensitizer conjugates, liposomes, and polymeric micelles (Sortino et al. 2006, Rijcken et al. 2007, Peng et al. 2008, 2009b, Nishiyama et al. 2009). These nanocarriers have been developed mainly because they show tumor-selective accumulation, due to the enhanced, microvascular permeability and impaired lymphatic drainage. However, the aggregation of photosensitizer molecules within the drug carrier is expected to reduce formation of ROS, due to light-quenching effects. Therefore, the release of the photosensitizer is needed for ensuring maximum therapeutic efficiency. The situation then resembles in many ways the design and properties requirements for drug delivery nanocarriers. Some illustrative examples of the application of polymeric nanoparticles in PDT are discussed in the following.

Thermosensitive mPEG-*b*-p(HPMAm-Lac2) micelles loaded with a hydrophobic solketal-substituted phthalocyanine (Si(sol)$_2$Pc) photosensitizer were studied by Hennink and coworkers (Rijcken et al. 2007). It was shown that the phthalocyanine molecule could be loaded efficiently in the micelles up to a concentration of ca. 2 mg/mL. The resulting nanoparticles had a diameter of ca. 75 nm. UV/Vis and fluorescence spectroscopy measurements indicated that at low concentrations ($\leq$0.05 μM, 0.45 mg/mL polymer), the photosensitizer molecules were molecularly dissolved within the micellar core, whereas Si(sol)$_2$Pc aggregates were formed at higher photosensitizer concentrations. *In vitro* studies with B16F10 and 14C cells showed that the photocytotoxicity of Si(sol)2Pc-loaded micelles (at a photosensitizer molar concentration of 0.05 μM) was similar to free Si(sol)$_2$Pc (IC50 values of 3.0 ± 0.2 nM were obtained in solutions containing 10% serum). The cellular uptake of the highly loaded micelles (with a 10 μM Si(sol)$_2$Pc concentration) was low and independent of the serum concentration (Figure 5.32). The nanoaggregates of Si(sol)$_2$Pc loaded in the micellar core could be released only during hydrolysis-induced micellar dissociation, which was observed after 5 h at pH 8.7 and at body temperature. The authors did not present *in vivo* studies with the particular nanoparticles, but the stability of the highly loaded micellar Si(sol)$_2$Pc formulation in the presence of serum,

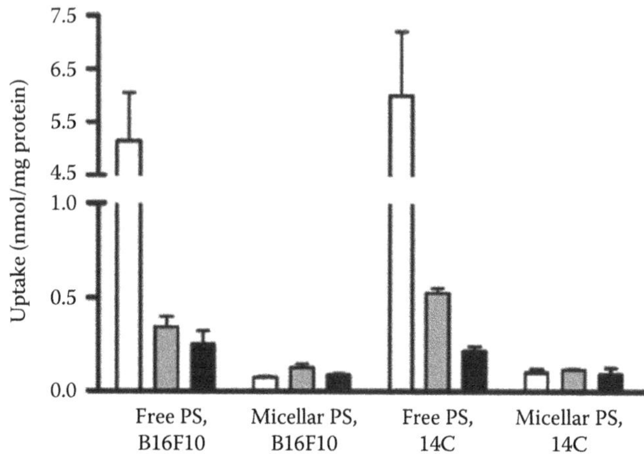

FIGURE 5.32 Cellular uptake of $Si(sol)_2Pc$ (nmol/mg protein) by B16F10 and 14C cells after 6 h of incubation, either administered as free drug or encapsulated in mPEG-*b*-p(HPMAm-Lac_2) micelles (10 μM $Si(sol)_2Pc$, 0.45 mg/mL polymer). White bars represent the uptake from medium without FBS, gray bars represent the uptake from medium supplemented with 10% FBS for B16F10 cells and 5% FBS for 14C cells, and the black bars show the uptake from medium supplemented with 50% or 25% FBS for B16F10 and 14C cells, respectively. (Reprinted from *J. Control. Release*, 124, Rijcken, C.J.F., Hofman, J.-W., van Zeeland, F., Hennink, W.E., and van Nostrum, C.F., 144–153, Copyright 2007, with permission from Elsevier.)

the controlled release of the photosensitizer molecules upon micellar disintegration, and the high photodynamic activity of $Si(sol)_2Pc$ make these micelles interesting candidates for PDT protocols.

Polymeric micelles encapsulating dendrimer phthalocyanine (DPc) as the photosensitizer were developed by Kataoka and coworkers (Nishiyama et al. 2009). These micelles were based on poly(ethylene glycol-*b*-L-lysine) (PEG-*b*-PLL) block copolymer carrying cationic charges and a second generation aryl ether dendrimer with a Zn(II) phthalocyanine center and 32 carboxylic groups on its periphery. Electrostatic complexation between the two building blocks results in complex polyion micelles with the DPc photosensitizer in the core (complexes with PLL blocks of the copolymers) and PEG coronas. The DPc-loaded micelles induced efficient and very rapid light-induced cell death, accompanied by characteristic morphological changes of the tumor cells, such as blebbing of cell membranes, when the cells were photoirradiated using a low-power halogen lamp or a high-power diode laser (Figure 5.33). Fluorescent microscopy observations, using organelle-specific dyes, demonstrated that, following internalization by endocytosis, DPc/PEG-*b*-PLL nanoparticles accumulate in the endo-/lysosomes. However, upon photo-irradiation, DPc/PEG-*b*-PLL is translocated to the cytoplasm and induces photodamage to the mitochondria, which may account for the enhanced photocytotoxicity of DPc/PEG-*b*-PLL nanoparticles. Additional studies demonstrated that DPc/PEG-*b*-PLL showed significantly higher *in vivo* PDT efficacy than the clinically used Photofrin® (polyhematoporphyrin esters, PHEs) in mice bearing human lung adenocarcinoma A549 cells. Furthermore, unlike the PHE-treated mice, the DPc/PEG-*b*-PLL-treated

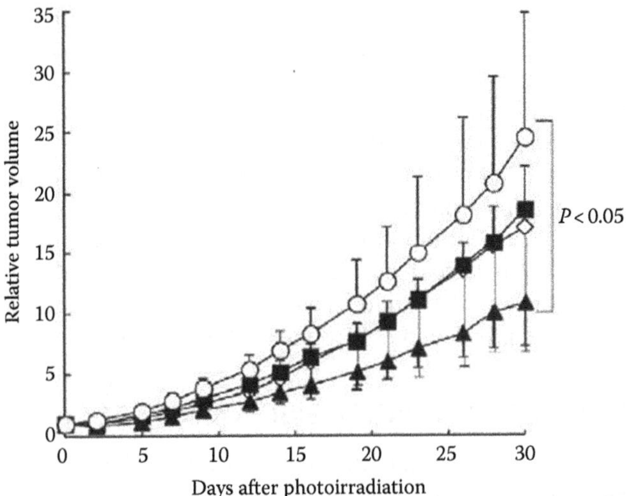

FIGURE 5.33 Growth curves of subcutaneous A549 tumors in control mice (open circle) and mice administered with 0.37 μmol/kg DPc (closed square), 0.37 μmol/kg DPc/m (closed triangle), and 2.7 μmol/kg Photofrin® (PHE) (open diamond) ($n = 6$). Twenty-four hours after administration of photosensitizing agents, the tumors were photoirradiated using a diode laser (fluence: 100 J/cm^2). (Reprinted from *J. Control. Release*, 133, Nishiyama, N., Nakagishi, Y., Morimoto, Y., Lai, P.-S., Miyazaki, K., and Urano, K., 245–251, Copyright 2009, with permission from Elsevier.)

mice showed no sign of skin phototoxicity, under the tested conditions. These results strongly suggest that the DPc/PEG-*b*-PLL nanosystem is expected to serve as an innovative photosensitizer formulation for improved effectiveness and safety of current PDT protocols.

Mixed polymeric micelles based on non-pH-sensitive and pH-sensitive graft copolymers, namely, (poly(*N*-vinylcaprolactam)-g-poly(D,L-lactide) (PVCLm-g-PDLLA) and poly(*N*-vinylcaprolactam-*co*-*N*-vinyl imidazole)-g-poly(D,L-lactide)) (P(VCLm-*co*-NVIM)-g-PDLLA), and PEG-*b*-PDLLA diblock copolymers, with or without fluorescent groups on the corona, were utilized for the encapsulation of protoporphyrin IX (PPIX) for *in vitro* and *in vivo* PDT-oriented studies by Tsai and coworkers (Tsai et al. 2012). Photochemical internalization was utilized to study the localization of pH- and non-pH-sensitive micelles uptake in the lysosome. After nontoxic light treatment, confocal microscopy observations indicated that PPIX molecules encapsulated in pH-sensitive micelles were found in the nucleus, while PPIX encapsulated in the non-pH-sensitive micelles was still localized in the lysosomal compartments. Formation of singlet oxygen was observed for both the block and graft copolymer micelles. Differences in the cell viability were ascribed to the damage occurring at the region where the PPIX was located. *In vivo* studies revealed that PPIX-loaded graft and diblock copolymer micelles presented prolonged blood circulation and enhanced tumor-targeting ability, especially in the case of micelles with targeting functionalities. The PPIX released from g-CIM micelles on tumor sites was also observed by *ex vivo* confocal imaging techniques. Mice treated with non-pH-sensitive micelle/PPIX nanoparticles showed a better repression of tumor growth than mice treated with PPIX alone. The latter result was attributed to the larger amount of photosensitizer localized in the tumor region still exhibiting therapeutic effects. Effective PDT-induced inhibition of tumor growth was found in mice treated with pH-sensitive micelle/PPIX.

Shan and coworkers presented the preparation and properties of hybrid nanoparticles comprised of tetraphenylporphyrin inorganic upconverting phosphors encapsulated within poly(lactic acid)-*b*-poly(ethylene oxide) block copolymers utilizing the flash nanoprecipitation method (Shan et al. 2011). The nanoassemblies were found to be stable under physiological conditions and demonstrated strong cancer cell deactivating abilities via the production of ROS under 980 nm illumination.

5.5 MULTIFUNCTIONAL POLYMERIC NANOPARTICLES

In recent years, the concept of multifunctional block copolymer micelles or polymeric nanoparticles for simultaneous therapy and diagnostics is gaining increased interest among researchers in the field. The basic idea behind these efforts concerns the incorporation into the polymeric nanoassembly or nanoparticle several functionalities/modalities that can be used for the delivery of therapeutic agents and other functional entities that are useful for imaging and diagnostics. This requires the combination of several design principles already discussed in previous sections and makes the construction of such nanodevices more difficult but at the same time more interesting. Some works exemplifying developments in this new rapidly increasing research area are presented as follows.

A nonviral nanoparticle gene carrier was developed by Liu et al. (2011d), and its efficiency for siRNA delivery and transfection was validated both *in vitro* and *in vivo*. The nanocarrier was constructed by a core of IO nanoparticles and a shell of alkylated PEI of 2000 kDa molecular weight (alkyl-PEI2k), abbreviated as alkyl-PEI2k-IO. The hybrid nanocarrier was able to bind with siRNA, resulting in well-dispersed nanoparticles with controlled structure and narrow size distribution. Electrophoresis studies show that the alkyl-PEI2k-IOs could retard siRNA completely at N:P ratios (i.e., PEI nitrogen to nucleic acid phosphate) above 10, protect siRNA from enzymatic degradation in serum, and release complexed siRNA efficiently in the presence of polyanionic heparin. The knockdown efficiency of the siRNA-loaded nanocarriers was tested with 4T1 cells stably expressing luciferase (fluc-4T1) and further with a fluc-4T1 xenograft model. Significant downregulation of luciferase was observed, and unlike high molecular weight analogues, the alkyl-PEI2k-coated IOs demonstrated good biocompatibility, highly efficient delivery of siRNA, and an innocuous toxic profile, making them a potential nanocarrier for gene therapy.

Cao and coworkers described the development of multifunctional polymeric micelles based on poly(ethylene glycol)-*b*-poly(D,L-lactide) amphiphilic block copolymers (Nasongkla et al. 2006). The PEG surface of the micelles was functionalized with cRGD-type peptide with targeting capability toward $\alpha_v\beta_3$-integrins, for controlled drug delivery of DOX to cancer cells. To this nanoconstruct, narrow distribution hydrophobically modified SPIONs were also encapsulated for achieving efficient MRI contrast characteristics. DOX and the SPIONs clusters were loaded successfully

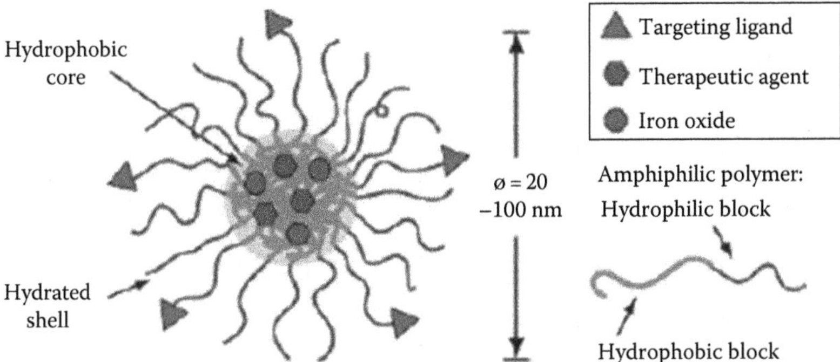

FIGURE 5.34 Multifunctional polymeric nanoparticle platform for targeted drug delivery and MRI. (Reprinted with permission from Nasongkla, N., Bey, E., Ren, J., Ai, H., Khemtong, C., and Guthi, J.S., *Nano Lett.*, 6, 2427–2432, 2006. Copyright 2006, American Chemical Society.)

inside the PDLLA micelle core utilizing a solvent-evaporation methodology, which allowed for a loading of ca. 7 w/w% for the superparamagnetic iron oxide (SPIO) particles and ca. 3 w/w% for DOX, respectively (Figure 5.34). Micellar sizes were in the range 40–50 nm for loaded and empty micelles. The presence of cRGD targeting peptide on the micelle surface resulted in the selective incorporation of the multifunctional micelles and the targeted drug delivery to SLK tumor endothelial cells as evidenced by flow cytometry and CLSM. A 2.5-fold increase of micellar uptake by cells, in comparison to nonfunctionalized micelles, was attributed to the presence of the targeting peptide. *In vitro* MRI and cytotoxicity studies demonstrated the ultrasensitive MRI (down to nanomolar levels) and the $\alpha_v\beta_3$-specific cytotoxic response of these multifunctional hybrid polymeric micelles (Figure 5.35). Longer incubation times were found to lead to an overall decrease of cancer cell viability for all of the micelle samples investigated.

Micellar hybrid nanoparticles for simultaneous magnetofluorescent imaging and anticancer drug delivery were reported by Sailor and coworkers (Park et al. 2008). This group utilized a solvent-evaporation method for the encapsulation of spherical OA-coated IO magnetic nanoparticles of ca. 11 nm diameter and elongated TOP-coated quantum dots (QDs) into PEG-lipid micellar nanoassemblies. The SPIO/QD ratio within the individual micelles could be adjusted by changing the mass ratio of magnetic nanoparticles to QDs during the preparation step. TEM imaging demonstrated the successful construction of the hybrid nanoassemblies. Imaging of the multifunctional nanoassemblies was possible even in the NIR region and in subnanomolar concentrations, despite the observed fluorescent quenching and adsorption due to the presence of the magnetic nanoparticles in close proximity to the QDs.

The ability of the novel multifunctional micelles to target and image in dual-mode (via MRI and fluorescence) tumor cells was tested against MDA-MB-435 human cancer cells. Specific targeting to tumor cells was made possible by chemical attachment of the targeting ligand F3 (a peptide known to target cell-surface nucleolin in endothelial cells in tumor blood vessels and in tumor cells and to become internalized in these cells) on the surface of the micelles (via chemical bond formation with

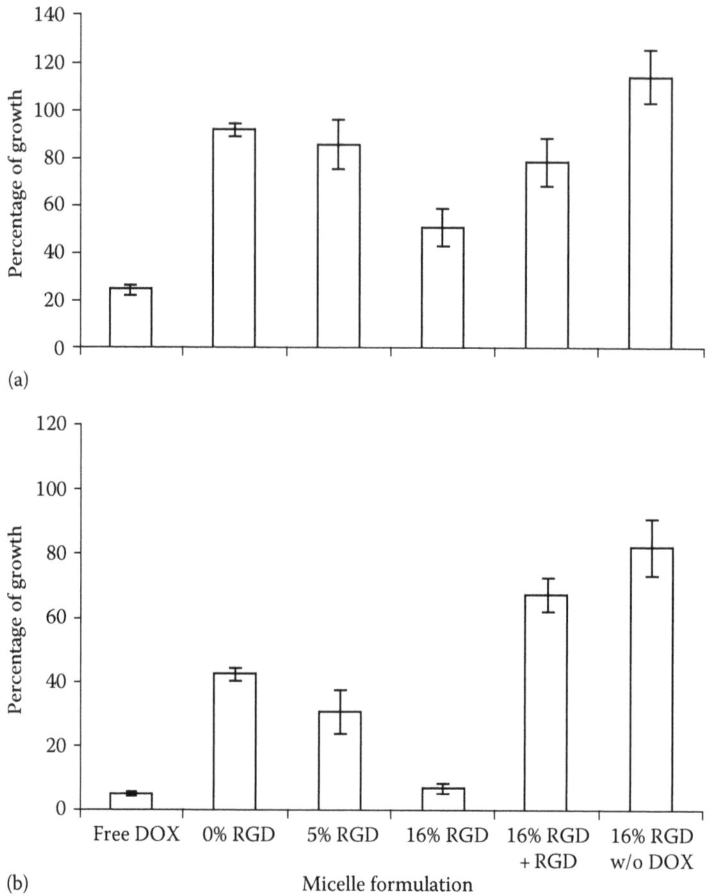

FIGURE 5.35 Inhibition of SLK cell growth in the presence of different formulations of SPIO-loaded micelles with or without 1 μM DOXO concentration after (a) 1 h or (b) 4 h incubation times. The percent inhibition of cell growth was calculated as the ratio of cell number in the treated sample divided by that in the untreated control. (Reprinted with permission from Nasongkla, N., Bey, E., Ren, J., Ai, H., Khemtong, C., and Guthi, J.S., *Nano Lett.*, 6, 2427–2432, 2006. Copyright 2006, American Chemical Society.)

the terminus of the PEG corona chains). Simultaneous imaging and drug delivery were demonstrated with the anticancer drug DOX, which was also loaded into the MHNs during preparation of the hybrid nanoassemblies (up to a DOX/micelle ratio of 0.093:1 w/w). The intrinsic fluorescence of DOX enabled the independent imaging of both DOX and QDs contained in the hybrid micelles. The intact hybrid micelles were observed to colocalize in some areas of MDA-MB-435 cells *in vitro* upon incubation for 2 h. It is interesting to note that although they were composed of relatively toxic QDs, no significant toxicity of the nanoassemblies was observed. On the other hand, F3-functionalized micelles in which DOX was encapsulated displayed significantly greater cytotoxicity than that of equivalent quantities of free DOX or hybrid micelles containing DOX without the targeting ligand. Biodistribution measurements on mice indicated that the multifunctional micelles were accumulated mainly in the liver, with no significant quantities being observed in other organs.

The preparation of organic–inorganic hybrid micelles by the comicellization of poly(ε-caprolactone)-*b*-poly(glycerol monomethacrylate) (PCL-*b*-PGMA) and poly(ε-caprolactone)-*b*-poly(oligoethylene glycol methacrylate-*co*-oligoethylene glycol methacrylate folic acid conjugate) (PCL-*b*-P(OEGMA-*co*-FA)) amphiphilic block copolymers, which physically encapsulate hydrophobic drugs within their micellar cores and with SPIONs embedded within their hydrophilic coronas, has been reported by Liu and coworkers (Hu et al. 2012). These nanoparticles possess integrated functions for chemotherapeutic drug delivery and MRI contrast enhancement. The model hydrophobic anticancer drug, PTX, and 4 nm SPIONs were loaded into micellar cores and hydrophilic coronas, respectively, taking advantage of the hydrophobicity of micellar cores and strong affinity between 1,2-diol moieties in PGMA and Fe atoms at the surface of SPIONs. A drug-loading content of ca. 8.5 w/w% was possible. Controlled and sustained release of PTX from the hybrid micelles was achieved, exhibiting a cumulative release of ca. 61% of encapsulated drug over a period of 130 h. The clustering of magnetic nanoparticles within the coronas led to considerably enhanced T_2 relaxivity, strongly suggesting that the particular hybrid micelles can serve as a T_2-weighted MRI contrast enhancer with improved performance. Preliminary *in vivo* MRI experiments indicated that this type of hybrid multifunctional polymeric micellar nanoparticles can act as a new nanoplatform integrating targeted drug delivery, controlled release, and disease diagnostic functions.

The same group also reported on the fabrication of an analogous mixed block copolymer micellar system composed of two types of amphiphilic diblock copolymers, namely, PCL-*b*-P(OEGMA-FA) and PCL-*b*-P(OEGMA-Gd), consisting of a hydrophobic PCL block and a hydrophilic poly(oligo(ethylene glycol) monomethyl ether methacrylate) (POEGMA) block, to which some segments carried covalently attached FA moieties and DOTA-Gd (Gd) side groups (Liu et al. 2012b). FA groups and Gd complexes provide synergistic functions for targeted delivery and MRI contrast enhancement. Loading of hydrophobic chemotherapeutic drugs, like PTX, could be achieved within the hydrophobic PCL cores of the mixed micelles. The as-prepared nanosized mixed micelles were capable of physically encapsulating PTX at a loading content of ca. 5.0 w/w%, exhibiting controlled release of up to ca. 60% of loaded drug and for a time period of almost 130 h. *In vitro* cell viability assays indicated that drug-free mixed micelles are almost noncytotoxic up to a concentration of 0.2 g/L, whereas PTX-loaded micelles could effectively kill HeLa cells at the same concentration. *In vitro* MRI experiments revealed an increased T_1 relaxivity (almost sevenfold increase) for mixed micelles compared to that of small molecular weight complex counterpart, alkynyl-DOTA-Gd. Additional, *in vivo* MRI experiments in rabbits indicated considerably enhanced signal intensity, prominent positive contrast enhancement, improved accumulation and retention, and extended blood circulation duration for FA-labeled mixed micellar nanoparticles within the rabbit liver, as compared to those for FA-free mixed micelles and alkynyl-DOTA-Gd complexes.

Micelles from fluorine-containing amphiphilic poly(hexafluorobutyl methacrylate-g-oligoethylene glycol methacrylate) (PHFMA-g-PEGMA) graft copolymers have been also utilized for the encapsulation of magnetic nanoparticles and a hydrophobic drug simultaneously (Li et al. 2012b). Encapsulated OA-modified magnetite (Fe_3O_4) nanoparticles were found to form clusters in the PHFMA-g-PEGMA micellar cores with a mean diameter of 100 nm, and the hybrid micelles showed high stability in aqueous media due to the highly hydrophobic fluorine segments in the graft copolymers that enhance the stability of the micelles. The hybrid micelles showed good cytocompatibility based on MTT cytotoxicity assay, and they possessed paramagnetic properties with saturation magnetization of ca. 17.14 emu/g. The hydrophobic drug 5-fluorouracil could be loaded into the hybrid micellar nanoparticles, using an emulsion method, with a loading efficiency of ca. 21 wt%. Controlled release of the drug was also achieved. The magnetic micelles had satisfactory transverse relaxivity rates and exhibited high efficacy as a negative MRI agent in T_2-weighted imaging. *In vivo* MRI studies demonstrated that the contrast between liver and spleen was enhanced by the magnetic copolymer micelles.

An interesting hybrid star poly(*N*-(2-hydroxypropyl)methacrylamide) (PHPMA)-based nanoplatform for simultaneous therapy and diagnosis was presented by Liu and coworkers (Liu et al. 2012c). pH-disintegrable micellar nanoparticles were fabricated from asymmetrically functionalized β-cyclodextrin (β-CD) core PHPMA star copolymers covalently conjugated with DOX, FA, and DOTA-Gd moieties for integrated cancer cell-targeted drug delivery and MRI contrast enhancement. The functionalized β-CD cores with 7 azide functionalities and 14 α-bromopropionate moieties, $(N_3)_7$-CD-$(Br)_{14}$, were used for the ATRP polymerization of HPMA through the bromide moieties and the click reaction with alkynyl-(DOTA-Gd) complex. DOX and FA groups were covalently bonded to the hydroxyl groups of HPMA segments. These reactions resulted in the formation of a (DOTA-Gd)$_7$-CD-(PHPMA-FA-DOX)$_{14}$ star copolymer possessing 7 DOTA-Gd complex moieties and 14 PHPMA arms carrying DOX and FA functionalities via acid-labile carbamate linkages and ester bonds, respectively. The covalent fixation of DOX molecules onto the PHPMA star copolymer arms (with ca. 14 wt% drug-loading content) makes the initially hydrophilic polymer amphiphilic, leading to the formation of micellar nanoparticles with sizes of several tens of nanometers in aqueous solution and at pH 7.4. The *in vitro* obtained DOX release profile was highly pH dependent (Figure 5.36). For a time period of 42 h, cumulative releases of 10%, 53%, and 89% conjugated DOX at pH 7.4, 5.0, and 4.0, respectively, were observed. Additionally, the pH-modulated release of conjugated DOX from micellar nanoparticles is accompanied with micelle disintegration due to the loss of the hydrophobic component in the star copolymer molecule. *In vitro* cell viability assays revealed that the (DOTA-Gd)$_7$-CD-(PHPMA$_{15}$)$_{14}$ star copolymer was almost noncytotoxic up to a concentration of 0.5 g/L, whereas DOX-conjugated micellar nanoparticles of (DOTA-Gd)$_7$-CD-(PHPMA-FA-DOX)$_{14}$ could effectively enter and kill HeLa cells at a concentration higher than 80 mg/L. *In vitro* MRI experiments showed an enhanced T_1 relaxivity for micellar nanoparticles compared to that for the neat alkynyl-DOTA-Gd complex. An *in vivo* MR imaging assay was also made in rats. The experiments revealed

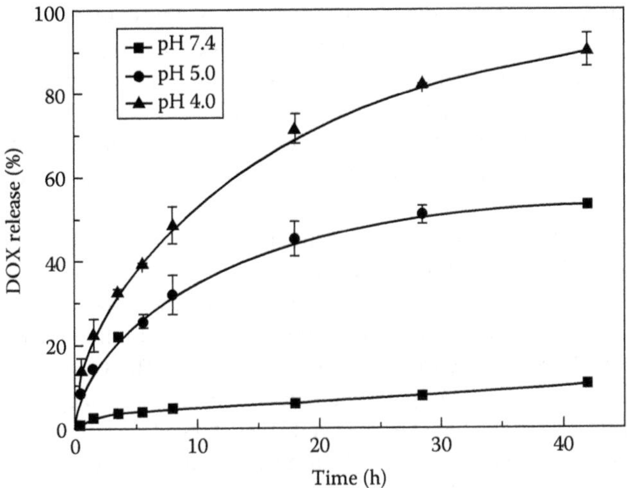

FIGURE 5.36 *In vitro* DOX release profiles at varying pH conditions (37°C, 20 mM buffer solution) from multifunctional micellar nanoparticles of (DOTA-Gd)7-CD-(PHPMA-FADOX) 14-star copolymers. (Reprinted from *Biomaterials*, 33, Liu, T., Li, X., Qian, Y., Hua, X., and Liu, S., 2521–2531, Copyright 2012c, with permission from Elsevier.)

an enhanced accumulation of hybrid copolymer nanoparticles within the liver and kidney of the animals and prominent positive contrast enhancement.

Novel hybrid magnetic/drug nanocarriers were prepared via *in situ* synthesis of magnetic IO nanocrystallites in the presence of poly(methacrylic acid)-graft-poly(ethylene glycol methacrylate) (p(MAA-*g*-EGMA)) able for interaction-induced self-assembly into hybrid micelles (Bakandritsos et al. 2012). The particular hybrid nanoparticles possessed bio-repellent properties due to the presence of EGMA macromonomer segments, pronounced magnetic response, and high loading capacity for the anticancer drug DOX, to an extent not observed before in such hybrid colloids. High magnetic responses were accomplished by engineering the size of the magnetic nanocrystallites (to ca. 13.5 nm) following an aqueous single-ferrous precursor route and through adjustment of the number of the cores in each colloidal assembly. The magnetic response of the hybrid nanocarriers was evaluated by conventional magnetometry and magnetophoretic experiments providing insight on the internal organization of the inorganic particles in respect to the organic/polymeric part and on their response to magnetic manipulation. The structural organization of the graft copolymer, locked on the surface of the IO nanocrystallites, was further probed by small-angle neutron scattering on single-cored colloids. Analysis showed that the MAA segments are selectively populating the area around the magnetic nanocrystallites, while the PEG-grafted chains are arranged as protrusions, pointing toward the aqueous environment, a structure similar to conventional linear block copolymer micelles. The nanocarriers were found to be stable at various pH and at highly salted aqueous media, as evidenced by DLS and electrophoretic light scattering measurements. Their colloidal stability was enhanced, as compared to uncoated nanocrystallites, presumably due to the presence of the external protective PEG canopy. The nanoparticles evaluated in terms of their bio-repellent properties, by assaying their stability using human blood plasma as the surrounding medium. DOX loading was achieved in the protonated form of the drug through physical adsorption onto the nanoparticles' core and reached 22 wt%. Initial experiments showed that DOX could be released in a sustainable manner in a number of aqueous media including blood serum.

A multifunctional nanostructure of nanometer dimensions self-assembled from a hydrophobic poly(DL-lactide-*co*-glycolide) (PLGA) core and a hydrophilic paramagnetic-folate-coated PEGylated lipid shell (PFPL; PEG) possessing simultaneous MRI and targeted therapeutics properties was also presented (Liao et al. 2011). The nanoassembly was found to have a well-defined core/shell structure as revealed by CLSM observations. The paramagnetic diethylenetriaminepentaacetic acid-gadolinium (DTPA-Gd) chelated to the shell layer showed significantly higher spin–lattice relaxivity (r_1) than the clinically used low molecular weight MRI contrast agent Magnevist. The PLGA core served as a nanocontainer to load and release the hydrophobic drugs. Drug-release studies indicated that the modification of the PLGA core with a polymeric liposome shell can reduce the drug-release rate. Folate functionalities enhanced cellular uptake of the nanocomplex compared to the non-functionalized analogue.

Shuai and coworkers presented the case of a multifunctional block copolymer micellar nanosystem for simultaneous tumor-targeted intracellular drug release and fluorescent imaging (Wang et al. 2012c). The micelles were formed by a triblock-like copolymer comprised of PEG, poly(*N*-(*N′*,*N′*-diisopropylaminoethyl) aspartamide), and a cholic acid hydrophobic relatively large end group (PEG-b-PAsp(DIP)-b-CA). The copolymer formed micelles in aqueous media, at pH 5.0, with a CA core. PAsp(DIP) forms a pH-sensitive interlayer that is expected to undergo hydrophobic/hydrophilic transition in the acidic lysosomes of cells. PTX and QDs were then incorporated into the copolymer nanoassembly. PTX (or fluorescein diacetate) was loaded due to hydrophobic interactions into the CA core, reaching a loading cargo of 2.6 wt%. The negatively charged QDs were incorporated in the positively charged PAsp(DIP) layer through electrostatic interaction (loading content 2.7%). The assembly had a size ca. 50 nm. At pH 7.4, the PAsp(DIP) layer was partially solvated and could act as an "on-off" pH switch controlling the release of the encapsulated drug. PTX release was very small at pH 7.4 (less than 5%), while at pH 5.0, a rapid release of the drug was observed. Folate groups were also incorporated at the PEG terminus for active targeting. *In vitro* experiments with Bel-7402 cells as well as in rats demonstrated the great potential of these hybrid micellar nanocarriers for anticancer therapy and imaging.

It is important to discuss the case of multifunctional nanocarriers able to deliver two different types of therapeutic agents (Figure 5.37). In this context, Yu and coworkers utilized pH-responsive block copolymer micelles from poly(2-(dimethylamino) ethyl methacrylate)-block-poly(2-(diisopropylamino)ethyl methacrylate) (PDMA-*b*-PDPA) diblock copolymers for the intracellular (endosomal) delivery of AmB and siRNA (Yu et al. 2011). AmB was loaded into the hydrophobic PDPA core at pH 7.4, and siRNA was complexed with the positively charged PDMA shell to form the micelleplexes. CLSM experiments showed the cellular uptake of the micellar

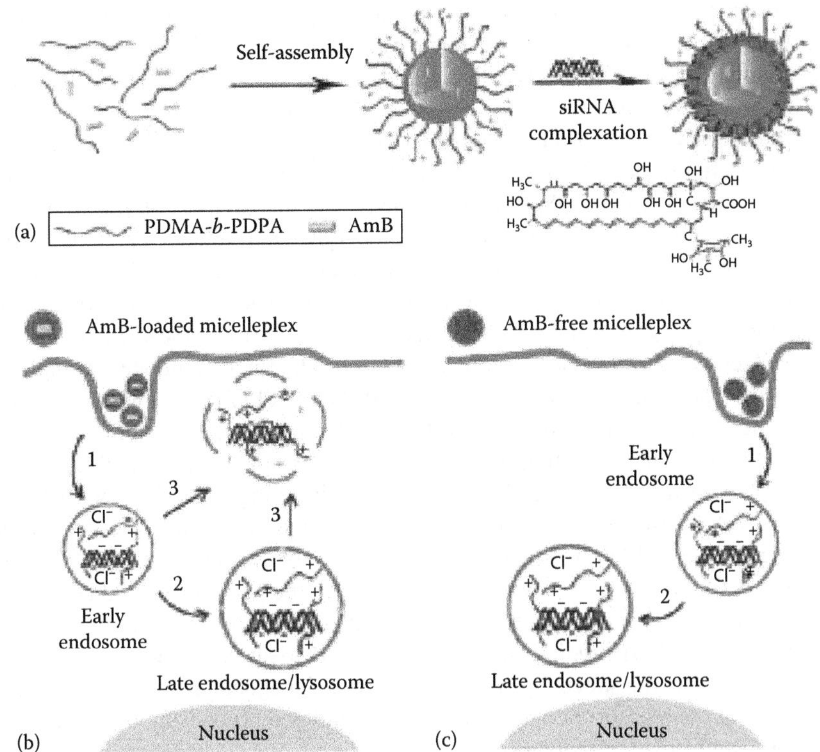

FIGURE 5.37 Preparation and action of AmB-loaded dual pH-responsive micelleplexes for siRNA delivery with enhanced siRNA endosomal escape ability. (a) Production of AmB-loaded PDMA-b-PDPA micelleplexes. AmB was loaded in the hydrophobic PDPA core, and siRNA was complexed with the PDMA corona shell. (b) AmB-facilitated endosome disruption and siRNA cytoplasmic release (1: AmB-loaded micelleplexes dissociated in early endosomes after cell uptake, and AmB molecules are inserted into endosomal membranes; 2: protonated PDMA-*b*-PDPA unimers complexed with siRNA and trafficked from early endosomes into late endosome/lysosomes, causing vesicle swelling; 3: AmB enhanced siRNA release from endosomes into cytoplasm via membrane destabilization). (c) In the case of AmB-free micelleplexes, polymer/siRNA complexes were entrapped in late endosomes or lysosomes without efficient cytoplasmic siRNA release. (Reprinted with permission from Yu, H., Zou, Y., Wang, Y., Huang, X., Huang, G., Sumer, B.D., and Boothman, D.A., and Gao, J., *ACS Nano*, 5, 9246–9255, 2011. Copyright 2011, American Chemical Society.)

nanoparticles. The PDMA-b-PDPA/siRNA micelleplexes were found to dissociate in early endosomes, and AmB was released from the nanocarriers. Live-cell imaging studies demonstrated that released AmB significantly increased the ability of siRNA to overcome the endosomal barrier, due to the ability of the drug to increase membrane permeability through the formation of transmembrane pores. Transfection studies also showed that AmB-loaded micelleplexes resulted in significant increase in luciferase knockdown efficiency over the AmB-free control. The enhanced luciferase knockdown efficiency was abolished by bafilomycin A1, a vacuolar ATPase inhibitor that inhibits the acidification of the endocytic organelles. The reported results support the basic hypothesis that membrane poration by AmB and increased endosomal swelling and membrane tension by a "proton sponge"–type polymer (like the particular PDMA-b-PDPA copolymer) provided a synergistic strategy to disrupt endosomes for improved intracellular delivery of siRNA.

A rather complex nanosystem composed of OA surface–modified magnetic nanoparticles, the cationic poly[2-(dimethylamino)ethyl methacrylate] end capped with a cholesterol moiety (Chol-PDMAEMA30), DNA, and the anionic poly[poly(ethylene glycol)methyl ether methacrylate]-*b*-poly(methacrylic acid) diblock carrying partial mercapto groups on the poly(methacrylic acid) block (PPEGMA-*b*-PMAASH), has been reported by Hao et al. (2012). The magnetic nanoparticles were first coated with the hydrophobically modified Chol-PDMAEMA30 polycation and then complexed with DNA chains through electrostatic interaction. The three component complexes were further physically binded with the brush-type PPEGMA-*b*-PMAASH polyanion. The resulting magnetic particle/DNA/polyion complexes could be further stabilized against disintegration by oxidizing the mercapto groups of PPEGMA-*b*-PMAASH to form a cross-linked PMAA shell with bridging disulfide bonds between the PPEGMA-*b*-PMAASH outer chains. It was shown that the presence of peripheral PPEGMA-*b*-PMAASH chains on the gene vector can reduce the nonspecific adsorption of blood proteins. The disulfide bonds on the periphery could be cleaved in a cell reducing environment enhancing the release of DNA. The combined results of zeta-potential, DNA-binding capacity, cytotoxicity assay, and *in vitro* transfection tests indicated that the specific gene vector possessed magnetic responsiveness, anti-nonspecific protein adsorption, low cytotoxicity, and good transfection efficiency on HEK 293T and HeLa cells.

A block copolymer–based nanosystem capable of synergistic chemotherapy and PDT toward cancer cells has been presented by Shieh and coworkers (Peng et al. 2008). Amphiphilic 4-armed star-shaped 5,10,15,20-tetrakis (4-aminophenyl)-21H,23H-chlorin (TAPC)-core block copolymers based on methoxy poly(ethylene glycol) (mPEG) and PCL were synthesized and studied. The photosensitizer-centered amphiphilic star-block copolymer formed micelles in aqueous media (CSBC micelles) that could encapsulate PTX in their hydrophobic inner PCL cores (PTX-loaded CSBC micelles denoted as PCSBC micelles). The star-block copolymer micelles exhibited efficient singlet oxygen generation, whereas the hydrophobic photosensitizer alone (without the diblock arms conjugation) failed, due to considerable aggregation in aqueous solution. The chlorin-core micelles alone exhibited obvious phototoxicity in MCF-7 breast cancer cells with 7 or 14 J/cm^2 light irradiation at a chlorin concentration of 125 mg/mL. After PTX loading, the size of the micelles increased from ca. 70 to 103 nm as a result of the incorporation of the hydrophobic drug. The PTX-loaded micelles were found to have an improved cytotoxicity toward MCF-7 cells after irradiation, most likely through a synergistic chemotherapy/PDT effect (Figure 5.38).

The same group also reported a star-block copolymer with a TAPC core and poly(ethylene glycol-*b*-(ε-caprolactone)) arms (CSBC), where the PCL blocks are directly connected to the core, that self-assembled into nanosized micelles, which acted as nanocarriers for the photosensitizing core moiety and the hydrophobic anticancer drug SN-38 (7-ethyl-10-hydroxy-camptothecin) (Peng et al. 2009b).

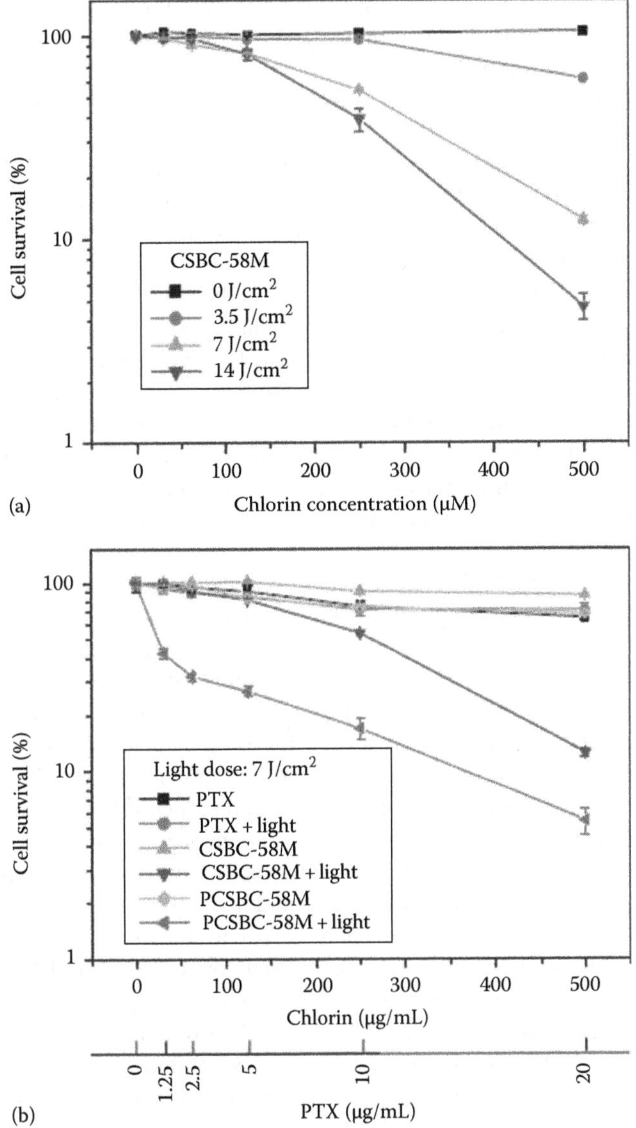

FIGURE 5.38 (a) Cytotoxicity of CSBC-58M under different irradiation and (b) PCSBC-58M in MCF-7 cells with 7 J/cm^2 irradiation. (Reprinted from *Biomaterials*, 29, Peng, C.-L., Shieh, M.-J., Tsai, M.-H., Chang, C.-C., and Lai, P.-S., 3599–3608, Copyright 2008, with permission from Elsevier.)

SN-38 was encapsulated into the star-block micelles by using a lyophilization–hydration protocol. Experiments showed a prolonged plasma residence time of SN-38/CSBC micelles as compared to free CPT-11, permitting the increased tumor accumulation and consequently improved antitumor activity. The combined effects of SN-38/CSBC-loaded micelles were evaluated in an HT-29 human colon cancer xenograft model. SN-38/CSBC-mediated simultaneous chemotherapy and PDT action successfully inhibited tumor growth, resulting in up to 60% complete regression of well-established tumors after three consecutive treatments. These treatments also decreased the microvessel density and cell proliferation within the subcutaneous tumors.

A similar system was also reported more recently by Zhang et al. (2012b). They synthesized a star-shaped PCL-*b*-PEO amphiphilic copolymer with a tetrakis-(4-aminophenyl)-terminated porphyrin core. PTX was *in situ* encapsulated into the star-block copolymer micelles. The fluorescent characteristic of the central porphyrin moiety allowed for monitoring the cellular uptake and biodistribution of the PTX-loaded micelles by fluorescent imaging in cell and live mice. It was concluded that the PTX-loaded micelles could be readily internalized by cancer cells and they showed a slightly higher cytotoxicity than the clinical PTX injection Taxol. *In vivo* real-time fluorescent imaging revealed that the micelles could accumulate at tumor sites via the blood circulation in tumor-bearing mice. *In vivo* antitumor efficacy examinations indicated that the PTX-loaded micelles had significantly superior efficacy in impeding tumor growth than Taxol and low toxicity to the living mice.

Micelles from amphiphilic star-block copolymers, having a hydrophobic hyperbranched core and amphiphilic fluoropolymer arms, were constructed as drug delivery agent assemblies as well as potential nanoparticulate MRI contrast agents (Du et al. 2008b). Several polymer structures were prepared from consecutive copolymerizations of 4-chloromethylstyrene with dodecyl acrylate and then 1,1,1-trifluoroethyl methacrylate with tBA, followed by acidolysis in order to produce the hydrophilic acrylic acid residues on the polymer chains. These polymer chains were labeled with cascade blue as a fluorescence reporter, allowing for a simultaneous optical imaging function of the nanoparticles. The series of materials differed primarily in the ratio of 1,1,1-trifluoroethyl methacrylate to acrylic acid units, to give differences in fluorine loading and hydrophobicity/hydrophilicity balance. DOX was used as a model hydrophobic drug to study the loading, release, and cytotoxicity of these polymer micellar constructs on an U87-MG-EGFRvIII-CBR cell line. The micelles (with TEM-measured diameters ranging 5–9 nm and DLS-measured hydrodynamic diameters 20–30 nm) had rather low loading capacities of ca. 4 wt% of DOX. The DOX-loaded micelles exhibited cytotoxicity with cell viabilities of 60%–25% at 1.0 μg/mL effective DOX concentrations, depending upon the polymer composition, as determined by MTT assays. These cell viability values are comparable to that of free DOX, suggesting an effective release of the cargo and delivery to the cell nuclei, which was further confirmed by fluorescence microscopy of the cells. ^{19}F-NMR spectroscopy indicated a partial degradation of the surface-available trifluoroethyl ester linkages of the micelles, which may have caused acceleration of the release of DOX. ^{19}F-NMR spectroscopy was also employed to confirm and to quantify the cell uptake of the drug-loaded micelles. Evidently, these dual-fluorescent and ^{19}F-labeled and chemically functional micelle carriers may be used in a variety of applications, such as cell labeling, imaging, and therapeutic delivery.

REFERENCES

Abdelhady, H. G., Allen, S., Davies, M. C., Roberts, C. J., Tendler, S. J., Williams, P. M. 2003. *Nucleic Acids Res*. 31(14):4001–4005.

Ahmed, M., Narain, R. 2011. *Biomaterials* 32(22):5279–5290.

Ahmed, M., Narain, R. 2012. *Biomaterials* 33(15):3990–4001.

Akagi, T., Baba, M., Akashi, M. 2007. *Polymer* 48:6729–6747.

Akbulut, M., Ginart, P., Gindy, M. E., Theriault, C., Chin, K. H., Soboyejo, W., Prud'homme, R. K. 2009. *Adv. Funct. Mater*. 19:718–725.

Alani, A. W., Bae, Y., Rao, D. A., Kwon, G. S. 2010. *Biomaterials* 31:1765–1772.

Aldawsari, H., Edrada-Ebel, R., Blatchford, D. R., Tate, R. J., Tetley, L., Dufes, C. 2011. *Biomaterials* 32(25):5889–5899.

Anderson, K., Sizovs, A., Cortez, M., Waldron, C., Haddleton, D. M., Reineke, T. M. 2012. *Biomacromolecules* 13(8):2229–2239.

Ando, S., Putnam, D., Pack, D. W., Langer, R. 1999. *J. Pharm. Sci*. 88(1):126–130.

Arifin, D. R., Palmer, A. F. 2005. *Biomacromolecules* 6:2172.

Arima, H., Tsutsumi, T., Yoshimatsu, A., Ikeda, H., Motoyama, K., Higashi, T., Hirayama, F., Uekama, K. 2011. *Eur. J. Pharm. Sci*. 44(3):375–384.

Ayame, H., Morimoto, N., Akiyoshi, K. 2008. *Bioconj. Chem*. 19:882–890.

Bae, Y., Fukushima, S., Harada, A., Kataoka, K. 2003. *Angew. Chem. Int. Ed*. 42:4640–4646.

Bakandritsos, A., Papagiannopoulos, A., Anagnostou, E. N., Avgoustakis, K., Zboril, R., Pispas, S., Tucek, J., Ryukhtin, V., Bouropoulos, N., Kolokithas-Ntoukas, A., Steriotis, T. A., Keiderling, U., Winnefeld, F. 2012. *Small* 8:2381–2393.

Barton, N. W., et al. 1991. *N. Engl. J. Med*. 324:1464–1470.

Bastakoti, B. P., Guragain, S., Yoneda, A., Yokoyama, Y., Yusab, S., Nakashima, K. 2010. *Polym. Chem*. 1:347–353.

Bauhuber, S., Lieble, R., Tomasetti, L., Rachel, R., Goepferich, A., Breunig, M. 2012. *J. Control. Release* 162(2):446–455.

Bergen, J. M., Pun, S. H. 2005. *MRS Bull*. 30(9):663–667.

Bisht, H. S., Manickam, D. S., You, Y. Z., Oupicky, D. 2006. *Biomacromolecules* 7:1169–1178.

Boyce, N. 2001. *Nature* 414:677.

Bremner, K. H., Seymour, L. W., Logan, A., Read, M. L. 2004. *Bioconjug. Chem*. 15(1):152–161.

Brunner, S., Sauer, T., Carotta, S., Cotton, M., Saltik, M., Wagner, E. 2000. *Gene Ther*. 7(5):401–407.

Budijono, S. J., Shan, J., Yao, N., Miura, Y., Hoye, T., Austin, R. H., Ju, Y., Prud'homme, R. K. 2010. *Chem. Mater*. 22:311–318.

Bulmus, V., Woodward, M., Lin, L., Murthy, N., Stayton, P., Hoffman, A. 2003. *J. Control. Release* 93:105–120.

Cabane, E., Malinova, V., Menon, S., Palivan, C. G., Meier, W. 2011. *Soft Matter* 7(19):9167–9176.

Check, E. 2002. *Nature* 420:116–118.

Chen, T.-H., Bae, Y., Furgeson, D. Y., Kwon, G. S. 2012a. *Int. J. Pharm*. 427(1):105–112.

Chen, Q., Osada, K., Ishii, T., Oba, M., Uchida, S., Tockary, T. A., Endo, T., Ge, Z., Kinoh, H., Kano, M. R., Itaka, K., Kataoka, K. 2012b. *Biomaterials* 33(18):4722–4730.

Chen, Z., Yu, D., Liu, C., Yang, X., Zhang, N., Ma, C. H., Song, J. B. Lu, Z. J. 2011. *J. Drug Targeting* 19:657–665.

Cheng, C., Convertine, A. J., Stayton, P. S., Bryers, J. D. 2012. *Biomaterials* 33(28):6868–6876.

Cho, H. Y., Srinivasan, A., Hong, J., Hsu, E., Liu, S., Arun Shrivats, A., Kwak, D. K., Bohaty, A. K., Paik, H., Jeffrey, O. Hollinger, J. O., Matyjaszewski, K. 2011. *Biomacromolecules* 12:3478–3486.

Choi, J. H., Joung, Y. K., Bae, J. W., Choi, J. W., Park, K. D. 2011. *Macromol. Res.* 19(2):180–188.

Christian, D. A., Cai, S., Bowen, D. M., Kim, Y., Pajerowski, J. D., Discher, D. E. 2009. *Eur. J. Pharm. Biopharm.* 71:463–474.

Chytil, P., Etrych, T., Kostka, L., Ulbrich, K. 2012. *Macromol. Chem. Phys.* 213:858–867.

Conner, S., Schmid, S. 2003. *Nature* 422(6927):37–44.

Convertine, A. J., Diab, C., Prieve, M., Paschal, A., Hoffman, A. S., Johnson, P. H., Stayton, P. S. 2010. *Biomacromolecules* 11:2904–2911.

Dash, B. C., Mahor, S., Carroll, O., Mathew, A., Wang, W., Woodhouse, K. A., Pandit, A. 2011. *J. Control. Release* 152(3):382–392.

Dauty, E., Verkman, A. S. 2005. *J. Biol. Chem.* 280(9):7823–7828.

Debus, H., Baumhof, P., Probst, J., Kissel, T. 2011. *J. Control. Release* 148(3):334–343.

De Duve, C., Wattiaux, R. 1966. *Annu. Rev. Physiol.* 28:435–492.

Dolmans, D., Fukumura, D., Jain, R. K. 2003. *Nat. Rev. Cancer* 3:380–387.

van Dongen, S. F. M., Verdurmen, W. P. R., Peters, R. J. R. W., Nolte, R. J. M., Brock, R., van Hest, J. C. M. 2010. *Angew. Chem. Int. Ed.* 49:7213–7216.

Du, F.-S., Wang, Y., Zhang, R., Li, Z.-C. 2010. *Soft Matter* 6:835–848.

Du, J., Jin, J., Yan, M., Lu, Y. 2012. *Curr. Drug Metabol.* 13:82–92.

Du, W., Nystrom, A. M., Zhang, L., Powell, K. T., Li, Y., Cheng, C., Wickline, S. A., Wooley, K. L. 2008a. *Biomacromolecules* 9:2826–2833.

Du, W., Xu, Z., Nystrom, A. M., Zhang, K., Leonard, J. R., Wooley, K. L. 2008b. *Bioconj. Chem.* 19:2492–2498.

van Duijnhoven, F. H., Aalbers, R., Rovers, J. P., Terpstra, O. T., Kuppen, P. J. K. 2003. *Immunobiology* 207:105–113.

El-Andaloussi, S., Holm, T., Langel, U. 2005. *Curr. Pharm. Des.* 11(28):3597–3611.

Falamarzian, A., Lavasanifar, A. 2010a. *Macromol. Biosci.* 10:648–656.

Falamarzian, A., Lavasanifar, A. 2010b. *Colloids Surf. B Biointerfaces* 81:313–320.

Farokhzad, O. C., Jon, S., Khademhosseini, A., Tran, T. N., Lavan, D. A, Langer, R. 2004. *Cancer Res.* 64:7668–7675.

Fu, K., Pack, D. W., Klibanov, A. M., Langer, R. 2000. *Pharm. Res.* 17(1):100–106.

Gao, G. H., Heo, H., Lee, J. H., Lee, D. S. 2010. *J. Mater. Chem.* 20:5454–5461.

Gao, G. H., Lee, J. W., Nguyen, M. K., Im, G. H., Yang, J., Heo, H., Jeon. P., Park, T. G., Lee, J. H., Lee, D. S. 2011b. *J. Control. Release* 155:11–17.

Gao, H., Lu, X., Ma, Y., Yang, Y., Li, J., Wu, G., Wang, Y., Fan, Y., Ma, J. 2011a. *Soft Matter* 7(19):9239–9247.

Gardikis, K., Dimas, K., Georgopoulos, A., Kaditi, E., Pispas, S., Demetzos, C. 2010. *Curr. Nanosci.* 6:277–284.

Gebhart, C. L., Kabanov, A. V. 2001. *J. Control. Release* 73(2–3):401–416.

von Gersdorff, K., Sanders, N., Vandenbroucke, R., De Smedt, S., Wagner, E., Ogris, M. 2006. *Mol. Ther.* 14(5):745–753.

Ghoroghchian, P. P., Frail, P. R., Susumu, K., Park, T.-H., Wu, S. P., Uyeda, H. T., Hammer, D. A., Therien, M. J. 2005. *J. Am. Chem. Soc.* 127:15388–15390.

Giannotti, M. I., Esteban, O., Oliva, M., Garcia-Parajo, M. F., Sanz, F. 2011. *Biomacromolecules* 12(7):2524–2533.

Gillies, E. R., Frechet, J. M. J. 2005. *Bioconj. Chem.* 16:361–366.

Grenha, A., Gomes, M. E., Rodrigues, M., Santo, V. E., Mano, J. F., Neves, N. M., Reis, R. L. 2010. *J. Biomed. Mater. Res. Part A* 92(4):1265–1272.

Grosse, S., Thevenot, G., Monsigny, M., Fajac, I. 2006. *J. Gene Med.* 8(7):845–851.

Haberland, A., Bottger, M. 2005. *Biotechnol. Appl. Biochem.* 42(Pt 2):97–106.

Han, L., Zhang, A., Wang, H., Pu, P., Kang, C., Chang, J. 2011. *J. Appl. Polym. Sci.* 121(6):3446–3454.

Hao, Y., Zhang, M., He, J., Ni, P. 2012. *Langmuir* 28:6448–6460.

Harada, A., Kataoka, K. 1998. *Macromolecules* 31:288.
Harada, Y., Yamamoto, T., Sakai, M., Saiki, T., Kawano, K., Maitani, Y., Yokoyama, M. 2011. *Int. J. Pharm.* 404:271–280.
Hardingham, T. E., Fosang, A. J. 1992. *FASEB J.* 6(3):861–870.
He, C., Yin, L., Tang, C., Yin, C. 2012. *Biomaterials* 33(33):8569–8578.
Hedley, M. Gene delivery using poly(lactide-co-glycolide) microspheres. In: Amiji M. M., ed. *Polymeric Gene Delivery: Principles and Applications.* Boca Raton, FL: CRC Press; 2005, pp. 451–466.
Hemp, S. T., Allen Jr., M. H., Green, M. D., Long, T. E. 2012a. *Biomacromolecules* 13(1):231–238.
Hemp, S. T., Smith, A. E., Bryson, J. M., Allen, M. H., Long, T. E. 2012b. *Biomacromolecules* 13(8):2439–2445.
Hood, E., Simone, E., Wattamwar, P., Dziubla, T., Muzykantov, V. 2011. *Nanomedicine* 6(7):157–172.
Horgan, A. M., Moore, J. D., Noble, J. E., Worsley, G. J. 2010. *Trends Biotechnol.* 28:485–494.
Hsu, J., Northrup, L., Bhowmick, T., Muro, S. 2012. *Nanomed. Nanotechnol. Biol. Med.* 8(5):731–739.
Hsu, J., Serrano, D., Bhowmick, T., Kumar, K., Shen, Y., Kuo, Y. C., Garnacho, C., Muro, S. 2011. *J. Control. Release* 149(3):323–331.
Hu, J., Qian, Y., Wang, X., Liu, T., Liu, S. 2012. *Langmuir* 28:2073–2082.
Islam, M. A., Shin, J.-Y., Firdous, J., Park, T.-E., Choi, Y.-J., Cho, M.-H., Yun, C.-H., Cho, C.-S. 2012. *Biomaterials* 33(34):8868–8880.
Jang, J. H., Shea, L. D. 2006. *J. Control. Release* 112(1):120–128.
Jeong, Y., Jin, G.-W., Choi, E., Jung, J. H., Park, J.-S. 2011. *Int. J. Pharm.* 420(2):366–370.
Ji, W., Panus, D., Palumbo, R. N., Tang, R., Wang, C. 2011. *Biomacromolecules* 12(12):4373–4385.
Jiang, Q., Shi, P., Li, C., Wang, Q., Xu, F., Yang, W., Tang, G. 2011b. *Macromol. Biosci.* 11(3):435–444.
Jiang, X., Sha, X., Xin, H., Chen, L., Gao, X., Wang, X., Law, K., Gu, J., Chen, Y., Jiang, Y., Ren, X., Ren, Q., Fang, X. 2011a. *Biomaterials* 32:9457–9469.
Jin, Y., Song, Y., Zhu, X., Zhou, D., Chen, C., Zhang, Z., Huang, Y. 2012. *Biomaterials* 33(5):1573–1582.
Johnson, R. N., Chu, D. S. H., Shi, J., Schellinger, J. G., Carlson, P. M., Pun, S. H. 2011. *J. Control. Release* 155(2):303–311.
Johnson, R. P., Jeong, Y.-I., Choi, E., Chung, C.-W., Kang, D. H., Oh, S.-O., Suh, H., Kim, I. 2012. *Adv. Funct. Mater.* 22:1058–1068.
Jooss, K., Yang, Y., Fisher, K. J., Wilson, J. M. 1998. *J. Virol.* 72:4212–4223.
Jule, E., Nagasaki, Y., Kataoka, K. 2003. *Bioconj. Chem.* 14:177–186.
Kabanov, A., Zhu, J., Alakhov, V. 2005. *Adv. Genet.* 53:231–261.
Kaouass, M., Beaulieu, R., Balicki, D. 2006. *J. Control. Release* 113(3):245–254.
Kasturi, S. P., Sachaphibulkij, K., Roy, K. 2005. *Biomaterials* 26(32):6375–6385.
Kataoka, K., Matsumoto, T., Yokoyama, M., Okano, T., Sakurai, Y., Fukushima, S. 2000. *J. Control. Release* 64:143–151.
Kedar, U., Phutane, P., Shidhaye, S., Kadam, V. 2010. *Nanomed. Nanotechnol. Biol. Med.* 6:714–729.
Kesharwani, P., Gajbhiye, V., Jain, N. K. 2012. *Biomaterials* 33(29):7138–7150.
Khalil, I., Kogure, K., Akita, H., Harashima, H. 2006. *Pharmacol. Rev.* 58(1):32–45.
Kichler, A., Mason, A. J., Bechinger, B. 2006. *Biochim. Biophys. Acta* 1758(3):301–307.
Kichler, A., Pages, J. C., Leborgne, C., Druillennec, S., Lenoir, C., Coulaud, D., Delain, E., Le Cam, E., Roques, B. P., Danos, O. 2000. *J. Virol.* 74(12):5424–5431.
Klink, D. T., Chao, S., Glick, M. C., Scanlin, T. F. 2001. *Mol. Ther.* 3(6):831–841.

Klutz, K., Willhauck, M. J., Dohmen, C., Wunderlich, N., Knoop, K., Zach, C., Senekowitsch-Schmidtke, R., Gildehaus, F.-J., Ziegler, S., Furst, S., Goke, B., Wagner, E., Ogris, M., Spitzweg, C. 2011. *Hum. Gene Therapy* 22(12):1563–1574.

Koo, H., Huh, M. S., Sun, I.-C., Yuk, S. H., Choi, K., Kim, K., Kwon, I. C. 2011. *Acc. Chem. Res.* 44:1018–1028.

Krammer, B. 2001. *Anticancer Res.* 21:4271–4277.

Kurisawa, M., Yokoyama, M., Okano, T. 2000. *J. Control. Release* 69:127–137.

Kwon, G., Suwa, S., Yokoyama, M., Okano, T., Sakura, Y., Kataoka, K. 1994. *J. Control. Release* 29:17–23.

Kwon, G. S., Forrest, M. L. 2006. *Drug Dev. Res.* 67:15–22.

Kwon, G. S., Naito, M., Yokoyama, M., Okano, T., Sakurai, Y., Kataoka, K. 1995. *Pharmacol. Res.* 12:192–195.

Lai, T. C., Kataoka, K., Kwon, G. S. 2012. *Colloids Surf. B: Biointerfaces* 99:27–37.

Lai, W.-F., Tang, G.-P., Wang, X., Li, G., Yao, H., Shen, Z., Lu, G., Poon, W. S., Kung, H.-F., Lin, M. C. M. 2011. *Bionanoscience* 1(3):89–96.

Langer, R. 1998. *Nature* 392:5–10.

Lavasanifar, A., Samuel, J., Kwon, G. S. 2002b. *J. Control. Release* 79:165–172.

Lavasanifar, A., Samuel, J., Sattari, S., Kwon, G. S. 2002a. *Pharm. Res.* 19:418–422.

Lavigne, M. D., Pennadam, S. S., Ellis, J., Alexander, C., Gorecki, D. C. 2007. *J. Gene Med.* 9:44–54.

Lechardeur, D., Sohn, K. J., Haardt, M., Joshi, P. B., Monck, M., Graham, R. W., Beatty, B., Squire, J., O'Brodovich, H., Lukacs, G. L. 1999. *Gene Ther.* 6(4):482–497.

Lee, E. S., Na, K., Bae, Y. H. 2005. *Nano Lett.* 5:325–329.

Lee, E. S., Shin, H. J., Na, K., Bae, Y. H. 2003. *J. Control. Release* 90:363–368.

Lee, J. C. M., Bermudez, H., Discher, B. M., Sheehan, M. A., Won, Y. Y., Bates, F. S., Discher, D. E. 2001. *Biotechnol. Bioeng.* 73:135.

Lee, S., Saito, K., Lee, H.-R., Lee, M. J., Shibasaki, Y., Oishi, Y., Kim, B.-S. 2012. *Biomacromolecules* 13:1190–1196.

Lee, Y., Bae, Y., Hiki, S., Ishii, T., Kataoka, K. 2007. *J. Am. Chem. Soc.* 129:5362.

Lee, Y., Ishii, T., Kim, H. J., Nishiyama, N., Hayakawa, Y., Itaka, K., Kataoka, K. 2010. *Angew. Chem. Int. Ed.* 49:2552–2555.

Levchenco, T. S., Mountrichas, G., Torchilin, V. P., Pispas, S. 2007. *Proceedings of the International Conference on Nanomedicine*, Chalkidiki, Greece, 2007, pp. 160–163.

Li, W., Chen, L., Huang, Z., Wu, X., Zhang, Y., Hu, Q., Wang, Y. 2011. *Org. Biomol. Chem.* 9(22):7799–7806.

Li, W., Wang, Y., Chen, L., Huang, Z., Hu, Q., Ji, J. 2012a. *Chem. Commun.* 48(81):10126–10128.

Li, X., Li, H., Liu, G., Deng, Z., Wu, S., Li, P., Xu, Z., Xu, H., Chu, P. K. 2012b. *Biomaterials* 33:3013–3024.

Li, Y., Ikeda, S., Nakashima, K., Nakamura, H. 2003. *Colloid Polym. Sci.* 281:562–568.

Li, Y., Pei, Y., Zhang, X., Gu, Z., Zhou, Z., Yuan, W., Zhou, J., Zhu, J., Gao, X. 2001. *J. Control. Release* 71:203–211.

Li, Y. Y., Dong, H. Q., Wang, K., Shi, D. L., Zhang, X. Z., Zhuo, R. X. 2010. *Sci. China Chem.* 53:447–457.

Liang, W., Gong, H., Yin, D., Lu, S., Fu, Q. 2011. *Chem. Pharm. Bull.* 59(9):1094–1101.

Liao, Z., Wang, H., Wang, X., Zhao, P., Wang, S., Su, W., Chang, J. 2011. *Adv. Funct. Mater.* 21:1179–1186.

Licciardi, M., Giammona, G., Du, J., Armes, S. P., Tang, Y., Lewis, A. L. 2006. *Polymer* 47:2946–2955.

Lin, C., Engbersen, J. F. J. 2011b. *Mater. Sci. Eng. C* 31(7):1330–1337.

Lin, L. Y., Lee, N. S., Zhu, J., Nystrom, A. M., Pochan, D. J., Dorshow, R. B., Wooley, K. L. 2011a. *J. Control. Release* 152:37–48.

Liu, G., Xie, J., Zhang, F., Wang, Z., Luo, K., Zhu, L., Quan, Q., Niu, G., Lee, S., Ai, H., Chen, X. 2011d. *Small* 7:2742–2749.

Liu, L., Zheng, M., Renette, T., Kissel, T. 2012a. *Bioconj. Chem.* 23(6):1211–1220.

Liu, S. Q., Wiradharma, N., Gao, S. J., Tong, Y. W., Yang, Y. Y. 2007. *Biomaterials* 28:1423–1428.

Liu, T., Li, X., Qian, Y., Hua, X., Liu, S. 2012c. *Biomaterials* 33:2521–2531.

Liu, T., Qian, Y., Hu, X., Ge, Z., Liu, S. 2012b. *J. Mater. Chem.* 22:5020–5030.

Liu, W.-M., Xue, Y.-N., Peng, N., He, W.-T., Zhou, R.-X., Huang, S.-W. 2011c. *J. Mater. Chem.* 21(35):13306–13315.

Liu, Y., Samsonova, O., Sproat, B., Merkel, O., Kissel, T. 2011b. *J. Control. Release* 153(3):262–268.

Liu, Z., Zheng, M., Meng, F., Zhong, Z. 2011a. *Biomaterials* 32(34):9109–9119.

Loh, X. J., Zhang, Z.-X., Mya, K. Y., Wu, Y.-L., He, C. B., Li, J. 2011. *J. Mater. Chem.* 20(47):10634–10642.

Lukacs, G. L., Haggie, P., Seksek, O., Lechardeur, D., Freedman, N., Verkman, A. S. 2000. *J. Biol. Chem.* 275(3):1625–1629.

Mahmud, A., Patel, S., Molavi, O., Choi, P., Samuel, J., Lavasanifar, A. 2009. *Biomacromolecules* 10:471–478.

Mahor, S., Dash, B. C., O'Connor, S., Pandit, A. 2012. *Bioconj. Chem.* 23(6):1138–1148.

Marshall, E. 1999. *Science* 286:2244–2245.

Mastrobattista, E., Hennink, W. E. 2012. *Nat. Mater.* 11(1):10–12.

Meyer, D. E., Chilkoti, A. 1999. *Nat. Biotechnol.* 17:1112–1115.

Mintzer, M. A., Simanek, E. E. 2009. *Chem. Rev.* 109(2):259–302.

Mishra, B., Patel, B.B., Tiwari, S. 2010. *Nanomed. Nanotechnol. Biol. Med.* 6:9–24.

Mislick, K. A., Baldeschwieler, J. D. 1996. *Proc. Natl. Acad. Sci. USA* 93(22):12349–12354.

Miyata, K., Christie, R. J., Kataoka, K. 2011. *React. Funct. Polym.* 71:227–234.

Moffatt, S., Wiehle, S., Cristiano, R. J. 2006. *Gene Ther.* 13(21):1512–1523.

Molavi, O., Ma, Z., Mahmud, A., Alshamsan, A., Samuel, J., Lai, R., Kwon, G. S., Lavasanifar, A. 2008. *Int. J. Pharm.* 347:118–127.

Monsigny, M., Rondanino, C., Duverger, E., Fajac, I., Roche, A. C. 2004. *Biochim. Biophys. Acta* 1673(1–2):94–103.

Munier, S., Messai, I., Delair, T., Verrier, B., Ataman-Onal, Y. 2005. *Colloids Surf B: Biointerfaces* 43(3–4):163–173.

Murata, H., Futami, J., Kitazoe, M., Yonehara, T., Nakanishi, H., Kosaka, M., Tada, H., Sakaguchi, M., Yagi, Y., Seno, M., Huh, N.-H., Yamada, H. 2008. *J. Biochem.* 144:447–455.

Nagasaki, Y., Yasugi, K., Yamamoto, Y., Harada, A., Kataoka, K. 2001. *Biomacromolecules* 2:1067–1070.

Nakamura, E., Makino, K., Okano, T., Yamamoto, T., Yokoyama, M. 2006. *J. Control. Release* 114:325–333.

Nam, H. Y., Kim, J., Kim, S., Yockman, J. W., Kim, S. W., Bull, D. A. 2011. *Biomaterials* 32(22):5213–5222.

Nasongkla, N., Bey, E., Ren, J., Ai, H., Khemtong, C., Guthi, J. S. 2006. *Nano Lett.* 6:2427–2432.

Nasongkla, N., Shuai, X., Ai, H., Weinberg, B. D., Pink, J., Boothman, D. A., Gao, J. M. 2004. *Angew. Chem. Int. Ed.* 43:6323–6327.

Nishiyama, N., Nakagishi, Y., Morimoto, Y., Lai, P.-S., Miyazaki, K., Urano, K. 2009. *J. Control. Release* 133:245–251.

Noga, M., Edinger, D., Rodl, W., Wagner, E., Winter, G., Besheer, A. 2012. *J. Control. Release* 159(1):92–103.

Nurmi, L., Peng, H., Seppala, J., Haddleton, D. M., Blakey, I., Whittaker, A. K. 2010. *Polym. Chem.* 1:1039–1047.

Nystrom, A. M., Bartels, J. W., Du, W., Wooley, K. L. 2009. *J. Polym. Sci. Part A: Polym. Chem.* 47:1023–1037.

O'Hagan, D. T., Singh, M., Ulmer, J. B. 2006. *Methods* 40(1):10–19.

Ornelas-Megiatto, C., Wich, P. R., Frechet, J. M. J. 2012. *J. Am. Chem. Soc.* 134(4):1902–1905.

O'Rorke, S., Keeney, M., Pandit, A. 2010. *Prog. Polym. Sci.* 35:441–458.

Osada, K., Shiotani, T., Tockary, T. A., Kobayashi, D., Oshima, H., Ikeda, S., Christie, R. J., Itaka, K., Kataoka, K. 2012. *Biomaterials* 33(1):325–332.

Pack, D. W., Hoffman, A. S., Pun, S., Stayton, P. S. 2005. *Nat. Rev. Drug Discov.* 4:581–593.

Palumbo, R. N., Zhong, X., Panus, D., Han, W., Ji, W., Wang, C. 2012. *J. Control. Release* 159(2):232–239.

Pang, Z. Q., Lu, W., Gao, H. L., Hu, K. L., Chen, J., Zhang, C. L., Gao, X. L., Jiang, X. G., Zhu, C. Q. 2008. *J. Control. Release* 128:120.

Pansare, V. J., Hejazi, S., Faenza, W. J., Prud'homme, R. K. 2012. *Chem. Mater.* 24:812–827.

Pante, N., Kann, M. 2002. *Mol. Biol. Cell* 13(2):425–434.

Panyam, J., Labhasetwar, V. 2003. *Adv. Drug Deliv. Rev.* 55(3):329–347.

Parelkar, S. S., Chan-Seng, D., Emrick, T. 2011. *Biomaterials* 32(9):2432–2444.

Park, E. K., Kim, S. Y., Lee, S. B., Lee, Y. M. 2005. *J. Control. Release* 109:158–168.

Park, J.-H., von Maltzahn, G., Sangeeta, E. R., Michael, N. B., Sailor, J. 2008. *Angew. Chem. Int. Ed.* 47:7284–7289.

Parker, A. L., Eckley, L., Singh, S., Preece, J. A., Collins, L., Fabre, J. W. 2007. *Biochim. Biophys. Acta* 1770:1331–1337.

Parveen, S., Misra, R., Sanjeeb, K. Sahoo, S. K. 2012. *Nanomed. Nanotechnol. Biol. Med.* 8:147–166.

Patil, M. L., Zhang, M., Minko, T. 2011. *ACS Nano* 22:1877–1887.

Peng, C.-L., Lai, P.-S., Lin, F.-H., Yueh-Hsiu Wu, S., Shieh, M.-J. 2009b. *Biomaterials* 30:3614–3625.

Peng, C.-L., Shieh, M.-J., Tsai, M.-H., Chang, C.-C., Lai, P.-S. 2008. *Biomaterials* 29:3599–3608.

Peng, H., Blakey, I., Dargaville, B., Rasoul, F., Rose, S., Whittaker, A. K. 2009a. *Biomacromolecules* 10:374–381.

Perche, F., Benvegnu, T., Berchel, M., Lebegue, L., Pichon, C., Jaffres, P.-A., Midoux, P. 2011a. *Nanomed. Nanotechnol. Biol. Med.* 7(4):445–453.

Perche, F., Gosset, D., Mevel, M., Miramon, M. L., Yaouanc, J.-J., Pichon, C., Benvegnu, T., Jaffres, P.-A., Midoux, P. 2011b. *J. Drug Targeting* 19(5):315–325.

Piest, M., Engbersen, J. F. J. 2011. *J. Control. Release* 155(2):331–340.

Ping, Y., Liu, C., Zhang, Z., Liu, K. L., Chen, J., Li, J. 2011. *Biomaterials* 32(32):8328–8341.

Pogue, B. W,, Pitts, J. D., Mycek, M. A., Sloboda, R. D., Wilmot, C. M., Brandsema, J. F. 2001. *Photochem. Photobiol.* 74:817–824.

Ponta, A., Bae, Y. 2010. *Pharm. Res.* 27:2330–2342.

Rameez, S., Alosta, H., Palmer, A. F. 2008. *Bioconj. Chem.* 19:1025.

Reilly, M. J., Larsen, J. D., Sullivan, M. O. 2012. *Mol. Pharm.* 9(5):1031–1040.

Rejman, J., Bragonzi, A., Conese, M. 2005. *Mol. Ther.* 12(3):468–474.

Rijcken, C. J. F., Hofman, J.-W., van Zeeland, F., Hennink, W. E., van Nostrum, C. F. 2007. *J. Control. Release* 124:144–153.

Rohrbach, M., Clarke, J. T. R. 2007. *Drugs* 67:2697–2716.

Sabeqzadeh, E., Rahbarizadeh, F., Ahmadvand, D., Rasaee, M. J., Parhamifar, L., Moghimi, S. M. 2011. *J. Control. Release* 156(1):85–91.

Sakuma, S., Suzuki, N., Kikuchi, H., Hiwatari, K., Arikawa, K., Kishida, A. 1997. *Int. J. Pharm.* 149:93–106.

Sakuma, S., Suzuki, N., Sudo, R., Hiwatari, K., Kishida, A., Akashi, M. 2002. *Int. J. Pharm.* 239:185–195.

Sant, V. P., Smith, D., Leroux, J. C. 2004. *J. Control. Release* 97:301–312.

Santander-Ortega, M. J., Bastos-Gonzalez, D., Ortega-Vinuesa, J. L., Alonso, M. J., 2009. *J. Biomed. Nanotechnol.* 5(1):45–53.

Santander-Ortega, M. J., Csaba, N., Gonzalez, L., Bastos-Gonzalez, D., Ortega-Vinuesa, J. L., Alonso, M. J. 2010. *Colloid Polym. Sci.* 288:141–150.

Santos, C. A., Freedman, B. D., Leach, K. J., Press, D. L., Scarpulla, M., Mathiowitz, E. 1999. *J. Control. Release* 60(1):11–22.

Schafer, A., Pahnke, A., Schaffert, D., Van Weerden, W. M., De Ridder, C. M. A., Rodl, W., Vetter, A., Spitzweg, C., Kraaij, R., Wagner, E., Ogris, M. 2011. *Hum. Gene Therapy* 22(12):1463–1473.

Schaffer, D. V., Lauffenburger, D. A. 1998. *J. Biol. Chem.* 273(43):28004–28009.

Sezer, A., D., Kazak, H., Oner, E. T. 2011. *Carbohydr. Polym.* 84(1):358–363.

Shahin, M., Lavasanifar, A. 2010. *Int. J. Pharm.* 389:213–222.

Shan, J., Budijono, S. J., Hu, G., Yao, N., Kang, Y., Ju, Y., Prud'homme, R. K. 2011. *Adv. Funct. Mater.* 21:2488–2495.

Shan, Y., Luo, T., Peng, C., Sheng, R., Cao, A., Cao, X., Shen, M., Guo, R., Tomas, H., Shi, X. 2012. *Biomaterials* 33(10):3025–3035.

Sharma, A., Soliman, G. M., Al-Hajaj, N., Sharma, R., Maysinger, D., Kakkar, A. 2012. *Biomacromolecules* 13:239–252.

Shen, Z., Shi, B., Zhang, H., Bi, J., Dai, S. 2012. *Soft Matter* 8(5):1385–1394.

Shi, J., Johnson, R. N., Schellinger, J. G., Carlson, P. M., Pun, S. H. 2012. *Int. J. Pharm.* 427(1):113–122.

Shiraishi, K., Kawano, K., Minowa, T., Maitani, Y., Yokoyama, M. 2009. *J. Control. Release* 136:14–20.

Simone, E. A., Dziubla, T. D., Colon-Gonzalez, F., Discher, D. E., Muzykantov, V. R. 2007. *Biomacromolecules* 8:3914–3921.

Simone, E. A., Dziuba, T. D., Discher, D. E., Muzykantov, V. R. 2009. *Biomacromolecules* 10(6):1324–1330.

Smith, A. E., Sizovs, A., Grandinetti, G., Xue, L., Reineke, T. M. 2011. *Biomacromolecules* 12(8):3015–3022.

Solaro, R. 2008. *J. Polym. Sci. Part A: Polym. Chem.* 46:1–11.

Soliman, G. M., Winnik, F. M. 2008. *Int. J. Pharmacol.* 356:248–258.

Son, S., Hwang, D. W., Singha, K., Jeong, J. H., Park, T. G., Lee, D. S., Kim, W. J. 2011. *J. Control. Release* 155(1):18–25.

Soppimath, K. S., Aminabhavi, T. M., Kulkarni, A. R., Rudzinski, W. E. 2001. *J. Control. Release* 70:1–20.

Soppimath, K. S., Liu, L. H., Seow, W. Y., Liu, S. Q., Chan, R. P. P. Y. 2007. *Adv. Funct. Mater.* 17:355–361.

Soppimath, K. S., Tan, D. C. W., Yang, Y. Y. 2005. *Adv. Mater.* 17:318–323.

Sortino, S., Mazzaglia, A., Monsu Scolaro, L., Marino Merlo, F., Valveri, V., Sciortino, M. T. 2006. *Biomaterials* 27:4256–4265.

Sulistio, A., Lowenthal, J., Blencowe, A., Bongiovanni, M. N., Ong, L., Gras, S. L., Zhang, X., Qiao, G. G. 2011. *Biomacromolecules* 12:3469–3477.

Sun, S. J., Liu, W. G., Cheng, N., Zhang, B. Q., Cao, Z. Q., Yao, K. D., Liang, D. C., Zuo, A. J., Guo, G., Zhang, J. Y. 2005. *Bioconj. Chem.* 16:972–980.

Sun, S.-L., Lo, Y.-L., Chen, H.-Y., Wang, L.-F. 2012. *Langmuir* 28 (7):3542–3552.

Takigawa, D. Y., Tirrell, D. A. 1985. *Macromolecules* 18(3):338–342.

Talelli, M., Iman, M., Varkouhi, A. K., Rijcken, C. J. F., Schiffelers, R. M., Etrych, T., Ulbrich, K., van Nostrum, C. F., Lammers, T., Storm, G., Hennink, W. E. 2010. *Biomaterials* 31:7797–7804.

Talelli, M., Rijcken, C. J. F., Lammers, T., Seevinck, P. R., Storm, G., van Nostrum, C. F., Hennink, W. E. 2009. *Langmuir* 25:2060–2067.

Talelli, M., Rijcken, C. J. F., Oliveira, S., van der Meel, R., van Bergen en Henegouwen, P. M. P., Lammers, T., van Nostrum, C. F., Storm, G., Hennink, W. E. 2011. *J. Control. Release* 151:183–192.

Talsma, S. S., Babensee, J. E., Murthy, N., Williams, I. R. 2006. *J. Control. Release* 112(2):271–279.

Tamboli, V., Mishra, G. P., Mitra, A. K. 2011. *Ther. Deliv.* 2(4):523–536.

Tan, J. P. K., Kim, S. H., Nederberg, F., Appel, E. A., Waymouth, R. M., Zhang, Y., Hedrick, J. L., Yang, Y. Y. 2009. *Small* 5:1504–1507.

Tan, J. P. K., Kim, S. H., Nederberg, F., Fukushima, K., Coady, D. J., Nelson, A., Yang, Y. Y., Hedrick, J. L. 2010. *Macromol. Rapid Commun.* 31:1187–1192.

Thomas, M., Klibanov, A. M. 2002. *Proc. Natl. Acad. Sci. USA* 99(23):14640–14645.

Tomar, L. K., Tyagi, C., Lahiri, S. S., Singh, H. 2011. *Polym. Adv. Technol.* 22:1760–1767.

Tsai, H. C., Tsai, C.-H., Lin, S.-L., Jhang, C.-R., Chiang, Y.-C., Hsiue, G.-H. 2012. *Biomaterials* 33:1827–1837.

Tu, S., Chen, Y.-W., Qiu, Y.-B., Zhu, K., Luo, X.-L, 2011. *Macromol. Biosci.* 11:1416–1425.

Turk, M., Dincer, S., Piskin, E. 2007. *J. Tissue Eng. Regen. Med.* 1:377–388.

Tyrrell, Z. L., Shen, Y., Radosza, M. 2010. *Prog. Polym. Sci.* 35:1128–1143.

Uchida, S., Itaka, K., Chen, Q., Osada, K., Ishii, T., Shibata, M.-A., Harada-Shiba, M., Kataoka, K. 2012. *Mol. Therapy* 20(6):1196–1203.

Ujiie, K., Kanayama, N., Asai, K., Kishimoto, M., Ohara, Y., Akashi, Y., Yamada, K., Hashimoto, S., Oda, T., Ohkohchi, N., Yanagihara, H., Kita, E., Yamaguchi, M., Fujii, H., Nagasaki, Y. 2011. *Colloids Surf. B* 88:771–778.

Urry, D. W. J. 1997. *Phys. Chem. B* 101:11007–11028.

Varkouhi, A. K., Mountrichas, G., Schiffelers, R. M., Lammers, T., Storm, G., Pispas, S., Hennink, W. E. 2012. *Eur. J. Pharm. Sci.* 45:459–466.

Vega, J., Ke, S., Fan, Z., Wallace, S., Charsangavej, C., Li, C. 2003. *Pharm. Res.* 20:826–832.

Velasco, D., Collin, E., San Roman, J., Pandit, A., Elvira, C. 2011. *Eur. J. Pharm. Biopharm.* 79(3):485–494.

Vercauteren, D., Piest, M., van der Aa, L. J., Al Soraj, M., Jones, A. T., Engbersen, J. F. J., De Smedt, S. C. Braeckmans, K. 2011. *Biomaterials* 32(11):3072–3084.

Vila, A., Sanchez, A., Janes, K., Behrens, I., Kissel, T., Vila Jato, J. L, Alonso, M. J. 2004. *Eur. J. Pharm. Biopharm.* 57:123–131.

Vila, A., Sanchez, A., Tobio, M., Calvo, P., Alonso, M. J. 2002. *J. Control. Rel.* 78:15–24.

Vinogradov, S., Batrakova, E., Li, S., Kabanov, A. 1999. *Bioconj. Chem.* 10:851–860.

Wagner, E. 2012. *Acc. Chem. Res.* 45:1005–1013.

Walter, E., Moelling, K., Pavlovic, J., Merkle, H. P. 1999. *J. Control. Release* 61(3):361–374.

Wang, B., Zhang, S., Cui, S., Yang, B., Zhao, Y., Chen, H., Hao, X., Shen, Q., Zhou, J. 2012b. *Biotechnol. Lett.* 34(1):19–28.

Wang, C., Ge, Q., Ting, D., Nguyen, D., Shen, H. R., Chen, J., Eisen, H. N., Heller, J., Langer, R., Putnam, D. 2004. *Nat. Mater.* 3(3):190–196.

Wang, C., Luo, X., Zhao, Y., Han, L., Zeng, X., Feng, M., Pan, S., Wu, C. 2012a. *Acta Biomater.* 8(8):3014–3026.

Wang, D., Robinson, D. R., Kwon, G. S., Samuel, J. 1999. *J. Control. Release* 57(1):9–18.

Wang, W., Cheng, D., Gong, F., Miao, X., Shuai, X. 2012c. *Adv. Mater.* 24:115–120.

Wang, Y., Xu, Z., Zhang, R., Li, W., Yang, L., Hu, Q. 2011. *Colloids Surf. B: Biointerfaces* 84(1):259–266.

Wang, Y. C., Li, Y., Sun, T. M., Xiong, M. H., Wu, J., Yang, Y. Y., Wang, J. 2010. *Macromol. Rapid Commun.* 31:1201–1206.

Wiradharma, N., Zhang, Y., Venkataraman, S., Hedrick, J. L., Yang, Y. Y. 2009. *Nano Today* 4:302–317.

Wong, S. Y., Pelet, J. M., Putnam, D. 2007. *Prog. Polym. Sci.* 32:799–837.

Wu, D. Q., Lu, B., Chang, C., Chen, C. S., Wang, T., Zhang, Y. Y. 2009. *Biomaterials* 30:1363–1369.

Wu, J., Yamanouchi, D., Liu, B., Chu, C.-C. 2012. *J. Mater. Chem.* 22(36):18983–18991.

Xiong, X. B., Falamarzian, A., Garg, S. M., Lavasanifar, A. 2011. *J. Control. Release* 155:248–261.

Xiong, X. B., Mahmud, A., Uludag, H., Lavasanifar, A. 2007. *Biomacromolecules* 8:874–884.

Xiong, X. B., Mahmud, A., Uludag, H., Lavasanifar, A. 2008. *Pharm. Res.* 25:2555–2566.

Xu, H., Meng, F., Zhong, Z. 2009. *J. Mater. Chem.* 19:4183–4190.

Xu, Y.-C., Dong, C.-M. 2012. *J. Polym. Sci. Part A: Polym. Chem.* 50, 1216–1225.

Yadav, S. Ch., Kumari, A., Yadav, R. 2011. *Peptides* 32:173–187.

Yanagishita, M., Hascall, V. C. 1992. *J. Biol. Chem.* 267(14):9451–9454.

Yang, X., Grailer, J. J., Rowland, I. J., Javadi, A., Hurley, S. A., Matson, V. Z., Steeber, D. A., Gong, S. 2010. *ACS Nano* 4:6805–6817.

Yant, S. R., Meuse, L., Chiu, W., Ivics, Z., Izsvak, Z., Kay, M. A. 2000. *Nat. Genet.* 25(1):35–41.

Yoo, H. S, Park, T. G. 2004. *J. Control. Release* 100:247–256.

Yu, H., Zou, Y., Wang, Y., Huang, X., Huang, G., Sumer, B. D., Boothman, D. A., Gao, J. 2011. *ACS Nano* 5:9246–9255.

Yuan, X. F., Harada, A., Yamasaki, Y., Kataoka, K. 2005. *Langmuir* 21:2668.

Zhang, G., Liu, J., Yang, Q., Zhuo, R., Jiang, X. 2012a. *Bioconj. Chem.* 23(6):1290–1299.

Zhang, L., Lin, Y., Zhang, Y., Chen, R., Zhu, Z., Wu, W., Jiang, X. 2012b. *Macromol. Biosci.* 12:83–92.

Zhang, R., Wang, Y., Du, F.-S., Wang, Y.-L., Tan, Y.-X., Ji, S.-P., Li, Z.-C. 2011. *Macromol. Biosci.* 11(10):1393–406.

Zhang, X., Cheng, J., Wang, Q., Zhong, Z., Zhuo, R. 2010. *Macromolecules* 43:6671–6677.

Zhao, H., Duong, H. H. P., Yung, L. Y. L. 2010. *Macromol. Rapid Commun.* 31:1163–1169.

Zheng, M., Liu, Y., Samsonova, O., Endres, T. Merkel, O., Kissel, T. 2012a. *Int. J. Pharm.* 427(1):80–87.

Zheng, M., Zhong, Y., Meng, F., Peng, R., Zhong, Z. 2011. *Mol. Pharm.* 8(6):2432–2443.

Zheng, S. W., Liu, G., Hong, R. Y., Li, H. Z., Li, Y. G., Wei, D. G. 2012b. *Appl. Surf. Sci.* 259:201–207.

Zhong, Z., Zheng, M., Zhong, Z., Zhou, L., Meng, F., Peng, R. 2012. *Biomacromolecules* 13(3):881–888.

Zhu, C., Zheng, M., Meng, F., Mickler, F. M., Ruthardt, N., Zhu, X., Zhong, Z. 2012. *Biomacromolecules* 13(3):769–778.

Zhu, Y., Sheng, R., Luo, T., Li, H., Sun, W., Li, Y., Cao, A. 2011. *Macromol. Biosci.* 11(2):147–186.

Zintchenko, A., Ogris, M., Wagner, E. 2006. *Bioconj. Chem.* 17:766–772.

Index

A

B

I

L

M

O

P

Q

R

T

U

W

Milton Keynes UK
Ingram Content Group UK Ltd.
UKHW021910071024
449327UK00022B/1647